Lecture Notes in Computer Science   1064

Edited by G. Goos, J. Hartmanis and J. van Leeuwen

Advisory Board: W. Brauer   D. Gries   J. Stoer

Springer
Berlin
Heidelberg
New York
Barcelona
Budapest
Hong Kong
London
Milan
Paris
Santa Clara
Singapore
Tokyo

Bernard Buxton Roberto Cipolla (Eds.)

Computer Vision – ECCV '96

4th European Conference
on Computer Vision
Cambridge, UK, April 15-18, 1996
Proceedings, Volume I

Springer

Series Editors

Gerhard Goos, Karlsruhe University, Germany

Juris Hartmanis, Cornell University, NY, USA

Jan van Leeuwen, Utrecht University, The Netherlands

Volume Editors

Bernard Buxton
University College London, Department of Computer Science
Gower Street, London WC1E 6BT, United Kingdom

Roberto Cipolla
University of Cambridge, Department of Engineering
Cambridge CB2 1PZ, United Kingdom

Cataloging-in-Publication data applied for

Die Deutsche Bibliothek - CIP-Einheitsaufnahme

Computer vision : proceedings / ECCV '96, Fourth European Conference on Computer
Vision, Cambridge, UK, April 1996. Bernard Buxton ; Roberto Cipolla (ed.). - Berlin ;
Heidelberg ; New York ; Barcelona ; Budapest ; Hong Kong ; London ; Milan ; Paris ;
Santa Clara ; Singapore ; Tokyo : Springer.

NE: Buxton, Bernard [Hrsg.]; ECCV <4, 1996, Cambridge>

Vol. 1. - 1996
  (Lecture notes in computer science ; Vol. 1064)
  ISBN 3-540-61122-3

NE: GT

CR Subject Classification (1991): I.3.5, I.5, I.2.9-10, I.4

ISBN 3-540-61122-3 Springer-Verlag Berlin Heidelberg New York

Typesetting: Camera-ready by author
SPIN 10512740   06/3142 – 5 4 3 2 1 0   Printed on acid-free paper

Preface

Following the highly successful conferences held in Antibes (ECCV'90), Santa Margherita Ligure (ECCV'92) and Stockholm (ECCV'94), the European Conference on Computer Vision has established itself as one of the major international events in the exciting and active research discipline of computer vision. It has been an honour and pleasure to organise the Fourth European Conference on Computer Vision, held in the University of Cambridge, 15–18 April 1996.

These proceedings collect the papers accepted for presentation at the conference. They were selected from 328 contributions describing original and previously unpublished research. As with the previous conferences each paper was reviewed double blind by three reviewers selected from the Conference Board and Programme Committee. Final decisions were reached at a committee meeting in London where 43 papers were selected for podium presentation and 80 for presentation in poster sessions. The decision to keep ECCV'96 a single track conference led to a competitive selection process and it is entirely likely that good submissions were not included in the final programme.

We wish to thank all the members of the Programme Committee and the additional reviewers who each did a tremendous job in reviewing over 25 papers papers in less than six weeks. We are also extremely grateful to the chairmen of previous conferences, Olivier Faugeras, Giulio Sandini and Jan-Olof Eklundh for their help in the preparations of the conference and to Roberto's colleagues and research students in the Department of Engineering for their patience and support. The conference was sponsored by the European Vision Society (EVS) and the British Machine Vision Association (BMVA). We are grateful to the Chairman of the BMVA and the executive committee for advice and financial support throughout the the organisation of the conference.

Cambridge, January 1996 Bernard Buxton and Roberto Cipolla

Conference Chairs

Bernard Buxton	University College London
Roberto Cipolla	University of Cambridge

Conference Board and Programme Committee

N. Ayache	INRIA, Sophia Antipolis
R. Bajcsy	University of Pennsylvania
A. Blake	University of Oxford
P. Bouthemy	IRISA/INRIA, Rennes
M. Brady	University of Oxford
H. Burkhardt	Technical University of Hamburg-Harburg
H. Buxton	University of Sussex
S. Carlsson	KTH, Stockholm
H. Christensen	Aalborg University
J. Crowley	INPG,Grenoble
R. Deriche	INRIA, Sophia Antipolis
M. Dhome	Blaise Pascal University of Clermont-Fd
E. Dickmanns	Universität der Bundeswehr, Munich
J.-O. Eklundh	KTH, Stockholm
W. Enkelmann	IITB, Karlsruhe
O. Faugeras	INRIA, Sophia Antipolis
G. Granlund	Linköping University, Sweden
W. Förstner	Bonn University
L. Van Gool	Katholieke Universiteit Leuven
D. Hogg	University of Leeds
R. Horaud	INPG,Grenoble
J.J. Koenderink	Universiteit Utrecht
H. Knutsson	Linköping University, Sweden
O. Kübler	ETH, Zurich
J. Malik	University of California at Berkeley
J. Mayhew	Sheffield University
R. Mohr	INPG, Grenoble
H. Nagel	IITB, Karlsruhe
B. Neumann	Hamburg University
S. Peleg	The Hebrew University of Jerusalem
B. M. ter Haar Romeny	Universiteit Utrecht
G. Sandini	University of Genova
W. von Seelen	Ruhr University, Bochum
F. Solina	Ljubljana University, Slovenia
G. Sparr	Lund University, Sweden
G. Sullivan	University of Reading
M. Tistarelli	University of Genova
V. Torre	University of Genova
S. Tsuji	Wakayama University, Japan
D. Vernon	Maynooth College, Ireland
A. Verri	University of Genova
J.J. Villanueva	Autonomous University of Barcelona
D. Weinshall	The Hebrew University of Jerusalem
A. Zisserman	University of Oxford
S. Zucker	McGill University, Canada

Additional Reviewers

Frank Ade
P. Anandan
Mats Andersson
Tal Arbel
Martin Armstrong
Minoru Asada
Kalle Astrom
Jonas August
Dominique Barba
Eric Bardinet
Benedicte Bascle
Adam Baumberg
Paul Beardsley
Marie-Odile Berger
Rikard Berthilsson
Jørgen Bjørnstrup
Jerome Blanc
Philippe Bobet
Luca Bogoni
Magnus Borga
Kadi Bouatouch
Pierre Breton
Joachim Buhmann
Andrew Bulpitt
Hans Burkhardt
Franco Callari
Nikolaos Canterakis
Carla Capurro
Geert de Ceulaer
Francois Chaumette
Jerome Declerck
Herve Delingette
Bernard Delyon
Jean-Marc Dinten
Leonie Dreschler-Fischer
Christian Drewniok
Nick Efford
Bernard Espiau
Hanny Farid
Jacques Feldmar
Peter Fiddelaers
David Forsyth
Volker Gengenbach
Lewis Griffin
Eric Grimson
Patrick Gros
Etienne Grossberg
Enrico Grosso
Keith Hanna
Mike Hanson
Friedrich Heitger
Fabrice Heitz

Marcel Hendrickx
Olof Henricsson
Anders Heyden
Richard Howarth
Steve Hsu
S. Iouleff
Michal Irani
Hiroshi Ishiguro
Peter Jürgensen
Frédéric Jurie
Ioannis Kakadiaris
Jørgen Karlholm
Hans-Joerg Klock
Lars Knudsen
Jana Kosecka
Steen Kristensen
Wolfgang Krüger
Rakesh Kumar
Claude Labit
Bart Lamiroy
Tomas Landelius
Jean-Thierry Lapreste
Ole Larsen
Fabio Lavagetto
Jean-Marc Lavest
Jean-Pierre Leduc
Chil-Woo Lee
Aleš Leonardis
Jean-Michel Létang
Tony Lindeberg
Oliver Ludwig
An Luo
Robert Maas
Brian Madden
Claus Madsen
Twan Maintz
Gregoire Malandain
Stephen Maybank
Phil McLauchlan
Etienne Memin
Baerbel Mertsching
Dimitri Metaxas
Max Mintz
Theo Moons
Luce Morin
Jann Nielsen
Mads Nielsen
Wiro Niessen
Alison Noble
Klas Nordberg
Eric Pauwels

Marcello Pelillo
Xavier Pennec
Patrick Perez
Bernard Peuchot
Nic Pillow
Paolo Pirjanian
Rich Pito
Marc Pollefeys
Marc Proesmans
Long Quan
Paolo Questa
Veronique Rebuffel
Ian Reid
Gérard Rives
Luc Robert
Karl Rohr
Bart ter Haar Romeny
Charlie Rothwell
Paul Sajda
Alfons Salden
Josè Santos-Victor
Gilbert Saucy
Harpreet Sawhney
Ralph Schiller
Cordelia Schmid
Christoph Schnoerr
Carsten Schroeder
Ulf Cah von Seelen
Eero Simmoncelli
Sven Spanne
Rainer Sprengel
H. Siegfried Stiehl
Kent Strahlén
Peter Sturm
Gerard Subsol
Gabor Szekely
Tieniu Tan
Hiromi Tanaka
Jean-Philippe Thirion
Phil Torr
Bill Triggs
Morgan Ulvklo
Dorin Ungureanu
Peter Vanroose
Jean-Marc Vezien
Uwe Weidner
Carl-Fredrik Westin
Johan Wiklund
Lambert Wixon
Gang Xu
Zhengyou Zhang

Contents of Volume I

Texture and Features

Recognition/Matching/Segmentation

Tracking (1)

Grouping and Segmentation

Stereo

Structure from Motion (2)

Contents of Volume II

Colour, Vision and Shading

Contents of Volume II

Medical Applications

Tracking (2)

Applications and Recognition

Calibration/Focus/Optics

Tracking (3)

Applications

Structure from Motion (3)

Author Index

Structure From Motion (I)

Structure From Motion (1)

Self-Calibration from Image Triplets

Martin Armstrong[1], Andrew Zisserman[1] and Richard Hartley[2]

[1] Robotics Research Group, Department of Engineering Science, Oxford University, England
[2] The General Electric Company, Research and Development Laboratory, Schenectady, NY, USA.

Abstract. We describe a method for determining affine and metric calibration of a camera with unchanging internal parameters undergoing
planar motion. It is shown that affine calibration is recovered uniquely
and metric calibration up to a two fold ambiguity.

The novelty of this work is first, that the distinguished objects
of affine geometry are fixed entities in the image, second, showing
that these fixed entities can be computed uniquely via the trifocal tensor
between image triplets; third, a robust and automatic implementation of
the method.

Results are included of affine and metric calibration and structure recovery from images of real scenes.

1   Introduction

From an image sequence acquired with an unchanging camera, scene structure can be recovered up to a projectivity [1, 6, 12]. However, if the camera
is constrained to have unchanging internal parameters then this ambiguity is
reduced to metric by calibrating the camera using only image correspondences. This is termed "self-calibration" [2, 11].

The various attempts to make use of the constraint for image pairs have generated
a set of polynomial equations that are solved numerically; continuation [?] or iteratively [8, 17] even so far. In this paper we demonstrate the advantages of
working in the projective directly for the case of planar motion, both in the method
and the results of the method, and in the presence of noise.

To reduce the ambiguity of projection we use two projections to affine: it is not
necessary to identify the plane at infinity, and to reduce further to metric we
ambiguity the absolute conic [?] ... once it is identified [4, 2]. Both fixed
and ??? are fixed entities under projective motions of 3-space. The key idea in
this paper is that these fixed entities can be recovered via fixed entities (points,
lines, conics) in the image.

To determine the fixed image entities we utilize the geometric relations between
lines that are independent of the 3-dimensional structure. The fundamental
geometric relation between two views of a line under planar motion is represented by
the fundamental matrix [4][?] this provides a mapping from points in one image
to lines in the other and consequently it is a suitable mapping for determining
fixed entities directly. However, before we show the fundamental geometric

Self-Calibration from Image Triplets

Martin Armstrong[1], Andrew Zisserman[1] and Richard Hartley[2]

[1] Robotics Research Group, Department of Engineering Science, Oxford University,
England
[2] The General Electric Corporate Research and Development Laboratory,
Schenectady, NY, USA.

Abstract. We describe a method for determining affine and metric calibration of a camera with unchanging internal parameters undergoing planar motion. It is shown that affine calibration is recovered uniquely, and metric calibration up to a two fold ambiguity.

The novel aspects of this work are: first, relating the distinguished objects of 3D Euclidean geometry to fixed entities in the image; second, showing that these fixed entities can be computed uniquely via the trifocal tensor between image triplets; third, a robust and automatic implementation of the method.

Results are included of affine and metric calibration and structure recovery using images of real scenes.

1 Introduction

From an image sequence acquired with an uncalibrated camera, structure of 3-space can be recovered up to a projective ambiguity [5, 7]. However, if the camera is constrained to have unchanging internal parameters then this ambiguity can be reduced to metric by calibrating the camera using only image correspondences (no calibration grid). This process is termed "self-calibration" [5, 11]. Previous attempts to make use of the constraint for image *pairs* have generated sets of polynomial equations that are solved by homotopy continuation [8] or iteratively [8, 17] over a sequence. In this paper we demonstrate the advantages of utilizing image *triplets* directly in the case of planar motion, both in the reduced complexity of the equations, and in a practical and robust implementation.

To reduce the ambiguity of reconstruction from projective to affine it is necessary to identify the plane at infinity, π_∞, and to reduce further to a metric ambiguity the absolute conic Ω_∞ on π_∞ must also be identified [4, 13]. Both π_∞ and Ω_∞ are fixed entities under Euclidean motions of 3-space. The key idea in this paper is that these fixed entities can be *accessed* via fixed entities (points, lines, conics) in the image.

To determine the fixed image entities we utilise geometric relations between images that are independent of three dimensional structure. The fundamental geometric relation between two views is the epipolar geometry, represented by the fundamental matrix [3]. This provides a mapping from points in one image to lines in the other, and consequently is not a suitable mapping for determining fixed entities directly. However, between three views the fundamental geometric relation is the trifocal tensor [6, 14, 15], which provides a mapping of points to points, and lines to lines. It is therefore possible to solve directly for fixed image entities as fixed points and lines under transfer by the trifocal tensor.

In the following we obtain these fixed image entities, and thence the camera calibration, from a triplet of images acquired by a camera with unchanging internal parameters undergoing "planar motion". Planar motion is the typical motion undergone by a vehicle moving on a plane — the camera translates in a plane and rotates about an axis perpendicular to that plane. This extends the work of Moons *et al.* who showed that affine structure can be obtained in the case of purely translational motion [12]. We show that

1. Affine structure is computed uniquely.
2. Metric structure can be computed up to a one parameter family, and this ambiguity resolved using additional constraints.

Section 2, describes the fixed image entities and their relation to π_∞ and Ω_∞, and describes how these are related to affine and metric structure recovery. Section 3 gives an algorithm for computing the image fixed points and lines uniquely using the trifocal tensor. Section 4.1 describes results of an implementation of this algorithm, and section 4.2 results for affine and metric structure recovery based on these fixed points from image triplets. All results are for real image sequences.

Notation We will not distinguish between the Euclidean and similarity cases, both will be loosely referred to as metric. Generally vectors will be denoted by $\mathbf{x}$, matrices as $\mathtt{H}$, and tensors as $\mathtt{T}_i^{jk}$. Image coordinates are lower case 3-vectors, e.g. $\mathbf{x}$, world coordinates are upper case 4-vectors, e.g. $\mathbf{X}$. For homogeneous quantities, $=$ indicates equality up to a non-zero scale factor.

2 Fixed Image Entities for Image Triplets

Planar Motion Any rigid transformation of space may be interpreted as a rotation about a *screw axis* and a simultaneous translation in the direction of the axis [2]. There are two special cases – pure translation and pure rotation. In this paper we consider the latter case. A planar motion of a camera consists of a rotation and a translation perpendicular to the rotation axis. This is equivalent to a pure rotation about a screw axis parallel to the rotation axis, but not in general passing through the camera centre. The plane through the camera centre and perpendicular to the rotation axis is the plane of motion of the camera. We consider sequences of planar motions of a camera, by which we mean a sequence of rotations about parallel but generally distinct rotation axes. The plane of motion is common to all the motions. For visualisation, we assume the plane of motion is horizontal and the rotation axes vertical.

3D fixed entities The plane at infinity and absolute conic are invariant under *all* Euclidean actions. These are the entities that we desire to find in order to compute respectively affine or metric structure. These entities can not be observed directly, however, so we attempt to find them indirectly. To this end we consider fixed points of a sequence of planar motions. A single planar motion has additional fixed entities, the screw axis (fixed pointwise), and the plane of motion (fixed setwise). In fact, any plane parallel to the plane of motion is fixed. The intersection of this pencil of planes with the plane at infinity is a line (fixed setwise). Although this line is fixed only as a set, its intersection with the absolute conic, Ω_∞, consisting of two points, is fixed pointwise by the motion. These two points are known as the *circular points*, denoted I and J, and lie on every plane parallel to the plane of motion. Knowledge of these two circular points is equivalent to knowing the metric structure in each of these planes ([13]).

The two circular points are fixed for all motions in a sequence of planar motions with common plane of motion. This is not true of the fixed screw axes, since we assume in general that the screw axis is not the same for all motions. However, since the screw axes are parallel, they all intersect at the plane at infinity at a point which we shall denote by V. The points I, J and V and their relation to Ω_∞ is shown in figure 1. They are fixed by all motions in the sequence.

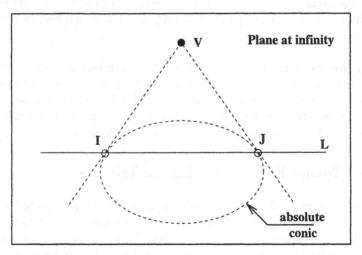

Fig. 1. The fixed entities on π_∞ of a sequence of Euclidean planar motions of 3-space. **V** is the ideal point of the screw axis, and **L** the ideal line of the pencil of planes, orthogonal to the screw axis. **I** and **J** are the circular points for these planes, defined by the intersection of **L** with Ω_∞. **V** and **L** are pole and polar with respect to the absolute conic.

If by some means we are able to find the locations of the points I, J and V in space, then we are able to determine the plane at infinity π_∞ as the unique plane passing through all three of them. This is equivalent to determining the affine structure of space. Although we do not know Ω_∞, and hence can not determine metric structure, we at least know two points on this absolute conic, and hence know the Euclidean geometry in every plane parallel to the plane of motion.

Fixed image entities Our goal is to find the three points I, J and V. Since they are fixed by the sequence of motions, their images will appear at the same location in all images taken by the moving camera (assuming fixed internal calibration). We are led to inquire which points are fixed in all images of a sequence. A fixed point in a pair of images is the image of a point in space that appears at the same location in the two images. It will be seen that apart from the images of I, J and V there are other fixed image entities. We will be led to consider both fixed points and lines.

The locus of all points in space that map to the same point in two images is known as the horopter curve. Generally this is a twisted cubic curve in 3-space passing through the two camera centres [10]. One can find the image of the horopter using the fundamental matrix of the pair of images, since a point on the horopter satisfies the equation $\mathbf{x}^\mathsf{T}\mathbf{F}\mathbf{x} = 0$. Hence, the image of the horopter is a conic defined by the symmetric part of $\mathbf{F}$, namely $\mathbf{F}_s = \mathbf{F} + \mathbf{F}^\mathsf{T}$.

In the case of planar motion, the horopter degenerates to a conic in the plane of motion, plus the screw axis. The conic passes through (and is hence defined by): the two camera centres, the two circular points, and the intersection of the screw axis with the plane of motion. It can be shown that for planar motion $\mathbf{F}_s$ is rank 2 [10], and the conic $\mathbf{x}^\mathsf{T}\mathbf{F}\mathbf{x} = \mathbf{x}^\mathsf{T}\mathbf{F}_s\mathbf{x} = 0$, which is the image of the horopter, degenerates to two lines. These lines are the image of the screw axis and the image of the plane of motion — the horizon line in the image [1]. The epipoles and imaged circular points lie on this horizon line. The apex (the vanishing point of the rotation axis) lies on the imaged screw axis. These points are shown in figure 2a. Although the lines can be computed from $\mathbf{F}_s$, and the imaged circular points and apex lie on these lines, we have not yet explained how to recover these points.

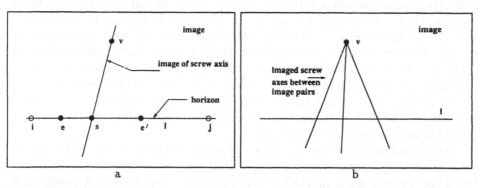

Fig. 2. Fixed image entities for planar motion. (a) For two views the imaged screw axis is a line of fixed points in the image under the motion. The horizon is a fixed line under the motion. (b) The relation between the fixed lines obtained pairwise for three images under planar motion. The image horizon lines for each pair are coincident, and the imaged screw axes for each pair intersect in the apex. All the epipoles lie on the horizon line.

We now consider fixed points in three views connected via planar motions. To do this, we need to consider the intersection of the horopter for cameras 1 and 2 with that for cameras 2 and 3. Each horopter consists of a conic in the plane of motion, plus the vertical axis. The two vertical axes, supposed distinct, meet at infinity at the point V. The two conics meet in 4 points, namely the circular points I and J, the centre of the second camera, plus one further point that is fixed in all three views. The horopter for cameras 1 and 3 will not pass through the second camera centre. Thus we are left with 4 points that are fixed in all three images. These are the circular points I and J, a third point X lying on the plane of motion, and the ideal point V.

Any two fixed points define a fixed line, the line passing through them. Since three of the points, namely the images of points I, J and the third point are collinear, there are just 4 fixed lines. There can be no others, since the intersection of two fixed lines must be a fixed point. We have sketched a geometric proof of the following theorem.

Theorem For three views from a camera with fixed internal parameters undergoing general planar motion, there are four fixed points, three of which are collinear:

1. *The vanishing point of the rotation axes, **v** (the apex).*
2. *Two complex points, the images of the two circular points I, J on the horizon line.*
3. *A third point **x** on the horizon line and peculiar to the image triplet.*

There are four fixed lines passing through pairs of fixed points.

3D Structure Determination A method for determining affine and metric structure is as follows. One determines the fixed points in the three images using the trifocal tensor as described in the following section. The third real collinear fixed point **x** can be distinguished from the complex circular points, the images of I and J. This third point is discarded. The 3-D points I, J and V corresponding to these fixed image points may be reconstructed. These three points define the plane at infinity, and hence affine structure. Planar metric structure is determined by the circular points I and J. Thus, in the absence of other constraints, 3D structure is determined up to a Euclidean transformation in planes parallel to the plane of motion, and up to a one dimensional affine transformation perpendicular to the plane of motion.

Following Luong and Vieville [9] an additional constraint is provided by assuming the skew parameter is zero i.e. that the image axes are perpendicular. This is a very good approximation in practice. This constraint results in a quadratic polynomial giving two solutions for the internal calibration matrix, and hence for the recovery of metric structure. Alternatively, an assumption of equal scale factors in the two coordinate axis directions will allow for unique metric reconstruction.

We have now described the structure ambiguity once the fixed image entities are identified. The next section describes a method of identifying the fixed image entities using the trifocal tensor.

3 Fixed image entities via the trifocal tensor

Suppose the 3×4 camera projection matrices for the three views are P, P$'$ and P$''$. Let a line in space be mapped to lines l, l$'$ and l$''$ in three images. A trilinear relationship exists between the coordinates of the three lines, as follows :

$$l_i = l'_j l''_k T_i^{jk} \tag{1}$$

where T_i^{jk} is the trifocal tensor [6]. Here and elsewhere we observe the convention that indices repeated in the contravariant and covariant positions imply summation over the range $(1, \ldots, 3)$ of the index. A similar relationship holds between coordinates of corresponding points in three images.

The trifocal tensor can be computed directly from point and line matches over three views. It can also be directly constructed from the camera projection matrices P, P$'$ and P$''$ as follows. Assuming that P $= [I \mid 0]$, we have the formula

$$T_i^{jk} = p_i'^j p_4''^k - p_4'^j p_i''^k \tag{2}$$

where $p_j'^i$ and $p_j''^i$ are the (ij)-th entry of the respective camera matrices, index i being the contravariant (row) index and j being the covariant (column) index.

Now in order to find fixed lines, we seek solutions l to the equations (from (1))

$$l_i = l_j l_k T_i^{jk} \tag{3}$$

In (1) as well as (3) the equality sign represents equality up to a non-zero scale factor. We may remove the unknown scale factor in (3) by taking the cross product of the two sides and equating the result to the zero vector. This results in three simultaneous homogeneous cubic equations for the components of l. In the following we discuss methods for obtaining the solutions to these cubics. First we describe the general case, and then show that this can be transformed to a special case where the solution reduces to a single cubic in one variable. The transformation required is a plane projective transformation of the images. Finally, we arrive at a two step algorithm, tailored to real images, for determining the fixed image points and lines.

3.1 General Planar Motion

We consider three views taken by a camera undergoing planar motion. Without loss of generality, we may assume that the camera is moving in the plane $Y = 0$. The rotation axes are perpendicular to this plane, and meet at the point at infinity $(0, 1, 0, 0)^\top$. We assume that the camera has fixed, but unknown calibration. The origin of coordinates may be chosen at the location of the first camera, which means that the camera has matrix P $= H[I \mid 0]$, for some matrix H. The other two camera differ by a planar motion from this first camera, which means that the three camera have the form

$$P = H[I \mid 0] \quad P' = H[R' \mid t'] \quad P'' = H[R'' \mid t''] \tag{4}$$

where R′ and R″ are rotations about the Y axis, and $\mathbf{t}'$ and $\mathbf{t}''$ are translations in the plane $Y = 0$.

One may solve for the fixed lines using the trifocal tensor T_i^{jk}. Denoting a fixed line $\mathbf{l}$ for convenience as $\mathbf{l} = (x, y, z)$ instead of (l_1, l_2, l_3), the fixed line equation $l_i = l_j l_k T_i^{jk}$ may be written as

$$
\begin{pmatrix} x \\ y \\ z \end{pmatrix} \approx \begin{pmatrix} h^{(2)}(x, y, z) \\ h'^{(2)}(x, y, z) \\ h''^{(2)}(x, y, z) \end{pmatrix} \tag{5}
$$

where the superscript (2) denotes the degree of the polynomial. Setting the cross-product of the two sides of this equation to zero, one obtains a set of three cubic equations in x, y and z. By the discussion of section 2, there should be four fixed lines as solutions to this set of equations.

The first thing to note, however, is that the three equations derived from (5) are not linearly independent. There are just two linearly independent cubics. Inevitably, for a trifocal tensor computed from real image correspondences, the solutions obtained depend on just which pair of the three equations one chooses. Furthermore, if there is noise present in the image measurements, then the number of solutions to these equations increases. In general, two simultaneous cubics can have up to 9 solutions. What happens is that one obtains a number of different solutions close to the four ideal solutions. Thus, for instance, there are a number of solutions close to the ideal horizon line. Generally speaking, proceeding in this way will lead into a mire of unpleasant numerical computation.

3.2 Normalized Planar Motion

One can simplify the problem by applying a projective transformation to each image before attempting to find the fixed lines. The transformation to be applied will be the same for each of the images, and hence will map the fixed lines to fixed lines of the transformed images. The transformation that we apply will have the effect of mapping the apex point $\mathbf{v}$ to the point at infinity $(0, 1, 0)^\mathsf{T}$ in the direction of the y-axis. In addition, it will map the horizon line to the x-axis, which has coordinates $(0, 1, 0)^\mathsf{T}$. The transformed images will correspond to camera matrices

$$
\tilde{P} = GH[I \mid 0] \qquad \tilde{P}' = GH[R' \mid \mathbf{t}'] \qquad \tilde{P}'' = GH[R'' \mid \mathbf{t}'']
$$

where G represents the applied image transformation. We considering now the first camera matrix $\tilde{P}$. This matrix maps $(0, 1, 0, 0)^\mathsf{T}$, the vanishing point of the Y axis, to the apex $(0, 1, 0)^\mathsf{T}$ in the image. Furthermore the plane $Y = 0$ with coordinates $(0, 1, 0, 0)$ is mapped to the horizon line $(0, 1, 0)^\mathsf{T}$ as required. This constrains the camera matrix $\tilde{P} = [GH \mid 0]$ to be of the form

$$
\tilde{P} = [GH \mid 0] = \begin{bmatrix} \times & 0 & \times & 0 \\ 0 & \times & 0 & 0 \\ \times & 0 & \times & 0 \end{bmatrix} \tag{6}
$$

where 0 represents a zero entry and × represents a non-zero entry.

Consider now the other camera matrices $\tilde{P}'$ and $\tilde{P}''$. Since R' and R'' are rotations about the Y axis, and t' and t'' are translations in the plane $Y = 0$, both $[R' \mid t']$ and $[R'' \mid t'']$ are of the form [16]

$$
\begin{bmatrix}
\times & 0 & \times & \times \\
0 & \times & 0 & 0 \\
\times & 0 & \times & \times
\end{bmatrix}
\tag{7}
$$

Premultiplying by GH, we find that both $\tilde{P}'$ and $\tilde{P}''$ are of the same form (7). This particularly simple form of the camera matrices allows us to find a simple form for the trifocal tensor as well. In order to apply formula (2), we require matrix P to be of the form $P = [I \mid 0]$. This can be achieved by right multiplication of all the camera matrices by the 3D transformation matrix $\begin{bmatrix} (GH)^{-1} & 0 \\ 0 & 1 \end{bmatrix}$. It may be observed that this multiplication does not change the format (7) of the matrices $\tilde{P}'$ and $\tilde{P}''$. Now, for $i = 1$ or 3, we see that $\tilde{p}'_i$ and $\tilde{p}''_i$ are of the form $(\times, 0, \times)^{\mathsf{T}}$, whereas for $i = 2$, they are of the form $(0, \times, 0)^{\mathsf{T}}$. Further, $\tilde{p}'_4$ is of the form $(\times, 0, \times)^{\mathsf{T}}$. One easily computes the following form for $\tilde{T}_i^{\cdot\cdot}$.

$$
\tilde{T}_i^{\cdot\cdot} = \begin{bmatrix}
\times & 0 & \times \\
0 & 0 & 0 \\
\times & 0 & \times
\end{bmatrix}
\quad \text{for } i = 1, 3 \qquad
\tilde{T}_2^{\cdot\cdot} = \begin{bmatrix}
0 & \times & 0 \\
\times & 0 & \times \\
0 & \times & 0
\end{bmatrix}
\tag{8}
$$

Using this special form of the trifocal tensor, we see that (3) may be written as

$$
\begin{pmatrix} x \\ y \\ z \end{pmatrix} = \begin{pmatrix} a_1 x^2 + b_1 xz + c_1 z^2 \\ d_2 xy + e_2 yz \\ a_3 x^2 + b_3 xz + c_3 z^2 \end{pmatrix}
\tag{9}
$$

where $\mathbf{1} = (x, y, z)^{\mathsf{T}}$ represents a fixed line. This set of equations has eight parameters $\{a_1 \ldots c_3\}$. The fixed lines may be found by solving this system of equations. One fixed point in the three views is the apex, $\mathbf{v} = (0, 1, 0)^{\mathsf{T}}$. Let us consider only lines passing through the apex fixed in all three views. Such a line has coordinates $(x, 0, z)$. Thus, we may assume that $y = 0$. The equations (9) now reduce to the form

$$
\begin{pmatrix} x \\ z \end{pmatrix} \approx \begin{pmatrix} a_1 x^2 + b_1 xz + c_1 z^2 \\ a_3 x^2 + b_3 xz + c_3 z^2 \end{pmatrix}
\tag{10}
$$

Cross-multiplying reduces this to a single equation

$$
z(a_1 x^2 + b_1 xz + c_1 z^2) = x(a_3 x^2 + b_3 xz + c_3 z^2)
\tag{11}
$$

This is a homogeneous cubic, and may be easily solved for the ratio $x : z$. The solutions to this cubic are the three lines passing through the apex joining it to three points lying on the horizon line. These three points are the images of the two circular points, and the third fixed point. The third fixed point may be distinguished by the fact that it is a real solution, whereas the two circular points are a pair of complex conjugate solutions. The third fixed point is of no special interest, and is discarded.

This analysis is an example of a generally useful technique of applying geometric transformations to simplify algebraic computation.

3.3 Algorithm Outline

We now put the parts of the algorithm together. The following algorithm determines the fixed lines in three views, and hence the apex and circular points on the horizon line. The first four steps reduce to the case of normalized planar motion. The fixed points and lines are then computed in steps 5 to 7, and the last step relates the fixed points back to the fixed points in the original images.

1. Compute the fundamental matrix for all pairs of image, and obtain the epipoles.
2. Find the orthogonal regression line fit to the epipoles. This is the horizon line $\mathbf{l}$.
3. Decompose the symmetric part of the $\mathbf{F}$'s into two lines, this generates the image of the screw axis for each pair. Find the intersection of the imaged screw axes, or in the presence of noise, the point with minimum squared distance to all the imaged screw axes. This determines the apex $\mathbf{v}$.
4. Find a projective transformation $\mathbf{G}$ taking the horizon line to the line $(0, 1, 0)$ and the apex to the point $(0, 1, 0)$. Apply this projective transform to all images.
5. For three views, compute the trifocal tensor from point and line matches, enforcing the constraint that it be of the form described in (8).
6. Compute the cubic polynomial defined in (9) and (11), and solve for the ratio $x : z$. There will be two imaginary and one real solution. Discard the real solution. The imaginary solutions will be lines with coordinates $(1, 0, z)$ and $(1, 0, \bar{z})$ passing through the apex and the two circular points on the horizon line.
7. Compute the intersection of the horizon line $(0, 1, 0)$ and the line $(1, 0, z)$. This is the point $(-z, 0, 1)$. Do the same for the other solution $(1, 0, \bar{z})$.
8. Apply the inverse transform $\mathbf{G}^{-1}$ to the two circular points to find the image of the two circular points in the original images.

4 Results

Numerical results are improved significantly by enforcing that both $\mathbf{F}$ and $\mathbf{F}_s$ are rank 2 during the minimization to compute the fundamental matrix. Implementation details of the algorithms are given in [18].

4.1 Fixed image points and lines

In this section we describe the results of obtaining the fixed points/lines over image triplets. These points are used for affine and metric calibration, which is described in section 4.2.

The image sequences used are shown in figure 3 (sequence I) and figure 4 (sequence II). The first sequence is acquired by a camera mounted on an Adept robot, the second by a different camera mounted on an AGV. The latter sequence has considerably more camera shake, and consequently is not perfect planar motion.

Figure 5 shows the two view fixed lines obtained from the seven sequential image pairs from sequence I. The trifocal tensor is computed for the six sequential image triplets, and the circular points computed from the fixed points of the tensor. The results are given in table 1a. The circular points are certainly stable, but it is difficult to quantify their accuracy directly because they are complex. In the next section the circular points are used to upgrade projective structure to affine and metric. The accuracy of the circular points is hence measured indirectly by the accuracy of the recovered structure. For comparison, the estimated circular points, based on approximate internal parameters, are $(400 \pm 1100i, -255, 1)^\top$.

The camera undergoes a smaller rotation in sequence II and the images are noisier due to camera shake. Superior results are obtained by using fundamental matrices from image pairs which are separated by 2 time steps (i.e. pairs $\{1,4\},\{1,5\},\{2,4\},...$), rather than sequential image pairs. Figure 6 shows these results. The tensor is calculated using image triplets separated by one time step (i.e. triplets $\{1,3,5\},\{2,4,6\},...$). The circular points computed are shown in table 1b. The estimated circular points, based on approximate internal parameters, are $(257 \pm 800i, 196 \pm 45i, 1)^\top$.

4.2 Structure Recovery

Section 4.1 obtained the 3 fixed points of image triplets using the algorithm of section 3.3. These points define the position of the plane at infinity π_∞, which allows affine structure to be recovered. In this section we describe the results of an implementation of affine and metric structure recovery, and assess the accuracy by comparing with ground truth.

Affine Structure Using the image sequence in figure 3, and the circular points listed in table 1a, affine structure is recovered. We can quantify the accuracy of

Fig. 3. Image sequence I: four images from an eight image sequence acquired by a camera mounted on an Adept robot arm. Planar motion with the rotation axis at approximately $25°$ to the image y axis and perpendicular to the image x axis. 149 corners are automatically matched and tracked through the 8 images. The Tsai grids are used only to provide ground truth, not to calibrate.

Fig. 4. Image sequence II: four images from a nine image sequence. Planar motion with the rotation axis approximately aligned with the image y axis. 75 corners are automatically matched across the 9 images. The sequence was acquired by a camera mounted on an AGV.

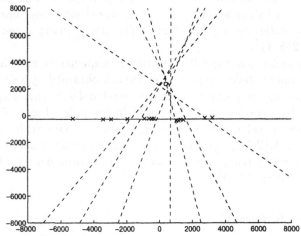

Fig. 5. The fixed points and lines obtained from all 7 possible sequential image pairs from sequence I (figure 3), with axes in pixels. × are the epipoles, dashed and solid lines are the screw axes and the horizon respectively, o is the apex at (394,2370). The horizon line is at $y = -255$.

the affine structure by comparing the values of affine invariants measured on the recovered structure to their veridical values. The affine invariant used is the ratio of line segment lengths on parallel lines. The lines in the scene are defined by the corners on the partially obscured calibration grid shown in figure 3. The veridical value of these ratios is 1.0, and the results obtained are shown in table 2. Clearly, the projective skewing has been largely removed.

Metric Structure Metric structure in planes parallel to the motion plane (the ground plane here) is recovered for the sequence in figure 3. The accuracy of the metric structure is measured by computing an angle in the recovered structure with a known veridical value. We compute two angles for each image triplet. First is the angle between the planes of the calibration grid. We fit planes to 23 and 18 points on the left and right faces respectively and compare the interplane angle to a veridical value of 90°. Second is the angle between the three computed camera centres for each image triplet which is known from the robot motion. The

Image Triplet	Circular points
123	$(415 \pm 1217i, -255, 1)^{\top}$
234	$(393 \pm 1229i, -255, 1)^{\top}$
345	$(407 \pm 1225i, -255, 1)^{\top}$
456	$(445 \pm 1222i, -255, 1)^{\top}$
567	$(372 \pm 1274i, -255, 1)^{\top}$
678	$(371 \pm 1232i, -255, 1)^{\top}$

a

Image Triplet	Circular points
135	$(301 \pm 698i, 199 \pm 39i, 1)^{\top}$
246	$(244 \pm 866i, 196 \pm 48i, 1)^{\top}$
357	$(265 \pm 714i, 197 \pm 40i, 1)^{\top}$
468	$(583 \pm 747i, 214 \pm 42, 1)^{\top}$
579	$(302 \pm 691i, 199 \pm 39, 1)^{\top}$

b

Table 1. The circular points obtained (complex conjugates) for (a) sequence I (figure 3) and (b) sequence II (figure 4). Note, the stability of the points estimated from different triplets.

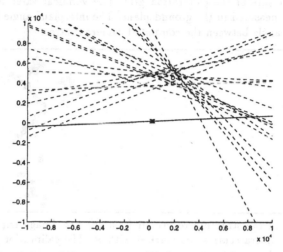

Fig. 6. The fixed points and lines sequence II, figure 4, with axes in pixels. × are the epipoles, dashed and solid lines are the screw axes and the horizon line respectively, o is the image apex at (1430,4675). The horizon line passes through (0,182).

camera centres are computed from the camera projection matrices.

Table 2 shows the computed angles for the 6 image triplets from sequence I, while figure 7 shows a plan view of the recovered metric structure from the first image triplet.

5 Conclusions and extensions

We have demonstrated the geometric importance of fixed points and lines in an image sequence as calibration tools. These fixed entities have been measured, and used to recover affine and partial metric structure from image sequences of real scenes. There are a number of outstanding questions, both numerical and theoretical:

1. We have demonstrated that estimates of the plane at infinity and camera internal parameters can be computed from image triplets. It now remains to derive the variance of these quantities. Then a recursive estimator can

Image Triplet	Affine Invariants				Metric Invariants		
	max	min	average	standard deviation	Plane angle $(90° \pm 1°)$	Motion Angle Actual	Motion Angle Computed
123	1.156	0.938	1.034	0.070	86.6	63.8	52.4
234	1.113	0.896	0.994	0.072	90.7	139.1	137.8
345	1.080	0.872	0.980	0.070	92.3	75.0	81.4
456	1.112	0.948	1.015	0.049	85.7	101.7	92.3
567	1.072	0.938	1.010	0.040	89.1	33.0	28.5
678	1.100	0.976	1.022	0.037	88.7	76.1	66.0

Table 2. *Affine Invariants* The ratio of lengths of parallel lines measured on the recovered affine structure of the calibration grid. The veridical value is unity. *Metric Invariants* Angles measured in the ground plane. The interplane angle for the calibration grid, and the angle between the computed camera centres.

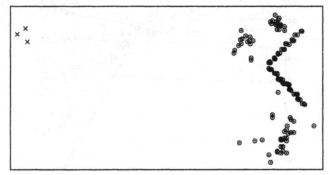

Fig. 7. Plan view of the structure recovered from the first image triplet of sequence I. The first three camera centres are marked with ×. The calibration grid and other objects are clearly shown.

be built, such as an Extended Kalman Filter, which updates the plane at infinity and camera calibration throughout an image sequence.

2. The image of the absolute conic is a fixed entity over all images with unchanging internal parameters. The study of fixed image entities opens up the possibility of solving for this directly as the fixed conic of a sequence.

Acknowledgements

Financial support for this work was provided by EU ACTS Project VANGUARD, the EPSRC and GE Research and Development.

References

1. Beardsley, P. and Zisserman, A. Affine calibration of mobile vehicles. In Mohr, R. and Chengke, W., editors, *Europe-China workshop on Geometrical Modelling and Invariants for Computer Vision.* Xi'an, China, 1995.
2. Bottema, O. and Roth, B. *Theoretical Kinematics.* Dover, New York, 1979.

3. Faugeras, O. What can be seen in three dimensions with an uncalibrated stereo rig? In *Proc. ECCV*, LNCS 588, pages 563–578. Springer-Verlag, 1992.

4. Faugeras, O. Stratification of three-dimensional vision: projective, affine, and metric representation. *J. Opt. Soc. Am.*, A12:465–484, 1995.

5. Faugeras, O., Luong, Q., and Maybank, S. Camera self-calibration: theory and experiments. In *Proc. ECCV*, LNCS 588, pages 321–334. Springer-Verlag, 1992.

6. Hartley, R. A linear method for reconstruction from lines and points. In *Proc. ICCV*, pages 882–887, 1995.

7. Hartley, R., Gupta, R., and Chang, T. Stereo from uncalibrated cameras. In *Proc. CVPR*, 1992.

8. Luong, Q. *Matrice Fondamentale et Autocalibration en Vision par Ordinateur.* PhD thesis, Université de Paris-Sud, France, 1992.

9. Luong, Q. and Vieville, T. Canonic representations for the geometries of multiple projective views. In *Proc. ECCV*, pages 589–597. Springer-Verlag, 1994.

10. Maybank, S. *Theory of reconstruction from image motion.* Springer-Verlag, Berlin, 1993.

11. Maybank, S. and Faugeras, O. A theory of self-calibration of a moving camera. *International Journal of Computer Vision*, 8:123–151, 1992.

12. Moons, T., Van Gool, L., Van Diest, M., and Pauwels, E. Affine reconstruction from perspective image pairs. In *Applications of Invariance in Computer Vision*, LNCS 825. Springer-Verlag, 1994.

13. Semple, J. and Kneebone, G. *Algebraic Projective Geometry.* Oxford University Press, 1979.

14. Shashua, A. Trilinearity in visual recognition by alignment. In *Proc. ECCV*, 1994.

15. Spetsakis, M.E. and Aloimonos, J. Structure from motion using line correspondences. *International Journal of Computer Vision*, pages 171–183, 1990.

16. Wiles, C and Brady, J.M. Ground plane motion camera models. In *Proc. ECCV*, 1996.

17. Zeller, C. and Faugeras, O. Camera self-calibration from video sequences: the kruppa equations revisited. Technical report, INRIA, 1995.

18. Zisserman, A., Armstrong, M., and Hartley, R. Self-calibration from image triplets. Technical report, Dept. of Engineering Science, Oxford University, 1996.

Parallax Geometry of Pairs of Points for 3D Scene Analysis

Michal Irani and P. Anandan

David Sarnoff Research Center, CN5300, Princeton, NJ 08543-5300, USA

Abstract. We present a geometric relationship between the image motion of *pairs* of points over multiple frames. This relationship is based on the *parallax* displacements of points with respect to an arbitrary planar surface, and does not involve epipolar geometry. A constraint is derived over two frames for any pair of points, relating their projective structure (with respect to the plane) based only on their image coordinates and their parallax displacements. Similarly, a 3D-rigidity constraint between pairs of points over multiple frames is derived. We show applications of these parallax-based constraints to solving three important problems in 3D scene analysis: (i) the recovery of 3D scene structure, (ii) the detection of moving objects in the presence of camera induced motion, and (iii) the synthesis of new camera views based on a given set of views. Moreover, we show that this approach can handle difficult situations for 3D scene analysis, e.g., where there is only a small set of parallax vectors, and in the presence of independently moving objects.

1 Introduction

The analysis of three dimensional scenes from image sequences has a number of goals. These include (but are not limited to): (i) the recovery of 3D scene structure, (ii) the detection of moving objects in the presence of camera induced motion, and (iii) the synthesis of new camera views based on a given set of views. The traditional approach to these types of problems has been to first recover the epipolar geometry between pairs of frames and then apply that information to achieve the abovementioned goals. However, this approach is plagued with the difficulties associated with recovering the epipolar geometry [24].

Recent approaches to 3D scene analysis have overcome some of the difficulties in recovering the epipolar geometry by decomposing the motion into a combination of a planar homography and residual parallax [11, 15, 17]. However, they still require the explicit estimation of the epipole itself, which can be difficult in many cases.

More recently, progress has been made towards deriving constraints directly based on collections of points in multiple views. Examples are the trilinearity constraints [18, 16] which eliminate the scene structure in favor of the camera geometries; the dual-shape tensor [23] which eliminates the camera motion in favor of scene structure; the more general framework of multipoint multiview geometry [6, 3]; and the work on multiple view invariants without requiring the recovery of the epipolar geometry [24]. In its current form, this class of methods does not address the problem of shape recovery in *dynamic* scenes, in particular when the amount of image motion due to independent moving object is not negligible.

In this paper we develop geometric relationships between the residual (planar) parallax displacements of *pairs* of points. These geometric relationships address the problem of 3D scene analysis even in *difficult* conditions, i.e., when the epipole estimation is ill-conditioned, when there is a small number of parallax vectors, and in the presence of moving objects. We show how these relationships can be applied to each of the three problems outlined at the beginning of this section. Moreover, the use of the parallax constraints derived here provides a continuum between "2D algorithms" and the "3D algorithms" for each of the problems mentioned above.

In Section 2 a *parallax-based structure constraint* is derived, which relates the *projective structure* of two points to their image positions and their *parallax displacements* alone. By eliminating the relative projective structure of a pair of points between *three* frames, we arrive at a constraint on the parallax displacements of two points *moving as a rigid object* over those frames. We refer to this as the *parallax-based rigidity constraint*.

In Section 3 an alternative way of deriving the *parallax-based rigidity constraint* is presented, this time geometrically rather than algebraically. This leads to a simple and intuitive geometric interpretation of the multiframe rigidity constraint and to the derivation of a *dual point* to the epipole.

In Section 4 the pairwise parallax-based constraints are applied to solving three important problems in 3D scene analysis, even in the abovementioned difficult scenarios: (i) the recovery of 3D scene structure, (ii) the detection of moving objects in the presence of camera induced motion, and (iii) the synthesis of new camera views based on a given set of views.

2 Parallax-Based Constraints on Pairs of Points

In this section we derive a constraint on the parallax motion of pairs of points between two frames. We show how this constraint can be used to recover relative 3D structure of two points from their parallax vectors alone, without any additional information, and in particular, without requiring the recovery of the camera epipoles. The parallax constraint is then extended to *multiple* frames and to multiple points to obtain *rigidity* constraints on image points based on their parallax displacements alone, without involving any scene or camera geometry.

2.1 The Planar Parallax Decomposition

To derive the parallax constraint, we first briefly describe the decomposition of the image motion into a homography (i.e., the image motion of an arbitrary planar surface) and residual parallax displacements. This decomposition has been previously derived and used in [11, 15, 17]. For a more detailed derivation see also [7].

Fig. 1 provides a geometric interpretation of the planar parallax. Let $\mathbf{P} = (X, Y, Z)^T$ and $\mathbf{P}' = (X', Y', Z')^T$ denote the Cartesian coordinates of a scene point with respect to two different camera views, respectively. Let $\mathbf{p} = (x, y)^T$

and $\mathbf{p}' = (x', y')^T$ respectively denote the corresponding coordinates of the corresponding image points in the two image frames. Let $\mathbf{T} = (T_x, T_y, T_z)$ denote

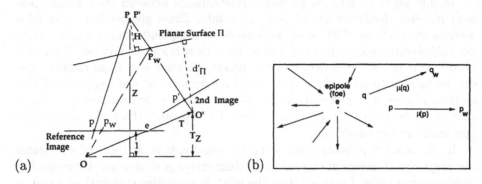

Fig. 1. *The plane+parallax decomposition. (a) The geometric interpretation. (b) The epipolar field of the residual parallax displacements.*

the camera translation between the two views. Let Π denote an arbitrary (*real* or *virtual*) planar surface in the scene, and let A' denote the homography that aligns the planar surface Π between the second and first frame (i.e., for all points $\mathbf{P} \in \Pi, \mathbf{P} = A'\mathbf{P}'$). It can be shown (see [11, 9, 15, 17]) that the 2D image displacement of the point $\mathbf{P}$ can be written as $\mathbf{u} = (\mathbf{p}' - \mathbf{p}) = \mathbf{u}_\pi + \mu$, where $\mathbf{u}_\pi$ denotes the *planar* part of the $2D$ image motion (the homography due to Π), and μ denotes the residual *planar parallax* 2D motion. The homography due to Π results in an image motion field that can be modeled as a 2D *projective transformation*. When $T_z \neq 0$:

$$\mathbf{u}_\pi = (\mathbf{p}' - \mathbf{p_w}) \; ; \; \mu = \gamma \frac{T_z}{d'_\pi}(\mathbf{e} - \mathbf{p_w}) \tag{1}$$

where $\mathbf{p_w}$ denotes the image point in the first frame which results from warping the corresponding point $\mathbf{p}'$ in the second image by the 2D parametric transformation of the plane Π. The $2D$ image coordinates of the epipole (or the *focus-of-expansion*, FOE) in the first frame are denoted by $\mathbf{e}$, and d'_π is the perpendicular distance from the second camera center to the reference plane (see Fig. 1). γ is a measure of the 3D shape of the point $\mathbf{P}$. In particular, $\gamma = \frac{H}{Z}$, where H is the perpendicular distance from the $\mathbf{P}$ to the reference plane, and Z is the "range" (or "depth") of the point $\mathbf{P}$ with respect to the first camera. We refer to γ as the projective 3D structure of the point $\mathbf{P}$. In the case when $T_z = 0$, the parallax motion μ has a slightly different form: $\mu = \frac{\gamma}{d'_\pi}t$, where $t = (T_x, T_y)^T$.

2.2 The Parallax Based Structure Constraint

Theorem 1: Given the planar-parallax displacement vectors μ_1 and μ_2 of two points that belong to the static background scene, their *relative 3D projective structure* $\frac{\gamma_2}{\gamma_1}$ is given by:

$$\frac{\gamma_2}{\gamma_1} = \frac{\mu_2{}^T(\Delta\mathbf{p_w})_\perp}{\mu_1{}^T(\Delta\mathbf{p_w})_\perp}, \tag{2}$$

where, as shown in Fig. 2.a, $\mathbf{p_1}$ and $\mathbf{p_2}$ are the image locations (in the reference frame) of two points that are part of the static scene, $\Delta\mathbf{p_w} = \mathbf{p_{w_2}} - \mathbf{p_{w_1}}$, the vector connecting the "warped" locations of the corresponding second frame points (as in Eq. (1)), and $\mathbf{v}_\perp$ signifies a vector perpendicular to $\mathbf{v}$.

Proof: From Eq. (1), we know that $\mu_1 = \gamma_1 \frac{T_z}{d'_\pi}(\mathbf{e} - \mathbf{p_{w_1}})$ and $\mu_2 = \gamma_2 \frac{T_z}{d'_\pi}(\mathbf{e} - \mathbf{p_{w_2}})$. Therefore,

$$\mu_1\gamma_2 - \mu_2\gamma_1 = \gamma_1\gamma_2 \frac{T_Z}{d'}(\mathbf{p_{w_2}} - \mathbf{p_{w_1}}) \tag{3}$$

This last step eliminated the epipole e. Eq. (3) entails that the vectors on both sides of the equation are parallel. Since $\gamma_1\gamma_2\frac{T_z}{d'}$ is a scalar, we get:

$$(\mu_1\gamma_2 - \mu_2\gamma_1) \parallel \Delta\mathbf{p_w} \Rightarrow (\mu_1\gamma_2 - \mu_2\gamma_1)^T(\Delta\mathbf{p_w})_\perp = 0 \Rightarrow \frac{\gamma_2}{\gamma_1} = \frac{\mu_2{}^T(\Delta\mathbf{p_w})_\perp}{\mu_1{}^T(\Delta\mathbf{p_w})_\perp}, \tag{4}$$

which is the *pairwise parallax constraint*. When $T_Z = 0$, a constraint stronger than Eq. (4) can be derived: $(\mu_1\frac{\gamma_2}{\gamma_1} - \mu_2) = 0$. However, Eq. (4), still holds. This is important, as we do not have a-priori knowledge of T_Z to distinguish between the two cases. ∎

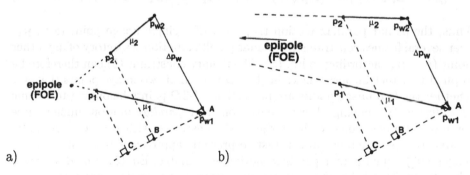

Fig. 2. The relative structure constraint. *(a) This figure geometrically illustrates the relative structure constraint (Eq. 2):* $\frac{\gamma_2}{\gamma_1} = \frac{\mu_2{}^T(\Delta\mathbf{p_w})_\perp}{\mu_1{}^T(\Delta\mathbf{p_w})_\perp} = \frac{AB}{AC}$. *(b) When the parallax vectors are nearly parallel, the epipole estimation is unreliable. However, the relative structure* $\frac{AB}{AC}$ *can be reliably computed even in this case.*

Fig. 2.a displays the constraint geometrically. The fact that relative structure of one point with respect to another can be obtained using only the two parallax vectors is not surprising: *In principle*, one could use the two parallax vectors to recover the epipole (the intersection point of the two vectors), and then use the magnitudes and distances of the points from the computed epipole to estimate their relative projective structure. The benefit of the constraint (2) is that it provides this information *directly* from the positions and parallax vectors of the two points, without the need to go through the computation of the epipole, using as much information as one point can give on another. Fig. 2.b graphically shows an example of a configuration in which estimating the epipole is very unreliable, whereas estimating the relative structure *directly* from Eq. (2) *is* reliable.

2.3 The Parallax-Based Rigidity Constraint

In this section we extend the parallax-based structure constraint to *multiple frames* and to *multiple points*, to obtain *rigidity* constraints that are based only on parallax displacements of the image points, and involve neither *structure* parameters nor *camera geometry*.

Rigidity Over Multiple Frames: Let p_1 and p_2 be two *image* points in the first (reference) frame. Let μ_1^j, μ_2^j be the parallax displacement displacements of the two points between the reference frame and the jth frame, and μ_1^k, μ_2^k be the parallax displacements between the reference frame and the kth frame. Let $(\Delta p_w)^j, (\Delta p_w)^k$ be the corresponding vectors connecting the warped points as in Eq. (2) and Fig. 2. Using the relative structure constraint (2), for any two frames j and k we get: $\frac{\gamma_2}{\gamma_1} = \frac{\mu_2^{j^T}(\Delta p_w)_\perp^j}{\mu_1^{j^T}(\Delta p_w)_\perp^j} = \frac{\mu_2^{k^T}(\Delta p_w)_\perp^k}{\mu_1^{k^T}(\Delta p_w)_\perp^k}$.

Multiplying by the denominators yields the *rigidity constraint* of the two points over three frames (reference frame, frame j, and frame k):

$$(\mu_1^{k^T}(\Delta p_w)_\perp^k)(\mu_2^{j^T}(\Delta p_w)_\perp^j) - (\mu_1^{j^T}(\Delta p_w)_\perp^j)(\mu_2^{k^T}(\Delta p_w)_\perp^k) = 0. \qquad (5)$$

Thus, the planar parallax motion trajectory of a single image point (e.g., p_1) over several frames constrains the planar parallax motion trajectory of any other point (e.g., p_2) according to Eq. (5). The rigidity constraint (5) can therefore be applied to detect inconsistencies in the $3D$ motion of two image points (i.e., say whether the two image points are projections of $3D$ points belonging to a same or different $3D$ moving objects) based on their *parallax* motion among three (or more) frames alone, without the need to estimate either *camera geometry* or *structure* parameters. In contrast to previous approaches (e.g., the trilinear tensor [16]), when planar parallax motion is available, Eq. (5) provides certain advantages: (i) it is based on the parallax motion of a *single* image point, (ii) it does not require any numerical estimation (e.g., unlike [16], it does not require estimation of tensor parameters), (iii) it does not involve, *explicitly or implicitly*, any shape or camera geometry information other than that already implicit in the planar parallax motion itself.

Rigidity Over Multiple Points: Instead of considering pairs of points over multiple frames, we can consider multiple points over pairs of frames to obtain a different form of the rigidity constraint.

Let p_1, p_2, and p_3 be three image points in the first (reference) frame. Let μ_i denote the $2D$ planar parallax motion of p_i from the first frame to another frame ($i = 1, 2, 3$).

Using the shape invariance constraint (2):

$$\frac{\gamma_2}{\gamma_1} = \frac{\mu_2{}^T(\Delta p_{w_{2,1}})_\perp}{\mu_1{}^T(\Delta p_{w_{2,1}})_\perp} \; ; \; \frac{\gamma_3}{\gamma_2} = \frac{\mu_3{}^T(\Delta p_{w_{3,2}})_\perp}{\mu_2{}^T(\Delta p_{w_{3,2}})_\perp} \; ; \; \frac{\gamma_3}{\gamma_1} = \frac{\mu_3{}^T(\Delta p_{w_{3,1}})_\perp}{\mu_1{}^T(\Delta p_{w_{3,1}})_\perp}.$$

Equating $\frac{\gamma_3}{\gamma_1} = \frac{\gamma_3}{\gamma_2}\frac{\gamma_2}{\gamma_1}$, and multiplying by the denominators, we get the rigidity constraint for three points over a pair of frames:

$$(\mu_3{}^T \Delta p_{w_{3,2}}{}_\perp)(\mu_2{}^T \Delta p_{w_{2,1}}{}_\perp)(\mu_1{}^T \Delta p_{w_{3,1}}{}_\perp) = (\mu_2{}^T \Delta p_{w_{3,2}}{}_\perp)(\mu_1{}^T \Delta p_{w_{2,1}}{}_\perp)(\mu_3{}^T \Delta p_{w_{3,1}}{}_\perp). \tag{6}$$

The fact that three points in two frames form a rigidity constraint is not surprising: *In principle*, one could use two of the three parallax vectors to obtain the epipole (the intersection point of the two vectors). 3D rigidity will constrain the parallax vector of the third point to lie on the epipolar line emerging from the computed epipole through the third point. The benefit of the rigidity constraint (6) is in the fact that it provides this information directly from the positions and parallax vectors of the three points, without the need to go through the unstable computation of the epipole, using as much information as two point can give on the third.

2.4 The Generalized Parallax Constraint

The pairwise-parallax constraint (Eq. (2)) can be extended to handle full image motion (as opposed to *parallax* motion), even when the homography is unknown. Eq. (1) can be rewritten (in homogeneous coordinates) as [11, 15, 17]:

$$p = p_w + \gamma\frac{1}{d'_\pi}(T_Z p_w - T) = \frac{A'p'}{a'_3{}^T p'} + \gamma\frac{1}{d'_\pi}(T_Z\frac{A'p'}{a'_3{}^T p'} - T), \tag{7}$$

where A' is an *unknown* homography from frame2 to the first frame, and a'_3 is the third row of the 3×3 matrix A'. It could relate to *any* planar surface in the scene, in particular a *virtual* plane.

Given two points p_1 and p_2, we can eliminate T in a manner similar to that done in Theorem 1. This yields the *generalized parallax constraint* in terms the relative projective structure $\frac{\gamma_2}{\gamma_1}$:

$$(\frac{\gamma_2}{\gamma_1}(p_1 - \frac{A'p'_1}{a'_3{}^T p'_1}) - (p_2 - \frac{A'p'_2}{a'_3{}^T p'_2}))^T (\frac{A'p'_1}{a'_3{}^T p'_1} - \frac{A'p'_2}{a'_3{}^T p'_2})_\perp = 0. \tag{8}$$

The generalized parallax constraint suggests a new implicit representation of *general* 2D image motion: Rather than looking at the representation of 2D image motion in terms of: *homography + epipole + projective structure* [11, 15, 17] it suggests an implicit representation of 2D image motion in terms of: *homography + relative projective structure* of pairs of points. Since this representation does not contain the epipole, it can be easily extended to multiple frames. In a similar manner, the *rigidity constraints* (5) and (6) can also be generalized to handle full image motion with unknown homography. For more details see [7].

3   Parallax Geometry and an Epipole Dual

In this section, we present a geometric view of the parallax-based rigidity constraint. This leads to derivation of a *dual point* to the epipole.

The 3D geometric structure associated with the planar parallax of pairs of points between two frames is illustrated in Fig. 3.a. In this figure, Π is the planar surface, and $\mathbf{P}$ and $\mathbf{Q}$ are the two scene points. As in the case of Fig. 1, $\mathbf{P_w}$ and $\mathbf{Q_w}$ are the intersections of rays $\mathbf{O'P}$ and $\mathbf{O'Q}$ with the plane Π. The points $\mathbf{p_w}$ and $\mathbf{q_w}$ on the reference image are the projections of $\mathbf{P_w}$ and $\mathbf{Q_w}$, and are therefore the points to which the planar homography transforms $\mathbf{p'}$ and $\mathbf{q'}$ respectively. Below, we refer to $\mathbf{p_w}$ and $\mathbf{q_w}$ as "warped points".

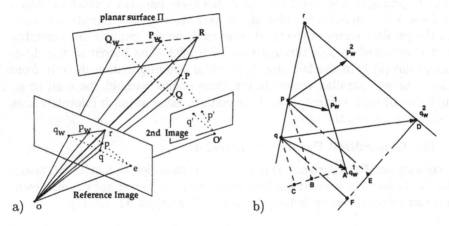

Fig. 3.   The dual of the epipole. *(a) The line connecting points* $\mathbf{p}, \mathbf{q}$ *in the reference image intersects the line connecting the warped points* $\mathbf{p_w}, \mathbf{q_w}$ *at* $\mathbf{r}$. *The point* $\mathbf{r}$ *is the image of the point* $\mathbf{R}$ *which is the intersection of the line* $\mathbf{PQ}$ *on the planar surface* Π. *(b) The line connecting points* $\mathbf{p}, \mathbf{q}$ *in the reference image and the lines connecting the corresponding warped points* $\mathbf{fp_w^1}, \mathbf{q_w^1}$ *and* $\mathbf{p_w^2}, \mathbf{q_w^2}$ *from other frames all intersect at* $\mathbf{r}$, *the dual of the epipole.*

Let $\mathbf{R}$ be the intersection of the line connecting $\mathbf{P}$ and $\mathbf{Q}$ with the plane Π, and $\mathbf{r}$ be its projection on the reference image plane. Since $\mathbf{P}, \mathbf{Q}, \mathbf{R}$ are colinear and $\mathbf{P_w}, \mathbf{Q_w}, \mathbf{R}$ are colinear, therefore $\mathbf{p_w}, \mathbf{q_w}, \mathbf{r}$ and colinear and $\mathbf{p}, \mathbf{q}, \mathbf{r}$ are colinear. *In other words, the line connecting* $\mathbf{p_w}$ *and* $\mathbf{q_w}$ *and the line connecting* $\mathbf{p}$ *and* $\mathbf{q}$ *intersect at* $\mathbf{r}$, *the image of the point* $\mathbf{R}$. (See also [24] for the same observation.)

Note that the point $\mathbf{r}$ does not depend on the second camera view. Therefore, if multiple views are considered, then the lines connecting the warped points $\mathbf{p_w^j}$ and $\mathbf{q_w^j}$ (for any frame j), meet at $\mathbf{r}$ for all such views.

The convergence of the lines is illustrated in Fig. 3.b. Referring to that figure, since the lines $\mathbf{qC}, \mathbf{pB}$ and $\mathbf{rA}$ are parallel to each other and intersect the lines $\mathbf{qpr}$ and $\mathbf{CAB}$: $\frac{\mathbf{qr}}{\mathbf{pr}} = \frac{\mathbf{CA}}{\mathbf{BA}} = \frac{\gamma_q}{\gamma_p}$. Similarly, $\frac{\mathbf{qr}}{\mathbf{pr}} = \frac{\mathbf{FD}}{\mathbf{ED}} = \frac{\gamma_q}{\gamma_p}$. Hence $\frac{\mathbf{qr}}{\mathbf{pr}} = \frac{\mathbf{CA}}{\mathbf{BA}} = \frac{\mathbf{FD}}{\mathbf{ED}}$. This is the same as the rigidity constraint of a pair of points over multiple frames derived in Section 2. Note, however, the rigidity constraint itself does not require the estimation of the point of convergence **r**, just as it does not require the estimation of the epipole.

The point **r** is the dual of the epipole: the epipole is the point of intersection of multiple parallax vectors between a pair of frames, i.e., the point of intersection of all lines connecting each image point with its warped point between a pair of frames. Whereas the dual point **r** is the point of intersection of all lines connecting a pair of points in the reference image and the corresponding pair of warped points from all other frames.

4 Applications of Pairwise Parallax Geometry

In this section we show how parallax geometry in its various forms, which was introduced in the previous sections, provides an approach to handling some well-known problems in $3D$ scene analysis, in particular: (i) Moving object detection, (ii) Shape recovery, (ii) New view generation.

An extensive literature exists on methods for solving the above mentioned problems. They can be roughly classified into two main categories: (i) *2D methods* (e.g.,[8, 2, 13]): These methods assume that the image motion of the scene can be described using a $2D$ parametric transformation. They handle *dynamic* scenarios, but are limited to planar scenes or to very small camera translations. These fail in the presence of parallax motion. (ii) *3D methods* (e.g., [11, 15, 17, 5, 14, 23, 6, 3]): These methods handle general $3D$ scenes, but are (in their current form) limited to static scenarios or to scenarios where the parallax is both dense and of significant magnitude in order to overcome "noise" due to moving objects.

The use of the parallax constraints derived here provides a continuum between "$2D$ algorithms" and the "$3D$ algorithms". The need for bridging this gap exists in realistic image sequences, because it is not possible to predict in advance which situation would occur. Moreover, both types of scenarios can occur within the same sequence, with gradual transitions between them.

Estimating Planar Parallax Motion: The estimation of the planar parallax motion used for performing the experiments presented in this section was done using two successive computational steps: (i) $2D$ image alignment to compensate for a detected planar motion (i.e., the homography) in the form of a $2D$ parametric transformation, and, (ii) estimation of residual image displacements between the aligned images (i.e, the parallax).

We use previously developed methods [1, 8] in order to compute the $2D$ *parametric* image motion of a single $3D$ planar surface in the scene. These techniques lock onto a "dominant" planar motion in an image pair, even in the presence of other differently moving objects in the field of view. The estimated parametric motion is used to warp the second image frame to the first. The *residual* image displacements (e.g., parallax vectors) are then estimated using the optical flow estimation technique described in [1].

4.1 Moving Object Detection

A number of techniques exist to handle multiple motions analysis in the simpler $2D$ case, where motions of independent moving objects are modeled by $2D$ parametric transformation [8, 2, 13]. These methods, however, would also detect points with planar parallax motion as moving objects, as they have a different $2D$ image motion than the planar part of the background scene.

In the general $3D$ case, the moving object detection problem is much more complex, since it requires detecting $3D$ motion inconsistencies. Typically, this is done by recovering the epipolar geometry. Trying to estimate epipolar geometry (i.e., camera motion) in the presence of multiple moving objects, with no prior segmentation, is extremely difficult. This problem becomes even more acute when there exists only sparse parallax information. A careful treatment of the issues and problems associated with moving object detection in 3D scenes is given in [19]. Methods have been proposed for recovering camera geometry in the presence of moving objects [12, 21] in cases when the available parallax was *dense* enough and the independent motion was *sparse* enough to be treated as *noise*.

Fig. 4.a graphically displays an example of a configuration in which estimating the epipole in presence of multiple moving objects can be very erroneous, even when using clustering techniques in the epipole domain as suggested by [12, 20]. Relying on the epipole computation to detect inconsistencies in $3D$ motion fails in detecting moving objects in such cases.

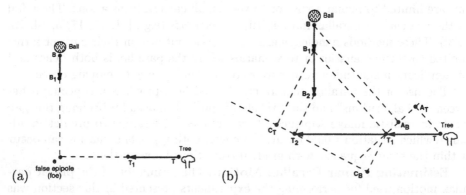

Fig. 4. Reliable detection of 3D motion inconsistency with sparse parallax information. *(a) Camera is translating to the right. The only static object with pure parallax motion is that of the tree. Ball is falling independently. The incorrectly estimated epipole e is consistent with both motions. (b) The rigidity constraint applied to this scenario detects 3D inconsistency over three frames, since $\frac{T_1A_B}{T_1A_T} \neq \frac{T_2C_B}{-T_2C_T}$. In this case, even the signs do not match.*

The parallax rigidity constraint(Eq. (5)) can be applied to detect inconsistencies in the $3D$ motion of one image point relative to another directly from their "parallax" vectors over multiple (three or more) frames, without the need to estimate either *camera geometry* or *shape* parameters. This provides a useful mechanism for clustering (or segmenting) the "parallax" vectors (i.e., the residual motion after planar registration) into consistent groups belonging to consistently $3D$ moving objects, even in cases such as in Fig. 4.a, where the parallax information is minimal, and the independent motion is not negligible. Fig. 4.b graphically explains how the rigidity constraint (5) detects the $3D$ inconsistency of Fig. 4.a over three frames.

Fig. 5 shows an example of using the rigidity constraint (5) to detect $3D$ inconsistencies.

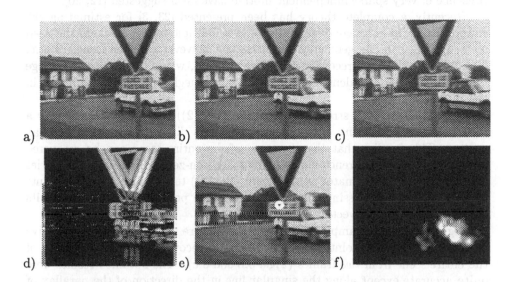

a)   b)   c)

d)   e)   f)

Fig. 5. Moving object detection relying on a single parallax vector.
(a,b,c) Three image frames from a sequence obtained by a camera translating from left to right, inducing parallax motion of different magnitudes on the house, road, and road-sign. The car moves independently from left to right. The middle frame (Fig. 5.b) was chosen as the frame of reference. (d) Differences taken after 2D image registration. The detected 2D planar motion was that of the house, and is canceled by the 2D registration. All other scene parts that have different 2D motions (i.e., parallax motion or independent motion) are misregistered. (e) The selected point of reference (on the road-sign) highlighted by a white circle. (f) The measure of 3D-inconsistency of all points in the image with respect to the road-sign point. Bright regions indicate violations in 3D rigidity detected over three frames. These regions correspond to the car. Regions close to the image boundary were ignored. All other regions of the image appear to move 3D-consistently with the road-sign point.

In [16] a rigidity constraint between three frames in the form of the trilinear tensor has been presented using regular image displacements. However, it requires having a collection of a set of image points which is known *a priori* to belong to the single $3D$ moving object. Selecting an inconsistent set of points will lead to an erroneous tensor. In [22] the trilinear constraint was used to segment and group multiple moving objects using robust techniques.

4.2 Shape Recovery

Numerous methods for recovering 3D depth from multiple views of calibrated cameras have been suggested. More recently, methods have been developed for recovering *projective structure* or *affine structure* from *uncalibrated* cameras [5, 14]. Under this category, methods for recovering structure using *planar parallax* motion have been proposed [10, 11, 15, 17]. Those methods rely on prior estimation of partial or full camera geometry. In particular, they rely on estimating the camera epipole. Some methods for recovering camera or scene geometry in the presence of very sparse independent motion have been suggested [12, 20].

Recently, a complete theory has been presented [23, 3] for estimating the shape *directly* from image displacements over multiple frames, *without* the need to recover the epipolar geometry. These assume, however, that the scene is static. The problem of shape recovery in dynamic scenes, where the amount of image motion due to independent moving object is not negligible, is not addressed in these works.

The parallax-based structure constraint (Eq. (2)) can be used to recover a the 3D relative structure between pairs of points directly from their parallax vectors. This implies that the structure of the entire scene can be recovered relative to a *single* reference image point (with non-zero parallax). Singularities occur when the denominator of constraint (Eq. (2)) tends to zero, i.e., for points that lie on the line passing through the reference point in the direction of its parallax displacement vector.

Fig. 6 shows an example of recovering structure of an entire scene relative to a *single* reference point. Fig. 6.f shows the recovered relative structure of the entire scene from *two* frames (Figs. 6.b and 6.c). The obtained results were quite accurate except along the singular line in the direction of the parallax of the reference point. The singular line is evident in Fig. 6.f.

The singularities can be removed and the quality of the computed structure can be improved either by using multiple frames (when their epipoles are non-colinear) or by using multiple reference points (An additional reference point should be chosen so that it does *not* lie on the singular line of the first reference point, and should be first verified to move consistently with the first reference point through the rigidity constraint (5) over a few frames). Of course, combinations of multiple reference points over multiple frames can also be used. Fig. 6.g shows an example of recovering structure of an entire scene from *three* frames relative to the same *single* reference point as in Fig. 6.f. The singular line in Fig. 6.f has disappeared.

The ability of obtain relatively good structure information even with respect to a *single* point has several important virtues: (i) Like [23, 3], it does not require the estimation of the epipole, and therefore does not require dense parallax information. (ii) Unlike previous methods for recovering structure, (including [23, 3]), it provides the capability to handle dynamic scenes, as it does not require having a collection of image points which is known *a priori* to belong to the single 3D moving object. (iii) Since it relies on a single parallax vector, it provides a natural continuous way to bridge the gap between 2D cases, that assume only planar motion exists, and 3D cases that rely on having parallax data.

4.3 New View Generation

The parallax rigidity constraint (5) can be used for generating novel views using a set of "model" views, without requiring any epipolar geometry or shape estimation. This work is still in early stages, and therefore no experimental results are provided.

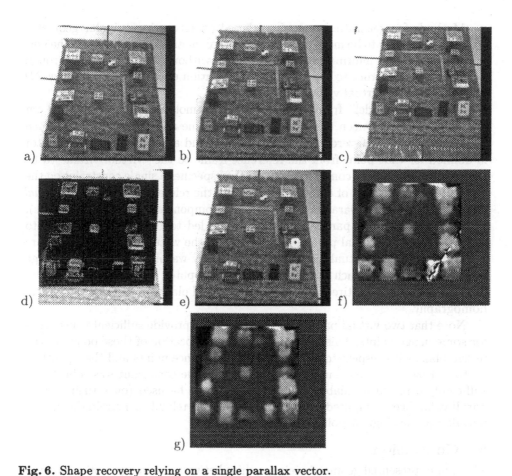

Fig. 6. Shape recovery relying on a single parallax vector.
(a,b,c) Three image frames from a sequence obtained by a hand-held camera of a rug covered with toy cars and boxes. The middle frame (Fig. 6.b) was chosen as the frame of reference. (d) Differences taken after 2D image registration. The detected 2D planar motion was that of the rug, and is canceled by the 2D registration. All other scene parts (i.e., toys and boxes) are misregistered. (e) The selected point of reference (a point on one of the boxes in the bottom right) highlighted by a white circle. (f) The recovered relative structure of the entire scene from two frames (Figs. 6.b and 6.c) relative to the selected point of reference. Regions close to the image boundary were ignored. The interpreted relative heights were quite accurate except along the singular line in the direction of the parallax displacement of the reference point. (g) The recovered relative structure of the entire scene using all three frames with respect to the selected point of reference. Regions close to the image boundary were ignored. The singular line has disappeared, providing more accurate shape.

Methods for generating new views based on recovering epipolar geometry (e.g., [4]) are likely to be more noise sensitive than methods that generate the new view based on $2D$ information alone [16], i.e., without going from $2D$ through a $3D$ medium in order to reproject the information once again onto a new $2D$ image plane (the virtual view).

Given two "model" frames, planar parallax motion can be computed for all image points between the first (reference) frame and the second frame. An image point with non-zero parallax is selected, and a "virtual" parallax vector is defined for that point from the reference frame to the "virtual" frame to be generated. The rigidity constraint (Eq. 5) then specifies a single constraint on the virtual parallax motion of all other points from the reference frame to the *virtual* frame. Since each $2D$ parallax vector has two components (i.e., two unknowns), at least two "virtual" parallax vectors are needed to be specified in order to solve for all other virtual parallax vectors. Once the virtual parallax vectors are computed, the new virtual view can be created by warping the reference image twice: First, warping each image point by its computed virtual parallax. Then, globally warping the entire frame with a $2D$ virtual planar motion for the virtual homography.

Note that two virtual parallax vectors may not provide sufficient constraints for some image points. This is due to unfavorable location of those points in the image plane with respect to the two selected reference points and their parallax vectors. However, other image points, for whom the constraint was robust and sufficient to produce reliable virtual parallax, can be used (once their virtual parallax has been computed) as additional points to reliably constrain the virtual parallax of the singular points.

5 Conclusion

This paper presented geometric relationships between the image motion of *pairs* of points over multiple frames. This relationship is based on the *parallax* displacements of points with respect to an arbitrary planar surface, and does not involve epipolar geometry. We derived constraint over two frames relating the projective structure (with respect to the plane) of any pair of points, based only on their image coordinates and their parallax motion. We also derived a $3D$-rigidity constraint between pairs of points over multiple frames.

We showed applications of these parallax-based constraints to the recovery of $3D$ scene structure, to the detection of moving objects in the presence of camera induced motion, and to "new view generation". Our approach can handle difficult situations for $3D$ scene analysis, e.g., where there is only a small set of parallax vectors, and in the presence of independently moving objects. The use of the parallax constraints derived here provides a continuum between "$2D$ algorithms" and the "$3D$ algorithms" for each of the problems mentioned above.

Finally, we outlined the generalization of our parallax based constraints to full image motion (as opposed to *parallax* motion), even when the homography is unknown. This is useful for handling scenes that do not contain a physical planar surface.

References

1. J.R. Bergen, P. Anandan, K.J. Hanna, and R. Hingorani. Hierarchical model-based motion estimation. In *ECCV*, pages 237–252, Santa Margarita Ligure, May 1992.
2. J.R. Bergen, P.J. Burt, R. Hingorani, and S. Peleg. A three-frame algorithm for estimating two-component image motion. *PAMI*, 14:886–895, September 1992.
3. S. Carlsson. Duality of reconstruction and positioning from projective views. In *Workshop on Representations of Visual Scenes*, 1995.
4. O. Faugeras and L. Robert. What can two images tell us about a third one? In *ECCV*, pages 485–492, May 1994.
5. O.D. Faugeras. What can be seen in three dimensions with an uncalibrated stereo rig? In *ECCV*, pages 563–578, Santa Margarita Ligure, May 1992.
6. O.D. Faugeras and B. Mourrain. On the geometry and algebra of the point and line correspondences between n images. In *ICCV*, pages 951–956, June 1995.
7. M. Irani and P. Anandan. Parallax geometry of pairs of points for 3d scene analysis. Technical report, David Sarnoff Research Center, October 1995.
8. M. Irani, B. Rousso, and S. Peleg. Computing occluding and transparent motions. *IJCV*, 12(1):5–16, January 1994.
9. M. Irani, B. Rousso, and S. Peleg. Recovery of ego-motion using image stabilization. In *CVPR*, pages 454–460, Seattle, Wa., June 1994.
10. J.J. Koenderink and A.J. van Doorn. Representation of local geometry in the visual system. *Biol. Cybern.*, 55:367 – 375, 1987.
11. R. Kumar, P. Anandan, and K. Hanna. Shape recovery from multiple views: a parallax based approach. In *DARPA IU Workshop*, November 1994.
12. J.M. Lawn and R. Cipolla. Robust egomotion estimation from affine motion parallax. In *ECCV*, pages 205–210, May 1994.
13. F. Meyer and P. Bouthemy. Region-based tracking in image sequences. In *ECCV*, pages 476–484, Santa Margarita Ligure, May 1992.
14. T. Chang R. Hartley, R. Gupta. Stereo from uncalibrated cameras. In *CVPR-92*.
15. Harpreet Sawhney. 3d geometry from planar parallax. In *CVPR*, June 1994.
16. A. Shashua. Algebraic functions for recognition. *PAMI*, 17:779–789, 1995.
17. A. Shashua and N. Navab. Relative affine structure: Theory and application to 3d reconstruction from perspective views. In *CVPR*, pages 483–489, 1994.
18. M. E. Spetsakis and J. Aloimonos. A unified theory of structure from motion. In *DARPA IU Workshop*, pages 271–283, 1990.
19. W.B. Thompson and T.C. Pong. Detecting moving objects. *IJCV*, 4:29–57, 1990.
20. P.H.S. Torr and D.W. Murray. Stochastic motion clustering. In *ECCV*, 1994.
21. P.H.S. Torr, A. Zisserman, and S.J. Maybank. Robust detection of degenerate configurations for the fundamental matrix. In *ICCV*, pages 1037–1042, 1995.
22. P.H.S. Torr, A. Zisserman, and D.W. Murray. Motion clustering using the trilinear constraint over three views. In *Workshop on Geometric Modeling and Invariants for Computer Vision*, 1995.
23. D. Weinshall, M.Werman, and A. Shashua. Shape descriptors: Bilinear, trilinear and quadlinear relations for multi-point geometry, and linear projective reconstruction algorithms. In *Workshop on Representations of Visual Scenes*, 1995.
24. Andrew Zisserman. A case against epipolar geometry. In *Applications of Invariance in Computer Vision*, pages 35–50, Ponta Delgada, Azores, October 1993.

Euclidean 3D reconstruction from image sequences with variable focal lengths

Marc Pollefeys*, Luc Van Gool, Marc Proesmans**

Katholieke Universiteit Leuven, E.S.A.T. / MI2
Kard. Mercierlaan 94, B-3001 Leuven, BELGIUM
Marc.Pollefeys, Luc.VanGool, Marc.Proesmans@esat.kuleuven.ac.be

Abstract. One of the main problems to obtain a Euclidean 3D reconstruction from multiple views is the calibration of the camera. Explicit calibration is not always practical and has to be repeated regularly. Sometimes it is even impossible (i.e. for pictures taken by an unknown camera of an unknown scene). The second possibility is to do auto-calibration. Here the rigidity of the scene is used to obtain constraints on the camera parameters. Existing approaches of this second strand impose that the camera parameters stay exactly the same between different views. This can be very limiting since it excludes changing the focal length to zoom or focus. The paper describes a reconstruction method that allows to vary the focal length. Instead of using one camera one can also use a stereo rig following similar principles, and in which case also reconstruction from a moving rig becomes possible even for pure translation. Synthetic data were used to see how resistant the algorithm is to noise. The results are satisfactory. Also results for a real scene were convincing.

1 Introduction

Given a general set of images of the same scene one can only build a *projective* reconstruction [4, 6, 15, 16]. Reconstruction up to a smaller transformation group (i.e. *affine* or *Euclidean*) requires additional constraints. In this article only methods requiring no a priori scene knowledge will be discussed. Existing methods assume that all internal camera parameters stay exactly the same for the different views. Hartley [7] proposed a method to obtain a Euclidean reconstruction from three images. The method needs a non-linear optimisation step wich is not guaranteed to converge. Having an affine reconstruction eliminates this problem. Moons *et al* [11] described a method to obtain an affine reconstruction when the camera movement is a pure translation. Armstrong *et al* [1] combined both methods [11, 7] to obtain a Euclidean reconstruction from three images with a translation between the first two views.

In contrast to the 2D case, where viewpoint independent shape analysis and the use of uncalibrated cameras go hand in hand, the 3D case is more subtle. The

---

* IWT fellow (Flemish Inst. for the Promotion of Scient.-Techn. Research in Industry)
** IWT post-doctoral researcher

precision of reconstruction depends on the level of calibration, be it in the form of information on camera or scene parameters. Thus, uncalibrated operation comes at a cost and it becomes important to carefully consider the pro's and con's of needing knowledge on the different internal and external camera parameters.

As an example, the state-of-the-art strategy to keep all internal camera parameters unknown but fixed, means that one is not allowed to zoom or adapt focus. This can be a serious limitation in practical situations. It stands to reason that the ability to keep the object of interest sharp and at an appropriate resolution would be advantageous. Also being allowed to zoom in on details that require a higher level of precision in the reconstruction can save much trouble. To the best of our knowledge no method using uncalibrated cameras for Euclidean reconstruction allows this (i.e. the focal length has to stay constant).

This paper describes a method to obtain a *Euclidean reconstruction* from images taken with an *uncalibrated camera* with a *variable focal length*. In fact, it is an adaptation of the methods of Hartley [7], Moons *et al* [11] and Armstrong *et al* [1], to which it adds an initial step to determine the position of the principal point. Thus, a mild degree of camera calibration is introduced in exchange for the freedom to change the focal length between views used for reconstruction. The very ability to change the focal length allows one to recover the principal point in a straight-forward way. From there, the method starts with an affine reconstruction from two views with a translation in between. A third view allows an upgrade to Euclidean structure. The focal length can be different for each of the three views. In addition the algorithm yields the relative changes in focal length between views.

Recently, methods for the Euclidean calibration of a fixed stereo rig from two views taken with the rig have been propounded [19, 3]. The stereo rig must rotate between views. It is shown here that also in this case the flexibility of variable focal length can be provided for and that reconstruction is also possible after pure translation, once the principal points of the cameras are determined.

2 Camera model

In this paper a pinhole camera model will be used. Central projection forms an image on a light-sensitive plane, perpendicular to the optical axis. Changes in focal length move the optical center along the axis, leaving the principal point[3] unchanged. This assumption is fulfilled to a sufficient extent in practice [9]. The following equation expresses the relation between image points and world points.

$$\lambda_{ij} m_{ij} = \mathbf{P}_j M_i \qquad (1)$$

Here P_j is a 3x4 camera matrix, m_{ij} and M_i are column vectors containing the homogeneous coordinates of the image points resp. world points, λ_{ij} expresses

---

[3] The principal point is defined as the intersection point of the optical axis and the image plane

the equivalence up to a scale factor. If P_j represents a Euclidean camera, it can be put in the following form [7]:

$$P_j = K_j [R_j | - R_j t_j] \qquad (2)$$

where R_j and t_j represent the Euclidean orientation and position of this camera with respect to a world frame, and K_j is the calibration matrix of the jth camera:

$$K_j = \begin{bmatrix} r_x^{-1} & -r_x^{-1} \cos \theta & f_j^{-1} u_x \\ & r_y^{-1} & f_j^{-1} u_y \\ & & f_j^{-1} \end{bmatrix} \qquad (3)$$

In this equation r_x and r_y represent the pixel width and height, θ is the angle between the image axes, u_x and u_y are the coordinates of the principal point, and f_j is the focal length. Notice that the calibration matrix is only defined up to scale. In order to highlight the effect of changing the focal length the calibration matrix K_j will be decomposed in two parts:

$$K_j = K_{f_j} K = \begin{bmatrix} 1 & 0 & (f_1/f_j - 1)u_x \\ & 1 & (f_1/f_j - 1)u_y \\ & & f_1/f_j \end{bmatrix} \cdot \begin{bmatrix} r_x^{-1} & -r_x^{-1} \cos \theta & f_1^{-1} u_x \\ & r_y^{-1} & f_1^{-1} u_y \\ & & f_1^{-1} \end{bmatrix} \qquad (4)$$

The second part K is equal to the calibration matrix K_1 for view 1, whereas K_{f_j} models the effect of changes in focal length (i.e. zooming and focusing). From equation (4) it follows that once the principal point u is known, K_{f_j} is known for any given value of f_j/f_1. Therefore, finding the principal point is the first step of the reconstruction method. Then, if the change in focal length between two views can be retrieved, its effect is canceled by multiplying the image coordinates to the left by $K_{f_j}^{-1}$.

The first thing to do is to retrieve the principal point u. Fortunately this is easy for a camera equiped with a zoom. Upon changing the focal length (without moving the camera or the scene), each image point according to the pinhole camera model will move on a line passing through the principal point. By taking two or more images with a different focal length and by fitting lines through the corresponding points, the principal point can be retrieved as the common intersection of all these lines. In practice these lines will not intersect precisely and a least squares approach is used to determine the principal point. This method has been used by others [18, 8, 9].

For the sake of simplicity we will assume $R_1 = I, t_1 = 0$ and $f_1 = 1$ in the remainder of this paper. Because the reconstruction is up to scaled Euclidean (i.e. similarity) and K_j is only defined up to scale this is not a restriction.

3 Affine structure from translation

It is possible to recover the affine structure of a scene from images taken by a translating camera [11]. This result can also be obtained when the focal length is not constant. Consider two perspective images with a camera translation and possibly a change in focal length in between.

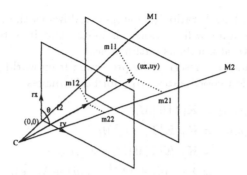

Fig. 1. *Illustration of the camera and zoom model. The focal lengths f_1 and f_2 are different, the other parameters ($r_x, r_y, u_x, u_y, \theta$) are identical.*

3.1   recovering focal length for translation

There are different methods to recover the change in focal length. We first give a straightforward method based on the movement of the epipoles[4]. The performance of this method degrades very fast with noise on the image correspondences. Therefore an alternative, non-linear method was developed, that uses all available constraints. This method gives good results even in the presence of noise.

The first method is based on the fact that epipoles move on a line passing through the principal point u when the focal length is changed while the camera is translated and that this movement is related to the magnitude of this change. The following equation follows from the camera model:

$$\lambda_{e_{21}} e_{21} = -\lambda_{e_{12}}(e_{12} + (f_2^{-1} - 1)u) \tag{5}$$

where e_{21} is the epipole in the second image and e_{12} the epipole in the first image. e_{21}, e_{12} and u are column vectors of the form $[x\,y\,1]^\top$. From equation (5) f_2^{-1} can be solved in a linear way. This method is suboptimal in the sense that it does not take advantage of the translational camera motion.

Determining the epipoles for two arbitrary images is a problem with 7 degrees of freedom. In the case of a translation (without changing the focal length) between two views, the epipolar geometry is the same for both images and the image points lie on their own epipolar lines. This means that the epipolar geometry is completely determined by knowing the position of the unique epipole (2 degrees of freedom). Adding changes in focal length between the images adds one degree of freedom when the principal point is known.

Given three points in the two views, one know that a scaling equal to the focal length ratio should bring them in position such that the lines through corresponding points intersect in the epipole. This immediately yields a quadratic

---

[4] an epipole is the projection of the optical center of one camera in the image plane of the other camera. The epipoles can be retrieved from at least 7 point correspondences between the two images [5].

equation in the focal length ratio. The epipole follows as the resulting intersection. In practice the data will be noisy, however, and it is better to consider information from several points as outlined next.

The following equations describe the projection from world point coordinates M_i to image projection coordinates $m_{i1,2}$ for both images

$$\lambda_{i1}m_{i1} = \mathbf{K}[\mathbf{I}\,|\,0]M_i \tag{6}$$

$$\begin{aligned}
\lambda_{i2}m_{i2} &= \mathbf{K}_{f_2}\mathbf{K}[\mathbf{I}\,|-t_2]M_i \\
&= \mathbf{K}_{f_2}\mathbf{K}[\mathbf{I}\,|\,0]M_i + \lambda_{e_{21}}e_{21} \\
&= \lambda_{i1}\left(m_{i1} + (f_2^{-1} - 1)u\right) + \lambda_{e_{21}}e_{21}
\end{aligned} \tag{7}$$

where m_{i1}, m_{i2}, u and e_{21} are column vectors of the form $[x\,y\,1]^\mathsf{T}$. Equation (7) gives 3 constraints for every point. If f_2 is known this gives a linear set of equations in $2n + 3$ unknowns ($\lambda_{i1}, \lambda_{i2}, \lambda_{e_{21}}e_{211}, \lambda_{e_{21}}e_{212}, \lambda_{e_{21}}e_{213}$). Because all unknowns of equation (7) (except f_2) comprise a scale factor, n must satisfy $3n \geq 2n + 3 - 1$ to have enough equations. To also solve for f_2 one needs at least one more equation, which means that at least 3 point correspondences are needed to find the relative focal length f_2 (remember that $f_1 = 1$).

For every value of f_2 one could try to solve the set of equations (7) by taking the singular value decomposition of the corresponding matrix. If f_2 has the correct value there will be a solution and the smallest singular value should be zero. With noisy data there will not be an exact solution anymore, but the value of f_2 which yields the smallest singular value will be the best solution in a least squares sense. For this paper the Decker-Brent algorithm was used to minimise the smallest singular value with respect to the relative focal length f_2. This gives very good results. Thanks to the fact that a non-linear optimisation algorithm in only one variable was used no convergence problems were encountered.

3.2 affine reconstruction

knowing the focal length, one can start the actual affine reconstruction. Notice that it follows from equation (6) that the scene points M_i are related to $\begin{bmatrix} \lambda_{i1}m_{i1} \\ 1 \end{bmatrix}$ by the affine transformation $\begin{bmatrix} \mathbf{K} & 0 \\ 0 & 1 \end{bmatrix}$. So it suffices to recover the λ_{i1} from equation (7) to have an affine reconstruction of the scene.

4 Euclidean structure from affine structure and supplementary camera motion

In this section the upgrade of the reconstruction from affine to Euclidean by using a supplementary image taken with a different orientation is discussed. Once an affine reconstruction of the scene is known the same constraints as in [7, 1, 19] can be used. Here they are less easy to use, because the focal length also appears in these constraints. Therefore one first has to find the relative change in focal length.

4.1 recovering focal length for a supplementary view

In this paragraph a method will be explained that allows to recover the relative focal length of a camera in any position and with any orientation (relative to the focal length of the camera at the beginning). This can be done by starting from an affine reconstruction. Choosing the first camera matrix to be $[\mathbf{I}\,|\,\mathbf{0}]$ the second camera matrix associated to our affine reconstruction is uniquely defined up to scale [15]. In the following equations the relationship between the affine camera matrixes $\mathbf{P}_{1A}, \mathbf{P}_{3A}$ and the Euclidean ones is given:

$$\mathbf{P}_{1A} = [\mathbf{I}\,|\,\mathbf{0}] = \mathbf{K}[\mathbf{I}\,|\,\mathbf{0}]\begin{bmatrix} \mathbf{K}^{-1} & \mathbf{0} \\ \mathbf{0} & 1 \end{bmatrix}$$

$$\mathbf{P}_{3A} \equiv [\tilde{\mathbf{P}}_{3A}|\,.\,] = \lambda_{\mathbf{P}_3}\mathbf{K}_{f_3}\mathbf{K}[\mathbf{R}_3|-\mathbf{R}_3 t_3]\begin{bmatrix} \mathbf{K}^{-1} & \mathbf{0} \\ \mathbf{0} & 1 \end{bmatrix}$$

$$= \lambda_{\mathbf{P}_3}\mathbf{K}_{f_3}[\mathbf{K}\mathbf{R}_3\mathbf{K}^{-1}|\,.\,] \tag{8}$$

By definition $\mathbf{K}\mathbf{R}_3\mathbf{K}^{-1}$ is conjugated to $\mathbf{R}_3$ and hence will have the same eigenvalues which for a rotation matrix all have modulus 1 (one of them is real and both others are complex conjugated or real). This will be called the *modulus* constraint in the remainder of this paper. From equation (8) it follows that $\tilde{\mathbf{P}}_{3A}$ is related to $\mathbf{K}\mathbf{R}_3\mathbf{K}^{-1}$ in the following way:

$$\mathbf{K}_{f_3}^{-1}\tilde{\mathbf{P}}_{3A} = \lambda_{\mathbf{P}_3}\mathbf{K}\mathbf{R}_3\mathbf{K}^{-1} \tag{9}$$

with

$$\mathbf{K}_{f_3}^{-1} = \begin{bmatrix} 1 & 0 & (f_3-1)u_x \\ & 1 & (f_3-1)u_y \\ & & f_3 \end{bmatrix}$$

The characteristic equation of $\mathbf{K}_{f_3}^{-1}\tilde{\mathbf{P}}_{3A}$ is as follows:

$$\det\left(\mathbf{K}_{f_3}^{-1}\tilde{\mathbf{P}}_{3A} - \lambda\mathbf{I}\right) \equiv a\lambda^3 + b\lambda^2 + c\lambda + d = 0 \tag{10}$$

The *modulus* constraint imposes $|\lambda_1| = |\lambda_2| = |\lambda_3| \,(= \lambda_{\mathbf{P}_3})$. From this one gets the following constraint:

$$ac^3 = b^3 d \tag{11}$$

Substituting the left hand side of equation (9) in equation (10), yields first order polynomials in f_3 for a, b, c, d. Substituting these in equation (11), one obtains a 4th order polynomial in f_3.

$$a_4 f_3^4 + a_3 f_3^3 + a_2 f_3^2 + a_1 f_3 + a_0 = 0 \tag{12}$$

This gives 4 possible solutions. One can see that if f_3 is a real solution, then $-f_3$ must also be a solution[5]. Filling this in in equation (12) one gets the following

---

[5] This is because the only constraint imposed is the *modulus* constraint (same modulus for all eigenvalues). If the real part of λ_2 and λ_3 have opposite sign then $\mathbf{K}_{f_3}^{-1}\tilde{\mathbf{P}}_{3A}$ does not represent a rotation but a rotation and a mirroring. Changing the sign of f_3 has the same effect.

result after som algebraic manipulations.

$$f_3 = \pm\sqrt{\frac{a_1}{a_3}} \qquad (13)$$

where the sign is dependent on camera geometry and is known[6]. One can conclude this paragraph by stating that the relative focal length f of any view with respect to a reference view can be recovered for any Euclidean motion.

4.2 Euclidean reconstruction

To upgrade the reconstruction to Euclidean the camera calibration matrix $\mathbf{K}$ is needed. This is equivalent to knowing the image $\mathbf{B}$ of the dual of the absolute conic for the first camera, since $\mathbf{B} = \mathbf{KK}^{\mathsf{T}}$. Images are constrained in the following way:

$$\kappa_{13}\mathbf{B}_3 = \mathbf{H}_{13\infty}\mathbf{BH}_{13\infty}^{\mathsf{T}} \qquad (14)$$

with $\mathbf{B}_3 = \mathbf{K}_3\mathbf{K}_3^{\mathsf{T}}$ the inverse of the image of the absolute conic in the third image and $\mathbf{H}_{13\infty}$ the infinity homography[7] between the two images. This would be a set of linear equations in the coefficients of $\mathbf{B}$ if κ_{13} was known. This can be achieved by imposing equal determinants for the left and right hand side of equation (14). But before doing this it is interesting to decompose $\mathbf{B}_3$:

$$\mathbf{B}_3 = \mathbf{K}_3\mathbf{K}_3^{\mathsf{T}} = \mathbf{K}_{f_3}\mathbf{KK}^{\mathsf{T}}\mathbf{K}_{f_3}^{\mathsf{T}} = \mathbf{K}_{f_3}\mathbf{BK}_{f_3}^{\mathsf{T}} \qquad (15)$$

From equation (15) one finds an equation for the determinant of $\mathbf{B}_3$ and by imposing the equality with the determinant of the right hand side of equation (14), the following equation is obtained.

$$\det \mathbf{B}_3 = (\det \mathbf{H}_{13\infty})^2 \det \mathbf{B} = (\det \mathbf{K}_{f_3})^2 \det \mathbf{B} \qquad (16)$$

Equation (16) will hold if the following equation holds:

$$\det \mathbf{H}_{13\infty} = \det \mathbf{K}_{f_3} \equiv f_3^{-1}, \qquad (17)$$

when f_3 has been obtained following the principles outlined in section 4.1. This constraint can easily be imposed because $\mathbf{H}_{13\infty}$ is only determined up to scale. The following equations (derived from equations (14) and (15)) together with the knowledge of u and f_3 then allows to calculate $\mathbf{B}$ (and $\mathbf{K}$ by cholesky factorisation).

$$\mathbf{K}_{f_3}\mathbf{BK}_{f_3}^{\mathsf{T}} = \mathbf{H}_{13\infty}\mathbf{BH}_{13\infty}^{\mathsf{T}} \qquad (18)$$

This approach could be simplified by assuming that the camera rows and columns are perpendicular ($\theta = 90^\circ$)[19]. In that case equation (18) boils down to an overdetermined system of linear equations in r_x^{-2} and r_y^{-2} which gives more stable results. r_x and r_y being the only unknowns left, one will also have $\mathbf{K}$. Finally the affine reconstruction can be upgraded to Euclidean by applying the following transformation

$$\mathbf{T}_{AE} = \begin{bmatrix} \mathbf{K}^{-1} & 0 \\ 0 & 1 \end{bmatrix} \qquad (19)$$

---

[6] for a non-mirrored image the sign must be positive.

[7] The infinity homography, which is a plane projective transformation, maps vanishing points from one image to the corresponding points in another image.

5 Euclidean calibration of a fixed stereo rig

The auto-calibration techniques proposed by Zisserman [19] and Devernay [3] for two views taken with a rotating fixed stereo rig can also be generalised to allow changes in focal lengths for both cameras independently and purely translational motions. In fact the method is easier than for a single camera.

For a fixed stereo rig the epipoles are fixed as long as one doesn't change the focal length. In this case the movement of the epipole in one camera is in direct relation with the change of its focal length. This is illustrated in figure 2. Knowing the relative change in focal length and the principal points allows to remove the effect of this change from the images. From then on the techniques of Zisserman [19] or Devernay [3] can be applied.

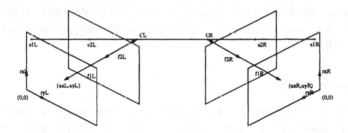

Fig. 2. *this figure illustrates how the epipoles will move in function of a change in focal length.*

By first extracting the principal points -i.e. mildly calibrating the camera - one can then also get a Euclidean reconstruction even for a translating stereo rig[12], which was not possible with earlier methods [19, 3]. Between any pair of cameras i and j we have the following constraints:

$$\kappa_{ij}\mathbf{B}_j = \mathbf{H}_{ij\infty}\mathbf{B}_i\mathbf{H}_{ij\infty}^{\mathsf{T}} \tag{20}$$

For two views with a fixed stereo rig there are 3 different constraints of the type of equation (20): for the left camera (between view 1 and 2), for the right camera (between view 1 and 2) and between the left and the right camera. For a translation $\mathbf{H}_{12\infty} = \mathbf{I}$ which means that the two first constraints become trivial. The constraint between the left and the right camera in general gives 6 independent equations[8]. This is not enough to solve for r_{Lx}, r_{Ly}, u_{Lx}, u_{Ly}, θ_L, r_{Rx}, r_{Ry}, u_{Rx}, u_{Ry} and θ_R. Knowing the principal points restricts the number of unknowns to 6, which could be solved from the available constraints. Assuming perpendicular images axes [19] one can solve for the 4 remaining unknowns in a linear way [13]. In practical cases this is very important because with the earlier techniques any movement close to translation gives unstable results which isn't the case anymore for this technique.

---

[8] the cameras of the stereo rig should not have the same orientation.

It is also useful to note that in the case of a translational motion of the rig, the epipolar geometry can be obtained with as few as 3 points seen in all 4 views. Superimposing their projections in the focal length corrected second views onto the first, it is as if one observes two translated copies of the points. Choosing two of the three points, one obtains four coplanar points from the two copies (coplanarity derives from the fact that the rig translates). Together with projections of the third point, this suffices to apply the algorithm propound in [2]. Needing as few as 3 points clearly is advantagous to detect e.g. independent motions using RANSAC strategies [17]

6 Results

In this section some results obtained with the single camera algortithm are presented. First an analysis of the noise resistance of the algorithm is given based on synthetic data with variable levels of noise. A reconstruction of a real scene is given as well.

6.1 synthetic data

Altough synthetic data were used to perform the experiments in this paragraph, due attention has been paid to mimic real data. A simple house shape was chosen as scene and the "camera" was given realistic parameter values. From this a sequence of 320x320 disparity maps was generated. These maps were altered with different amounts of noise to see how robust the method is. The focal length change between the first two images (translation) can be recovered very accurately. The non-linear method was used to obtain f_2/f_1. The focal length for the third image is much more sensitive to noise, but this doesn't seem to influence too much the calculation of r_x^{-1} or r_y^{-1}. This is probably due to the fact that the set of equations (18) gives us 6 independent equations for only 2 unknowns. The influence of a bad localisation of the principal point u was also analysed. The errors on the estimated parameters f_2, f_3, r_x^{-1} and r_y^{-1} came out to be of the same order as the error on u, which in practice was small when determined from zooming. From these experiments one sees that the Euclidean calibration of the camera and hence also the reconstruction degrades gracefully in the presence of noise. This indicates that the presented method is usable in practical circumstances.

6.2 real images

Here some results obtained from a real scene are presented. The scene consisted of a cornflakes boxes, a lego box and a cup. The images that were used can be seen in figure 3. The scene was chosen to allow a good qualitative evaluation of the Euclidean reconstruction. The boxes have right angles and the cup is cylindrical. These characteristics must be preserved by a *Euclidean* reconstruction, but will in general not be preserved by an *affine* or *projective* reconstruction.

Fig. 3. *The 3 images that were used to build a Euclidean reconstruction. The camera was translated between the first two views (the zoom was used to keep the size more or less constant). For the third image the camera was also rotated.*

To build a reconstruction one first needs correspondences between the images. Zhang's corner matcher[5] was used to extract these correspondences. A total of 99 correspondences were obtained between the first two images. From these the affine calibration of the camera was computed. From the first up to the third image 34 correspondences could be tracked. The corresponding scene points were reconstructed from the first two images which allowed to find an affine calibration for the camera in the third position. From this the method described in section 5 to find the Euclidean calibration of the camera was used. Subsequently, the output of an algorithm to compute dense point correspondences[14] was used to generate a more complete reconstruction. This algorithm yields a pointwise correspondence and confidence level. Only points with a confidence level above a fixed threshold were used for the reconstruction.

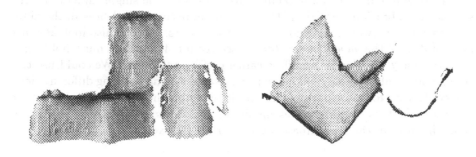

Fig. 4. *front and top view of the reconstruction.*

Figure 4 shows two views of the reconstructed scene. The left image is a front view while the right image is a top view. Note , especially from the top view, that 90° angles are preserved and that the cup keeps its cylindrical form which is an indication of the quality of the *Euclidean* reconstruction. Figure 5 shows a further view, both shaded and texture mapped to indicate the consistency with the original image texture.

Fig. 5. *side views of the reconstructed scene (with shading and with texture).*

7 Conclusions and future work

In this paper the possibility to obtain the auto-calibration of both a single moving camera and a moving stereo-rig was demonstrated and this without the need of keeping all internal parameters constant. The complete method for a single camera was described. From the experiments one can conclude that this method is relatively stable in the presence of noise. This makes it suitable for practical use. Also a method for auto-calibration of a fixed stereo rig with independently zooming cameras was briefly presented. An additional advantage of this method is that it is even suitable for a *purely translating* stereo rig, whereas previous methods required a rotational motion component.

We plan to enhance the implementations of both the single camera and the stereo rig calibration algorithm. The input of more correspondences in the auto-calibration stage would certainly yield better results. We will also look at other possibilities of the *modulus constraint* (see section 4.1) which must hold for a camera in any position in an affine camera reference frame[9]. We could use this constraint to calculate any unknown parameter of a camera. Using different views one could solve for more parameters, like for example the *affine* calibration of the matrix itself. This could be intresting because it allows to work in a 3 parameter space in stead of the 8 parameters that Hartley [7] had to solve for at once.

Acknowledgement

Marc Pollefeys and Marc Proesmans acknowledge a specialisation resp. research grant from the Flemish Institute for Scientific Research in Industry (IWT). Financial support from the EU ACTS project AC074 'VANGUARD' and from the IUAP-50 project of the Belgian OSTC is also gratefully acknowledged.

---

[9] it must be the same affinely calibrated camera for all views and the camera matrix must be [I|0] for the first view.

References

1. M. Armstrong, A. Zisserman, and P. Beardsley, Euclidean structure from uncalibrated images, *Proc. BMVC'94*, 1994,
2. B.Boufama, R. Mohr, Epipole and fundamental matrix estimation using virtual parallax, *Proc. ICCV'95*, pp.1030-1036, 1995
3. F. Devernay and O. Faugeras, From Projective to Euclidean Reconstruction, *INSIGHT meeting Leuven*, 1995,
4. O. Faugeras, What can be seen in three dimensions with an uncalibrated stereo rig, *Proc. ECCV'92*, pp.321-334, 1992.
5. R. Deriche, Z. Zhang, Q.-T. Luong, O. Faugeras, Robust Recovery of the Epipolar Geometry for an Uncalibrated Stereo Rig *Proc. ECCV'94*, pp.567-576, 1994.
6. R. Hartley, Estimation of relative camera positions for uncalibrated cameras, *Proc. ECCV'92*, pp.579-587, 1992.
7. R. Hartley, Euclidean reconstruction from uncalibrated views, in: J.L. Mundy, A. Zisserman, and D. Forsyth (eds.), *Applications of invariance in Computer Vision*, Lecture Notes in Computer Science **825**, pp. 237-256, Springer, 1994.
8. J.M. Lavest, G. Rives, and M. Dhome. 3D reconstruction by zooming. *IEEE Robotics and Automation*, 1993
9. M. Li, Camera Calibration of a Head-Eye System for Active Vision *Proc. ECCV'94*, pp.543-554, 1994.
10. Q.T. Luong and T. Vieville. Canonic representations for the geometries of multiple projective views. *Proc. ECCV'94*, pp.589-597, 1994.
11. T. Moons, L. Van Gool, M. Van Diest, and E. Pauwels, Affine reconstruction from perspective image pairs, Proc. Workshop on Applications of Invariance in Computer Vision II, pp.249-266, 1993.
12. M. Pollefeys, L. Van Gool, and T. Moons, Euclidean 3D Reconstruction from Stereo Sequences with Variable Focal Lengths *Recent Developments in Computer Vision*, Lecture Notes in Computer Science, pp.405-414, Springer-Verlag, 1996.
13. M. Pollefeys, L. Van Gool, and M. Proesmans, Euclidean 3D reconstruction from image sequences with variable focal lengths, Tech.Rep.KUL/ESAT/MI2/9508, 1995.
14. M. Proesmans, L. Van Gool, and A. Oosterlinck, Determination of optical flow and its discontinuities using non-linear diffusion, *Proc. ECCV'94*, pp. 295-304, 1994.
15. C. Rothwell, G. Csurka, and O.D. Faugeras, A comparison of projective reconstruction methods for pairs of views, *Proc. ICCV'95*, pp.932-937, 1995.
16. M. Spetsakis, Y.Aloimonos, A Multi-frame Approach to Visual Motion Perception *International Journal of Computer Vision*, 6:3, 245-255, 1991.
17. P.H.S. Torr, Motion Segmentation and Outlier Detection, *Ph.D.Thesis*, Oxford 1995.
18. R.Y. Tsai. A versatile camera calibration technique for high-accuracy 3D machine vision using off-the-shelf TV cameras and lenses. *IEEE Journal of Robotics and Automation*, RA-3(4):323-331, August 1987.
19. A. Zisserman, P.A.Beardsley, and I.D. Reid, Metric calibration of a stereo rig. *Proc. Workshop on Visual Scene Representation*, Boston, 1995.

Recognition (1)

Eigenfaces vs. Fisherfaces: Recognition Using Class Specific Linear Projection

Peter N. Belhumeur   João P. Hespanha*   David J. Kriegman**

Department of Electrical Engineering, Yale University, New Haven CT 06520-8267

Abstract. We develop a face recognition algorithm which is insensitive to gross variation in lighting direction and facial expression. Taking a pattern classification approach, we consider each pixel in an image as a coordinate in a high-dimensional space. We take advantage of the observation that the images of a particular face, under varying illumination but fixed pose, lie in a 3-D linear subspace of the high dimensional feature space — if the face is a Lambertian surface without self-shadowing. However, since faces are not truly Lambertian surfaces and do indeed produce self-shadowing, images will deviate from this linear subspace. Rather than explicitly modeling this deviation, we project the image into a subspace in a manner which discounts those regions of the face with large deviation. Our projection method is based on Fisher's Linear Discriminant and produces well separated classes in a low-dimensional subspace, even under severe variation in lighting and facial expressions. The Eigenface technique, another method based on linearly projecting the image space to a low dimensional subspace, has similar computational requirements. Yet, extensive experimental results demonstrate that the proposed "Fisherface" method has error rates that are significantly lower than those of the Eigenface technique for tests on the same database.

1   Introduction

Within the last several years, numerous algorithms have been proposed for face recognition; for detailed surveys see [1, 25]. While much progress has been made toward recognizing faces under small variations in lighting, facial expression and pose, reliable techniques for recognition under more extreme variations have proven elusive.

In this paper, we outline a new approach for face recognition — one that is insensitive to extreme variations in lighting and facial expressions. Note that lighting variability includes not only intensity, but also direction and number of light sources. As seen in Fig. 1, the same person, with the same facial expression, seen from the same viewpoint, can appear dramatically different when light sources illuminate the face from different directions.

Our approach to face recognition exploits two observations:

1. For a Lambertian surface without shadowing, all of the images of a particular face from a fixed viewpoint will lie in a 3-D linear subspace of the high-dimensional image space [30].

---

* J. Hespanha was supported by NSF Grant ECS-9206021, AFOSR Grant F49620-94-1-0181, and ARO Grant DAAH04-95-1-0114.

** D. Kriegman was supported by NSF under an NYI, IRI-9257990 and by ONR N00014-93-1-0305.

Eigenfaces vs. Fisherfaces:
Recognition Using Class Specific Linear Projection

Peter N. Belhumeur   João P. Hespanha*   David J. Kriegman**

Dept. of Electrical Engineering, Yale University, New Haven, CT 06520-8267

Abstract. We develop a face recognition algorithm which is insensitive to gross variation in lighting direction and facial expression. Taking a pattern classification approach, we consider each pixel in an image as a coordinate in a high-dimensional space. We take advantage of the observation that the images of a particular face under varying illumination direction lie in a 3-D linear subspace of the high dimensional feature space – if the face is a Lambertian surface without self-shadowing. However, since faces are not truly Lambertian surfaces and do indeed produce self-shadowing, images will deviate from this linear subspace. Rather than explicitly modeling this deviation, we project the image into a subspace in a manner which discounts those regions of the face with large deviation. Our projection method is based on Fisher's Linear Discriminant and produces well separated classes in a low-dimensional subspace even under severe variation in lighting and facial expressions. The Eigenface technique, another method based on linearly projecting the image space to a low dimensional subspace, has similar computational requirements. Yet, extensive experimental results demonstrate that the proposed "Fisherface" method has error rates that are significantly lower than those of the Eigenface technique when tested on the same database.

1 Introduction

Within the last several years, numerous algorithms have been proposed for face recognition; for detailed surveys see [4, 24]. While much progress has been made toward recognizing faces under small variations in lighting, facial expression and pose, reliable techniques for recognition under more extreme variations have proven elusive.

In this paper we outline a new approach for face recognition – one that is insensitive to extreme variations in lighting and facial expressions. Note that lighting variability includes not only intensity, but also direction and number of light sources. As seen in Fig. 1, the same person, with the same facial expression, seen from the same viewpoint can appear dramatically different when light sources illuminate the face from different directions.

Our approach to face recognition exploits two observations:
1. For a Lambertian surface without self-shadowing, all of the images of a particular face from a fixed viewpoint will lie in a 3-D linear subspace of the high-dimensional image space [25].

* J. Hespanha was supported by NSF Grant ECS-9206021, AFOSR Grant F49620-94-1-0181, and ARO Grant DAAH04-95-1-0114.
** D. Kriegman was supported by NSF under an NYI, IRI-9257990 and by ONR N00014-93-1-0305

| Subset 1 | Subset 2 | Subset 3 | Subset 4 | Subset 5 |

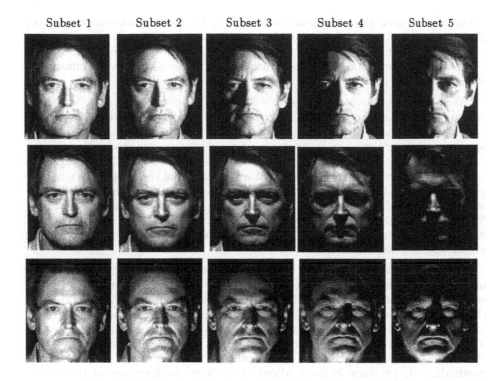

Fig. 1. The same person seen under varying lighting conditions can appear dramatically different. These images are taken from the Harvard database which is described in Section 3.1.

2. Because of expressions, regions of self-shadowing and specularity, the above observation does not exactly apply to faces. In practice, certain regions of the face may have a variability from image to image that often deviates drastically from the linear subspace and, consequently, are less reliable for recognition.

We make use of these observations by finding a linear projection of the faces from the high-dimensional image space to a significantly lower dimensional feature space which is insensitive both to variation in lighting direction and facial expression. We choose projection directions that are nearly orthogonal to the within-class scatter, projecting away variations in lighting and facial expression while maintaining discriminability. Our method Fisherfaces, a derivative of Fisher's Linear Discriminant (FLD) [9, 10], maximizes the ratio of between-class scatter to that of within-class scatter.

The Eigenface method is also based on linearly projecting the image space to a low dimensional feature space [27, 28, 29]. However, the Eigenface method, which uses principal components analysis (PCA) for dimensionality reduction, yields projection directions that maximize the total scatter across all classes, i.e. all images of all faces. In choosing the projection which maximizes total scatter, PCA retains some of the unwanted variations due to lighting and facial expression. As illustrated in Fig. 1 and stated by Moses, Adini, and Ullman, "the variations between the images of the same face due to illumination and viewing direction are almost always larger than image variations due to change in face identity" [21]. Thus, while the PCA projections are optimal for reconstruction from a low dimensional basis, they may not be optimal from a discrimination standpoint.

We should point out that Fisher's Linear Discriminant [10] is a "classical" technique in pattern recognition [9] that was developed by Robert Fisher in 1936 for taxonomic classification. Depending upon the features being used, it has been applied in different ways in computer vision and even in face recognition. Cheng *et al.* presented a method that used Fisher's discriminator for face recognition where features were obtained by a polar quantization of the shape [6]. Contemporaneous with our work [15], Cui, Swets, and Weng applied Fisher's discriminator (using different terminology, they call it the Most Discriminating Feature – MDF) in a method for recognizing hand gestures [8]. Though no implementation is reported, they also suggest that the method can be applied to face recognition under variable illumination.

In the sections to follow, we will compare four methods for face recognition under variation in lighting and facial expression: correlation, a variant of the linear subspace method suggested by [25], the Eigenface method [27, 28, 29], and the Fisherface method developed here. The comparisons are done on a database of 500 images created externally by Hallinan [13, 14] and a database of 176 images created at Yale. The results of the tests on both databases shows that the Fisherface method performs significantly better than any of the other three methods. Yet, no claim is made about the relative performance of these algorithms on much larger databases.

We should also point out that we have made no attempt to deal with variation in pose. An appearance-based method such as ours can be easily extended to handle limited pose variation using either a multiple-view representation such as Pentland, Moghaddam, and Starner's View-based Eigenspace [23] or Murase and Nayar's Appearance Manifolds [22]. Other approaches to face recognition that accommodate pose variation include [2, 11]. Furthermore, we assume that the face has been located and aligned within the image, as there are numerous methods for finding faces in scenes [5, 7, 17, 18, 19, 20, 28].

2 Methods

The problem can be simply stated: Given a set of face images labeled with the person's identity (*the learning set*) and an unlabeled set of face images from the same group of people (*the test set*), identify the name of each person in the test images.

In this section, we examine four pattern classification techniques for solving the face recognition problem, comparing methods that have become quite popular in the face recognition literature, i.e. correlation [3] and Eigenface methods [27, 28, 29], with alternative methods developed by the authors. We approach this problem within the pattern classification paradigm, considering each of the pixel values in a sample image as a coordinate in a high-dimensional space (*the image space*).

2.1 Correlation

Perhaps, the simplest classification scheme is a nearest neighbor classifier in the image space [3]. Under this scheme, an image in the test set is recognized by assigning to it the label of the closest point in the learning set, where distances are measured in the image space. If all of the images have been normalized to be zero mean and have unit variance, then this procedure is equivalent to choosing the image in the learning set that best correlates with the test image. Because of the normalization process, the result is independent of light source intensity and the effects of a video camera's automatic gain control.

This procedure, which will subsequently be referred to as correlation, has several well-known disadvantages. First, if the images in the learning set and test set are gathered under varying lighting conditions, then the corresponding points in the image space will not be tightly clustered. So in order for this method to work reliably under variations in lighting, we would need a learning set which densely sampled the continuum of possible lighting conditions. Second, correlation is computationally expensive. For recognition, we must correlate the image of the test face with each image in the learning set; in an effort to reduce the computation time, implementors [12] of the algorithm described in [3] developed special purpose VLSI hardware. Third, it requires large amounts of storage – the learning set must contain numerous images of each person.

2.2 Eigenfaces

As correlation methods are computationally expensive and require great amounts of storage, it is natural to pursue dimensionality reduction schemes. A technique now commonly used for dimensionality reduction in computer vision – particularly in face recognition – is principal components analysis (PCA) [13, 22, 27, 28, 29]. PCA techniques, also known as Karhunen-Loeve methods, choose a dimensionality reducing linear projection that maximizes the scatter of all projected samples.

More formally, let us consider a set of N sample images $\{x_1, x_2, \ldots, x_N\}$ taking values in an n-dimensional feature space, and assume that each image belongs to one of c classes $\{\chi_1, \chi_2, \ldots, \chi_c\}$. Let us also consider a linear transformation mapping the original n-dimensional feature space into an m-dimensional feature space, where $m < n$. Denoting by $W \in \mathbb{R}^{n \times m}$ a matrix with orthonormal columns, the new feature vectors $y_k \in \mathbb{R}^m$ are defined by the following linear transformation:

$$y_k = W^T x_k, \qquad k = 1, 2, \ldots, N.$$

Let the total scatter matrix S_T be defined as

$$S_T = \sum_{k=1}^{N} (x_k - \mu)(x_k - \mu)^T$$

where $\mu \in \mathbb{R}^n$ is the mean image of all samples.

Note that after applying the linear transformation, the scatter of the transformed feature vectors $\{y_1, y_2, \ldots, y_N\}$ is $W^T S W$. In PCA, the optimal projection W_{opt} is chosen to maximize the determinant of the total scatter matrix of the projected samples, i.e.

$$W_{opt} = \arg\max_{W} |W^T S_T W| = [w_1 \ w_2 \ \ldots \ w_m] \qquad (1)$$

where $\{w_i \mid i = 1, 2, \ldots, m\}$ is the set of n-dimensional eigenvectors of S_T corresponding to the set of decreasing eigenvalues. Since these eigenvectors have the same dimension as the original images, they are referred to as Eigenpictures in [27] and Eigenfaces in [28, 29].

A drawback of this approach is that the scatter being maximized is not only due to the between-class scatter that is useful for classification, but also the within-class scatter that, for classification purposes, is unwanted information. Recall the comment by Moses, Adini and Ullman [21]: Much of the variation from one image to the next is due to illumination changes. Thus if PCA is presented with images of faces under varying illumination, the projection matrix W_{opt} will contain principal components (i.e. Eigenfaces) which retain, in the projected feature space, the variation due lighting. Consequently, the points in projected space will not be well clustered, and worse, the classes may be smeared together.

It has been suggested that by throwing out the first several principal components, the variation due to lighting is reduced. The hope is that if the first principal components capture the variation due to lighting, then better clustering of projected samples is achieved by ignoring them. Yet it is unlikely that the first several principal components correspond solely to variation in lighting; as a consequence, information that is useful for discrimination may be lost.

2.3 Linear Subspaces

Both correlation and the Eigenface method are expected to suffer under variation in lighting direction. Neither method exploits the observation that for a Lambertian surface without self-shadowing, the images of a particular face lie in a 3-D linear subspace.

Consider a point p in a Lambertian surface and a collimated light source characterized by a vector $s \in \mathbb{R}^3$, such that the direction of s gives the direction of the light rays and $\|s\|$ gives the intensity of the light source. The irradiance at the point p is given by

$$E(p) = a(p) < n(p), s > \tag{2}$$

where $n(p)$ is the unit inward normal vector to the surface at the point p, and $a(p)$ is the albedo of the surface at p [16]. This shows that the irradiance at the point p, and hence the gray level seen by a camera, is linear on $s \in \mathbb{R}^3$. Therefore, in the absence of self-shadowing, given three images of a Lambertian surface from the same viewpoint taken under three known, linearly independent light source directions, the albedo and surface normal can be recovered; this is the well known method of photometric stereo [26, 30]. Alternatively, one can reconstruct the image of the surface under an arbitrary lighting direction by a linear combination of the three original images, see [25].

For classification, this fact has great importance: It shows that for a fixed viewpoint, all images of a Lambertian surface lie in a 3-D linear subspace embedded in the high-dimensional image space. This observation suggests a simple classification algorithm to recognize Lambertian surfaces – invariant under lighting conditions.

For each face, use three or more images taken under different lighting directions to construct a 3-D basis for the linear subspace. Note that the three basis vectors have the same dimensionality as the training images and can be thought of as basis images. To perform recognition, we simply compute the distance of a new image to each linear subspace and choose the face corresponding to the shortest distance. We call this recognition scheme the Linear Subspace method. We should point out that this method is a variant of the photometric alignment method proposed in [25] and, although it is not yet in press, the Linear Subspace method can be thought of as special case of the more elaborate recognition method described in [14].

If there is no noise or self-shadowing, the Linear Subspace algorithm would achieve error free classification under *any* lighting conditions, provided the surfaces obey the Lambertian reflectance model. Nevertheless, there are several compelling reasons to look elsewhere. First, due to self-shadowing, specularities, and facial expressions, some regions of the face have variability that does not agree with the linear subspace model. Given enough images of faces, we should be able to learn which regions are good for recognition and which regions are not. Second, to recognize a test image we must measure the distance to the linear subspace for each person. While this in an improvement over a correlation scheme that needs a large number of images for each class, it is still too computationally expensive. Finally, from a storage standpoint, the Linear Subspace algorithm must keep three images in memory for every person.

2.4 Fisherfaces

The Linear Subspace algorithm takes advantage of the fact that under ideal conditions the classes are linearly separable. Yet, one can perform dimensionality reduction using linear projection and still preserve linear separability; error free classification under any lighting conditions is still possible in the lower dimensional feature space using linear decision boundaries. This is a strong argument in favor of using linear methods for dimensionality reduction in the face recognition problem, at least when one seeks insensitivity to lighting conditions.

Here we argue that using class specific linear methods for dimensionality reduction and simple classifiers in the reduced feature space one gets better recognition rates in substantially less time than with the Linear Subspace method. Since the learning set is labeled, it makes sense to use this information to build a more reliable method for reducing the dimensionality of the feature space. Fisher's Linear Discriminant (FLD) [10] is an example of a *class specific method*, in the sense that it tries to "shape" the scatter in order to make it more reliable for classification. This method selects W in such a way that the ratio of the between-class scatter and the within-class scatter is maximized. Let the between-class scatter matrix be defined as

$$S_B = \sum_{i=1}^{c} |\chi_i| \, (\mu_i - \mu)(\mu_i - \mu)^T$$

and the within-class scatter matrix be defined as

$$S_W = \sum_{i=1}^{c} \sum_{x_k \in \chi_i} (x_k - \mu_i)(x_k - \mu_i)^T$$

where μ_i is the mean image of class χ_i, and $|\chi_i|$ is the number of samples in class χ_i. If S_W is nonsingular, the optimal projection W_{opt} is chosen as that which maximizes the ratio of the determinant of the between-class scatter matrix of

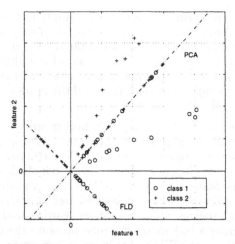

Fig. 2. A comparison of principal component analysis (PCA) and Fisher's linear discriminant (FLD) for a two class problem where data for each class lies near a linear subspace.

the projected samples to the determinant of the within-class scatter matrix of the projected samples, i.e.

$$W_{opt} = \arg\max_W \frac{|W^T S_B W|}{|W^T S_W W|} = [\,w_1 \; w_2 \ldots w_m\,] \tag{3}$$

where $\{w_i \mid i = 1, 2, \ldots, m\}$ is the set of generalized eigenvectors of S_B and S_W corresponding to set of decreasing generalized eigenvalues $\{\lambda_i \mid i = 1, 2, \ldots, m\}$, i.e.

$$S_B w_i = \lambda_i S_W w_i\,, \qquad i = 1, 2, \ldots, m.$$

Note that an upper bound on m is $c - 1$ where c is the number of classes. See [9].

To illustrate the benefits of the class specific linear projections, we constructed a low dimensional analogue to the classification problem in which the samples from each class lie near a linear subspace. Figure 2 is a comparison of PCA and FLD for a two-class problem in which the samples from each class are randomly perturbed in a direction perpendicular to the linear subspace. For this example $N = 20$, $n = 2$, and $m = 1$. So the samples from each class lie near a line in the 2-D feature space. Both PCA and FLD have been used to project the points from 2-D down to 1-D. Comparing the two projections in the figure, PCA actually smears the classes together so that they are no longer linearly separable in the projected space. It is clear that although PCA achieves larger total scatter, FLD achieves greater between-class scatter, and consequently classification becomes easier.

In the face recognition problem one is confronted with the difficulty that the within-class scatter matrix $S_W \in \mathbb{R}^{n \times n}$ is always singular. This stems from the fact that the rank of S_W is less than $N - c$, and in general, the number of pixels in each image (n) is much larger than the number of images in the learning set (N). This means that it is possible to chose the matrix W such that the within-class scatter of the projected samples can be made exactly zero.

In order to overcome the complication of a singular S_W, we propose an alternative to the criterion in Eq. 3. This method, which we call Fisherfaces, avoids this problem by projecting the image set to a lower dimensional space so that the resulting within-class scatter matrix S_W is nonsingular. This is achieved by using PCA to reduce the dimension of the feature space to $N - c$ and then, applying the standard FLD defined by Eq. 3 to reduce the dimension to $c - 1$. More formally, W_{opt} is given by

$$W_{opt} = W_{fld} W_{pca} \tag{4}$$

where

$$W_{pca} = \arg \max_W |W^T S_T W|$$

$$W_{fld} = \arg \max_W \frac{|W^T W_{pca}^T S_B W_{pca} W|}{|W^T W_{pca}^T S_W W_{pca} W|}.$$

Note that in computing W_{pca} we have thrown away only the smallest c principal components.

There are certainly other ways of reducing the within-class scatter while preserving between-class scatter. For example, a second method which we are currently investigating chooses W to maximize the between-class scatter of the projected samples after having first reduced the within-class scatter. Taken to an extreme, we can maximize the between-class scatter of the projected samples subject to the constraint that the within-class scatter is zero, i.e.

$$W_{opt} = \arg \max_{W \in \mathcal{W}} |W^T S_B W| \tag{5}$$

where $\mathcal{W}$ is the set of $n \times m$ matrices contained in the kernel of S_W.

3 Experimental Results

In this section we will present and discuss each of the aforementioned face recognition techniques using two different databases. Because of the specific hypotheses that we wanted to test about the relative performance of the considered algorithms, many of the standard databases were inappropriate. So we have used a database from the Harvard Robotics Laboratory in which lighting has been systematically varied. Secondly, we have constructed a database at Yale that includes variation in both facial expression and lighting.[3]

3.1 Variation in Lighting

The first experiment was designed to test the hypothesis that under variable illumination, face recognition algorithms will perform better if they exploit the fact that images of a Lambertian surface lie in a linear subspace. More specifically, the recognition error rates for all four algorithms described in Section 2 will be compared using an image database constructed by Hallinan at the Harvard Robotics Laboratory [13, 14]. In each image in this database, a subject held his/her head steady while being illuminated by a dominant light source. The space of light source directions, which can be parameterized by spherical angles, was then sampled in 15° increments. From a subset of 225 images of five people in this database, we extracted five subsets to quantify the effects of varying lighting. Sample images from each subset are shown in Fig. 1.

---

[3] The Yale database is available by anonymous ftp from daneel.eng.yale.edu.

53

Subset 1 contains 30 images for which both of the longitudinal and latitudinal angles of light source direction are within 15° of the camera axis.

Subset 2 contains 45 images for which the greater of the longitudinal and latitudinal angles of light source direction are 30° from the camera axis.

Subset 3 contains 65 images for which the greater of the longitudinal and latitudinal angles of light source direction are 45° from the camera axis.

Subset 4 contains 85 images for which the greater of the longitudinal and latitudinal angles of light source direction are 60° from the camera axis.

Subset 5 contains 105 images for which the greater of the longitudinal and latitudinal angles of light source direction are 75° from the camera axis.

For all experiments, classification was performed using a nearest neighbor classifier. All training images of an individual were projected into the feature space. The images were cropped within the face so that the contour of the head was excluded.[4] For the Eigenface and correlation tests, the images were normalized to have zero mean and unit variance, as this improved the performance of these methods. For the Eigenface method, results are shown when ten principal components are used. Since it has been suggested that the first three principal components are primarily due to lighting variation and that recognition rates can be improved by eliminating them, error rates are also presented using principal components four through thirteen. Since there are 30 images in the training set, correlation is equivalent to the Eigenface method using 29 principal components.

We performed two experiments on the Harvard Database: extrapolation and interpolation. In the extrapolation experiment, each method was trained on samples from Subset 1 and then tested using samples from Subsets 1, 2 and 3.[5] Figure 3 shows the result from this experiment.

In the interpolation experiment, each method was trained on Subsets 1 and 5 and then tested the methods on Subsets 2, 3 and 4. Figure 4 shows the result from this experiment.

These two experiments reveal a number of interesting points:

1. All of the algorithms perform perfectly when lighting is nearly frontal. However as lighting is moved off axis, there is a significant performance difference between the two class-specific methods and the Eigenface method.
2. It has also been noted that the Eigenface method is equivalent to correlation when the number of Eigenfaces equals the size of the training set [22], and since performance increases with the dimension of the Eigenspace, the Eigenface method should do no better than correlation [3]. This is empirically demonstrated as well.
3. In the Eigenface method, removing the first three principal components results in better performance under variable lighting conditions.
4. While the Linear Subspace method has error rates that are competitive with the Fisherface method, it requires storing more than three times as much information and takes three times as long.

---

[4] We have observed that the error rates are reduced for all methods when the contour is included and the subject is in front of a uniform background. However, all methods performed worse when the background varies.

[5] To test the methods with an image from Subset 1, that image was removed from the training set, i.e. we employed the "leaving-one-out" strategy [9].

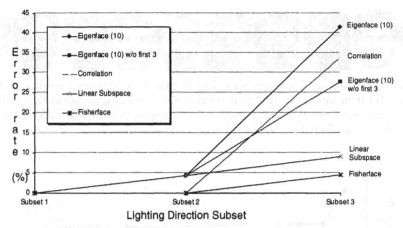

Fig. 3. Extrapolation: When each of the methods is trained on images with near frontal illumination (Subset 1), the graph and corresponding table show their relative performance under extreme light source conditions.

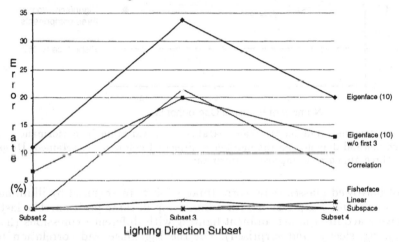

Fig. 4. Interpolation: When each of the methods is trained on images from both near frontal and extreme lighting (Subsets 1 and 5), the graph and corresponding table show the methods' relative performance under intermediate lighting conditions.

5. The Fisherface method had error rates many times lower than the Eigenface method and required less computation time.

3.2 Variation in Facial Expression, Eyewear, and Lighting

Using a second database constructed at the Yale Vision and Robotics Lab, we constructed tests to determine how the methods compared under a different range of conditions. For sixteen subjects, ten images were acquired during one session in front of a simple background. Subjects included females and males (some with facial hair), and some wore glasses. Figure 5 shows ten images of one subject. The first image was taken under ambient lighting in a neutral facial expression, and the person wore his/her glasses when appropriate. In the second image, the glasses were removed if glasses were not normally worn; otherwise,

Fig. 5. The Yale database contains 160 frontal face images covering sixteen individuals taken under ten different conditions: A normal image under ambient lighting, one with or without glasses, three images taken with different point light sources, and five different facial expressions.

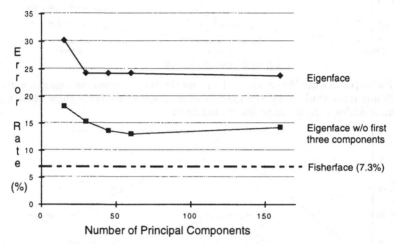

Fig. 6. As demonstrated on the Yale Database, the variation in performance of the Eigenface method depends on the number of principal components retained. Dropping the first three appears to improve performance.

a pair of borrowed glasses were worn. Images 3-5 were acquired by illuminating the face in a neutral expression with a Luxolamp in three position. The last five images were acquired under ambient lighting with different expressions (happy, sad, winking, sleepy, and surprised). For the Eigenface and correlation tests, the images were normalized to have zero mean and unit variance, as this improved the performance of these methods. The images were manually centered and cropped to two different scales: The larger images included the *full face* and part of the background while the *closely cropped* ones included internal structures such as the brow, eyes, nose, mouth and chin but did not extend to the occluding contour.

In this test, error rates were determined by the "leaving-one-out" strategy [9]: To classify an image of a person, that image was removed from the data set and the dimensionality reduction matrix W was computed. All images in the database, excluding the test image, were then projected down into the reduced space to be used for classification. Recognition was performed using a nearest neighbor classier. Note that for this test, each person in the learning set is represented by the projection of ten images, except for the test person who is represented by only nine.

In general, the performance of the Eigenface method varies with the number of principal components. So, before comparing the Linear Subspace and Fisher-

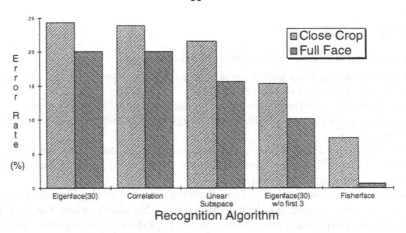

Fig. 7. The graph and corresponding table show the relative performance of the algorithms when applied to the Yale database which contains variation in facial expression and lighting.

face methods with the Eigenface method, we first performed an experiment to determine the number of principal components that results in the lowest error rate. Figure 6 shows a plot of error rate vs. the number of principal components when the initial three principal components were retained and when they were dropped.

The relative performance of the algorithms is self evident in Fig. 7. The Fisherface method had error rates that were better than half that of any other method. It seems that the Fisherface method learns the set of projections which performs well over a range of lighting variation, facial expression variation, and presence of glasses.

Note that the Linear Subspace method faired comparatively worse in this experiment than in the lighting experiments in the previous section. Because of variation in facial expression, the images no longer lie in a linear subspace. Since the Fisherface method tends to discount those portions of the image that are not significant for recognizing an individual, the resulting projections W tend to mask the regions of the face that are highly variable. For example, the area around the mouth is discounted since it varies quite a bit for different facial expressions. On the other hand, the nose, cheeks and brow are stable over the within-class variation and are more significant for recognition. Thus, we conjecture that Fisherface methods, which tend to reduce within-class scatter for all classes, should produce projection directions that are good for recognizing other faces besides the training set. Thus, once the projection directions are determined, each person can be modeled by a single image.

All of the algorithms performed better on the images of the full face. Note that there is a dramatic improvement in the Fisherface method where the error rate was reduced from 7.3% to 0.6%. When the method is trained on the entire face, the pixels corresponding to the occluding contour of the face are chosen as good features for discriminating between individuals. i.e., the overall shape of the face is a powerful feature in face identification. As a practical note however, it is expected that recognition rates would have been much lower for the full face images if the background or hairstyles had varied and may even have been worse than the closely cropped images.

4 Conclusion

The experiments suggest a number of conclusions:

1. All methods perform well if presented with an image in the test set which is similar to an image in the training set.
2. The Fisherface method appears to be the best at extrapolating and interpolating over variation in lighting, although the Linear Subspace method is a close second.
3. Removing the initial three principal components does improve the performance of the Eigenface method in the presence of lighting variation, but does not alleviate the problem.
4. In the limit as more principal components are used in the Eigenface method, performance approaches that of correlation. Similarly, when the first three principal components have been removed, performance improves as the dimensionality of the feature space is increased. Note however, that performance seems to level off at about 45 principal components. Sirovitch and Kirby found a similar point of diminishing returns when using Eigenfaces to represent face images [27].
5. The Fisherface method appears to be the best at simultaneously handling variation in lighting and expression. As expected, the Linear Subspace method suffers when confronted with variation in facial expression.

Even with this extensive experimentation, interesting questions remain: How well does the Fisherface method extend to large databases. Can variation in lighting conditions be accommodated if some of the individuals are only observed under one lighting condition? i.e., how can information about the class of faces be exploited?

Additionally, current face detection methods are likely to break down under extreme lighting conditions such as Subsets 4 and 5 in Fig. 1, and so new detection methods will be needed to support this algorithm. Finally, when shadowing dominates, performance degrades for all of the presented recognition methods, and techniques that either model or mask the shadowed regions may be needed. We are currently investigating models for representing the set of images of an object under *all possible* illumination conditions; details will appear in [1].

Acknowledgements We would like to thank Peter Hallinan for providing the Harvard Database, and Alan Yuille and David Mumford for many useful discussions.

References

1. P. Belhumeur and D. Kriegman. What is the set of images of an object under all possible lighting conditions? In *IEEE Proc. Conf. Computer Vision and Pattern Recognition*, 1996.
2. D. Beymer. Face recognition under varying pose. In *Proc. Conf. Computer Vision and Pattern Recognition*, pages 756–761, 1994.
3. R. Brunelli and T. Poggio. Face recognition: Features vs templates. *IEEE Trans. Pattern Anal. Mach. Intelligence*, 15(10):1042–1053, 1993.
4. R. Chellappa, C. Wilson, and S. Sirohey. Human and machine recognition of faces: A survey. *Proceedings of the IEEE*, 83(5):705–740, 1995.
5. Q. Chen, H. Wu, and M. Yachida. Face detection by fuzzy pattern matching. In *Int. Conf. on Computer Vision*, pages 591–596, 1995.

6. Y. Cheng, K. Liu, J. Yang, Y. Zhuang, and N. Gu. Human face recognition method based on the statistical model of small sample size. In *SPIE Proc.: Intelligent Robots and Computer Vision X: Algorithms and Techn.*, pages 85–95, 1991.
7. I. Craw, D. Tock, and A. Bennet. Finding face features. In *Proc. European Conf. on Computer Vision*, pages 92–96, 1992.
8. Y. Cui, D. Swets, and J. Weng. Learning-based hand sign recognition using SHOSLIF-M. In *Int. Conf. on Computer Vision*, pages 631–636, 1995.
9. R. Duda and P. Hart. *Pattern Classification and Scene Analysis*. Wiley, New York, 1973.
10. R. Fisher. The use of multiple measures in taxonomic problems. *Ann. Eugenics*, 7:179–188, 1936.
11. A. Gee and R. Cipolla. Determining the gaze of faces in images. *Image and Vision Computing*, 12:639–648, 1994.
12. J. Gilbert and W. Yang. A Real–Time Face Recognition System Using Custom VLSI Hardware. In *Proceedings of IEEE Workshop on Computer Architectures for Machine Perception*, pages 58–66, 1993.
13. P. Hallinan. A low-dimensional representation of human faces for arbitrary lighting conditions. In *Proc. IEEE Conf. on Comp. Vision and Patt. Recog.*, pages 995–999, 1994.
14. P. Hallinan. *A Deformable Model for Face Recognition Under Arbitrary Lighting Conditions*. PhD thesis, Harvard University, 1995.
15. J. Hespanha, P. Belhumeur, and D. Kriegman. Eigenfaces vs. Fisherfaces: Recognition using class specific linear projection. Center for Systems Science 9506, Yale University, PO Box 208267, New Haven, CT 06520, May 1995.
16. B. Horn. *Computer Vision*. MIT Press, Cambridge, Mass., 1986.
17. A. Lanitis, C. Taylor, and T. Cootes. A unified approach to coding and interpreting face images. In *Int. Conf. on Computer Vision*, pages 368–373, 1995.
18. T. Leung, M. Burl, and P. Perona. Finding faces in cluttered scenes using labeled random graph matching. In *Int. Conf. on Computer Vision*, pages 637–644, 1995.
19. K. Matsuno, C. Lee, S. Kimura, and S. Tsuji. Automatic recognition of human facial expressions. In *Int. Conf. on Computer Vision*, pages 352–359, 1995.
20. Moghaddam and Pentland. Probabilistic visual learning for object detection. In *Int. Conf. on Computer Vision*, pages 786–793, 1995.
21. Y. Moses, Y. Adini, and S. Ullman. Face recognition: The problem of compensating for changes in illumination direction. In *European Conf. on Computer Vision*, pages 286–296, 1994.
22. H. Murase and S. Nayar. Visual learning and recognition of 3-D objects from appearence. *Int. J. Computer Vision*, 14(5–24), 1995.
23. A. Pentland, B. Moghaddam, and Starner. View-based and modular eigenspaces for face recognition. In *Proc. Conf. Computer Vision and Pattern Recognition*, pages 84–91, 1994.
24. A. Samal and P. Iyengar. Automatic recognition and analysis of human faces and facial expressions: A survey. *Pattern Recognition*, 25:65–77, 1992.
25. A. Shashua. *Geometry and Photometry in 3D Visual Recognition*. PhD thesis, MIT, 1992.
26. W. Silver. *Determining Shape and Reflectance Using Multiple Images*. PhD thesis, MIT, Cambridge, MA, 1980.
27. Sirovitch, L. and Kirby, M. Low-dimensional procedure for the characterization of human faces. *J. Optical Soc. of America A*, 2:519–524, 1987.
28. M. Turk and A. Pentland. Eigenfaces for recognition. *J. of Cognitive Neuroscience*, 3(1), 1991.
29. M. Turk and A. Pentland. Face recognition using eigenfaces. In *Proc. IEEE Conf. on Comp. Vision and Patt. Recog.*, pages 586–591, 1991.
30. R. Woodham. Analysing images of curved surfaces. *Artificial Intelligence*, 17:117–140, 1981.

Rapid Object Indexing and Recognition Using Enhanced Geometric Hashing

Bart Lamiroy and Patrick Gros*

GRAVIR – IMAG & INRIA Rhône-Alpes,
46 avenue Félix Viallet, F-38031 Grenoble, France

Abstract. We address the problem of 3D object recognition from a single 2D image using a model database. We develop a new method called *enhanced geometric hashing*. This approach allows us to solve for the indexing and the matching problem in one pass with linear complexity. Use of quasi-invariants allows us to index images in a new type of geometric hashing table. They include topological information of the observed objects inducing a high numerical stability.

We also introduce a more robust Hough transform based voting method, and thus obtain a fast and robust recognition algorithm that allows us to index images by their content. The method recognizes objects in the presence of noise and partial occlusion and we show that 3D objects can be recognized from any viewpoint if only a limited number of key views are available in the model database.

1 Introduction

In this paper we address the problem of recognizing 3D objects from a single 2D image. When addressing this problem and when the objects to be recognized are stored as models in a database, the questions *"Which one of the models in the database best fits the unknown image?"* and *"What is the relationship between the image and the proposed model(s)?"* need to be answered. The first question can be considered as an indexing problem, the second as a matching problem. Indexing consists in calculating a key from a set of image features. Normally similar keys correspond to similar images while different keys correspond to different images. The complexity of comparing images is reduced to the comparison of their simpler indexing keys. Comparing two keys gives a measure of global similarity between the images, but does not establish a one-to-one correspondence between the image features they were calculated from [12]. Although indexing techniques are fast, the lack of quantitative correspondences, makes this approach unsuited for recognizing multiple or partially occluded objects in a noisy image.

Since the indexing result only establishes a weak link between the unknown image and a model, a second operation is needed to determine the quantitative relationship between the two. Common used relationships are the exact location

---

* This work was performed within a joint research programme between (in alphabetical order) CNRS, INPG, INRIA, UJF

of the model in the image, the viewpoint form which the image of the model was taken and/or aspect of the model the unknown image corresponds to. They can be calculated by solving the matching problem. I.e. establishing a correspondence between the image and model features. Solving the matching problem, however, is an inherently combinatorial problem, since any feature of the unknown image can, a priori, be matched to any feature in the model.

Several authors have proposed ways to solve the recognition problem. Systems like the one proposed in [4] use a prediction-verification approach. They rely on a rigidity constraint and their performance generally depends on the optimisation of the hypothesis tree exploration. They potentially evaluate matches between every feature of every model and every feature of the unknown image. Lamdan and Wolfson [11] first suggested the geometric hashing technique, which was later extended in several ways [8, 5, 6, 1]. It assumes rigidity as well, but the search is implemented using hashing. This reduced a part the complexity. Its main advantage is that it is independent of the number of models, although it still potentially matches every feature of every model. It has several other drawbacks however. Other approaches include subgraph matching, but are of a high complexity because the rigidity constraint is relaxed. They rely on topological properties which are not robust in the presence of noise. As a result, hashing techniques were developed using topological properties [10], without much success. Stochastic approaches containing geometric elements [2, 14], or based on Markov models generally demand very strong modeling, and lack flexibility when constructing the model database. Signal based recognition systems [13] usually give a yes-no response, and do not allow a direct implementation of a matching algorithm.

Our approach builds on geometric hashing type of techniques. Classical geometric hashing solves for the indexing problem, but not for the matching problem however. The new method that we propose solves simultaneously for indexing and matching. It is therefore able to rapidly select a few candidates in the database and establish a feature to feature correspondence between the image and the related candidates. It is able to deal with noisy images and with partially occluded objects. Moreover, our method has reduced complexity with respect to other approaches such as tree search, geometric hashing, subgraph matching, etc.

The method, called *enhanced geometric hashing*, introduces a way of indexing a richer set of geometric invariants that have a stronger topological meaning, and considerably reduce the complexity of the indexing problem. They serve as a key to a multi-dimensional hash table, and allow a vote in a Hough space. The use of this Hough transform based vote renders our system robust, even when the number of collisions in the hash table bins is high, or when models in the model base present a high similarity.

In the following section we shall describe the background of our approach. We shall explain the different choices we made and situate them in the light of previous work. Section 3 gives a brief overview of our recognition algorithm. Sections 4, 5 and 6 detail the different parts of our algorithm while Sect. 7 contains our experimental results. Finally, we shall discuss the interest of our approach, as well as future extensions in the last section.

2 Background and Justification

Our aim is to develop a recognition system based on the matching technique proposed in [8, 9]. We shall further detail this technique in section 4. It is a 2D-2D matching algorithm that extensively uses quasi-invariants [3]. This induces that our recognition and indexing algorithm will be restricted to 2D-2D matching and recognition. We can easily introduce 3D information however, by adding a layer to our model database[2]. Instead of directly interpreting 3D information, we can store different 2D aspects of a 3D model, and do the recognition on the aspects. Once the image has been identified, it is easy to backtrack to the 3D information.

Since our base method uses geometric invariants to model the images, and since we want to index these images in a model base, the geometric hashing algorithm by Lamdan and Wolfson [11] seems an appropriate choice.

The advantage of this method is that it develops a way of accessing an image database with a complexity $\mathcal{O}(n)$ that depends uniquely on the size n of the unknown image. Multiple inconvenients exist however. They are the main reason for which we developed a new method we call *enhanced geometric hashing*. In our method we kept the idea of an invariant indexed hash table and the principle of voting for one or several models. The similarity stops there however.

The classical geometric hashing has proved to contain several weaknesses. Our approach solves them on many points.

- Grimson, Huttenlocher and Jacobs showed in [7] that the 4-point affine invariant causes fundamental mismatches due to the impossibility to incorporate a correct error model. Another account for false matches is that the 4-point invariant, used in the classical geometric hashing method is far too generic, and thus, by taking every possible configuration all topological information of the objects in the image is lost. We reduce this incertainty and loss of topological information by using the quasi-invariants proposed in [3, 9]. Instead of just extracting interest points, and combining them to form affine frames, we use connected segments to calculate the needed quasi-invariants. The segments are more robust to noise and the connectedness constraint reduces the possibility of false matches, since it is based on a topological reality in the image.
- The new type of quasi-invariants and the connectedness constraint add another advantage to our method. For a given image containing n interest points, the number of invariants calculated by Lamdan and Wolfson is about $\mathcal{O}(n^4)$. In our method, the number of quasi-invariants mainly varies linearly

---

[2] Model database will also be referred to as model base.

with the number of interest points and lines. We therefore considerably reduce the total complexity of the problem.

- It is known that the performance of the classical geometric hashing algorithm decreases when the number of collisions between models increases. The main reason for this is that the voting process only takes into account the quantitative information the possible matches offer. There is no qualitative measure that would allow votes to be classified as coherent or incoherent. We introduce a measure of coherence between votes by using the apparent motion between the unknown image and the possible corresponding model that is defined by the match (cf. Sec. 4). Basically, the matched quasi-invariants define a geometric transform. This transform corresponds to an n-dimensional point in a corresponding Hough space. Coherent votes will form a cluster, while incoherent votes will be spread out throughout the whole transformation space. This enhanced voting method will allow us a greater robustness during the voting process.
- Our quasi-invariants are particularly well suited to describe aspects of 3D objects, since they vary in a controllable way with a change of viewpoint. It is therefore easy to model a 3D object by its 2D aspects. Storing these aspects in our model base will allow us to identify an object from any viewpoint.

3 The Recognition Algorithm

In this section we present the global lay-out of our recognition system. We shall briefly address its different parts. More detailed information will be given in the following sections. From now on, we shall refer to the images in the model base as "models", while the unknown images to recognize will be simply referred to as "images". It is to be noted, however, that there is no a priori structural difference between the two. The classification only reflects the fact that "models" correspond to what we know, and what is stored in the model base, while "images" correspond to what is unknown what we want to identify.

The recognition algorithm can be separated in four steps. The first step can be considered "off-line". This does not mean, however, that, once the model base constructed it cannot be modified. New models can be added without affecting the performances of the recognition algorithm, and without modification of the model base structure.

1. **Setting up the Model Base** and Vote Space. For all the models we want to be able to recognize, we proceed in the following way: we extract interest points and segments from the model image. These points and segments will provide the configurations needed for the calculation of the quasi-invariants. Once these invariants calculated we label them with the model from which they originated, and add them to the model base. Each model added to the model base also defines a transformation space, ready to receive votes.
2. **Extracting the necessary invariants** from the unknown image.
3. **Confronting Invariants** with the Model Base. We calculate the needed invariants from an unknown image, and need to find the models which are

likely to correspond to it. As in the classical geometric hashing technique, the complexity of this operation is completely independent of the number of models or invariants present in the current model base. Every invariant of the image having been confronted to the model base, we obtain an output list containing the possible matches between image and the model features. The information contained in one element the output list consists of an invariant of the unknown image I_i, its possible corresponding invariant in one of the models J_{jm}, and the model to which J_{jm} belongs: M_m.

4. **Voting** in the Transformation Space. The output list obtained from the previous step, contains enough information to calculate an apparent motion between the unknown image and the corresponding model for each found match. This transform defines a point in the transformation space, and due to the properties of the quasi-invariants, we know that coherent votes will form clusters, while unrelated votes will spread out in the vote space. The best model will be the one with the highest density cluster in its transformation space.

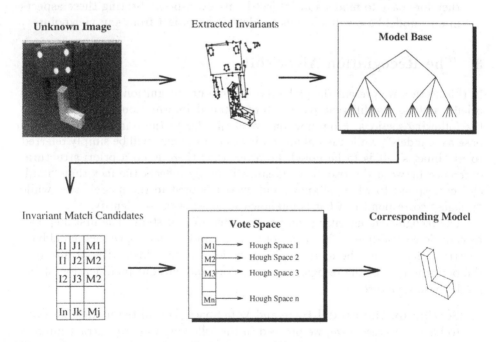

Fig. 1. Object Recognition Algorithm: We extract the invariants of the unknown image and feed them to the model base. The model base returns possible matches, which are used to calculate the transforms in the vote space. These transforms are added in the corresponding transformation space. The model presenting the best density cluster is returned as a match for the image.

4 Matching Invariants

In this section we address the matching algorithm which our system is built on [8]. It was initially designed for image to image matching, but as we will show, it can easily be used to match an image to a model base. The principle is to calculate the invariants of n models, as well as those of an unknown image i. By comparing the invariants of the n models to those of i, and by using a voting technique, we can determine the model image that most resembles i.

4.1 Single Image to Image Matching

The base algorithm calculates quasi-invariants from features in an image. Quasi-invariants vary in a controllable way under projective transforms and are explicitly described in [3]. Since they vary only slightly with a small change in the viewpoint, they are well suited as a similarity measure between two configurations.

As shown in [3], the couple (θ, ρ) formed by the angle θ and the length ratio ρ defined by two intersecting segments, form a quasi-invariant. Moreover, these values are invariant under any similarity transform of the image. Gros [9] showed that it is valid to approximate the apparent motion between two images of a same object by such a similarity transform[3] in order to obtain a match between the two.

If we consider the configurations of intersecting segments in two images of the same object (taken from different viewpoints, but with a camera movement that remains within reasonable bounds) the invariants of two corresponding configurations should not differ too much. Matching these configurations can be done is a two-stage process.

We consider any pair of invariants that are not too different from one another. Hence, the corresponding configurations are similar, and form a possible match. This gives us a list of possible matches between the two images. In order to determine which of them correspond to correct matches, we approximate the apparent motion between the two images by a similarity transform. Each possible match defines such a transform, and allows us to calculate its parameters [8]. If we present each calculated similarity as a point in the appropriate transformation space, a correct match should define a similarity close to those defined by the other correct ones, while incorrect matches will uniformly spread out in the transformation space. In the second stage, we therefore only need to search the four dimensional Hough space. We refer to this Hough space as "transformation space". The point with the highest density cluster wields the best candidate for the apparent motion between the two images. The invariant pairs having contributed to this cluster are those that give the correct matches between configurations. (A weight can be affected to the votes, depending on the confidence and/or robustness accorded to both configurations.)

---

[3] By taking a different form of invariants, one can approximate the apparent motion by an affine transform [9]. We shall be using the similarity case, keeping in mind that the affine case could be applied to our method without loss of generality.

4.2 Single Image to Multiple Image Matching

In order to match the invariants of one image i to those of the models in a model base, we could proceed by matching i to every model in the model base. As we have mentioned in the introduction, this operation is of too high a complexity.

Since we aim at rendering the comparison phase independent of the number of models, we need to store our invariants in such a way that comparison only depends on the number of invariants in i, and not on the number of stored models in our model base. By using the values of the invariants as indexing keys in a multi-dimensional hash table we achieve the requested result.

When we compare our invariants to the model base (as will be described in Sect. 5) the list of possible matches will be similar to the one obtained in the single image matching case. The only difference is that the found matches will no longer refer to one model, but to various models present in the model base. By assigning a transformation space to each of the found models, we can vote in an appropriate transformation space for every possible match.

Unlike Lamdan and Wolfson, we do not select the model having obtained the highest number of votes, but the one having the highest density cluster. By doing this we found an elegant way of filtering the incoherent votes from the correct ones, and we obtain a very robust voting system. Furthermore, the density cluster allows us to solve the matching problem, since it contains the invariant pairs that correctly contributed to the vote.

5 Storing Model Invariants and Accessing the Model Base

Now that we have defined the matching context, we can take a look at the way of easily store model invariants and allow a fast comparison with the image invariants.

It has been shown [1, 6] that hash tables behave in an optimal manner when the keys are uniformly spread over the indexing space. Given a segment (on which we specify an origin $\mathcal{O}$), however, a random second segment (with $\mathcal{O}$ as origin) will give a distribution of quasi-invariants (θ, ρ) that is not uniform. In order to obtain an optimal use of the table entries and minimize the risk of losing precision within certain zones, we need to transform our initial invariant values to obtain a uniform probability.

The distribution of the first member of the invariant, the angle θ, is uniform, but it is not practical to compare two values in the $[0, 2\pi[$ range since they are defined modulo π. By systematically taking the inner (i.e. smaller) angle between two segments, we obtain values within the $[0, \pi[$ range. They conserve the property of uniform distribution and make comparison easier.

We can solve the problem of the non-uniform distribution of the length ratio ρ in a similar fashion. Instead of considering the value $\rho = \frac{l_1}{l_2}$, which has a non uniform distribution, we can use $\rho' = ln(\rho) = ln(l_1) - ln(l_2)$ which behaves correctly as indexing key.

Now that we have solved the indexing problem, we can put the model invariants in our hash table. Since invariants contain noise and since their values are only quasi-invariant to the projective transforms that model the 3D-2D mapping, it is necessary to compare the invariants of the image to those in the model base up to an ε. To compare an unknown invariant to those in the model base, we can proceed in any of the two following manners:

- Since ε is known at construction time, multiple occurrences of one invariant can be stored in the model base. To be assured that an invariant spans the whole ε-radius incertainty circle[4] we can store a copy of it in each of the neighbouring bins. At recognition time we only need to check one bin to find all invariants within an ε radius.
- We can obtain the same result by storing only a single copy of the invariants. At recognition time however, we need to check the neighbouring bins in order to recover all invariants within the ε scope.

6 Voting Complexity

Voting in the vote space is linear in the number of found initial match candidates. The number candidates depends on the number of hits found by the model base. Their quantity is a function of the number of bins in the model base, and the number of models contained therein. Given an invariant of the unknown image, the probability to hit an invariant of a model in the model base varies linearly with the number of models. This probability also depends inversely on the number of bins and the distribution of the models in the model base. Basically, the complexity of the voting process is $\mathcal{O}(n \times m)$. Where n is the number of invariants in the image and m the number of models in the model base.

7 Experimental Results

In this section we present the results of some of the experiments we conducted on our method. The first test, using CAD data, shows that our method allows us to correctly identify the aspect of a 3D object. A second test with real, noisy images, validates our approach for detecting and recognizing different objects in a same image.

7.1 Synthetic Data: Aspect Recognition

We present here an example of 3D object recognition using aspects. We placed a virtual camera on a viewing sphere, centered in an L shaped CAD model,

---

[4] As a matter of fact ε is a multidimensional vector, defining an uncertainty ellipsoid. For convenience we'll assume that it is a scalar defining "ε-radius incertainty circle". This affects in no way the generality of our results.

and took about 200 images from different angles. These images were used to automatically calculate 34 aspects that were fed into the model base. We then presented each of the 200 initial images to our recognition system[5]. Since we have the program that generated the aspects at our disposal, we could easily check for the pertinence of the results given by our system.

The results we obtained can be categorized in three groups: *direct hits*, *indirect hits* and *failed hits*. *Direct hits* are the images our system found the corresponding aspect for. *Failed hits* are those the system misinterpreted and attributed a wrong aspect. The *indirect hits* need some more attention and will be defined below.

In order to easily understand why *indirect hits* occur, we need to detail how the algorithm we used, proceeds in calculating the different aspects. The program uses a dissimilarity measure between images. In function of a dynamic threshold it constructs different classes of similar images. It is important to note that each image can belong to only one unique class. Once the different classes obtained, the algorithm synthesises all of the members of a class into a new "mean" view. This view is then used as the representative of this class.

It is the combined effect of having an image belong to only one class, and synthetically creating a representative of a class that causes the *indirect hits*. It is possible that some of the members of other classes are similar to this synthesized view, although they were too dissimilar to some of the class members to be accepted in that class. One of the cases where this is situation is common, is when the algorithm decides to "cut" and start a new class. Two images on either side of the edge will be very similar, while their class representatives can differ allot. *indirect hits* do not necessarily occur systematically on boundaries however. The example in Fig. 2 (top) shows this clearly.

We therefore define these hits as *indirect* since they refer to a correct aspect (in the sense that it is the one that corresponds to the best similarity transform), although it is not the one initially expected. They are considered as valid recognition results.

As for the *failed hits*, we have noted that most come from degenerate views (cf. example in Fig. 2 bottom). Since we took a high number of images from all over the viewing sphere we necessarily retrieved some views where some distortion occurred. In the cited example, for instance, the "correct" model is the middle one. Our recognition algorithm proposed the model on the right for a simple reason: the distorted face of both the right model and the image are completely equivalent and differ only by some rotation, and the calculated "correct" model is a complete degenerate view that contains almost zero information.

We obtained the following results for the recognition of our images:

---

[5] In a first time we presented the 34 aspects to the model base to check whether the system was coherent. We got a 100% recognition rate. Moreover, this result has been observed in all cases where the images our system was to recognize, had an identical copy in the model base. This result is not fundamental, but proves that the implementation of the different parts is sound.

Example of an *indirect hit*: the unknown image should have been assigned to the first model. Our system matched it with the right model. As a matter of fact, the right model and the image only differ by a 114° rotation which is a better similarity transform than the one that approximates the apparent notion between the image and the other aspect.

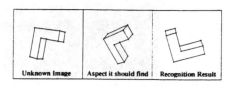

Example of an *failed hit*: the unknown image should have been assigned to the first model. Our system matched it with the right model. The reason for this mismatch is due to the lack of information in the middle model, and the similarity between the two distorted lower faces.

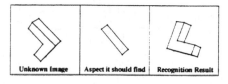

Fig. 2. Examples of recognition results.

# Models	# Images	Direct	Indirect	Failed	Recognition Rate
34	191	151 (79.0%)	26 (13.6%)	14 (7.3%)	92.6%
Without Hough vote: 36 correct hits					18.8%

7.2 Real Data: Object Recognition

We shall now present the tests using real images containing noise. We took the image database of 21 objects from [10]. We then presented the a series of real images to the model base, containing one or more objects. Note that the image base is incomplete for our standards, since we don't have a complete series of aspects for the objects at our disposal. This will allow us to verify the behaviour of our method when unknown aspects or objects are present in an image. Some of the 16 images we used to test recognition algorithm can be found in Fig. 3.

We conducted two types of tests: first we tried to detect one object per image for all images; in a second stage we asked our system to recognize the number of objects we knew were present in the image. We obtained a 94% recognition rate for the first kind of tests (first three results in Fig. 3). The second series show that there are three types of images: those showing known objects in a known aspect, images showing known objects in an unknown aspect, and those featuring unknown objects.

For the objects showing a known aspect we obtain a very good recognition rate (9 out of 11 objects are recognized). This confirms the effectiveness of our method. For the image containing no known objects we get excellent results as

	Known objects in a known aspect	Unknown objects	Known objects in an unknown aspect
Images			
Invariants			
Output			

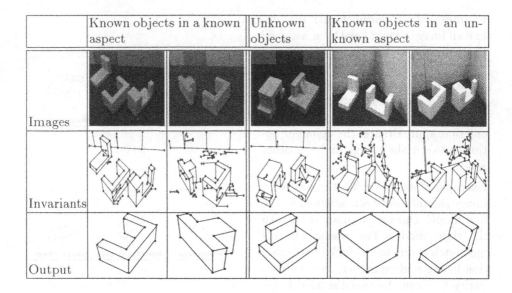

Fig. 3. Some of the 16 images used for recognition.

well: 6 models were matched ex aequo with a very low vote weight compared to the previous results.

However, none of the matched objects with an unknown aspect was correct (last two results Fig. 3). Although this result may seem deceptive, we note that the weight of the votes resulting in the second match are significantly lower than those identifying a correct aspect. Furthermore, our system does not use any 3D information whatsoever. Recognition of unknown aspects from 3D objects does not fall within its scope, which we showed with this experiment.

8 Discussion and Future Extensions

In this paper we presented a new solution to the problem of object recognition in the presence of a model database. We have developed an indexing technique we called *enhanced geometric hashing*. Our approach allows us to solve for the indexing and matching problem in one pass in a way similar to the geometric hashing approach and is able to detect and recognize multiple objects in an image. Other contributions consists in using invariants that contain more topological information and are more robust than the simple 4 point configuration, enhancing the voting phase by introducing a richer and more discriminating vote, and reducing the complexity of the problem.

We validated our approach on both CAD and real image data, and found that our approach is very robust and efficient. Although it is based on 2D image information, we have showed that we can recognize 3D models through their aspects.

Medium term evolutions of our system may consist of: automatically finding multiple objects in one image, matching n images to each other instead of a single image to m models (This may prove extremely useful for the aspect calculating clustering algorithm), optimizing the voting process by precalculating parts of the affine transform or by using techniques of sparse distributed memory, extending the scope of invariants used, etc.

References

1. G. Bebis, M. Georgiopoulos, and N. da Vitoria Lobo. Learning geometric hashing functions for model-based object recognition. In *Proceedings of the 5th International Conference on Computer Vision, Cambridge, Massachusetts, USA*, pages 543–548. IEEE, June 1995.
2. J. Ben-Arie. The probabilistic peaking effect of viewed angles and distances with application to 3-D object recognition. IEEE *Transactions on Pattern Analysis and Machine Intelligence*, 12(8):760–774, August 1990.
3. T.O. Binford and T.S. Levitt. Quasi-invariants: Theory and exploitation. In *Proceedings of* DARPA *Image Understanding Workshop*, pages 819–829, 1993.
4. R.C. Bolles and R. Horaud. 3DPO : A three-Dimensional Part Orientation system. *The International Journal of Robotics Research*, 5(3):3–26, 1986.
5. A. Califano and R. Mohan. Multidimensional indexing for recognizing visual shapes. IEEE *Transactions on Pattern Analysis and Machine Intelligence*, 16(4):373–392, April 1994.
6. L. Grewe and A.C. Kak. Interactive learning of a multiple-attribute hash table classifier for fast object recognition. *Computer Vision and Image Understanding*, 61(3):387–416, May 1995.
7. W.E.L. Grimson, D.P. Huttenlocher, and D.W. Jacobs. A study of affine matching with bounded sensor error. In *Proceedings of the 2nd European Conference on Computer Vision, Santa Margherita Ligure, Italy*, pages 291–306, May 1992.
8. P. Gros. Using quasi-invariants for automatic model building and object recognition: an overview. In *Proceedings of the NSF-ARPA Workshop on Object Representations in Computer Vision, New York, USA*, December 1994.
9. P. Gros. Matching and clustering: Two steps towards object modelling in computer vision. *The International Journal of Robotics Research*, 14(5), October 1995.
10. R. Horaud and H. Sossa. Polyhedral object recognition by indexing. *Pattern Recognition*, 28(12):1855–1870, 1995.
11. Y. Lamdan and H.J. Wolfson. Geometric hashing: a general and efficient model-based recognition scheme. In *Proceedings of the 2nd International Conference on Computer Vision, Tampa, Florida, USA*, pages 238–249, 1988.
12. H. Murase and S.K. Nayar. Visual learning and recognition of 3D objects from appearance. *International Journal of Computer Vision*, 14:5–24, 1995.
13. R.P.N. Rao and D.H. Ballard. Object indexing using an iconic sparse distributed memory. In *Proceedings of the 5th International Conference on Computer Vision, Cambridge, Massachusetts, USA*, pages 24–31, 1995.
14. I. Shimshoni and J. Ponce. Probabilistic 3D object recognition. In *Proceedings of the 5th International Conference on Computer Vision, Cambridge, Massachusetts, USA*, pages 488–493, 1995.

Recognition of Geons by Parametric Deformable Contour Models

Maurizio Pilu* and Robert B. Fisher

Department of Artificial Intelligence, University of Edinburgh,
5 Forrest Hill, Edinburgh EH1 2QL - SCOTLAND (UK)

Abstract. This paper presents a novel approach to the detection and recognition of qualitative parts like geons from real 2D intensity images. Previous works relied on semi-local properties of either line drawings or good region segmentation. Here, in the framework of Model-Based Optimisation, whole geons or substantial sub-parts are recognised by fitting parametric deformable contour models to the edge image by means of a Maximum A Posteriori estimation performed by Adaptive Simulated Annealing, accounting for image clutter and limited occlusions. A number of experiments, carried out both on synthetic and real edge images, are presented.

1 Introduction

In Computer Vision, the task of detecting and recognising general 3D objects in static scenes is still very far from a solution.

One of the most relevant and early recognised approaches to overcome the limitations of traditional CAD-model based vision is the *recognition by parts* theory, pioneered by Marr and Binford [3], which provides both a paradigm and a computational model for computer vision as well as for human vision.

Early attempts to define part models notably included polyhedra, which turned out to be too simple, and generalised cylinders which turned out to be too general and not easily detectable from real images.

Building on Lowe's work, Biederman's Recognition-by-Components (RBC) theory [2] provided a link between studies on human perception and computational vision by proposing a novel part classification scheme based on four significant *non-accidental properties*, cross-section shape, symmetry, sweeping rule and shape of axis, which were then used as a perceptual basis for the generation of a set of 36 components that he called geometrical ions, or geons (examples of geons are shown in Fig. 1-right).

There have been very few works that aimed at the detection and recognition of geons. Bergevin [1] exploited precise properties of line drawings (such as T-junctions, corners, faces, etc.) and so did Hummel and Biederman [7] by mean of their elegant neural network implementation. In the approach by Dickinson *et al.* [4], aspects of geons were pre-compiled and stored in a hierarchical structure

---

* Partially supported by SGS-THOMSON Microelectronics.

and recognition performed by a probabilistic graph matching technique. Later, Metaxas *et al.* [13] built upon that work and both improved the segmentation and integrated stereo information to fit superquadrics models. The main drawback was that a good initial segmentation was required.

This paper describes a new approach to the detection of qualitative parts such as geons in 2D images. The approach matches deformable contours of whole geons to the edge image of a scene. In contrast to other methods, ours is truly global, in the sense that the whole geon is holistically considered as a single entity to be extracted from the image. A new parametric model is introduced that allow us to represent compactly and efficiently the various geon outlines. Then, within the framework of Model-Based Optimisation (MBO), a cost function that represent the quality of fit between model and image is maximised in the parameter space by using Adaptive Simulated Annealing (ASA) and the best image-model match is found. The cost function expresses in Bayesian terms the probability that a certain number of edgels match the geon contour model and some others do not, which has also an analogy in information theoretical terms.

At this stage of the research, the initialisation of the optimisation procedure is performed semi-automatically by manually selecting out the image regions corresponding to parts and by computing the principal moments of the thresholded regions.

The structure of the paper is as follows. First we describe the construction of geon contour models. Next we present the design of the objective function and how we perform its optimisation by Adaptive Simulated Annealing; some experimental results are given that support the validity of this approach. We conclude by proposing possible future extensions.

2 Parametrically Deformable Contour Model of Geons

Within the framework of Model-Based Optimisation, the recognition of geons from 2D images needs a model that can describe in a compact way their contour. As the geon models are computed inside the innermost loop of the optimisation process, this must be done as efficiently as possible.

Following [14] and much recent work on part decomposition from range data (e.g. [16, 17, 15, 13]), we extend the use of superquadrics (SQ) to the 2D case by using their projected visible contour (outline) as a 2D model, thus creating a geon *parametrically deformable contour model* (henceforth PDCM).

However, since a direct computation of SQ outlines is extremely expensive, we have pragmatically created an efficient approximate model that is suitable for qualitative geon PDCMs. Starting with a cylinder centred on the z axis with superelliptical cross-section, centred on the origin (Fig. 1-left) we apply deformations and rotations and find the outline by simple geometric considerations. The initial superelliptical cylinder C of height $2 \cdot a_z$ and semi-axes a_x and a_y can be expressed as

$$\mathbf{C} = \begin{bmatrix} \mathbf{x}(\eta) \\ \mathbf{y}(\eta) \\ \mathbf{z} \end{bmatrix} = \begin{bmatrix} a_x \cos(\eta)^\epsilon \\ a_y \sin(\eta)^\epsilon \\ \mathbf{z} \end{bmatrix} \qquad \begin{matrix} -\pi \leq \eta \leq \pi \\ -a_z \leq \mathbf{z} \leq a_z \end{matrix}, \qquad (1)$$

where $0 \leq \epsilon \leq 1$ controls the degree of squareness of the cross-section from a virtual block for $\epsilon \to 0$ to a cylinder for $\epsilon \to 1$.

Any curve lying on this cylinder can be variously deformed. We use three types of deformations, *tapering* ($\mathcal{T}$), *bending* ($\mathcal{B}$) and *swelling* ($\mathcal{S}$) along the principal axis.

Let us indicate by $\mathbf{x}$, $\mathbf{y}$, $\mathbf{z}$ and $\mathbf{X}$, $\mathbf{Y}$, $\mathbf{Z}$ the vector of shape points before and after the deformations, respectively. Then we have:

$\mathcal{T}(\mathbf{C}, K_x, K_y) =$	$\mathcal{B}(\mathbf{C}, c) =$	$\mathcal{S}(\mathbf{C}, s) =$
$\begin{cases} X = (\frac{K_x}{a_z}z + 1)x \\ Y = (\frac{K_y}{a_z}z + 1)y \\ Z = z \end{cases}$	$\begin{cases} X = x + sign(c)(R'' - r) \\ Y = y \\ Z = \sin(\gamma)(\kappa^{-1} - R'') \end{cases}$	$\begin{cases} X = x + sign(x)(R' \cos\alpha - (R' - \sigma)) \\ Y = y + sign(y)(R' \cos\alpha - (R' - \sigma)) \\ Z = R' \sin\alpha \end{cases}$

where $r = sign(c)\cos(\beta)\sqrt{x^2 + y^2}$, $\beta = \arctan \frac{y}{x}$, $R'' = \kappa^{-1} - \cos(\gamma)(\kappa^{-1} - r)$,
$\gamma = \frac{z}{\kappa^{-1}}$, $\kappa^{-1} = \frac{a_z}{|c|}$, $\sigma = a_x s$, $R' = \frac{(a_x^2 - \sigma^2)}{2\sigma}$, $\alpha = \arctan \frac{z}{(R' - \sigma)}$ and $0 \leq s \leq 1$,
$-1 \leq c \leq 1$ and $-1 \leq K_x = K_y \leq 1$ are, respectively, the swelling, bending and tapering control parameters.

The tapering and bending deformations have been derived from [16]; the latter has been slightly modified by normalising the bending control parameter to a_z, which has improved the stability of its estimation, and to allow bending on both sides. The swelling deformation, however, has been introduced here to represent the geons' "expanding and contracting" sweeping rule [2].

Once deformed, the shape is rotated in space (by θ_{pan}, θ_{tilt}), projected (P) and roto-translated in the image plane (by t_x and t_z and θ_{opt}). The whole chain of transformations of the initial 3D shape $\mathbf{C}$ to its full projection onto the image plane z-x $\mathbf{C}' = \begin{bmatrix} \mathbf{x}' \\ \mathbf{z}' \end{bmatrix}$ is then:

$$\mathbf{C}' = T(R_y(P(R_z(R_x(\mathcal{B}(\mathcal{S}(\mathcal{T}(\mathbf{C}, K_x, K_x), s), c), \theta_{pan}), \theta_{tilt})), \theta_{opt}), t_x, t_z) \qquad (2)$$

Now we are ready to describe the construction of the geons' PDCM. The knottiest problem is to determine the occluding contour. For doing this, we employed the following approximation.

We apply the transformations in Eqn. (2) to the two bases of the superelliptical cylinder and take the four outermost points $P1'_a$, $P1'_b$ and $P2'_a$, $P2'_b$ (small circles in Fig. 1-left-B) and find the two corresponding points in the original undeformed superellipses (small circles in Fig. 1-left-A). These two points are linked by two 3D straight lines $L1$ and $L2$, as shown in Fig. 1-left-A, and successively deformed according to Eqn. (2); the resulting $L1'$ and $L2'$ (Fig. 1-left-B) will then be used as the two lateral parts of the occluding contour.

By checking the projection of the normals $\mathbf{n}_a$ and $\mathbf{n}_b$ to the superelliptical ends on the image plane, we can then determine whether each of the two ends

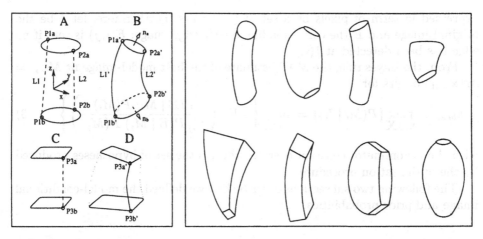

Fig. 1. Determination of occluding contour (left) and some examples of PDCM (right).

are visible or not: if visible, the whole superellipse contour will be added in the geon PDCM; otherwise only its outermost part between $P1'_a$ ($P1'_b$) and $P2'_a$ ($P2'_b$) will be included in the final PDCM.

In the case we have a geon with squared cross-section (say $\epsilon < 0.5$), the central edge is determined by joining the two corners $P3_a$ and $P3_b$ (Fig. 1-left-C) from the undeformed superelliptical bases occurring at $\eta = \pi/4$ in Eqn. (1) by a 3D straight line and then deforming it by Eqn. (2); the resulting 2D curve is shown in Fig. 1-left-D.

Some examples of geon PDCMs produced by this method can be seen in Fig. 1-right; it takes less than 1ms of a SPARC 10 machine to create a model instance, over two order of magnitudes faster that any other method that would use raster scan techniques or surface normals.

This superquadric-inspired model can, as in [16, 17], represent the 12 geon classes which have cross-section symmetry [2] with a sufficient level of accuracy. However, by virtue of the employed geometric construction of the outline, they do not work properly with large bending or under certain viewing directions (about $|\theta_{tilt}| > \frac{\pi}{3}$), where the occluding contour construction we employed is no longer applicable; despite this, it is a good trade-off between accuracy and speed.

3 Objective Function Design

Within the framework of Model-Based Optimisation, the matching of a PDCM to a geon image involves the minimisation of an objective (or cost) function that expresses the quality of image-model fit.

Seeking the best interpretation of the available data by the model, we express this cost function in Bayesian terms.

Let $\mathbf{x}_i = [a_x\ a_y\ a_z\ \epsilon\ K_x\ s\ c\ \theta_{pan}\ \theta_{tilt}\ \theta_{opt}\ P_x\ P_z]^T$ be the vector of the PDCM parameters and $M_i = M(\mathbf{x}_i)$ be a geon PDCM instance built as in Sec. 2, expressed in term of pixels by a set of (i, j) pairs. Furthermore, let I be the original image and E the corresponding binary edge image: $E(i, j)$ is zero if no edge has been detected at $I(i, j)$.

From the Bayes rule, the MAP estimate of the best model-image fit $M_{best} = M(\mathbf{x}_{best})$ occurs for

$$\mathbf{x}_{best} : \max_{\mathbf{x}_i \in \mathbf{X}} \{P(M_i \mid E)\} = \min_{\mathbf{x}_i \in \mathbf{X}} \left\{ -\log\left(\frac{P(E \mid M_i) \, P(M_i)}{\sum_j P(E \mid M_j) \, P(M_j)} \right) \right\}, \quad (3)$$

with the denominator constant over $\mathbf{X}$, which is the set of hypotheses produced by the optimisation procedure.

The following two subsections describe how we defined the model-conditional image and prior probabilities.

3.1 Model-Conditional Image Probability

In Eqn. (3), $P(E \mid M_i)$ expresses the conditional probability of having particular image evidence in the presence of the model.

Earlier experiments using various techniques based on sum of distances between model and edges (as used by Lowe [12] and others) had problems of unwanted shrinking or expansions so we concluded that they cannot properly cope with the use of deformable models and sparse unsegmented data. Our approach is to find the model that best accounts for the image in terms of matched and unmatched edgels.
Let

$$E_m(M_i) = \{(k, l) \; : \; \mid (i, j) - (k, l) \mid \le d \, , \; (i, j) \in M_i\}$$

be the d-neighbourhood of the model contour M_i, and $E_b(M_i) = E - E_m(M_i)$ the rest of the edge image which is not covered by it (background); henceforth we drop the M_i arguments wherever there cannot be ambiguities.
Let N_{b1} and N_{b0} be the number of pixels in the background E_b that are edge ("1") or non-edge ("0"). On the other hand, in the model neighbourhood E_m, and N_{m1} the number of *edgels* whose direction is locally consistent with that of the model M_i at that point (we set a threshold of $\pi/10$); furthermore let N_{m0} be the number of pixels in E_m that are *not* matching the model, either because no edge has been detected there or because of directional inconsistency.

The model conditional probability $P(E \mid M_i)$ globally expresses the probability of the presence or absence of edge points or consistent edgels in E_b and E_m. Since E_b and E_m can be regarded, to a first approximation, as realizations of two independent binary ergodic processes, whose probabilities are given by the product of single local outcome probabilities p_{b1} and p_{m1}, respectively, we have:

$$P(E \mid M_i) = P(E_b \mid M_i) \cdot P(E_m \mid M_i) = \left[p_{b1}^{N_{b1}} (1 - p_{b1})^{N_{b0}} \right] \cdot \left[p_{m1}^{N_{m1}} (1 - p_{m1})^{N_{m0}} \right].$$

The value of p_{b1} is given by the ratio between edge locations and the number of pixels in the image (typical values: 0.02-0.06); p_{m1} ranges from 0.6 to 0.9,

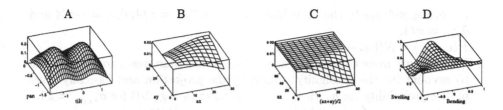

Fig. 2. Model prior probabilities.

depending on the neighbourhood dimension d and how good the edge detection of object contour is expected to be. By taking the logarithms on both sides and expanding we obtain:

$$\log(P(E|M_i)) = \log(P(E_b|M_i)) + \log(P(E_m|M_i)) = \qquad (4)$$

$$K + [N_{m1}\log(p_{m1}) + N_{m0}\log(1 - p_{m1})] - [N_{m1}\log(p_{b1}) + N_{m0}\log(1 - p_{b1})],$$

where K is a constant term that can be dropped in the MAP estimation.

The terms of this equation can be also reinterpreted as the number of bits "saved" by representing E by the model M_i bringing all into an information theoretical framework (see, e.g., [6, 11]).

3.2 Model A Priori Probability

Within a Bayesian framework we can assign an occurrence probability $P(M_i)$ of a certain PDCM, called the *model a priori probability*. The reasons for introducing it are essentially three: *i)* some parameter configuration are unlikely to occur (such as a bent and swollen object); *ii)* Certain configurations of parameters arise from a weird view point that would make detection impossible; and *iii)* It biases the fitting to more perceptually likely shapes. These considerations are both practical and also correspond to sensible assumptions to reduce the quantitative shape ambiguities caused by the projection.

We have defined a sensible heuristic to represent these loose constraints. The probability of each PDCM is expressed by overlapping (multiplying) marginal densities of parameter values or combination of them, tacitly assuming statistical inter-independence. The parameters we took into considerations are a_x, a_y and a_z, swelling, bending and the viewpoint angles; the tapering and roundness parameters have not been taken into account and thus are given uniform probabilities. Below we describe these probabilities, whose non-normalised p.d.f. are also displayed in Fig. 2.

- $P(\theta_{tilt}, \theta_{pan}) = N(\theta_{tilt} - \frac{\pi}{4}, \sigma_{\theta_{tilt}}) \cdot N(\theta_{tilt} + \frac{\pi}{4}, \sigma_{\theta_{tilt}}) \cdot N(\theta_{pan} - \frac{\pi}{4}, \sigma_{\theta_{pan}})$
  In our perception [2] there is a bias towards objects or surfaces in canonical stable positions. $P(\theta_{tilt}, \theta_{pan})$ has therefore probability density like the one shown in Fig. 2-A for $\sigma_{\theta_{tilt}} = \sigma_{\theta_{pan}} = \pi/6$, which has two Gaussian bells

in correspondence to the most likely values of $\theta_{tilt} = \pi/4$, $\theta_{tilt} = -\pi/4$ and $\theta_{pan} = \pi/4$. [2]

- $P(a_x, a_y) = N(\mid a_x - a_y \mid, \sigma_{a_x,a_y})$

  This favours more compact cross-sections rather than weird rotation angles to account for the ambiguity caused by the projection and basically gives higher probability for a_x close to a_y, as shown in Fig. 2-B for $\sigma_{a_x,a_y} = 20$.

- $P(a_z \mid a_x, a_y) = \begin{cases} N(\tau \frac{\mid a_x + a_y \mid}{2} - a_z, \sigma_{a_z}) \text{ if } a_z \leq \tau \frac{\mid a_x + a_y \mid}{2} \\ \frac{1}{\sqrt{2\pi} \cdot \sigma_{a_z}} \qquad\qquad \text{otherwise} \end{cases}$

  This favours elongated geons by giving lower and lower probability if its length is smaller than the average of cross-section dimensions by a factor τ, as shown in Fig. 2-C for $\tau = 1.5$ and $\sigma_{a_z} = 20$

- $P(c, s) = N(\mid c \mid \cdot s, \sigma_{c,s})$.

  This expresses the perceptual incompatibility between high swelling and bending; the p.d.f. is shown in Fig. 2-D for $\sigma_{c,s} = 0.3$.

Now that we have all the non-normalised p.d.f. and given the assumption of prior inter-independence between parameters, we just multiply them together to obtain the (non-normalised) *a priori* p.d.f. of the model:

$$\log(P(M_i)) = H + \log(P(\theta_{tilt}, \theta_{pan})) + \log(P(a_x, a_y)) +$$
$$\log(P(a_z \mid a_x, a_y)) + \log(P(c, s)) \qquad (5)$$

The normalisation constant H is unnecessary because it does not affect the MAP estimate.

We wish to remark that there are other possible ways of defining the model prior probability and that we could also incorporate more detailed specific domain-dependent knowledge about the scene structure.

4 Model Fitting Procedure

The model fitting is obtained, from (3), by the minimisation of

$$- \log(P(M_i \mid E)) = - \log(P(E \mid M_i)) - \log(P(M_i)), \qquad (6)$$

where the two terms are given by Equations (5) and (4). This minimisation is, however, rather difficult to achieve, since it is extremely irregular and presents many shallow and/or narrow minima. By trying to minimise Eqn. (6) alone, we also found that sometime the optimisation got stuck in local minima because of the step-like nature of the model-conditional probability of Eqn. (4) (remember we used a binary "belonging to the model" criteria). For overcoming this problem we have added to Eqn. (6) a small smoothing term representing the average minimal distance between contour model and image edge points (we used a minimal distance transform computed off-line). The smoothing term does not

---

[2] θ_{tilt} and θ_{pan} affect the topology of the PDCM and these values can be regarded also as giving the high disambiguation distance between visual events [9].

	a_x	a_y	a_z	ϵ	K_x	s	c	θ_{pan}	θ_{tilt}	θ_{opt}	P_x	P_z
Lower	-5%	-5%	-5%	.01	-1.0	0.0	-0.8	$(\bar\theta_{pan} - 0.3)$	$(\bar\theta_{tilt} - 0.3)$	$(\bar\theta_{opt} - 0.4)$	$\frac{-N}{10}\%$	$\frac{-N}{10}\%$
Upper	+5%	+5%	+5%	.99	1.0	1.0	0.8	$(\bar\theta_{pan} + 0.3)$	$(\bar\theta_{tilt} + 0.3)$	$(\bar\theta_{opt} + 0.4)$	$\frac{+N}{10}\%$	$\frac{+N}{10}\%$
Δ	1.0	1.0	1.0	0.2	0.2	0.2	0.1	0.05	0.05	0.05	1	1

Table 1. Ranges and gradient steps used in the optimisation procedure.

affect the MAP estimate but just helps convergence in cases where image and model are much displaced and the numerical computation of the gradient become meaningless due to the low number of edge points falling inside the model neighbourhood. This term can then be seen as "telling the optimiser where to go" in absence of other information.

In early stages of the work we used a Levenberg-Marquandt method with added random perturbations as used in [16] and other works, but this approach had difficult convergence. Our choice fell then to Simulated Annealing [10], which is a powerful optimisation tool that efficiently combines gradient descent and controlled random perturbations to perform the minimisation of non-convex functions. We used a recent publicly available version of Simulated Annealing called Adaptive Simulated Annealing (ASA) [8].

The initialisation and set-up of the ASA algorithm is discussed below.

INITIALISATION   The initialisation part is concerned with estimating initial coarse part hypotheses (sometime called *frames*) that comprise position, orientation of the major axis and dimensions. This initialisation need not be precise and the degree of allowed inaccuracy depends upon the power of the optimisation procedure. Since no method is currently available for segmenting qualitative parts from 2D real cluttered images, we have used a semi-automatic procedure. Regions corresponding to parts are first selected out manually (see Fig. 4) and an adaptive thresholding technique is used to yield blob-like regions; orientation and size are found by computing the first and second moments and principal directions of the blobs, which are assigned to $\bar{P}_x$, $\bar{P}_z$, $\bar{a}_z$, $\bar{a}_x = \bar{a}_y$ and $\bar\theta_{opt}$, respectively. The values of $\bar{a}_x$ and $\bar{a}_y$ are set to be equal because we do not have prior information on the aspect-ratio of the part cross-section. Moreover tapering, bending and swelling are all initially set to zero, $\bar\theta_{pan} = 0$ and $\bar\theta_{tilt}$ experimentally set to $\pi/6$.

OPTIMISATION SET-UP   The bounds of the hyper-rectangular search space and the step for the computation of the gradient, needed by the ASA optimiser greatly affect the nature of the convergence. Table 1 summarises the values we found reasonably good through our experiments. (The values expressed in percentage are relative to their respective initial values and N is the image resolution in pixels)

The two parameters specifying the annealing schedule, the Temperature Ratio Scale and Temperature Annealing Scale [8] have also been experimentally set to $5 \cdot 10^{-5}$ and 50, respectively, and the Cost Precision to 0.0001. Finally, the number of iterations has been set to 2000, which we found to be a good trade-off between speed and good convergence. The neighborood distance d was set to 6 pixels for 512x512 images and to 2 pixels for 128x128 images. Each ASA running time ranged from 5 to 15 secs on a SPARC 10 machine, according to image resolution.

5 Experimental Results

In this section we present two sets of experiments that show the validiy of the approach.

In the first set, shown in Fig. 3, a 512x512 image of six plasticine isolated geons was taken (Image C) and, after the application of a Canny Edge detector, no post-processing was carried out (Image C-left). From it, we created two synthetic edge images with the same geons but with roundish (Image A) and squared (Image B) cross-sections. The initializations (unique for each geon across Images A,B and C and overlapped to the edge images in the left figures) are rather crude; for the synthetic images the right aspect topology was imposed to each instance whereas it was set free for Image C. It can be seen that with synthetic images the results (displayed in the right column) are rather good in both examples. The results with real images are reasonably good too, considering that we intentionally left a high cluttering level; geon #1,#4 and #6, however, have been slightly misfit due to intollerably high noise within the geon body that, in particular, caused a change in topology.

In the second set, shown in Fig. 4, two real 128x128 images of an handset, a mug and a banana were used, and their edge images produced by a Canny edge detector and simple filtering (top); initializations are performed as outlined in Sec. 4 and are displayed in the middle row. In the case of the handset, three good estimates were produced although the edge image is cluttered and incomplete. Except for the mouth piece, which is slightly over-swollen, the essential qualitative features of the geons (such as cross-section and curvature) are extracted. The experiment with the banana is succesfull too, but, of course, because of the long shading edge running along its body, the fitting yielded squared cross-section. This spurious effect can be overcome only by integrating further information. The mug example was, as expected, a failure because both the initialization and edge image were excessively poor, undistinguishable even to the human eye.

6 Conclusion and Future Work

In this paper we have presented a novel method for the fitting of generic geon-like parts to 2D edge image. We developed a new efficent approximated model of superquadrics contour that, in the framework of Model-Based Optimisation, is fitted to the image by a MAP estimation procedure which seeks the best interpretation of the available data in terms of the model. Some experiments have been described that show the validity of the approach.

Single parts are currently semi-automatically selected out from the image, since no methods is available for authomatically doing it from real images, but we

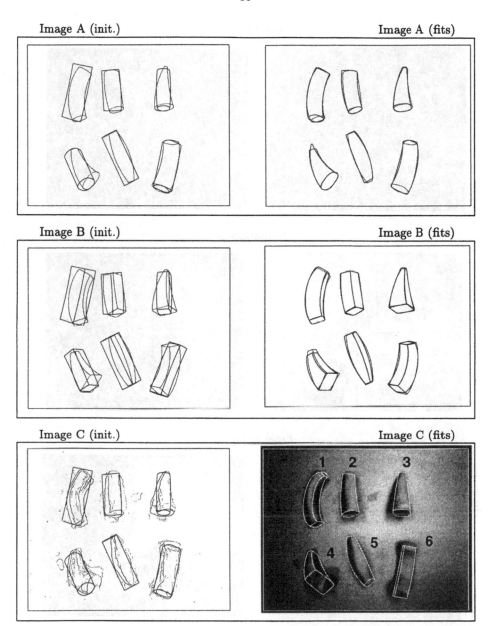

Fig. 3. First set of experiments designed to assess the convergence of the method (see text for details).

Edge Image

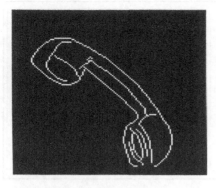
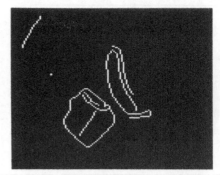

Initialisation

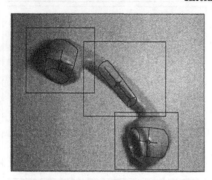
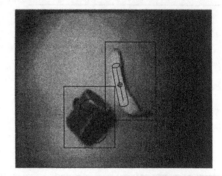

Final estimates

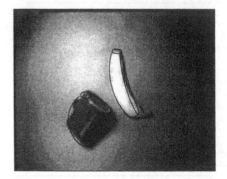

Fig. 4. Second set of experiments with semi-automatic initialization (see text for details).

are actively investigating into the problem. Furthermore, early experiments using different aspect topologies [5], have shown a substantial increase in robustness. We are also planning to integrate other information, such as shading. to further increase stability and convergence.

Acknowledgements: Thanks to A. Fitzigibbon for useful suggestions.

References

1. R. Bergevin. Primal access recognition of visual objects. Technical Report TR-CIM-90-5, Mc Gill University, Canada, February 1990.
2. I. Biederman. Recognition-by-components: A theory of human image understanding. *Psychological Review*, 94:115–147, 1987.
3. T.O. Binford. Visual perception by computer. In *Proceedings of the IEEE System Science and Cybernetic Conference*, Miami, December 1971.
4. S.J. Dickinson, A.P. Pentland, and A. Rosenfeld. 3-D Shape Recovery Using Distributed Aspect Matching. *IEEE PAMI*, 14(2):130–154, 1992.
5. D. Eggert, L. Stark, and K. Bowyer. Aspect graphs and their use in object recognition. *Annals of Mathematics and Artificial Intelligence*, 13:347–375, 1995.
6. P. Fua and A.J. Hanson. Objective functions for feature discrimination: Applications to semiautomated and automated feature extraction. In *DARPA Image Understanding Workshop*, pages 676–694, 1989.
7. J.E. Hummel and I. Biederman. Dynamic binding in a neural net model for shape recognition. *Psychological Review*, 99:480–517, 1992.
8. L. Ingber. *Adaptive Simulated Annealing*. Lester Ingber Research, Mc Lean, VA, 1993. [ftp.alumni.caltech.edu./pub/ingber/ASA.tar.zip].
9. J.R. Kender and D.G. Freudenstain. What is a "degenerate" view? In *DARPA Image Understanding Workshops*, pages 589–598, 1987.
10. S. Kirkpatrick, C.D. Gelatt, and M.P. Vecchi. Optimization by simulated annealing. *Science*, 220:671–680, 1983.
11. Y.G. Leclerc. Constructing simple stable description for image partitioning. *International Journal of Computer Vision*, 3:73–102, 1989.
12. D. Lowe. Fitting parametrized 3D models to images. *PAMI*, 13(5):441–450, May 1991.
13. D. Metaxas, S.J. Dickinson, R.C., Munck-Fairwood, and L. Du. Integration of quantitative and qualitative techniques for deformable model fitting from orthographic, perspective and stereo projection. In *Fourth International Conference on Computer Vision*, pages 364–371, 1993.
14. A.P. Pentland. Perceptual organization and the representation of natural form. *Artificial Intelligence*, 28:293–331, 1986.
15. N.S. Raja and A.K. Jain. Obtaining generic parts from range data using a multiview representation. In *Appl. Artif. Intell. X: Machine Vision and Robotics, Proc. SPIE 1708*, pages 602–613, Orlando, FL, April 1992.
16. F. Solina and R. Bajcsy. Recovery of parametric models from range images: The case of superquadrics with global deformations. *IEEE PAMI*, 12(2):131–147, February 1990.
17. K. Wu and M.D. Levine. Recovering of parametric geons from multiview range data. In *IEEE Conference on Computer Vision and Pattern Recognition*, Seattle, WA, 1994.

Geometry and Stereo

Automatic Extraction of Generic House Roofs from High Resolution Aerial Imagery *

Frank Bignone, Olof Henricsson, Pascal Fua+ and Markus Stricker

Communications Technology Laboratory
Swiss Federal Institute of Technology ETH
CH-8092 Zurich, Switzerland
+SRI International, Menlo Park, CA 94025, USA

Abstract. We present a technique to extract complex suburban roofs from sets of aerial images. Because we combine 2-D edge information, photometric and chromatic attributes and 3-D information, we can deal with complex houses. Neither do we assume the roofs to be flat or rectilinear nor do we require parameterized building models. From only one image, 2-D edges and their corresponding attributes and relations are extracted. Using a segment stereo matching based on all available images, the 3-D location of these edges are computed. The 3-D segments are then grouped into planes and 2-D enclosures are extracted, thereby allowing to infer adjoining 3-D patches describing roofs of houses. To achieve this, we have developed a hierarchical procedure that effectively pools the information while keeping the combinatorics under control. Of particular importance is the tight coupling of 2-D and 3-D analysis.

1   Introduction

The extraction of instances of 3-D models of buildings and other man-made objects is currently a very active research area and an issue of high importance to many users of geo-information systems, including urban planners, geographers, and architects.

Here, we present an approach to extract complex suburban roofs from sets of aerial images. Such roofs can neither be assumed to be flat nor to have simple rectangular shapes. In fact, their edges may not even form ninety degrees angles. They do tend, however, to lie on planes. This specific problem is a typical example of the general Image Understanding task of extracting instances of generic object classes that are too complex to be handled by purely image-based approaches and for which no specific template exists.

Because low-level methods typically fail to extract all relevant features and often find spurious ones, existing approaches use models to constrain the problem [15]. Traditional approaches rely almost exclusively on the use of edge-based features and their 2-D or 3-D geometry. Although 3-D information alleviates the problem, instantiating the models is combinatorially explosive. This difficulty

* We acknowledge the support given to this research by ETH under project 13-1993-4.

is typically handled by using very constrained models, such as flat rectilinear roofs or a parameterized building model, to reduce the size of the search space. These models may be appropriate for industrial buildings with flat roofs and perpendicular walls but not for the complicated suburban houses that can be found in scenes such as the one of Fig. 1.

It has been shown, however, that combining photometric and chromatic region attributes with edges leads to vastly improved results over the use of either alone [6, 11]. The houses of Fig. 1 require more flexible models than the standard ones. We define a very generic and free-form roof primitive: we take it to be a 3-D patch that is roughly planar and encloses a compact polygonal area with consistent chromatic and luminance attributes. We therefore propose an approach that combines 2-D and 3-D edge geometry with region attributes. This is not easy to implement because the complexity of the approach is likely to increase rapidly with the number of information sources. Furthermore, these sources of information should be as robust as possible but none of them can be expected to be error-free and this must be taken into account by the data-fusion mechanism.

Figure 1 Two of the four registered 1800×1800 images that are part of our residential dataset (Courtesy of IGP at ETH Zürich).

To solve this problem, we have developed a procedure that relies on hierarchical hypothesis generation, see Fig. 2. The procedure starts with a multi-image coverage of a site, extracts 2-D edges from a source image, computes corresponding photometric and chromatic attributes, and their similarity relationships. Using both geometry and photometry, it then computes the 3-D location of these edges and groups them to infinite planes. In addition, 2-D enclosures are extracted and combined with the 3-D planes to instances of our roof primitive, that is 3-D patches. All extracted hypotheses of 3-D patches are ranked according to their geometric quality. Finally, the best set of 3-D patches that are mutually consistent are retained, thus defining a scene parse. This procedure has proven powerful enough so that, in contrast to other approaches to generic roof extraction (e.g. [14, 6, 4, 13, 7, 12]), we need not assume the roofs to be flat or rectilinear or use a parameterized building model.

Note that, even though geometric regularity is the key to the recognition of man-made structures, imposing constraints that are too tight, such as requiring that edges on a roof form ninety degrees angles, would prevent the detection of many structures that do not satisfy them perfectly. Conversely, constraints that are too loose will lead to combinatorial explosion. Here we avoid both problems by working in 2-D and 3-D, grouping only edges that satisfy loose coplanarity constraints, weak 2-D geometric and similarity constraints on their photometric and chromatic attributes. None of these constraints is very tight but, because we pool a lot of information from multiple images, we are able to retain only valid object candidates.

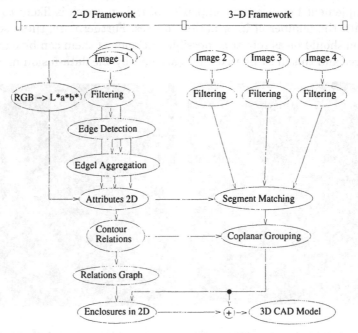

Figure 2 Our hierarchical framework, a feed-forward scheme, where several components in the 2-D scheme mutually exchange data and aggregates with the 3-D modules.

We view the contribution of our approach as the ability to robustly combine information derived from edges, photometric and chromatic area properties, geometry and stereo, to generate well organized 3-D data structures describing complex objects while keeping the combinatorics under control. Of particular importance is the tight coupling of 2-D and 3-D analysis.

For our experiments, we use a state-of-the-art dataset produced by the Institute of Geodesy and Photogrammetry at ETH Zürich. It consists of a residential and an industrial scene with the following characteristics: 1:5,000 image scale vertical aerial photography, four-way image overlap, color imagery, geometrically accurate film scanning with 15 microns pixel size, precise sensor orientation, and accurate ground truth including DTM and manually measured building CAD-models. The latter are important to quantitatively evaluate our results.

Our hierarchical parsing procedure is depicted in Fig. 2. Below we describe each of its components: 2-D edge extraction, computation of photometric and chromatic attributes, definition of similarity relationships among 2-D contours, 3-D edge matching and coplanar grouping, extraction of 2-D enclosures, and finally, generation and selection of candidate 3-D object models. Last, we present and discuss our results.

2 Attributed Contours and their Relations

2.1 Edge Detection and Edgel Aggregation

Our approach is based on grouping contour segments. The presented work does not require a particular edge detector, however, we believe it is wise to use the best operator available to obtain the best possible results. For this reason we use the SE energy operator (suppression and enhancement) recently presented in [8]. The operator produces a more accurate representation of edges and lines in images of outdoor scenes than traditional edge detectors due to its superior handling of interferences between edges and lines, for example at sharp corners.

The edge and line pixels are then aggregated to coherent contour segments by using the algorithm described in [10]. The result is a graph representation of contours and vertices, as shown in Fig. 3B. Each contour has geometric attributes such as its coarse shape, that is *straight, curved* or *closed*.

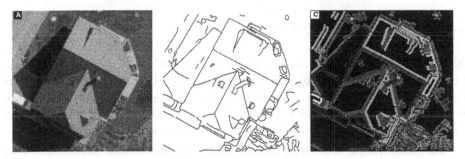

Figure 3 (A) a cut-out 350 × 350 from the dataset in Fig.1, (B) The resulting attributed graph with all its contours and vertices, (C) the flanking regions with their corresponding median luminance attributes. The background is black.

2.2 Photometric and Chromatic Contour Attributes

The contour graph contains only basic information about geometry and connectivity. To increase its usefulness, image attributes are assigned to each contour and vertex. The attributes reflect either properties along the actual contour (e.g. integrated gradient magnitude) or region properties on either side, such as chromatic or photometric homogeneity.

Since we are dealing with fairly straight contours the construction of the flanking regions is particularly simple. The flanking region is constructed by a translation of the original contour in the direction of its normal. We define a flanking region on each side of the contour. When neighboring contours interfere with the constructed region, a truncation mechanism is applied. In Fig. 3C we display all flanking regions. For more details we refer to [9].

To establish robust photometric and chromatic properties of the flanking regions, we need a color model that accurately represents colors under a variety of illumination conditions. We chose to work with HVC color spaces since they separate the luminant and chromatic components of color. The photometric attributes are computed by analyzing the value component, whereas the chromatic attributes are derived from the hue and chroma components. As underlying color space we use the CIE(L*a*b*) color space because of its well based psychophysical foundation; it was created to measure *perceptual* color differences [16].

Since each flanking region is assumed to be fairly homogeneous (due to the way it is constructed), the data points contained in each region tend to concentrate in a small region of the color space. As we deal with images of aerial scenes where disturbances like chimneys, bushes, shadows, or regular roof texture are likely to be within the defined regions, the computation of region properties must take outliers into account. Following the approach in [11] we represent photometric attributes with the median luminance and the interquartile range (IQR), see Fig. 3C. The chromatic region properties are computed analogously from the CIE(a*b*) components and are represented by the center of the chromatic cluster and the corresponding spreads.

2.3 Contour Similarity Relations

Although geometric regularity is a major component in the recognition of man-made structures, neglecting other sources of information that can assert the relatedness among straight contours imposes unnecessary restrictions on the approach. We propose to form a measure that relate contours based on similarity in position, orientation, and photometric and chromatic properties.

For each straight contour segment we define two directional contours pointing in opposite directions. Two such directional contours form a *contour relation* with a defined logical interior. For each contour relation we compute four scores based on similarity in *luminance*, *chromaticity*, *proximity*, and *orientation* and combine them to a single similarity score by summation.

Three consecutive selection procedures are applied, retaining only the best non-conflicting interpretations. The first selection involves only two contours (resp. four directional contours) and aims at reducing the eight possible interpretations to less or equal to four. The second selection procedure removes short-cuts among three directed contours. The final selection is highly data-driven and aims at reducing the number of contour relations from each directed contour to only include the *locally* best ones. All three selection procedures are based on analysis of the contour similarity scores. Due to lack of space we refer to [11] for more details.

3 Segment Stereo Matching

Many methods for edge-based stereo matching rely on extracting straight 2-D edges from images and then matching them [1]. These methods, although fast and reliable, have one drawback: if an edge extracted from one image is occluded or only partially defined in one of the other images, it may not be matched. In outdoor scenes, this happens often, for example when shadows cut edges. Another class of methods [2] consists of moving a template along the epipolar line to find correspondences. It is much closer to correlation-based stereo and avoids the problem described above. We propose a variant of the latter approach for segment matching that can cope with noise and ambiguities. Edges are extracted from *only one* image (the source image) and are matched in the other images by maximizing an "edginess measure" along the epipolar line. The source image is the nadir (most top-view) image because it is assumed to contain few (if any) self-occluded roof parts. Geometric and photometric constraints are used to reduce the number of 3-D interpretations of each 2-D edge. We outline this approach below and refer the interested reader to [3] for further details.

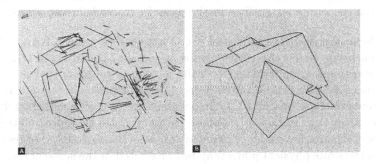

Figure 4 (A) Matched 3-D segments. Notice the false matches.
(B) Manually measured 3-D CAD model.

For a given edge in the source image we want to find the location of its correspondences in the other image. A segment is described by the position of its middle point, its orientation and length. We use the epipolar geometry to constrain the location in the second image so that only 2 parameters are required to describe its counterpart: s_m, the position along the epipolar line, and θ the orientation. The length l, in the other images, is predicted by using (s_m, θ) and the epipolar geometry. For a given s_m and θ, we evaluate its probability of being correct by measuring the *edginess* f. It is a function of the image gradient:

$$f(s_m, \theta) = \sum_{r=-\frac{l}{2}}^{r=\frac{l}{2}} \|G(r)\| \cdot e^{-\frac{(\theta - \theta(r))^2}{2\sigma^2}}$$

where $G(r)$ is the image gradient at r, $\theta(r)$ its orientation. The function f is maximum when (A, B) lies on a straight edge and decreases quickly, when (A, B)

is not an edge or is poorly located. Further, f can be large even if if the edge is only partially visible in the image, that is occluded or broken.

The search for the most likely counterparts for the source edge now reduces to finding the maxima of f by discretizing θ and s_m and performing a 2-D search. In the presence of parallel structures, the edginess typically has several maxima that cannot be distinguished using only two images. However, using more than two images, we can reduce the number of matches and only keep the very best by checking for consistency across image pairs.

We can further reduce the hypothesis set by using the photometric edge attributes of section 2.2 after photometric equalization of the images. We compute the 2-D projections of each candidate 3-D edge into all the images. The image photometry in areas that pertains to at least one side of the 2-D edges should be similar across images. Figure 4 shows all matched 3-D segments as well as the manually measured CAD model for the house in Fig. 3A.

4 Coplanar Grouping of 3-D Segments

To group 3-D segments into infinite planes, we propose a simple and deterministic method that accounts for outliers in the data. It proceeds in two steps:

- **Explore**: We first find an initial set of hypotheses using a RANSAC style approach [5]: Given the relationships of section 2.3 and the 3-D geometry of the segments, we fit planes to pairs of related contours that are roughly coplanar. We then extend the support of those planes by iteratively including segments that are related to the hypothesis and that are close enough to the plane. After each iteration the plane parameters are re-approximated.
- **Merge**: We now have a set of plane hypotheses. Because all the edges belonging to the same physical plane may not be related in the sense of section 2.3, this plane may give rise to several hypotheses that must be merged. This is done by performing an F-test on pairs of parallel planar hypotheses to check whether or not they describe the same plane.

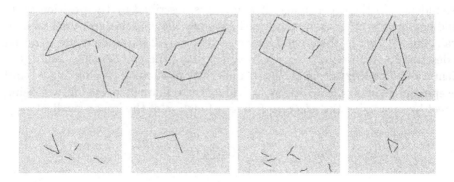

Figure 5 Selection of planes extracted from the 3-D segments of the house in Fig. 3A.

Each plane in Fig. 5 consists of a number of 3-D segments, some of which are correctly matched and do belong to a planar object part. However, quite a few 3-D segments are incorrectly matched and accidentally lie on the plane and other segments, such as the contours on the ground aligned with the roof plane, are correctly matched but do not belong to the object part.

5 Extracting and Selecting 2-D Enclosures

In the preceding section we presented an approach to group 3-D segments into *infinite* planes. However, only a subset of all segments on each plane actually belong to valid 3-D patches, see Fig. 5. To obtain an ordered list of 2-D contours describing a 3-D patch, we propose to group contours in 2-D where *more complete* data is available and subsequently merge the extracted enclosures with the corresponding planes. The tight coupling between the 2-D and 3-D processes plays an important role; the extracted planes that are *not* vertical initialize the enclosure finding algorithm. We therefore do not need to find *all* possible 2-D enclosures, only those that overlap with non-vertical planes.

We use the edge and region based approach described in [11] since it allows to group contours on other grounds than geometric regularity. The method consists of defining contour similarity relations (section 2.3), which are then used to build a relations graph, in which each cycle define a 2-D enclosure. At last, all extracted 2-D enclosures are ranked according to simple geometric shape criteria.

5.1 Extracting 2-D Enclosures

Instances of 2-D roof-primitives can be found by grouping related contours to polygonal shaped structures. A computationally attractive approach is to build a relations graph and use it to find these structures [6, 12]. By construction each cycle in the relations graph describe an enclosure. Each contour relation define a node in the graph and two nodes are linked together if they have a compatibly directed contour. We use a standard depth-first search algorithm to find cycles in the *directed* relations graph.

The procedure work as follows: select a not already used node *that belongs to the plane* and find all *valid* cycles in the graph given this start node. Pick the next not already used node *on the same plane* and iterate the procedure until there are no more nodes left. A *valid* cycle is a set of directed contours that have a boundary length not exceeding a large value and that does not form a self-loop; the boundary of the enclosure must be compact.

5.2 Selecting 2-D Enclosures

The above algorithm produces for each plane a set of 2-D enclosure hypotheses. To alleviate the fusion of enclosures and planes, we rank the enclosures within each plane according to simple geometrical shape criteria. We assume that each roof part has a compact and simple polygonal shape. In addition we require a large overlap between the contours in the 2-D enclosure and the corresponding 3-D segments of the plane. We propose the following criteria:

Shape simplicity Shape simplicity is defined as number of straight contours required to represent the enclosure boundary (including missing links). Given an error tolerance, we use a standard polygon approximation algorithm to compute the required number of straight lines. The simpler the description of a 2-D enclosure is, the more likely it is that it will describe a roof part.

Shape compactness Compactness is defined as the squared length of the boundary of the enclosure divided by the enclosed area.

3-D completeness The 3-D completeness is defined as the ratio of the length of the 3-D contours that lie on the enclosure boundary and on the plane, with respect to the total length of the enclosure boundary. This measure will be high whenever a large portion of the extracted 2-D contours have correctly matched 3-D segments that lie on the same infinite plane.

Figure 6 shows a few representative 2-D enclosures for the larger planes of the house in Fig. 3A. Two thresholds are applied, one for shape simplicity (≤ 10) and one for 3-D completeness (≥ 0.4). Together with the 3-D patch consistency test in next section these thresholds preclude highly unlikely hypotheses of 2-D enclosures *before* fusing them with planes to hypotheses of 3-D patches.

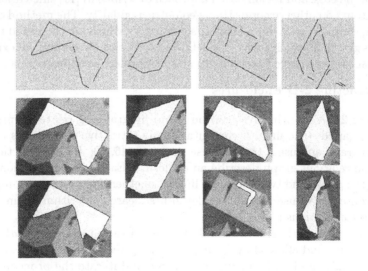

Figure 6 A few representative 2-D enclosures for the larger planes.

6 Finding Coherent 3-D Patches

Each 2-D enclosure describes a possible boundary description of the corresponding 3-D plane. It is reasonable to assume that roofs are usually constructed of adjoining planes. For this reason, only hypotheses of 3-D patches that mutually adjoin with other 3-D patches *along their boundaries* are retained. In addition we require that the 2-D contours belonging to the adjoining boundary of the 3-D patches are collinear in 2-D. Those 3-D patches that fulfill these constraints are consistent.

The iterative procedure initially selects a subset of 3-D patches and verifies the mutual consistency along the boundaries. If one or more 3-D patches do not fulfill this consistency, they are rejected and a new subset of 3-D patches is selected. Moreover the subset of 3-D patches should be maximally consistent, i.e. maximum number of mutually consistent boundaries. The order of selection is initially based on shape simplicity, and in a second step on the product of the normalized compactness and 3-D completeness. To obtain the 3-D coordinates of those contours that are contained in the 2-D enclosure but not on the plane, we project their endpoints onto the plane. The result is a complete 3-D boundary for each plane that is likely to describe a roof.

7 Results

We use the presented framework to extract complex roofs of houses in suburban scenes, see Fig. 1. The process is initialized by selecting a rectangular window enclosing the same house in all four images. It has been demonstrated [7] that this initialization procedure can be automated by locating elevation blobs in the digital surface model. After this initialization, the roof is *automatically extracted*.

The roof depicted in Fig. 7A is complex because it consists of several adjoining non-planar and non-rectangular shapes. The feature extraction finds 171 straight 2-D edges. The segment stereo matching produces 170 3-D segments, and the coplanar grouping extracts 33 infinite planes of which 7 are mutually adjoining and non-vertical. Given these 7 non-vertical planes, the algorithm finds 373 valid 2-D enclosures (resp. 3-D patches). The five 3-D patches in Fig. 7B are finally selected since they maximizes the geometrical shape score and mutual consistency among all 3-D patches. The result is a 3-D CAD model with 3-D segments, 3-D planes and their topology. This procedure yields the main parts of the roof, however, 3-D patches that are not mutually consistent with the final set of 3-D patches, but nevertheless belongs to the house, are not included. One such example is the 3-D patch describing the dormer window in Fig. 7A.

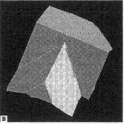

Figure 7 (A) cut-out from the aerial image in Fig.1A, (B) the reconstructed house roof in 3-D.

In Fig. 8 we demonstrate the performance of our approach on the entire scene in Fig. 1. To the automatically extracted CAD models of the roofs we add artificial vertical walls. The height of the vertical walls is estimated through the available digital terrain model (DTM). Ten of the twelve house roofs are extracted, nine of them with a high degree of accuracy and completeness. The marked house to the right is not complete, since the algorithm fails to extract the two triangular shaped planes, however, the corresponding 2-D enclosures are correctly extracted. The algorithm fails to extract the two upper left houses. The lower of the two is under construction and should not be included in the performance analysis. Even manual measuring this house is troublesome. The upper house is complicated because a bunch of trees cast large shadows on the right roof part. Because of these shadows the algorithm fails to find the corresponding plane, however, the left roof part is correctly reconstructed.

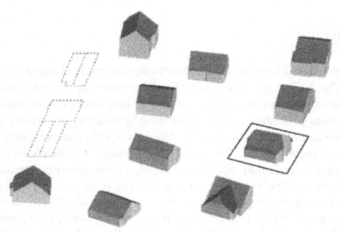

Figure 8 The result of running the algorithm on all houses in the scene of Fig.1. The artificial vertical walls are added and projected down to the ground. The ground height is estimated through the digital terrain model (DTM).

8 Conclusions and Future Work

We have addressed the problem of delineating complex objects in real images for which no specific template exists. We have implemented a working framework for the purpose of extracting complicated and generic roofs from sets of aerial images. The framework successfully performs the following tasks: hierarchical integration of multiple data sources, generic shape extraction, and efficient hypothesis generation and selection.

We combine geometry, photometry, chromaticity and stereo information about edges and their flanking regions. We thus make effective use of much of the available 2-D and 3-D information present in several images of a given site. As a result, our procedure is more robust than one that uses only partial information. We further use weak coplanarity constraint together with a generic extraction of 2-D enclosures allow us to effectively find instances of 3-D patches. The polygonal shape of these 3-D patches can be arbitrarily complex.

Future work will concentrate on: improvement of each individual module (whenever possible), better exploitation of existing knowledge, e.g. DSMs and

96

color, better interaction between 2-D and 3-D information, further tests with other data, and improving the selection of 2-D enclosures by also modelling the color homogeneity of the enclosed area with for example MDL encoding, see [6].

Bibliography

[1] N. Ayache and F. Lustman. Fast and reliable passive trinocular stereo vision. In *International Conference on Computer Vision*, June 1987.

[2] E. Baltsavias. *Multiphoto Geometrically Constrained Matching*. PhD thesis, Institute for Geodesy and Photogrammetry, ETH Zurich, December 1991.

[3] F. Bignone. Segment Stereo Matching and Coplanar Grouping. Technical Report BIWI-TR-165, Institute for Communications Technology, Image Science Lab, ETH, Zürich, Switzerland, 1995.

[4] R.T. Collins, A.R. Hanson, M.R. Riseman, and H. Schultz. Automatic Extraction of Buildings and Terrain from Aerial Images. In *Automatic Extraction of Man-Made Objects from Aerial and Space Images*, pages 169–178. Birkhäuser Verlag,1995.

[5] M.A Fischler and R.C Bolles. Random Sample Consensus: A Paradigm for Model Fitting with Applications to Image Analysis and Automated Cartography. *Communications ACM*, 24(6):381–395, 1981.

[6] P. Fua and A.J. Hanson. An Optimization Framework for Feature Extraction. *Machine Vision and Applications*, 4:59–87, 1991.

[7] N. Haala and M. Hahn. Data fusion for the detection and reconstruction of buildings. In *Automatic Extraction of Man-Made Objects from Aerial and Space Images*, pages 211–220. Birkhäuser Verlag, Basel, 1995.

[8] F. Heitger. Feature Detection using Suppression and Enhancement. Technical Report TR-163, Image Science Lab, ETH-Z, Switzerland, 1995.

[9] O. Henricsson. Inferring Homogeneous Regions from Rich Image Attributes. In *Automatic Extraction of Man-Made Objects from Aerial and Space Images*, pages 13–22. Birkhäuser Verlag, Basel, 1995.

[10] O. Henricsson and F. Heitger. The Role of Key-Points in Finding Contours. In J.O. Eklundh, editor, *Computer Vision - ECCV'94*, volume II, pages 371–383. Springer Verlag, Berlin, 1994.

[11] O. Henricsson and M. Stricker. Exploiting Photometric and Chromatic Attributes in a Perceptual Organization Framework. In *ACCV'95, Second Asian Conference on Computer Vision*, pages 258–262, 1995.

[12] T. Kim and J.P. Muller. Building Extraction and Verification from Spaceborne and Aerial Imagery using Image Understanding Fusion Techniques. In *Automatic Extraction of Man-Made Objects from Aerial and Space Images*, pages 221–230. Birkhäuser Verlag, Basel, 1995.

[13] C. Lin, A. Huertas, and R. Nevatia. Detection of Buildings from Monocular Images. In *Automatic Extraction of Man-Made Objects from Aerial and Space Images*, pages 125–134. Birkhäuser Verlag, Basel, 1995.

[14] D.M. McKeown, W.A. Harvey, and J. J. McDermott. Rule-Based Interpretation of Aerial Images. *IEEE Transactions on Pattern Analysis and Machine Intelligence*, 7(5):570–585, 1985.

[15] P. Suetens, P. Fua, and A. Hanson. Computational Strategies for Object Recognition. *ACM Computing Surveys*, 24(1):5–61, 1992.

[16] G. Wyszecki and W.S. Stiles. *Color Science*. Wiley, New York, 1982.

Three Dimensional Object Modeling via Minimal Surfaces

Vicent Caselles[1], Ron Kimmel[2], Guillermo Sapiro[3], and Catalina Sbert[1]

[1] Dept. of Mathematics and Informatics, University of Illes Balears,
07071 Palma de Mallorca, Spain
[2] Mail-stop 50A-2129, LBL, UC Berkeley, CA 94720, USA
[3] Hewlett-Packard Labs, 1501 Page Mill Road, Palo Alto, CA 94304

Abstract. A novel geometric approach for $3D$ object segmentation and representation is presented. The scheme is based on geometric deformable surfaces moving towards the objects to be detected. We show that this model is equivalent to the computation of surfaces of minimal area, better known as 'minimal surfaces,' in a Riemannian space. This space is defined by a metric induced from the $3D$ image (volumetric data) in which the objects are to be detected. The model shows the relation between classical deformable surfaces obtained via energy minimization, and geometric ones derived from curvature based flows. The new approach is stable, robust, and automatically handles changes in the surface topology during the deformation. Based on an efficient numerical algorithm for surface evolution, we present examples of object detection in real and synthetic images.

1 Introduction

One of the basic problems in image analysis is object detection. This can be associated with the problem of boundary detection, when boundary is roughly defined as a curve or surface separating homogeneous regions. "Snakes," or active contours, were proposed by Kass *et al.* in [17] to solve this problem, and were later extended to $3D$ surfaces. The classical snakes and $3D$ deformable surfaces approach are based on deforming an initial contour or surface towards the boundary of the object to be detected. The deformation is obtained by minimizing a functional designed such that its (local) minima is obtained at the boundary of the object [3, 36]. The energy usually involves two terms, one controlling the smoothness of the surface and another one attracting it to the boundary. These energy models are not capable of changing its topology when direct implementations are performed. The topology of the final surface will in general be as that of the initial one, unless special procedures are used for detecting possible splitting and merging [24, 34]. This approach is also non intrinsic, i.e., the energy functional depends on the parameterization. See for example [23, 39] for comments on advantages and disadvantages of energy approaches for deforming surfaces.

Recently, geometric models of deformable contours/surfaces were simultaneously proposed by Caselles *et al.* [4] and by Malladi *et al.* [23]. In these models, the curve/surface is propagating by an implicit velocity that also contains two terms, one related to the regularity of the deforming shape and the other attracting it to the boundary. The model is given by a geometric flow (PDE), based on mean curvature motion, and is not a result of minimizing an energy functional. This model automatically handles changes in topology when implemented using the level-sets numerical algorithm [26]. Thereby, several objects can be detected simultaneously, without previous knowledge of their exact number in the scene, and without special tracking procedures.

In [5], we have shown the formal mathematical relation between these two approaches for $2D$ object detection. We also extended them proposing what we named "geodesic active contours." The geodesic active contours model has the following main properties: 1- It connects energy and curve evolution approaches of active contours. 2- Presents the snake problem as a geodesic computation one. 3- Improves existing models as a result of the geodesic formulation. 4- Allows simultaneous detection of interior and exterior boundaries of several objects without special contour tracking procedures. 5- Holds formal existence, uniqueness, and stability results. 6- Stops automatically.

In this paper we extend the results in [5] to $3D$ object detection. The obtained geometric flow is based on geometric deformable surfaces. We show that the desired boundary is given by a minimal surface in a Riemannian space defined by the image. In other words, segmentation is achieved via the computation of surfaces of minimal area, where the area is defined in a non-Euclidean space. The obtained flow has the same advantages over other $3D$ deformable models, similar to the advantages of the geodesic active contours over previous $2D$ approaches.

We note that the deformable surfaces model is related to a number of previously or simultaneously developed results. It is of course closely related to the works of Terzopoulos and colleagues on energy based deformable surfaces, and the works by Caselles et $al.$ and Malladi et $al.$ [4, 23]. It is an extension of the $2D$ model derived in [5]. The basic equations in this paper, as well as the corresponding $2D$ ones in [5], were simultaneously developed in [18, 33]. Similar $3D$ models are studied in [37, 38] as well. Extensions to [4, 23] are presented also in [35]. The similitude and differences with those approaches will be presented after describing the basic principles of the model.

2   Energy and Geometry based approaches of deformable surfaces

The $3D$ extension of the $2D$ snakes, known as the deformable surface model, was introduced by Terzopoulos et $al.$ [36]. It was extended for $3D$ segmentation by many others (e.g. [9, 10, 11]). In the $3D$ case, a parameterized surface $v(r,s) = (x(r,s), y(r,s), z(r,s))$ $(r,s) \in [0,1] \times [0,1]$, is considered, and the energy functional is given by

$$E(v) = \int_{\Omega} \left[\omega_{10} \left| \frac{\partial v}{\partial r} \right|^2 + \omega_{01} \left| \frac{\partial v}{\partial s} \right|^2 + \omega_{11} \left| \frac{\partial^2 v}{\partial r \partial s} \right|^2 + \omega_{20} \left| \frac{\partial^2 v}{\partial r^2} \right|^2 + \omega_{02} \left| \frac{\partial^2 v}{\partial s^2} \right|^2 + P \right] drds,$$

where $P := - \| \nabla I \|^2$, or any related decreasing function of the gradient, where I is the image. The first terms are related to the smoothness of the surface, while the last one is responsible of attracting it to the object. The algorithm starts with an initial surface S_0, generally near the desired $3D$ boundary O, and tries to move S_0 towards a local minimum of E.

The geometric models proposed in [4, 23] can easily be extended to $3D$ object detection. Let $Q =: [0, a] \times [0, b] \times [0, c]$ and $I : Q \to I\!\!R^+$ be a given $3D$ data image. Let $g(I) = 1/(1 + |\nabla \tilde{I}|^p)$, where $\tilde{I}$ a regularized version of I, and $p = 1$ or 2. $g(I)$ acts as an edge detector so that the object we are looking for is ideally given by the equation $g = 0$. Our initial active surface S_0 will be embedded as a level set of a function $u_0 : Q \to I\!\!R^+$, say $S_0 = \{x : u_0(x) = 0\}$ with u_0 being positive in the exterior and negative in the interior of S_0. The evolving active surface is defined by $S_t = \{x : u(t, x) = 0\}$ where $u(t, x)$ is the solution of

$$\frac{\partial u}{\partial t} = g(I)|\nabla u|\mathrm{div}\left(\frac{\nabla u}{|\nabla u|}\right) + \nu g(I)|\nabla u| = g(I)(\nu + \mathbf{H})|\nabla u|, \qquad (1)$$

with initial condition $u(0, x) = u_0(x)$ and Neumann boundary conditions. Here $\mathbf{H} = \mathrm{div}\left(\frac{\nabla u}{|\nabla u|}\right)$ is the sum of the two principal curvatures of the level sets S (twice its mean curvature,) and ν is a positive real constant. The $2D$ version of this model was heuristically justified in [4, 23]. It contains: 1. A smoothing term: Twice the mean curvature in the case of (1). More efficient smoothing velocities as those proposed in [2, 7, 25] can be used instead of $\mathbf{H}$.[4]. 2. A constant balloon-type force ($\nu|\nabla u|$). 3. A stopping factor ($g(I)$). The sign conventions here are adapted to inwards propagating active contours. For active contours evolving from the inside outwards, we take $\nu < 0$. This is a drawback of this model: the active contours cannot go in both directions (see also [35]). Moreover, we always need to select $\nu \neq 0$ even if the surface is close to the object's boundary.

Our goal will be to define a $3D$ geometric model (with level set formulation) corresponding the minimization of a meaningful and intrinsic energy functional. It is motivated by the extension of $2D$ geometric model to the geodesic active contours as done in [5].

3 Three dimensional deformable models as minimal surfaces

In [5], a model for $2D$ object detection based on the computation of geodesics in a given Riemannian space was presented. This means that we are computing paths or curves of minimal (weighted) length. This idea may be extended to $3D$ surfaces by computing surfaces of minimal area, where the area is defined in a given Riemannian space. In the case of surfaces, arc length is replaced by surface area $A := \int \int da$, and weighted arc length by "weighted" area

$$A_R := \int \int g(I) da, \qquad (2)$$

---

[4] Although curvature flows smooth $2D$ curves [15, 16, 30, 31], a $3D$ geometric flow that smoothes all possible surfaces was not found [25]. Frequently used are mean curvature or the positive part of the Gaussian curvature flows [2, 7].

where da is the (Euclidean) element of area. Surfaces minimizing A are denoted as *minimal surfaces* [27]. In the same manner, we will denote by minimal surfaces those surfaces that minimize (2). The area element da is given by the classical area element in Euclidean space, while the area element da_r is given by $g(I)da$. Observe that da_r corresponds to the area element induced on a surface of $I\!\!R^3$ by the metric of $I\!\!R^3$ given by $g_{ij}\,dx_i dx_j$ with $g_{ij} = g(I)^2 \delta_{ij}$. This is the 3D analogue of the metric used in [5] to construct the geodesic active contour model. The energy A_R can be formally derived from the original energy formulation using basic principles of dynamical systems [5], further justifying this model. The basic element of our deformable model will be given by minimizing (2) by means of an evolution equation obtained from its Euler-Lagrange. Let us point out the basic characteristics of this flow.

The Euler-Lagrange of A is given by the mean curvature $\mathbf{H}$, resulting a curvature (steepest descent) flow $\frac{\partial S}{\partial t} = \mathbf{H}\mathcal{N}$, where S is the 3D surface and $\mathcal{N}$ its inner unit normal. With the sign conventions explained above, the corresponding level set [26] formulation is $u_t = |\nabla u|\text{div}\left(\frac{\nabla u}{|\nabla u|}\right) = |\nabla u|\mathbf{H}$. Therefore, the mean curvature motion provides a flow that computes (local) minimal surfaces [8]. Computing the Euler-Lagrange of A_R, we get

$$S_t = (g\mathbf{H} - \nabla g \cdot \mathcal{N})\mathcal{N}. \tag{3}$$

This is the basic weighted minimal surface flow. Taking a level set representation, the steepest descent method to minimize (2), yields

$$\frac{\partial u}{\partial t} = |\nabla u|\text{div}\left(g(I)\frac{\nabla u}{|\nabla u|}\right) = g(I)|\nabla u|\text{div}\left(\frac{\nabla u}{|\nabla u|}\right) + \nabla g(I) \cdot \nabla u. \tag{4}$$

We note that comparing with previous geometric surface evolution approaches for 3D object detection, the minimal surfaces model includes a new term, $\nabla g \cdot \nabla u$. This term is fundamental for detecting boundaries with fluctuations in their gradient; see [5] for details.

As in the 2D case, we can add a constant force to the minimization problem (minimizing volume), obtaining the general *minimal surfaces model* for object detection:

$$\frac{\partial u}{\partial t} = |\nabla u|\text{div}\left(g(I)\frac{\nabla u}{|\nabla u|}\right) + \nu g(I)|\nabla u|. \tag{5}$$

This is the flow we will further analyze and use for 3D object detection. It has the same properties and geometric characteristics as the geodesic active contours, leading to accurate numerical implementations and topology free object segmentation. The following results can be proved for this flow

Theorem 1 ([6]). *Assume that $g \geq 0$ is sufficiently smooth. Then, for any Lipschitz initial condition u_0, there exists a unique viscosity solution $u(t, x)$ of (5) with $u(0, x) = u_0(x)$.*

In practice, we choose an initial condition u_0 with $\{x : u_0(x) \leq 0\}$ containing the desired object and we let it evolve according to (5). The active surface $\mathcal{S}(t)$ is the boundary of the set $\{x : u(t, x) \leq 0\}$. One can show [6] the independence of the evolution from the particular function u_0 used to define the initial active surface. Finally, the model (5) enables us to show the correctness of the geometric formulation in some special yet important cases. We have

Theorem 2 ([6]). *Assume that $\mathcal{S} = \{x : g(x) = 0\}$ is a compact connected smooth surface embedded in $\mathbb{R}^3$ which is unknotted. Then, if the constant ν is sufficiently large, then $\mathcal{S}(t) \to \mathcal{S}$ in the Hausdorff distance as $t \to \infty$. The same result can be proved for all compact smooth surfaces which can be unknotted by adding them a finite number of handles. And also for finite unions of surfaces in that class.*

This covers a large class of surfaces which can be found in practice. Several questions arise concerning this theorem: 1- how large should the constant ν be? It can be seen from the proof in [6] that ν should be larger than the mean curvature of the evolving surfaces. A reasonable assumption is that ν should be larger than the curvature of the desired surface. On the other hand, for initial condition of a surface close to the desired object, one can choose $\nu = 0$. In practice, convergence can also be obtained for $\nu = 0$ if obstacles do not stop the active surface, yet the process is slower. 2- The presence of noise may disturb the convergence. This can be avoided by preprocessing the original image I. In practice, if the noise is not dominant and is not structured along a surface, it will not stop the active surface. 3- The above theorem assumes that the desired surface is given by $g(x) = 0$. We do not give a proof for the general case in which $g(x) > 0$ along the desired surface. In that case the equilibrium position should be along the local minimum and a balance of the forces yields the result.

In [13] it was shown that the curvature along the $2D$ geodesics minimizing the weighted arclength may be bounded by $|\kappa| \leq \sup_{p \in [0,a] \times [0,b]} \left\{ \frac{|\nabla g(I(p))|}{g(I(p))} \right\}$. This result is obtained directly form the Euler-Lagrange equation of the weighted arclength integral. It is easy to see that there is no need for the geodesic itself for limiting the curvature values. In [13], motivated by [21, 22], this bound helped in the construction of different potential functions.

A straightforward generalization of this result to our three dimensional model yields the bound over the mean curvature $\mathbf{H}$. From the equations above, it is clear that for a steady state (*i.e.* $\mathcal{S}_t = 0$) the mean curvature along the surface $\mathcal{S}$ is given by $\mathbf{H} = \frac{\nabla g \cdot \mathcal{N}}{g} - \nu$. We readily obtain the following upper bound for the mean curvature magnitude along the final surface $|\mathbf{H}| \leq \sup \left\{ \frac{|\nabla g|}{g} \right\} + |\nu|$, where the sup operation is taken over all the $3D$ domain. The above bound gives an estimation of the allowed gaps in the edges of the object to be detected as a function of ν. A pure gap is defined as a part of the object boundary at which, for some reason, $g = $ constant$\neq 0$ in a large enough neighborhood. At these locations $|\mathbf{H}| = |\nu|$. Therefore, pure gaps of radius larger than $1/\nu$ will cause the propagating surface to penetrate into the segmented object. It is also clear that

$\nu = 0$ allows the detection of gaps of any given size, and the boundary at such places will be detected as the minimal surface 'gluing' the gaps boundaries.

The basic equations for $3D$ segmentation here described, and those for $2D$ in [5], were recently independently proposed by Kichenassamy et al. [18, 19] based on a slightly different initial approach. Shah [33] also recently presented a $2D$ active contours formulation as the one in [5], which is the $2D$ analogue of the model here described. Although the works in [18, 33] also present the problem of $2D$ active contours as geodesic computations, they do not show the connections between energy models and curve evolution ones. Actually, to the best of our knowledge, non of the previous works on curve/surface evolution for object segmentation show the mathematical relation between those models and classical energy approaches, as done in [5] for the $2D$ case and extended in this paper and in [6] for the $3D$ one. Actually, in general the two approaches are considered independent. In [5, 6] and here we show that they are mathematically connected, and one can enjoy the advantages of both of them in the same model. Although the extension from the $2D$ model to the $3D$ one is easy, no $3D$ examples are presented in [18, 33]. Also, not all the theoretical results here quoted [6] can be found in [18, 19, 33] (in [19] the authors do show a number of very important theoretical results as those in [6] and quoted here). Three dimensional examples are given in [37], where similar equations as the presented are proposed. The equations there are obtained by extending the flows in [4, 23], again without showing that they can be obtained in a natural fashion from a re-interpretation of energy based snakes via minimal surfaces. In [35], motivated by work reported in [20], the authors based their work on the models in [4, 23]. One of the key ideas there, motivated by the shape theory of shocks developed by Kimia et al., is to perform multiple initializations. A normalized version of A was derived in [14] from a different point of view, giving as well different flows for $2D$ active contours. Extension of that model to $3D$ was presented in [12].

4 Experimental results

We now present some examples of our minimal surfaces deformable model (5). The numerical implementation is based on the algorithm for surface evolution via level sets [26]. It allows the evolving surface to change topology without monitoring the deformation. Using new results in [1], the algorithm can be made to converge very fast. In the numerical implementation of Eq. (5) we have chosen central difference approximation in space and forward difference approximation in time. This simple selection is possible due to the stable nature of the equation, however, when the coefficient ν is taken to be of high value or when the gradient term is dominant, more sophisticated approximations are required [26].

In our examples, the initialization is in general given by a surface surrounding all the possible objects in the scene. In the case of outward flows [5], a surface is initialized inside each object. The first example of the minimal surfaces deformable model is presented in Figure 1. A 'knotted surface' composed of two tori forming a 'chain' is detected. The initial surface is an ellipsoid surrounding the two tori (top left). Note how the model manages to change its topology and detect the final surface (bottom right).

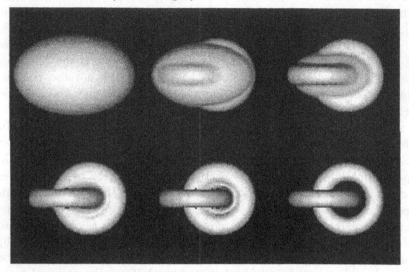

Fig. 1. Detection of two linked tori.

Figure 2 presents the $3D$ detection of a tumor in an MRI image. The initial surface is presented in the first row on the left followed by 3 evolution steps. The final surface, the 'weighted minimal surface', is presented at the lower right frame. Figure 3 shows slices of the $3D$ detection painted on the corresponding MRI data.

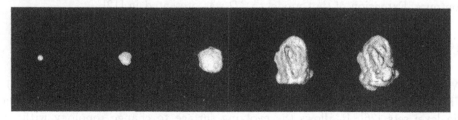

Fig. 2. Detection of a tumor in MRI.

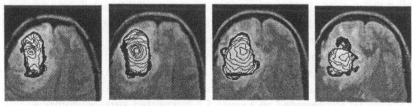

Fig. 3. Slices of the $3D$ detection in Figure 2.

Figure 4 presents the segmentation of the interior and exterior of a $3D$ MRI data of a bone. The two slices show the process of locating the outer and inner parts. Two views of the final segmentation of the inner and outer parts are presented in upper and lower rows. This figure also demonstrate the power of the the proposed technique in accurate analysis of medical images.

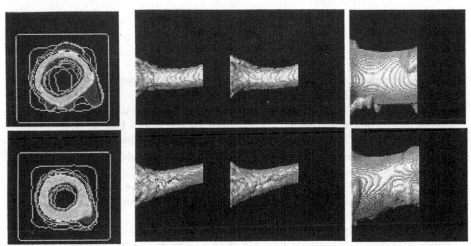

Fig. 4. Two slices and two orthographic views of $3D$ detection of the inner and outer parts of a bone in an MRI image.

5   Concluding remarks

In this paper we presented a novel formulation of deformable surfaces for $3D$ object detection, extending our previous $2D$ work [5]. We proposed a solution to deformable surfaces approach for boundary detection. It is given as a minimal surface in a Riemannian space defined by a metric derived from the given image. This means that detecting the object is equivalent to finding a surface of minimal weighted area. This approach allowed to relate classical energy based models with new surface evolution ones. The minimal surfaces formulation introduced a new term that attracts the deforming surface to the boundary, improving the detection of boundaries with large differences in their gradient. This new term also frees the model from the need to estimate critical parameters. Therefore, the minimal surfaces formulation not only connects previous models, but also improves them. Results regarding existence, uniqueness, stability, and correctness of the solution obtained by our model were summarized and will be reported elsewhere.

Experiments for different kind of images were presented. These experiments demonstrate the ability to detect several objects, as well as the power to simultaneously detect interior and exterior boundaries. The sub-pixel accuracy intrinsic to the algorithm allows to perform accurate measurements after the object is detected [28].

Acknowledgments

The authors thank Prof. J. Blat, Prof. P.L. Lions, and Prof. J.M. Morel for interesting discussions and their constant support. The work of RK was partially supported by the Applied Mathematical Science subprogram of the Office of Energy Research, U.S. Department of Energy, under Contract Number DE-AC03-76SF00098.

References

1. D. Adalsteinsson and J. A. Sethian, "A fast level set method for propagating interfaces," *J. of Comp. Phys.*, 118:269–277, 1995.
2. L. Alvarez, F. Guichard, P. L. Lions, and J. M. Morel, "Axioms and fundamental equations of image processing," *Arch. Rational Mechanics* 123, 1993.
3. A. Blake and A. Zisserman, *Visual Reconstruction*, MIT Press, Cambridge, 1987.
4. V. Caselles, F. Catte, T. Coll, F. Dibos, "A geometric model for active contours," *Numerische Mathematik* 66, pp. 1-31, 1993.
5. V. Caselles, R. Kimmel, and G. Sapiro, "Geodesic active contours," to appear *International Journal of Computer Vision*. A short version appears at *ICCV'95*, Cambridge, June 1995.
6. V. Caselles, R. Kimmel, G. Sapiro and C. Sbert, "Minimal surfaces: A three-dimensional segmentation approach," *Technion Technical Report* 973, June 1995.
7. V. Caselles and C. Sbert, "What is the best causal scale-space for 3D images?," *SIAM J. on Applied Math*, to appear.
8. D. Chopp, "Computing minimal surfaces via level set curvature flows," *J. of Comp. Phys.*, 106(1):77–91, 1993.
9. P. Cinquin, "Un modele pour la representation d'images medicales 3d," *Proc. Euromedicine, Sauramps Medical*, 86, pp 57-61, 1986.
10. L. D. Cohen, "On active contour models and balloons," *CVGIP: Image Understanding* 53, pp. 211-218, 1991.
11. I. Cohen, L. D. Cohen, and N. Ayache, "Using deformable surfaces to segment 3D images and infer differential structure," *CVGIP: Image Understanding* 56, pp. 242-263, 1992.
12. L. D. Cohen and I. Cohen, "Finite element methods for active contour models and ballons for 2D and 3D images," *IEEE Tran. on PAMI* 15(11), November, 1993.
13. L. D. Cohen, and R. Kimmel, "Minimal geodesics of active contour models for edge integration and segmentation," TR-INRIA, 1995.
14. P. Fua and Y. G. Leclerc, "Model driven edge detection," *Machine Vision and Applications,* 3, pp. 45-56, 1990.
15. M. Gage and R. S. Hamilton, "The heat equation shrinking convex plane curves," *J. Differential Geometry* 23, pp. 69-96, 1986.
16. M. Grayson, "The heat equation shrinks embedded plane curves to round points," *J. Differential Geometry* 26, pp. 285-314, 1987.
17. M. Kass, A. Witkin, and D. Terzopoulos, "Snakes: Active contour models," *International Journal of Computer Vision* 1, pp. 321-331, 1988.
18. S. Kichenassamy, A. Kumar, P.Olver, A. Tannenbaum, and A. Yezzi, "Gradient flows and geometric active contour models," *Proc. ICCV*, Cambridge, June 1995.

19. S. Kichenassamy, A. Kumar, P.Olver, A. Tannenbaum, and A. Yezzi, "Conformal curvature flows: from phase transitions to active vision," to appear *Archive for Rational Mechanics and Analysis.*

20. B. B. Kimia, A. Tannenbaum, and S. W. Zucker, "Shapes, shocks, and deformations, I," *International Journal of Computer Vision* 15, pp. 189-224, 1995.

21. R. Kimmel, A. Amir, A. M. Bruckstein, "Finding shortest paths on surfaces using level sets propagation," *IEEE–PAMI*, 17(6):635–640, 1995.

22. R. Kimmel, N. Kiryati, A. M. Bruckstein, "Distance maps and weighted distance transforms," *Journal of Mathematical Imaging and Vision*, Special Issue on Topology and Geometry in Computer Vision, to appear.

23. R. Malladi, J. A. Sethian and B. C. Vemuri, "Shape modeling with front propagation: A level set approach," *IEEE Trans. on PAMI*, January 1995.

24. T. McInerney and D. Terzopoulos, "Topologically adaptable snakes," *Proc. ICCV*, Cambridge, June 1995.

25. P. J. Olver, G. Sapiro, and A. Tannenbaum, "Invariant geometric evolutions˙ of surfaces and volumetric smoothing," *SIAM J. of Appl. Math.*, to appear.

26. S. J. Osher and J. A. Sethian, "Fronts propagation with curvature dependent speed: Algorithms based on Hamilton-Jacobi formulations," *Journal of Computational Physics* 79, pp. 12-49, 1988.

27. R. Osserman, *Survey of Minimal Surfaces*, Dover, 1986.

28. G. Sapiro, R. Kimmel, and V. Caselles, "Object detection and measurements in medical images via geodesic active contours," SPIE-Vision Geometry, San Diego, July 1995.

29. G. Sapiro, R. Kimmel, D. Shaked, B. B. Kimia, and A. M. Bruckstein, "Implementing continuous-scale morphology via curve evolution," *Pattern Recog.* 26:9, pp. 1363-1372, 1993.

30. G. Sapiro and A. Tannenbaum, "On affine plane curve evolution," *Journal of Functional Analysis* 119:1, pp. 79-120, 1994.

31. G. Sapiro and A. Tannenbaum, "Affine invariant scale-space," *International Journal of Computer Vision* 11:1, pp. 25-44, 1993.

32. H.M. Soner, "Motion of a set by the curvature of its boundary," *J. of Diff. Equations* 101, pp. 313-372, 1993.

33. J. Shah, "Recovery of shapes by evolution of zero-crossings," Technical Report, Math. Dept. Northeastern Univ. Boston MA, 1995.

34. R. Szeliski, D. Tonnesen, and D. Terzopoulos, "Modeling surfaces of arbitrary topology with dynamic particles," *Proc. CVPR*, pp. 82-87, 1993.

35. H. Tek and B. B. Kimia, "Image segmentation by reaction-diffusion bubbles," *Proc. ICCV*, Cambridge, June 1995.

36. D. Terzopoulos, A. Witkin, and M. Kass, "Constraints on deformable models: Recovering 3D shape and nonrigid motions," *Artificial Intelligence* 36, pp. 91-123, 1988.

37. R. T. Whitaker, "Volumetric deformable models: Active blobs," *ECRC TR* 94-25, 1994.

38. R. T. Whitaker, "Algorithms for implicit deformable models," *Proc. ICCV'95*, Cambridge, June 1995.

39. S. C. Zhu, T. S. Lee, and A. L. Yuille, "Region competition: Unifying snakes, region growing, energy/Bayes/MDL for multi-band image segmentation," *Proc. ICCV*, Cambridge, June 1995.

Class Based Reconstruction Techniques Using Singular Apparent Contours

G.J.Fletcher and P.J.Giblin

Dept. of Mathematics, University of Liverpool, PO BOX 147, LIVERPOOL, L69 3BX, U.K.

Abstract. We present methods for the *global* reconstruction of some classes of special surfaces. The contour ending (cusp on the apparent contour) is tracked under a dynamic monocular perspective observer. The classes of surfaces considered are surfaces of revolution (SOR), canal surfaces and ruled surfaces. This paper presents theoretical methods for surface reconstruction and error analysis of reconstruction under noise. We find the techniques used exhibit stability even under large noise.
This work has added to the accumulating body of work that has arisen in the computer vision community, concerning the differential geometric aspects of special surface classes.

1 Introduction.

We present methods for the *global* reconstruction of some classes of special surfaces by tracking contour endings (cusps) [1] of the apparent contour (also known as the profile and the occluding contour) under a known dynamic monocular perspective observer.

There has been considerable interest in the vision community concerning families of surfaces ([1], [6], [7], [10], [13], [14] for instance). Much of the literature exploits *rich image features*, such as inflections, bitangents, the symmetry set, to aid reconstruction and viewpoint-invariant representation. While there have been some theoretical results concerning the cusps of apparent contours on special surfaces [11],[12], there has been little exploitation of the geometry with regard to reconstruction.

We examine surfaces of revolution (SOR), canal surfaces (piped) and ruled surfaces. Each of these special types of surface is generated in a special way by a moving curve. For example the ruled surface is generated by a sweeping line, the SOR generated by a varying radius circle centred on a straight line and the canal surface is described by sweeping a circle along a space curve, keeping it in the normal plane. (The canal surface can also be considered as an envelope of spheres of constant radius centred along a space curve). If we can recover the generator curve then we recover part of the original surface, even resulting in areas that are unseen and beyond the *frontier* [5]. We recommend [9] for an introduction to the differential geometry of the above classes of surfaces.

---

[1] We use the terms 'contour ending' and 'cusp' synonymously. A cusp is observed in the image for a transparent surface and for brevity we often refer to 'cusps'.

For each class of surface we provide simulated experiments that illustrate the technique and demonstrate the stability of the reconstruction under extremely noisy data. This simulates the uncertainty in the detection of the contour ending that is present in any practical situation. A contour ending can be seen in Figure 1 where it appears as a dark blob, but the observed location is subject to error.

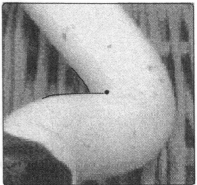

Fig. 1. Canal surface with a T-junction and a contour ending magnified on the right.

2 Background.

For smooth curved surfaces an important image feature is the apparent contour or profile. This is the projection of the locus of points on the surface which separates the visible from occluded parts. Under perspective projection this locus—the contour generator or critical set—can be constructed as the set of points on the surface where rays through the projection centre c are tangent to the surface. Each viewpoint will generate a different contour generator with the contour generators 'slipping' over the visible surface under viewer motion. This is the familiar situation of [2].

The projection of the contour generator in the image sphere or image plane gives a curve called the apparent contour or profile. This paper is concerned with the situation when the apparent contour ends, or cusps. It is well known (e.g. [8, p.422]) that the apparent contour cusps if and only if the viewing direction is along an *asymptotic direction* and this is equivalent to the view direction being tangent to the contour generator. Recall that a hyperbolic surface patch has two asymptotic directions, and if we extend these directions indefinitely we expect the lines to fill a region of space. Thus any camera position in that region will lie on some asymptotic direction and hence we expect to see cusps in the image.

We recall some results concerning the tracking of cusps from [3].

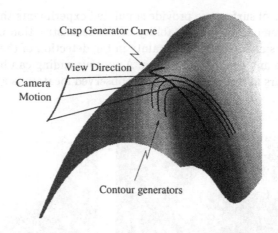

Fig. 2. Cusp generator on a smooth surface.

2.0.1 Definition: *If a point r is on the contour generator and projects in the image to a cusp point then r is said to be a **cusp generator point**. As the camera moves the locus of cusp points in the image is the **cusp locus**, and the locus of cusp generator points on the surface is the **cusp generator curve**. See Figure 2.*

The following proposition informs us that by tracking cusps we can reconstruct a surface strip (i.e. a curve with surface normals) complete with the second fundamental form at the cusp generator points. The main results of this paper concern the extending of this surface strip globally, in a certain sense, for certain classes of surfaces.

As in [3] we consider the camera measurements to be in unrotated *world* coordinates. Using an image sphere with centre **c** for mathematical convenience, **c** + **p** is the position in world coordinates of a point on the apparent contour. The unit vector **p** runs from the centre of the sphere to the apparent contour.

2.0.2 Proposition:[3] *If the camera motion is* **c**(t), *the cusp generator curve* **r**(t), *the cusp locus* **p**(t), *the surface normal (equal to the apparent contour normal) at the cusp point* **n**, *with* **r** = **c** + λ**p** *then,*

$$\lambda = -\frac{\mathbf{c}_t.\mathbf{n}}{\mathbf{p}_t.\mathbf{n}}$$

$$K = \frac{-(\mathbf{p}_t.\mathbf{n})^4}{[\mathbf{p},\mathbf{c}_t,\mathbf{p}_t]^2}$$

$$H = \frac{\mathbf{p}_t.\mathbf{n}(\mathbf{c}_{tt}.\mathbf{n}\ \mathbf{p}_t.\mathbf{n} - \mathbf{c}_t.\mathbf{n}\ \mathbf{p}_{tt}.\mathbf{n} - 2\mathbf{p}.\mathbf{c}_t(\mathbf{p}_t.\mathbf{n})^2)}{2[\mathbf{p},\mathbf{c}_t,\mathbf{p}_t]^2}$$

where λ *is the depth,* K *is the Gauss curvature and* H *is the Mean curvature, and the suffix denotes differentiation with respect to t.*

A major practical concern is the actual detection of contour endings. Recognising this we provide simulations using large uncertainties in the position in the image of the contour ending, typically ten pixels. We have reported elsewhere [4] of the stability of the formulas in Proposition 2.0.2 under image uncertainty, and the stability is high even under extreme noise.

3 Surfaces of Revolution.

3.1 Theory.

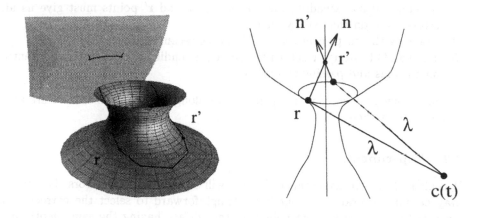

Fig. 3. Left figure: apparent contour (projected on to image sphere) and contour generator (on surface) shown for a particular value of t. Observe the cusp generator pair r and r' on the same parallel. Right figure: cusp generator pair r and r' on a surface of revolution with surface normals n and n' intersecting on the axis.

We now show that by tracking a cusp pair on a surface of revolution (SOR) global information about the surface can be found. More specifically, we assert that by tracking the cusp pair over parallels of our SOR we can reconstruct those parallels. A parallel (section) of a SOR is a plane section of a SOR perpendicular to the axis of rotation. It is a plane circle.

We shall need the following facts.

3.1.1 Fact: *If the point r generates an ordinary cusp for a certain camera position then there exists another point r' on the same parallel with the same depth to the camera that also generates a cusp. Note that the surface at r is congruent to that at r', in particular the Gauss curvatures are equal. See Figure 3.*

This fact tells us that cusps always appear in the image in pairs, and resulting from the same parallel. The following facts will be used in the reconstruction process.

3.1.2 Fact: *The normals to a SOR along a parallel all intersect at a point on the axis, see Figure 3.*

The basic reconstruction technique is as follows.

1. We observe a cusp pair in the image. Note that cusp pairs from the same parallel on the SOR have equal depths and Gauss curvatures by Fact 3.1.1, so we can easily verify from the image which of the cusps we observe do in fact arise from the same parallel.
2. We reconstruct the depth using Proposition 2.0.2 to get two points $\mathbf{r}, \mathbf{r}'$ these are the so-called cusp-generator points.
3. The surface normals are preserved under perspective projection to an image sphere, and so extending the normals at $\mathbf{r}$ and $\mathbf{r}'$ points must give us an intersection on the axis by Fact 3.1.2.
4. Tracking the cusps over time gives us the reconstructed axis.
5. The parallel through $\mathbf{r}$ is then the circle perpendicular to the axis with centre on the axis and passing through $\mathbf{r}$.

As our camera moves the cusp pair sweeps along the parallels and we are able to reconstruct them.

3.2 Experiment.

It is clear that in practice this technique will be susceptible to errors. The image may contain several cusps but it is straightforward to select the correct pair since these cusps arise from points on the surface having the same depth and Gaussian curvature (see Proposition 2.0.2). This provides a consistency check.

In practice when we reconstruct the cusp generator points and extend the normals we find they do not quite intersect. We take the nearest point in this instance and fit an axis to the noisy points.

The reconstruction technique was tested for different amounts of error in the observation of the cusp images. An error of x degrees means that up to x degrees of Gaussian noise was added to the cusp locus on the image sphere to give a noisy locus. For a camera with a focal length of 20mm and pixel density of 500pixels per 5mm, we find that an angular separation of 0.03 degrees is about 1 pixel. For the following SOR experiments errors of 0.3 degrees (10 pixels) and 0.6 degrees (20 pixels) were used.

We now produce some simulated examples which demonstrate the reconstruction technique. The surface used was the following,

$$\mathbf{r}(s,\theta) = ((1+s^2)\cos\theta, (1+s^2)\sin\theta, s)$$

and the camera motion,

$$\mathbf{c}(t) = (10 + 2t, 0.3t + 0.1t^2, -5 + 4t).$$

Note that the axis of the SOR is the z-axis $(0,0,1)$. We observed a cusp pair at discrete times and added some Gaussian noise of various amounts. This was then smoothed with a cubic curve via a least-squares method to give the observed cusp loci. The depth was calculated and then the nearest intersection point to the normals was calculated. This gave points on the SOR axis, and a straight axis was fitted. The parallels could then be generated resulting in a radius function that could then be smoothed giving a complete SOR.

We now illustrate some of the results for an error of 0.3 degrees and 0.6 degrees. The reconstructed axis for an error of 0.3 degrees was calculated as $[.051 - .001u, -.013 + .003u, -1.198 + .999u]$, recall that the actual axis is $[0, 0, u]$. The axis for 0.6 degrees is $[.158 - .012u, -.074 + .013u, -1.142 + .999u]$. See Figure 4.

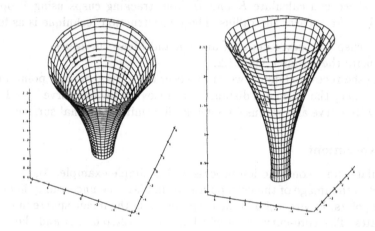

Fig. 4. Actual surface (cut away) compared with recovered surface (0.3degs) (left). Actual surface (cut away) compared with reconstructed surface (0.6 degs) (right).

4 Canal Surfaces.

4.1 Theory

Let $\gamma(t)$ be a space curve and $N(t)$ be its principal normal and $B(t)$ be the binormal. Then the standard parameterisation for a canal surface is the following:

$$\mathbf{r}(t, \theta) = \gamma(t) + r(N(t) \cos \theta + B(t) \sin \theta).$$

We can also think of the canal surface as an envelope of a family of spheres of radius r centred on $\gamma(t)$.

4.1.1 Definitions. *The space-curve γ is the* **core curve***, the factor r is the (constant)* **radius** *of the canal surface. The circle $\gamma(t_0) + r(N(t_0) \cos \theta + B(t_0) \sin \theta)$ is the* **characteristic circle** *for $t = t_0$.*

We assert that by tracking a single cusp along the canal surface we can reconstruct the characteristic circles (and hence the complete surface) as the cusp sweeps along the surface. We note that this reconstruction technique works with incomplete viewer information, such as when only one 'side' of the canal surface is visible. We shall need the following fact.

4.1.2 Fact:

1. *The radius of a canal surface can be expressed in terms of the Gaussian curvature K, and the Mean curvature H, as $r = \frac{H - \sqrt{H^2 - K}}{K}$.*
2. *The normal to a canal surface at a point p passes through the centre of the characteristic circle of p.*

Recall that we can calculate K and H from tracking cusps using Proposition 2.0.2 and so can recover the radius. The reconstruction technique is as follows,

1. Track cusp to recover depth, Gaussian and Mean curvature.
2. Calculate the radius r via 4.1.2.
3. Using the recovered depth we can recover the cusp generator point and then move along the normal a distance r to recover the core curve by 4.1.2.
4. The core curve and radius completely determine the canal surface.

4.2   Experiment.

We simulate the reconstruction process with a simple example. Again noise will be added to the image of the cusp points to simulate the uncertainty in detecting the cusp points. Figure 5 shows the cusp points on the image sphere in theta/phi coordinates. The core-curve used will be, $\gamma(t) = (2t, 0.6t^2, 0)$ and the radius 1. Note that the core curve of this canal surface is planar; this is just to simplify the calculations and does not imply a restriction inherent in the technique used. An error of 0.5 degrees was added in this example and the recovered radius was 0.973. It is difficult to quantify the error in the core curve, but Figure 6 shows the actual and recovered core curves. The recovered and actual surfaces are shown in Figure 6. It is unclear how best to empirically measure the 'success' of the reconstruction, other than simply a visual inspection. Figure 7 shows a series of experiments performed on different canal surfaces all with radius one, and varying camera motions. The horizontal axis indicates increasing noise added, and the vertical axis shows the recovered radius. A deviation from a radius equal to one, shows the effect of the noise. We don't expect this relationship to be simple since the radius depends on second derivatives of the cusp locus (Fact 4.1.2). We merely wish to assess the stability under large noise.

5   Ruled surfaces.

5.1   Theory.

We now consider tracking a cusp on a ruled surface. As the cusp sweeps across the rulings we find that we are able to reconstruct the rulings and hence the

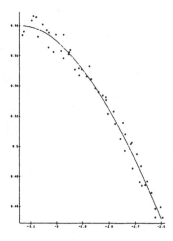

Fig. 5. Image of noisy cusp points with a cubic curve fitted.

whole surface. The crucial observation is contained in the following fact from [8, p.361].

5.1.1 Fact: *If the angle between asymptotic directions at a hyperbolic point of a surface is ϕ, then $\tan \phi = \frac{\sqrt{-K}}{H}$.*

By tracking cusps we can recover K, H, the depth, the surface normal and one asymptotic direction (namely the view direction). Recall that for a ruled surface one asymptotic direction is always along the ruling and $K \leq 0$. The ingredients are now all present along with Fact 5.1.1, and the recipe is now given.

1. Track cusp and recover the depth, K and H.
2. The view direction is one asymptotic direction and the other is the ruling. Calculate the angle between them by Fact 5.1.1 and since we know that the ruling lies in the tangent plane this constrains it.
3. This gives the direction of the ruling, and it passes through the cusp generator point which can be recovered with knowledge of the depth.

Figure 7 shows the result of a reconstruction experiment on a ruled surface where the maximum Gaussian error in observed cusp points was 0.3 degrees.

6 Conclusion

We have shown that the cusps on the apparent contours of certain classes of smooth surfaces give enough information to enable the *complete* reconstruction of the surface by tracking cusps alone. This work has built on the theory developed in [3].

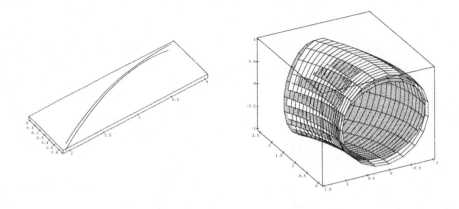

Fig. 6. Actual and recovered core curves (left), and actual and recovered surfaces (right). Note that the scale and orientation are different on left and right.

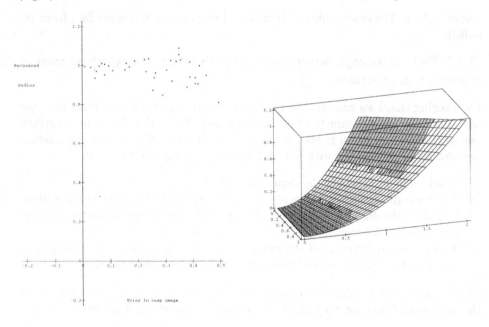

Fig. 7. Left figure: increase in maximum angular error of cusp points (horizontal) with recovered radius of canal surface (vertical). Right figure: actual and reconstructed ruled surface with an error of 0.3 degrees.

Recognising that cusps are hard to detect in real images we have given an error analysis that demonstrates the stability of the reconstruction even under large image perturbation.

Future work will include extending the methods to other classes of special surfaces, and the analysis of real image data.

Acknowledgements: we thank Dr.R.Morris for his software 'The Liverpool Software Modelling Package' that produced Figure 2 and Figure 3 and Dr.A.Zisserman for Figure 1.

We also acknowledge support through grant GR/H59855 from the British research council EPSRC (formerly SERC). The first author is supported by EPSRC.

References

1. J.W.Bruce, P.J.Giblin, F.Tari, 'Special surfaces and their duals; preliminary version', *Liverpool University Preprint.* (1995).
2. R.Cipolla and A.Blake, 'Surface shape from deformation of apparent contours', *Int. J. of Computer Vision* 9 (1992), 83-112.
3. R.Cipolla, G.Fletcher, P.Giblin, 'Surface geometry from cusps of apparent contours', *Proc. Fifth Int. Conf. on Computer Vision*, Cambridge, Mass, June 1995, pp.858-863.
4. R.Cipolla, G.Fletcher, P.Giblin, 'Following Cusps', *to appear in Int.J.Computer Vision.*
5. R.Cipolla, K.E.Åström and P.J.Giblin, 'Motion from the frontier of curved surfaces', *Proc. Fifth Int. Conf. on Computer Vision*, Cambridge, Mass, June 1995, pp.269-275
6. D.A.Forsyth, J.L.Mundy, A.Zisserman, C.A.Rothwell, 'Recognising rotationally symmetric surfaces from their outlines', *Proc. Second European Conf. on Computer Vision*, Santa Margherita Ligure (Italy), pp.639-647, May 1992.
7. R.Glachet, M.Dhome, J.T.Lapreste, 'Finding the pose of an object of revolution', *Proc. Second European Conf. on Computer Vision*, Santa Margherita Ligure (Italy), pp.681-686, May 1992.
8. J.J.Koenderink, *Solid Shape*, M.I.T.Press 1990.
9. B. O'Neill, *Elementary Differential Geometry*, Academic Press 1966.
10. J.Ponce, 'Invariant properties of straight homogeneous generalized cylinders'. *IEEE PAMI*, vol.11,no.9,pp.951-965,1989.
11. J.Ponce, D.Chelberg, 'Finding the limbs and cusps of Generalized Cylinders.' *Int. J.Comp. Vision 1*, (1987) 195-210.
12. J.H.Rieger,'Projections of generic surfaces of revolution', *Geometriae Dedicata*, 1993, Vol.48, No.20, pp.211-230.
13. A.Zisserman, J.Mundy, D.Forsyth, J.Liu, N.Pillow, C.Rothwell and S.Utcke, 'Class-Based Groupings in Perspective Images', *Proc. Fifth Int. Conf. on Computer Vision*,Cambridge, Mass, June 1995.
14. M.Zerroug, R.Nevatia, 'Segmentation and recovery of SHGCs from a real intensity image.', *Proc. Third European Conf. on Computer Vision*, Stockholm, Sweden Vol.I pp.319-330, May 1994.

Reliable Surface Reconstruction from Multiple Range Images *

A. Hilton, A.J.Stoddart, J. Illingworth and T.Windeatt

VSSP Group, Department of Electronic and Electrical Engineering
University of Surrey, Guildford. GU2 5XH. U.K.
a.hilton@ee.surrey.ac.uk

Abstract. This paper addresses the problem of reconstructing an integrated 3D model from multiple 2.5D range images. A novel integration algorithm is presented based on a continuous implicit surface representation. This is the first reconstruction algorithm to use operations in 3D space only. The algorithm is guaranteed to reconstruct the correct topology of surface features larger than the range image sampling resolution. Reconstruction of triangulated models from multi-image data sets is demonstrated for complex objects. Performance characterization of existing range image integration algorithms is addressed in the second part of this paper. This comparison defines the relative computational complexity and geometric limitations of existing integration algorithms.

1 Introduction

Recent research has resulted in the independent publication of several algorithms that reconstruct triangulated 3D surface models of complex objects [4, 6, 7, 9]. The goal of surface reconstruction is to estimate a manifold surface, S', that approximates an unknown object surface, S, using a sample of points, $\mathbf{x} = (x, y, z)$, in 3D Euclidean space, $X = \{\mathbf{x}_0, ..., \mathbf{x}_{N-1}\}$, combined with knowledge about the sampling resolution, $\Delta\mathbf{x}$, measurement error, ε, and measurement confidence, $p(S'/\mathbf{x}_i)$. Given a method we want to be able to specify conditions on the original surface, S, and sample, X, that allow S' to be a reliable model.

Hoppe et al. [4] presented a general method for constructing an implicit surface representation from unstructured 3D points. Polygonal models were then generated using a 'marching cubes' approach [1]. The algorithm is 'static' in the sense that all the image data is required prior to the polygonisation process. Soucy et al. [7] integrated range images using canonic subsets of the Venn diagram. The canonic subsets each represent the overlap between a subset of the 2.5D range images and are associated with a 2D viewpoint reference frame. The 2D reference frames are used to eliminate redundant data and merge intersecting regions. Soucy et al. [8] extended this algorithm to be 'dynamic' allowing the sequential integration of new range images. Turk et al. [9] integrated range image triangulations using a dynamic mesh 'zippering' approach. Overlapping regions of meshes are eroded and the boundary correspondence found by operations in

---

* Supported by EPSRC GR/K04569. 'Finite Element Snakes for Depth Data Fusion'

3D space. A local 2D constrained triangulation is then used to join overlapping mesh boundaries to form a single mesh. Boissonnat [2] and Rutishauser et al. [6] retriangulate two overlapping meshes using local 2D constraints on triangle shape. The four integration algorithms are completely different in their approach to constructing a single triangulated model. This results in different complexity, limitations and failure modes.

A new range image integration algorithm is presented in section 2. This is based on a continuous implicit surface representation which combines geometry and topology information from individual range images. This is the first integration algorithm which uses operations entirely in 3D space. Unlike all previous approaches the method does not require either local or global projection to 2D sub-planes. This eliminates limitations on local surface geometry inherent in previous approaches. Performance characterization of existing integration methods is presented in section 3. This comparison defines the relative computational complexity and geometric limitations for reliable reconstruction.

2 New Integration Algorithm

The new range image integration algorithm based on a 'continuous implicit surface' is presented in section 2.1. A triangulated model is construted using a standard implicit surface polygonisation algorithm, section 2.2. Results for the reconstruction of complex objects are presented in section 2.3. The computational complexity and limitations of this approach are discussed in section 3.

2.1 Continuous Implicit Surface Construction

An implicit surface is defined as the zero-set of a scalar field function $f(\mathbf{x}) = 0$. The aim of representing a set of surface measurements, $X = \{\mathbf{x}_0....\mathbf{x}_{N-1}\}$, as an implicit surface is to construct a smooth field function, $f(\mathbf{x})$, such that the zero-set approximates the data X as closely as possible. A piecewise continuous implicit surface function for multiple range images is presented in this section. This is based on the integration of multiple overlapping 2.5D triangulations. Accurate representation of surface geometry is achieved by combining overlapping measurements according to their confidence. Boundary information is integrated from multiple meshes to obtain an explicit representation of the measured surface topology.

Meshes, M_i, are initially constructed from each range image using a step discontinuity constrained triangulation, [6, 7, 9]. Range images are triangulated in the 2D image plane using a constant distance threshold, $t_d = n\Delta\mathbf{x}$. This defines the local surface continuity for each range image.

For a single mesh M the field function, $f(\mathbf{x})$, is constructed as the signed distance to the nearest mesh point, $\mathbf{p}(x, y, z)$. A binary function, $b(\mathbf{x}) = [0, 1]$, is used to explicitly label field function values, $f(\mathbf{x})$, with nearest points on the mesh boundary. If the nearest point, $\mathbf{p}$, is not on the mesh boundary, $b(\mathbf{x}) = 0$, the signed distance is the dot product of the vector to the nearest point, $(\mathbf{x} - \mathbf{p})$, with the surface normal, $\mathbf{n}_p$, at the nearest point: $f(\mathbf{x}) = (\mathbf{x} - \mathbf{p}).\mathbf{n}_p$. Alternatively, if the nearest point, $\mathbf{p_b}$, is on the mesh boundary, $b(\mathbf{x}) = 1$, the signed distance function is evaluated as the sign of the dot product of the vector to the nearest points, $(\mathbf{x} - \mathbf{p_b})$, with the nearest point normal, $\mathbf{n}_{p_b}$, multiplied by the Euclidean distance: $f(\mathbf{x}) = sign[(\mathbf{x} - \mathbf{p_b}).\mathbf{n}_{p_b}] \times |(\mathbf{x} - \mathbf{p_b})|$. The zero-set of the field function, $f(\mathbf{x})$, for a single mesh, M, is thus a piecewise continuous function with the same topology as the mesh. The implicit surface representation for a single mesh is illustrated in Figure 1(a) for a 2D cross section.

Integration of multiple range images requires the construction of an implicit surface function based on multiple overlapping meshes M_k where $k = 0...M - 1$. A field function $f_k(\mathbf{x})$ can be implemented as described above for a single mesh. The problem is then to integrate the individual field functions into a single continuous surface, $f(\mathbf{x})$, and boundary label, $b(\mathbf{x})$. This is achieved by first evaluating $f_k(\mathbf{x})$ for each individual mesh and then integrating them using a simple set of rules based on local surface geometry:

i) Evaluate the signed field function, $f_k(\mathbf{x})$, and boundary function, $b_k(\mathbf{x})$, for each mesh M_k for $k = 0...m - 1$.
ii) Find the nearest non-boundary mesh point, $b_k(\mathbf{x}) = 0$, from the set $f_k(\mathbf{x})$, $f_{min}(\mathbf{x})$.
iii) If a non-boundary point, $f_{min}(\mathbf{x})$, does not exist return the nearest boundary point, $b_k(\mathbf{x}) = 1$: $f(\mathbf{x}) = f_{min_{bound}}(\mathbf{x})$ and $b(\mathbf{x}) = 1$.
iv) Else find all non-boundary points, $b_k(\mathbf{x}) = 0$, with the same orientation as the nearest point, $f_{min}(\mathbf{x})$: $F = \{f_{same_i}(\mathbf{x})\}$ $i = 0...N_{same}$, where $n_{min}.n_{same_i} > 0$
v) Find the nearest non-boundary point with opposite orientation, $f_{opposite}(\mathbf{x})$, where $n_{min}.n_{opposite} < 0$.
vi) Eliminate all points in F, where $f_{same_i}(\mathbf{x}) > f_{opposite}(\mathbf{x})$.
vii) Evaluate the nearest point as a weighted average of all points in F:
$f(\mathbf{x}) = \sum_k w_k f_k(\mathbf{x})$ where $\sum w_k = 1$ and $b(\mathbf{x}) = 0$

This set of rules enables the integration of overlapping meshes to define a continuous zero-set of the field function, $f(\mathbf{x}) = 0$, in all regions where a mesh is continuous and non-zero elsewhere. The rules account for the special cases of two overlapping surfaces with different orientations and multiple overlapping surfaces. The field function evaluation according to the rules given above is illustrated schematically in Figure 1 for the different cases of overlapping surface. Step (iii) explicitly defines mesh boundaries, $b(\mathbf{x}) = 1$, providing an integrated representation of the local surface topology. Steps (iv—vi) eliminate ambiguity if there are multiple overlapping meshes corresponding to different surface regions. This enables correct representation for surface regions of high curvature and different surfaces in close proximity, section 3.2. The weighted average of nearest points, step (vii), enables smooth integration of overlapping meshes using estimates of measurement confidence or blending functions, [3].

2.2 Implicit Surface Polygonisation

Polygonisation of implicit surfaces has received considerable interest for visulisation in medical imaging and computer graphics. The 'Marching Cubes' algorithm

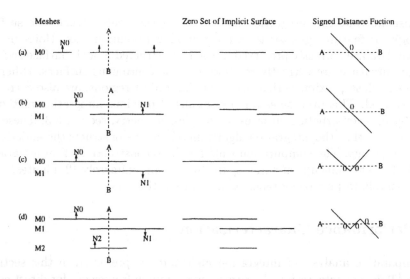

Fig. 1. Continuous Implicit Surface Function: (a) Single mesh, (b) Two meshes from same surface, (c) Two meshes from different adjacent surface with opposite orientation, (d) Three meshes from three different adjacent surfaces.

uses a uniform subdivision of 3D space to reconstruct a triangulated model of a manifold surface without boundaries, [1]. Extension of marching cubes to the polygonisation of a bounded implicit surface was addressed by Hoppe et al. [4]. The modified marching cubes algorithm only reconstructs the implicit surface for cubes which do not intersect the implicit surface boundary. Application of a modified marching cubes algorithm to the mesh based implicit surface enables a 3D triangulated model to be constructed. Explicit representation of the surface boundary, $b(\mathbf{x}) = 1$, together with the implicit surface, $f(\mathbf{x}) = 0$, enables guaranteed reconstruction of a polygonal model with the same topology as the surface measurements. To ensure correct reconstruction of surface features the spatial subdivision used in the marching cubes algorithm must be less than the sampling resolution, $\Delta\mathbf{x}$. Previous implicit surface based approaches [4] did not use a continuous representation, resulting in reconstruction errors, section 3.2.

2.3 Results

Results of the reconstruction process for four objects are shown in Figure 2. All models were constructed from 8—10 range images. The bunny [2] and telephone[2] model consist of approximately 25000 triangular elements. This data was previously used to demonstrate the mesh zippering algorithm, [9]. The teapot [3] and soldier[3] consist of 15000 and 80000 triangular elements. This data was previously used to demonstrate the Canonic subsets algorithm, [8]. The results

---

[2] Cyberware scanner range data registered using ICP [9]
[3] NRCC scanner range data [5] registered using InnovMetric software [8]

demonstrate that the integration algorithm correctly reconstructs the surface topology for features greater than the sampling resolution, Δx. Holes in the original data are correctly preserved in the reconstructed model. Surface regions of high curvature are correctly reconstructed as continuous surfaces. Different surfaces in close proximity, that occur for thin object regions, are also correctly reconstructed. This overcomes the limitations of previous integration methods, [4, 6, 9], in reconstructing surfaces of complex geometry, section 3.3. These results indicate that the integration algorithm reliably reconstructs the underlying surface topology. The computation time for the reconstruction of an integrated model of 25000 elements on a Sun sparc 10 was approximately 12 minutes. This is comparable to previously reported integration times [8].

3   Performance Characterization

A comparative analysis of integration algorithms is presented in this section. Hilton [3] defines integration algorithms in a common framework for direct comparison. Implementation is also discussed including requirements that were previously undocumented. The comparison of integration algorithms presented here focuses on two principal issues. Firstly, the computational complexity of each of the algorithms, section 3.1. Secondly, identification of inherent limitations in each of the integration methods, section 3.2. The comparative analysis considers the following algorithms:

   I: Point-Normal Implicit Surface (Hoppe et al. [4]).
   II: Mesh Implicit Surface (this paper).
   III: Canonic Views (Soucy et al. [7]).
   IV: Mesh Zippering (Turk et al. [9]).
   V: Mesh Growing (Rutishauser et al. [6]).

3.1   Time Complexity

Defining a general form for the computational complexity of each of the integration algorithms is not possible as it is a function of the particular image set. This depends on the number of images, m, the number of points in each image, N_k, the proportion of redundancy between images and the length of the boundary between overlapping images. A qualitative comparison of the worst-case computational complexity of each algorithm is given, for m images of N points, by approximating the cost of each stage of the integration process and deriving the overall order of complexity. To enable quantitative comparison of the computational complexity we consider a special case: the cost of integrating two images of N points with 50% overlap. Results of the complexity analysis are summarised in Table 1.

Nearest point search is common to all integration algorithms. Implemented as a brute force search the time complexity is $O(N)$, where N is the number of points. Uniform subdivision of the 3D space facilitates a local search for the nearest point, [4, 9]. Subdivision of the 3D space into voxels of approximately

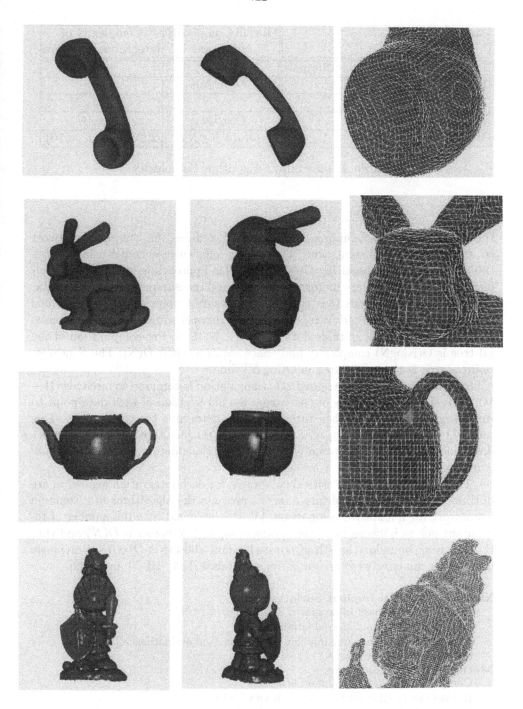

Fig. 2. Reconstructed 3D Models

Method	Overall Complexity m Images	Complexity of Integrating 2 images
I Implicit Surface Points+Normals	$O(m^2 N)$	$6N$
II Implicit Surface Mesh	$O(m^2 N)$	$12N$
III Canonic Views	$O(m^2 N)$	$4N + 2\sqrt{N}$
IV Mesh Zippering	$O(m^2 N)$	$N + \sqrt{N}$
V Mesh Growing	$O(mN \log N)$	$20N \log N (K = 10)$

Table 1. Integration Algorithm Complexity

the same size as the sampling resolution, Δx, reduces the computational cost to approximately constant time. This can be efficiently encoded using a hash table representation requiring O(N) space. This approach assumes that the individual range images are uniformly sampled and measurement error, $\varepsilon \ll \Delta x$. Throughout this analysis it is assumed that nearest point search is performed in constant time. Method V requires the neighbourhood of K nearest points. A k-D tree provides a suitable data structure, [2, 4, 6]. Pre-computation of the kd-tree is O(NlogN) complexity and space requirement is O(N). The K nearest point search is then computed in $O(\log N)$ time.

Step discontinuity constrained 2D triangulation is common to methods (II—V). This operation consists of comparing the 3D position of each data point to its 8-neighbourhood and then thresholding to determine the connectivity. This requires on average 3 comparisons and Euclidean distance computations per data point. The step discontinuity constrained triangulation has computational complexity, $O(N)$.

Time complexity of the critical sub-stages for each integration algorithm are outlined below. This complexity analysis presents the algorithms in a common framework for qualitative comparison, [3]. It is assumed that the number of redundant points between two overlapping images of N points is $O(N)$ and that the resulting boundary length of non-redundant subsets is $O(\sqrt{N})$. Previously complexity analyses were presented for methods I [1, 4], III [8] and V [2].

Methods I and II: Implicit Surfaces
    Computation of hash table spatial subdivision: O(mN)
    Implicit surface function evaluation: $0(m)$
    Number of marching cube implicit surface function evaluations: $0(mN)$
    Overall complexity: $0(m^2 N)$
Method III: Canonic views
    Computation of Venn Diagram: $0(m^2 N)$
    Reparameterisation into canonic sub-views: $O(m^2 N)$
    Retriangulation of all redundant canonic subsets: $O(m^2 N)$
    Elimination of redundancy in canonic subsets: $O(m^2 N)$
    Retriangulation to build model: $O(m\sqrt{N})$
    Overall Complexity: $O(m^2 N)$

Method IV: Mesh Zippering
  Redundancy test using nearest point: $O(m)$
  Elimination of redundant mesh elements: $0(m^2 N)$
  Clipping of mesh boundary: $0(m\sqrt{N})$
  Overall complexity: $O(m^2 N)$
Method V: Mesh Growing
  Computation of the k-D tree: $O(mN \log N)$
  Search for K-Nearest Neighbours: $O(K \log N) = O(\log N)$
  Surface retriangulation: $O(KmN \log N) = O(mN \log N)$
  Overall Complexity: $O(mN \log N)$

3.2 Geometric Limitations

This section identifies limitations inherent in each of the integration algorithms
for reliable surface reconstruction. The comparison focuses on limitations on
correct reconstruction of surface topology and geometry. Restrictions are identified by considering three cases: minimum hole size, maximum surface curvature
and minimum surface separation (Figure 3). Results are summarised in Table
2. Table 2 also includes general characteristics which relate to the integration
performance for a particular application. Computation type is specified as 2D or
3D according to the requirement for local or global projection to 2D sub-planes
which imposes limitations on the local geometry. Static computation requires
all data to be present prior to integration, conversely dynamic computation allows sequential addition of new data. Additionally some integration algorithms
impose inherent limitations on the sampling process, restricting integration of
range images at a single resolution. Further details are given in [3].

	2D/3D	Static/ Dynamic	Sampling	Minimum Feature Size	Minimum Crease Angle	Minimum Surface Separation
I	3D	static	uniform	$> 3n\Delta\mathbf{x}$	$140°$	$n\Delta\mathbf{x}$
II	3D	static	non-uniform	$n\Delta\mathbf{x}$	$30°$	ε_{max}
III	2D	semi-dynamic	non-uniform	$n\Delta\mathbf{x}$	$30°$	ε_{max}
IV	2D/3D	dynamic	uniform	$2n\Delta\mathbf{x}$	$90°$	$n\Delta\mathbf{x}$
V	3D	dynamic	non-uniform	$2n\Delta\mathbf{x}$	$90°$	ε_{max}

Table 2. Integration Algorithm Limitations for Reliable Reconstruction

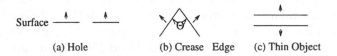

Surface —— ——   (a) Hole   (b) Crease Edge   (c) Thin Object

Fig. 3. Limitations on Surface Geometry

Minimum Feature Size: This section defines the minimum hole size that is guaranteed to be reconstructed. **Method I** does not explicitly represent mesh boundaries and therefore requires a threshold to be set for the maximum distance from the mesh, $\sqrt{3}n\Delta\mathbf{x}$. This results in invalid boundary extension of order $\sqrt{3}n\Delta\mathbf{x}$ which will fill any holes less than $2\sqrt{3}n\Delta\mathbf{x}$ in size. **Methods II-IV** use the local mesh connectivity for model reconstruction. The aim is to preserve holes that are identified in the step discontinuity constrained triangulation. The constrained triangulation imposes a lower limit on the surface feature size, $n\Delta\mathbf{x}$, for which the local surface connectivity will be reliably reconstructed. However, **method IV** may fail due to ambiguities between overlapping mesh elements that arise in the zippering algorithm. Holes less than twice the size of the individual mesh elements may redundantly overlap. This limits the minimum feature size for reliable reconstruction to $2n\Delta x$. **Method V** does not use the local mesh continuity to constrain the topology of the triangulation. The maximum size of an element added to the triangulation is limited explicitly by a distance threshold. This threshold limits the minimum feature size for which the topology will be correctly reconstructed to $2n\Delta x$.

Minimum Crease Angle: This section defines limitations on the reliability of the integration process for regions of high surface curvature and crease edges. This is quantified as the minimum crease angle, θ_{min}, between two intersecting planes across which the local topology is guaranteed to be correctly reconstructed. **Methods II—IV** may fail in regions of high curvature if the constrained triangulation does not correctly reconstruct the connectivity due to the increased distance between adjacent samples. The step discontinuity threshold, $t_d = n\Delta\mathbf{x}$, imposes a limit on the minimum crease angle, $\theta_{min} = 2tan^{-1}(\Delta\mathbf{x}/n\Delta\mathbf{x})$. Typically $n \approx 4$ thus the sampling resolution imposes a minimum crease angle $\theta_{min} \approx 30°$. **Methods IV and V** both require explicit thresholds to be set for the maximum difference in orientation between adjacent surface elements, $> 90°$. This imposes a hard limit on the minimum crease angle that is reliably reconstructed, $\theta_{min} \approx 90°$. **Method I** will fail if an incorrect tangent plane is used for the implicit surface distance function evaluation, $f(\mathbf{x})$. This results in spurious mesh artifacts in regions of high curvature. In practice this imposes a minimum crease angle for reliable reconstruction, $\theta_{min} \approx 140°$.

Minimum Surface Separation: The third geometric structure that causes failure of the integration process occurs where different surface regions are in close proximity (i.e. thin object parts). This can be quantified as the minimum separation between different surface regions. Reconstruction will fail if the algorithm does not use the local connectivity and orientation information. **Methods II,III and V** use the local surface orientation to define overlapping regions. The lower limit for reliable reconstruction for these methods is therefore surface separated by the maximum measurement error ε_{max}. **Method I** does not use the local surface orientation and will therefore fail for surface regions separated by less than the sampling resolution, $n\Delta\mathbf{x}$. **Method IV** relies on a nearest point search to define the boundary intersection between meshes. This test may fail if meshes from different surfaces are closer than the sampling resolution, $n\Delta\mathbf{x}$.

4 Conclusion

A novel registered range image integration algorithm has been presented. This uses a mesh based implicit surface function to define the object surface as the zero-set of a field function defined at any point in 3D space. The aim of this algorithm is to estimate the underlying surface topological type from the measured data by reconstructing a triangulated model with the same topology. This is the first integration algorithm that operates entirely in 3D space. Reliable reconstruction is demonstrated for complex objects.

Performance characterization of existing integration algorithms is presented. Qualitative analysis of the computational complexity demonstrates that most algorithms are $O(m^2 N)$ for integration of m images of N points. Quantitative analysis in specific cases indicates large difference in the constant time complexity associated with each method. Geometric limitations for reliable reconstruction are identified. This analysis demonstrates that reliable reconstruction is only guaranteed for the integration algorithms based on Canonic sub-views [8] and the mesh based implicit surface (this paper).

Future development of the integration algorithm presented in this paper should focus on dynamic data integration and computational efficiency. Further work on performance characterization is required to benchmark the relative computational cost and validate the geometric limitations identified.

References

1. J. Bloomenthal. An implicit surface polygonizer. *Graphics Gems ed. Heckbert,P.S.*, 4:324—350, 1994.
2. J.D. Boissonnat. Geometric structures for three-dimensional shape representation. *ACM Transactions on Graphics*, 3(4):266—286, 1984.
3. A. Hilton. On reliable surface reconstruction from multiple range images. In *VSSP-TR-5-95*. ftp://ftp.ee.surrey.ac.uk/pub/vision/papers/hilton-vssp-tr-5-95.ps.Z, 1995.
4. H. Hoppe, T. DeRose, T. Duchamp, J. McDonald, and W. Stuetzle. Surface reconstruction from unorganised points. *Computer Graphics*, 26(2):71—77, 1992.
5. M. Rioux. Laser range finder based on synchronized scanners. *Applied Optics*, 23(21):3837—3844, 1984.
6. M. Rutishauser, M. Stricker, and M. Trobina. Merging range images of arbitrarily shaped objects. In *Proceedings of IEEE Conference on Computer Vision and Pattern Recognition*, pages 573—580, 1994.
7. M. Soucy and D. Laurendeau. Multi-resolution surface modelling from multiple range images. In *Proceedings of IEEE Conference on Computer Vision and Pattern Recognition*, pages 348–353, 1992.
8. M. Soucy and D. Laurendeau. A general surface approach to the integration of a set of range views. *IEEE Trans. Pattern Analysis and Machine Intelligence*, 14(4):344–358, 1995.
9. G. Turk and M. Levoy. Zippered polygon meshes from range images. In *Computer Graphics Proceedings, SIGGRAPH*, 1994.

Shape from Appearance:

A Statistical Approach to Surface Shape Estimation*

Darrell R. Hougen and Narendra Ahuja

Beckman Institute and Coordinated Sciences Laboratory
University of Illinois, Urbana, Illinois 61801, USA

Abstract. This paper is concerned with surface shape estimation by a method in which an empirically determined associative model relating appearance to surface shape is used. Significantly, the estimated model is more accurate than the algorithm that generates the examples. The method presented here is a generalization of shape from shading methods that does not rely upon idealized models of the image formation process. As a relative of shape from shading, this method more accurately recovers small surface detail than is possible with methods such as stereo and motion. The present approach is a continuous analogue of pattern recognition and is closely related to methods of joint space learning used in robotics. Experiments on real scenes are used to illustrate the concepts involved.

1 Introduction

This paper describes a method of surface shape estimation that involves automatic generation of an *associative model* that relates surface shape to appearance. It is shown that through a scale change and the use of a smoothness requirement, the estimated model can be made to be more accurate than the algorithm that produced the examples. The performance increase is key to the utility of this method and sets it apart from the approach of Lehky and Sejnowski [7]

Associative modelling techniques are considered by the authors to be important because of the generality and precision made possible by such techniques. The shape estimation procedure described below is a generalization of physics based methods and embodies many of the advantages of such methods with few of the disadvantages. Physics based methods generally rely heavily on idealized models of the image formation process which do not capture the complexity of real scenes [2, 3, 4]. In addition, such models often contain hard to estimate parameters [3, 4]. However, shading information, in particular, is useful for recovering small surface detail.

* This research was supported by the Advanced Research Projects Agency and the National Science Foundation under grant IRI-89-02728 and by the Army Advance Construction Technology Center under grant DAAL 03-87-K-0006.

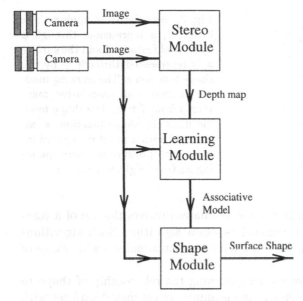

Fig. 1. A stereo module produces coarse shape estimates. A learning module produces an associative model of the relationship of shape to appearance. The shape module estimates the surface shape of novel objects.

In comparison, methods that rely upon the coincidence or correlation of features in two or more images, such as stereo and motion based methods, cannot be used to recover small surface detail due to the sparseness of discriminable image features [5] and the fact that the accuracy of such techniques drops with the square of depth [5]. In addition, such methods are only reliable in highly textured regions or in the presence of well defined image features. However, such methods are based on relatively weak assumptions and are therefore useful for recovering coarse or sparse depth estimates.

Recent papers by Leclerc and Bobick [6] and Hougen and Ahuja [3, 4] discuss integrated methods which combine the strengths of the above methods while avoiding many of the limitations. However, despite the increase in generality, such methods are still dependent upon highly restrictive idealized models of the image formation process.

In order to escape such restrictions, it should be noted that the relationship between local appearance and corresponding surface shape can always be captured in the form of a probability density function. The density estimation problem encountered here is the continuous analogue of the pattern recognition problem [1] and is closely related to function learning problems encountered in robotics [8]. Indeed, in the presence of nonlocal, contextual information, it may be possible to simplify the problem and treat the local relationship of shape to appearance as a functional relationship. This approach is explored in the following sections.

2 Algorithm Overview

The three major components of the local shape-from-appearance estimation procedure are illustrated in figure 1 including, (1) generation of coarse surface shape estimates for statistical modelling, (2) estimation of the associative model governing the statistical relationship of shape to appearance, and (3) surface shape estimation through application of the statistical model to the desired image.

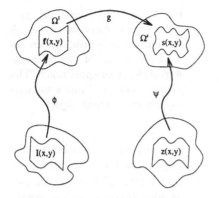

Fig. 2. Associative mapping. The image, I, is represented through ϕ by a set of features f and the surface z is represented through ψ by the shape function s. The learning module estimates an associative mapping g from f to s. The shape module finds the shape function, s, associated with a novel image and inverts ψ to produce the corresponding surface height function, z.

Generation of the coarse surface shape estimates involves the use of a standard vision algorithm such as a stereo or motion algorithm. Such algorithms are capable of producing coarse yet reasonable estimates under a wide range of conditions.

Estimation of the associative model governing the relationship of shape to appearance is accomplished by associating examples of estimated surfaces with corresponding examples from images as illustrated in figure 2. Each example of surface shape and appearance is represented by one or more mathematical features which are designed to capture important information about the examples. The relationship between corresponding examples is represented by a statistical model and serves as a substitute for the derivation of an idealized model.

Physics based methods are generally based upon restrictive assumptions. The present method requires only that the conditional density of shape given appearance be sufficiently informative. If the relationship is particularly simple, it may be possible to replace estimation of the probability density with estimation of a mapping function. In this case the model is referred to as an *associative map*. More detail appears in the following sections.

Surface shape estimation is accomplished through application of the previously estimated associative model to the desired image. The features calculated at a particular image location serve as input vectors to the associative model, allowing the statistically expected shape parameters to be computed at that location. A more recognizable surface description is obtained by inverting the feature calculation process as described in section 5.

3 Associative Modelling

Let $z : A \to \Re^{-}$ be a surface height function and $I : A \to [0, I_{\max}]$ be an image brightness function defined on $A \subset \Re^2$. The functions z and I may be viewed as members of the ensembles $\mathcal{Z}$ and $\mathcal{I}$ respectively where $\mathcal{Z}$ is the set of all viewable height functions and $\mathcal{I}$ is the set of all images of the height functions in $\mathcal{Z}$.

For a single surface height function, variable lighting conditions and surface marking patterns make possible a wide variety of possible images. Conversely, a single image may correspond to more than one surface height function. However, some height functions are more likely than others to have produced a given image. The problem considered here is to find the surface *most likely* to have produced a given image.

3.1 Choosing Features

Important criteria for choosing a surface representation are simplicity, symmetry, completeness, and learnability. An operator, ψ, that at least partially satisfies these criteria is the Laplacian of a Gaussian smoothing filter. It is simple, symmetric, and complete; z can be reconstructed from $s(x, y; \sigma) = \nabla^2 G(\sigma) * z$. It is also more apparent locally, and therefore more learnable, than, for example, surface height or slope.

The most important criterion for choosing an image representation is that it be locally informative. The image irradiance at a point is uninformative but higher order functions of the local image irradiance are statistically related to surface shape and therefore useful. In general, the image, I, is represented by an image feature vector, $\boldsymbol{f}(\sigma)$, where $f_j(\sigma) = \phi_j(\sigma)I$, $j = 1, \ldots, k$. Here, polynomial coefficients are used.

3.2 Conditional Density

At a point, (x, y), the conditional density of shape given appearance in the local neighborhood of (x, y) is written $p_{A(\sigma)}(s(x, y; \sigma) | \boldsymbol{f}(x, y; \sigma))$ and is scale dependent. If the image feature vector is given on $A \subset \Re^2$ and the surface feature vector at a particular point depends only on the image feature vector at that point then the maximum log-likelihood solution for surface shape given the information about appearance is given by

$$z(x, y) = \max_{z(x, y)} \int \int_A \log(p_{A(\sigma)}(s(x, y; \sigma) | \boldsymbol{f}(x, y; \sigma))) dx dy$$

where the maximization is over all functions $z : A \to \Re^-$.

The above formulation can be simplified by using the marginal probability density in which the dependency upon σ is removed. This is a reasonable simplification given the fact that most objects that appear in the world are observed at many ranges and therefore at many scales making the conditional density nearly independent of scale. More formally, it is assumed that, $p_{A(\sigma)} \approx p_A$.

3.3 Scaling to Increase Performance

The fact that a surface curve or bend or marking has previously been seen up close, allowing its shape to be accurately determined, means that the shape can be accurately determined later when the surface is far away. The statistical association between surface shape and appearance determined at a coarse scale is stored in the form of a conditional probability density and used later to estimate the surface shape at a finer scale. If the probability density is scale independent, it is not necessary to know the change in scale to reconstruct the surface.

As a preliminary, it should be noted that scaling to increase power can only work if the new shape estimation procedure has higher performance than the example generator. As an example, the resolution and accuracy are higher for shape from shading methods than for stereo, motion, or focus methods [5]. The present method is a relative of shape from shading and has similar performance characteristics, making possible an increase in performance.

Although the precise increase in performance is the subject of ongoing research, the following considerations are relevant. The probability density, p_Λ, should be estimated using examples given at a scale σ_e that is chosen to obtain the maximum performance from the example producing algorithm. If the resultant surface is estimated at scale a σ_r that is chosen optimally or suboptimally with $\sigma_r \geq \sigma_{\text{opt}}$. Then, if $\sigma_r < \sigma_e$ the resolution and hence the performance is increased by a factor related to σ_e/σ_r.

4 Associative Mapping

In many instances considered in computer vision, the probability density may be simple enough to be well approximated by a sum of normal variates. The correspondence of a smooth surface to a smoothly varying image or of a long narrow specularity to a surface with a convex or concave bend are examples of unimodal or bimodal distributions. If the mode is assumed known, the maximization of the log-likelihood reduces to the solution of a least squares problem. This is not unreasonable in cases in which a single choice is required for an entire region as is the case that the surface curvature has constant sign in a region of interest.

Let $g : \Omega^I \to \Omega^z$ be a function from the image feature space, Ω^I, into the surface feature space, Ω^z, such that $s(x,y;\sigma) = g(f(x,y;\sigma)) + \epsilon$ where ϵ is zero mean Gaussian white noise. If the conditional density, p_Λ, is Gaussian white noise with mean, $s(x,y;\sigma)$, then g is guaranteed to exist and is given by $g(f(\sigma)) = \langle s(\sigma)|f(\sigma)\rangle$ where $\langle \cdot \rangle$ denotes expectation. Thus, estimation of g is a regression problem. Note that the mapping function, g, is assumed to be independent of the the scale factor σ. If the mapping is scale dependent, then $g(f(\sigma))$ may be written $g(f(\sigma);\sigma)$.

Let $(f_i(\sigma_e), s_i(\sigma_e))$, $i = 1, \ldots, N$, be pairs of image feature vectors and corresponding surface shape estimates generated by the stereo program. In order to obtain the least squares estimate of g a criterion function Q is defined by

$$Q(\theta) = \frac{1}{N} \sum_{i=1}^{N} (s_i(\sigma_e) - g(f_i(\sigma_e); \theta))^2$$

where g is parameterized by $\theta = (\theta_1, \ldots, \theta_m)$.

If g is modelled by a linear sum of basis functions, $(B_1, \ldots, B_m)$, with coefficients $(\theta_1, \ldots, \theta_m)$. Then $\hat{g}$ may be written, $\hat{g}(f_i(\sigma_e)) = \sum_{j=1}^{m} \theta_j B_j(f_i(\sigma_e))$. In this case, the regression reduces to an ordinary linear least squares problem.

Figure 2 illustrates the relationships between $I, \phi(\sigma), f(\sigma), z, \psi(\sigma), s(\sigma)$ and g. The feature operators, $\phi(\sigma)$ and $\psi(\sigma)$ transform the functions $I(x,y)$ and $z(x,y)$ into $f(x,y;\sigma)$ and $s(x,y;\sigma)$ respectively. The function g maps each point of $f(x,y;\sigma)$ to a point in the surface feature space that differs from $s(x,y;\sigma)$ by an amount ϵ.

5 Surface Estimation

In the preceding analysis, g was treated as a random variable with mean $s(\sigma_e)$. For purposes of surface estimation, $s(\sigma_r)$ is identified as a random variable with mean g. If $s(\sigma_r)|f(\sigma_r)$ has a normal distribution with mean $g(f(\sigma_r))$ and variance ρ^2 then $\log(p_\Lambda(s(\sigma_r)|f(\sigma_r))) = (s(\sigma_r) - g(f(\sigma_r)))^2/(2\rho^2) - \kappa$ where $\kappa = \frac{1}{2}\log(2\pi\rho^2)$ is a constant. Therefore, the maximum likelihood estimate of the surface is found by maximizing the criterion function

$$D(z) = \iint_A (s(x,y;\sigma_r) - g(f(x,y;\sigma_r)))^2 \, dxdy$$

over all surfaces $z : A \to \Re^-$.

Let $Z, H(\sigma_r), S(\sigma_r)$ and $\Psi(\sigma_r)$ be the Fourier transforms of $z, g(f(\sigma_r)), s(\sigma_r)$ and $\psi(\sigma_r)$ respectively. Then, by Parseval's theorem,

$$D(z) = \iint |S(\omega_1,\omega_2;\sigma_r) - H(\omega_1,\omega_2;\sigma_r)|^2 \, d\omega_1 d\omega$$

The minimum integrated squared error is achieved by the function that minimizes the error at each point of the domain. Since $S(\sigma_r) = \Psi(\sigma_r)Z$, that minimum is achieved by setting, $\hat{Z} = H(\sigma_r)/\Psi(\sigma_r)$. The solution surface, $\hat{z}$, is the inverse transform of $\hat{Z}$.

Although $\hat{z}$ is the maximum likelihood estimate in the absence of noise, a better estimate in the presence of noise is found by Wiener filtering. The resulting optimal estimate is found by setting

$$\hat{Z} = \frac{H(\sigma_r)\Psi(\sigma_r)}{\Psi^2(\sigma_r) + K^2}$$

where $K^2(\omega_1,\omega_2) = \langle \eta^2(\omega_1,\omega_2)/Z^2(\omega_1,\omega_2)\rangle$ is the variance of the noise divided by the expected power spectrum of the surface. Under the assumption of white noise, the noise term reduces to a constant. If the surface is assumed to be fractal Brownian, the final value of K^2 is given by $K^2 = \eta^2(\omega_1^2 + \omega_2^2)^2$.

6 Experimental Results

The algorithm described in section 2 can be thought of as operating in two major modes, the model estimation mode and the surface estimation mode. In the model estimation mode, the input images are used by the stereo module to produce surface shape estimates which serve as examples to be used in estimating the associative model. In the surface estimation mode, the model is used to estimate the shape of a previously unseen surface. The experiments explained in this section are designed to illustrate both major operational modes.

6.1 Model Estimation Results

Figure 3 shows one of four images of an oriented ridge surface with its stereo depth map and corresponding level curves. The image is the left image of a stereo pair of images taken at a depth of about 10cm with a baseline of about 1cm. Once the surface estimates have been computed by the stereo program, regions from each image and corresponding surface are selected to act as input data for the model estimation procedure.

The original data regions are converted to data points, $(\boldsymbol{f}_i(\sigma_e), s_i(\sigma_e)), i = 1, \ldots N$ through the action of the feature operators, $\phi(\sigma_e)$ and $\psi(\sigma_e)$. For the experiments reported here, the image feature operators were defined to be local, second degree polynomial fits and the output features were the polynomial coefficients. Figure 4 shows a plot of the surface data projected onto a two-dimensional subspace of the image feature space along with two projections of the mapping function, g, which is represented by a low degree polynomial. The size of each dot shows its magnitude. The clear trend in the data suggests that a low order model should account for a large percentage of the variance.

6.2 Surface Estimation Results

This test is designed to show that the system can recover the shape of a previously unseen surface that is very different from the surfaces used in the model estimation phase. The shape estimation procedure is conducted using the model estimated from the ridge surfaces of the previous section.

Figures 5 and 6 illustrate the shape recovery process applied to the image of a clay face. The first step of the surface recovery procedure is calculation of the image feature vector at every image location. The middle image in figure 5 shows one of six feature arrays produced using the local polynomial fit method described above using a scale factor σ_r. Each feature can be thought of as encoding a particular type of information about the local image structure. Once the feature arrays have been computed, the mapping function g is evaluated at each point producing a surface map, $s(\sigma_r)$, shown at the right of figure 5.

The final step of the surface estimation procedure is deconvolution of the surface map using the kernel, $\psi(\sigma_r)$, to obtain a surface height map, z. The resultant depth map, corresponding level curves, and reconstructed image are shown in figure 6. Note that the current method does not depend upon stereo during the surface estimation procedure and therefore produces a smooth surface. Another example is shown in figure 7 which shows a picture of a human subject followed by the corresponding depth map, level curves, and shaded depth map.

7 Conclusions

The shape from appearance method is a new method for estimating surface shape based on a learned associative model. The model is generated by associating examples of local surface shape with corresponding image features. The model may be a probability density or an associative map. The associative map is easier to use but can only be used in the presence of sufficient contextual information.

The experiments described in this paper involve the use of a stereo module as a source of local shape examples. It has been shown that through a scale change, the associative model can be made more accurate than the stereo algorithm. As a consequence, it is possible to recover the shape of an unknown surface more accurately than is possible with the stereo algorithm.

In general, the method reported here is more accurate than stereo, motion, or focus based methods. Shading based methods are also quite accurate, but they are based on strong and often unrealistic assumptions about reflectance, lighting, shadowing and other scene characteristics. The method described here is based on much weaker assumptions. There is still much theoretical development to be done. However, its generality promises to make it useful under a wider range of conditions than existing methods.

References

1. R. O. Duda and P. E Hart, *Pattern Classification and Scene Analysis.* New York: John Wiley & Sons, 1973.
2. D. Forsyth and A. Zisserman, "Shape from shading in the light of mutual illumination," *Image Vision Comput.,* vol. 8, no. 1, pp 42–49, 1990.
3. D. R. Hougen and N. Ahuja, "Estimation of the Light Source Distribution and its Use in Shape Recovery from Stereo and Shading," in *Fourth Intl Conf. Comput. Vis.,* pp. 148–155, May 1993.
4. D. R. Hougen and N. Ahuja, "Adaptive Polynomial Modelling of the Reflectance for Shape Estimation from Stereo and Shading," in *Proc. IEEE Conf. Comput. Vis. Patt. Recog.,,* pp. 991–994, June 1994.
5. D. R. Hougen and N. Ahuja, "Resolution and Accuracy of Stereo, Motion, and Shading Methods," submitted to *Comput. Vis. Patt. Recog.,,* 1996.
6. Y. G. Leclerc and A. F. Bobick, "The direct computation of height from shading," *Proc. Comput. Vis. Patt. Recog.,* pp. 552–558, June 1991.
7. S. R. Lehky and T. J. Sejnowski, "Neural network model of visual cortex for determining surface curvature from images of shaded surfaces," *Proc. R. Soc. Lond. B,* vol. 240, pp. 251–278, 1990.
8. S. Omohundro, "Geometric learning algorithms," *Physica D,* vol. 42 pp. 307–321, 1990, and in *Emergent Computation,* ed. Stephanie Forrest, MIT Press 1991.

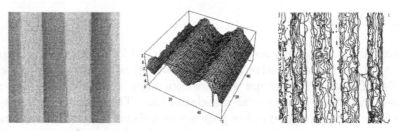

Fig. 3. One of four oriented ridge surface images with stereo depth map and corresponding level curves. The stereo program extracts a coarse estimate of the surface for use as input to the model building program.

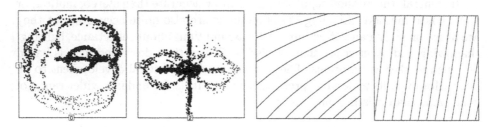

Fig. 4. Data collected from ridge images and resultant model projected onto a two-dimensional subspace of the image feature space. The size of each dot corresponds to value of the LOG surface feature. The model is a low order polynomial.

Fig. 5. Image of clay cherub figure, one of six feature maps, and shaded surface map, $s(x,y) = g(f(x,y))$. Each feature is one coefficient of second degree polynomial fit to the local gray level surface.

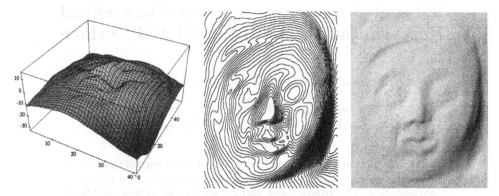

Fig. 6. Depth map corresponding to cherub image, level curves, and an image produced by shading the depth map from $(-1, 1, 1)/\sqrt{3}$.

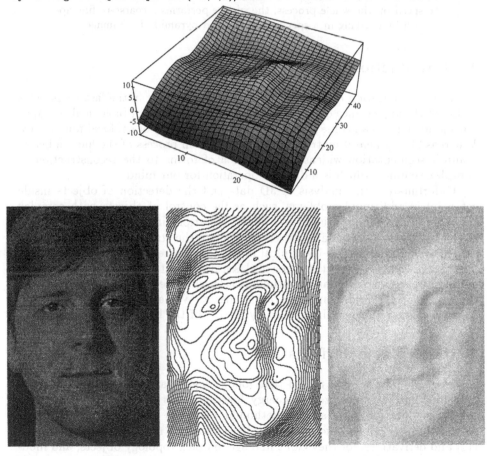

Fig. 7. Image of one author, depth map, level curves, and an image produced by shading the depth map from $(-1, 1, 1)/\sqrt{3}$.

Volumic Segmentation using Hierarchical Representation and Triangulated Surface

Jacques-Olivier Lachaud and Annick Montanvert

LIP, ENS-Lyon, 46, allée d'Italie, 69364 Lyon Cedex 7, France
e-mail: (jolachau, montanv)@lip.ens-lyon.fr

Abstract. This article presents a new algorithm for segmenting 3D images. It is based on a dynamic triangulated surface and on a pyramidal representation. The triangulated surface, which can as well modify its geometry as its topology, segments images into their components by altering its shape according to internal and external constraints. In order to speed up the whole process, the surface performs a coarse-to-fine approach by evolving in a specifically designed pyramid of 3D images.

1 Introduction

Volumic segmentation has become a major research topic in the last years. This is due to the appearance of 3D data in medical, geological or biological domains. This kind of data can come from either MR, tomography or confocal microscopy. Whereas 2D segmentation tries to mimic the vision process of the human being, volumic segmentation widens the detection of forms to the reconstruction of complex volumes, which is a difficult operation for our mind.

Unfortunately the analysis of 3D data and the detection of objects inside set a lot of additional problems, such as the control of objects with complex topology or the computational cost of the operations in a volumic space.

Hence the first purpose has been to evaluate the state of the art in order to develop a deformable and dynamical model that possesses a variable topology. Then, in order to overstep the scope of classical segmentation and to speed up the process, we associate a scalar continuous field with the image and we introduce the notion of pyramids composed of 3D images. We have tested our model and we have measured the savings given by the use of the hierarchical approach.

2 Deformable Surfaces

A deformable model is a model that follows a general principle: the object is deformed until it minimizes an energy function. This can be done either by a direct computation (least-square method for instance) or by the repeated application of constraints on the model. We have discarded models based on quadrics [12] and derivatives, for they can't manage complex topology objects, and models based on implicit surfaces (*blobs* [14] or front propagation [9]), because their computationnal cost is often heavy.

We prefer instead models of deformable meshes, such as cubic splines [7] or triangulated surfaces [10] which carry out segmentation by constraining the model on its vertices.

Considering the potentiality of the combination of meshes with some characteristics of bi-dimensionnal *snakes* [4], such as local minimization and inner constraints, we have chosen the last approach, the triangulated mesh, in order to segment 3D images. Quickness ($\mathcal{O}(N^2)$ vertices for a volumic image with edges of N voxels), rendering, are its main advantages. To these points we can add the opportunity of extracting features of the object, such as the area and the volume defined by the object, moments and topological informations.

3 Description of the Model

We define our surface as a closed oriented triangulated mesh. With this definition the surface always represents the boundary of a real volume. Our surface may thus be composed of several connected components, all closed and oriented.

Physical Aspect. The mesh is assimilated to a dynamic system of particles [11], which are the vertices of the triangles. Practically, interactions between particles occur only between direct neighbours. This neighbourhood allows us to define two internal constraints, the surface tension $\mathcal{F}_c$ and the surface elasticity $\mathcal{F}_e$, which follow the action/reaction principle. We transform the discrete image into a continuous scalar field (see section 4.1), in order to express the image influence through two external constraints: the force $\mathcal{F}_i$ which search for an isopotential surface and the force $\mathcal{F}_{di}$ which is a classical gradient descent.

The algorithm carrying out the displacement of the surface can be summarized into an iteration of the following steps:

1. Computing of internal and external forces for all vertices.
2. Re-sampling of the time scale to limit the vertex displacements.
3. Application of the Dynamic Fundamental Law for each vertex.
4. Effective displacement of the vertices.

We emphasize that this process expresses only the geometrical displacement of the surface and not the intrinsic topological modifications.

Geometrical Aspect. Meshes tend to intersect when they evolve. [5] introduces a global invariant δ which bounds the minimal and maximal sizes of each edge. By this way, local geometrical modifications are made easier and we can detect collisions by tests over vertex distance. Let us recall the two geometrical constraints induced:

$$\forall (U,V), \text{ if neighbours: } \delta \leq \|\overrightarrow{UV}\| \leq 2.5 \, \delta, \text{ and if not: } \frac{2.5}{\sqrt{3}} \delta \leq \|\overrightarrow{UV}\| \quad (1)$$

Topological breaks are controlled via Euler-Poincaré's characteristic χ [3]. Note that the different topology changes of the surface correspond to a modification of 2 or -2 of χ. Additionnal informations may be found in [6].

Initialization of the triangulated mesh. The triangulated surface is initialized with one icosahedron embracing the volumic image in real coordinates or with several icosahedra scattered in the image. The surface is then globally divided (see section 5.1) until it follows our geometrical constraints. After that, the surface is free to evolve according to its dynamic and geometrical rules.

4 Image Workspace and Pyramids

4.1 Transformation toward a Continuous Scalar Field

Let $I(i, j, k)$ be our discrete image of size $M \times N \times P$. Let μ, ν, π be its real size: it corresponds to the volume really occupied by the image in space. We determine the continuous potential function $\Pi_I(x, y, z)$, $(x \in [0, \mu[,\ y \in [0, \nu[,\ z \in [0, \pi[)$, by a first degree interpolation of $I()$ from space M, N, P to space μ, ν, π.

In this way we obtain a continuous scalar field that allows the computation of $\mathcal{F}_i$, but which is not derivable everywhere. Therefore $\mathcal{F}_{di}$ is computed by interpolating the discrete gradient deduced from the image. The computed scalar field can be interpreted as a stack of isopotential surfaces. Note that the interpolation degree has no influence about this fact and first degree is sufficient for an isopotential surface tracking.

4.2 Multi-scale Approach with 3-D Pyramids

Direct approach of image segmentation is not totally satisfactory. The influence of the image is indeed localized around vertices and makes sense only if the mesh has the same preciseness than the resolution of the 3D image. Common solutions either use a mesh with a refinement comparable to the one of the image, or take an interest in a wider area around each vertex. Hence the computational cost is deeply proportionnal to the image size. Therefore, our approach is to compute once and for all the influence of the image areas at different scales. This mixed solution can be done by the computation of a 3D image pyramid.

We take a particular interest in pyramids of frequency decomposition [1]: they provide a set of images at decreasing resolutions and details. These pyramids create no wrong contours, hence the model can exploit the results obtained at coarse levels in order to start the calculation on a finer level with more efficiency.

Their successive levels are computed by the convolution of a Gaussian kernel of side 5 voxels. It guarantees a low cost filtering without phase translation linked to a reduction factor of two for each image dimension [2]. Let G_0 be the initial 3D image and the base of the pyramid. The computation of G_{h+1} according to G_h (image of level h in the pyramid) is given by the discrete convolution formula:

$$G_{h+1}(i', j', k') = \sum_{m=-2}^{2} \sum_{n=-2}^{2} \sum_{p=-2}^{2} \omega(m, n, p) \cdot G_h(2i' + m, 2j' + n, 2k' + p) \ (2)$$

where ω is a Gaussian convolution kernel of size 5 voxels: $(\frac{1}{16}[1\ 4\ 6\ 4\ 1])^3$.

4.3 3-D Image Pyramids of any Reduction Factor

The previous formulation (2) is not usable as it is, because we have to take into account two major constraints within our context:
- voxels are not bound to be cubic (sampling frequencies are highly dependent of the acquisition means and are not identical in the general case),
- the reduction factor of the re-sampling must be coherent to the surface representation.

We will realize the convolution operations efficiently, by defining a real work-space corresponding to the discrete structure including the initial data.

Our goal is to determine a list of discrete images, denoted $G_0, \ldots, G_{max}$, and which represents the pyramid. G_0 is the initial image (given for segmentation) of sizes M, N and P. It is the image that possesses the greatest amount of informations. G_{max} will be the image that includes only the lowest frequencies. Let M_h, N_h and P_h be the sizes of the discrete image G_h. Their values are still unknown. Let $I_h = M_h \times N_h \times P_h$. With these definitions a discrete image G_h is a function from I_h toward $[0, 1]$.

We denote I_R the space defined by the real image of size $[0, \mu[\times[0, \nu[\times[0, \pi[$. Because all images G_h represent at different scales the same real image, all of them have a real size of μ, ν, π. The immersion of a voxel (i, j, k) of G_h into the real image space I_R is given by the transformation T_h as follows:

$$
\begin{aligned}
T_h : \quad I_h \quad &\longrightarrow I_R \\
(i, j, k) &\longrightarrow \left(i\frac{\mu}{M_h}, j\frac{\nu}{N_h}, k\frac{\pi}{P_h}\right)
\end{aligned}
\tag{3}
$$

We call *unit* of the real space the value $U_h = \min(\mu/M_h, \nu/N_h, \pi/P_h)$. It's the smallest distance between the immersions of two voxels in the real image. In the case of an isotropic image, we got $U_h = \mu/M_h = \nu/N_h = \pi/P_h$.

A reduction factor is needed in order to build the successive pyramid levels. Being for the moment unpredictable (see section 5.1), our pyramid construction must authorize any reduction factor. Unlike purely discrete formulations, the transformation into a continuous image (see section 4.1) associated with our immersion process allows us to build pyramids of any factor. The chosen kernel is of side 5, therefore the reduction factor T must be less than two.

Let V_0 be the base of our pyramid of real images. V_0 is given by the immersion then by the interpolation of the discrete data. As a matter of fact $V_0 = \Pi_{G_0}$. Let V_h be the level h of the real image pyramid. V_{h+1} is calculated from V_h. The number and the localization (in I_R) of the points to be calculated are determined by the reduction factor T, and the values are obtained after convolution of some points of V_h. Their storing after computation on the real image space is of course done in an array of voxels, assimilated to the discrete pyramid (G_i) at level $h+1$.

The discrete sizes M_h, N_h, P_h and the measure unit U_h correspond to a real image V_h. Its characteristics are defined recursively with:

$$
\begin{aligned}
M_0 &= M & N_0 &= N & P_0 &= P & U_0 &= \min(\mu/M, \nu/N, \pi/P) \\
M_{h+1} &= \left\lfloor \tfrac{M_h}{T} \right\rfloor & N_{h+1} &= \left\lfloor \tfrac{N_h}{T} \right\rfloor & P_{h+1} &= \left\lfloor \tfrac{P_h}{T} \right\rfloor & U_{h+1} &= U_h \cdot T
\end{aligned}
\tag{4}
$$

Let $R(i, j, k)$ be a voxel of the discrete data of G_{h+1}. Its immersion R_V in the real image V_{h+1} has coordinates of $T_{h+1}(i, j, k)$ (see figure 1a). In order to establish the value of R, the convolution operation is defined over points of V_h. The central point has the same position in V_h and in V_{h+1}. The localization of the other points involved in the convolution is determined via the use of the unit U_h to discretize V_h around the point R_V (see figure 1b).

G_h is known, V_h is defined by Π_{G_h}. We obtain:

$$
G_{h+1}(i, j, k) = \sum_{m=-2}^{2} \sum_{n=-2}^{2} \sum_{p=-2}^{2} \omega(m, n, p) \, V_h[T_{h+1}(i, j, k) + (mU_h, nU_h, pU_h)] \tag{5}
$$

G_{h+1} defined then V_{h+1} implicitly $(V_{h+1} = \Pi_{G_{h+1}})$.

Because of the unknown reduction factor, the 5^3 points involved in the convolution do not coincide with given points of G_h in the general case (see figure 1c). Moreover there usually won't be any cover between points involved in two neighbouring convolutions. Besides, each point of V_h compulsory for the convolution computation is interpolated from 8 data points of V_h (so stored in G_h) which form the parallelepiped containing this point (see section 4.1 and figure 1c too).

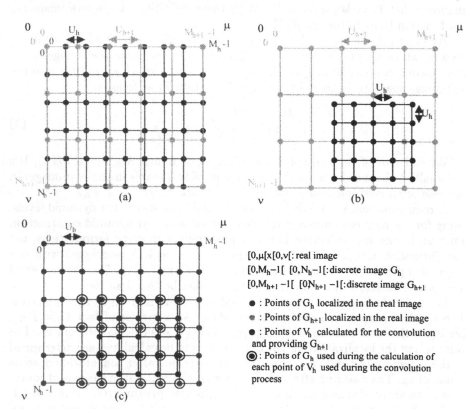

[0,μ[x[0,ν[: real image
[0,M_h−1[[0,N_h−1[: discrete image G_h
[0,M_{h+1}−1[[0,N_{h+1}−1[: discrete image G_{h+1}

● : Points of G_h localized in the real image
❋ : Points of G_{h+1} localized in the real image
● : Points of V_h calculated for the convolution and providing G_{h+1}
◉ : Points of G_h used during the calculation of each point of V_h used during the convolution process

Fig. 1. View of a convolution computation in 2-D: (a) The two superposed levels G_h and G_{h+1}, (b) computation of level G_{h+1} together with the localization of the convolution mask applied on one point, (c) application of the convolution mask over level G_h and visualization of the discrete points of G_h involved in the computation.

The Gaussian convolution kernel (of size 5^3) is applied successively along the three dimensions because of its separability. We can estimate the optimization savings by using the following notations and equations: Let T be the reduction factor, t_1 the execution time of a classical algorithm, t_2 the execution time of the optimized algorithm. A short calculation gives:

$$\frac{t_2}{t_1} = \frac{1}{25}(T^2 + T + 1) \quad \text{if the Gaussian kernel size is } 5^3 \tag{6}$$

The optimized algorithm is thus faster for $0 \le T \le \frac{-1+\sqrt{97}}{2}$ (≈ 4.42).

5 Segmentation

5.1 Image/Model Appropriateness

In section 4.2 we have chosen to express surface-image interaction with constraints locally computed around each vertex. In order to obtain a correct appropriateness between the surface and the pyramid images, we must first examine the relations linking the model preciseness to the resolution of an image, and secondly, we must establish an algorithm of surface refinement, which ensures the appropriateness surface-image during the whole coarse-to-fine process.

Surface-Image Relationships. According to (1) we have $d_{min} = \delta$ and $d_{max} = 2.5\ \delta$. The refinement can so be defined entirely with the invariant δ. Let δ_h be the invariant δ at level h of the pyramid. d_{min}^h and d_{max}^h are defined as same.

The image resolution is linked to the unit U_h (see section 4.3). We can deduce the relationships between δ_h and U_h with the help of the following considerations:

1. an edge may represent a contour formed by two 6-connected voxels (distant of U_h), so $d_{min}^h \leq U_h$,
2. an edge may represent a contour formed by two stricly 26-connected voxels (distant of $\sqrt{3}\ U_h$), so $d_{max}^h \geq \sqrt{3}\ U_h$.

$$\text{Hence,} \quad \frac{\sqrt{3}}{2.5} \leq \frac{U_h}{\delta_h} \leq 1 \quad \text{with } U_h = \min\left(\frac{\mu}{M_h}, \frac{\nu}{N_h}, \frac{\pi}{P_h}\right) \tag{7}$$

Surface Refinement. The surface works in an image pyramid and, consequently, must refine its mesh every time it goes down a level of the pyramid. We propose a process refining triangulated surface with a factor K. This factor will determine the reduction factor T of the pyramid.

Refinement process (or global surface division) (see figure 2):

1. in a first scan, a new vertex is created in the center of each facet of the model; the vertex is connected to the three vertices that delimit its facet,
2. in a second scan, the edges, which link together the old vertices (those which were not created during the first pass), are reversed in order to regularize edge lengths in a systematic way.

Such an algorithm reduces the average edge length to $1/\sqrt{3}$ of the old one. We may so apply to the invariant a reduction factor whose value is also $\sqrt{3}$, so $K = \sqrt{3}$. In order that inegality (7) be respected at the initialization moment and during all successive pyramid levels, an identical reduction factor is chosen for the pyramid construction:

$$T = K = \sqrt{3} \text{ and } \forall(h = 0 \ldots h_{max} - 1), \begin{cases} \delta_{h_{max}} = \delta_{init}, \ \delta_h = \delta_{h+1}/K \\ U_0 = U, \ U_{h+1} = U_h \cdot T \\ \text{thus we got } \frac{\sqrt{3}}{2.5} \leq \frac{U_h}{\delta_h} \leq 1 \end{cases} \tag{8}$$

At the initialization moment, a bubble or a set of bubbles, whose invariant δ_{init} is consistent with (7) at level h_{max}, are created. After that, the recursive process described by (8) guarantees a correct surface-image appropriateness, whichever are the iteration or the current level in the pyramid.

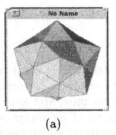 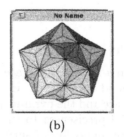 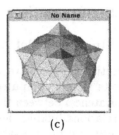

(a)   (b)   (c)

Fig. 2. Example of a global division process over a polyhedron with sixty facets: (a) before global division, (b) after first pass, (c) after second pass.

5.2 Anisotropy during segmentation

During the segmentation process, the edges have to keep their meaning with regards to a voxel space deformed by a possible anisotropy. So we make the surface evolve in a real space derived from the space (μ, ν, π) by affine transformation such that it has the same proportions than the discrete image it interpolates (its sizes are thus $(\lambda M_h, \lambda N_h, \lambda P_h)$. The internal forces have a slightly different behaviour than the one they would have in the real physical space. An alternative was to equip the real space with an anisotropic metrics in order to obtain a real physical behaviour, but such a work would not have been relevant in the context of segmentation.

6 Results

We first test our model on a volumic image of a human skull [1] of discrete sizes $256 \times 256 \times 68$ and of real sizes $1.0 \times 1.0 \times 1.0625$. The model segments the image by searching an isopotential surface. We give the surface some inner constraints to smooth the result. We provide our process with a full reliable heuristic that quickens the treatment of motionless or quasi-motionless vertices. The segmentation process is run two times for comparison purposes:

- First a direct processing on the image without any pyramid is shown on figure 3: the surface slowly sticks on the outer part of the skull and then goes inside to segment its inner part (orbits of the eyes, brain cavity, ...). The surface needs more than 400 iterations to meet equilibrium.
- We run after the process by making use of the pyramid built up from this volumic image with a reduction factor of $\sqrt{3}$. The process waits for its complete convergence at one pyramid level before going down one level. Figure 4 represents the coarse-to-fine evolution of the surface in the image pyramid.

Figure 5 analyzes the behaviour of both algorithms. The one of the classical segmentation algorithm is quite simple: the kinetic energy curve shows the slow segmentation convergence (see figure 5b), the number of vertex (see figure 5d) and the average edge length (see figure 5c) are subject to few changes, the time cost (see figure 5a) slowly decreases but only because of the use of the heuristic.

---

[1] Thanks to Yves Usson (C.H.U. Grenoble) for the volumic database.

144

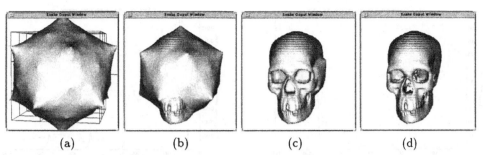

(a)   (b)   (c)   (d)

Fig. 3. Surface evolution during a segmentation without pyramid: (a) iteration 0 on image G_1, (b) iter. 100 on image G_1, (c) iter. 250 on image G_1, (d) iter. 700 on image G_1.

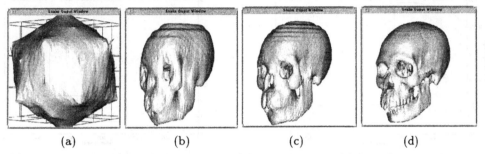

(a)   (b)   (c)   (d)

Fig. 4. Surface evolution during a pyramidal segmentation: (a) iteration 0 on image G_3, (b) iter. 225 on image G_3, (c) iter. 333 on image G_2, (d) iter. 461 on image G_1.

The behaviour of the pyramidal segmentation algorithm displays the four used pyramid levels: the number of vertex (see figure 5d) and the average edge length (see figure 5c) show that the triangulated mesh has gone down a level at the iterations 226, 334 and 462. The kinetic energy evolution (see figure 5b) highlights the convergences at each step and the duration graph of each iteration (see figure 5a) explains the time savings provided by a coarse-to-fine process. The final surface has 1 connected component and 23 topological holes.

We then test our model in a more problematic case: a phase contrast MR angiographic image [2]. Its discrete sizes are $256 \times 256 \times 124$. We can observe that such images are mainly composed of vessels whose proportions are not suited for a pyramidal representation. Within this context, the top image represents nearly nothing and the initialization of the process is very bad. Nevertheless the model succeeds in following the vessels to recover the forgotten ones as shown by figure 6, but each level needs more time to converge than for the first example.

7 Conclusion

We have designed and developed an efficient volumic segmentation algorithm by means of a deformable triangulated surface evolving in a 3D image pyramid. The obtained results show that their quality is identical to other similar algorithms;

[2] Acknowledgements to the UMDS Image Processing Group, London, for the angiographic image.

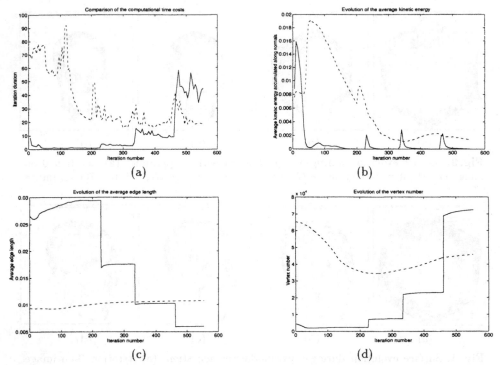

Fig. 5. Statistics over a skull segmentation process and comparison between the approach without a pyramid (in dotted line) and the pyramidal approach (in solid line): (a) duration of each iteration, (b) evolution of the average kinetic energy accumulated along the surface normals, (c) average edge length according to the iteration, (d) number of vertex of the surface according to the iteration.

the tests demonstrate the efficiency and the quickness of our algorithm. Several points may be nevertheless explored:

- The first convergence step can be greatly speeded up if the surface initialization contains more information. A marching-cube process [8] applied on the highest level of the pyramid would give a better first approximation. However surfaces obtained by this process are generally not closed.
- The convergence may be also guided by some new constraints, for instance by introducing attractive forces generated either by *sure* points of the object or by particular edges detected during a pre-treatment.
- The overall speed can be improved by accelerating the detection of topological breaks. For instance the use of some efficient algorithms [13] would reduce the cost of self-collision detection, which is for the moment within $\mathcal{O}(n \cdot \log(n))$ if n is the number of vertex.

References

1. P. J. BURT. "Fast filter transforms for image processing". *Computer Graphics and Image Processing*, 16:20–51, January 1981.

146

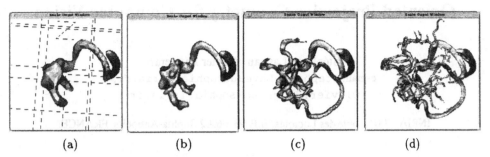

(a)   (b)   (c)   (d)

Fig. 6. Surface evolution during the segmentation of an angiographic image with pyramid: (a) After convergence on image G_3, (b) After convergence on image G_2,(c) After convergence on image G_1, (d) Final result

2. A. CHÉHIKIAN. "Algorithmes optimaux pour la génération de pyramides passes-bas et laplaciennes". *Traitement du Signal*, 9:297–308, January 1992.
3. H.B. GRIFFITHS. *"Surfaces"*. Cambridge University Press, January 1976.
4. M. KASS, A. WITKIN, and D. TERZOPOULOS. "Snakes: active contour models". In *1st Conference on Computer Vision*, Londres, June 1987.
5. J.O. LACHAUD and A. MONTANVERT. "Volumic Segmentation using Hierarchical Representation and Triangulated Surface". Research Report 95-37, LIP - ENS Lyon, France, November 1995.
6. J.O. LACHAUD and A. MONTANVERT. "Segmentation tridimensionnelle hiérarchique par triangulation de surface". In *10ème Congrès Reconnaissance des Formes et Intelligence Artificielle*, January 1996.
7. F. LEITNER and P. CINQUIN. "Complex topology 3D objects segmentation". In *Advances in Intelligent Robotics Systems*, volume 1609 of *SPIE*, Boston, November 1991.
8. W. E. LORENSEN and H. E. CLINE. "Marching Cubes: A High Resolution 3D Surface Construction Algorithm". *Computer Graphics*, 21:163–169, January 1987.
9. R. MALLADI, J. A. SETHIAN, and B. C. VEMURI. "Shape Modelling with Front Propagation: A Level Set Approach". *IEEE Transactions on Pattern Analysis and Machine Intelligence*, 17(2):158–174, February 1995.
10. J.V. MILLER, D.E. BREEN, W.E. LORENSEN, R.M. O'BARNES, and M.J. WOZNY. "Geometrically deformed models: A method for extracting closed geometric models from volume data". *Computer Graphics*, 25(4), July 1991.
11. R. SZELISKI and D. TONNESEN. "Surface Modeling with oriented Particle Systems". Technical Report CRL-91-14, DEC Cambridge Research Lab., December 1991.
12. D. TERZOPOULOS and A. WITKIN. "Deformable Models: Physically based models with rigid and deformable components". *IEEE Computer Graphics and Applications*, 8(6):41–51, November 1988.
13. P. VOLINO and Thalmann N. MAGNENAT. "Efficient self-collision detection on smoothly discretized surface animations using geometrical shape regularity". In *Eurographics'94*, volume 13(3), September 1994.
14. R.T. WHITAKER. "Volumetric deformable models: active blobs". In *VBC*, volume 2359 of *SPIE*, pages 122–134, March 1994.

Oriented Projective Geometry for Computer Vision

Stéphane Laveau   Olivier Faugeras
e-mail: Stephane.Laveau@sophia.inria.fr,
Olivier.Faugeras@sophia.inria.fr

INRIA. 2004, route des Lucioles. B.P. 93. 06902 Sophia-Antipolis. FRANCE.

Abstract. We present an extension of the usual projective geometric framework for computer vision which can nicely take into account an information that was previously not used, i.e. the fact that the pixels in an image correspond to points which lie in front of the camera. This framework, called the oriented projective geometry, retains all the advantages of the unoriented projective geometry, namely its simplicity for expressing the viewing geometry of a system of cameras, while extending its adequation to model realistic situations.
We discuss the mathematical and practical issues raised by this new framework for a number of computer vision algorithms. We present different experiments where this new tool clearly helps.

1 Introduction

Projective geometry is now established as the correct and most convenient way to describe the geometry of systems of cameras and the geometry of the scene they record. The reason for this is that a pinhole camera, a very reasonable model for most cameras, is really a projective (in the sense of projective geometry) engine projecting (in the usual sense) the real world onto the retinal plane. Therefore we gain a lot in simplicity if we represent the real world as a part of a projective 3-D space and the retina as a part of a projective 2-D space.

But in using such a representation, we apparently loose information: we are used to think of the applications of computer vision as requiring a Euclidean space and this notion is lacking in the projective space. We are thus led to explore two interesting avenues. The first is the understanding of the relationship between the projective structure of, say, the environment and the usual affine and Euclidean structures, of what kind of measurements are possible within each of these three contexts and how can we use image measurements and/or a priori information to move from one structure to the next. This has been addressed in recent papers [7, 2]. The second is the exploration of the requirements of specific applications in terms of geometry. A typical question is, can this application be solved with projective information only, affine, or Euclidean. Answers to some of these questions for specific examples in robotics, image synthesis, and scene modelling are described in [11, 3, 4], respectively.

In this article we propose to add a significant feature to the projective framework, namely the possibility to take into account the fact that for a pinhole camera, both sides of the retinal plane are very different: one side corresponds to what is in front of the camera, one side to what is behind! The idea of visible points, i.e. of points located in front of the camera, is central in vision and the problem of enforcing the visibility of reconstructed points in stereo, motion or shape from X has not received a satisfactory answer as of today. A very interesting step in the direction of a possible solution has been taken by Hartley [6] with the idea of Cheirality invariants. We believe that our way of extending the framework of projective geometry goes significantly further.

Thus the key idea developed in this article is that even though a pinhole camera is indeed a projective engine, it is slightly more than that in the sense that we know for sure that all 3-D points whose images are recorded by the camera are in front of the camera. Hence the imaging process provides a way to tell apart both sides of the retinal plane. Our observation is that the mathematical framework for elaborating this idea already exists, it is the oriented projective geometry which has recently been proposed by Stolfi in his book [13].

2 A short introduction to oriented projective geometry

An n-dimensional projective space, $\mathcal{P}^n$ can be thought of as arising from an $n+1$ dimensional vector space in which we define the following relation between non zero vectors. To help guide the reader's intuition, it is useful to think of a non zero vector as defining a line through the origin. We say that two such vectors x and y are equivalent if and only if they define the same line. It is easily verified that this defines an equivalence relation on the vector space minus the zero vector. It is sometimes also useful to picture the projective space as the set of points of the unit sphere $\mathcal{S}^n$ of $\mathbb{R}^{n+1}$ with antipodal points identified. A point in that space is called a projective point; it is an equivalence class of vectors and can therefore be represented by any vector in the class. If x is such a vector, then λx, $\lambda \neq 0$ is also in the class and represents the same projective point.

In order to go from projective geometry to oriented projective geometry we only have to change the definition of the equivalence relation slightly:

$$\exists \lambda > 0 \text{ such that } y = \lambda x \tag{1}$$

where we now impose that the scalar λ be positive. The equivalence class of a vector now becomes the half-line defined by this vector. The set of equivalence classes is the oriented projective space $\mathcal{T}^n$ which can also be thought of as $\mathcal{S}^n$ but without the identification of antipodal points. A more useful representation, perhaps, is Stolfi's straight model [13] which describes $\mathcal{T}^n$ as two copies of $\mathbb{R}^n$, and an infinity point for every direction of $\mathbb{R}^{n+1}$, i.e. a sphere of points at infinity, each copy of $\mathbb{R}^n$ being the central projection of half of $\mathcal{S}^n$ onto the hyperplane of $\mathbb{R}^{n+1}$ of equation $x_1 = 1$. These two halves are referred to as the front range ($x_1 > 0$) and the back range ($x_1 < 0$) and we can think of the front half as the set of "real" points and the back half as the set of "phantom" points, or vice versa. Thus, given a point x of $\mathcal{T}^n$ of coordinate vector x, the point represented by $-x$ is different from x, it is called its antipode and noted $\neg x$.

The nice thing about $\mathcal{T}^n$ is that because it is homeomorphic to $\mathcal{S}^n$ (as opposed to $\mathcal{P}^n$ which is homeomorphic to $\mathcal{S}^n$ where the antipodal points have been identified), it is orientable. It is then possible to define a coherent orientation over the whole of $\mathcal{T}^n$: if we imagine moving a direct basis of the front range across the sphere at infinity into the back range and then back to the starting point of the front range, the final basis will have the same orientation as the initial one which is the definition of orientability. Note that this is not possible for $\mathcal{P}^n$ for even values of n.

3 Oriented projective geometry for computer vision

We now use the previous formalism to solve the problem of determining "front" from "back" for an arbitrary number of weakly calibrated pinhole cameras, i.e. cameras for which only image correspondences are known. We know that in this case only the projective structure of the scene can be recovered in general [1, 5]. We show that in fact a lot more can be recovered.

As usual, a camera is modeled as a linear mapping from $\mathcal{P}^3$ to $\mathcal{P}^2$ defined by a matrix P called the perspective projection matrix. The relation between a 3-D point M and its image m is $m \simeq PM$ where $\simeq$ denotes projective equality. By modeling the environment as $\mathcal{T}^3$, instead of $\mathcal{P}^3$, and by using the fact that the imaged points are in front of the retinal plane Π_f of the camera, we can orient that retinal plane in a natural way. In the case of two cameras, we can perform this orientation coherently and in fact extend it to the epipolar lines and the fundamental matrix. This applies also to any number of cameras. We call the process of orienting the focal plane of a camera orienting the camera.

3.1 Orienting the camera

The orienting of a camera is a relatively simple operation. It is enough to know the projective coordinates of a visible point. We say that this point is *in the front range* of the camera. By choosing one of the two points of $\mathcal{T}^3$ associated with this point of $\mathcal{P}^3$, we identify the front and the back range relatively to the focal plane of the camera.

The existence of such an information (the coordinates of a point in space and in the image) is verified in all practical cases. If the scene is a calibration grid, its space and image coordinates are known. In the case of *weak calibration*, a projective reconstruction is easily obtained by triangulation.

In order to define the orientation of the camera, we also need the assumption that the points we are considering do not project onto the line at infinity in the image plane. This is also verified in all practical cases, because no camera can see points on its focal plane.

Let us write the perspective projection matrix P as

$$P = \begin{pmatrix} l_1^T \\ l_2^T \\ l_3^T \end{pmatrix}$$

where l_i, $i = 1, 2, 3$ are 4×1 vectors defined up to scale. We know that the optical center is the point C verifying $PC = 0$ and that l_3 represents the focal plane Π_f of the camera. Π_f is a plane of $\mathcal{T}^3$ which we can orient by defining its positive side as being the front range of the camera and its negative side as being the back range of the camera. This is equivalent to choosing l_3 such that, for example, the image of the points in front of the camera have their last coordinate positive, and negative for the points behind the camera. Again, this is just a convention, not a restriction.

The last coordinate of m is simply $l_3.M$. According to our conventions, this expression must be positive when M ($M \in \mathcal{T}^3$) is *in front* of the focal plane. This determines the sign of l_3 and consequently the sign of P. P is then defined up to a *positive* scale factor.

Hence we have a clear example of the application of the oriented projective geometry framework: the retinal plane is a plane of $\mathcal{T}^3$ represented by two copies of $\mathbb{R}^2$ its front and back ranges in the terminology of section 1, which are two affine planes, and a circle of points at infinity. The front range "sees" the points in the front range of the camera, the back range "sees" the points in the back range of the camera.

The sign of P determines the orientation of the camera without ambiguity. A camera with an opposite orientation will look in the exact opposite direction, with the same projection characteristics. It is reassuring that these two different cameras are represented by two different mathematical objects. For clarity, in the Figure 1, consider that we are working in a plane containing M and C. The scene is then T^2 that we represent as a sphere, whereas the focal plane appears as a great circle. We know that the scene point will be between C and ¬C.

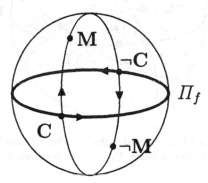

Fig. 1. A plane passing through M and C is represented as $\mathcal{S}^2$. The trace of the focal plane is an oriented line. It is oriented so that the front appears on the left when moving along the line.

3.2 Orienting the epipolar geometry

In this section, we consider a pair of cameras whose perspective projection matrices are P_1 and P_2. The epipoles are noted e_{12} and e_{21}.

The orienting of the cameras is not without having an incidence on the other attributes the stereo rig. The epipoles which are defined as the projections of the optical centers are oriented in the same fashion as the other points: if the optical center of the second camera is in front of (resp. behind) the focal plane of the first camera, the epipole has a positive (resp. negative) orientation, that is to say, a positive (resp. negative) last coordinate. This is achieved from the oriented pair of perspective projection matrices. The projective coordinates of the optical center C_2 are computed from P_2 by solving $P_2 C_2 = 0$. From this information only we cannot decide which of the T^3 objects corresponding to our projective points are the optical centers. Geometrically, this means that both C_2 and $\neg C_2$ are possible representations of the optical center. Of course, only one of them is correct. This is shown in Figure 2. If we choose the wrong optical center, the only incidence is that the orientation of all the epipoles and epipolar lines is going to be reversed.

The epipoles are then computed as images of the optical centers. In our case, $e = PC'$. The epipolar lines can then be computed using $l_{m'} = e \times m$. As we can see, there are only two possible orientations for the set of epipolar lines. Setting the orientation of one of them imposes the orientation of all epipolar lines in the image. This ambiguity also goes away if we know the affine structure of the scene. The plane at infinity splits the sphere in two halves. We can choose its orientation so that M is on the positive side. Because C_1 and C_2 are *real* points, they also are on the positive side of the plane at infinity. This will discriminate between C_2 and $\neg C_2$.

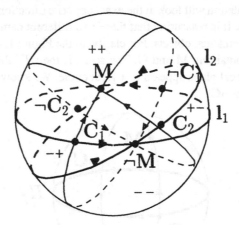

Fig. 2. The epipolar plane of M represented as the T^2 sphere. l_1 and l_2 represent the intersection of the focal planes with the epipolar plane. The cameras are oriented so that M is in front of the cameras.

4 Applications

In this section, we show some applications of the oriented projective geometry in problems in computer vision involving weakly calibrated cameras.

4.1 Possible and impossible reconstructions in stereo

In this section, in order to keep the demonstrations clear and concise, we consider only two cameras. The extension to any given number of cameras is easy. It is a fact that for a stereo system, all reconstructed points must be in front of both cameras. This allows us to eliminate false matches which generate impossible reconstructions. This will be characterized by a scene crossing the focal plane of one or several cameras. The focal planes of the two cameras divide the space into four zones as shown in Figure 3. The reconstruction must lie in zone ++ only.

When we reconstruct the points from the images, we obtain 3-D points in $\mathcal{P}^3$ which are pairs of points in T^3. We have no way of deciding which of the antipodal points is a real point. Therefore, we choose the point in T^3 to be in front of the first camera. We are then left with points possibly lying in ++ and +−. The points in +− are impossible points.

From this we see that we do not have a way to discriminate against points which are in the −− zone. These points can be real (i.e. in the front range of T^3), but they will always have an antipodal point in ++. On the other hand, points in −+ have their antipodal point in +− and can always be removed.

We can eliminate the points in −− only if we know where is the front range of T^3. This is equivalent to knowing the plane at infinty ant its orientation or equivalently the affine structure. Once we know the plane at infinity, we also know its orientation since

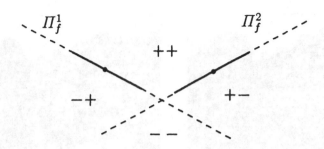

Fig. 3. Division of T^3 in 4 zones.

our first point of reference used to orient the cameras must be *in front*. We can now constrain every reconstructed point to be in the front range of T^3. This enables us to choose which point in T^3 corresponds to the point in $\mathcal{P}^3$. The points appearing in $--$ can be removed, their antipodal point being an impossible reconstruction also.

We are not implying that we can detect this way all of the false matches, but only that this inexpensive step[1] can improve the results at very little additional expense. It should be used in conjunction with other outlier detection methods like [14] and [15].

The method is simple. From our correspondences, we compute a fundamental matrix as in [8] for example. From this fundamental matrix, we obtain two perspective projection matrices [1, 5], up to an unknown projective transformation. We then orient, perhaps arbitrarily, each of the two cameras. The reconstruction of the image points yields a cloud of pairs of points which can lie in the four zones.

In general one of the zones contains the majority of points because it corresponds to the real scene[2]. The points which are reconstructed in the other zones are then marked as incorrect and the cameras can be properly oriented so that the scene lies in front of them.

The pair of images in figure 4 has been taken with a conventional CCD camera. The correspondences were computed using correlation, then relaxation. An outlier rejection method was used to get rid of the matches which did not fulfill the epipolar constraints. Most outliers were detected using the techniques described in [14] and [15]. Still, these methods are unable to detect false matches which are consistent with the epipolar geometry. Using orientation, we discovered two other false matches which are marked as points 41 and 251. This is not a great improvement because most outliers have already been detected at previous steps, but these particular false matches could not have been detected using any other method.

4.2 Hidden surface removal

The problem is the following: given two points M^a and M^b in a scene which are both visible in image 1 but project to the same image point in image 2, we want to be able

---

[1] A projective reconstruction can be shown to be equivalent to a least-squares problem, that is to say a singular value decomposition.

[2] We are making the (usually) safe assumption that the correct matches outnumber the incorrect ones.

Fig. 4. Outliers detected with orientation only.

to decide from the two images only and their epipolar geometry which of the two scene points is actually visible in image 2 (see Figure 5). This problem is central in view transfer and image compression [3, 9].

It is not possible to identify the point using the epipolar geometry alone, because both points belong to the same epipolar line. We must identify the closest 3-D point to the optical center on the optical ray. It is of course the first object point when the ray is followed from the optical center towards infinity in the front range of the camera. It is a false assumption that it will necessarily be the closest point to the epipole in the image. In fact, the situation changes whenever the optical center of one camera crosses the focal plane of the other camera. This can be seen in the top part of Figure 5 where the closest point to the epipole switches from m_1^a to m_2^a when C_2 crosses Π_{f1} and becomes C_2' with the effect that e_{12} becomes e_{12}'.

We can use oriented projective geometry in order to solve this problem in a simple and elegant fashion. We have seen in the previous section that every point of the physical space projects onto the images with a sign describing its position with respect to the focal plane. We have also seen that the epipolar lines were oriented in a coherent fashion, namely from the epipole to the point. When the epipole is in the front range of the retinal plane as for e_{12} (right part of the bottom part of Figure 5), when we start from e_{12} and follow the orientation of the epipolar line, the first point we meet is m_1^a which is correct. When the epipole is in the back range of the retinal plane as for e_{12}' (left part of the bottom part of Figure 5), when we start from e_{12}' and follow the orientation of the epipolar line, we first go out to infinity and come back on the other side to meet m_1^a which is again correct!

Hence we have a way of detecting occlusion even if we use only projective information. The choice of which representant of C_2 we use will determine a possible orientation. But what happens when the chosen orientation is incorrect? In order to understand the problem better, we synthesized two views of the same object, using the same projection matrices, but with two different orientations. This is shown in Figure 6. The erroneous left view appears as seen "from the other side". The geometric interpretation

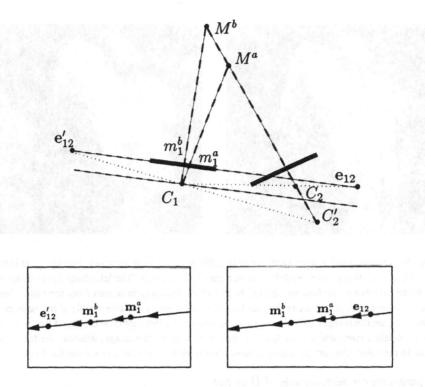

Fig. 5. Change of orientation when C_2 crosses the focal plane of the first camera.

is simple: The wrongly oriented camera looks in the opposite direction, but far enough to go through the sphere at infinity and come back to the other side of the object. Please note that without the use of oriented projective geometry, we would have a very large number of images, namely two possible orientations for each pixel.

4.3 Convex hulls

Another application of oriented projective geometry is the ability to build convex hulls of objects from at least two images. A method for computing the 3-D convex hull of an object from two views has been proposed before by Robert and Faugeras [12]. However, it is clear that the method fails if any point or optical center crosses any focal plane, as noted by the authors. Their approach is based on the homographies relating points in the images when their correspondents lie on a plane. Our approach makes full use of the oriented projective geometry framework to deal with the cases where their method fails. It consists in using once again the fact that the physical space is modelled as T^3 and that in order to compute the convex hull of a set of 3-D points we only need to be able to compare the relative positions of these points with respect to any plane, i.e. to decide whether two points are or are not on the same side of a plane.

This is possible in the framework of oriented projective geometry. Let Π be a plane represented by the vector u which we do not assume to be oriented. Given two scene points represented by the vectors M and M' in the same zone (++ for example), comparing the signs of the dot products $u \cdot M$ and $u \cdot M'$ allows us to decide whether the

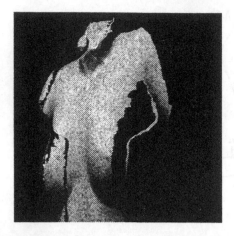

Fig. 6. Two synthesized images with cameras differing only in orientation. The image is incomplete because the source images did not cover the object entirely. The left image presents anomalies on the side due to the fact that the left breast of the mannequin is seen from the back. Hence, we see the back part of the breast first and there is a discontinuity line visible at the edge of the object in the source images. What we are seeing first are the last points on the ray. If the object was a head modelled completely, only the hair would be seen in one image, whereas the face would appear in the other. One image would appear seen from the back, and one from the front.

two points are on the same side of Π or not.

The usual algorithms for convex hull building can then be applied. There are of several sorts, a good reference being [10]. The results are of course identical to those of the method of Robert and Faugeras. The interested reader is then referred to [12] for details and results.

5 Conclusion

We have presented an extension of the usual projective geometric framework which can nicely take into account an information that was previously not used, i.e. the fact that we know that the pixels in an image correspond to points which lie in front of the camera. This framework, called the oriented projective geometry, retains all the advantages of the unoriented projective geometry, namely its simplicity for expressing the viewing geometry of a system of cameras, while extending its adequation to model realistic situations.

References

1. Olivier Faugeras. What can be seen in three dimensions with an uncalibrated stereo rig. In G. Sandini, editor, *Proceedings of the 2nd European Conference on Computer Vision*, volume 588 of *Lecture Notes in Computer Science*, pages 563–578, Santa Margherita Ligure, Italy, May 1992. Springer-Verlag.
2. Olivier Faugeras. Stratification of 3-d vision: projective, affine, and metric representations. *Journal of the Optical Society of America A*, 12(3):465–484, March 1995.
3. Olivier Faugeras and Stéphane Laveau. Representing three-dimensional data as a collection of images and fundamental matrices for image synthesis. In *Proceedings of the International Conference on Pattern Recognition*, pages 689–691, Jerusalem, Israel, October 1994. Computer Society Press.

4. Olivier Faugeras, Stéphane Laveau, Luc Robert, Cyril Zeller, and Gabriella Csurka. 3-d reconstruction of urban scenes from sequences of images. In A. Gruen, O. Kuebler, and P. Agouris, editors, *Automatic Extraction of Man-Made Objects from Aerial and Space Images*, pages 145-168, Ascona, Switzerland, April 1995. ETH, Birkhauser Verlag. also INRIA Technical Report 2572.

5. Richard Hartley, Rajiv Gupta, and Tom Chang. Stereo from uncalibrated cameras. In *Proceedings of the International Conference on Computer Vision and Pattern Recognition*, pages 761-764, Urbana Champaign, IL, June 1992. IEEE.

6. Richard I. Hartley. Cheirality invariants. In *Proceedings of the ARPA Image Understanding Workshop*, pages 745-753, Washington, DC, April 1993. Defense Advanced Research Projects Agency, Morgan Kaufmann Publishers, Inc.

7. Q.-T. Luong and T. Viéville. Canonical representations for the geometries of multiple projective views. Technical Report UCB/CSD 93-772, Berkeley, 1993. Oct. 1993, revised July 1994.

8. Quang-Tuan Luong. *Matrice Fondamentale et Calibration Visuelle sur l'Environnement- Vers une plus grande autonomie des systèmes robotiques*. PhD thesis, Université de Paris-Sud, Centre d'Orsay, December 1992.

9. Leonard McMillan and Gary Bishop. Plenoptic modeling: An image-based rendering system. In *SIGGRAPH*, Los Angeles, CA, August 1995.

10. F. Preparata and M. Shamos. *Computational Geometry*. Springer-Verlag, New-York, 1985.

11. L. Robert, C. Zeller, O. Faugeras, and M. Hébert. Applications of non-metric vision to some visually-guided robotics tasks. In Y. Aloimonos, editor, *Visual Navigation: From Biological Systems to Unmanned Ground Vehicles*, chapter ? Lawrence Erlbaum Associates, 1996. to appear, also INRIA Technical Report 2584.

12. Luc Robert and Olivier Faugeras. Relative 3-D positioning and 3-D convex hull computation from a weakly calibrated stereo pair. *Image and Vision Computing*, 13(3):189-197, 1995. also INRIA Technical Report 2349.

13. Jorge Stolfi. *Oriented Projective Geometry, A Framework for Geometric Computations*. Academic Press, Inc., 1250 Sixth Avenue, San Diego, CA, 1991.

14. Philip Torr. *Motion Segmentation and Outlier Detection*. PhD thesis, Department of Engineering Science, University of Oxford, 1995.

15. Zhengyou Zhang, Rachid Deriche, Olivier Faugeras, and Quang-Tuan Luong. A robust technique for matching two uncalibrated images through the recovery of the unknown epipolar geometry. *Artificial Intelligence Journal*, 1994. to appear, also INRIA Research Report No.2273, May 1994.

Fast Computation of the Fundamental Matrix for an Active Stereo Vision System

Fuxing Li, Michael Brady, Charles Wiles

Department of Engineering Science, University of Oxford, Oxford OX1 3PJ

Abstract. This paper investigates the problem of computing the fundamental matrix for a class of active stereo vision system, namely with common elevation platform. The fundamental matrix is derived for such a system, and a number of methods are proposed to simplify its computation. Experimental results validate the feasibility of the different methods. These methods are then used in a real application to validate the correctness of the fundamental matrix form for an active stereo system. We demonstrate that typical variations in camera intrinsic parameters do not much affect the epipolar geometry in the image. This motivates us to calibrate the camera intrinsic parameters approximately and then to use the calibration results to compute the epipolar geometry directly in real time.

1 Introduction

Active stereo vision systems based on a common elevation platform have been widely used for research on the visual control and navigation of mobile robots [1]. Usually, this kind of vision system has four degrees of freedom. The two cameras have independent vergence joints and use a common elevation platform. A pan joint is used to increase the field of view of the system. There is a platform for each of the cameras so that their optical centres can be adjusted to a position near the intersection of the elevation axis and the vergence axes to ensure that the elevation and vergence of the head approximately causes a pure rotation of the cameras. Let P be the fixation point of the two cameras, θ_e the elevation angle of the system, and θ_l and θ_r the two camera angles. Clearly, the epipolar geometry depends only on the two vergence joint angles. When the cameras verge, the epipolar geometry changes dynamically.

Such a stereo head has been mounted on the mobile robot at Oxford University to provide visual information for navigation. With the LICA(*Locally Intelligent Control Agent*)[3] based distributed architecture, parallel implementation of a number of algorithms using on-board transputers makes real-time visual guidance practical. Using such a robot head, the vision system is able to actively change its geometry to adapt to the task requirements and to plan and execute accurate camera motions. We have demonstrated real-time target tracking and following using this method[1].

In order to solve the feature-based correspondence problem in real time, a fast method must be employed to compute the epipolar geometry, hence to limit the search from two dimensions to one for correspondence. Furthermore, during the active verging process, all camera-related parameters are unchanged apart from the camera verge angles, which are available from the head state feedback.

In this paper, we investigate specific forms of the fundamental matrix, the algebraic representation of the epipolar geometry, for this important kind of stereo system. As a result, a number of methods are proposed to compute the fundamental matrix quickly and reliably. From extensive experimental results of applying these methods, epipolar geometry related camera intrinsic parameters are calibrated approximately. We demonstrate with these calibration results and the head state feedback, it is possible to update the epipolar geometry directly with sufficient accuracy. This will be applied to revisit the parallel implementation of PMF[5] in the near future and the result used to actively control the head.

2 The Common Elevation Fundamental Matrix

A stereo head with common elevation platform can be regarded as two cameras related by a translation $\mathbf{T}$ along the baseline and a rotation $\mathbf{R}$ about the Y axis. The fundamental matrix[8] is:

$$\mathbf{F} = \mathbf{C'}^\mathsf{T}\mathbf{R}[\mathbf{T}]_\mathbf{x}\mathbf{C}^{-1} = \mathbf{C'}^{-\mathsf{T}}\mathbf{R}\mathbf{C}^\mathsf{T}[\mathbf{CT}]_\mathbf{x},$$

where

$$\mathbf{C} = \begin{bmatrix} fk_u & 0 & u_0 \\ 0 & -fk_v & v_0 \\ 0 & 0 & 1 \end{bmatrix}, \quad \mathbf{T} = \begin{bmatrix} -B\sin\theta_l \\ 0 \\ -B\cos\theta_l \end{bmatrix}, \quad \mathbf{R} = \begin{bmatrix} \cos\theta_v & 0 & \sin\theta_v \\ 0 & 1 & 0 \\ -\sin\theta_v & 0 & \cos\theta_v \end{bmatrix},$$

The matrix $\mathbf{C'}$ is similar to $\mathbf{C}$. $\mathbf{C}$ and $\mathbf{C'}$ are the *camera calibration matrices* for the left and right cameras respectively, and represent the transformation from camera coordinates to image coordinates. B is the length of the stereo baseline; θ_l, θ_r are the left and right camera angles respectively; and $\theta_v = \theta_r - \theta_l$ is called the camera vergence angle.

We find that in this case $\mathbf{F}$ can be written in the following form:

$$\mathbf{F} = \begin{bmatrix} 0 & a & ab \\ c & 0 & d \\ b'c & e & b'd + be \end{bmatrix}, \tag{1}$$

where

$$a = \frac{fk_u}{fk'_u}\cos\theta_r, \quad b = -v_0, \quad b' = -v'_0, \quad c = -\frac{fk_v}{fk'_v}\cos\theta_l,$$

$$d = \frac{fk_v}{fk'_v}(fk_u\sin\theta_l + u_0\cos\theta_l), \quad e = -(fk_u\sin\theta_r + \frac{fk_u}{fk'_u}u'_0\cos\theta_r).$$

We will call this the common elevation fundamental matrix in the remainder of the paper.

The epipoles $\mathbf{e} = (e_x, e_y)^\top$ in the left image and $\mathbf{e}' = (e'_x, e'_y)^\top$ in the right image are:

$$e_x = -\frac{d}{c} = f k_u \tan\theta_l + u_0, \quad e_y = -b = v_0,$$

$$e'_x = -\frac{e}{a} = f k'_u \tan\theta_r + u'_0, \quad e'_y = -b' = v'_0.$$

Of the entries of the common elevation fundamental matrix:

$$d = \frac{f k_v}{f k'_v} sqrt(f k_u^2 + u_0^2) \cos(\theta_l - \phi),$$

where $\tan\phi = f k_u / u_0$.

For d to be zero, $\cos(\theta_l - \phi) = 0$. This implies $\theta_l = \phi + \frac{2n+1}{2}\pi$; but $\phi \neq 0$ and $0^o < \theta_l \leq 90^0$, so d is highly unlikely to be zero.

Since the fundamental matrix is only significant up to scale, using the property that d is unlikely to be zero, we set $d = 1$ in the following to simplify computations.

3 Computation

From equation 1, it can be seen that the main difference between the general fundamental matrix and that for a stereo head is that the latter has two zero entries which reflect the constraint of common elevation. That means we can compute the other entries with a simplified algorithm that is better conditioned and that can easily operate in real time.

8-point algorithm: Generally, each correspondence point pair generates one constraint on the fundamental matrix $\mathbf{F}$:

$$[x'_i \ y'_i \ 1] \begin{bmatrix} f_1 & f_2 & f_3 \\ f_4 & f_5 & f_6 \\ f_7 & f_8 & f_9 \end{bmatrix} \begin{bmatrix} x_i \\ y_i \\ 1 \end{bmatrix} = 0.$$

This can be rearranged as $\mathbf{Mf} = 0$ where $\mathbf{M}$ is a $n \times 9$ measurement matrix, and $\mathbf{f}$ is the fundamental matrix represented as a 9-vector:

$$\begin{bmatrix} x'_1 x_1 & x'_1 y_1 & x'_1 & y'_1 x_1 & y'_1 y_1 & y'_1 & x_1 & y_1 & 1 \\ \vdots & \vdots & \vdots & \vdots & \vdots & \vdots & \vdots & \vdots & \vdots \\ x'_n x_n & x'_n y_n & x'_n & y'_n x_n & y'_n y_n & y'_n & x_n & y_n & 1 \end{bmatrix} \begin{bmatrix} f_1 \\ \vdots \\ f_9 \end{bmatrix} = 0.$$

If the correspondence point pairs are reliable, 7 pairs suffice to solve $\mathbf{F}$ up to scale. To solve the problem linearly, it is customary to use 8 points to estimate f_i, $i = 1, \cdots, 9$, first and then enforce the zero determinant constraint afterwards[2].

6-point algorithm (linear method): After reparameterising the common elevation fundamental matrix from equation 1, we get

$$[x'_i \ y'_i \ 1] \begin{bmatrix} 0 & f_2 & f_3 \\ f_4 & 0 & 1 \\ f_7 & f_8 & f_9 \end{bmatrix} \begin{bmatrix} x_i \\ y_i \\ 1 \end{bmatrix} = 0.$$

This can be rewritten and solved using Singular Value Decomposition(SVD):

$$
\begin{bmatrix}
x'_1 y_1 & x'_1 & y'_1 x_1 & x_1 & y_1 & 1 \\
\vdots & \vdots & \vdots & \vdots & \vdots & \vdots \\
x'_n y_n & x'_n & y'_n x_n & x_n & y_n & 1
\end{bmatrix}
\begin{bmatrix}
f_2 \\ f_3 \\ f_4 \\ f_7 \\ f_8 \\ f_9
\end{bmatrix}
=
\begin{bmatrix}
-y'_1 \\ \vdots \\ -y'_n
\end{bmatrix}.
\tag{2}
$$

However, this does not ensure $det\mathbf{F} = 0$, and we have to enforce this constraint subsequently.

A convenient way to do this is to correct the matrix $\mathbf{F}$ found by equation 2. $\mathbf{F}$ is replaced by the matrix $\mathbf{F}'$ that minimises the Frobenius norm $||\mathbf{F} - \mathbf{F}'||$ subject to the condition $det\mathbf{F}' = 0$. This method was suggested by Tsai and Huang [7], adopted by Hartley [2], and has been proven to minimise the Frobenius norm $||\mathbf{F} - \mathbf{F}'||$, as required. Please consult [4] for details.

This kind of linear algorithm is normally not stable and is sensitive to noisy data. As Hartley argued in [2], normalising the data improves the performance of the eight point algorithm, which is used to compute the general form fundamental matrix. Our experimental results support this argument.

5-point algorithm (nonlinear method): Equation 1 can be written as:

$$
\mathbf{F} =
\begin{bmatrix}
0 & a & ab \\
c & 0 & 1 \\
b'c & e & b' + be
\end{bmatrix}
=
\begin{bmatrix}
1 & 0 & 0 \\
0 & 1 & 0 \\
0 & b' & 1
\end{bmatrix}
\begin{bmatrix}
0 & a & 0 \\
c & 0 & 1 \\
0 & e & 0
\end{bmatrix}
\begin{bmatrix}
1 & 0 & 0 \\
0 & 1 & b \\
0 & 0 & 1
\end{bmatrix}.
$$

Let

$$
\mathbf{F}' =
\begin{bmatrix}
0 & a & 0 \\
c & 0 & 1 \\
0 & e & 0
\end{bmatrix}, \quad
\mathbf{T}' =
\begin{bmatrix}
1 & 0 & 0 \\
0 & 1 & b' \\
0 & 0 & 1
\end{bmatrix}, \quad
\mathbf{T} =
\begin{bmatrix}
1 & 0 & 0 \\
0 & 1 & b \\
0 & 0 & 1
\end{bmatrix},
$$

then

$$
\mathbf{x}'^{\mathsf{T}} \mathbf{F} \mathbf{x} = \mathbf{x}'^{\mathsf{T}} \mathbf{T}'^{\mathsf{T}} \mathbf{F}' \mathbf{T} \mathbf{x} = (\mathbf{T}'\mathbf{x}')^{\mathsf{T}} \mathbf{F}' \mathbf{T} \mathbf{x} = 0,
$$

ie.

$$
\begin{bmatrix} x'_i & y'_i + b' & 1 \end{bmatrix}
\begin{bmatrix}
0 & a & 0 \\
c & 0 & 1 \\
0 & e & 0
\end{bmatrix}
\begin{bmatrix}
x_i \\ y_i + b \\ 1
\end{bmatrix}
= 0.
$$

Now we have five free parameters. In theory, given five correspondence point pairs, we should be able to solve for them. However this is a nonlinear problem.

If b and b' are unknown, the problem can be formed as a nonlinear optimisation problem to find b and b' to minimise

$$
\sum (x'_i(y_i + b)a + x_i(y'_i + b')c + (y_i + b)e + (y'_i + b'))^2,
$$

where a, c, e are computed to produce the sum of squares using SVD, given a set of b and b', as follows:

$$
x'_i(y_i + b)a + x_i(y'_i + b')c + (y_i + b)e = -(y'_i + b').
$$

3-point algorithm A (linear method with known b and b'): If the cameras are partially calibrated, ie. b and b' are known, then from the previous section, we can omit the nonlinear process and use SVD to solve a, c, e directly as follows:

$$\left[x_i'(y_i + b) \; x_i(y_i' + b') \; (y_i + b) \right] \begin{bmatrix} a \\ c \\ e \end{bmatrix} = - (y_i' + b')$$

4-point algorithm: If (as is often reasonable in practice) we assume that the two cameras forming the stereo head have approximately the same intrinsic parameters, the fundamental matrix can be further simplified:

$$\mathbf{F} = \begin{bmatrix} 0 & cos\theta_r & -v_0 cos\theta_r \\ -cos\theta_l & 0 & fk_u sin\theta_l + u_0 cos\theta_l \\ v_0 cos\theta_l & -(fk_u sin\theta_r + u_0 cos\theta_r) & u_0 v_0 (cos\theta_r - cos\theta_l) + fk_u v_0 (sin\theta_r - sin\theta_l) \end{bmatrix}$$

$$= \begin{bmatrix} 0 & a & ab \\ c & 0 & d \\ bc & e & b(d+e) \end{bmatrix},$$

where

$$a = cos\theta_r, \quad b = -v_0, \quad c = -cos\theta_l,$$
$$d = fk_u sin\theta_l + u_0 cos\theta_l, \quad e = -fk_u sin\theta_r - u_0 cos\theta_r.$$

As before, we can set $d = 1$.

If we further set $r = \frac{a}{c}$, the fundamental matrix becomes:

$$\mathbf{F} = \begin{bmatrix} 0 & rc & rcb \\ c & 0 & 1 \\ bc & e & b(1+e) \end{bmatrix}. \tag{3}$$

There are four free parameters b, c, e and r. In theory, with four correspondence point pairs, we should be able to solve for them. This is also a nonlinear problem. Compared to the situation when the two cameras have different intrinsic parameters, the nonlinearity is caused by just one parameter b. We can derive the nonlinear equation of b given four correspondence point pairs, then a regression method can be used to solve for b. Please consult [4] for details.

3-point algorithm B (camera angles are known): For an active vision system, we can usually acquire the head state, hence the camera angles θ_l and θ_r. In this case, $r = \frac{a}{c} = \frac{cos(\theta_r)}{-cos(\theta_l)}$ is known since it relies only on the camera angles whatever scale the fundamental matrix uses. We can then further simplify the 4-point algorithm to be the 3-point algorithm B.

Given three correspondence point pairs, we can derive a quadratic equation for b from the definition of the fundamental matrix. It is very easy to determine its closed form solution and then compute parameters c, e and a. Please consult [4] for details.

From the formula for the fundamental matrix, we can see that although we set $d = 1$, which scales the parameters which form the fundamental matrix, this does not affect the value of b, so b remains equal to $-v_o$. We can guess its initial value, and this can be used to select one of the two solutions of the quadratic.

Random sampling [6] is used to choose the best three point pairs to compute the fundamental matrix with the above algorithm. Another benefit of using the 4-point algorithm and 3-point algorithm B discussed in this section is they can be used to distinguish inliers and outliers of correspondence pairs.

4 Results

We have performed extensive experiments on synthetic and real data, [4] gives details. We present results here for a typical experiment on real data.

An image pair of a calibration grid standing in our laboratory was taken from our active stereo head with camera angles $\theta_l = 75^o$ and $\theta_r = 105^o$. The camera vergence angle is 30° which is usually the maximum value in real applications.

The number of the inliers among the point pairs used in the computation is one of the criteria to evaluate the quality of the computed fundamental matrix. If $d_i = \frac{d(p'_i, Fp_i) + d(p_i, F^T p'_i)}{2} < d_t$, we say that the computed fundamental matrix fits for the point pair (p_i, p'_i), and point pair (p_i, p'_i) is a inlier. Here (p_i, p'_i), $i = 1, \cdots, n$, are the n correspondence point pairs and $d(\cdot, \cdot)$ is the point-to-line Euclidean distance expressed in pixels. d_t is the threshold value which reflects the quality of the computed fundamental matrix as another criterion.

The fundamental matrices computed by the different algorithms are listed in Table 1, and we give the numbers of inliers in the table explicitly, given $d_t < 1$. We do not compute results using the 3-point A algorithm, since we do not know the true values of b and b'.

The results show that the first two diagonal elements are indeed close to zero when the 8-point and 6-point algorithms are used. The small difference from zero in these elements may be due to small manufacturing errors and slight misalignment in the common elevation geometry. However, we observe that the epipolar geometry computed by the 8-point and 6-point algorithms (see Figure 1) is in fact consistent with the plane to plane homography of the grid in the foreground of the scene. We conclude that, in this example, the point set used to compute the epipolar geometry must be close to approximating a plane and hence the computation of epipolar geometry with the 8-point and 6-point algorithms is underconstrained, resulting in a number of equally valid geometries close to that computed above. The 5-point algorithm enforces the leading diagonal elements to equal zero and hence limits the epipolar geometry to a unique solution. In fact the epipolar geometry computed with the 5-point algorithm more acurately approximates the true epipolar geometry of the stereo rig.

The results for the 4-point and 3-point B algorithms are poor, suggesting that the assumption of identical intrinsic parameters is not valid in this case (thus we do not show their epipolar lines in Figure 1).

Note that although the epipoles computed by the different algorithms are quite different, the large variation in b and b' does not significantly affect the

Table 1. *The results of the epipolar geometry computed by different algorithms for a stereo image pair. The first column is the algorithm used, the second the computed F matrix, the third the epipoles, and the fourth the number of inliers and total number of points considered. The corresponding epipolar lines are shown in Figure 1.*

8-pt	$F = \begin{bmatrix} 0.000049 & -0.000707 & 0.058852 \\ -0.000720 & -0.000045 & 1.000000 \\ 0.122604 & -0.809891 & -23.999954 \end{bmatrix}$	$e = \begin{bmatrix} 1389.71 \\ 83.2074 \end{bmatrix}, e' = \begin{bmatrix} -1145.06 \\ 170.384 \end{bmatrix}$	20/20
6-pt	$F = \begin{bmatrix} -0.000043 & -0.000241 & 0.063241 \\ -0.001126 & -0.000000 & 1.000000 \\ 0.129292 & -0.879739 & -23.914293 \end{bmatrix}$	$e = \begin{bmatrix} 888.455 \\ 262.196 \end{bmatrix}, e' = \begin{bmatrix} -3647.36 \\ 114.87 \end{bmatrix}$	20/20
5-pt	$F = \begin{bmatrix} 0 & -0.000620 & 0.080674 \\ -0.000804 & 0 & 1 \\ 0.108096 & -0.830662 & -26.2668 \end{bmatrix}$	$e = \begin{bmatrix} 1242.7 \\ 130.094 \end{bmatrix} e' = \begin{bmatrix} -1339.51 \\ 134.331 \end{bmatrix}$	20/20
4-pt	$F = \begin{bmatrix} 0 & -0.002125 & 0.918267 \\ 0.002209 & 0 & 1 \\ -0.954708 & -0.717975 & -121.866 \end{bmatrix}$	$e = \begin{bmatrix} -452.60952 \\ 432.109 \end{bmatrix}, e' = \begin{bmatrix} -337.85787 \\ 432.109 \end{bmatrix}$	12/20
3-pt B	$F = \begin{bmatrix} 0 & -0.000766 & 0.128619 \\ -0.000766 & 0 & 1 \\ 0.128619 & -0.786932 & -35.7733 \end{bmatrix}$	$e = \begin{bmatrix} 1305.3757 \\ 167.896 \end{bmatrix}, e' = \begin{bmatrix} -1027.2419 \\ 167.896 \end{bmatrix}$	9/20

epipolar lines in the image (since they are far from the image center). We discuss this further in section 5.

The analysis presented above assumes that the common elevation stereo head has been manufactured perfectly; but in practice this is never the case. The upshot of this is that intrinsic parameters that should be identical are measured to be different, and even though manufacturing errors may be slight the corresponding errors in the intrinsic parameters can be huge. In related work [4], we have shown how the fundamental matrix can be used to identify (small) manufacturing errors and correct for them in software automatically. Here, for purposes of clarity in exposition, we assume that slight manufacturing errors have been identified and corrected for. Please consult [4] for details.

5 Calibration

Our experiments show that the variation of b and b' does not much affect the epipolar lines in the image. If this is true for all the intrinsic parameters, we can then calibrate them using algorithms for computing the common elevation fundamental matrix and use the calibration results to compute the fundamental matrix directly according to its theoretical form.

We have studied the effects of intrinsic parameter variation on the epipolar geometry. Table 2 shows a typical result, corresponding to variations in the estimation of u_0. Here $(e_x, e_y)^{\top}$ is the epipole in the left image, $(e'_x, e'_y)^{\top}$ is the epipole in the right image, d_{cl} represents the corner-to-epipolar line distance in

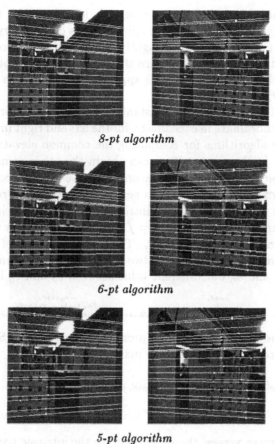

8-pt algorithm

6-pt algorithm

5-pt algorithm

Fig. 1. *Epipolar lines computed by various algorithms for an image pair. The corresponding fundamental matrices are shown in Table 1.*

the left image, d'_{cl} represents the corner-to-epipolar line distance in the right image, s represents the slope of the epipolar line in the left image, s' represents the slope of the epipolar line in the right image, E and σ represents the mean value and the standard deviation of the correspondence variable on the table. 500 randomly selected stereo match pairs are used for the statistical computation.

Table 2. *The error mean and covariance about epipolar geometry when u_o changes.*

Variation	10%				30%			
Vergence	10°		30°		10°		30°	
	E	σ^2	E	σ^2	E	σ^2	E	σ^2
e_x	12	0	12	0	36	0	36	0
e_y	0	0	0	0	0	0	0	0
e'_x	12	0	12	0	36	0	36	0
e'_y	0	0	0	0	0	0	0	0
d_{cl}	0.46104	0.119537	1.01775	0.504995	1.03841	0.542331	2.83341	3.64044
d'_{cl}	0.460199	0.11874	1.02727	0.510776	1.04803	0.546761	2.96384	3.94416
s	0.229193	1.11552	0.479116	8.44234	0.747675	1.98968	1.37208	11.4518
s'	-0.0305518	0.788074	-0.313323	6.92041	0.0892043	0.917001	-0.655558	7.11436

It is clear that when the vergence angle is less than 30°, the epipoles are far from the image planes, so that the epipolar lines change very little when the camera intrinsic parameters change slightly. In particular, the corner-to-epipolar line distance, which is used to measure the accuracy of the epipolar geometry, changes very little. This enables us to approximate calibration to compute the epipolar geometry efficiently.

We are able to get the nearly constant b and b', which represent the image centre's vertical coordinate in each image, for the left and right images of different frames using the algorithms for computing the common elevation fundamental matrix proposed in this paper. If we can obtain the other camera intrinsic parameters based on which the epipolar geometry can be computed, we can update the epipolar geometry using the stereo head state feedback in real time.

If we assume that the remaining intrinsic parameters are indeed equal between the two cameras such that $fk_u = fk'_u$ and $u_0 = u'_0$, then we can compute them given the camera angles θ_l and θ_r. The theoretical form of the common elevation fundamental matrix when the two cameras have the same value between fk_u and fk'_u, u_0 and u'_0 but different value between v_0 and v'_0 is:

$$\mathbf{F} = \begin{bmatrix} 0 & cos\theta_r & -v_0 cos\theta_r \\ -cos\theta_l & 0 & fk_u sin\theta_l + u_0 cos\theta_l \\ v'_0 cos\theta_l & -(fk_u sin\theta_r + u_0 cos\theta_r) & v_0(fk_u sin\theta_r + u_0 cos\theta_r) - v'_0(fk_u sin\theta_l + u_0 cos\theta_l) \end{bmatrix}. \quad (4)$$

Given the camera angles and v_0, v'_0 approximately, we can use SVD to compute fk_u and u_0 according to known stereo match pairs:

$$\left[sin(\theta_l)(y' - v'_0) - sin(\theta_r)(y - v_0) \quad cos(\theta_l)(y' - v'_0) - cos(\theta_r)(y - v0) \right] \begin{bmatrix} fk_u \\ u_0 \end{bmatrix}$$

$$= (y' - v'_0)x * cos(\theta_l) - (y - v_0)x' cos(\theta_r).$$

The more the camera verges, the more accurate the intrinsic parameters we can compute from the fundamental matrix according to the stereo match pairs. We use stereo match pairs from the image pair with vergence angle 30°.

The calibration results are then used to compute the fundamental matrix according to equation 4 for real image pairs taken from our active stereo head with different vergence angles. The computed fundamental matrices are sufficiently accurate for stereo correspondence search. [4] gives details of these results. This means that using this approximate calibration results, we can then compute the fundamental matrix in real time with the head state feedback. The calibration algorithm is listed in Figure 2, and a typical result is shown in Table 3.

References

1. F. Du and J.M. Brady. A four degree-of-freedom robot head for active vision. *International Journal of Pattern Recognition and Artificial Intelligence*, 8(6) 1994.
2. R. Hartley. In defence of the 8-point algorithm. In *Procceding of ICCV'95 International Conference on Computer Vision*, pages 1064–1070, 1995.
3. H. Hu, M. Brady, J. Grothusen, F. Li, and P. Probert. Licas: A modular architecture for intelligent control of mobile robots. In *Procceding of IROS'95 International Conference on Intelligent Robots and Systems*, 1995.

```
Input       I_l(0),I_r(0)      : Image pair with parallel geometry;
            I_l(30),I_r(30)    : Image pair verged 30°;
Output      fk_u, u_0, v_0, v_0'  : Camera intrinsic parameters.

1. Obtain the stereo match pairs for each image pair using the 8-point algorithm.
```

$$F = \begin{bmatrix} f_1 & f_2 & f_3 \\ f_4 & f_5 & f_6 \\ f_7 & f_8 & f_9 \end{bmatrix};$$

2. Identify and correct small manufacturing errors[4] so that both fundamental matrices have the form:

$$F = \begin{bmatrix} 0 & a & ab \\ c & 0 & 1 \\ b'c & e & b' + be \end{bmatrix},$$

where $b' = b + \delta b$. δb is obtained in this step by identify small manufacturing errors[4];

3. Compute $v_0 = -b$ and $v_0' = -b' = -(b + \delta b)$ using the 5-point algorithm from:

$$F(30) = \begin{bmatrix} 0 & a & ab \\ c & 0 & 1 \\ b'c & e & b' + be \end{bmatrix};$$

4. Compute fk_u and u_0 using SVD from:

$$\left[\sin(\theta_l)(y' - v_0') - \sin(\theta_r)(y - v_0) \quad \cos(\theta_l)(y' - v_0') - \cos(\theta_r)(y - v0) \right] \begin{bmatrix} fk_u \\ u_0 \end{bmatrix}$$
$$= (y' - v_0')x * \cos(\theta_l) - (y - v_0)x' \cos(\theta_r),$$

here $\theta_l = 75°$, $\theta_r = 105°$. v_0 and v_0' are computed from the above step.

Fig. 2. *The algorithm of calibrating the camera intrinsic parameters for computing the epipolar geometry.*

Table 3. *The results of the epipolar geometry for different image pairs using the calibration parameters and camera angles. The first column is the vergence angle, the second the computed F matrix, the third the epipoles, and the fourth the number of inliers and total number of points considered.*

Parallel image pair	$F = \begin{bmatrix} -0.000000 & -0.000000 & 0.024713 \\ -0.000000 & 0.000000 & 1.000000 \\ 0.001243 & -1.000000 & -4.720147 \end{bmatrix}$	$e = \begin{bmatrix} 8.2e+07 \\ 7.1e+07 \end{bmatrix}, e' = \begin{bmatrix} -2.9e+09 \\ 102441 \end{bmatrix}$	19/20
20°	$F = \begin{bmatrix} -0.000011 & -0.000479 & 0.078777 \\ -0.000479 & 0.000011 & 1.000000 \\ 0.058886 & -0.879146 & -21.207220 \end{bmatrix}$	$e = \begin{bmatrix} 2088.92 \\ 164.534 \end{bmatrix}, e' = \begin{bmatrix} -1836.19 \\ 123.009 \end{bmatrix}$	20/20

4. F. Li, J.M. Brady, and C. Wiles. Calibrating the camera intrinsic parameters for epipolar geometry computation. Technical report, OUEL 2075/1995, 1995.
5. Stephen B Pollard, John Porrill, John E W Mayhew, and John P Frisby. Disparity gradient, lipschitz continuity, and computing binouclar correspondences. In John E W Mayhew and John P Frisby, editors, *3D Model Recognition From Stereoscopic Cues*, pages 25–32. MIT, 1991.
6. P. Torr. *Motion Segmentation and Outlier Detection*. PhD thesis, University of Oxford, 1995.
7. R. Y. Tsai and T. S. Huang. Uniqueness and estimation of three dimensional motion parameters of rigid objects with curved surfaces. *IEEE Trans. Patt. Anal. Machine Intell.*, PAMI-6:13–27, 1984.
8. A. Zisserman. The geometry of multiple views. Seminar note for computer vision, 1995.

On Binocularly Viewed Occlusion Junctions

Jitendra Malik

Computer Science Division, Department of EECS
University of California at Berkeley, Berkeley, CA 94720
email: malik@cs.berkeley.edu

Abstract. Under binocular viewing of a scene, there are inevitably regions seen only in one eye or camera. Normally, this is a source of trouble for stereopsis algorithms which must deal with these regions on non-correspondence. This paper points out that in fact half-occlusion can be a source of valuable information. This is done by deriving a formula relating the displacement of an occlusion junction in the two eyes' images and the depth difference between the scene edges that comprise the occlusion junction. This paper represents the first quantitative result on the cue of half-occlusion in stereopsis.

1 Introduction

Under binocular viewing of a scene, there are inevitably regions seen only in one eye. Interesting phenomena related to this "half-occlusion" have been pointed out by Lawson and Gulick[10], Nakayama and Shimojo[13], Anderson and Nakayama[3], Anderson[1] and Anderson and Julesz[2].

Early computational models of stereopsis ignored half-occlusion altogether. Two recent models that have taken the phenomena seriously are those due to Jones and Malik[9] and Belhumeur and Mumford[4]. Jones and Malik find depth discontinuities and half-occlusion zones in an iterative framework where initial estimates of disparity from corresponding points are used to estimate the locations of occlusion contours, leading to a re-estimation of disparity and so on. Belhumeur and Mumford use dynamic programming to optimize a functional which explicitly allows for a suspension of the smoothness constraint at disparity discontinuities. However, even in these models half-occlusion is treated only as something that must be coped with rather than as a cue to depth discontinuities.

So far, there is no *quantitative* theory of the cue to depth discontinuity that is available in these regions of half-occlusion. Perhaps the most detailed attempt at a *qualitative* understanding is that due to Anderson and Julesz[2]. They point out that as a consequence of half-occlusion, the image locations of the junctions can have both a horizontal as well as vertical disparity between the two views. This is illustrated in the stereo pair in Figure 2 and the zoom of a junction in Figure 3. They then seek to understand the phenomena by performing a case analysis of a variety of different junctions seen stereoscopically in terms of features and contour segments that are seen only in the left eye or only in the right eye.

The goal of this paper is to develop a quantitative model of the depth discontinuity cue available in half-occlusion. It turns out to be surprisingly easy. The key point to focus on is the displacement of the image location of the junction between the two views of the scene. Generically, there are only two kinds of occlusion junctions in images of natural scenes–T junctions and X junctions. The hypothesis is that the visual system is able to localize these features independently in the two monocular images and then use the 2D displacement in making inferences. In section 2, I derive the main result of this paper – a formula relating this 2D displacement, *pseudo-disparity*, to the true disparity and the orientations of the occluding and occluded contours. Section 3 aims to quantify the magnitude of this effect under ecological viewing conditions. Section 4 discusses how the cue of pseudo disparity could be incorporated in computational models of stereopsis. The Appendix contains a generalization of the result derived in Section 2.

In this paper, we will analyze occlusion in the context of binocular stereopsis. These ideas apply, with minor modifications, to motion as will be reported in a companion paper.

2 The concept of pseudo-disparity

The cue of interposition for signalling depth has been known to artists for centuries. Helmholtz drew attention to T-junctions as indicating depth relationships. In computational vision, Guzman[7], Huffman[8], Clowes[6], Malik[11] have developed catalogs listing different image junctions and interpreting them in terms of possible scene events.

In order to understand what happens additionally under binocular viewing, it is sufficient to consider two classes of junctions–occlusion junctions and 3–D intersection junctions. Generically, under ecologically valid conditions, we have only two junctions associated with occlusion events: T and X junctions arising at occlusion by opaque and transparent surfaces respectively. All the rest of the junctions – L , arrow, Y, curvature-L, three tangent etc.– correspond to the projections of 3D intersections of scene contours. Note that T and X junctions do not *necessarily* signal occlusion–under certain conditions they could correspond just to surface markings.

Unlike most[1] 3D intersection junctions, an occlusion junction viewed from two different eye positions, does not correspond to the same physical point. We will now calculate the displacement of the location of such a junction between the two images.

First some notation: we use (x_l, y_l) to denote the coordinates in the left eye, and (x_r, y_r) to denote coordinates in the right eye, in each case measured with

---

[1] Exceptions are those 3D intersection junctions that involve a self-occlusion contour on a curved surface where the line of sight is tangent to the surface. Here slightly different points are being viewed in the two eyes because the occluding contour is now in a slightly different position on the object. This is well understood geometrically: Cipolla and Blake have used this cue to estimate surface curvature.

respect to the fixation point being the origin of the coordinate system. Then retinal disparity is classically defined to be $(x_r - x_l, y_r - y_l)$.

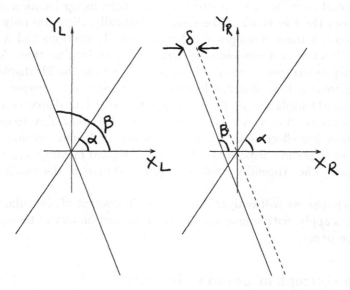

Fig. 1. A binocularly viewed occlusion junction.

Consider (Figure 1) two edges e_α and e_β in an occlusion relationship, either a T–junction (for an opaque occluding surface) or a X–junction (when the occluding surface is transparent). For simplicity, assume[2] that both edges lie in frontoparallel planes, so that there is no orientation disparity in the views in the two eyes. Let the edges e_α, e_β make angles α, β with respect to the positive x axis ($\alpha \neq \beta$). Let e_α, the occluding edge, be at zero disparity and e_β, the occluded edge, have a disparity $(\delta, 0)$.

Without loss of generality assume that the occlusion junction is at the origin in the left eye. We will compute its location in the right eye.

In the left eye the equations of the two edges are given by

$$x_l \sin \alpha - y_l \cos \alpha = 0 \tag{1}$$

$$x_l \sin \beta - y_l \cos \beta = 0 \tag{2}$$

which intersect at $(0, 0)$. In the right eye, the equations are

$$x_r \sin \alpha - y_r \cos \alpha = 0 \tag{3}$$

$$(x_r - \delta) \sin \beta - y_r \cos \beta = 0 \tag{4}$$

We solve (3) and (4) simultaneously to find the location of the junction in the right eye. Multiplying Eqn (3) by $\cos \beta$ and (4) by $\cos \alpha$ and subtracting, we eliminate y_r to get

$$x_r \sin \alpha \cos \beta - (x_r - \delta) \cos \alpha \sin \beta = 0$$

---

[2] This assumption is relaxed in the formula for the general case derived in the Appendix

giving

$$x_r = \frac{-\delta \cos \alpha \sin \beta}{\sin(\alpha - \beta)} \qquad (5)$$

Similarly, we can eliminate x_r between (3) and (4) to get,

$$y_r = \frac{-\delta \sin \alpha \sin \beta}{\sin(\alpha - \beta)} \qquad (6)$$

We can now compute the displacement of the occlusion junction in the right eye relative to the left eye as

$$(h_o, v_o) = \left(\frac{-\delta \cos \alpha \sin \beta}{\sin(\alpha - \beta)}, \frac{-\delta \sin \alpha \sin \beta}{\sin(\alpha - \beta)} \right) \qquad (7)$$

We will refer to (h_o, v_o) as the *pseudo-disparity* of the occlusion junction. The pseudo– prefix is intended to be a reminder that the junctions in the two eyes do not correspond to the same physical point.

What we have just established is that pseudo-disparity is uniquely determined by the horizontal disparity difference δ and the orientations α and β of the contours at the occlusion junction. We may think of pseudo-disparity as an 'emergent' cue at occlusion junctions, additional to the disparity of the defining contours. An analogy to orientation disparity or spatial frequency disparity is appropriate. In principle, these cues are just geometric consequences of positional disparity. However, the visual system may find it convenient to measure and exploit these cues directly.

As an illustration of this phenomenon, the reader is invited to examine Figure 2 where the epipolar lines are horizontal but the location of the T-junction is displaced vertically as well as horizontally. The zoom in Figure 3 makes this even more evident. The utility of eqn 7 is that it provides the first quantitative explanation of this phenomenon.

3 Magnitude of pseudo-disparity

The pseudo-disparity can be considerably larger than the horizontal disparity difference δ and typically has a significant vertical component v_o. In Figure 4, the magnitude of v_o is plotted as a function of α and β. Note that larger values are obtained for contours that have small orientation difference $(\alpha - \beta)$ and for nearly vertical contours $(\alpha \approx \pi/2)$.

I performed Monte Carlo simulations to estimate the probability distribution of the magnitude of v_o assuming that the orientations α and β are drawn independently from a uniform distribution on $[0, \pi]$. The median value of v_o is about 0.50δ, 5% of the v_o values are above 6.06δ, and 1% are even greater than 32.7δ. If one allows a slightly higher fraction of near vertical orientations, the fraction of large vertical pseudo-disparities goes up slightly.

One motivation for using pseudo-disparity can be seen as follows. If a vision system could make use of measurements of pseudo-disparity, its ability to identify

Fig. 2. An example of a junction showing pseudo-disparity at an occlusion junction. Zoom in Figure 3.

which of two edges in an occlusion relationship are closer would be much better than if it relied on horizontal disparity alone. For a human observer looking at a shrub 50 m away with branches that differ in depth by 1 m, the horizontal disparity is just 5″, at the limits of stereoacuity. However if the branches are nearly vertical and differ in orientation by less than 5°, the vertical component of the pseudo-disparity v_o is as much as 1′. It should be noted that the motivation for using the pseudo-disparity cue under natural viewing conditions may be even stronger–the sensitivity to disparity increments/decrements decreases rapidly away from the horopter.

4   Subjective occluding contours

Anderson[1] and Anderson and Julesz[2] have shown how oriented subjective contours are formed in the presence of half-occlusion. Eqn 7 provides a quantitative prediction of the results of the experiment in Fig. 3 in Anderson. Here the visual system knows the disparity δ, the pseudo-disparity (h_o, v_o) , and the orientation β of the occluded contour. The subjective occluding contour appears at the (unique) orientation α such that the formula for pseudo-disparity (Eqn 7) is satisfied. Anderson tested only for the sign of the slope of the occluding contour. I predict that in fact the orientation of the occluding contour is determined quantitatively. This could be tested psychophysically.

Fig 17 in Anderson and Julesz[2] shows the tradeoff between orientation and depth of the occluding surface associated with a given difference in vertical half-occlusion. Again, this can be made quantitative from (Eqn 7). For a vertical half-occluded edge, $\beta = \pi/2$, and the pseudo-disparity is $(\delta, \delta \tan \alpha)$. A given

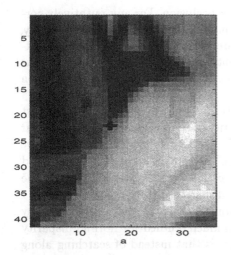
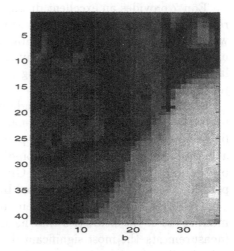

Fig. 3. An example of a junction showing pseudo-disparity at an occlusion junction. Detail of a junction from Figure 2.

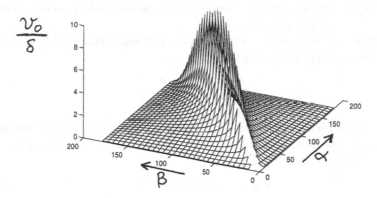

Fig. 4. Distribution of vertical component of pseudo-disparity.

vertical pseudo-disparity $\delta \tan \alpha$ can result from small disparity δ and large slope $\tan \alpha$ or *vice versa*.

5 Applications

How the constraint captured in equation 7 may be profitably exploited in vision systems remains a topic for future research. I sketch some ideas here.

Eqn 7 provides an excellent diagnostic for occlusion. For any putatively corresponding image junctions where more than one orientation is present locally, we can determine if the two orientations are associated with an occlusion situation or if they correspond to physically intersecting contours. We could thus label each image junction as being an occlusion junction or a 3D intersection junction.

In models for finding corresponding features, typically a search for the best match is necessary. For the traditional disparity cue, a 1–D search along epipolar lines is performed. To find "corresponding" occlusion junctions, we again have a 1–D search; here the "pseudo-epipolar line" line in the right image is defined by the locus of $(x_l + h_o, y_l + v_o)$. Given α, β (measured in the left image), the pseudo-epipolar line is parametrized by the disparity step δ.

There is a false target problem in both contexts. It can be resolved by making use of the heuristics that have been much studied for normal disparity measurements–the most significant difference is that instead of searching along epipolar lines $(x_l + \delta, y_l)$, one searches along $(x_l + h_o(\delta), y_l + v_o(\delta))$, where $h_o(\delta)$, $v_o(\delta)$ are the functions defined in Eqn 7.

We thus have the outlines of a computational mechanism adequate for a theory of stereopsis where occlusion and disparity are treated as equally powerful cues for identifying depth discontinuities:

- Direct inference from an occlusion junction with measured pseudo-disparity. Cooperative interactions among the occluding contours could be used to complete surface boundaries[1]. Associated with each occluding contour would be a zone of "non-correspondence".
- Indirect inference from surface interpolation and segmentation[12] at locations where disparity changes 'a lot'.

Appendix

We develop here the pseudo-disparity formula when the edges e_α and e_β are not necessarily in frontoparallel planes. As before, in the left eye the equations of the two edges can be expressed as

$$x_l \sin \alpha - y_l \cos \alpha = 0 \tag{8}$$
$$x_l \sin \beta - y_l \cos \beta = 0 \tag{9}$$

which intersect at $(0, 0)$.

To find the equations of the edges in the right eye coordinate system, we first need to model the changing horizontal disparity as a function of the image y coordinate. To do this, let the horizontal disparity along e_α, the occluding edge, be $\delta'_\alpha y_r$ and e_β, the occluded edge, be $\delta'_\beta y_r + \delta$. Here δ of e_β at the x–axis, same as before, and δ'_α, δ'_β are used to model the rate of change of disparity as a function of y. We can now write the equations for the right eye as

$$(x_r - \delta'_\alpha y_r) \sin \alpha - y_r \cos \alpha = 0 \tag{10}$$
$$(x_r - \delta'_\beta y_r - \delta) \sin \beta - y_r \cos \beta = 0 \tag{11}$$

We solve (10) and (11) simultaneously to find the location of the junction in the right eye.

$$h_o = \frac{-\delta \sin \beta (\cos \alpha + \delta'_\alpha \sin \alpha)}{\sin(\alpha - \beta) + (\delta'_\beta - \delta'_\alpha) \sin \alpha \sin \beta} \tag{12}$$

$$v_o = \frac{-\delta \sin \alpha \sin \beta}{\sin(\alpha - \beta) + (\delta'_\beta - \delta'_\alpha) \sin \alpha \sin \beta} \tag{13}$$

References

1. B.L. Anderson. The role of partial occlusion in stereopsis. *Nature*, 367:365–368, 1994.
2. B.L. Anderson and B.Julesz. A Theoretical Analysis of Illusory Contour Formation in Stereopsis. *Psych Review*, 1995 (in press).
3. B.L. Anderson and K. Nakayama. Toward a general theory of stereopsis: Binocular matching, occluding contours, and fusion. *Psych Review*, 101, pp. 414-445, 1994
4. P.N. Belhumeur and D. Mumford. A Bayesian treatment of the stereo correspondence problem using half-occluded regions. *Proc. of IEEE CVPR*, Urbana,IL, June 1992.
5. R. Cipolla and A. Blake. Surface shape from the deformation of apparent contours. *Int. Journ. of Computer Vision*, 9(2): 83-112, 1992.
6. M.B. Clowes, "On seeing things," *Artificial Intelligence* **2** ,pp. 79-116, 1971.
7. A. Guzman, "Computer Recognition of three–dimensional objects in a scene," MIT Tech. Rept. MAC-TR-59 ,1968.
8. D.A. Huffman, "Impossible objects as nonsense sentences," *Machine Intelligence* **6** ,pp. 295-323, 1971.
9. D.G. Jones and J. Malik. Computational framework for determining stereo correspondence from a set of linear spatial filters. *Image and Vision Computing*, 10: 699-708, 1992.
10. R.B. Lawson and W.L. Gulick. Stereopsis and Anomalous Contour. *Vision Research*, 7: 271-297.
11. J. Malik. Interpreting Line Drawings of Curved Objects. *Int. Journ. of Computer Vision*, 1(1): 73-103, 1987.
12. G. Mitchison. Planarity and segmentation in stereoscopic matching. *Perception*, 17:753-782, 1988.
13. K. Nakayama and S. Shimojo. DaVinci Stereopsis: Depth and subjective occluding contours from unpaired image points. *Vision Research*, 30(11): 1811-1825, 1990.

Understanding the Shape Properties of Trihedral Polyhedra

Charlie Rothwell and Julien Stern

INRIA, 2004, Route des Lucioles, Sophia Antipolis, 06902 CEDEX, France

Abstract. This paper presents a general framework for the computation of projective invariants of arbitrary degree of freedom (dof) trihedral polyhedra. We show that high dof. figures can be broken down into sets of connected four dof. polyhedra, for which known invariants exist. Although the more general shapes do not possess projective properties as a whole (when viewed by a single camera), each subpart does yield a projective description which is based on the *butterfly invariant*. Furthermore, planar projective invariants can be measured which link together the subparts, and so we can develop a local-global description for general trihedral polyhedra. We demonstrate the recovery of polyhedral shape descriptions from images by exploiting the local-global nature of the invariants.

1 Introduction

In this article we introduce a general scheme for understanding the shape properties of trihedral polyhedra. Trihedral polyhedra are solid polyhedra made up of planes in arbitrary positions, and as such, no special constraints exist between the planes. The nomenclature trihedral derives from the fact that the vertices of the polyhedra are only ever defined by triples of planes: points in space need at least three planes to assert their locations, but any more would provide excess constraint and hence would not be generic (and stably realisable). The results in this paper are a summary of those given in [9].

In all, we generalise the result in [8] which showed how a projectively invariant description can be computed for four degree of freedom (dof) polyhedra from a single view. In turn, [8] was a extension of the work of Sugihara [11]. The latter dealt with scaled orthographic projection and the calibrated perspective cases, whereas the former demonstrated the projective equivalence of all members of the family of four dof. polyhedra generating a set of scene measurements using an uncalibrated camera. We show in this paper that the approach of [8] can be extended to include all trihedral polyhedra.

We also build on some recent work for computing the invariants of minimal point configurations in three-dimensional space. Being able to compute measures for small local feature groups provides robustness to occlusion. More global descriptions can be built up using the local-global nature of many shape descriptions [3, 10]. We derive a part-whole decomposition by drawing the invariant description of [8], and the invariants based on the *butterfly configuration* of Mundy [13] together. The butterfly invariant is a geometric description of a special six-point configuration.

Our interest in the butterfly invariant was promoted by the recent paper of Sugimoto [12]. This paper discusses an invariant very similar to the original butterfly invariant, but suggests an algebraic rather than a geometric formulation. However, Sugimoto suggested that the invariants in [12] in some way replace the invariants described by [8]. In fact, these two types of invariant can be taken hand-in-hand and are exactly complementary. This we show partly in this paper, and in more detail in [9].

The contributions of this paper are three-fold: in Section 2 we discuss how the original invariant description of [8] can be decomposed into a set of three independent butterfly invariants. Then we show in Section 3 how to reduce a five dof. figure into sets of

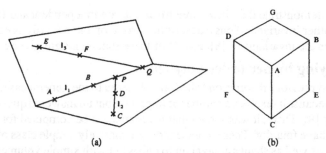

Fig. 1. *(a) the butterfly consists of two sets of four coplanar points yielding a unique cross ratio. (b) the labelling of a cube used for the butterfly invariant computation.*

figures with four dof. For the simplest case of such an object (two cube-like structures stuck together using a common plane), we can recover three invariants for each half of the object, and a further eight joint-invariants based on planar measures between the two halves. We also describe how higher dof. polyhedra can be broken down into four dof. objects related through pairwise planar constraints. Finally, in Section 4 we report on algorithms for the extraction of the polyhedral descriptions and their image invariants.

2 The butterfly invariant

The butterfly is the simplest known invariant for a set of points in 3D space.[1] The configuration is composed of six points in space broken up into two four point groups, $\{A, B, C, D\}$ and $\{A, B, E, F\}$. Each four point group is coplanar, and two points are shared between the groups. This is shown in Fig. 1a. The invariant for the butterfly is measured through the construction of a cross ratio. As can be seen in Fig. 1a, it is possible to form a set of four collinear points and hence the invariant cross ratio [2]. These are the points $\{A, B, P, Q\}$ from which we define the cross ratio $\tau = \{A, B; P, Q\}$. In fact, we show in [9] that the same butterfly invariant actually takes the algebraic form:

$$\tau = \frac{|M_{ACD}|}{|M_{AEF}|} \cdot \frac{|M_{BEF}|}{|M_{BCD}|}, \tag{1}$$

which allows the direct computation of the invariant values from image data. Here M_{ijk} is the 3×3 matrix whose columns are the points i, j and k and $|M|$ is the determinant of M. Although this invariant is very similar to that derived by Sugimoto [12] the interested reader will note that it can be derived more simply (shown in [9]). It is also worth noting that the form above was derived by Carlsson [1] using a more difficult, but in fact more general (and so elegant) approach based on the double algebra.

2.1 Computing polyhedral invariants with the butterfly

The projective invariants for polyhedral figures can be computed using the butterfly. Here we consider how to compute the invariants for a cube-like structure such as that shown in Fig. 1b. In fact, we see later that understanding the invariants for this type of figure is as far as we need to go to comprehend the invariants for all trihedral polyhedra. For a cube-like polyhedron, there are three independent invariants, for instance:

$$\tau_1 = \frac{|M_{ACE}|}{|M_{ADG}|} \cdot \frac{|M_{BDG}|}{|M_{BCE}|}, \qquad \tau_2 = \frac{|M_{ADF}|}{|M_{ABE}|} \cdot \frac{|M_{CBE}|}{|M_{CDF}|}, \qquad \tau_3 = \frac{|M_{ABG}|}{|M_{ACF}|} \cdot \frac{|M_{DCF}|}{|M_{DBG}|}. \tag{2}$$

---

[1] That is for a configuration which satisfies no single geometric constraint.

It is simple to demonstrate that these three invariants are independent and form a basis for all of the other invariants. Thus a description based on the butterfly is equivalent to one which uses the invariants of [8]. Proofs of these statements are given in [9].

3 Simplifying higher order polyhedra

We now extend previous theories on trihedral polyhedra to the general case. This result is significant because it resolves a number of important but unanswered questions which were posed by [8]. There, it was shown that invariants can be computed for polyhedral figures which have four dof. These represent only a relatively simple class of polyhedra such as cubes, as well as shapes equivalent to cubes but with simple volumes cut-out of them. Note that the number of dof. of a polyhedra represent the dimensionality of the space of reconstructions which arise out of the image constraints.

In the following paragraphs we demonstrate that figures with more than four dof. can be decomposed into sets of related polyhedra which have only four dof. The first example we consider is the decomposition of five dof. figures into sets of four dof. figures. More complicated figures can be decomposed similarly. The simplification process means that the invariants we have seen previously for four dof. polyhedra (such as the butterfly invariants) can be employed for each subpart of a more complicated structure, and so informs us that the basic invariants for general polyhedral structures are based on those for four dof. figures. However, no projective invariants of the complete structures exist as each four dof. subpart maintains a certain amount of (non-projective) independence from the other parts. We discuss this claim in Section 3.3.

Additionally, we can compute joint-invariants between the pairs of adjacent four dof. figures in the subpart hierarchy; these joint-invariants provide the glue which holds the subparts together. Therefore, although there are no *general* projective constraints between the subparts of a polyhedron, the subparts are not reconstructed in space with total liberty, but are placed in related frames. Usually, each related pair of subparts in a figure share a common plane in which *plane projective invariants* enforce constraints.

These two sets of invariants are all those which may be computed for an arbitrary trihedral polyhedron. We now proceed with proofs and explications of the statements given above. Due to limitations of space we are unable to repeat the theory given in [8], but only highlight the key points. However, we do make use of the same notation.

3.1 Reconstructing polyhedral figures

The assumptions we make about the polyhedra we treat are that they are made up of trihedral vertices and thus satisfy genericity conditions. Additionally, they are bounded solids which, due to self-occlusion under imaging, can be considered *not* to have interior volumes or surfaces. Thus, any *complete set* of adjacent planes forms a solid polyhedron. A complete polyhedron has three planes meeting at every vertex, and two planes at every edge. We can always render a set of adjacent planes complete using the trihedral constraint by defining additional planes passing through the edges and vertices.

We use a 3D coordinate system (X, Y, Z) and so planes satisfy: $a_j X_i + b_j Y_i + c_j Z_i + 1 = 0$, $j \in \{1, ..., n\}$ where n is the number of planes on the object. Each of the (X_i, Y_i, Z_i) represent the polyhedral vertices. Under the pinhole projection model, projection onto the plane $Z = 1$ maps the point (X_i, Y_i, Z_i) to (x_i, y_i) where: $x_i = X_i/Z_i$ and $y_i = Y_i/Z_i$. Substituting these into the plane constraint equation, dividing through by Z_i and setting $t_i = 1/Z_i$ yields:

$$a_j x_i + b_j y_i + c_j + t_i = 0. \tag{3}$$

Here we have assumed a camera calibration; as stated in [8], this need not be the case. The trihedral assumption means that each point lies on three planes, say $j \in \{\alpha, \beta, \gamma\}$. Eliminating the t_i between pairs of constraints yields two linear equations per point:

$$(a_\beta - a_\alpha)x_p + (b_\beta - b_\alpha)y_p + (c_\beta - c_\alpha) = 0,$$
$$(a_\gamma - a_\alpha)x_p + (b_\gamma - b_\alpha)y_p + (c_\gamma - c_\alpha) = 0. \tag{4}$$

Now, from the m image vertices observed in each image, we form $\mathtt{A}(\mathbf{x}, \mathbf{y}) \, \mathbf{w} = 0$ which is a matrix equation of $2m$ constraints in $3n$ unknowns. Here $\mathbf{x} = (x_1, \ldots, x_m)^\top$, $\mathbf{y} = (y_1, \ldots, y_m)^\top$, $\mathtt{A}(\mathbf{x}, \mathbf{y})$ is a $2m \times 3n$ matrix, and $\mathbf{w} = (a_1, b_1, c_1, a_2, \ldots, c_n)^\top$. We now state the following theorem:

Theorem 3.1 *The kernel of the matrix* $\mathtt{A}$ *represents all of the possible solutions to the reconstruction of a trihedral polyhedron imaged by a single pinhole camera.*

Proof See [8] and more completely [7]. □

The kernel of $\mathtt{A}$ can be represented as $\mathbf{w} = \sum_{i=1}^{d} \lambda_i \mathbf{b}_i$. We can orient the basis for $\mathbf{w}$ so that $\mathbf{b}_i = (\delta_1, \delta_2, \delta_3, \ldots, \delta_1, \delta_2, \delta_3)^\top$ for $1 \leq i \leq 3$ and where $\delta_i = 1$, and $\delta_{mod_3(i+1)} = \delta_{mod_3(i+2)} = 0$. We may also fix the other basis vectors so that:

$$\mathbf{b}_i = (0, 0, 0, (\mathbf{r}_i)_2^\top, \ldots, (\mathbf{r}_i)_n^\top)^\top,$$

for $4 \leq i \leq d$. This means that the solution for $\mathbf{w}$ with $\lambda_i = 0$ for $i > 3$, represents a simple planar configuration (strictly at least one of λ_i should satisfy $\lambda_i \neq 0$ for $i \leq 3$). The initial three elements of $\mathbf{b}_i$ being zero means that the first plane of the reconstructed polyhedron has coordinates $(\lambda_1, \lambda_2, \lambda_3, 1)^\top$.

3.2 Four dof. polyhedron

It was shown in [7] that for $d = 4$ the solutions represented by $\mathbf{w}$ are all projectively equivalent to each other. In this case we thus need consider only one solution from $\mathbf{w}$ for the computation of the three-dimensional projective invariants on the shape.

3.3 Five dof. polyhedron

Consider the case for $d = 5$. To simplify the discussion we represent $\mathbf{b}_4$ and $\mathbf{b}_5$ by:

$$\mathbf{b}_4 = (0, 0, 0, \mathbf{r}_2^\top, \ldots, \mathbf{r}_n^\top)^\top \quad \text{and} \quad \mathbf{b}_5 = (0, 0, 0, \mathbf{s}_2^\top, \ldots, \mathbf{s}_n^\top)^\top.$$

Separating out the first four dof. figure

We return to the form of the constraints given in eqn (4). Consider the first plane, and another plane i which shares an edge with it; the endpoints of the edge are the points p and q. First we construct a set of planes S_1 which initially contains the first and the ith planes. Consider writing $\alpha = (\lambda_1, \lambda_2, \lambda_3)^\top$. Then, making use of the fact that both $\mathbf{r}_1$ and $\mathbf{s}_1$ are null vectors means that $(a_1, b_1, c_1)^\top = \alpha$ and $(a_i, b_i, c_i)^\top = \alpha + \lambda_4 \mathbf{r}_i + \lambda_5 \mathbf{s}_i$. Subsequently, eqn (4) reduces to the following vectorial constraints:

$$(x_p \ y_p \ 1)^\top \cdot (\lambda_4 \mathbf{r}_i + \lambda_5 \mathbf{s}_i) = 0, \quad \text{and} \quad (x_q \ y_q \ 1)^\top \cdot (\lambda_4 \mathbf{r}_i + \lambda_5 \mathbf{s}_i) = 0. \tag{5}$$

These define $(\lambda_4 \mathbf{r}_i + \lambda_5 \mathbf{s}_i)$ uniquely up to a scale as $\mathbf{x}_p$ and $\mathbf{x}_q$ are distinct points. This solution exists for all λ_4 and λ_5, and so either $\mathbf{r}_i$ is parallel to $\mathbf{s}_i$, or one of $\mathbf{r}_i$ and $\mathbf{s}_i$ is null (but not both). Now, within the bounds of the geometric constraints used so far, we are free to orient b_4 and b_5 as we desire. In order to simplify the overall description, we make a further change of basis. Assuming that r_i and s_i are parallel, a change of basis is effected so that s_i becomes a null vector, and r_i is non-zero. Should either of the two vectors have already been null we can make a change of basis (perhaps trivially) to ensure that r_i is non-zero and s_i is null. Once done, neither the first plane, nor plane i, will depend on λ_5. By reapplying the reasoning, we may extend the process and provide independence to λ_5 for all of the other planes which have r_j parallel to s_j (or are null vectors). All planes of this type are included in the set S_1.

It is certain that a number of planes will be included in S_1. Geometrically we see that providing independence to λ_5 for other planes is based on the trihedral assumption underlying our manipulations. Consider any third plane which shares an edge with both the first and i^{th} planes; the third plane is precisely defined by the positions of the other two. The first plane is independent of both λ_4 and λ_5, and the ith one depends only on λ_4. Thus the new plane depends only on λ_4 (and so its s must be null). This process can be continued for all other planes which have two edges in common with a pair of planes from S_1. One will thus build up a set of planes constituting S_1 which depend only on λ_4, and which consequently represent a four dof. figure.

The second four dof. figure

Once the first four dof. subpart has been extracted we can be sure that no planes remain outside S_1 which are in contact with more than one plane in S_1. This is because if a plane not in S_1 were attached to two planes in S_1, then it would have to be fixed in space. We label the complementary set to S_1 by S. We may now make the following statements:
1. The planes in S_1 use up four of the five dof. for the polyhedron. This is because the planes in S_1 are defined by the λ_i, $i \le 4$.
2. S_1 does not represent the entire polyhedra (S is not empty). If it did, then b_5 would be null and we would have only a four dof. figure.
3. There must be at least one plane in S which is adjacent to a plane in S_1. If this were not the case, then the planes in S would make up a figure with at least three more dof. These extra three dof. would lead to the entire figure having too many dof.

We exploit the third of these statements to base the rest of the reconstruction around the plane in S_1 which is adjacent to S. To do this, we again change basis so that this plane is accounted for by the first three parameters of w. For simplicity of argument we also reorder the rest of planes so that those dependent on only λ_4 appear first in the basis vectors. The mth plane (which is the first plane that does not belong to S_1) is chosen so that is shares an edge with the first plane. Thus we have:

$$b_4 = (0,0,0,r_2^\top,\ldots,r_{m-1}^\top,r_m^\top,\ldots,r_n^\top)^\top,$$
$$b_5 = (0,0,0,\ldots\ldots,0,0,0,s_m^\top,\ldots,s_n^\top)^\top.$$

We now repeat the process used previously for the elimination of the components of s_i from b_5 for $i < m$, though this time we eliminate the r_i. Given that the first plane is represented by the zero vector, and that the mth plane shares an edge with it, we know that r_m and s_m are parallel (cf. eqn (5)). We may now eliminate either of r_m or s_m to yield dependence only on one of λ_4 or λ_5. Clearly this plane cannot depend on λ_4 otherwise it would already have been included within S_1. Therefore, we are able to eliminate r_m. In the same way that we added planes to S_1 we continue to build up planes dependent only on λ_5. We place all of these planes in the set S_2. Once all possible r_i have been set to zero, we are left with basis vectors of the form:

$$\mathbf{b}_4 = (0, 0, 0, \mathbf{r}_2^\top, \ldots, \mathbf{r}_{m-1}^\top, 0, 0, 0, \ldots, 0, 0, 0, \mathbf{r}_k^\top, \ldots, \mathbf{r}_n^\top)^\top,$$
$$\mathbf{b}_5 = (0, 0, 0, \ldots \ldots, 0, 0, 0, \mathbf{s}_m^\top, \ldots \ldots, \mathbf{s}_{k-1}^\top, \mathbf{s}_k^\top, \ldots, \mathbf{s}_n^\top)^\top. \tag{6}$$

Such a basis represents a decomposition of the polyhedron into two four dof. complete solid polyhedra built around the first plane (the first polyhedron of planes 1 to $(m-1)$, and the second of plane 1, and planes m to $(k-1)$). The first plane is called the *common plane*. In addition to these sub-figures, there may be a set of planes k to n which we represent by the set S_3. S_3 is empty for simple polyhedra, and so $k = n + 1$. However, this is not the general case and S_3 can well contain further polyhedral subparts. Note that the planes in this subsidiary set are still constrained completely by λ_4 and λ_5.

S_3 at first has the appearance of being a more complex figure than the four dof. subparts which we have already extracted, though this is not really the case. S_3 will always be composed of other four dof. polyhedra. The simplest way to understand this is to re-order the planes so that one of the planes in S_1 adjacent to S_3 is set to the first plane of w (such a plane must exist). We then repeat the above elimination process. Doing this would mean that a subset of planes in S_3 would be dependent only on the new λ_5 (S_1 is still parametrized by λ_4) and so would have only four dof. By progressive choices of common planes between the extracted subparts of S_3 and their complement in S_3, we can gradually demonstrate that S_3 is composed only of four dof. figures. The trick is to see that any two polyhedra sharing a common plane can always be demonstrated to have four dof. We thus extend our current understanding of trihedral polyhedra via:

Theorem 3.2 *Five dof. trihedral polyhedra are made up of simple building blocks consisting only of constrained four dof. polyhedra. As each subpart has only four dof., their own local equivalence classes are projective.*

Proof As discussed above, and more completely in [9]. □

Projective inequivalence of five dof. figures
Unlike the four dof. figures discussed previously in [8], two different reconstructions of a five dof. figure need not necessarily be projectively equivalent. Formally:

Theorem 3.3 *The different reconstructions of a five dof. non-trivial polyhedron imaged by an uncalibrated pinhole camera are not in the same projective equivalence class.*

Proof Given in [9]. □

3.4 Invariants for five dof. figures
There are two sets of invariants which can be measured for five dof. polyhedral figures. The first are based on the invariants of each of the subparts, and so are the same as the four dof. invariants given in [8]. As stated in Section 2.1, these can be measured using the butterfly invariants. We can also measure a number of invariants in the common plane from the coplanar edges of each adjacent sub-polyhedra. Generically the number of invariants of this form depends on the structure of the polyhedra: if there are m_i edges in each of the j common planes between the subparts, $i \in \{1, \ldots, j\}$, then there are $\sum_{i=1}^{j} (2m_i - 8)$ computable planar invariants. It can be shown that the butterfly and planar invariants represent all of the invariants which can be computed for such a figure.

3.5 Higher degree of freedom figures

We can use the results of Section 3.3 to complete the description paradigm for polyhedra of six or more dof. We proceed with higher dof. figures in a similar way to five dof. polyhedra. First, extract a four dof. subpart and define the planes which it has in common with the rest of the polyhedron. Then, continue to examine the rest of the polyhedron by ignoring the planes in the first subpart, extracting further four dof. polyhedra. In this manner the whole polyhedron can be decomposed into a set of four dof. subparts. The computation of the invariants for the entire object is then straightforward:

- Compute projective invariants for the planes within each subpart.
- Compute planar invariants between the edges in all the common planes.

It follows obviously from the result for five dof. polyhedra that the families of reconstructions for higher dof. figures are not projectively equivalent.

4 Finding polyhedra in images

In this section we report on a pair of approaches which show how polyhedral descriptions can be extracted from real images. As always, we are plagued by the difficulty of extracting accurate segmentations from real images. We discuss two different methods and provide demonstrations of them working on relatively simple images containing well-defined polyhedral objects. The solutions are:

- Use an edgel detector to initialize a polyhedral snake on sets of image features. Then compute invariants for the snake.
- Search for pairs of adjacent closed regions in the image suitable for estimating butterfly invariants. These invariants are used to index into a model base and subsequently to provide hypotheses suggesting which object might be present in the scene. Hypotheses are combined post-indexing to derive richer object descriptions.

Both of these approaches require edgel and line extraction. The edgel detector we use is described in [6] and ensures good connectivity around image junctions. Straight lines are fitted to edgel-chains using orthogonal regression and a topological (connectivity) structure composed of connected straight lines and edgel-chain segments is produced. The actual data structures we use are built around a *vertex-edge-face* topology hierarchy [5]. Vertices are typically used to represent junctions and the interfaces between pairs of lines or edgel-chains. Edges link vertices and are instantiated geometrically by lines and edgel-chains. Faces represent closed cycles of edges (we also have 1-chains which represent non-cyclic edge chains). This topology hierarchy allows the straightforward extraction of closed regions from the image as each *face* is a closed region.

4.1 Polyhedral snakes

The primary goal of the snake extraction approach is to recover structures which have the same topological forms as the projections of the polyhedra which we wish to recognize. Once these have been extracted, they can be matched to snake models, and then the snakes develop in the image by interaction with the image intensity surface. After a number of iterations we can measure the invariants of the snake (which are similar to projected polyhedra), and thus hypothesize the identity of the object in the scene.

The first phase of processing involves the extraction of edges, as demonstrated in Fig. 2. Notice that we have been able to recover a fairly good description of the polyhedron's topology with the edge detector. Principally the edge description consists of three closed regions each of which match projected polyhedral faces. Being able to find these image regions means that we can reject scene clutter (non-polyhedral regions) rapidly

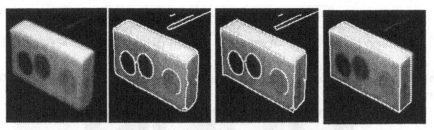

Fig. 2. *First we show a polyhedron and then the output of the edgel detector superimposed. Note that the basic topological structure of the edges is correct. Next are the lines fitted by orthogonal regression and finally the polyhedral snake.*

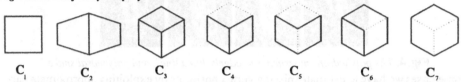

C_1 C_2 C_3 C_4 C_5 C_6 C_7

Fig. 3. *The topological descriptions stored for the projections of a six-plane polyhedron.*

during subsequent processing. We then fit lines and throughout ensure the maintenance of the topological description. After fitting, we extract faces from the image. For Fig. 2 there are three basic regions which are suitable, plus a number of others which include combinations of these regions. Subsequent analysis is focused on the faces and on the lines which they contain. However, over-segmentation can arise due to the edge detector recovering small noisy features. A number of these are shown in Fig. 2. These features disrupt the line fitting process and cause breaks to be inserted between pairs of line segments. We therefore traverse the boundary of each face and test whether a pair of adjacent lines would be better represented by a single straight line segment. If so, a single line segment is substituted. Once the line merging process is complete, we count the number of lines in each face to see whether it remains of interest. Polyhedral faces must contain at least three line segments, and so any with less can be discarded. For this application we desire at least four lines. Consecutive line segment endpoints should also be reasonably close as polyhedral faces are generally polygonal. Given this, we can discard the edgel-chains between lines and reduce the face representations to ideal polygons.

The penultimate stage of processing involves the matching of the adjacency graph of the faces to topological models which we have for the polyhedra in the model base. The topological descriptions of the lines and the snake are consistent in Fig. 2. The complete set of descriptions from which we form snakes is given in Fig. 3. Given a match, such as that shown, we initialise a snake which is allowed to relax onto the contrast boundaries in the original image. The snakes are a crude implementation of those described in [4]. The final position of the snake yields sufficient structure for invariant computation. In Fig. 2, all of the butterfly invariants are unity because the polyhedron is projectively equivalent to a cube (which is composed of three of sets of parallel planes). Table 1 shows how the three invariants for the shape change over the iterations of the snake.

A similar example of polyhedron extraction using snakes is in Fig. 4. In Fig. 5 we have a harder case. The edgel detector failed to recover the polyhedron's vertical internal boundary, and so we must use a simplified model with which to initialize the snake. In this case we have a description equivalent to C_6 (in Fig. 3) and so we are able to hypothesize the location of the missing edge by joining up a pair of vertices. We return to the complete snake description (C_3) once the missing edge has been inserted. In

iteration	invariant 1	invariant 2	invariant 3
0	-0.920797	-0.897423	-1.34495
4	-0.956112	-1.02075	-1.04808
8	-0.964451	-1.01821	-1.02506

Table 1. *The three invariants for the polyhedral snake of Fig. 2 for the first eight iterations. The invariant values should be all converge to unity.*

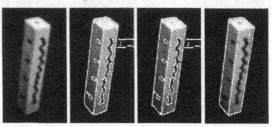

Fig. 4. *The polyhedron, superimposed edgels, fitted lines, and polyhedral snake.*

some cases we have to estimate missing vertex positions by exploiting approximate parallelism; the action of the snake removes the effects of the approximation and usually yields suitable fits if the correct model hypothesis has been made. In practice we have found that either two quadrilateral faces or a single hexagon are required to initialize the snake model. One finding a lone quadrilateral, such as C_1, we terminate processing.

4.2 Finding polyhedra using local butterfly invariants

The above approach is global; we now describe a local method which uses the butterfly invariant. The initial processing is the same in that we extract edgels and lines, and then recover *faces* from the image which represent closed regions. These faces are processed to merge lines and are then represented as polygons if suitable. All polygons which are not quadrilateral are rejected as they are not appropriate for use with the butterfly invariant. Faces which approximately (or exactly) share edges are paired to produce butterflies. Vertices of the common edge are adjusted so that they lie suitably between the two faces (by averaging the image coordinates). From this merged configuration we can immediately compute a butterfly invariant. The combined face-pair feature is accepted depending on whether the invariant value belongs to a model in the model base (potentially this is evaluated through indexing). Each accepted invariant value can be used to form a local hypothesis. Larger hypotheses can be formed by combining the local hypotheses using the hypothesis extension process described in [10]. Briefly, two hypotheses are consistent if they represent different parts of the same model, and if the feature correspondences between the image and the model are also consistent. The final results of the process are the four dof. descriptions of the polyhedra in the images. In

Fig. 5. *First the polyhedron and the edgels. The complete topological structure of the polyhedron has not been recovered. Next are the lines and finally the polyhedral snake.*

Fig. 6. *After segmentation and closed region detection we extract three butterflies with invariants approximately equal to unity. These butterflies can be merged into a single polyhedral description.*

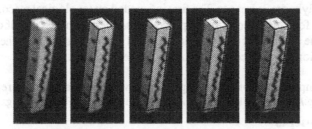

Fig. 7. *We recover three butterflies for this imaged polyhedron.*

principal we can re-apply the hypothesis extension process between the four dof. figures by using the planar projective invariants in the common planes as tests of consistency.

Extraction of the butterflies is shown in Fig. 6. We show the three different extracted butterflies for the image section in the figure. The butterflies are composed of three image faces each containing four principal line segments, and have invariant values of 1.007, 1.022, and 1.000 respectively. All of these match butterfly invariants for objects which have projective equivalence to a cube. They can thus be merged into the single polyhedral description shown on the right which is a single four dof. polyhedron.

Another example is in Fig. 7. There the three butterflies have invariants equal to 1.000, 1.000, and 1.029. However, in Fig. 8 we show an example in which only a single butterfly can be recovered (with an invariant of 1.000) due to the failure of the edgel detector to recover more than two of the polyhedral faces as closed regions. In this case we have only weak support for a polyhedral hypothesis as one matching invariant carries less weight than three. Nevertheless, the precision of the single invariant suggests that the features probably do correspond to a polyhedron with similarity to a cube.

5 Conclusions

We have made a number of different contributions in this paper. The first two are at a theoretical level, and the third expresses a pair of practical approaches for the extraction of polyhedral descriptions from images. The first theoretical study presented a very

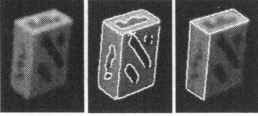

Fig. 8. *For this polyhedron we recover only two polyhedral faces as closed regions in the edgel description. Thus we can find only a single butterfly configuration. However, the invariant value for the butterfly matches precisely that which we would expect to find for such a polyhedron.*

simple algebraic formulation for the butterfly invariant and showed how it is related to the polyhedral invariants of [8].

The second theoretical contribution is a completion of the projective invariant description of [8] to trihedral polyhedra of arbitrary degrees of freedom. We demonstrated that all higher order polyhedra can be broken down into sets of four dof. polyhedra associated using plane projective invariants. This yields a complete description paradigm for arbitrary trihedral polyhedra.

Finally, we developed two different implementations of algorithms intended for the detection of polyhedra in images. The first is founded on topological reasoning which is used as a basis for a polyhedral snake process. The second makes use of closed region detection for the creation of local hypotheses, and then the use of hypothesis extension for the generation of more complete polyhedral descriptions.

A more complete version of this paper [9] is available as a technical report from the URL: <URL ftp://ftp.inria.fr/INRIA/publication/publi-ps-gz/RR/RR-2661.ps.gz>.

Acknowledgments

We acknowledge the input of a number of people in this work, principally: Olivier Faugeras, David Forsyth, Bill Hoffman, Joe Mundy, and Andrew Zisserman. Thanks to Pippa Hook for proof-reading this article. CR is funded by a Human Capital and Mobility grant from the European Community. JS was visiting INRIA from the Ecole Normale Supérieure de Lyon.

References

[1] S. Carlsson. Multiple image invariance using the double algebra. In *Applications of Invariance in Computer Vision*, volume 825 of *LNCS*, p.145–164. Springer-Verlag, 1994.

[2] R. Duda and P. Hart. *Pattern Classification and Scene Analysis*. Wiley, 1973.

[3] G. Ettinger. Large hierarchical object recognition using libraries of parameterized model sub-parts. *Proc. CVPR*, p.32–41, 1988.

[4] M. Kass, A. Witkin, and D. Terzopoulos. Snakes: Active contour models. *Proc. ICCV*, p.259–268, 1987.

[5] C. Rothwell, J. Mundy, and W. Hoffman. Representing objects using topology. In preparation, 1996.

[6] C. Rothwell, J. Mundy, W. Hoffman, and V.-D. Nguyen. Driving vision by topology. *Proc. IEEE International Symposium on Computer Vision*, p.395–400, 1995.

[7] C. Rothwell, D. Forsyth, A. Zisserman, and J. Mundy. Extracting projective information from single views of 3D point sets. TR 1927/92, Oxford Univ. Dept Eng. Sci., 1992.

[8] C. Rothwell, D. Forsyth, A. Zisserman, and J. Mundy. Extracting projective structure from single perspective views of 3D point sets. *Proc. ICCV*, p.573–582 1993.

[9] C. Rothwell and J. Stern. Understanding the shape properties of trihedral polyhedra. TR 2661, INRIA, 1995.

[10] C. Rothwell. *Object recognition through invariant indexing*. Oxford University Press, 1995.

[11] K. Sugihara. *Machine Interpretation of Line Drawings*. MIT Press, 1986.

[12] A. Sugimoto. Geometric invariant of noncoplanar lines in a single view. *Proc. ICPR*, p.190–195, 1994.

[13] A. Zisserman, D. Forsyth, J. Mundy, C. Rothwell, J. Liu, and N. Pillow. 3D object recognition using invariance. TR 2027/94, Oxford Univ. Dept Eng. Sci., 1994, to appear the AI Journal.

Texture and Features

Nonlinear Scale-Space from n-Dimensional Sieves

J. Andrew Bangham, Richard Harvey, Paul D. Ling, and Richard V. Aldridge

School of Information Systems, University of East Anglia, Norwich, NR4 7TJ, UK

Abstract. The one-dimensional image analysis method known as the
sieve[1] is extended to new finite dimensional image. It preserves all the
useful scale-space properties but has some additional forms that we
believe make it more attractive than the diffusion-based methods. We
present some simple examples of how it might be used.

1 Introduction

The use of scale-space for the analysis of images is well established and there is
particular interest in incorporating scale-space processing as part of biologically or human
vision tasks.

Scale-space vision is usually associated with diffusion-based systems [2, 3, 4]
in which the image is subjected to initial conditions for a discrete implementation of the diffusion
equation $\partial_t \nabla^2 = \nabla_t$, In these scale-space processors the diffusion is controlled by
D ... a constant and, although this system has the desirable property that it does
not introduce new extrema at scale increases, it also drives scale-space smoothing of
edges the loss of objects in the scene, since edges are thought to be important
for most high-level vision applications, account in high D is made so any non-
linear function of $\|\nabla I\|$ is sometimes used.

It is said that scale-space should be "scale-space finally meaningful" for which
there properties are the most contentions [5]. Broadly, (although sometimes hard
driven and preventing smoothing. The second is a requirement that the "region
boundaries should be sharp and coincide with structurally meaningful boundaries" that resolution. And the third property is that flatter/other smoothing
should ideally preferentially smoothes low-contrast blobs.

To meet these several scale-space concerns are often required. Firstly [6] The operations should be scale-calibrated. At a point that scale-space should so a nature of
only that scale. This allows shapes to be measured accurately. (ii) The scale-
space should be non-negative. One should be able to process the image in the
scale-space domain and reconstruct it to produce an enhanced image.

This paper presents a system that has all the desirable properties of scale-
space and also some additional very powerful features. The system has
similarities with morphological linear-space systems.

Mathematical morphology [7] as known for the analysis of shape and has
developed separately. In ways that have several recent development has been made
first to unify some of scale in the scale-space and morphology [8 9, 10, 11].
We continue this to end with an insight into and experimental study of a type of
decomposition, called a sieve, resulting in two or more dimensions.

Nonlinear Scale-Space from n-Dimensional Sieves

J. Andrew Bangham, Richard Harvey, Paul D.Ling and Richard V. Aldridge

School of Information Systems, University of East Anglia, Norwich, NR4 7TJ, UK.

Abstract. The one-dimensional image analysis method know as the *sieve*[1] is extended to any finite dimensional image. It preserves all the usual scale-space properties but has some additional features that, we believe, make it more attractive than the diffusion-based methods. We present some simple examples of how it might be used.

1   Introduction

The use of scale-space for the analysis of images is well established and there is an interest in incorporating scale-space processors as part of high-level computer vision tasks.

Scale-space vision is usually associated with diffusion based systems [2, 3, 4] in which the image forms the initial conditions for a discretization of the diffusion equation $\nabla \cdot (D \nabla I) = I_t$. In linear scale-space processors the diffusion parameter, D, is a constant and, although the system has the desirable property that it does not introduce new extrema as scale increases (preserves *scale-space causality*), it blurs the edges of objects in the scene. Since edges are thought to be important for most high-level vision operations, a variant, in which D is made to vary as a function of $\|\nabla I\|$, is sometimes used.

It is said that scale-space should be "semantically meaningful," for which three properties have been enunciated [5], *causality* (above), *immediate localisation* and *piecewise smoothing*. The second is a requirement that the "region boundaries should be sharp and coincide with semantically meaningful boundaries at that resolution" and the third property is that "intra-region smoothing should occur preferentially over inter-region smoothing."

To these properties one might add some practical requirements: (i) The system should be *scale-calibrated*. At a particular scale one should see features of only that scale. This allows shapes to be measured accurately. (ii) The scale-space should be *manipulable*. One should be able to process the image in the scale-space domain and reconstruct it to produce an enhanced image.

This paper presents a system that has all the desirable properties of scale-space and also some additional, extremely powerful, features. The system has similarities with morphological image processing systems.

Mathematical morphology [6, 7] is based on the analysis of shape and has developed separately. However, a recent welcome development has been the effort to achieve some unification of scale-space and morphology [8, 9, 1, 10, 11]. We continue this trend with an analytical and experimental study of a type of decomposition called a *sieve* operating in two or more dimensions.

2 Properties of sieves

2.1 Definitions

An arbitrary array of pixels, or voxels, can be described by a connected graph [12] $G = (V, E)$ where V is the set of vertices and E is the set of pairs that describe the edges.

Definition 1. When G is a graph and $r \geq 1$, we let $C_r(G)$ denote the set of connected subsets of G with r elements. When $x \in V$, we let $C_r(G, x) = \{\, \xi \in C_r(G) \mid x \in \xi \,\}$.

All subsequent operations take place over such sets. The structure of an element of C_r is determined by the adjacency of image pixels. In three dimensions the connected subsets of G would normally be six-connected.

Definition 1 allows a compact definition of an *opening*, ψ_r, and *closing*, γ_r, of size r.

Definition 2. For each integer, $r \geq 1$, the operators, $\psi_r, \gamma_r, \mathcal{M}^r, \mathcal{N}^r \colon Z^V \to Z^V$ are defined by

$$\psi_r f(x) = \max_{\xi \in C_r(G,x)} \min_{u \in \xi} f(u), \; \gamma_r f(x) = \min_{\xi \in C_r(G,x)} \max_{u \in \xi} f(u),$$

and $\mathcal{M}^r = \gamma_r \psi_r$, $\mathcal{N}^r = \psi_r \gamma_r$. ($\psi_r$ and γ_r are well defined since, for each $x \in V$, $C_r(G, x)$ is nonempty [13].)

Thus $\mathcal{M}^r$ is an opening followed by a closing, both of size r and in any finite dimensional space.

We can now define the M- and N-sieves of a function, $f \in Z^V$.

Definition 3. Suppose that $f \in Z^V$.
(a) The M-sieve of f is the sequence, $(f_r)_{r=1}^{\infty}$, given by

$$f_1 = \mathcal{M}^1 f = f, \qquad f_{r+1} = \mathcal{M}^{r+1} f_r, \quad \text{for integers, } r \geq 1.$$

(b) The N-sieve of f is the sequence, $(f_r)_{r=1}^{\infty}$, given by

$$f_1 = \mathcal{N}^1 f = f, \qquad f_{r+1} = \mathcal{N}^{r+1} f_r, \quad \text{for integers, } r \geq 1.$$

M- and N-sieves are one type of alternating sequential filter [7]. We use the term sieve, partly for brevity, and partly because the properties we concentrate on are not unique to alternating sequential filters but are also to be found with, for example, a sequence of recursive median filters [14]. In addition, not all alternating sequential filters have the properties studied here.

Theorem 4. *If $(f_r)_{r=1}^{\infty}$ is the M- or the N-sieve of an $f \in Z^V$, then, for each integer, $r \geq 1$, f_r is r-clean.*

Definition 5. For each integer, $r \geq 1$, we let $M^r = \mathcal{M}^r \cdots \mathcal{M}^2 \mathcal{M}^1$, and $N^r = \mathcal{N}^r \cdots \mathcal{N}^2 \mathcal{N}^1$. So the M-sieve (resp. N-sieve) of an $f \in \mathbf{Z}^V$ is $(M^r f)_{r=1}^{\infty}$ (resp. $(N^r f)_{r=1}^{\infty}$).

The term r-clean means that extremal level connected sets have r or more pixels [13]. The sieve has the effect of locating intensity extrema and "slicing off peaks and troughs" to produce *flat zones* [15] of r or more pixels hence r-clean, a function from which the flat zones have been "cleaned."

Since all the pixels within each extremal connected set have the same intensity, a simple graph reduction at each stage can lead to a fast algorithm [16]. The one-dimensional equivalent has already been reported [17]. It has order complexity p, where p is the number of pixels, for a complete decomposition ($r = p+1$).

In one-dimension the flat zones created by the sieve have a length r. This means that the scale parameter can be used for the precise measurement of features [18]. For a two-dimensional image with regular pixelation the flat zones created by the sieve have a defined area and we refer to the sieve as an area decomposition c.f. [19]. In three dimensions we have a decomposition by volume.

2.2 Properties

One important property of the M- and N-sieves is that, for a particular edge, $\{x,y\} \in E$, as r increases, the change in image intensity, $f_r(y) - f_r(x)$, does not change sign and its absolute value never increases. In particular, if it vanishes for some r, it is then zero for all larger r. Formally, we have the following result.

Theorem 6. *Suppose that $(f_r)_{r=1}^{\infty}$ is the M- or the N-sieve of an $f \in \mathbf{Z}^V$, that $\{x,y\} \in E$ and put $\delta_r = f_r(y) - f_r(x)$, for each, r. Then $\delta_1 \geq \delta_2 \geq \cdots \geq 0$, or $\delta_1 \leq \delta_2 \leq \cdots \leq 0$.*

In other words n-dimensional sieves preserve scale-space causality.

At each stage of the M- or N-sieve one can examine either the filtered image (usually the case) or the difference between successive stages. These differences are called *granules* and are defined as follows.

Definition 7. When $f \in \mathbf{Z}^V$, the M-granule decomposition of f is the sequence of functions, $(d_r)_{r=1}^{\infty}$, defined by $d_r = f_r - f_{r+1}$, for each $r \geq 1$, and $\{d\}$ is the granularity domain. There is an equivalent definition for the N-granule decomposition.

Theorem 8 (Invertibility). *Suppose that $(d_r)_{r=1}^{\infty}$ is the M-granule decomposition of an $f \in \mathbf{Z}^V$ with finite support then the original image can be rebuilt from the granularity domain $f = \sum_{r=1}^{\infty} d_r$. Suppose that $(\hat{d}_r)_{r=1}^{\infty} \subseteq \mathbf{Z}^V$ are such that, for each integer, $r \geq 1$, $\hat{d}_r = 0$ on $[d_r = 0]$, $\hat{d}_r$ is constant and nonnegative on each positive granule of d_r and $\hat{d}_r$ is constant and nonpositive on each negative granule of d_r and let $\hat{f} = \sum_{r=1}^{\infty} \hat{d}_r$. Then $(\hat{d}_r)_{r=1}^{\infty}$ is the M-granule decomposition of $\hat{f}$.*

This last property is vital if one wishes to manipulate an image in the granularity domain. For example, a pattern recognising filter can be produced by: decomposition using M^r or N^r, selectively removing granules (filtering in the granularity domain); and rebuilding. Such a filter is idempotent.

3 Results

Figure 1(A) shows an image of a doll manually segmented from a larger image.

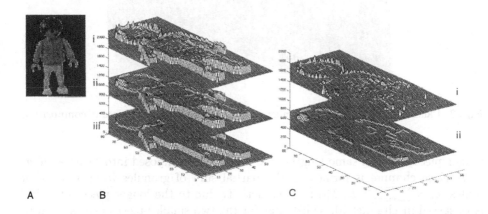

Fig. 1. (A) Shows the original image I; (B)(i) shows, in relief, the original image, B(ii) shows $M_4(I)$ and B(iii) shows $M_{25}(I)$. (C) shows the differences between B(i), B(ii) and B(iii)

It is shown as a topological relief in Figure 1(B). The spiky detail, particularly evident around the head, represents small scale regional extremes of area that are removed by sieving to scale 4, as in Figure 1(B)ii. The small extrema that are removed are shown in Figure 1(C)i (actually $M^4(I) - I$). Likewise the pale highlight on the (left side) of the hair, is a feature represented by a larger scale extremum that is removed with a larger, scale 25, area sieve. The differences between Figure 1(B)ii and (B)iii are shown in Figure 1(C)ii. No new features can be seen in Figure 1(B)iii and what remains is a set of large flat zones with edges in the same places as those in the original. This is consistent with Theorem 6 and Theorem 4. This localisation of features is an important advantage over, say, Gaussian scale-space which blurs the edges.

The process is essentially regional and is not equivalent to redefining the intensity map over the whole image. Nor does it alter the corners of objects in the way that a morphological or median filter with a rigid structuring element would be expected to.

However, there is a problem. In Figure 1(A) there are over 5,000 pixels and so, potentially, 5,000 different area scales. In practice, this is often far too many. Figure 2 (bottom right panel) shows an image of four pieces of string. The two

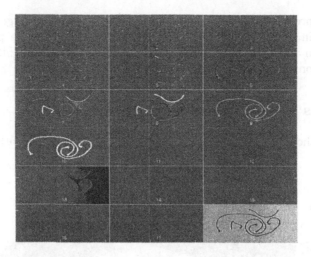

Fig. 2. Panel 18 shows the original image and panels 1–17 show an area decomposition

longer pieces are the same length. The image is decomposed into 17 area *channels*. Each channel is formed as the partial sum of granules from a range of scales, $c_i = \sum_{r=r_i}^{r=r_{i+1}} d_r$. Most of the activity due to the longer pieces of string is observed in channel 10, whilst that for the two smaller pieces can be seen in channels 7 and 8. Of course, since the pieces of string are not of uniform intensity, and there are shadows, activity is not confined to a single channel, rather it peaks at the appropriate scale.

This result would be difficult to achieve using a filter with a structuring element (for example, an alternating sequential filter by reconstruction [20] with square structuring element) because, a structuring element chosen to encompass the small bits of string would remove the more coiled of the long strings. The area sieve does not rely on a match between the shape of a structuring element and an object. The strings can be arranged in any shape (although, if the string forms loops it might be necessary to use a sieve that takes advantage of the sign of the local extremum, cf. [19]).

Figure 3 shows a more complicated example: a human face. Note the nostrils in channel 5, the eyes in channels 6 and 7, the mouth in channels 9 and 10 and the entire face in channel 15. The larger scale channels can be used for segmentation, for example, channel 15 could be used as a mask to pick out facial features from the background for, by Theorem 6, it accurately represents features in the original image. This image is one of a movie sequence in which the activity associated with the mouth moves from a peak in channel 9 to a peak in channel 10/11 and back.

Figure 4 shows a set of transverse scans of a human head generated by X-ray tomography. The sections start at the level of the nose and go up to the top of the head. A guide frame, which is opaque to X-rays, is attached to the head. The white, calcified bony tissue of the skull is readily identified and could be

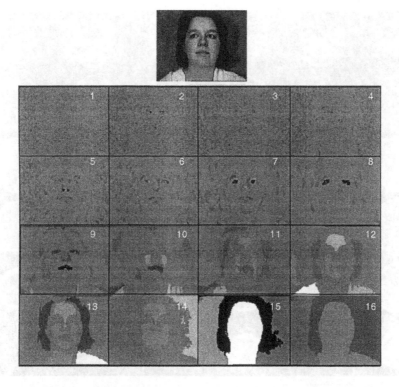

Fig. 3. An area decomposition of an image (top) into a number of channels

segmented out by simple thresholding, however, we have chosen to use a very large scale volume sieve.

Figure 5(A) shows a rendered reconstruction of the volume channel containing between 25,600 and 51,200 voxels. The left cheekbone can be seen looping out at the bottom of the image, and the upper part of the nose is visible at the bottom left. The guide frame fastenings protrude from the skull.

The third ventricle is a complex shaped void within the brain. It appears as a darkened region circled in panel 16 of Figure 4 and runs up through the brain, bifurcating to form two "wings" that come together at the top. The primary segmentation is made by taking a volume channel encompassing 800 to 1600 voxels. This first step achieves a near perfect segmentation but some other features of the same volume, are also present. Since they do not penetrate to the same depth, they occupy a larger area of those slices in which they are present and so can be removed by area sieving each slice to remove areas larger than 100 pixels. This leaves some small regions, which are no longer connected through the slices, that can be removed by re-sieving through the original volume sieve. From Theorem 8 this leaves the third ventricle unchanged.

The final example shows how the sieve might be used to create compact feature vectors from an image. An isolated object in an image can be represented as a hierarchy of nested connected subsets of G. The pixels that form the graph

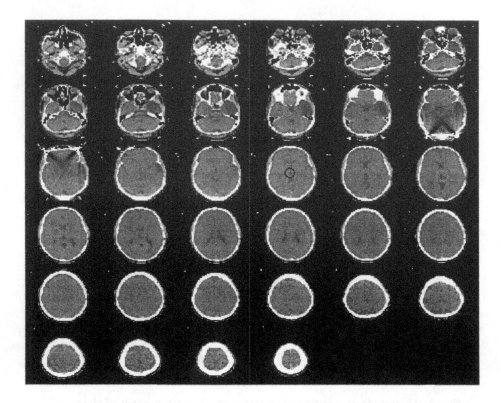

Fig. 4. Thirty-five Computer tomography scans through a human head. The third ventricle is circled in the sixteenth slice. Note thin connecting filaments are not rendered.

of a large granule may themselves be elements of smaller graphs that represent the finer scale features of that object. This is illustrated in Figure 6

In Figure 6 the top sketch shows an apparently meaningless collection of objects but in the lower sketch the shaded areas immediately identify two faces. The bar charts represent a count of the number granules at a particular scale enclosed within a particular object as a proportion of the total number of granules within the object. Both the faces in (B) contain the same distribution of area-granules, and so this feature vector can be used for rotation-independent pattern recognition; even though, in this case, the shaded regions have different shapes.

Figure 7(B) shows an area-channel obtained from the image in Figure 7(A) (a QuickTake image taken at a "freshers" party with a flash). Notice that the regions labelled 4, 8 and 11 are readily visible. This is because they form maxima relative to the background in their region.

Each of the regions in Figure 7(B) are well defined, for the edges correspond directly to intensity changes in the original image (a consequence of Theorem 6, but the faces are not necessarily identifiable from the outlines because the illumination is very uneven and the face yielding region 11 is partly occluded. The

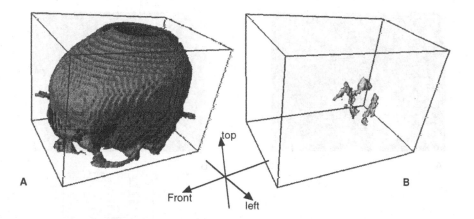

Fig. 5. (A) shows a rendered 3D image of the skull. (B) shows the third ventricle on the same co-ordinated system as (A)

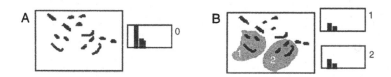

Fig. 6. (A) Shows a collection of shapes (black connected sets). In (B) the same shapes that have been grouped to make sense. Panel 0 shows granule density (described in the text) of A, Panels 1 and 2 show the density for features within sets 1 and 2.

set of segments do, however, form a starting point for developing a heuristic for finding the faces. Here we use the new approach, indicated in Figure 6, for obtaining a set of rotation-independent features for each region.

Each region is taken in turn, Figure 7(C), and examined to find the number of smaller area regions it contains at each scale. This is plotted on the ordinate, as a fraction of the total number of smaller area segments within each segment, as a function of scale, abscissa. Each of the plots is an intensity and rotation-independent signature of the regions in Figure 7(B) and, to the extent that plots 4, 8 and 11 are similar to each other and differ from the others, they represent a way of distinguishing faces from the background. In this experiment only regions 1 and 6 might be confused with faces on this criterion alone. A better feature vector might include other components (measurements made with lower dimensional sieves for example) that would allow more reliable recognition. Here, our purpose is only to illustrate the principle of using area decomposition to aid rotation independent recognition.

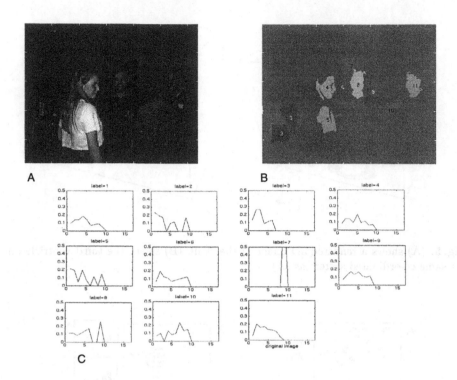

C

Fig. 7. (A) original image; (B) labelled single channel of original image; (C) granule densities for each of the regions in (B)

4 Conclusion

A nonlinear method of generating scale-space has been presented. It has the following features: it preserves scale-space causality; sharp-edged objects in the original image remain precisely localised; it operates on images defined in any finite dimension; it is scale-calibrated; the image may be processed in the scale-space domain; the decomposition appears to offer semantically meaningful feature vectors.

The method presented here is unusual, and may be unique, in having all these properties. Although the theory has been presented for M- and N-sieves our experimental evidence suggests that one can form an n-dimensional sieve using recursive median filters and it has the above properties. This has been proved in the one-dimensional case and such sieves have been shown to have good performance in noise [14]. The empirical evidence suggests that the n-dimensional sieve is also robust.

The one-dimensional version is extremely quick to compute and the n-dimensional version may also be efficiently coded although the order complexity has yet to be presented. Initial studies show that the n-dimensional sieve can form a valuable part of computer vision systems.

References

1. J. Andrew Bangham, Paul Ling, and Richard Harvey. Scale-space from nonlinear filters. In *Proc. First International Conference on Computer Vision*, pages 163–168, 1995.
2. J.J.Koenderink. The structure of images. *Biological Cybernetics*, 50:363–370, 1984.
3. Tony Lindeberg. *Scale-space theory in computer vision*. Kluwer Academic, Dordrecht, Netherlands, 1994.
4. Bart M. ter Harr Romeny, editor. *Geometry-driven diffusion in Computer vision*. Kluwer Academic, Dordrecht, Netherlands, 1994. ISBN 0-7923-3087-0.
5. Pietro Perona and Jitendra Malik. Scale-space and edge detection using anisotropic diffusion. *IEEE Trans. Patt. Anal. Mach. Intell.*, 12(7):629–639, July 1990.
6. G.Matheron. *Random sets and integral geometry*. Wiley, 1975.
7. J.Serra. *Image analysis and mathematical morphology Volume 2: Theoretical Advances*, volume 2. Academic Press, London, 1988. ISBN 0-12-637241-1.
8. Paul T. Jackway and Mohamed Deriche. Scale-space properties of the multiscale morphological dilation-erosion. In *Proc. 11th IAPR Conference on Pattern Recognition*, 1992.
9. R. van den Boomgaard and A. Smeulders. The morphological structure of images: the differential equations of morphological scale-space. *IEEE Trans. Patt. Anal. Mach. Intell.*, 16(11):1101–1113, November 1994.
10. Corinne Vachier and Fernand Meyer. Extinction value: a new measurement of persistence. In Ionas Pitas, editor, *Proc. 1995 IEEE Workshop on nonlinear signal and image processing*, volume 1, pages 254–257, JUNE 1995.
11. M. H. Chen and P. F. Yan. A multiscale approach based upon morphological filtering. *IEEE Trans. Patt. Anal. Mach. Intell.*, 11:694–700, 1989.
12. H.J.A.M.Heijmans, P.Nacken, A.Toet, and L.Vincent. Graph morphology. *Journal of Visual Computing and Image Representation*, 3(1):24–38, March 1992.
13. J. Andrew Bangham, Paul Ling, and Richard Harvey. Nonlinear scale-space in many dimensions. Internal report, University of East Anglia, 1995.
14. J. Andrew Bangham, Paul Ling, and Robert Young. Multiscale recursive medians, scale-space and transforms with applications to image processing. *IEEE Trans. Image Processing*, pages –, January 1996. Under review.
15. J.Serra and P.Salembier. Connected operators and pyramids. In *Proc. SPIE*, volume 2030, pages 65–76, 1994.
16. Luc Vincent. Morphological grayscale reconstruction in image analysis: applications and efficient algorithms. *IEEE Trans. Image Processing*, 2(2):176–201, April 1993.
17. J. A. Bangham, S. J. Impey, and F. W. D. Woodhams. A fast 1d sieve transform for multiscale signal decomposition. In *EUSIPCO*, 1994.
18. J. A. Bangham, T. G. Campbell, and M. Gabbouj. The quality of edge preservation by non-linear filters. In *Proc. IEEE workshop on Visual Signal Processing and Communication*, pages 37–39, 1992.
19. Luc Vincent. Grayscale area openings and closings, their efficent implementation and applications. In Jean Serra and Phillipe Salembier, editors, *Proc. international workshop on mathematical morphology and its applications to signal processing*, pages 22–27, May 1993.
20. P. Salembier and M. Kunt. Size sensitive multiresolution decomposition of images with rank order based filters. *Signal Processing*, 27:205–241, 1992.

Hierarchical Curve Reconstruction.
Part I: Bifurcation Analysis and
Recovery of Smooth Curves. *

Stefano Casadei and Sanjoy Mitter

Massachusetts Institute of Technology, Cambridge Ma, 02139

Abstract. Conventional edge linking methods perform poorly when multiple responses to the same edge, bifurcations and nearby edges are present. We propose a scheme for curve inference where divergent bifurcations are initially suppressed so that the smooth parts of the curves can be computed more reliably. Recovery of curve singularities and gaps is deferred to a later stage, when more contextual information is available.

1 Introduction

The problem of curve inference from a brightness image is of fundamental importance for image analysis. Computing a curve representation is generally a difficult task since brightness data provides only uncertain and ambiguous information about curve location. Two sources of uncertainty are curve bifurcations (junctions) and "invisible curves" (e.g. the sides of the Kanisza triangle). Local information is not sufficient to deal with these problems and "global" information has to be used somehow. Methods based on optimization of a cost functional derived according to Bayesian, minimum description length, or energy-based principles [4, 9, 11, 12, 20] introduce global information by simply adding an appropriate term to the cost functional. These formulations are simple and compact but lead usually to computationally intractable problems. Moreover, it is often difficult or impossible to guarantee that the optimal solution of these cost functionals represents correctly all the desired features, such as junctions and invisible curves [16]. Curve evolution approaches, where the computed curves are defined to be the stationary solutions of some differential equation [7, 19], can be computationally efficient but usually require some external initialization in order to converge to the desired solution.

A way to exploit global information efficiently without the need of external initialization is to use a hierarchy of intermediate representations between the brightness data and the final representation. The complexity and spatial extent of the descriptors in these representations increase gradually as one moves up in this hierarchy. Global information is introduced gradually so that computation is always efficient.

* Research supported by US Army grant DAAL03-92-G-0115, Center for Intelligent Control Systems

An intermediate representation used by most hierarchical approaches is given by of a set of points or tangent vectors obtained by locating and thresholding the maxima of the brightness gradient [1, 6, 15]. Linking and recursive grouping techniques exist to connect these points into polygonal curves [8, 3]. This aggregation is based on "perceptual organization" criteria such as collinearity and proximity [10]. Iterative procedures, such as relaxation labeling, have also been proposed to infer a set of curves from a set of tangent vectors [13, 5].

To guarantee that computation is efficient and robust, the hierarchy of representations should be "smooth". That is any two consecutive levels of the hierarchy should be "close" so that each level contains all the information necessary to reconstruct the objects at the following level efficiently and robustly. Consider for instance the following hierarchy:

brightness data → tangent vectors
→ smooth curves
→ curves with corners and junctions
→ closed partly invisible curves
→ overlapping regions ordered by depth.

Notice that the first stage of this hierarchy does not recover curve singularities (corners and junctions) nor invisible curves. These are recovered later when information about the smooth portions of the curves is available. Resolving uncertainty in small steps allows the algorithm to make difficult decisions when more information is present. In fact, what is uncertain or ambiguous at some level of the hierarchy might become certain and unambiguous at a higher level when more global and contextual information is available. Also, to achieve robustness, it is important that uncertainties be not eliminated arbitrarily, as is done by many threshold-based methods. Rather, these uncertainties should be represented explicitly and propagated to the higher levels. Thus it is important to understand what can be computed reliably at any particular stage and what should instead be deferred until more contextual information is present.

To illustrate this point, observe the output of the Canny edge detector followed by conventional greedy edge-point linking (figure 4, third and fourth columns). Every edge-point is linked to the neighboring point which is best aligned with the local estimate of edge orientation. If the thresholds of the edge-point detection algorithm are set to low values then many edges are completely missed (third column). On the other hand, if these thresholds are set to higher values, then more edge points are present but conventional edge linking fails in the vicinity of curve singularity and when edges are close to each other (fourth column). The algorithm is trying to make too many difficult decisions at the same time.

Uncertainty in edge linking occurs at bifurcations, namely points which can be linked to more than one other point. Notice that the paths from a bifurcation remain close to each other in some cases while in other cases they diverge. The first kind of bifurcation will be called *stable* and the second type *divergent*.

Stable bifurcation are typically caused by multiple responses to the same edge or uncertainty in curve localization. Divergent bifurcations can be either due to the topology of the curves to be reconstructed (e.g. junctions) or to noise and interference from nearby edges. Conventional edge linking does not distinguish between these different types of bifurcations and decides how each bifurcation should be disambiguated in a single step based solely on local similarity properties. Instead, the approach described in this paper resolves ambiguity at bifurcations in more than one stage. The first stage, described in this paper, "disables" temporarily divergent bifurcations and recovers smooth curves by disambiguating stable bifurcations. Curve singularity (junctions and corners) are left for a subsequent stage.

There are three reasons why divergent bifurcations are initially disabled. First, this eliminates the risk that a spurious path created by noise disrupts tracking of the true curve (see for instance the fourth column of figure 4). Secondly, it simplifies the task of resolving stable bifurcations since tracking is reduced to a one dimensional problem. Thirdly, the lost information about curve singularity is usually inaccurate because local edge-point detectors are known to have poor performance near singular points.

2 Suppression of divergent bifurcations

Let P be a set of points in $\mathbf{R}^2$ which represent sampled estimates of the location of the curves (figure 1(b)). In the experiments shown here this set has been obtained by locating the maxima of the brightness gradient in the gradient direction to sub-pixel accuracy. The gradient and its derivative have been estimated by fitting linear and cubic polynomials to blocks of 9 to 12 pixels respectively. For each $p \in P$, $\theta(p)$ denotes the estimated orientation of the curve given by the orthogonal direction to the brightness gradient and $\phi(p)$ denotes the gradient magnitude.

Let S (figure 1(c)) be the set of segments obtained by connecting every pair of points in P estimated from adjacent blocks of pixels (each point is connected to 8 other points). Segments whose orientation differs from the estimated curve orientation at its endpoints by more than $\Theta = 40°$ are discarded.

As figure 1(c) illustrates the planar graph associated with the set of segments S contains both stable and divergent bifurcations. Roughly speaking, stable bifurcations are associated with *collateral* paths, namely paths which remain "close" to each other whereas divergent bifurcations occur when these paths diverge. The definition below makes precise the distinction between collateral and divergent paths. The first part of the algorithm (table 1) computes a divergence-free subgraph of S by detecting pairs of diverging segments and "suppressing" the "weakest" segment of each such pair (figures 1(d) and 1(e)).

For any polygonal path π whose segments are elements of S let $N_w(\pi)$ be the set of points in $\mathbf{R}^2$ with distance from π less or equal to some $w > 0$ ($w = 0.75$ in our experiments). The boundary of $N_w(\pi)$, $\beta_w(\pi)$, is decomposed into its left,

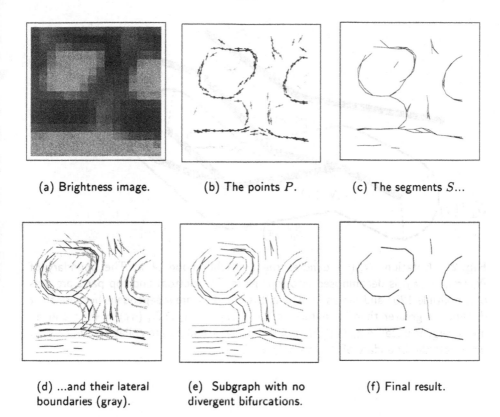

(a) Brightness image.　　(b) The points P.　　(c) The segments S...

(d) ...and their lateral　　(e) Subgraph with no　　(f) Final result.
boundaries (gray).　　divergent bifurcations.

Fig. 1. The steps of the proposed algorithm. A set of points P (b) is computed by locating the maxima of the brightness gradient. A planar graph (c) is obtained by connected neighboring points. Figure (d) shows the lateral boundaries $\beta_w^{\text{lat}}(s)$ (gray) of every segment in $s \in S$. Whenever the lateral boundary of some segment intersects another segment, the weaker of the two segments is suppressed. The resulting planar graph (e) does not contain divergent bifurcations. Finally, (f) shows the polygonal curves obtained by the procedure in table 2.

right, bottom and top parts (see figure 2): $\beta_w(\pi) = \beta_w^{\text{rig}}(\pi) \cup \beta_w^{\text{top}}(\pi) \cup \beta_w^{\text{left}}(\pi) \cup \beta_w^{\text{bot}}(\pi)$. Let also $\beta_w^{\text{lat}}(\pi) = \beta_w^{\text{left}}(\pi) \cup \beta_w^{\text{rig}}(\pi)$ (lateral boundary).

Definition 1 *Two paths π_1, π_2 (figure 2) are said to be* independent *if*

$$\pi_1 \cap N_w(\pi_2) = \pi_2 \cap N_w(\pi_1) = \emptyset \tag{1}$$

They are collateral *if they are not independent and*

$$\pi_1 \cap \beta_w^{\text{lat}}(\pi_2) = \pi_2 \cap \beta_w^{\text{lat}}(\pi_1) = \emptyset \tag{2}$$

Finally, π_1, π_2 are divergent *if*

$$\pi_1 \cap \beta_w^{\text{lat}}(\pi_2) \neq \emptyset \quad or \quad \pi_2 \cap \beta_w^{\text{lat}}(\pi_1) \neq \emptyset \tag{3}$$

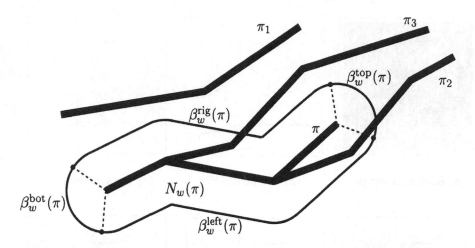

Fig. 2. Notation: $N_w(\pi)$ denotes the w-neighborhood of π. The boundary of $N_w(\pi)$, $\beta_w(\pi)$, is decomposed into four parts. The bottom and top parts are arcs of amplitude $120°$ and radius w. The paths π and π_1 are independent because their distance is greater than w, namely $\pi_1 \cap N_w(\pi_2) = \pi_2 \cap N_w(\pi_1) = \emptyset$. π and π_3 are divergent because $\pi_3 \cap \beta_w^{\mathrm{rig}}(\pi) \neq \emptyset$. Finally, π and π_2 are collateral because π_2 intersects the boundary of $N_w(\pi)$ only at $\beta_w^{\mathrm{top}}(\pi)$.

Then let $\delta_w(\pi_1, \pi_2)$ be the boolean variable which indicates whether two paths are divergent or not. Thus, $\delta_w(\pi_1, \pi_2) = 1$ if π_1 and π_2 are divergent and $\delta_w(\pi_1, \pi_2) = 0$ otherwise.

Definition 2 *A set of segments S is said to be* divergence-free *if it does not contain divergent regular paths, namely if for any pair of regular paths π_1, π_2 in S, $\delta_w(\pi_1, \pi_2) = 0$.*

For a definition of "regular" and a proof of the following results see [2]. The algorithm described in table 1, which extracts a divergence-free subset from S, hinges on the following proposition which "localizes" the notion of divergence.

Proposition 1 *Let S be a set of segments such that the orientation difference between two connected segments is never larger than $60°$. Then S is divergence-free if and only if $\delta_w(s_1, s_2) = 0$ for any $s_1, s_2 \in S$.*

This proposition ensures that the property of being divergence-free is equivalent to a local condition which involves only pairs of neighboring segments. Notice that it assumes that the orientation difference between two connected segments is $\leq 60°$. Thus, to apply this result one has to eliminate from the graph all the segments which violate this condition (lines 3-6 in table 1).

The non-maximum suppression procedure described in table 1 computes a divergence-free subset S^{DF} of S. For this purpose, a positive function $\phi : S \to \mathbf{R}^+$

```
1      For every s ∈ S
2          α(s) := 1
3      For every s₁, s₂ ∈ S, s₁, s₂ connected, T(s₁, s₂) ≥ 60°
4          If φ(s₁) ≤ φ(s₂)
5              α(s₁) := 0
6      S' = {s ∈ S : α(s) = 1}
7      For every s₁, s₂ ∈ S' such that δ_w(s₁, s₂) = 1
8          If φ(s₁) ≤ φ(s₂)
9              α(s₁) := 0
10     S^DF = {s ∈ S : α(s) = 1}
```

Table 1. The non-maximum suppression procedure to eliminate divergent bifurcations. $T(s_1, s_2)$ denotes the orientation difference between the segments s_1, s_2. The variable $\alpha(s)$ is initially set to 1 to indicate that every segment is initially "active". The loop in lines 3-5 ensures that S' does not contain pairs of connecting segments with orientation difference $> 60°$. This guarantees that proposition 1 can be applied to S'. The loop in lines 7-9 ensures that S^{DF} is divergence-free by disactivating the weaker segment of any pair of inconsistent (i.e. divergent) segments in S'.

is used to decide which of a pair of segments should be suppressed. The function ϕ used in our implementation is given by $\phi(s) = \min(\phi(p_1), \phi(p_2))$ where p_1 and p_2 are the end-points of s.

Proposition 2 *The graph S^{DF} defined in table 1 is divergence-free.*

If the length of the segments in S is bounded from above by some $L > 0$, then the two loops of the procedure need not be carried out over all pairs of segments. In fact, for each segment, it is enough to consider all the segments in a neighborhood of size $L/2 + w$ around its midpoint. If we further assume that the density of segments in the image is also bounded from above, then the complexity of the procedure is linear in the number of segments. The result of this procedure applied on the segments in figure 1(c) is shown in figure 1(e).

3   Computing the longest paths with minimum curvature

After eliminating the divergent bifurcations, the remaining ones, which are stable, can be disambiguated by computing a set of "maximally long" paths in S^{DF} which "cover" every possible path in S^{DF} (the notions of "maximally long" and "cover" are discussed more rigorously in [2]). The basic idea is as follows. Whenever a segment is connected to more than one other segment, the one which yields the "longest" path is selected. If this is not sufficient to remove all the ambiguities, namely there are two or more choices yielding "equally long" paths,

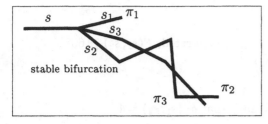
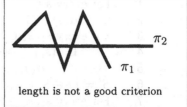

Fig. 3. Left: The segment s is connected to three other segments, s_1, s_2, s_3 which yield three different paths π_1, π_2, π_3. The algorithm in table 2 links s to s_3. Right: The length of π_1 is greater than the length of π_2. Yet, the path π_2 covers π_1 while π_1 does not cover π_2.

then the one which minimizes the total turn (that is the sum of the orientation differences of consecutive segments) is selected (see figure 3, left).

Notice that the length of a path is not the right measure to decide whether a path covers or is "longer" than another path (see figure 3, right). A more reliable way to check that a path π_1 covers π_2 is to test whether π_1 intersects $\beta_w^{\text{top}}(\pi_2)$. In fact, one can prove that if π_1, π_2 are two regular paths in S^{DF} having the same initial segment then $\pi_1 \cap \beta_w^{\text{top}}(\pi_2) = \emptyset$ implies that π_2 covers π_1 in a precise sense [2]. Conversely, $\pi_1 \cap \beta_w^{\text{top}}(\pi_2) \neq \emptyset$ implies that π_1 covers π_2.

Let $C(s)$ denote all the segments connected to $s \in S^{\text{DF}}$ which form an obtuse angle with it. The algorithm computes three variables for each segment:

- $\nu(s) \in C(s)$ is segment linked to s.
- $\lambda(s)$ is the last segment of $\pi^\star(s)$, where $\pi^\star(s)$ denotes the "best" path starting at s. Initially, $\lambda(s)$ is set equal to *nil*.
- $\tau(s)$ is the total turn of $\pi^\star(s)$.

The segments of the the path $\pi^\star(s)$ are $s, \nu(s), \nu(\nu(s)), \ldots, \lambda(s)$. The variables $\nu(s)$, $\lambda(s)$, $\tau(s)$ are computed by a procedure $link(s)$ which calls itself recursively on the elements of $C(s)$. After the procedure $link(s)$ returns, the variables $\nu(s)$, $\lambda(s)$, $\tau(s)$ have already converged to their final value because this problem is a special case of dynamic programming. This approach is similar to the saliency maximization technique used in [17, 18]. However, the method proposed here is more computationally efficient since the procedure is called exactly once on every segment and no multiple iterations of any sort are needed. A way to visualize the algorithm is to think that the procedure $link(s)$ explores all the possible paths originating from s but it prunes the search every times it hits a segment which has already been solved. The procedure is described in table 2 for the case where no close loops are present. The details to deal with close loops are explained elsewhere. Since each step can be done in constant time and the procedure is called exactly once for every segment (this is ensured by the condition at line 8) the procedure runs in linear time in the number of segments.

After the procedure $link$ has been called on all the segments, all the bifurcations have been disambiguated and every segment is linked to at most one

Definition of $link(s)$

```
1      If C(s) = ∅
2          λ(s) := s
3          τ(s) := 0
4          ν(s) := nil
5          return
6      else
7          For every s′ ∈ C(s)
8              If λ(s′) = nil
9                  link(s′)
10         C* = {s′ ∈ C(s)  :  β_w^top(λ(s′)) ∩ π*(s″) = ∅, s″ ∈ C(s)}
11         ν(s) = argmin_{s′∈C*} τ(s′)
12         τ(s) = τ(ν(s)) + T(s, ν(s))
13         λ(s) = λ(ν(s))
14         return
```

Table 2. The procedure $link(s)$. The set $C^\star$ contains all the segments $s' \in C(s)$ yielding a path $\pi^\star(s')$ which is "maximally long" among the paths originating from s. In fact, if $s' \in C^\star$, the top boundary of the last segment of $\pi^\star(s')$, $\lambda(s')$, is not "cut" (i.e. intersected) by any other other path $\pi^\star(s'')$ originating from s. Note that the variables of all the segments in $\pi^\star(s'')$, $s'' \in C(s)$ have already been computed before execution arrives at line 10. The test $\beta_w^{\mathrm{top}}(\lambda(s')) \cap \pi^\star(s'') = \emptyset$ can be done in constant time by detecting what segments intersect $\beta_w^{\mathrm{top}}(\lambda(s'))$ and then checking whether any of these segments has its variable λ equal to some $\lambda(s'')$, $s'' \in C(s)$.

segment. To extract the paths explicitly one has to identify a good set of initial points. The details of how this is done are explained in [2].

4   Experimental results

The result of the algorithm on four test images are shown in figure 4, second column. The parameter w has been set to 0.75 for all the experiments. The other parameters (used to compute P) have been kept constant also. The results are compared against Canny's algorithm with sub-pixel accuracy followed by greedy edge linking, as implemented in [14] (smoothing scale = 0.7)

5   Conclusions

To infer curves reliably and efficiently from a set of edge-points or tangent vectors it is necessary to deal with bifurcations appropriately and resolve uncertainties in the right context. We proposed a hierarchical scheme for curve reconstruction whose early stages are based on the distinction between stable and divergent

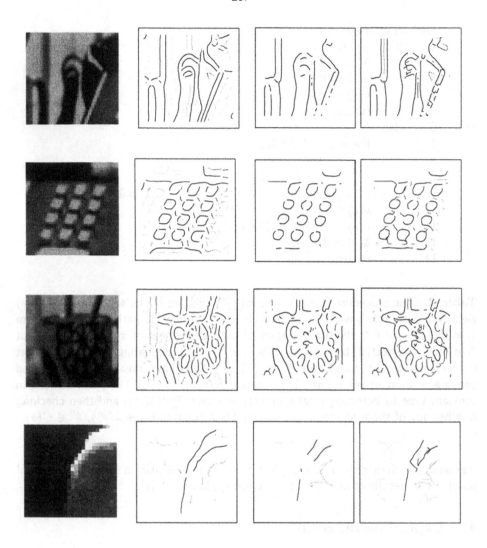

Fig. 4. The results obtained by the proposed algorithm are shown in the second column. The third and fourth column are the result of Canny's algorithm with two different sets of thresholds. The gray level of the edges is proportional to the magnitude of the brightness gradient.

bifurcations. The first stage of this hierarchy, described in this paper, recovers only the smooth portions of the curves. Work is currently being done to design and implement the following stages which combine the representation computed by the first stage with more global information to recover junctions, gaps and to construct a region-based description of the image. Recently, some theoretical results have been proved which guarantee robust performance of the algorithm in reconstructing a class of smooth curve models.

References

1. J. Canny. A computational approach to edge detection. *IEEE Transactions on Pattern Analysis and Machine Intelligence*, 8:679–698, 1986.
2. S. Casadei and S.K. Mitter. A hierarchial approach to high resolution edge contour reconstruction. Unpublished, October 1995.
3. J. Dolan and E. Riseman. Computing curvilinear structure by token-based grouping. In *Conference on Computer Vision and Pattern Recognition*. IEEE Computer Society, 1992.
4. S. Geman and D. Geman. Stochastic relaxation, gibbs distributions, and the bayesian restoration of images. *IEEE Transactions on PAMI*, 6:721–741, November 1984.
5. E.R. Hancock and J. Kittler. Edge-labeling using dictionary-based relaxation. *IEEE trans. Pattern Anal. Mach. Intell.*, 12:165–181, 1990.
6. R.M. Haralick. Digital step edges from zero crossing of second directional derivatives. *IEEE Transactions on Pattern Analysis and Machine Intelligence*, 6:58–68, 1984.
7. M. Kass, A. Witkin, and D. Terzopoulos. Snakes: Active contour models. *International Journal of Computer Vision*, 1:321–331, 1988.
8. D. Lowe. *Perceptual Organization and Visual Recognition*. Kluwer, 1985.
9. J. Marroquin, S. Mitter, and T. Poggio. Probabilistic solution of ill-posed problems in computational vision. *J. Am. Stat. Ass.*, 82(397):76–89, Mar. 1987.
10. Rakesh Mohan and Ramakant Nevatia. Perceptual organization for scene segmentation and description. *IEEE trans. Pattern Anal. Mach. Intell.*, 14, June 1992.
11. D. Mumford and J. Shah. Boundary detection by minimizing functionals. *Image Understanding 1989*, 1:19–43, 1989.
12. M. Nitzberg and D. Mumford. The 2.1-d sketch. In *Proceedings of the Third International Conference of Computer Vision*, pages 138–144, 1990.
13. Pierre Parent and Steven W. Zucker. Trace inference, curvature consistency, and curve detection. *IEEE trans. Pattern Anal. Mach. Intell.*, 11, August 1989.
14. P. Perona, 1995. Lecture notes, Caltech.
15. P. Perona and J. Malik. Detecting and localizing edges composed of steps, peaks and roofs. In *Proceedings of the Third International Conference of Computer Vision*, pages 52–57, Osaka, 1990. IEEE Computer Society.
16. T. J. Richardson. Scale independent piecewise smooth segmentation of images via variational methods. Technical Report LIDS-TH-1940, Laboratory for Information and Decision Systems, Massachusetts Institute of Technology, feb 1990.
17. A. Sha'ashua and S. Ullman. Structural saliency: the detection of globally salient structures using a locally connected network. In *Second International Conference on Computer Vision (Tampa,, FL, December 5-8, 1988)*, pages 321–327, Washington, DC,, 1988. Computer Society Press.
18. J.B. Subirana-Vilanova and K.K. Sung. Perceptual organization without edges. In *Image Understanding Workshop*, 1992.
19. A. Tannenbaum. Three snippets about curve evolution, 1995.
20. S.C. Zhu, T.S. Lee, and A.L. Yuille. Region competition: unifying snakes, region growing, and bayes/mdl for multi-band image segmentation. Technical Report CICS-P-454, Center for Intelligent Control Systems, march 1995.

Texture Feature Coding Method for Classification of Liver Sonography

Ming-Huwi Horng[1] , Yung-Nien Sun[1] and Xi-Zhang Lin[2]

[1]Institute of Information Engineering,[2]Department of Internal Medicine, National Cheng Kung University, Tainan, Taiwan, R.O.C.

Abstract   Liver sonography is a widely used noninvasive diagnostic tool. Analyzing histology changes in sonograms provides a means of diagnosing and monitoring chronic liver diseases. Nonetheless, conventional ultrasonography is still qualitative. To improve reliability of liver diagnosis, quantitative image analysis is highly desirable for the assessment of various liver states. In this paper, a novel approach, called Texture Feature Coding Method (TFCM) is presented for texture classification of liver sonography, more specifically, classification of normal liver, hepatitis and cirrhosis. TFCM is a texture analysis technique based on gray-level gradient variations in a 3x3 texture unit. It transforms an image into a texture feature image in which each pixel is represented by a texture feature number (*TFN*) coded by TFCM. The obtained texture feature numbers are then used to generate a *TFN* histogram and a *TFN* co-occurrence matrix which will produce texture feature descriptors. By coupling with a supervised maximum likelihood (ML) classifier, these descriptors form a classification system to discriminate the three above-mentioned liver classes. The TFCM-supervised ML system is trained by 30 liver samples proven by biopsy and tested on a set of 90 samples. The results show that the designed *TFN*-supervised ML system performs better than do existing techniques, and the correct classification rate can reach as high as 83.3%.

1. Introduction

In chronic liver diseases the severity of infected patients may range from healthy carrier to cirrhosis. In the past literature, several characteristics have been used to evaluate diffuse parenchyma liver diseases in past literature [2-4]. These include changes in echotexture, echogenicity, liver surface, inferior edge, diameter of hepatic duct and cystic vein. Of particular interest are the changes of echotexture. But the quality of the ultrasonic liver images is very poor since some images properties, such as edgeness, coarseness, etc., are severely affected by various noises. According to clinical experts, the liver histological changes from normal to cirrhosis can be described by the gray-level gradient variation of the echotexture in ultrasonic liver images. Thus, if we can establish correlation between gray-level gradient variations in echotexture and liver histology, using this information will be of great advantage for liver sonography analysis because it not only can assist physicians in diagnosis of liver diseases, but also provides assessment of progressive development of liver diseases.

Over the past several years a number of approaches have been proposed in natural texture classification. Among them are the gray-level co-occurrence matrix [1], statistical feature matrix [5] and texture spectrum [6] which are the most interesting texture analysis techniques that can be applied to liver images. However, these approaches do not produce satisfactory results. To improve their performance, a novel approach to texture analysis for liver sonography, called Texture Feature Coding Method (TFCM) is presented in this paper, particularly, for classification of three liver states, normal liver, hepatitis, cirrhosis. TFCM is a coding scheme which transforms an image into a texture feature image in which each pixel is encoded by TFCM into a texture feature number (*TFN*) which represents a certain type of local texture. In order to discriminate subtle details of a texture, the ideas of gray-level histogram and co-occurrence matrix are introduced into TFCM, called a texture feature number histogram and a texture feature number co-occurrence matrix respectively from which a set of texture feature number (*TFN*) descriptors are generated for texture classification. Finally, these *TFN* descriptors are coupled with a supervised maximum likelihood (ML) classifier to form a TFCM system consisting of *TFN* descriptors and a supervised ML classifier for liver sonography. The results show that the system outperforms existing methods and the correct classification rate can reach as high as 83.3%.

V_2	V_3	V_4
V_1	V_0	V_5
V_8	V_7	V_6

2	1	2
1	X	1
2	1	2

Figure1. Texture unit of texture spectrum   Figure 2. Texture unit of TFCM

2. Texture Feature Coding Method

In this section, we propose a novel approach to generating texture feature numbers, called Texture Feature Coding Method (TFCM). The design rationale of this method is based on gray-level gradient variations of a 3×3 texture unit.

2.1 Texture Feature Number Generation

2.1.1 Texture Unit
TFCM is a coding scheme which transforms an original image into a texture feature image whose pixels are represented by texture feature numbers coded by TFCM. The texture feature number of each pixel **X** is generated on the basis of gray-level changes of its 8 surrounding pixels, called a texture unit, a term was in He and Wang's work [6] described in Figure 1. Unlike He and Wang's texture spectrum, we consider the connectivity of a texture unit.

2.1.2. First-order and Second-order Connectivity of A Texture Unit
The 8 neighboring pixels in Figure 2 constitute the 8-connectivity of the texture unit which can be divided into the first-order 4-connectivity pixels and second-order 4-connectivity pixels. The four pixels labelled by 1 satisfy the first-order 4-connectivity of the texture unit because they are immediately adjacent to the pixel **X**. They will be denoted by first-order connectivity pixels. The other four pixels labelled by 2 satisfy the second-order 4-connectivity of the texture unit which

are diagonally adjacent to **X** and will be denoted by second-order connectivity pixels. In general, first-order connectivity pixels have higher correlation with pixel **X** than do second-order connectivity pixels.

2.1.3 Scan Lines of First-order and Second-order

In order to code pixel **X** in Figure 2, TFCM produces a pair of integers (α,β) where α and β represent gray-level gradient variations of three successive first-order connectivity and second-order connectivity pixels respectively. As shown in Figure 3, two scan lines along 0°-180° and 90°-270° directions produce two sets of three successive first-order connectivity pixels with pixel **X** in the middle which are horizontal line and vertical line denoted by "+". Similarly, two scan lines along diagonal direction 45°-225° and asymmetric diagonal direction 135°-315° denoted by "$\times$" also produce two sets of three successive second-order connectivity pixels as shown in Figure 4. The α and β will be used to indicate types of the gray-level gradient variations between pixels (*a* and *b*) and (*b* and *c*) described below in equation (1) or Figure 5.

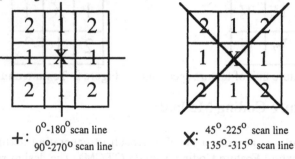

+: 0°-180° scan line
   90°-270° scan line

×: 45°-225° scan line
   135°-315° scan line

Figure 3. First-order 4-connectivity   Figure 4. Second-order 4-connectivity

In other words, each pixel in an original image will be coded by a pair of (α,β) based on types of gray-level gradient variations using the first-order connectivity and second-order connectivity pixels of its texture unit.

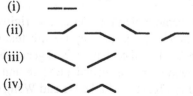

(i)

(ii)

(iii)

(iv)

Figure 5. Types of gray-level graphical structure gradient variation:

2.1.4 Types of Gray-Level Gradient Variation of a Texture Unit

Suppose that (G_a,G_b,G_c) corresponds the gray levels of three pixels (*a,b,c*) respectively. The two successive gray-level changes between two pairs (G_a,G_b) and (G_b,G_c) form four different types of variations. Let Δ be a tolerance of variation. Type (i) describes the case that the gray levels of *a*, *b* and *c* are very close within the tolerance Δ. Type (ii) is the case that one pair of gray levels are very close within

Δ, but the other pair is not and its gray-level gradient variation exceeds Δ. Type (iii) is the case that the gray levels of a,b,c are continuously decreasing or increasing with gray-level differences larger than Δ. Type (iv) is the most drastic case that either the gray-level variation is first decreasing, then increasing or first increasing, then decreasing where all the increments and decrements in this type exceed Δ.

(i) $if\ (\ |G_a - G_b| \leq \Delta) \cap (|G_b - G_c| \leq \Delta)$

(ii) $if\ [(|G_a - G_b| \leq \Delta) \cap (|G_b - G_c| \leq \Delta)] \cup [(|G_a - G_b| \geq \Delta) \cap (|G_b - G_c| \leq \Delta)]$

(iii) $if\ [(G_a - G_b > \Delta) \cap (G_b - G_c > \Delta)] \cup [(G_b - G_a > \Delta) \cap (G_c - G_b > \Delta)]$

(iv) $if\ [(G_a - G_b > \Delta) \cap (G_c - G_b > \Delta)] \cup [(G_b - G_a > \Delta) \cap (G_b - G_c > \Delta)]$ (1)

2.1.5 Types of Gray-Level Graphical Structure Gradient Variations

The four types of gray-level graphical structure gradient variations given by equation (1) can be graphed by the following gray-level graphical structure variations respectively.

First-order 4-connectivity

	(i)	(ii)	(iii)	(iv)
(i)	1	2	3	4
Second-order 4-connectivity(ii)	2	5	6	7
(iii)	3	6	8	9
(iv)	4	7	9	10

Figure 6. Texture feature number generation table

According to the degree of gray-level graphical structure variation. The higher the order of graphical structure variation, the more the changes of gray-level are the larger the gray-level gradient variation.

Figure 6 is symmetric where the column represents the horizontal scan line 0^0-180^0 for α and the diagonal line 45^0-225^0 for β, and the row represents the vertical scan line 90^0-270^0 for α and the asymmetric diagonal line 135^0-315^0 for β. Finally, the texture feature number of each pixel is generated by taking the product of α and β. More precisely, let the gray level of the pixel with spatial location (x,y) be denoted by $G(x,y)$ and the corresponding texture feature number by $TFN(x,y)$. Then

$$TFN(x, y) = \alpha(x, y)\beta(x, y)$$ (2)

where $\alpha(x,y)$ and $\beta(x,y)$ are values obtained by Figure 6 for the pixel at spatial location (x,y). In order to find an optimal gray-level gradient variation tolerance Δ, we adopt the following *TFN* entropy definition as the criterion for optimality,

213

$$H_{TEN}(\Delta) = -\sum_x \sum_y p_\Delta(TFN(x,y)) \log p_\Delta(TFN(x,y)). \tag{3}$$

where (x,y) is taken over all pixels in the image. The optimal choice for Δ is obtained by finding the maximum entropy of equation (3). That is, let Δ^* be the optimal choice for Δ, then Δ^* satisfies

$$\Delta^* = \arg\{\max_\Delta H_{TEN}(\Delta)\}. \tag{4}$$

According to experiments, we have found that $\Delta^* = 3$ yielded the maximum entropy of equation (4).

2.2 Texture Feature Number Histogram

According to equation (4) $TFN(x,y)$ can take on 100 values ranging from 1 to 100. However, we can compress 100 values to 42 values by removing unused texture feature numbers, for instance, all prime numbers are removed since they cannot be decomposed into a product of two integers as shown in equation (2). By relabelling we can assume that these 42 values take on values from 0 to 41, i.e., $\{0,1,2,L,41\}$. In this case, we can define a texture feature number histogram by

$$p_\Delta(n) = \frac{N_\Delta(n)}{N}, \quad n \in \{0,1,2L,41\} \tag{5}$$

where Δ is the gray-level gradient variation tolerance given in equation (1), $N_\Delta(n)$ is the frequency of occurrence of the texture feature number n and N is the total number of pixels in the feature image.

2.3 Texture Feature Number Co-occurrence Matrix

In the TFCM approach, we define a co-occurrence matrix on texture feature numbers of the feature image obtained by TFCM, called texture feature number co-occurrence matrix. In analogy with equation (1), a probability distribution of transitions between any pair of arbitrary two texture feature numbers can be defined similarly by

$$p_\Delta(i,j|d,\theta) = \frac{N_{\Delta d,\theta}(i,j)}{N_t}, \quad i,j \in \{0,1,2,L,41\} \tag{6}$$

where Δ is the gray-level gradient variation tolerance given in equation (3), $N_{\Delta d,\theta}(i,j)$ is defined similarly as equation (1) with the gray-level gradient variation tolerance Δ, i and j are texture feature numbers rather than gray levels as defined in equation (1) and N_t is the normalization factor which is the total number of TFN transitions.

3. Texture Feature Descriptors

In what follows, we derive 8 texture feature descriptors based on definitions of equation (5) and equation (6). Of particular interest is the last descriptor which not only takes care of the joint occurrence of two TFNs, but also includes the pixel's spatial location (x,y). Thus, we call it a second-order correlation descriptor because it accounts for *TFN*'s auto-correlation and *TFN*-spatial correlation.

1) Coarseness: $Coarse = \sum_{\Delta=0}^{R} p_{\Delta}(41)$ where R is a positive integer. A pixel

corresponding to *TFN* 41 represents a drastic change in its 8-connectivity neighborhood. So, the total number of these *TFN*s also provides a good indication of coarseness.

2) Homogeneity: $Hom = \sum_{\Delta=0}^{R} p_{\Delta}(0)$. A pixel corresponding to *TFN* 0 represents no

significant change in its 8-connectivity neighborhood. So, the total number of these *TFN*s provides a good indication of homogeneity.

3) Mean Convergence: $MC = \sum_{n=0}^{41} \frac{\left| n \cdot p_{\Delta}(n) - \mu_{\Delta} \right|}{\sigma_{\Delta}}$ This feature descriptor

indicates how close the texture approximates the mean.

4) Variance: $Var = \sum_{n=0}^{41} (n - \mu_n)^2 \cdot p_{\Delta}(n)$. The variance measures deviation of

*TFN*s from the mean.

5) Entropy: $Entropy = -\sum_{i=0}^{41} \sum_{j=0}^{41} p_{\Delta}(i, j | d, \theta) \log p_{\Delta}(i, j | d, \theta)$ where

$p_{\Delta}(i, j | d, \theta)$ is the (i, j)-th entry of the *TFN* co-occurrence matrix.

6) Run Length Density: $RLD = \sum_{i=1}^{41} p_{\Delta}^2(i, i | d, \theta)$ where $p_{\Delta}(i, j | d, \theta)$ is defined

above with $i = j$. This feature descriptor is used to calculate the density of run length of *TFN*s in its 8-connectivity neighborhood.

7) Regularity: $Regularity = \sum_{i=0}^{41} \sum_{j=0}^{41} \frac{p_{\Delta}(i, j | d, \theta)}{1 + (i - j)^2}$ This feature descriptor measures

the regularity of *TFN*s. The higher the regularity number, the more close the *TFN*s in its 8-connectivity neighborhood.

8) Gray-Level Resolution Similarity: $GLRS = \sum_{i=0}^{41} \sum_{j=0}^{41} \frac{p(i, j | x, y)}{1 + (i - j)^2}$. where

$p(i, j | x, y)$ is defined as the joint probability of *TFN* i of the pixel (x,y) in the *TFN* co-occurrence matrix with $\Delta = 0$ and *TFN* j of the same pixel (x,y) in the *TFN* co-occurrence matrix with $\Delta^* = 3$. This feature descriptor provides information about the probability of a pixel at (x,y) whose *TFN* is i at $\Delta = 0$ and *TFN* j at $\Delta^* = 3$. The higher the *GLRS*, the less the change in *TFN*s of the same pixel, thus, the less change the gray levels in this neighborhood. Unlike feature descriptor (7), this feature descriptor includes the information of changes in gray levels from $\Delta = 0$ to $\Delta^* = 3$.

4. Supervised Maximum Likelihood Classification

In Section III, we introduced TFCM to generate a set of 8 texture feature descriptors which will be used for texture classification. The classification to be presented in this section is the well-known maximum likelihood classification. Suppose that K is the number classes of interest into which a set of samples will be classified. Let $p(x|\omega_i)$ be the conditional class probability distribution, viz., the probability of x given that x belongs to the class ω_i. A feature classifier is a decision function which assigns each feature to its associated class according to a certain criterion for optimality. A maximum likelihood (ML) feature classifier is a feature classifier designed based on the conditional class probability distribution $p(x|\omega_i)$ so that a feature x will be assigned to class ω_i if $p(x|\omega_i)$ yields largest probability among all K classes. More precisely,

$$x \in \omega_i \quad \text{if} \quad p(x|\omega_i) = \max_{j \in \{0,1,L,K\}} p(x|\omega_j). \tag{7}$$

However, the ML classifier is *a priori* classification technique where the probability distribution $p(x|\omega_i)$ is assumed to be given. In many applications, $p(x|\omega_i)$ must be estimated by observations. As a result, a set of training samples is generally needed to obtain required information. An ML classifier is called supervised if it requires a set of training samples to estimate and determine the designed parameters prior to classification. For example, if the conditional class probability distribution is assumed to be Gaussian, the designed parameters for class ω_i will be the class mean μ_i and variance σ_i^2.

5. Experimental Results and Discussions

All ultrasonic images used in experiments were captured by a Toshiba Sonolayer SSA250A with PVE375 3.75 MHz (dynamic focusing) transducer at National Cheng Kung University Hospital. The images were transferred via a VFG frame grabber to a PC-AT PC and digitized into 256x512 pixels with 256 gray levels. The resulting digital images were then transmitted through NFS network system to a Sun Sparc II workstation where the proposed TFCM-supervised ML classification system was implemented with C program language. Figures 7(a-c) are sample images of normal liver, hepatitis and cirrhosis taken from intercoastal view, respectively. For each image an area of interest with 41x91 pixels is selected under the probe from 4.5-6.5 scale depth and a specific time-gain control. Each area of interest is chosed, if possible, to include solely liver tissue without major blood vessel or hepatic duct. The experiments were conducted based on 90 test images which will be classified into three liver disease classes. These liver images have proven by liver biopsy and equally divided into 3 groups, each of which has 30 cases. Four methods will be evaluated based on these 90 test images and compared in terms of classification rate and computing time.

We first analyze the classification rates of the four methods. Table 1 lists all the features selected by the forward sequential search algorithm for all the four methods and used in the classification procedures and their computing times. In Table 1 given below, the rows of tables represent the correct results proven by biopsy and the columns of tables are classification resulting from classified techniques. From this Table 1, TFCM is the best among the four methods. It is found that the TFCM completely extracts the gray level gradient variation in the ultrasonic liver images for classification. The second best is Method 2 using the co-occurrence matrix. The result reveals that the disease changes of ultrasonic liver images are related to the gray level changes in liver images. The TS and SFM mainly measure some image properties such as coarseness or some specific arrangement of image pattern in a small area. Form the Table 1, The SFM and TS are the worst and their performances are nearly the same. We are not surprised with the results since these features generated by TS and SFM are not e nough to overcome the poor image quality of a small area in the ultrasonic liver images.

As far as computing time is concerned, the Table 1 reveal that TS requires the least time 2.98 seconds and CM is the slowest method requiring 51.56 seconds. Contrary to CM, the TFCM is better to generating texture features and has the best performance among other methods. It is our opinion that based on the classification rate and computing time, we can conclude that the TFCM may be the best candidate system for liver texture classification among all the methods examined in our experiments. Unlike the texture spectrum, we consider the connectivity of each texture unit to generate the feature number. In this feature number coding method, the permutation of the first-order and second-order connectivity of a texture unit do not change the feature number, since the feature number of a small area will persist when the estimated image has few rotation. Futhermore, In the past studies, conventional methods suffered the poor quality of ultrasonic liver images. In TFCM method, we use the maximum TFN entropy criterion to find the optimal the tolerance of gray level variation. Thus, the influence of the noises can be suppressed effectively. In spite of the fact that the performance of CM method is less than the TFCM method, it is found that the gray level changes of the ultrasonic liver image is valuable for classification. Thus, It is very interesting work in the future to combine the TFCM method with CM method.

6.Conclusion

In this paper, a TFCM classification system was designed to classify three classes of liver diseases. A new texture feature coding method (TFCM) was introduced into the system which transforms a liver image into a feature image where each pixel in the feature image is represented by a texture feature number (*TFN*) coded by TFCM. Based on the texture feature numbers in the feature image, a texture feature number histogram and co-occurrence matrix can be generated from which a set of texture feature descriptors is derived for texture classification. As for classification, a supervised maximum likelihood classifier is implemented. The performance of the proposed TFCM-supervised ML classification system was compared to the gray-level co-occurrence matrix method, statistical feature method and texture spectrum method. The experimental results conducted based on 90 test liver images showed that the TFCM-supervised ML classification system outperformed all other methods and correct classification rate could achieve as high as 83.3%.

Table 1. Features selected by the forward sequential search algorithm [7], computing
time and correct classification rate.

Method	Feature Selectesd	Correct classificatio n rate	Comput- ing Time
CM	$SE(\theta = 90^0)$, $DE(\theta = 0^0)$, $COR(\theta = 90^0)$ $ASM(\theta = 90^0)$	75.7%	51.56
TS	GS, MDS2,BWS and MHS	57.78%	3.489
SFM	Contrast, Coarseness and Periodicity Measure	55.67%	2.986
TFCM	Var, GLRS, Entropy and Similarity	83.3%	5.899

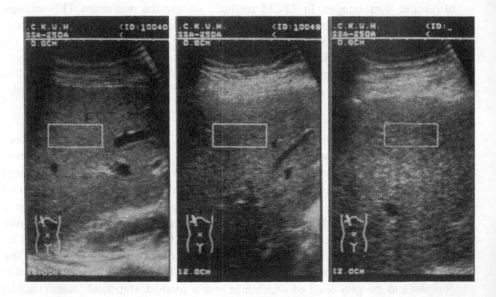

Figure 7(a). A case of normal liver (b). A case of liver hepatitis (c). A case of liver cirrhosis

7. Acknowledgment

The author would like to thank the National Science Council, R.O.C. under Grant No. NSC 84-2213-E-006-086 and the Department of Public Health under Grant No. DOH84-HR-416 for support of this work.

8. References

1. R.M. Haralick, K. Shanugan and I. Dinstein, "Texture features for image classification," *IEEE Trains. Syst. Man. Cybernet.* Vol 3, pp. 610-621, 1973.
2. A. Duerinckx, K. Rosenberg, D. Aufrichting, A Beuget, G. Kanel and S. Lottenberg."I"In vivio acoustic attention in liver correlation with blood tests and histology," Ultrasound in Med. and Biol., vol. 14, pp 405-413, 1988.
3. Y.N. Sun, H.T. Chiu and X.Z. Lin, "A computer system for the analysis of liver cirrhosis from ultrasonic images," *Chinese Journal of Medical and Biological Engineering,* Vol. 11, No. 2, pp. 119-135, 1991.
4. C.M. Wu, Y.C. Chen, and K.S. Hsieh, "Texture feature for classification of ultrasonic liver images," *IEEE Trans. Med. Imaging,* Vol. 11, No. 2, pp. 141-152, 1992.
5. C.M. Wu, and Y.C. Chen, "Statistical feature matrix for texture analysis," *Comput. Vision Graphic, Image Processing* , Vol. 54, No. 5, pp 407-419, 1992.
6. D.C. He and L. Wang. "T"Texture features based on texture spectrum," *Pattern Recognition,* Vol. 24, No. 5, pp.391-399, 1991.
7. D.C. He, L. Wang and J. Guibert, "Texture discrimination based on an optimal utilization of texture features," *Pattern Recognition,* Vol. 21, No. 2, pp141-146, 1988.

Local Appropriate Scale in Morphological Scale-Space

Ullrich Köthe

Fraunhofer Institute for Computer Graphics Rostock
Joachim-Jungius-Str. 9, 18059 Rostock, Germany
Email: koethe@egd.igd.fhg.de

Abstract. This paper discusses the problem of selecting appropriate scales for region detection *prior* to feature localization. We argue that an approach in morphological opening-closing scale-space is better than one in Gaussian scale-space. The proposed operator is based on a new shape decomposition method called morphological band-pass filter that decomposes an image into structures of different size *and* different curvature polarity. Local appropriate scale is then defined as the scale that maximizes the response of the band-pass filter at each point. This operator gives constant scale values in a region of constant width, and its zero-crossings coincide with local maxima of the gradient magnitudes. Its usefulness is demonstrated by some examples.

1 Introduction

Since their introduction by Witkin [14] scale-space representations have become a universal approach to a wide variety of computer vision tasks. They are based on the observation that real world objects and their projections onto images exist as meaningful entities only over certain ranges of scale. By making scale a parameter, an image can be transformed into a family of gradually simplyfied versions of itself. The scale parameter controls the amount of smoothing, thus the greater it is the more fine scale information is suppressed.

The most common implementation of this idea is the *Gaussian scale-space* which is defined by a convolution of the image $f(\mathbf{x})$ with a Gaussian kernel where the scale parameter determines the width of the Gaussian. The properties of this scale-space have been studied intensively by several researchers, see e.g. Lindeberg [8]. Since it is a "pure scale-space", i.e. it does not require any prior knowledge about the image content and treets all scales equally, an important question arises: If no scale is special in any way, how do we know at which scale level the interesting information can be found?

A very interesting answer to this question was given by Lindeberg [7]. He proposed to measure *local appropriate scales* which optimize the trade-off between smoothing and feature visibility. These measurements are then used to appropriately tune subsequent operators. Lindeberg defines the appropriate scale as the scale that maximizes the response of certain nonlinear operators w.r.t. scale. For example, a measure for the sizes of blobs and ridges, i.e. local extrema of image brightness, is obtained by maximizing (w.r.t. scale) the magnitudes of scale-normalized Laplaceans of Gaussian [7] or second directional derivatives of Gaussians [2], [4]. These operators give good results near the centers of blobs. However, near edges they reflect the sharpness of the edges rather than the blob sizes as is illustrated by figure 1 (center). Consequently, one has to localize blobs and edges *before* the results of these operators can be interpreted correctly.

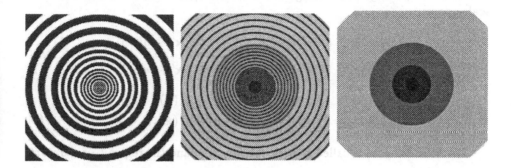

Fig. 1. Left: test image containing regions that have different constant widths, Center: appropriate scales measured by second directional derivatives are always small near edges, Right: magnitude of appropriate scales measured by our new method are constant within each region

In our opinion, appropriate scale measurements would be even more useful if they were available *prior* to feature detection. Hence we need an operator that works *uniformly* all over the image, regardless of what feature type a pixel belongs to. This could be achieved most naturally if appropriate scale were always associated with blob size, i.e. the edge response were supressed. In particlular all points in a region with constant width should have the same scale value - see the right image in figure 1. Two main problems must be solved:

1. A point may belong to regions at different scales simultaneously. These different regions must be identified, and the size of the most salient among them should determine the appropriate scale.
2. The width of a region must be defined and measured at every point without making unnecessary assumptions about possible region properties.

In this paper we propose to use *greyscale morphology* [12] to solve these problems. As opposed to convolution morphological operations are sensitive to geometrical shape. Morphological shape decomposition methods exploiting this fact have been developed by several researchers, e.g. [9], [11], [1], and [13]. Moreover, in [5] and [3] a solid theory of *morphological scale-space* is developed. This enables us to define local appropriate scales on the basis of morphological bandpass filters that will be defined as a generalisation of Wang et al. [13]. Due to space limitations all proofs have been omitted. Interested readers should refer to [6].

2 Morphological scale-space

2.1 Definition

Morphological scale-space has been developed into a coherent theory independently by Jackway [5] and van den Boomgaard and Smeulders [3]. Erosions and dilations or openings and closings are the basic operations needed to built the scale-space:

$$
\begin{aligned}
\text{Erosion:} \quad & (f \ominus g)(\mathbf{x}) = \inf_{\mathbf{x}' \in G} \left(f(\mathbf{x} + \mathbf{x}') - g(\mathbf{x}') \right) \\
\text{Dilation:} \quad & (f \oplus g)(\mathbf{x}) = \sup_{\mathbf{x}' \in G} \left(f(\mathbf{x} - \mathbf{x}') + g(\mathbf{x}') \right) \quad (1) \\
\text{Opening:} \quad & (f \circ g)(\mathbf{x}) = \left((f \ominus g) \oplus g \right)(\mathbf{x}) \\
\text{Closing:} \quad & (f \bullet g)(\mathbf{x}) = \left((f \oplus g) \ominus g \right)(\mathbf{x}) \quad (2)
\end{aligned}
$$

The function $g(\mathbf{x})$ is called *structuring function*, and the region G its *support*.

Definition 1. A morphological opening-closing scale-space is defined [5] as

$$
F(\mathbf{x}, s) = \begin{cases} (f \bullet g_s)(\mathbf{x}) & \text{if } s > 0 \\ f(\mathbf{x}) & \text{if } s = 0 \\ (f \circ g_{-s})(\mathbf{x}) & \text{if } s < 0 \end{cases} \quad (3)
$$

The unification of opening and closing in one single scale-space with positive and negative scale values is possible because the two operations are *non-self-dual* - the former operates on local maxima of $f(\mathbf{x})$, while the latter operates on minima. The subscript at g_s indicates that we are using a family of structuring functions scaled by s $(g_s(\mathbf{x}) = g(\mathbf{x}/|s|))$. If the structuring function $g(\mathbf{x})$ is anti-convex the resulting scale-space satisfies a causality theorem, i.e. no new detail is introduced by the smoothing operations [5].

2.2 Morphological low-pass and high-pass filters

As we want to decompose images w.r.t. to region size we must choose the structuring function accordingly. To make as few assumptions as possible about the regions we model them as blobs, i.e. local extrema of $f(\mathbf{x})$ and their neighborhood[1]. Shape and size of a region can then be measured by anlysing the iso-contour lines each point belongs to. Formally, we get:

Definition 2. A point $\mathbf{x}_0$ constitutes a light (dark) blob or ridge of size s if there exists a closed disk D_s of radius s that contains $\mathbf{x}_0$ so that $f(\mathbf{x}) \geq f(\mathbf{x}_0)$ $(f(\mathbf{x}) \leq f(\mathbf{x}_0))$ for every point in the disk, and no such disk exists for any $s' > s$. All points in the disk are said to lie inside a blob of size s.

---

[1] Regions where the local extrema property holds only in certain directions will be called ridges.

Consequently, a point may lie inside blobs of different sizes. To identify the most salient of all possible blob sizes for a point will become the idea behind appropriate scale identification.

Definition 3. A low-pass filter with respect to blob size is characterized by the following properties (s is the limiting blob size of the filter):

1. The filter is isotropic.
2. Blobs smaller than $|s|$ are not present in the filtered image.
3. Blobs larger than $|s|$ are not be affected by the filtering.

A special case of property 3 is that a blob of infinite size, e.g. a single step edge, must not be changed for any finite $|s|$. This leads to the following proposition:

Proposition 4. *The only isotropic, anti-convex structuring functions that do not change a blob of infinite size under opening or closing are the disks with radius s (proof see [6]):*

$$g_s(\mathbf{x}) := d_s(\mathbf{x}) = \begin{cases} 0 & \text{if } |\mathbf{x}| \leq |s| \\ -\infty & \text{otherwise} \end{cases} \tag{4}$$

The relationship between blobs as defined above and opening-closing with disk structuring functions is established by the following proposition:

Proposition 5. *Morphological opening and closing with disk structuring functions $d_s(\mathbf{x})$ are perfect low-pass filters w.r.t. blob size. (proof see [6])*

Note also that only "flat" structuring functions (like disks) ensure invariance of morphological operations w.r.t. brightness scaling (i.e. $(\lambda f \circ g) = \lambda(f \circ g)$, see [10]). Therefore we will use disk structuring functions throughout this paper. Now we define a *high-pass with respect to feature size* by the relationship:

$$H(\mathbf{x}, s) = f(\mathbf{x}) - F(\mathbf{x}, s) \tag{5}$$

which gives rise to the following proposition:

Proposition 6. *An image morphologically high-pass filtered according to (5) does not contain blobs of size s and larger. (proof see [6])*

2.3 Morphological band-pass filters

In the next step we combine low- and high-pass filters to define a *band-pass filter with respect to blob size*.

Definition 7. A band-pass filter w.r.t. blob size has the following properties (s_l, s_u are lower and upper limiting sizes, $|s_l| < |s_u|$, $s_u s_l > 0$):

1. The filter should act isotropically.
2. Blobs smaller than $|s_l|$ are not present in the filtered image $B_{s_l}^{s_u}(\mathbf{x})$.
3. Blobs larger than $|s_u|$ are not present in the filtered image.

Generalizing an idea from Wang et al. [13] to our scale-space definition we get the following recursive algorithm that alternately high- and low-pass filters the image starting with high-pass filtering at the coarsest scales:

Proposition 8. *A family of perfect morphological band-pass filters with limiting blob sizes* $-\infty = s_{-n-1} < s_{-n} < \ldots < s_0 = 0 < \ldots < s_n < s_{n+1} = \infty$ *is obtained by the following formula (s_n must be larger than the image diagonal):*

$$
\begin{aligned}
\text{for } s_k \geq 0: \quad & H_{s_{n+1}}(\mathbf{x}) = f(\mathbf{x}) \\
& B_{s_k}^{s_{k+1}}(\mathbf{x}) = (H_{s_{k+1}} \bullet d_{s_k})(\mathbf{x}) \\
& H_{s_k}(\mathbf{x}) = H_{s_{k+1}}(\mathbf{x}) - B_{s_k}^{s_{k+1}}(\mathbf{x})
\end{aligned}
$$

$$
\begin{aligned}
\text{and for } s_k \leq 0: \quad & H_{s_{-n-1}}(\mathbf{x}) = f(\mathbf{x}) \\
& B_{s_k}^{s_{k-1}}(\mathbf{x}) = (H_{s_{k-1}} \circ d_{s_k})(\mathbf{x}) \\
& H_{s_k}(\mathbf{x}) = H_{s_{k-1}}(\mathbf{x}) - B_{s_k}^{s_{k-1}}(\mathbf{x})
\end{aligned}
$$

(6)

where the resulting $B_{s_k}^{s_l}(\mathbf{x})$ represent a morphological decomposition of the image into bands of different blob sizes and curvature polarities (H_{s_k} are intermediate high-pass filtered images). The original image can be exactly reconstructed from both the positive and the negative parts of the decomposition (proof see [6]):

$$
\sum_{k=0}^{n} B_{s_k}^{s_{k+1}}(\mathbf{x}) = \sum_{k=-n}^{0} B_{s_k}^{s_{k-1}}(\mathbf{x}) = f(\mathbf{x})
$$

(7)

Figure 2 shows a family of band-pass filtered images using (6). s is sampled in octaves. It is clearly visible how the image is decomposed into different structure sizes by the filter family.

3 Appropriate scale measurements in opening-closing scale-space

Similar to the proposal of Lindeberg [7], we identify the local appropriate scale as the scale that maximizes the response of a normalized band-pass filter with respect to scale:

Definition 9. The local appropriate scale of a blob at $\mathbf{x}$ is defined as:

$$
s_A(\mathbf{x}) = \arg_{s_k} \left(\max_{\substack{s_k = s_{-n}, \ldots, s_{-1}, \\ s_1, \ldots, s_n}} \left| \frac{B_{s_{k\mp1}}^{s_k}(\mathbf{x})}{s_k - s_{k\mp1}} \right| \right)
$$

(8)

where s_{k-1} applies if $s_k > 0$ and s_{k+1} if $s_k < 0$. The expression $B_{s_{k\mp1}}^{s_k}(\mathbf{x})/(s_k - s_{k\mp1})$ will be called normalized band-pass filter.

Figure 1 (right) shows the result of this operator on a test image. Note that the scale values within a region of constant width are constant. Figure 3 illustrates the application of the new appropriate scale operator to a natural image. Again the scale values correspond to the width of the region a pixel belongs

Fig. 2. Decomposition of an image (center) with respect to structure sizes. Left from top to bottom: $s = 2, 4, 8$ - dark blobs and ridges. Right: $s = -2, -4, -8$ - light blobs and ridges.

to. Scales are positive if this region is darker than its surroundings, negative otherwise.

Another fact is, however, somewhat surprising: The borders between areas of positive and negative scale ("zero-crossings" of the appropriate scale) correspond to image edges (local maxima of the image gradient). Although this behavior has been justified experimentally on a large number of images, we do not yet have a full theoretical explanation in 2D. An analysis of the scale operator in 1D indicates, however, that maxima of the normalized bandpass $B_{s_{k\mp1}}^{s_k}(x)/(s_k - s_{k\mp1})$ are indeed correlated with local maxima of the gradient $f'(x)$.

Consider the function in figure 4. The effect of morphological opening is best visualized by "fitting the structuring function under the original function". Opening with $d_s(x)$ therefore replaces the function between $f(x_1)$ and $f(x_2)$ with a straight line (where $x_2 - x_1 = 2s$). Likewise, opening with $d_{s'}(x)$ results in the straight line between $f(x_1')$ and $f(x_2')$ (with $x_2' - x_1' = 2s'$). Now normalized band-pass filtering between x_1 and x_2 yields

$$\left| \frac{B_s^{s'}(x)}{s' - s} \right| = \frac{f(x_1) - f(x_1')}{s' - s} = \frac{f(x_2) - f(x_2')}{s' - s}$$

$|s_A| =$ *1 2 4 8 16 ≥32*

Fig. 3. Appropriate scale measurements. For better visibility positive and negative scales are decomposed into two images (top: positive scales, bottom: magnitude of negative scales).

If we expand the r.h.s. into a Taylor series we arrive at

$$\lim_{s' \to s} \left| \frac{B_s^{s'}(x)}{s' - s} \right| = 2 \left(\frac{1}{|f'(x_1)|} + \frac{1}{|f'(x_2)|} \right)^{-1}$$

Hence the result of the morphological band-pass filter is proportional to the harmonic mean of the gradients at the points where the structuring function touches the original function. The gradients are maximized at edge points, thus the appropriate scale is obtained when the structuring function just fits between two edge points.

The following proposition establishes that the appropriate scale is invariant under rotation, translation, and brightness scaling while it scales accordingly when the spatial coordinates are uniformly scaled. This result is a major prerequisite for practical applications of approproate scales (like tuning of scale dependent operators towards suitable scales).

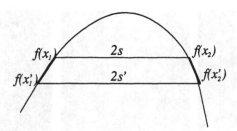

Fig. 4. Analysis of the appropriate scale operator in 1D (see text).

Proposition 10. *Let $\tilde{f}(\mathbf{x})$ be the image after similarity transformation and brightness rescaling, i.e. $\tilde{f}(\mu \mathbf{R}\mathbf{x} + \mathbf{x}_0) = \lambda f(\mathbf{x})$ where $\mathbf{R}$ denotes an arbitrary 2D rotation matrix, $\mathbf{x}_0$ an arbitrary translation, and $\lambda, \mu > 0$ are scalars. Then the appropriate scale $\tilde{s}_A$ of the transformed image $\tilde{f}(\mathbf{x})$ is given by (proof see [6]):*

$$\tilde{s}_A(\mu \mathbf{R}\mathbf{x} + \mathbf{x}_0) = \mu s_A(\mathbf{x}) \tag{9}$$

4 Applications

4.1 Appropriate scale ridges

In this section we demonstrate that our new appropriate scale operator can be combined with Gaussian scale-space to select appropriate scales for ridge detection. We measure the scale-dependent ridge strength in Gaussian scale-space by the eigenvalues of the Hessian matrix ($\Phi_{xx}(\mathbf{x}, s)$ etc. denote second derivatives at scale s):

$$\gamma_{1,2}(\mathbf{x}, s) = \frac{1}{2}\left(\Phi_{xx}(\mathbf{x}, s) + \Phi_{yy}(\mathbf{x}, s) \pm \sqrt{(\Phi_{xx}(\mathbf{x}, s) - \Phi_{yy}(\mathbf{x}, s))^2 + 4\Phi_{xy}(\mathbf{x}, s)^2}\right) \tag{10}$$

Now we use the appropriate scale $s_A(\mathbf{x})$ to select the correct scale, where the larger eigenvalue is taken for positive scales and the smaller for negative scales:

$$r(\mathbf{x}) = \begin{cases} \gamma_1(\mathbf{x}, s_A(\mathbf{x})) & (s_A(\mathbf{x}) > 0) \\ -\gamma_2(\mathbf{x}, -s_A(\mathbf{x})) & (s_A(\mathbf{x}) < 0) \end{cases} \tag{11}$$

Since the appropriate scale depends only on the width of the ridge and does not change near edges, the ridge strength operator $r(\mathbf{x})$ shows very good step edge supression as is illustrated by figure 5.

4.2 Parameter-free binarization

The fact that the zero-crossings of the appropriate scales often coincide with image edges suggests to use the sign of the scales as a simple means to binarize the image:

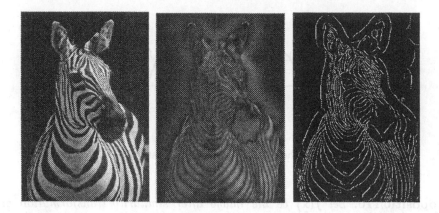

Fig. 5. Appropriate scale ridges. Center: ridge strength, Right: ridge location after non-maxima supression

$$b(\mathbf{x}) = \begin{cases} 1 & (s_A(\mathbf{x}) < 0) \\ 0 & (s_A(\mathbf{x}) > 0) \end{cases} \qquad (12)$$

This technique has two important advantages over thresholding: The binarization is invariant under brightness rescaling (Proposition 10) and the positions of the zero-crossings are insensitive to large scale shading. Figure 6 illustrates this by comparing classical thresholding with the new binarization method.

5 Conclusions

In this paper we discussed the problem of selecting appropriate scales *uniformly* at each pixel, regardless of the feature type the pixel belongs to. This led to a new appropriate scale operator built upon a morphological band-pass filter which proved more suitable to define uniform region based appropriate scale measurements than approaches based on Gaussian scale-space. In particular it has the following very interesting properties:

- The scale value obtained at any point does reflect the width of the most salient region the point belongs to.
- It is invariant w.r.t. brightness rescaling of the image.
- If the image undergoes a similarity transform it is scaled by the same amount as the spatial coordinates.
- The positions of zero-crossings of the appropriate scale seem to be correlated with maxima of the gradient magnitudes of the image.

Further investigation is needed to theoretically establish the last property in 2D.

Two applications illustrate that the new operator can indeed be used to improve other operators in making them invariant under similarity transformations and brightness scaling or tuning them towards the best scale of operation. We expect very interesting results from further research in this direction.

228

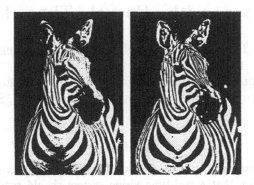

Fig. 6. Left: conventional thresholding with hand-tuned threshold, Right: parameter-free binarization

References

1. A. Bangham, P. Chardaire, P. Ling: "The Multiscale Morphology Decomposition Theorem", in: J. Serra, P. Soille (eds.): Mathematical Morphology and Its Applications to Image Processing, Proc. of ISMM '94, Kluwer 1994
2. B. Burns, K. Nishihara, S. Rosenschein: "Appropriate Scale Local Centers: a Foundation for Parts-based Recognition", Teleos Research Technical Report TR-94-05, 1994
3. R. van den Boomgaard, A. Smeulders: "The Morphological Structure of Images: The Differential Equations of Morphological Scale-Space", IEEE Trans. on Pattern Analysis and Machine Intelligence vol. 16, no. 11, pp 1101-1113, 1995
4. K. Dana, R. Wildes: "A Dynamic Energy Image with Applications", Proc. ARPA Image Understanding Workshop, 1994
5. P.T. Jackway: "Morphological Scale-Space With Application to Three-Dimensional Object Recognition", PhD-Theses, Queensland University of Technology, 1994
6. U. Köthe: "Local Appropriate Scale in Morphological Scale-Space", Fraunhofer Institute for Computer Graphics Technical Report no. 96i001-FEGD, January 1996
7. T. Lindeberg: "On Scale Selection for Differential Operators", Proc. 8th Scandinavian Conf. on Image Analysis, Tromsø, Norway, 1993
8. T. Lindeberg: "Scale-Space Theory in Computer Vision", Kluwer Academic Publishers, 1994
9. I. Pitas, A. Maglara: "Range Image Analysis by Using Morphological Signal Decomposition", Pattern Recognition, vol. 24, no. 2, pp. 165-181, 1991
10. J.-F. Rivest, J. Serra, P. Soille: "Dimensionality in Image Analysis", J. of Visual Communication and Image Representation, vol. 3, no. 2, pp 137-146, 1992
11. P. Salembier, M. Kunt: "Size-sensitive multiresolution decomposition of images with rank order based filters", Signal Processing, vol. 27, no. 2, pp. 205-241, 1992
12. S. Sternberg: "Grayscale Morphology", Computer Vision, Graphics, and Image Processing, vol. 35, pp 333-355, 1986
13. D. Wang, V. Haese-Coat, A. Bruno, J. Ronsin: "Texture Classification and Segmentation Based on Iterative Morphological Decomposition", J. of Visual Communication and Image Representation, vol. 4, no. 3, pp 197-214, 1993
14. A.P. Witkin: "Scale-Space Filtering", Proc. Intl. Joint Conf. on Artificial Intelligence, Karlsruhe, Germany, 1983

Scale-Space with Casual Time Direction

Tony Lindeberg and Daniel Fagerström

Computational Vision and Active Perception Laboratory (CVAP)*
KTH (Royal Institute of Technology), Stockholm, Sweden.

Abstract

This article presents a theory for multi-scale representation of *temporal data*. Assuming that a real-time vision system should represent the incoming data at different time scales, an additional causality constraint arises compared to traditional scale-space theory—we can only use what has occurred in the past for computing representations at coarser time scales. Based on a previously developed scale-space theory in terms of *non-creation of local maxima with increasing scale*, a complete classification is given of the scale-space kernels that satisfy this property of non-creation of structure and *respect the time direction as causal*. It is shown that the cases of continuous and discrete time are inherently different.

For continuous time, there is no non-trivial time-causal semi-group structure. Hence, the time-scale parameter *must* be discretized, and the only way to construct a linear multi-time-scale representation is by (cascade) convolution with truncated exponential functions having (possibly) different time constants. For discrete time, there is a canonical semi-group structure allowing for a continuous temporal scale parameter. It gives rise to a *Poisson-type temporal scale-space*. In addition, geometric moving average kernels and time-delayed generalized binomial kernels satisfy temporal causality and allow for highly efficient implementations.

It is shown that *temporal derivatives* and derivative approximations can be obtained directly as *linear combinations* of the temporal channels in the multi-time-scale representation. Hence, to maintain a representation of temporal derivatives at multiple time scales, there is no need for other time buffers than the temporal channels in the multi-time-scale representation.

The framework presented constitutes a useful basis for expressing a large class of algorithms for computer vision, image processing and coding.

1   Introduction

The notion of multi-scale representation is essential when dealing with measured data, such as images. Philosophically, this need arises from the fact that we perceive real-world structures as meaningful entities only over certain ranges of scale. Traditionally, multi-scale concepts such as pyramids (Burt 1981; Crowley 1981) and scale-space representation (Witkin 1983; Koenderink 1984; Yuille and Poggio 1986; Koenderink and van Doorn 1992; Florack 1993; Lindeberg 1994)

---

*The support from the Swedish Research Council for Engineering Sciences, TFR, and the Esprit-NSF collaboration Diffusion is gratefully acknowledged. *Address:* KTH, NADA, S-100 44 Stockholm, Sweden. *Email:* tony@bion.kth.se, danielf@bion.kth.se. http://www.bion.kth.se.

have been developed over a spatial domain, in which data are available in all directions. Most works have avoided the constraints arising from the fact that time runs in a special direction and a genuine real-time vision cannot access the future—only what has occurred in the past can be used for generating representations at different time scales. An early suggestion for how to treat time in a multi-scale context was given by (Koenderink 1988), who proposed to transform the time axis so as to map the present moment to the unreachable infinity. In the transformed domain, he then applied the traditional scale-space concept by Gaussian convolution. The subject of this article is to reconsider the problem of constructing a multi-time-scale representation from an axiomatic viewpoint.

2 Continuous and discrete scale-space kernels: Review

A fundamental requirement when constructing a multi-scale representation is that the transformation from a fine scale to a coarser scale should constitute a simplification in the sense that fine-scale image structures should be successively suppressed. In the literature on traditional (spatial) scale-space representation, this property has been formalized in different ways. A noteworthy coincidence is that several different ways of choosing *scale-space axioms* lead to the Gaussian kernel as the unique choice.

In this article, we shall follow the scale-space formulation in (Lindeberg 1990, 1994) based on non-creation of local extrema (zero-crossings) with increasing scale. As shown in the abovementioned references, the class of convolution operators satisfying this requirement can be completely classified based on classical results by (Schoenberg 1953) (see also (Karlin 1968)). Besides translation and rescaling, there are two primitive types of linear and shift-invariant smoothing transformations in the continuous case:

- convolution with *Gaussian kernels*,

$$h(\xi) = e^{-\gamma \xi^2}, \tag{1}$$

- convolution with *truncated exponential functions*,

$$h(\xi) = \begin{cases} e^{-\xi/|\mu|} & \xi \geq 0, \\ 0 & \xi < 0, \end{cases} \qquad h(\xi) = \begin{cases} e^{\xi/|\mu|} & \xi \leq 0, \\ 0 & \xi > 0, \end{cases} \tag{2}$$

Correspondingly, in the discrete case, there are besides rescaling and translation, three primitive types of smoothing transformations (where $f_{out} = h * f_{in}$):

- two-point weighted averaging or *generalized binomial smoothing*,

$$\begin{aligned} f_{out}(x) &= f_{in}(x) + \alpha_i f_{in}(x-1) & (\alpha_i \geq 0), \\ f_{out}(x) &= f_{in}(x) + \delta_i f_{in}(x+1) & (\delta_i \geq 0), \end{aligned} \tag{3}$$

- moving average or *first-order recursive filtering*,

$$\begin{aligned} f_{out}(x) &= f_{in}(x) + \beta_i f_{out}(x-1) & (0 \leq \beta_i < 1), \\ f_{out}(x) &= f_{in}(x) + \gamma_i f_{out}(x+1) & (0 \leq \gamma_i < 1), \end{aligned} \tag{4}$$

- *infinitesimal smoothing* described by the generating function

$$H_{semi-group}(z) = e^{t(az^{-1}+bz)}. \tag{5}$$

In the symmetric case, $a = b = \alpha/2$, this transformation corresponds to convolution with the *discrete analogue of the Gaussian kernel*,

$$T(n; \sigma^2) = e^{-\alpha\sigma^2} I_n(\alpha\sigma^2), \tag{6}$$

where I_n are the modified Bessel functions of integer order.

Among these *scale-space kernels*, we recognize the continuous Gaussian kernel $g(x; \sigma^2)$ and its discrete analogue $T(n; \sigma^2)$, which arise as unique symmetric choices if the scale parameter is required to be continuous and a semi-group structure is imposed (Lindeberg 1990, 1994). The generalized binomial kernels provide a natural basis for constructing pyramid representations (Burt 1981; Crowley 1981), whereas recursive filters can be used for efficient implementations of smoothing operations (Deriche 1987).

3 Time-causal scale-space kernels

The review in the previous section is general and does not take the specific nature of the time direction into account. For scale-space kernels treating the time direction as causal, an obvious requirement is that only function values in the past can be accessed. Hence, the kernels must satisfy $h(t) = 0$ when $t < 0$. Here, we shall analyse the implications of imposing this constraint on scale-space kernels in the continuous and discrete domains.

Continuous time. An immediate consequence of the classification of semi-groups of continuous scale-space kernels (the Gaussian kernel is unique) is that we cannot preserve a continuous semi-group structure with respect to the time-scale parameter if the time direction is to be treated as causal. Hence, the *only* choice is to discretize the time-scale parameter. The only primitive scale-space kernels with one-sided support are the truncated exponential functions. After normalization to unit L_1-norm they can be written

$$h_{exp}(t; \mu) = \frac{1}{\mu} e^{-t/\mu} \qquad (t > 0). \tag{7}$$

By varying μ, we obtain first-order filters having different time constants. The classification of continuous scale-space kernels implies that a kernel is a time-causal scale-space kernel if and only if it can be decomposed into a sequence of convolutions with such filters. Hence, the architecture on a time-scale representation imposed by this construction is a set of *first-order recursive filters in cascade*, each having a (possibly) different time constants μ_i. Such a filter has mean value $M(h_{composed}(\cdot; \mu)) = \sum_{i=1}^{\infty} \mu_i$, variance $\lambda = V(h_{composed}(\cdot; \mu)) = \sum_{i=1}^{\infty} \mu_i$, and a (bilateral) Laplace transform of the form

$$H_{composed}(s; \mu) = \int_{t=-\infty}^{\infty} (*_{i=1}^{\infty} h_{exp}(t; \mu_i)) e^{-st} dt = \prod_{i=1}^{\infty} \frac{1}{1 + \mu_i s}. \tag{8}$$

If we in analogy with a semi-group requirement, require the transformation from any fine-scale representation to any coarser-scale representation to be a scale-space transformation, then the only possibility is that all the (discrete) scale levels in the multi-time scale representation can be generated by a cascade of such truncated exponential filters.

Discrete time. For discrete time sampling, the discrete analogue of the truncated exponential filters are the first-order recursive filters (4). With normalization to unit l_1-norm, and $\mu = \beta/(1-\beta)$, their generating functions can be written

$$H_{geom}(z) = \frac{1}{1 - \mu(z-1)}, \tag{9}$$

Computationally, these filters are highly efficient, since only few arithmetic operations and no additional time buffering are required to compute the output at time $t+1$ given the output at time t. In normalized form, the recursive smoothing operation is

$$f_{out}(t) - f_{out}(t-1) = \frac{1}{1+\mu}(f_{in}(t) - f_{out}(t-1)). \tag{10}$$

In analogy with the case of continuous time, a natural way to combine these filters into a discrete multi-time-scale representation is by cascade coupling. The mean and variance of such a composed filter are $M(h_{geom}(\cdot; \mu)) = \sum_{i=1}^{\infty} \mu_i$ and $\lambda = V(h_{geom}(\cdot; \mu)) = \sum_{i=1}^{\infty} \mu_i^2 + \mu_i$. In the case of discrete time, we can also observe that the generalized binomial kernels (3) indeed satisfy temporal causality, if combined with a suitable time delay. In this respect, there are more degrees of freedom in the case of discrete time sampling.

Time-causal semi-group structure exists only for discrete time. The case of discrete time it also special in the sense that a semi-group structure is, indeed, compatible with temporal causality. If we let $q_{-1} = 0$ and $q_1 = \lambda$ in (5) and multiply by the normalization factor $\exp(-\lambda)$, we obtain a generating function of the form $P(z; \lambda) = e^{\lambda(z-1)}$ (Lindeberg 1996) with associated filter coefficients

$$p(n; \lambda) = e^{-\lambda}\frac{\lambda^n}{n!}. \tag{11}$$

This filter corresponds to a Poisson distribution and the kernel p will be referred to as the *Poisson kernel*. Intuitively, it can be interpreted as the limit case of repeated convolution of kernels of the form (9) with time constants $\mu = \lambda/m$:

$$\lim_{m \to \infty}\left(H_{geom}(z; \frac{\lambda}{m})\right)^m = \lim_{m \to \infty}\frac{1}{(1 - \frac{\lambda}{m}(z-1))^m} = P(z; \lambda). \tag{12}$$

Such a kernel has mean $M(p(\cdot; \lambda)) = \lambda$, and variance $V(p(\cdot; \lambda)) = \lambda$. From the ratio $\frac{p(n+1; \lambda)}{p(n; \lambda)} = \frac{\lambda}{n+1}$, it can be seen for $\lambda < 1$ the filter coefficients decrease monotonically for $n \geq 0$, while for $\lambda > 1$ there is a local maximum at the smallest integer less than λ: $n = [\lambda] > 0$. Similarly, there are two inflexion points at $n \approx \lambda + \frac{1}{2} \pm (\lambda + \frac{1}{4})^{1/2}$. Concerning the qualitative behaviour, it also well-known from statistics that the Poisson distribution approaches the normal distribution with increasing standard deviation (see figure 1).

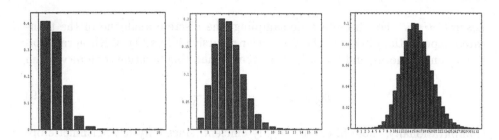

Figure 1: Graphs of the Poisson kernels for $\lambda = 0.9$, 3.9 and 15.9.

Under variations of λ, the Poisson kernel satisfies $\partial_\lambda p(n; \lambda) = -(p(n; \lambda) - p(n - 1; \lambda))$. Thus, if we define a multi-time-scale representation $L: \mathbb{R} \times \mathbb{R}_+ \to \mathbb{R}$ of a discrete signal $f: \mathbb{R} \to \mathbb{R}$, having a *continuous time-scale parameter*, by

$$L(t; \lambda) = \sum_{n=-\infty}^{\infty} p(n; \lambda) f(t - n), \tag{13}$$

this representation satisfies the first-order semi-differential equation $\partial_\lambda L = -\delta_- L$, where δ_- denotes the backward difference operator $\delta_- L(t; \lambda) = L(t; \lambda) - L(t - 1; \lambda)$. Hence, in contrast to multi-scale representations of the spatial domain, for which derivatives with respect to scale are related to second-order derivatives/differences in the spatial domain, temporal scale derivatives are here related to *first-order* temporal differences.

Note that a corresponding time-causal structure does not exist for continuous signals. If we apply the same way of reasoning and compute the limit case of primitive kernels of the form (8) for which all λ_i are equal, we obtain the trivial semi-group corresponding to translations of the time axis by a time delay λ.

4 Temporal scale-space and temporal derivatives

So far, we have shown how general constraints concerning non-creation of local extrema with increasing scale combined with temporal causality restrict the class of operations that can be used for generating multi-scale representations corresponding to temporal integration over different time scales. When to use these results in practice, an obvious issue concerns how to distribute a (finite) set of discrete scale levels over scales and how to compute temporal derivatives (or derivative approximations) at different time scales.

Distribution of scale levels. A useful property of the Poisson-type scale-space (13) is that there is no need for selecting scale levels *in advance*. If we have access to all data in the past, we can compute the temporal scale-space representation at any scale. Assuming that a vision system is to operate at a set of K temporal scales, a natural *a priori* distribution of these scale levels λ_k between some minimum scale λ_{min} and some maximum scale λ_{max} is according to a geometric series $\lambda_k = \gamma^k \lambda_{min}$ where $\gamma^K = \lambda_{max}/\lambda_{min}$.

Concerning the multi-time scale representations having a discrete time-scale parameter, let us assume that a *minimal design* is chosen, in the sense that the transformation between adjacent scales is always of the form (7) or (9). Since variances are additive under convolution, it follows that the time constants between adjacent scales should satisfy $\lambda_k = \lambda_{k-1} + \mu_k$ for continuous signals and $\lambda_k = \lambda_{k-1} + \mu_k + \mu_k^2$ for discrete signals.

Temporal scale-space derivatives in the continuous case. Given a continuous signal f, assume that a level k in a time-scale representation

$$L(\cdot; \lambda_k) = (*_{i=1}^{k} h_{exp}(t; \mu_i)) * f \tag{14}$$

has been computed at some temporal scale λ_k by cascade filtering with a set of k truncated exponential filters with time constants μ_i. From this representation, a *temporal scale-space derivative* of order r at scale λ_k is defined by

$$L_{t^r}(\cdot; \lambda_k) = \partial_{t^r} L(\cdot; \lambda_k) = (\partial_{t^r}(*_{i=1}^{k} h_{exp}(t; \mu_i))) * f, \tag{15}$$

and the Laplace transform of the composed (equivalent) derivative kernel is

$$H^{(r)}_{composed}(s; \lambda_k) = s^r \prod_{i=1}^{k} \frac{1}{1 + \mu_i s}. \tag{16}$$

For this kernel to have a net integration effect (well-posed derivative operators), an obvious requirement is that the total order of differentiation should not exceed the total order of integration. Thereby, $r < k$ is a necessary requirement. As a consequence, the transfer function must have finite L_2-norm.

A useful observation in this context is that these *temporal scale-space derivatives can be equivalently computed from differences between the temporal channels*. Assume, for simplicity, that all μ_i are different in (16). Then, a decomposition of $H^{(r)}_{composed}$ into a sum of r such transfer functions at finer scales

$$H^{(r)}_{composed}(s; \lambda_k) = \sum_{i=k-r}^{k} B_i H_{composed}(s; \lambda_i) \tag{17}$$

shows that the weights B_i are given as the solution of a triangular system of equations provided that the necessary condition $r < k$ is satisfied

$$\frac{(-1)^r}{\mu_i^r} \prod_{j=i+1}^{k} \frac{1}{(1 - \mu_j/\mu_i)} = B_i + \sum_{\nu=i+1}^{k} B_\nu \prod_{j=i+1}^{\nu} \frac{1}{(1 - \mu_j/\mu_i)} \qquad (k - r \le i \le k).$$

Hence, each temporal derivative can be computed as a linear combination of the representations at finer time scales. Moreover, the Laplace transforms of the equivalent derivative computation kernels satisfy the recurrence relation

$$H^{(r)}_{composed}(s;\ \lambda_k) = -\frac{1}{\mu_k}\left(H^{(r-1)}_{composed}(s;\ \lambda_k) - H^{(r-1)}_{composed}(s;\ \lambda_{k-1})\right). \quad (18)$$

In other words, higher-order temporal derivatives can be computed as finite differences of lower-order derivatives (analogous to finite difference operators in the spatial domain). Derivative computations will thus be highly efficient.

Temporal derivative approximations in the discrete case. In (Lindeberg and Fagerström 1996) it is shown that a corresponding structure holds in the discrete case, for multi-scale temporal derivative approximations obtained by applying (either symmetric or non-symmetric) central difference operators to the discrete

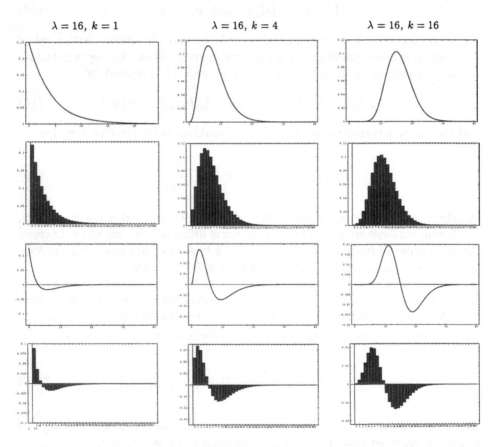

Figure 2: Graphs of equivalent smoothing kernels and first-order derivative (approximation) kernels in the continuous and discrete cases, respectively, for k cascade coupled smoothing steps in which all the primitive time constants μ_i are equal. (Here, μ_i have been determined from k such that the variance λ is the same for all smoothing kernels.)

multi-time-scale representation constructed by cascade convolution with first-order recursive filters of the form (10).

Temporal derivatives from linear combinations of temporal channels This special structure is highly useful for practical purposes, since it makes explicit construction of temporal derivative kernels unnecessary. In other words, no other time buffers are necessary for computing temporal scale-space derivatives than the actual channels in the multi-time-scale representation. An intuitive explanation of why this is possible is that the different temporal channels, which represent the incoming data at different time-scales, have different effective temporal delays. Besides the primary effect of producing an integrated representation over a certain time-scale, each such channel serves as a temporal buffer.

Kernel graphs and trade-off issues. Figure 2 shows graphs of equivalent (continuous and discrete) convolution kernels for a few combinations of the number of recursive filters in cascade, k, and the individual time constants, μ_i.

The parameter values have been chosen such that the variance λ of the smoothing kernel is the same for all filters. Hence, they represent different ways of computing the representation at a certain scale.

As can be seen, the kernels are discontinuous if $r \geq k - 1$, whereas the degree of smoothness increases with k. To guarantee a certain minimum degree of temporal smoothness at the finest temporal scale, it can therefore (depending on the external sampling conditions) be useful to precede the recursive temporal multi-scale representations by a common pre-smoothing step (such as a few steps of recursive filtering or time-delayed binomial smoothing). For a more detailed analysis, including frequency properties, see (Lindeberg and Fagerström 1996).

5 Spatio-temporal scale-space

When to combine these multi-time-scale representations with a spatial representation for dealing with time-varying images, let us first treat space and time as separable dimensions. This is a natural assumption in the absence of further information (such as velocity information). The spatio-temporal scale-space representation we then obtain is the Cartesian product of the spatial and temporal scale-space representations, and is parameterized by a spatial scale parameter σ^2 and a temporal scale parameter λ.

Depending on whether the spatial domain $\mathbb{S}$ is continuous or discrete, and correspondingly for the temporal domain $\mathbb{T}$ as well as the domains Σ and Λ of the spatial and temporal scale parameters, we then obtain one out of twelve possible types of spatio-temporal scale-space representations (see figure 3).

Denote the transfer function of the spatial smoothing kernel by $H_\mathbb{S}(u; \sigma^2)$ and the transfer function of the temporal smoothing kernel by $H_\mathbb{T}(v; \lambda)$. Then, the transfer function for mapping a spatio-temporal signal $f \colon \mathbb{S}^N \times \mathbb{T} \to \mathbb{R}$ to its *spatio-temporal scale-space representation* $L \colon \mathbb{S}^N \times \mathbb{T} \times \Sigma \times \Lambda \to \mathbb{R}$ is given by

$$H(u, v; \sigma^2, \lambda) = H_\mathbb{S}(u; \sigma^2) \, H_\mathbb{T}(v; \lambda). \tag{19}$$

		Spatial domain $\mathbb{S}$	
		Continuous	Discrete
Spatial scale Σ	Continuous	Continuous Gaussian	Discrete Gaussian
	Discrete	+ Trunc. exp.	+ Binom. and geom. averaging

		Temporal domain $\mathbb{T}$	
		Continuous	Discrete
Temporal scale Λ	Continuous	——	Poisson kernel
	Discrete	Trunc. exp.	+ Binom. and geom. averaging

Figure 3: Scale-space kernels satisfying non-creation of local extrema with increasing scale in the cases of a continuous/discrete domain, a continuous/discrete scale parameter, and a spatial/temporal domain without or with preferred direction.

When implementing this operation in practice, the linearity implies that the spatial and temporal smoothing operators commute. For time-recursive temporal smoothing, it will therefore be more efficient to compute the spatial scale-space representation at the finest temporal scale, and then apply subsequent temporal smoothing to each spatial scale layer in this representation. If there is a common temporal smoothing component for all temporal scales (such as time-delayed binomial smoothing to reduce temporal aliasing due to poor temporal sampling), it will be computationally more efficient to apply such filters before constructing the spatial scale-space representation. Concerning temporal derivatives, it was shown that these can be computed by linear combinations of the temporal channels at each spatial scale. Before or after this step, finite difference operators can be applied to compute spatial derivative approximations (see figure 4).

In summary, this spatio-temporal scale-space concept leads to a visual front-end model, which at every time moment outputs a set of spatio-temporal derivatives at different spatio-temporal scales. Concerning time buffering, there is essentially no need for the visual front-end to represent the past in any other ways than as the temporal channels in the multi-time-scale representation. Hence, for two-dimensional image data, we obtain a visual front-end, which over time maintains a four-dimensional representation of the current (delayed) moment. This data set constitutes one time slice of the five-dimensional spatio-temporal representation of the complete history of the visual observer.

Figure 5 shows an example of multi-scale spatio-temporal image descriptors computed in this way. It shows second-order temporal derivatives computed from an image sequence for a number of different values of the spatial and temporal scale parameters. Observe how qualitatively different types of responses are obtained at the different spatio-temporal scales.

A more extensive treatment of this subject is presented in (Lindeberg 1996), including scale-space properties, necessity results and the non-separable case.

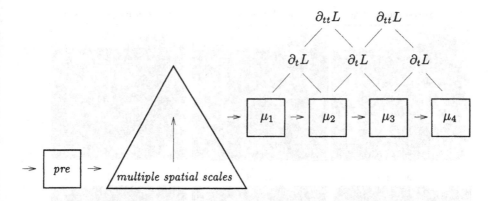

Figure 4: The composed architecture of the resulting spatio-temporal visual front-end consists of the following types of processing steps. (i) Optional temporal preprocessing. (ii) Spatial multi-scale representation, *e.g.* a pyramid or a scale-space representation. (iii) One set of recursive temporal smoothing stages associated with each spatial scale. (iv) Temporal derivatives from linear combinations of temporal channels. (v) Spatial derivative approximations from finite spatial differences (not shown in this figure).

6 Summary and discussion

We have presented a theory for how the linear scale-space concept can be extended to the temporal domain. The theory is complete in the sense that it provides a complete catalogue of all linear scale-space concepts that satisfy temporal causality in the cases of continuous *vs.* discrete time as well as continuous *vs.* discrete scale. Essentially, there are three main categories.

The construction started from similar scale-space axioms as have been used for deriving the uniqueness of the Gaussian kernel in the spatial domain, namely linearity, shift invariance, symmetry and non-creation of maxima (zero-crossings) with increasing scale. In the case of a continuous scale parameter, the latter assumption corresponds to a semi-group structure. Then, we replaced the symmetry condition by the essential requirement that the time direction should be treated as causal, and only what has occurred in the past may be used as input for computing representations at coarser time scales.

A kernel satisfying these properties was termed a time-causal scale-space kernel, and a complete classification was given for continuous and discrete time domains. For continuous time, the only primitive time-causal kernels are truncated exponential kernels corresponding to first-order integration over time. The discrete correspondences to these are geometric moving average kernels. In the discrete domain, also time-shifted binomial kernels satisfy temporal causality.

In the case of discrete time, and only in this case, there is a non-trivial time-causal semi-group structure. It corresponds to convolution with Poisson kernels, and can be regarded as the canonical model of a temporal scale-space, since it is the only time-causal scale-space having a continuous scale parameter.

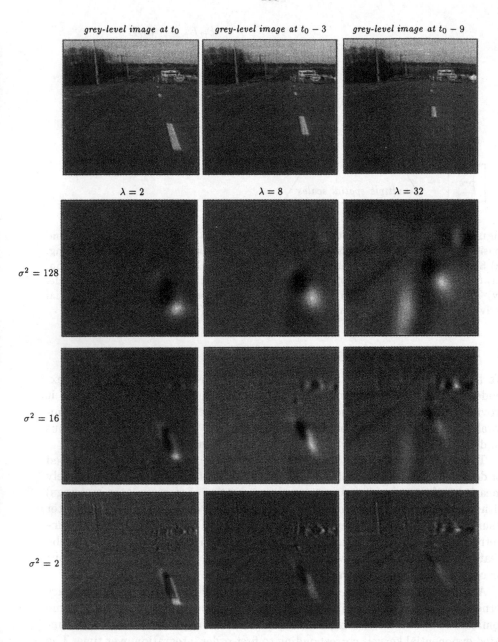

Figure 5: Second-order temporal scale-space derivatives computed for a few combinations of spatial scales and temporal scales. The top row shows three frames from the image sequence, whereas the following rows show spatio-temporal data for $\sigma^2 = 2$, 16 and 128 (from the bottom to the top) and $\lambda = 2$, 8 and 32 (from the left to the right).

We analysed derivative operators and derivative approximations with respect to their scale-space properties. Specifically, we made the important observation that the temporal channels themselves contain sufficient information for computing temporal derivatives at the current moment. Hence, there is no need for additional time buffering as would be needed if computing temporal derivative approximations by explicit finite differences.

More generally, the time recursive properties of the smoothing kernels corresponding to a discrete scale parameter imply that it is sufficient for the visual front-end to maintain a representation over time that corresponds to a four-dimensional slice of the entire five-dimensional spatio-temporal scale-space. This dimensionality reduction is of crucial importance, since it substantially reduces computational and hardware requirements.

An attractive property of the presented theory is that it leads to a conceptually very simple architecture (illustrated in figure 4) and allows for computationally highly efficient implementations. To update the temporal information to the next time moment (according to equation (10)) it is sufficient to perform one multiplication and two additions per pixel and spatio-temporal channel. Whereas recursive filters are common in signal processing and constitute a natural choice on an *ad hoc* basis, an important result of this treatment is that this design can be derived by necessity from first principles.

References

P. J. Burt. Fast filter transforms for image processing. *CVGIP*, 16:20–51, 1981.

J. L. Crowley. *A Representation for Visual Information*. PhD thesis. Carnegie-Mellon University, Robotics Institute, Pittsburgh, Pennsylvania, 1981.

R. Deriche. Using Canny's criteria to derive a recursively implemented optimal edge detector. *IJCV*, 1:167–187, 1987.

L. M. J. Florack. *The Syntactical Structure of Scalar Images*. PhD thesis. Dept. Med. Phys. Physics, Univ. Utrecht, NL-3508 Utrecht, Netherlands, 1993.

N. L. Johnson and S. Kotz. *Discrete Distributions*. Houghton Mifflin, Boston, 1969.

S. Karlin. *Total Positivity*. Stanford Univ. Press, 1968.

J. J. Koenderink and A. J. van Doorn. Generic neighborhood operators. *IEEE-PAMI*, 14(6):597–605, 1992.

J. J. Koenderink. The structure of images. *Biol. Cyb.*, 50:363–370, 1984.

J. J. Koenderink. Scale-time. *Biol. Cyb.*, 58:159–162, 1988.

T. Lindeberg. Scale-space for discrete signals. *IEEE-PAMI*, 12(3):234–254, 1990.

T. Lindeberg. *Scale-Space Theory in Computer Vision*. Kluwer, Netherlands, 1994.

T. Lindeberg. Linear spatio-temporal scale-space. in preparation, 1996.

T. Lindeberg and D. Fagerström. Scale-space with causal time direction. Tech. rep. ISRN KTH/NA/P–96/04–SE, NADA, KTH, Stockholm, Sweden, jan 1996.

I. J. Schoenberg. On smoothing operations and their generating functions. *Bull. Amer. Math. Soc.*, 59:199–230, 1953.

A. P. Witkin. Scale-space filtering. In *8th IJCAI*, pages 1019–1022, 1983.

A. L. Yuille and T. A. Poggio. Scaling theorems for zero-crossings. *IEEE-PAMI*, 8:15–25, 1986.

Tracing Crease Curves by Solving a System of Differential Equations

Antonio M. López, Joan Serrat

Computer Vision Center
Departamento de Informática, Edificio C
Universidad Autónoma de Barcelona,
08193 Bellaterra, Spain
tel: +34 3 5811863, Fax: +34 3 5811670
e–mail : {antonio | joans}@upisun1.uab.es

Abstract. Different kinds of digital images can be modelled as the sampling of a continuous surface, being described and analyzed through the extraction of geometric features from the underlying surface. Among them, ridges and valleys or, generically, creases, have deserved special interest. The computer vision community has been relying on different crease definitions, some of them equivalent. Although they are quite valuable in a number of applications, they usually do not correspond to the real creases of a topographic relief. These definitions give rise either to algorithms that label pixels as crease points, and then focus on the problem of grouping them into curves, or to operators whose outcome is a creaseness image. We draw our attention to the real crease definition for a landscape, due to Rudolf Rothe, which is based on the convergence of slopelines. They are computed by numerically solving a system of differential equations. Afterwards, we extract Rothe creases which are parts of slopelines where others converge, avoiding in such a way any pixel-grouping step. At the same time we compute a creaseness image according to this definition.

Key Words : ridge, valley, creaseness, slopelines, differential equation.

1 Introduction

In computer vision there is an image model underlying every image analysis method. One of these models considers images as the sampling of a manifold, such as a graphic surface in the case of two–dimensional images. Due to the fact that dealing with a surface as a whole is, in general, computationally infeasible, one must look for manifolds of lesser dimension confined in it having a visual meaning or serving as image descriptors. Among them, ridges and valleys or, generically, *creases*, have been widely used for several purposes. For instance, they are involved in the description and segmentation of medical images [3, 5], and digital elevation models [7, 14]. They have also been shown to be useful for 2D and 3D medical image registration [2, 13], since they correspond to anatomic features. Creases are also especially useful for range image processing [4, 11], because these images are true sampled surfaces.

In computer vision literature, we find diverse crease characterisations, some of them equivalent. In [1] we find a rather complete classification according to several criteria. In essence, we distinguish three main classes : implementations of definitions from differential geometry that 1) take into account or 2) not, the existence of a singled out direction, and 3) algorithmic or constructive definitions.

Even if all these definitions and their implementations are very useful in a number of applications, they share one or several of the following drawbacks :

1. None of them are the true creases in the sense of ridges and valleys of a topographic relief; that is, the loci of points where water gathers to run downhill, in the case of valleys, and similarly for ridges but with the same relief turned upside down.

2. Implementations usually do not compute curves, that is, sequences of adjacent pixel coordinates, but images where pixels are labelled as ridge, valley or background. Therefore, a further error prone process of grouping is necessary.

3. Crease operators give relatively high responses in the neighbourhood of a crease, so an additional decision rule (e.g. thresholding) must be applied, which commonly produces thick lines. These operators also tend to have high outputs at regions which exhibit high curvature but are not necessarily creases.

4. Some implementations rely on critical points like maxima, minima or saddle points which are quite dependent on noise. As a consequence, creases are unstable.

In this paper, we present a new method for crease extraction from two–dimensional images that overcomes all of the former problems. Moreover, this method is based on the true characterisation of creases in the topographic sense, due to Rudolf Rothe [12], which is discussed in detail in [8]. In order to compute Rothe creases, we first realize that curves lying on a surface can be formulated as the solution of a system of coupled Ordinary Differential Equations (ODEs). It is solvable by a standard numerical integration 'engine', which only needs to be 'fuelled' with values depending on surface derivatives. Some useful curves in computer vision such as parabolic curves, curves of mean curvature extrema, the classical ridge and valley curves and many others [9], can be expressed and computed in this way. Here, we shall focus on *slopelines*, because they converge towards Rothe creases. Therefore, we have devised a convergence assessment procedure for discrete curves.

This article is organized as follows. Section 2 reviews different crease definitions. In Sect. 3 we formulate curves on a parametric surface in $\mathbb{R}^3$ as the solution of a system of coupled ODEs. Next, in Sect. 4, we present an algorithm to extract Rothe creases from slopelines. Section 5 illustrates the results obtained. Finally, in Sect. 6 we discuss the conclusions and future work.

2 Different Crease Definitions

2.1 Definitions Independent of a Singled Out Direction

They are based on the principal curvatures of the surface which are independent of how the surface is embedded in the space. For instance, a crest line (ridge or valley) of a surface given in implicit form is defined in [13] as the extrema of the maximum principal curvature in absolute value k_{max}, in its principal direction v_{max}, that is, $\nabla k_{max} \cdot v_{max} = 0$ where ∇k_{max} is the gradient of k_{max}. Ridges are distinguished from valleys by the sign of the mean curvature k_M, which, conversely to the principal curvatures, is a non–directional quantity. It provides information about the concavity ($k_M > 0$, valley) or convexity ($k_M < 0$, ridge) of the surface. Actually, creases have been alternatively defined as the extrema of the mean curvature [1, 11].

These definitions are suitable for spaces where there is not a privileged direction. They have the advantage of rotational invariance because they are based on principal curvatures. Hence, they have also been used in the context of spaces with a privileged direction, e.g. for object recognition in range images [4, 11].

2.2 Creases Based on a Singled Out Direction

In this case we model images as a height function, being the height axis the singled out direction. In 2D images we have a function $I(x_1, x_2)$ where the singled out direction is most often the intensity axis. Now, we distinguish the crease characterisation due to De Saint–Venant from the one due to Rothe.

The De Saint–Venant condition. Several authors [2, 6] have taken *crease* points of a function $I(x_1, x_2)$ as the height extrema in the directions where their second directional derivative is also extreme. Creases have also been identified as the loci of curvature extrema of the level curves (isohypses, isophotes). This definition has given rise to several implementations [3, 7] and the same idea has been extended to three–dimensional implicit surfaces [13].

Actually, both characterisations are equivalent and correspond to the condition of creases given by De Saint–Venant [8] as the loci of extreme slope along a level curve. If we denote the first and second order partial derivatives of $I(x_1, x_2)$ by $I_{x_\alpha} = \partial I(x_1, x_2)/\partial x_\alpha$ and $I_{x_\alpha x_\beta} = \partial^2 I(x_1, x_2)/\partial x_\alpha \partial x_\beta$, respectively, the gradient of $I(x_1, x_2)$ will be $\nabla I = (I_{x_1}, I_{x_2})$, the vector orthogonal to it $\nabla_\perp I = (I_{x_2}, -I_{x_1})$ and the Hessian of $I(x_1, x_2)$

$$\nabla\nabla I = \begin{bmatrix} I_{x_1 x_1} & I_{x_1 x_2} \\ I_{x_1 x_2} & I_{x_2 x_2} \end{bmatrix} . \tag{1}$$

Then, taking the magnitude of the gradient as a slope measure, the De Saint–Venant condition is expressed as

$$\frac{\nabla I \cdot \nabla\nabla I \cdot \nabla_\perp I^T}{\| \nabla I \|^2} = 0 . \tag{2}$$

Rothe's characterisation. Koenderink and van Doorn saved from oblivion the fact that the former definition does not correspond to the intuitive notion of creases as steepest descent/ascent water paths in a landscape [8]. Breton de Champ proved that the only curves satisfying both conditions are confined to vertical planes. However, it is clear that in general this is not true. We owe to R. Rothe the right ridge and valley characterisation as parts of slopelines where other slopelines converge. These parts are also referred as special slopelines. The family of slopeline curves, also called creeplines or flowlines, is defined as being orthogonal to the family of level curves. That is to say, those following the gradient direction. We shall see in Sect. 4 that solving the slopeline differential equation

$$\nabla_\perp I \cdot d\mathbf{x} = 0 \tag{3}$$

for initial points spread all over the image, the computed curves converge to all the salient creases.

2.3 Algorithmic Definitions

They are based on critical points of the image seen as a landscape, namely, local maxima, minima and saddle points. Therefore, they also assume the existence of a privileged direction. Perhaps, the most well known algorithmic definition is the morphological watershed which computes regions of influence of local minima [14]. Each region is claimed to be a catch basin and their closed borders are identified as ridges or divide lines in the sense of a topographic relief. Another algorithmic definition consists in tracing curves that join critical points following the gradient direction [5]. These algorithms suffer from the instability of critical points, that may easily appear, disappear or move due to noise. In addition, they also fail to match the true creases, despite its undeniable usefulness.

In all these characterisations, crease points are classified as ridges or valleys, depending on the sign of the second directional derivative of the height function $I(x_1, x_2)$ in the direction orthogonal to the gradient :

$$I_{\nabla_\perp \nabla_\perp} = \frac{\nabla_\perp I \cdot \nabla\nabla I \cdot \nabla_\perp I^{\mathrm{T}}}{\| \nabla I \|^2} . \tag{4}$$

$I_{\nabla_\perp \nabla_\perp}$ is negative at convex regions (ridges) and positive at concave regions (valleys). As a matter of fact, there are convolution operators which approximate $I_{\nabla_\perp \nabla_\perp}$ at a given scale in such a way that points with an output magnitude above a certain threshold are considered creases [2], implementing in this way an approximation to the De Saint–Venant condition.

3 Curves on Surfaces as Coupled ODEs

Let us consider a parametric surface $s(x_1, x_2)$ in $\mathbb{R}^3$, and a curve $s(\mathbf{x}(t))$ lying on it, for $\mathbf{x}(t) = (x_1(t), x_2(t))$ on the plane of parametrisation $x_1 x_2$. $\mathbf{x}(t)$ is completely determined by a certain relationship between parameters x_1 and x_2. This relationship may take the form of an ODE

$$\mathbf{f} \cdot \mathbf{dx} = 0 \tag{5}$$

where $\mathbf{f} = (f_1(x_1, x_2), f_2(x_1, x_2))$, for certain functions f_1, f_2, and $\mathbf{dx} = (dx_1, dx_2)$. This is the case of (3), the slopeline ODE, where $f_1 = I_{x_2}$ and $f_2 = -I_{x_1}$.

To solve this equation means to find the integral curves of the vectorial field $\mathbf{w} = (f_2(x_1, x_2), -f_1(x_1, x_2))$, on the plane of parametrisation, orthogonal to the vectorial field $\mathbf{f}$. That is to say, to find curves $\mathbf{x}(t)$ which are the solution of the following system of coupled ODEs

$$d\mathbf{x}(t)/dt = (dx_1(t)/dt, dx_2(t)/dt) = \mathbf{w} \tag{6}$$

for *some unknown* parametrisation t. That is, the t that matches at each point $(x_1(t), x_2(t))$ the tangent of the curve with the direction, *and* the magnitude of $\mathbf{w}$. However, we know that two possible parametrisations are the arclength s of the curve $\mathbf{x}(t)$ lying on the $x_1 x_2$–plane and of the curve $\mathbf{s}(\mathbf{x}(t))$ lying on the surface. Then, we can write

$$\frac{d\mathbf{x}}{ds} = \pm \frac{\mathbf{w}}{\sqrt{\mathbf{w} \cdot \mathbf{G} \cdot \mathbf{w}^T}} \tag{7}$$

where $\mathbf{G}$ is in the first case the identity matrix and in the second case the matrix with the covariant components of the surface metric tensor, this is,

$$\mathbf{G} = \begin{bmatrix} 1 + I_{x_1} & I_{x_1} I_{x_2} \\ I_{x_1} I_{x_2} & 1 + I_{x_2} \end{bmatrix} . \tag{8}$$

We can solve (7) with a numerical integration method. In particular, we use the fourth order Runge–Kutta with adaptative step [10]. In order to start the numerical integration and obtain curve points, only an initial point and the component functions of $\mathbf{w}$, are needed.

In this paper, we are interested in curves on the plane of parametrisation because in two–dimensional images they are the projection of the corresponding curves on the surface. Hence, we should parametrise by the arclength of $\mathbf{x}(t)$, which has the advantage over the parametrisation by $\mathbf{s}(\mathbf{x}(t))$ that the solution curve runs faster because $\mathbf{w} \cdot \mathbf{w} \leq \mathbf{w} \cdot \mathbf{G} \cdot \mathbf{w}^T$. However, the integration method has a local error of order $\mathcal{O}(\Delta s^5)$ for an increment Δs of the parameter variable. Thus, a higher speed implies a greater local error in each step. This in turn may involve a computation overhead because the integration process can be compelled to try too many fractions of Δs in order not to exceed the allowed local error. Therefore, we have chosen the second parametrisation.

4 Extraction of Rothe Creases

We have seen that the characterisation of creases in a continuous landscape identifies them as (parts of) slopelines where other slopelines converge. In the discrete case, we consider that creases are formed by contiguous slopeline segments, where there is a high degree of convergence. More precisely, we shall consider that a set of sampled slopelines converge into a certain slopeline segment if all of them overlap in it. The more curves crowd together, the higher the convergence is. The central idea of the algorithm is to compute a large number of slopelines and store them as separate curves. Then for each one, the segments with high convergence are selected. Finally, redundant segments similar to and shorter than other selected segments are discarded. The algorithm has the following steps: 1) Slopeline extraction, 2) Formation of the creaseness image, and 3) Crease extraction.

Slopeline extraction. To determine how many slopelines have to be traced and from which starting points, we note that too few curves will cause a poor overlapping and therefore some creases will surely be missed, and that too many curves just slightly improves the result. Hence, we limit the number of computed curves to those which are needed in order to cover the whole image, making sure that at least one curve passes through each pixel. Thus, at any given time, a pixel is liable to be an initial point $\mathbf{x}(0)$ for (7) if no curve has visited it yet. By following this rule, we are able to find the main and wider creases in several kinds of images. However, in images with small details, sometimes narrow creases are missed. We have overcome this problem by working at subpixel resolution. This means that, if it is necessary, slopelines are sampled not at integer coordinates but at a finer resolution r, being for instance $r = 2$ for double resolution. Values of $r = 2, 3$ are sufficient to produce good results in these cases, as we shall see in Sect. 5.

To integrate the ODEs system of the slopelines numerically, the values of the first partial derivatives of the image are required at points with integer coordinates but also at points in between. We approximate the derivative of the image at a certain scale σ, at which creases are extracted, by the convolution with the sampled derivative of a bidimensional Gaussian of variance σ^2. For first order derivatives, $I_{x_\alpha}(x_1, x_2; \sigma) \approx I(x_1, x_2) * G_{x_\alpha}(x_1, x_2; \sigma)$, $\alpha = 1, 2$. Note that this equation applies only to pixel coordinates, that is, both x_1 and x_2 are integers. Elsewhere, we approximate subpixel derivatives evaluating the derivatives of an algebraic polinomial, obtained by a bicubic interpolation [10] from the four nearest pixels.

Formation of the creaseness image. The convergence of curves in the continuous space is interpreted as the overlapping of sampled curves and the discrete ones. Thus, we need to count, for each point of the image domain, how much slopelines pass through it. This process must be done at resolution r, therefore in a matrix r^2 times larger than the original image. This matrix has an interesting meaning: it is a *creaseness* image, analogously to images produced by crease operators like $I_{\nabla_{\perp}\nabla_{\perp}}$ (Fig. 1e), though corresponding to Rothe creases (Fig. 1d).

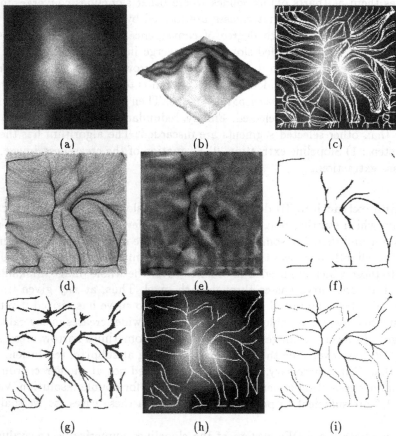

Fig. 1. (a) Heart scintigraphic image. (b) Image seen as a landscape. (c) Several slope-lines for $r = 3$ and $\sigma = 1.0$ pixels. (d) Creaseness image where darker grey–levels denote greater accumulation. (e) Operator $I_{\nabla_{\perp}\nabla_{\perp}}$ applied to (a), displaying convexity (dark) and concavity (bright). (f) Thresholding of (d) showing regions of greater accumulation. (g) Selected slopeline segments for $L = T = 4$. (h) Special slopelines for $P = 0.95$. (g) Classification into ridges (black) and valleys (grey) according to the sign of $I_{\nabla_{\perp}\nabla_{\perp}}$.

Crease extraction. Creases are segments of certain slopelines where there is a high convergence. Hence, we select those segments from the stored slopelines where accumulation at each point is greater than a threshold T. We impose a further condition in order to avoid small segments, most probably due to the sampling of slopelines which do not converge but just get closer than $1/r$ pixels and then run far away: those segments must be longer than rL, for another threshold parameter L.

Finally, if we look at the image of segments passing the two conditions, it displays bunches of segments, not isolated curves (Fig. 1g). The reason is precisely that they group because of their high accumulation one over the others, thus fulfilling the two conditions above. In order to discard redundant segments, we apply the following rule : given two segments, we eliminate the shorter one if it overlaps the longer one in more than a certain length fraction P.

The results are not very sensitive to parameters T, L and P. We mean that they are just minimum values intended to get rid of several abnormal cases like very short or very weak creases. Typical values are $T = 3, L = 3$ and $P = 0.9$. The increase of T and L, or the decrease of P, do not change substantially the final result.

5   Results

We now present the results obtained for Rothe's definition of ridge and valley in a landscape. We shall illustrate its utility in the context of two applications on different medical image modalities: coronary arteriography and brain MRI. In addition, results obtained for two range images are also presented.

In coronary arteriography the goal is to delineate its vessels. We have detected them searching the ridges of the image. In fact, thin vessels *are* ridges. Figure 2b shows the extracted ridges superimposed to the image at the same scale σ for which they were computed.

The second application is the extraction of salient features to be used in the registration of two MR or MR and CT brain images. In Fig. 3b we see the ridges and valleys that have been extracted by our algorithm. They can be used to calculate the geometric transform between a pair of images of the same slice, by means of some process of curve matching. Alternatively, it is possible to obtain the registration transform from the correlation of the creaseness images (Fig. 3c) of the two slices as it is done in [2] with the $I_{\nabla_\perp \nabla_\perp}$ operator (Fig. 3d).

On the other hand, we have experimented with our algorithm in some range images. Figure 4b shows the creases of the simple range image of a block. They perfectly match its roof borders. We have also tested the algorithm with a Digital Elevation Model image (Fig. 4c), which is a real topographic relief.

In the previous examples, such as in Fig. 2b, we observe short disconnected segments that could be joined in order to produce longer, perceptually better crease curves. This fragmentation is caused by the saddle points of the image, that, by definition, slopelines can not reach nor cross. Formally, there is nothing wrong with this, but in many applications it should be convenient to link these segments.

The computation times depend mainly on the image size and the resolution r. For instance, in a SPARC 10 computer, the creases of Fig. 1a which is 64×64 pixels, have been computed in 1 minute at a resolution of $r = 3$ whereas for $r = 1$ the time is 15 seconds.

6 Conclusions and Future Work

In this paper we proposed a new method to extract ridges and valleys or, generically, creases, of a landscape according to the correct definition by Rothe. We have developed an algorithm which obtains these creases, first covering the whole image with slopelines and then extracting from them the segments where they converge. In this way, we obtain creases not as a binary image but as a set of curves, avoiding a further grouping step. At the same time our algorithm extracts a creaseness image, according to this definition.

As a future work we will join Rothe creases through saddle points, because it is interesting for most applications. Finally, we will study the evolution of Rothe creases in scale space.

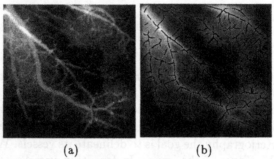

(a)   (b)

Fig. 2. (a) Coronary arteriography. (b) Ridges for $\sigma = 3.0$ pixels, $r = 1$

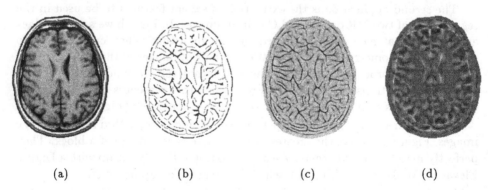

(a)   (b)   (c)   (d)

Fig. 3. (a) Brain MRI (axial slice). (b) Ridges (black) and valleys (grey) for $\sigma = 3.0$ pixels, $r = 1$ (c) Creaseness image from Rothe's definition. (d) Response to the $I_{\nabla_\perp \nabla_\perp}$ operator at the same scale.

References

1. D. Eberly, R. Gardner, B. Morse, S. Pizer, C. Scharlach. *Ridges for Image Analysis. Journal of Mathematical Imaging and Vision*, **4**, N. 4, p. 353–373, 1994
2. P. A. Van den Elsen, J. B. A. Maintz, E. D. Pol, M. A. Viergever. *Automatic Registration of CT and MR Brain Images Using Correlation of Geometrical Features. IEEE Trans. on Medical Imaging*, **14**, N. 2, p. 384–396, June 1995.

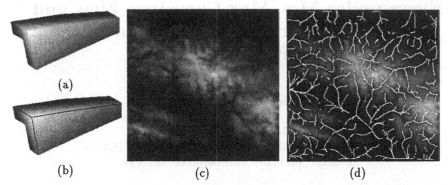

Fig. 4. (a) Range image of a block. (b) Rothe ridges for $\sigma = 1.5$ pixels, $r = 1$. (c) Digital Elevation Model fragment of a central region of Haiti. (d) Valleys for $\sigma = 1.0$ pixels, $r = 2$.

3. J. M. Gauch, S. M. Pizer. *Multiresolution analysis of ridges and valleys in grey-scale image. IEEE Trans. on Pattern Analysis and Machine Intelligence*, 15, N. 6, p. 635–646, June 1993.
4. G. G. Gordon. *Face Recognition based on Maps and Surface Curvature.* In SPIE **1570** Geometric Methods in Computer Vision, p. 234–247, 1991.
5. L. D. Griffin, A. C. F. Colchester, G. P. Robinson. *Scale and Segmentation of Grey-Level Images using Maximum Gradient Paths.* In Information Processing in Medical Imaging (IPMI), p. 256–272, July 1991. Lecture Notes in Computer Science, **511**, Springer-Verlag.
6. R. M. Haralik, L. G. Shapiro. *Computer and Robot Vision* (1). Addison-Wesley, 1992.
7. I. S. Kweon, T. Kanade. *Extracting Topographic Terrain Features from Elevation Maps.* CVGIP-Image Understanding, 59, N. 2, p. 171–182, March 1994.
8. J. J. Koenderink, A. J. van Doorn. *Local Features of smooth Shapes: Ridges and Courses.* In SPIE **2031** Geometric Methods in Computer Vision II, p. 2–13, 1993.
9. A. M. Lopez, J. Scrrat. *Image Analysis Through Surface Geometric Descriptors.* In Proceedings of the VI Spanish Symposium on Pattern Recognition and Image Analysis, p. 35–42, April 1995.
10. W. H. Press, S. A. Teukolsky, W. T. Vetterling, B. P. Flannery. *Numerical Recipes in C*, 2nd Edition Cambridge University Press, 1992.
11. S. Panikanti, C. Dorai, A. K. Jain. *Robust feature detection for 3D object recognition and matching.* In SPIE **2031** Geometric Methods in Compuetr Vision, p. 366–377, 1993.
12. R. Rothe. *Zum Problem des Talwegs. Sitzungsber. Berliner Math. Gesellschaft,* 14, p. 51–69, 1915.
13. J. P. Thirion, A. Gourdon. *Computing the Differential Characteristics of Isointensity Surfaces. CVGIP-Image Understanding*, 61, N. 2, p. 190–202, March 1995.
14. L. Vincent, P. Soille. *Watersheds in Digital Spaces: An Efficient Algorithm Based on Immersion Simulations. IEEE Trans. on Pattern Analysis and Machine Intelligence*, 13, N. 6, p. 583–598, June 1991.

Flows under Min/Max Curvature Flow and Mean Curvature: Applications in Image Processing*

R. Malladi** and J. A. Sethian

Lawrence Berkeley National Laboratory
University of California, Berkeley, CA 94720, USA
e-mail: {malladi,sethian}@csr.lbl.gov

Abstract. We present a class of PDE-based algorithms suitable for a wide range of image processing applications. The techniques are applicable to both salt-and-pepper grey-scale noise and full-image continuous noise present in black and white images, grey-scale images, texture images and color images. At the core, the techniques rely on a level set formulation of evolving curves and surfaces and the viscosity in profile evolution. Essentially, the method consists of moving the isointensity contours in a image under curvature dependent speed laws to achieve enhancement. Compared to existing techniques, our approach has several distinct advantages. First, it contains only one enhancement parameter, which in most cases is automatically chosen. Second, the scheme automatically stops smoothing at some optimal point; continued application of the scheme produces no further change. Third, the method is one of the fastest possible schemes based on a curvature-controlled approach.

1 INTRODUCTION

The essential idea in image smoothing is to filter noise present in the image signal without sacrificing the useful detail. In contrast, image enhancement focuses on preferentially highlighting certain image features. Together, they are precursors to many low level vision procedures such as edge finding [11, 2], shape segmentation, and shape representation [9, 10, 7]. In this paper, we present a method for image smoothing and enhancement which is a variant of the geometric heat equation. This technique is based on a min/max switch which controls the form of the application of the geometric heat equation, selecting either flow by the positive part of the curvature or the negative part, based on a local decision. This approach has several key virtues. First, it contains only one enhancement parameter, which it most cases is automatically chosen. Second, the scheme

---

* Supported in part by the Applied Mathematics Subprogram of the Office of Energy Research under DE-AC03-76SF00098, and the National Science Foundation DARPA under grant DMS-8919074.
** Supported in part by the NSF Postdoctoral Fellowship in Computational Science and Engineering

automatically picks the stopping criteria; continued application of the scheme produces no further change. Third, the method is one of the fastest possible schemes based on a curvature-controlled approach.

Traditionally, both 1-D and 2-D signals are smoothed by convolving them with a Gaussian kernel; the degree of blurring is controlled by the characteristic width of the Gaussian filter. Since the Gaussian kernel is an isotropic operator, it smooths across the region boundaries thereby compromising their spatial position. As an alternative, Perona and Malik [13] have used an anisotropic diffusion process which performs intraregion smoothing in preference to interregion smoothing. A significant advancement was made by Alvarez, Lions, and Morel (ALM) [1], who presented a comprehensive model for image smoothing.

The ALM model consists of solving an equation of the form

$$I_t = g(|\nabla G * I|) \, \kappa \, |\nabla I|, \quad \text{with} \quad I(x,y,t=0) = I_0(x,y), \tag{1}$$

where $G * I$ denotes the image convolved with a Gaussian filter. The geometric interpretation of the above diffusion equation is that the isointensity contours of the image move with speed $g(|\nabla G * I|)\kappa$, where $\kappa = \text{div} \frac{\nabla I}{|\nabla I|}$ is the local curvature. One variation of this scheme comes from replacing the curvature term with its affine invariant version (see Sapiro and Tannenbaum [15]). By flowing the isointensity contours normal to themselves, smoothing is performed perpendicular to edges thereby retaining edge definition. At the core of both numerical techniques is the Osher-Sethian level set algorithm for flowing the isointensity contours; this technique was also used in related work by Rudin, Osher and Fatemi [14].

In this work, we return to the original curvature flow equation, namely $I_t = F(\kappa) \mid \nabla I \mid$, and Osher-Sethian [12, 17] level set algorithm and build a numerical scheme for image enhancement based on a automatic switch function that controls the motion of the level sets in the following way. Diffusion is controlled by flowing under $\max(\kappa, 0)$ and $\min(\kappa, 0)$. The selection between these two types of flows is based on local intensity and gradient. The resulting technique is an automatic, extremely robust, computationally efficient, and a straightforward scheme.

To motivate this approach, we begin by discussing curvature motion, and then develop the complete model which includes image enhancement as well. The crucial ideas on min/max flows upon which this paper is based have been reported earlier by the authors in [5]; more details and applications in textured and color image denoising may be found in Malladi and Sethian [6]. The outline of this paper is as follows. First, in Section II, we study the motion of a curve moving under its curvature, and develop an automatic stopping criteria. Next, in Section III, we apply this technique to enhancing binary and grey-scale images that are corrupted with various kinds of noise.

2 MOTION OF CURVES UNDER CURVATURE

Consider a closed, nonintersecting curve in the plane moving with speed $F(\kappa)$ normal to itself. More precisely, let $\gamma(0)$ be a smooth, closed initial curve in R^2, and let $\gamma(t)$ be the one-parameter family of curves generated by moving $\gamma(0)$ along its normal vector field with speed $F(\kappa)$. Here, $F(\kappa)$ is a given scalar function of the curvature κ. Thus, $n \cdot x_t = F(\kappa)$, where x is the position vector of the curve, t is time, and n is the unit normal to the curve. For a specific speed function, namely $F(\kappa) = -\kappa$, it can be shown that an arbitrary closed curve (see Gage, [3] Grayson [4]) collapses to a single point.

2.1 The Min/Max flow

We now modify the above flow. Motivated by work on level set methods applied to grid generation [18] and shape recognition [7], we consider two flows, namely $F(\kappa) = \min(\kappa, 0.0)$ and $F(\kappa) = \max(\kappa, 0.0)$. As shown in Figure 1, the effect of flow under $F(\kappa) = \min(\kappa, 0.0)$ is to allow the inward concave fingers to grow outwards, while suppressing the motion of the outward convex regions. Thus, the motion halts as soon as the convex hull is obtained. Conversely, the effect of flow under $F(\kappa) = \max(\kappa, 0.0)$ is to allow the outward regions to grow inwards while suppressing the motion of the inward concave regions. However, once the shape becomes fully convex, the curvature is always positive and hence the flow becomes the same as regular curvature flow; hence the shape collapses to a point. We can summarize the above by saying that, for the above case, flow under $F = \min(\kappa, 0.0)$ preserves some of the structure of the curve, while flow under $F = \max(\kappa, 0.0)$ completely diffuses away all of the information.

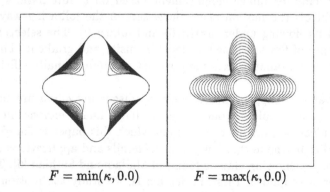

$$F = \min(\kappa, 0.0) \qquad F = \max(\kappa, 0.0)$$

Fig. 1. Motion of a curve under Min/Max flow

Here, we have evolved the curve using the Osher-Sethian level set method, see [12], which grew out of earlier by Sethian [16] on the mathematical formulation of curve and surface motion. Briefly, this technique works as follows. Given a moving closed hypersurface $\Gamma(t)$, that is, $\Gamma(t = 0) : [0, \infty) \to R^N$, we wish to produce an Eulerian formulation for the motion of the hypersurface propagating along its normal direction with speed F, where F can be a function of various arguments, including the curvature, normal direction, e.t.c. The main idea is to embed this propagating interface as the zero level set of a higher dimensional function ϕ. Let $\phi(x, t = 0)$, where $x \in R^N$ is defined by

$$\phi(x, t = 0) = \pm d \tag{2}$$

where d is the distance from x to $\Gamma(t = 0)$, and the plus (minus) sign is chosen if the point x is outside (inside) the initial hypersurface $\Gamma(t = 0)$. Thus, we have an initial function $\phi(x, t = 0) : R^N \to R$ with the property that

$$\Gamma(t = 0) = (x|\phi(x, t = 0) = 0) \tag{3}$$

It can easily be shown that the equation of motion given by

$$\phi_t + F|\nabla\phi| = 0 \tag{4}$$

$$\phi(x, t = 0) \quad given \tag{5}$$

is such that the evolution of the zero level set of ϕ always corresponds to the motion of the initial hypersurface under the given speed function F.

Consider now the square with notches on each side shown in Figure 2a. We let the color black correspond to the "inside" where $\phi < 0$ and the white correspond to the "outside" where $\phi > 0$. We imagine that the notches are one unit wide, where a unit most typically will correspond to a pixel width. Our goal is to use the above flow to somehow remove the notches which protrude out from the sides. In Figure 2b, we see the effect of curvature flow; the notches are removed, but the shape is fully diffused. In Figure 2c, we see the effect of flow with speed $F = \min(\kappa, 0.0)$; here, one set of notches are removed, but the other set have been replaced by their convex hull. If we run this flow forever, the figure will not change since the convex hull has been obtained, which does not move under this flow. Conversely, as shown in Figure 2d, obtained with speed $F = \max(\kappa, 0.0)$, the inner notches stay fixed and the front moves in around them, while the outer notches are diffused. Continual application of this flow causes the shape to shrink and collapse. If the roles of black and white in the figure are reversed, so are the effects of min and the max flow.

The problem is that in some places, the notch is "outwards", and in others, the notch is "inwards". Our goal is a flow which somehow chooses the correct choice of flows between $F = \max(\kappa, 0.0)$ and $F = \min(\kappa, 0.0)$. The solution lies in a switch function which determines the nature of the notch.

2.2 The switch

In this section, we present the switch function to flow the above shape. Our construction of a switch is motivated by the idea of comparing the value of a function with its value in a ball around the function. Thus, imagine the simplest case, namely that of a black and white image, in which black is given the

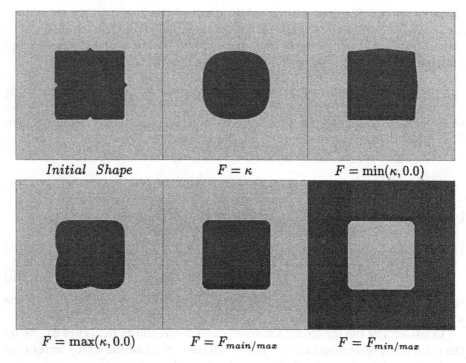

Fig. 2. Motion of notched region under various flows

value $\phi = -1$ and white given the value $\phi = 1$. We select between the two flows based on the sign of the deviation from the mean value theorem. Define $Average(x, y)$ as the average value of the image intensity $I(x, y)$ in a square centered around the point (x, y) with sidelength $(2. * StencilWidth + 1)$, where, for now $StencilWidth = 0$. Then, at any point (x, y), define the flow by

$$F_{min/max} = \begin{cases} \min(\kappa, 0) & \text{if } Average(x, y) < 0 \\ \max(\kappa, 0) & \text{otherwise} \end{cases} \qquad (6)$$

Here, we view 0 as the "threshold" value $T_{threshold}$; since it is halfway between the black value of -1 and the white value of 1. This flow can be seen to thus choose the "correct" flow between the min flow and the max flow. As a demonstration, in Figure 2e, we show the result of using the min/max flow given in Eqn. 6 on Figure 2a. To verify that our scheme is independent of the positioning of the colors, we reverse the initial colors and show the results of the *same* min/max flow in Figure 2f. What happens is that the small-scale "noise" is removed; once this happens, the boundary achieves a final state which does not change and preserves structures larger than the one-pixel wide noise.

We note that the level of noise removed is a function of the size of the stencil used in computing the switch in the min/max speed. What remains are structures than are not detected by our threshold stencil. Thus, the stencil size is the single parameter that determines the flow and hence the noise removal capabilities. We view this as a natural and automatic choice of the stencil, since it is given by the pixel refinement of the image. However, for a given pixel size, one can choose a larger stencil to exact noise removal on a larger scale; that is, we can choose to remove the next *larger* level of noise structures by increasing the size of our threshold stencil by computing the average $Average(x, y)$ over a larger square. We then use this larger stencil and continue the process by running the min/max flow. We have done this in Figure 3; we start with an initial shape in Figure 3a which has "noise" on the boundary. We then perform the min/max flow until steady-state is achieved with stencil size zero in Figure 3b; that is, the "average" consists only of the value of ϕ at the point (x, y) itself. We note that when we choose a stencil size of zero, nothing happens; see Malladi and Sethian [6] for details. In Figure 3c, we perform the min/max flow until steady-state is achieved with a stencil size of 1, and the continue min/max flow with a larger stencil until steady-state is again achieved in Figure 3d. As the stencil size is increased, larger and larger structures are removed. We can summarize our results as follows:

1. The single min/max flow selects the correct motion to diffuse the small-scale pixel notches into the boundary.

2. The larger, global properties of the shape are maintained.

3. Furthermore, and equally importantly, the flow stops once these notches are diffused into the main structure.

4. Edge definition is maintained, and, in some global sense, the area inside the boundary is roughly preserved up to the order of the smoothing.

5. The noise removal capabilities of the min/max flow is scale-dependent, and can be hierarchically adjusted.

6. The scheme requires only a nearest neighbor stencil evaluation.

The above min/max flow switch is, in fact, remarkably subtle in what it does. It works because of three reasons:

- First, the embedding of a front as a level set allows us to use information about neighboring level sets to determine whether to use the min flow or the max flow.

- Second, the level set method allows the construction of barrier masks to thwart motion of the level sets.

- Third, the discretization of the problem onto a grid allows one to select a natural scale to the problem.

Interested reader is referred to Malladi and Sethian [6] for a detailed explanation of the above issues.

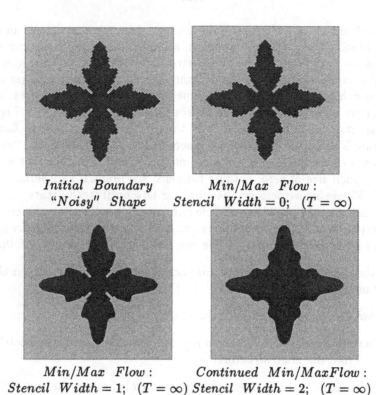

Initial Boundary
"Noisy" Shape

Min/Max Flow :
Stencil Width = 0; (T = ∞)

Min/Max Flow :
Stencil Width = 1; (T = ∞)

Continued Min/MaxFlow :
Stencil Width = 2; (T = ∞)

Fig. 3. Motion of a StarShaped region with noise under Min/Max flow at various stencil levels

3 APPLICATIONS

3.1 Application of Min/Max flows to binary images

We now apply our scheme given by Eqn. 6 to the problem of binary images with noise. Since we are looking at black and white images, where 0 corresponds to black and 255 to white, the threshold value $T_{threshold}$ is taken as 127.5 rather than 0. In Figures 5a & c, we add noise to a black and white image of a handwritten character. The noise is added as follows; 50% noise means that at 50% of the pixels, we replace the given value with a number chosen with uniform distribution between 0 and 255. Thus, a full spectrum of gray noise is added to the original binary image. In Figures 5b & d, we show the reconstructed images and stress that the results have converged and continued application of the scheme does not change anything.

3.2 Grey-scale images: Min/Max flows and scale-dependent noise removal

Imagine a grey-scale image; for example, two concentric rings of differing grey values. Choosing a threshold value of 127.5 is clearly inappropriate, since the value "between" the two rings may not straddle the value of 127.5, as it would it an original binary image. Instead, our goal is to locally construct an appropriate thresholding value. We follow the philosophy of the algorithm for binary images.

Imagine a grey scale image, such as the two concentric rings, in which the inner ring is slightly darker then the exterior ring; here, we interpret this as ϕ being more negative in the interior ring than the exterior. Furthermore, imagine a slight notch protruding outwards into the lighter ring, (see Figure 4). Our goal is to decide whether the area within the notch belongs to the lighter region, that is, whether it is a perturbation that should be suppressed and "reabsorbed" in to the appropriate background color. We determine this by first computing the average value of the intensity ϕ in the neighborhood around the point. We then must determine a comparison value which indicates the "background" value. We do so by computing a threshold $T_{threshold}$, defined as the average value of the intensity obtained in the direction perpendicular to the gradient direction. Note that since the direction perpendicular to the gradient is tangent to the isointensity contour through (x, y), the two points used to compute are either in the same region, or the point (x, y) is an inflection point, in which the curvature is in fact zero and the min/max flow will always yield zero.

Formally then,

$$F_{min/max} = \begin{cases} \max(\kappa, 0) \text{ if } Average(x,y) < T_{threshold} \\ \min(\kappa, 0) \text{ otherwise} \end{cases} \qquad (7)$$

This has the following effect. Imagine again our case of a grey disk on a lighter grey background, where the darker grey corresponds to a smaller value of ϕ than the lighter grey. When the threshold is larger than the average, the max is selected, and the level curves move in. However, as soon as the average becomes larger, the min switch takes over, and the flow stops. The arguments are similar to the ones given in the binary case.

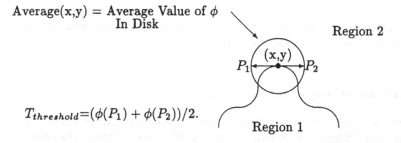

Average(x,y) = Average Value of ϕ In Disk

Region 2

$T_{threshold}=(\phi(P_1) + \phi(P_2))/2.$

Region 1

Fig. 4. Threshold test for Min/Max flow

Now we use this scheme to remove salt-and-pepper gray-scale noise from a grey-scale image. Once again, we add noise to the figure by replacing $X\%$ of the pixels with a new value, chosen from a uniform random distribution between 0 and 255, Our results are obtained as follows. Figure 5e shows an image where 25% of the pixels are corrupted with noise. We first use the min/max flow from Eqn.7 until a steady-state is reached (Figure 5f). This removes most of the noise. We then continue with a larger stencil for the threshold to remove further noise (Figure 5g). For the larger stencil, we compute the average $Average(x, y)$ over a larger disk, and compute the threshold value $T_{threshold}$ using a correspondingly longer tangent vector.

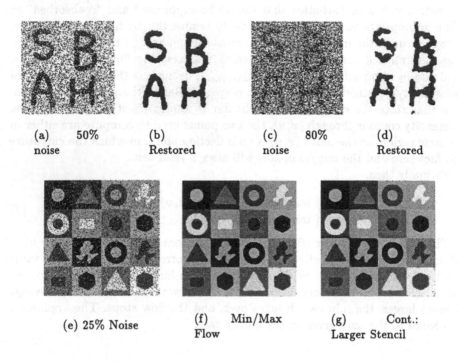

(a)   50%   (b)   (c)   80%   (d)
noise   Restored   noise   Restored

(e) 25% Noise   (f)   Min/Max   (g)   Cont.:
                      Flow   Larger Stencil

Fig. 5. Image restoration using min/max flow of binary and grey-scale images corrupted with grey-scale salt-and-pepper noise

3.3 Selective smoothing of medical images

In certain cases, one may want to remove some level of detail in an image; for example, in medical imaging, in which a low level of noise or image gradient is undesired, and the goal is enhancement of features delineated by large gradients. In this case, a simple modification of our min/max flow can achieve good results. We begin by defining the mean curvature of the image when viewed as a graph; that is, let

$$M = \frac{(1 + I_{xx})I_y^2 - 2I_x I_y I_{xy} + (1 + I_{yy})I_x^2}{(1 + I_x^2 + I_y^2)^{3/2}} \tag{8}$$

be the mean curvature. If we flow the image according to its mean curvature, i.e.,

$$I_t = M(1 + I_x^2 + I_y^2)^{1/2} \tag{9}$$

this will smooth the image. Thus, given a user-defined threshold $V_{gradient}$ based on the local gradient magnitude, we use the following flow to selectively smooth the image:

$$F_{min/max/smoothing} = \begin{cases} M & \text{if } | \nabla I | < V_{gradient} \\ \text{min/max flow} & \text{otherwise} \end{cases} \tag{10}$$

Thus, below a prescribed level based on the gradient, we smooth the image using flow by mean curvature; above that level, we use our standard min/max flow. Other choices for the smoothing flow include isotropic diffusion and curvature flow. We have had the most success with mean curvature flow; isotropic diffusion is too sensitive to variations in the threshold value $V_{gradient}$, since edges just below that value are diffused away, while edges are preserved in mean curvature flow. Our choice of mean curvature flow over standard curvature flow is because mean curvature flow seems to perform smoothing in the selected region somewhat faster. This is an empirical statement rather than one based on a strict proof.

In Figure 6, we show results of this scheme (Eqn.10) applied to a digital subtraction angiogram (DSA). In Figure 6a, we show the original image. In Figure 6b, we show the steady-state min/max flow image. In Figure 6c, we show the steady-state obtained with min/max flow coupled to mean curvature flow in the lower gradient range.

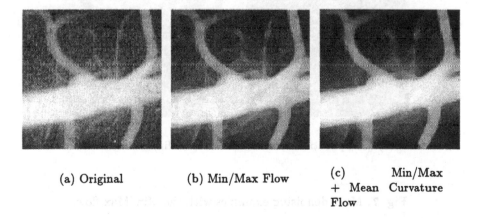

| (a) Original | (b) Min/Max Flow | (c)   Min/Max + Mean Curvature Flow |

Fig. 6. Min/Max flow with selective smoothing: The left image is the original. The center image is the steady-state of min/max flow. The right image is the steady-state of the min/max flow together with mean curvature flow in lower gradient range.

3.4 Additional examples

In this section, we present further images which are enhanced by means of our min/max flows. We begin with a medical image in Figure 7a; here, no noise is artificially added, and instead our goal is to enhance certain features within the given images and make them aminable to further processing like shape finding [9, 10, 8].

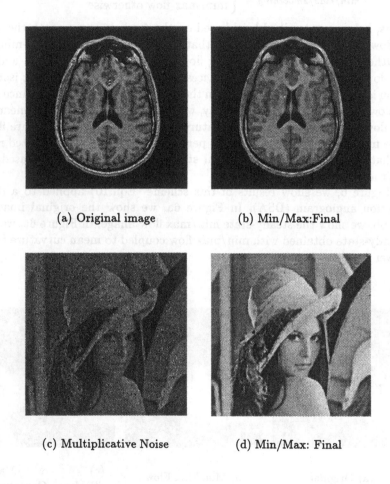

(a) Original image   (b) Min/Max:Final

(c) Multiplicative Noise   (d) Min/Max: Final

Fig. 7. More denoising examples with the Min/Max flow.

Next, we study the effect of our min/max scheme on multiplicative noise added to a grey-scale image. In Figure 7c & d, we show the reconstruction of an image with 15% multiplicative noise. Finally, interested reader is referred to Malladi and Sethian [6] for examples enhancing both gray scale and color images corrupted with Gaussian noise.

References

1. L. Alvarez, P. L. Lions, and J. M. Morel, "Image selective smoothing and edge detection by nonlinear diffusion. II," *SIAM Journal on Numerical Analysis*, Vol. 29(3), pp. 845–866, 1992.

2. J. Canny, "A computational approach to edge detection," *IEEE Trans. Pattern Analysis and Machine Intelligence*, Vol. PAMI-8, pp. 679–698, 1986.

3. M. Gage, "Curve shortening makes convex curves circular," *Inventiones Mathematica*, Vol. 76, pp. 357, 1984.

4. M. Grayson, "The heat equation shrinks embedded plane curves to round points," *J. Diff. Geom.*, Vol. 26, pp. 285–314, 1987.

5. R. Malladi and J. A. Sethian, "Image processing via level set curvature flow," *Proc. Natl. Acad. of Sci., USA*, Vol. 92, pp. 7046–7050, July 1995.

6. R. Malladi and J. A. Sethian, "Image processing: Flows under min/max curvature and mean curvature," to appear in *Graphical Models and Image Processing*, March 1996.

7. R. Malladi and J. A. Sethian, "A unified approach for shape segmentation, representation, and recognition," Report LBL-36069, Lawrence Berkeley Laboratory, University of California, Berkeley, August 1994.

8. R. Malladi, D. Adalsteinsson, and J. A. Sethian, "Fast method for 3D shape recovery using level sets," submitted.

9. R. Malladi, J. A. Sethian, and B. C. Vemuri, "Evolutionary fronts for topology-independent shape modeling and recovery," in *Proceedings of Third European Conference on Computer Vision*, LNCS Vol. 800, pp. 3–13, Stockholm, Sweden, May 1994.

10. R. Malladi, J. A. Sethian, and B. C. Vemuri, "Shape modeling with front propagation: A level set approach," *IEEE Trans. on Pattern Analysis and Machine Intelligence*, Vol. 17(2), pp. 158–175, Feb. 1995.

11. D. Marr and E. Hildreth, "A theory of edge detection," *Proc. of Royal Soc. (London)*, Vol. B207, pp. 187–217, 1980.

12. S. Osher and J. A. Sethian, "Fronts propagating with curvature dependent speed: Algorithms based on Hamilton-Jacobi formulation," *Journal of Computational Physics*, Vol. 79, pp. 12–49, 1988.

13. P. Perona and J. Malik, "Scale-space and edge detection using anisotropic diffusion," *IEEE Trans. Pattern Analysis and Machine Intelligence*, Vol. 12(7), pp. 629–639, July 1990.

14. L. Rudin, S. Osher, and E. Fatemi, "Nonlinear total variation based noise removal algorithms," *Modelisations Matematiques pour le traitement d'images*, INRIA, pp. 149–179, 1992.

15. G. Sapiro and A. Tannenbaum, "Image smoothing based on affine invariant flow," *Proc. of the Conference on Information Sciences and Systems*, Johns Hopkins University, March 1993.

16. J. A. Sethian, "Curvature and the evolution of fronts," *Commun. in Mathematical Physics*, Vol. 101, pp. 487–499, 1985.

17. J. A. Sethian, "Numerical algorithms for propagating interfaces: Hamilton-Jacobi equations and conservation laws," *Journal of Differential Geometry*, Vol. 31, pp. 131–161, 1990.

18. J. A. Sethian, "Curvature flow and entropy conditions applied to grid generation," *Journal of Computational Physics*, Vol. 115, No. 2, pp. 440–454, 1994.

Decomposition of the Hough Transform: Curve Detection with Efficient Error Propagation

Clark F. Olson

Department of Computer Science, Cornell University, Ithaca, NY 14853, USA

Abstract. This paper describes techniques to perform fast and accurate curve detection using a variant of the Hough transform. We show that the Hough transform can be decomposed into small subproblems that examine only a subset of the parameter space. Each subproblem considers only those curves that pass through some small subset of the data points. This property allows the efficient implementation of the Hough transform with respect to both time and space, and allows the careful propagation of the effects of localization error in the detection process. The use of randomization yields an $O(n)$ worst-case computational complexity for this method, where n is the number of data points, if we are only required to find curves that are significant with respect to the complexity of the data. In addition, this method requires little memory and can be easily parallelized.

1 Introduction

The Hough transform is a method to detect parameterized models (e.g. curves and surfaces) in data by mapping data features into manifolds in the parameter space [3, 5]. The models are detected by locating peaks in the parameter space (which is typically performed using multi-dimensional histograming). In this paper, we consider methods to improve curve detection by decomposing the Hough transform into many small subproblems. We use randomization to limit the number of subproblems that we must examine and we carefully propagate the effects of localization error in the subproblems that we do examine. While we concentrate on curve detection, similar Hough transform techniques can be applied to surface detection and a number of other problems.

We will use a modified version of the formal definition of the Hough transform given by Princen *et al.* [6]. Let $X = (x, y)$ be a point in the image space, $\Omega = (\omega_1, ..., \omega_N)$ be a point in an N-dimensional parameter space, and $f(X, \Omega) = 0$ be the function that parameterizes the set of curves. We will call the set of data points $\mathcal{E} = \{X_1, ..., X_n\}$.

Standard Hough transform implementations discretize the parameter space and maintain a counter for each cell. The counters record the number of data points that map to a manifold that intersects each of the cells. In the errorless case, each data point maps to an $N - 1$ dimensional manifold in the parameter space. Princen *et al.* denote a cell in parameter space centered at Ω by C_Ω. They define:

$$p(X, \Omega) = \begin{cases} 1, & \text{if } \{\Lambda : f(X, \Lambda) = 0\} \cap C_\Omega \neq \emptyset \\ 0, & \text{otherwise} \end{cases}$$

Thus, $p(X, \Omega)$ is 1 if any curve in the parameter space cell, C_Ω, passes through the point, X, in the image. If we assume that there is no localization error, the Hough transform can then be written:

$$H(\Omega) = \sum_{j=1}^{n} p(X_j, \Omega) \tag{1}$$

$H(\Omega)$ is now the number of data points that any curve in C_Ω passes through. In an ideal system, the discretization of the parameter space would not be important. Instead, we should consider the error in the localization of the image points. Let's assume that the true location of each data point lies within a bounded region, N_X, of the determined location, X. We can redefine $p(X, \Omega)$ as follows:

$$p(X, \Omega) = \begin{cases} 1, & \text{if } \{Y : f(Y, \Omega) = 0\} \cap N_X \neq \emptyset \\ 0, & \text{otherwise} \end{cases}$$

Now, $p(X, \Omega)$ is 1 if the curve represented by Ω passes through N_X. With this definition we can still use (1) to describe the Hough transform. This yields, for each curve, the number of data points that the curve passes through up to the localization error. Since discretization of the parameter space will not be important for the techniques that we present here, we will use the new definition.

2 Mapping Point Sets into the Parameter Space

A technique that has been recently introduced [1, 2, 4, 7] maps point sets rather that single points into the parameter space. Rather than considering each point separately, this method considers point sets of some cardinality, k. For each such set, the curves that pass through each point in the set (or their error boundaries) are determined and the parameter space is incremented accordingly. The benefit of this technique is that each mapping is to a smaller subset of the parameter space. If the curve has N parameters, then, in the errorless case, N non-degenerate data points map to a finite set of points in the parameter space. For the curves we examine here, this will be a single point. We thus need to increment only one bin in the parameter space for each set, rather than the bins covering an $N - 1$ dimensional manifold for each point. Of course, we don't need to use sets of size N, we could use any size, $k > 0$. If $k \leq N$, each non-degenerate set maps to an $N - k$ dimensional manifold in the parameter space. The disadvantage to methods that map point sets into the parameter space is that there are $\binom{n}{k}$ sets of image pixels with cardinality k to be considered.

An examination of how the technique of mapping point sets into the parameter space is related to the standard Hough transform is informative. Let's label the Hough transform technique that maps sets of k points into the parameter space $H^k(\Omega)$. An image curve (a point in the parameter space) now gets a vote only if it passes within the error boundary of each point in the set, so we have:

$$H^k(\Omega) = \sum_{\{g_1,\dots,g_k\} \in \binom{\mathcal{E}}{k}} p(X_{g_1}, \Omega) \cdot \dots \cdot p(X_{g_k}, \Omega)$$

where $\binom{\mathcal{E}}{k}$ is the set of all k-subsets of the data points, $\mathcal{E}$.

Consider this function at an arbitrary point in the parameter space. For some set of data points, $\{X_{g_1}, \dots, X_{g_k}\}$, the product, $p(X_{g_1}, \Omega) \cdot \dots \cdot p(X_{g_k}, \Omega)$, will be 1 if and only if each of the $p(X, \Omega)$ terms is 1 and otherwise it will be 0. If there are x points such that $p(X, \Omega)$ is 1 (these are the points that lie on Ω up to the localization error), then there are $\binom{x}{k}$ sets with cardinality k that contribute 1 to the sum. $H^k(\Omega)$ will thus be $\binom{x}{k}$. Since the standard Hough transform will yield $H(\Omega) = x$ in this case, we can express $H^k(\Omega)$ simply in terms of $H(\Omega)$:

$$H^k(\Omega) = \binom{H(\Omega)}{k}$$

If the standard Hough transform uses threshold $t \geq k$ to find peaks and the method of mapping point sets into the parameter space uses threshold $\binom{t}{k}$, these methods will find the same set of peaks according to the above analysis. Their accuracy is thus the same.

3 Decomposition into Subproblems

Let us now introduce a new technique, where we map only those point sets into the parameter space that share some set of j *distinguished points*, $\mathcal{D} = \{X_{d_1}, \dots, X_{d_j}\}$. We will still vary $k - j$ data points, $\mathcal{G} = \{X_{g_1}, \dots, X_{g_{k-j}}\}$, in these sets. The point sets we are mapping into parameter space are thus $\mathcal{D} \cup \mathcal{G}$. This yields:

$$H^{\mathcal{D},k}(\Omega) = \sum_{\mathcal{G} \in \binom{\mathcal{E} \setminus \mathcal{D}}{k-j}} \prod_{i=1}^{j} p(X_{d_i}, \Omega) \prod_{i=1}^{k-j} p(X_{g_i}, \Omega)$$

Consider this function at an arbitrary point in the parameter space. Since we aren't varying the distinguished points, $\{X_{d_1}, \dots, X_{d_j}\}$, the curve must pass through the error boundary of each of these to yield a non-zero response. If x points lie on a curve and we use a set of j of distinguished points on the curve, then $x - j$ of these points remain in $\mathcal{E} \setminus \mathcal{D}$. We thus have:

$$H^{\mathcal{D},k}(\Omega) = \begin{cases} \binom{H(\Omega) - j}{k - j}, & \text{if } \prod_{i=1}^{j} p(X_{d_i}, \Omega) = 1 \\ 0, & \text{otherwise} \end{cases}$$

We should thus use a threshold of $\binom{t-j}{k-j}$ in this case if the standard Hough transform used a threshold of t. We would then find those curves that are found by the standard Hough transform that pass through the distinguished points up to the localization error. We can formulate algorithms to recognize arbitrary curves by considering several subproblems, each of which examines a particular set of distinguished points, as above. A deterministic algorithm using these ideas would consider each possible set of distinguished points. This would guarantee that we would examine a correct set of distinguished points for each curve. If we are willing to allow a small probability of failure, we can use randomization to considerably reduce the number of sets of distinguished points that we must examine (see Section 5).

To gain the maximum decomposition of the problem, we want j, the number of distinguished points, to be as large as possible, but note that if we choose $j \geq k$, we will have $H^{\mathcal{D},k}(\Omega) = 0$ or 1 for all Ω. Our response will be 1 if Ω goes through the j points and otherwise 0, but it yields no other information. We thus want to have $j < k$. In addition, we want $k \leq N$ or else we will examine sets that are larger than necessary. The optimal choice is thus $j = k-1 = N-1$.

Note that considering sets of N data points that vary in only one point (i.e. when $j = k - 1 = N - 1$) constrains the transform to lie on a 1-dimensional manifold (a curve) in the parameter space. This can easily be seen since we have N variables (the curve parameters) and the $N-1$ distinguished points yield $N-1$ equations in them. Let's call this curve the *Hough curve*. When localization error is considered, the transform will no longer be constrained to lie on the Hough curve, but the transform points will remain close to this curve. This yields two useful properties. First, since the Hough curve is essentially 1-dimensional, it is much easier to search than the full parameter space. Second, it is now much easier to propagate localization error carefully. This will be accomplished by determining tight bounds on the range that a set of points can map to in parameter space.

4 Error Propagation

Let's now examine how to propagate the localization error in the curve detection process. We will first consider how error would be propagated in the ideal case. Each set of points maps to a subset of the parameter space under given error conditions. This subset consists of the curves that pass through the set of points up to the error criteria. Call this subset of the parameter space the *error cloud* of the set of points. Ideally, we would determine how many error clouds intersect at each point of the parameter space. This would tell us, for any curve, how many of the points the curve passes through up to the localization error. We do not do this since it is not practical, but for the subproblems we now examine, we can efficiently compute a good approximation.

Since the Hough curve is one-dimensional in the noiseless case, we can parameterize it in a single variable, t. Consider the projection of the error clouds onto the t-axis (see below for examples). The number of such projected error clouds that intersect at some point in this projection yields a bound on the number of error clouds that intersect on a corresponding hypersurface in the full space. Furthermore, since the error clouds do not vary far from the Hough curve,

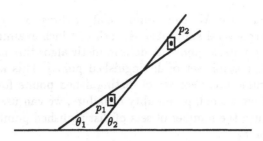

Fig. 1. For any two points, we can determine bounds on the range of θ by considering lines that pass through the boundaries of their possible localization error.

this yields a good approximation to the maximum number of intersecting error clouds, which is the information we want.

Once we have projected each of the sets that we consider in some subproblem onto the t-axis, we can find the peaks along the Hough curve in one of two ways. We could simply discretize t and perform voting by incrementing the bins consistent with each range in t that an error cloud projects to. This discretization can be done finely since it is only in one dimension. Alternatively, we could sort the minimal and maximal t points of each error cloud and use a sweep algorithm. This method would examine the extremal points in sorted order and keep a counter that is incremented each time we hit a minimal point and decremented each time we hit an maximal point. If the counter reaches a large enough value, then a line has been found which passes through (or close to) many points.

The following subsections describe how we can parameterize the Hough curve in t for the cases of lines and circles, and how we can project the error cloud for the point sets onto the t-axis for each case.

4.1 Lines

If we use the ρ-θ parameterization for lines (i.e. $x \cos \theta + y \sin \theta = \rho$), we can simply parameterize the Hough curve by θ, since ρ is a function of θ. To project the error cloud for a pair of points onto the θ-axis, we simply determine the minimal and maximal θ that a pair of points can yield. If we use square error boundaries, we need only consider the corners of the squares in determining these minimal and maximal θ values. See Fig. 1.

4.2 Circles

We can parameterize the space of circles by the coordinates of the center of the circle and the radius, so there are three parameters: (x_c, y_c, r). For this case, the optimal decomposition will use $j = N - 1 = 2$ distinguished points. We can parameterize the Hough curve by the distance of the center of the circle from the midpoint of the two distinguished points (which we will take to be positive when the center is to the right of the segment connecting the distinguished points, and negative otherwise).

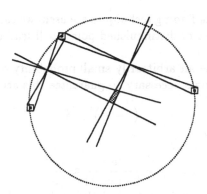

Fig. 2. We can determine bounds on the position of the center of a circle passing through 3 points (up to localization error) by examining the range of possible perpendicular bisectors for the segments between the points.

To project the error cloud onto the t-axis, we now want to determine error bounds on this distance given three points and their localization error boundaries. Recall that the center of the circle passing through three points is the point where the perpendicular bisectors of the segments between the points meet. We can thus determine bounds on the location of the center of the circle by examining the set of points in which two of the perpendicular bisectors of the segments can meet (see Fig. 2). The minimum and maximum distance from the center of the circle to the midpoint of the distinguished points can easily be determined by examining the extremal points of this set.

5 Computational Complexity

This section determines the computational complexity of the techniques described in this paper. Let's first determine how many of the sets of distinguished points we must examine to maintain a low rate of failure. We'll assume that we only need to find curves that comprise some fraction, ϵ, of the total number of data points. The probability that a single set of j random points lie on a particular such curve is then at least:

$$p_0 \geq \frac{\binom{\epsilon n}{j}}{\binom{n}{j}} \approx \frac{(\epsilon n)^j}{n^j} = \epsilon^j$$

since we must have $\binom{\epsilon n}{j}$ sets of distinguished points that lie on the curve among the $\binom{n}{j}$ possible sets of distinguished points. If we take t such trials, the probability that all of them will fail for a particular curve is no more than:

$$p \leq (1 - p_0)^t \approx (1 - \epsilon^j)^t$$

For each curve, we thus have a probability no larger than p that we will fail to examine a set of distinguished points that is a subset of the curve in t trials.

Since conservative peak finding techniques are used, we can assume that any trial examining a correct set of distinguished points will lead to the identification of the curve.

We can now choose an arbitrarily small probability of failure, δ, and determine the number of trials necessary to guarantee this accuracy:

$$\left(1 - \epsilon^j\right)^t \leq \delta$$

Solving for t yields:

$$t \geq \frac{\ln \delta}{\ln(1 - \epsilon^j)} \approx \frac{\ln \frac{1}{\delta}}{\epsilon^j}$$

For each trial, we now have to find the peaks on the Hough curve. Recall that we use $j = N - 1$ in our method to facilitate the propagation of error. If we use voting, the time is dependent on how finely the Hough curve is discretized. If there are α bins, we need to increment $O(\alpha)$ bins per trial per point, yielding $O(n\alpha)$ time per trial. The total time requirement is thus $O(\frac{n\alpha \log \delta}{\epsilon^{N-1}})$ or simply $O(n)$ when measured by the size of the input (α, δ, ϵ, and N are constants).

If we use the sweep algorithm, we must sort the $O(n)$ maximal and minimal points of the error clouds, requiring $O(n \log n)$ time per trial. Processing the sorted points requires $O(n)$ time. We thus require $O(\frac{n \log n \log \delta}{\epsilon^{N-1}})$ total time or $O(n \log n)$ when measured by the size of the input.

6   Results

These techniques have been applied to real images to test their efficacy. Figure 3 shows an image that was used to test the line detection techniques. The edges were determined to sub-pixel accuracy. For tests on this image, square error boundaries were used such that the true location of each point was assumed to be within 0.25 pixels of the measured location in both x and y. When a large threshold was used ($\epsilon = 0.01$), all of the long lines were found in the image, but short or curving lines were not found. When a lower threshold was used ($\epsilon = 0.004$), even short lines were found in the image.

Figure 4 shows an image that was used to test the circle detection techniques. This image is an engineering drawing that has been scanned. For this reason, it was not possible to determine the location of pixels to sub-pixel accuracy. In addition, the presence of small and dashed circles and the clutter in the image make this a difficult test case. While all of the large circles were found with $\epsilon = 0.04$, the small and dashed circles did not comprise a large enough fraction of the image to be found. With $\epsilon = .008$, the implementation finds several of circles, some of which are not perceptually obvious. Note that in each of the insalient circles, the pixels found overlay most of the perimeter of a circle and thus if we want to find small and/or broken circles it is difficult to rule out these circles without using additional information. In addition, the implementation has difficulty finding both of the dashed circles with the same center since they are so close together and are imperfect circles. The dashed circles shown in Fig. 4(c) consist of the top half of one of the circles and the bottom half of the other.

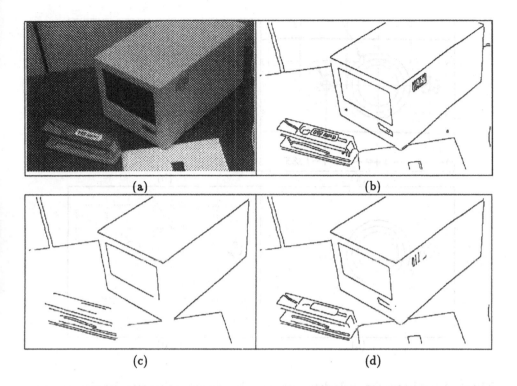

Fig. 3. A test image for line detection. (a) The original image. (b) The edges detected. (c) The lines found with $\epsilon = .01$. (d) The lines found with $\epsilon = .004$.

7 Ellipses and Other High-Order Curves

When applying these techniques to curves with several degrees of freedom, we must take special care, since the number of trials that are required can become large. Let's consider the detection of ellipses, which have five parameters. If the image is sparse or we can segment the image, then we should have no problems. For example, if we only need to detect ellipses that comprise 50% of the image pixels (or some subset after segmentation), then the number of trials required to achieve 99% accuracy is 74. On the other hand, if we wish to detect ellipses that comprise at least 10% of the image pixels using these techniques in a straightforward manner, then this would require 46,052 trials to achieve 99% accuracy.

When we wish to detect high-order curves in complex images there are additional techniques that we can use in order to perform curve detection quickly. One simple technique is to use more information at each curve pixel. For example, we can use the orientation of the curve at each pixel (as determined from the gradient, the curve normal, or the tangent). When we do this, we require fewer curve points to determine the position of the curve. We can determine the

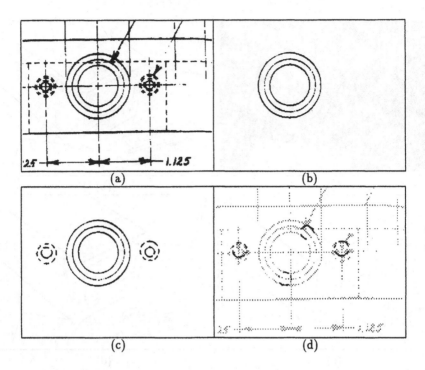

Fig. 4. A test image for circle detection. (a) The original engineering drawing. (b) The circles found with $\epsilon = 0.04$. (c) Perceptually salient circles found with $\epsilon = 0.008$. (d) Insalient circles found with $\epsilon = 0.008$.

position of an ellipse using three points with orientations rather than five unoriented points. We would thus use two, rather than four, distinguished points, and we would require many fewer trials to ensure that there is a low probability of not selecting a correct set of distinguished points. Of course, we do not need to restrict this technique to high-order curves. We can use two oriented points to determine the position of a circle, rather than the three unoriented points used in the previous sections.

An alternate technique that can detect high-order curves quickly is to use a two-step technique, where we first determine a subset of the curve parameters and then determine the remaining parameters. For examine, Yuen *et al.* [8] describe a method for detecting ellipse centers. They note that the center of an ellipse must lie on the line connecting the intersection of the tangents of two points on the ellipse with the midpoint of the segment between the two points. A point of intersection between several such lines yields a likely ellipse center. We can use decomposition techniques similar to those already described in this paper when this method is used to detect ellipse centers. Once the center of the ellipse has been detected, there are three remaining ellipse parameters. These can be detected using a three parameter Hough transform similar to the detection of circles.

8 Summary

We have considered efficient techniques to perform the Hough transform with careful propagation of localization error. To this end, we have modified a formal definition of the Hough transform to allow localization error to be analyzed appropriately. We then considered a new method where the Hough transform is decomposed into several subproblems, each of which examines a subset of the parameter space, by considering only those point sets that include some set of distinguished points. These subproblems allow us, first, to propagate the localization error efficiently and accurately in the parameter space, and second, to use randomization techniques to reduce the complexity while maintaining a low probability of missing an important curve. The overall complexity of the resulting algorithm is $O(n)$, where n is the number of data points in the image. Finally, we have given results of this system detecting straight lines and circles in cluttered and noisy real images and discussed the application of these techniques to curves with several parameters.

References

1. J. R. Bergen and H. Shvaytser. A probabilistic algorithm for computing Hough transforms. *Journal of Algorithms*, 12:639–656, 1991.
2. A. Califano, R. M. Bolle, and R. W. Taylor. Generalized neighborhoods: A new approach to complex parameter feature extraction. In *Proceedings of the IEEE Conference on Computer Vision and Pattern Recognition*, pages 192–199, 1989.
3. J. Illingworth and J. Kittler. A survey of the Hough transform. *Computer Vision, Graphics, and Image Processing*, 44:87–116, 1988.
4. V. F. Leavers. The dynamic generalized Hough transform: Its relationship to the probabilistic Hough transforms and an application to the concurrent detection of circles and ellipses. *CVGIP: Image Understanding*, 56(3):381–398, November 1992.
5. V. F. Leavers. Which Hough transform? *CVGIP: Image Understanding*, 58(2):250–264, September 1993.
6. J. Princen, J. Illingworth, and J. Kittler. A formal definition of the Hough transform: Properties and relationships. *Journal of Mathematical Imaging and Vision*, 1:153–168, 1992.
7. L. Xu, E. Oja, and P. Kultanen. A new curve detection method: Randomized Hough transform (RHT). *Pattern Recognition Letters*, 11:331–338, May 1990.
8. H. K. Yuen, J. Illingworth, and J. Kittler. Detecting partially occluded ellipses using the Hough transform. *Image and Vision Computing*, 7(1):31–37, 1989.

Image Retrieval Using Scale-Space Matching*

S. Ravela[1]   R. Manmatha[2]   E. M. Riseman[1]

[1] Computer Vision Research Laboratory
[2] Center for Intelligent Information Retrieval
University of Massachusetts at Amherst
{ravela, manmatha}@cs.umass.edu

Abstract. The retrieval of images from a large database of images is an important and emerging area of research. Here, a technique to retrieve images based on appearance that works effectively across large changes of scale is proposed. The database is initially filtered with derivatives of a Gaussian at several scales. A user defined template is then created from an image of an object similar to those being sought. The template is also filtered using Gaussian derivatives. The template is then matched with the filter outputs of the database images and the matches ranked according to the match score. Experiments demonstrate the technique on a number of images in a database. No prior segmentation of the images is required and the technique works with viewpoint changes up to 20 degrees and illumination changes.

1 Introduction

The advent of multi-media and large image collections in several different domains brings forth a necessity for image retrieval systems. These systems will respond to visual queries by retrieving images in a fast and effective manner. The application potential is enormous; ranging from database management in museums and medicine, architectural and interior design, image archiving, to constructing multi-media documents or presentations[3].

Simple image retrieval solutions have been proposed, one of which is to annotate images with text and then use a traditional text-based retrieval engine. While this solution is fast, it cannot however be effective over large collections of complex images. The variability and richness of interpretation is quite enormous as is the human effort required for annotation. To be effective an image retrieval system should exploit image attributes such as color distribution, motion, shape [1], structure, texture or perhaps user drawn sketches or even abstract token sets (such as points, lines etc.). Image retrieval can be viewed as an ordering of match scores that are obtained by searching through the database. The key challenges in building a retrieval system are the choice of attributes,

---
*   This work was supported in part by the Center for Intelligent Information Retrieval, NSF Grants IRI-9208920, CDA-8922572 and ARPA N66001-94-D-6054

(i) Original Image   (ii) Composite (Query)

Fig. 1. Construction of a query begins with a user marking regions of interest in an image, shown by the rectangles in (i).The regions of interest and their spatial relationships define a query, shown in (ii).

their representations, query specification methods, match metrics and indexing strategies.

In this paper a method for retrieving images based on appearance is presented. Without resorting to token feature extraction or segmentation, images are retrieved in the order of their *similarity in appearance* to a *query*.

Queries are constructed from raw images, as illustrated in Figure 1. The regions in Figure 1(ii) along with their spatial relationship are conjunctively called as the query[1]. Images are then retrieved from the database in the order of their similarity of appearance to the query. Similarity of appearance is defined as the similarity of shape under small view variations. The proposed definition constrains view variations, but does not constrain scale variations.

A measure of similarity of appearance is obtained by correlating filtered representations of images. In particular a *vector representation*(VR) of an image is obtained by associating each pixel with a vector of responses to Gaussian derivative filters of several different orders. To retrieve similar looking images under varying scale a representation over the scale parameter is required and scale-space representations [6] are a natural choice. Lists of VRs generated using banks of Gaussian derivative filters at several different scales form a scale-space representation [6] of the object. A match score for any pair of images is obtained by correlating their scale-space vector representations.

Thus, the entire process of retrieval can be viewed as the following three-step process. The first is an off-line computation step that generates VRs of database images for matching (described in Section 3). The second is construction of queries and their VRs (described in Section 5). The third is an ordering of images ranked by the correlation of their VRs with that of the query (described in Section 4). In Section 6 experiments with this procedure demonstrate retrieval of similar looking objects under varying scale.

While one is tempted to argue that retrieval and recognition problems have a lot in common, one should also note the sharp contrasts between the two paradigms. First, putting a user in the "loop" , shifts the burden of the deter-

---

[1] The retrieved images for this case are shown in Figure 3.

mination of feature saliency to the user. For example, only regions of the car in Figure 1(i) (namely, the wheels, side-view mirror and mid-section) considered salient by the user are highlighted. Second, user interaction can be used in a retrieval system of sufficient speed to evaluate the ordering of retrieved images and reformulate queries if necessary. Thus, in the approach presented in this paper, alternate regions could be marked if the retrieval is not satisfactory. Third, a hundred percent accuracy of retrieval is desirable but not at all critical (for comparison the best text-based retrieval engines have retrieval rates less than 50%). The user ultimately views and evaluates the results, allowing for tolerance to the few incorrect retrieval instances.

2   Related Work

A number of researchers have investigated the use of shape for retrieval [1, 9, 10]. However, unlike the technique presented in this paper, these methods all require prior segmentation of the object using knowledge of the contour or binary shape of the object.

It has been argued by Koenderink and van Doorn [5] and others that the structure of an image may be represented using Gaussian derivatives. Hancock et al [4] have shown that the principal components of a set of images containing natural structures may be modeled as the outputs of a Gaussian and its derivatives at several scales. That is, there is a natural decomposition of an image into Gaussian derivatives at several scales. Gaussians and their derivatives have, therefore, been successfully used for matching images of the same object under different viewpoints (see [12] for references). This paper is an extension to matching "similar" objects using Gaussian derivatives.

3   Matching Vector Representations

The key processing involves obtaining and matching vector-representations of a sample gray level image patch S and a candidate image C. The steps involved in doing this will now be described:

Consider a Gaussian described by it's coordinate $\mathbf{r}$ and scale σ

$$G\left(\mathbf{r},\sigma\right) = \frac{1}{\sqrt{2\pi}\sigma}e^{-\frac{r^2}{2\sigma^2}} \tag{1}$$

A vector-representation $\mathbf{I}$ of an image I is obtained by associating each pixel with a vector of responses to partial derivatives of the Gaussian at that location. Derivatives up to the second order are considered. More formally, $\mathbf{I}$ takes the form $\langle I_x, I_y, I_{xx}, I_{xy}, I_{yy}\rangle$ where I_x, I_y denote the the filter response of I to the first partial derivative of a Gaussian in direction x and y respectively. I_{xx}, I_{xy} and I_{yy} are the appropriate second derivative responses. The choice of first and second Gaussian derivatives is discussed in [12].

The correlation coefficient η between images $\mathbf{C}$ and $\mathbf{S}$ at location (m, n) in $\mathbf{C}$ is given by:

$$\eta\left(m,n\right) = \sum_{i,j} \hat{C_M}\left(i,j\right) \cdot \hat{S_M}\left(m-i, n-j\right) \tag{2}$$

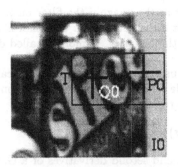

Fig. 2. I1 is half the size of I0. To match points p_0 with p_1, Image I_0 should be filtered at point p_0 by a Gaussian of a scale twice that of the Gaussian used to filter image I_1 (at p_1). To match a template from I_0 containing p_0 and q_0, an additional warping step is required. See text in Section 4.

where

$$\hat{S_M}(i,j) = \frac{\mathbf{S}(i,j) - S_M}{||\mathbf{S}(i,j) - S_M||}$$

and S_M is the mean of $\mathbf{S}(i,j)$ computed over S. $\hat{C_M}$ is computed similarly from $\mathbf{C}(i,j)$. The mean C_M is in this case computed at (m,n) over a neighborhood in C (the neighborhood is the same size as S).

Vector correlation performs well under small view variations. It is observed in [12] that typically for the experiments carried out with this method, in-plane rotations of up to $20°$, out-of plane rotation of up to 30^0 and scale changes of less than 1.2 can be tolerated. Similar results in terms of out-of-plane rotations were reported by [11].

4 Matching Across Scales

The database contains many objects imaged at several different scales. For example, the database used in our experiments has several diesel locomotives. The actual image size of these locomotives depends on the distance from which they are imaged and shows considerable variability in the database. The vector correlation technique described in Section 3 cannot handle large scale changes, and the matching technique, therefore, needs to be extended to handle large scale changes.

In Figure 2 image I_1 is half the size of image I_0 (otherwise the two images are identical). Thus,

$$I_0(\mathbf{r}) = I_1(s\mathbf{r}) \tag{3}$$

where $\mathbf{r}$ is any point in image I_0 and $s\mathbf{r}$ the corresponding point in I_1 and the scale change $s = 0.5$. In particular consider two corresponding points p_0 and p_1 and assume the image is Gaussian filtered at p_0. Then it can be shown that [7],

$$\int I_0(\mathbf{r})G(\mathbf{r} - \mathbf{p_0}, \sigma)dr = \int I_1(s\mathbf{r})G(s\mathbf{r} - \mathbf{p_1}), s\sigma)d(s\mathbf{r}) \tag{4}$$

In other words, the output of I_0 filtered with a Gaussian of scale σ at p_0 is equal to the output of I_1 filtered with a Gaussian of scale $s\sigma$ i.e. the Gaussian has to be stretched in the same manner as the image if the filter outputs are to be equal. This is not a surprising result if the output of a Gaussian filter is viewed as a Gaussian weighted average of the intensity. A more detailed derivation of this result is provided in [7].

The derivation above does not use an explicit value of the scale change s. Thus, equation 4 is valid for any scale change s. The form of equation 4 resembles a convolution and in fact it may be rewritten as a convolution

$$I_0(\mathbf{r}) \star G(.,\sigma) = I_1(s\mathbf{r}) \star G(.,s\sigma) \tag{5}$$

Similarly, filtering with the first and second derivatives of a Gaussian gives [8]

$$I_0 \star \mathbf{G}'(.,\sigma) = I_1 \star \mathbf{G}'(.,s\sigma) \tag{6}$$

and,

$$I_0 \star \mathbf{G}''(.,\sigma) = I_1 \star \mathbf{G}''(.,s\sigma) \tag{7}$$

where the normalized first derivative of a Gaussian is given by

$$\mathbf{G}'(\mathbf{r},s\sigma) = s\sigma \ dG(\mathbf{r},s\sigma)/d\mathbf{r} \tag{8}$$

and the normalized second derivative of a Gaussian is given by

$$\mathbf{G}''(\mathbf{r},s\sigma) = (s\sigma)^2 \ d^2 G(\mathbf{r},s\sigma)/d(\mathbf{r}\mathbf{r}^{\mathbf{T}}) \tag{9}$$

Note that the first derivative of a Gaussian is a vector and the second derivative of a Gaussian a 2 by 2 matrix.

The above equations are sufficient to match the filter outputs (in what follows assume only Gaussian filtering for simplicity) at corresponding points (for example at $\mathbf{p_0}$ and $\mathbf{p_1}$). A further complication is introduced if more than one point is to be matched while preserving the relative distances (structure) between the points. Consider for example the pair of corresponding points $\mathbf{p_0}, \mathbf{q_0}$ and $\mathbf{p_1}, \mathbf{q_1}$. The filter outputs at points $\mathbf{p_0}, \mathbf{q_0}$ may be visualized as a template and the task is to match this template with the filter outputs at points $\mathbf{p_1}, \mathbf{q_1}$. That is, the template is correlated with the filtered version of the image I_1 and a best match sought. However, since the distances between the points $\mathbf{p_1}, \mathbf{q_1}$ are different from those between $\mathbf{p_0}, \mathbf{q_0}$ the template cannot be matched correctly unless either the template is rescaled by a factor of $1/2$ or the image I_1 is rescaled by a factor of 2. The matching is, therefore, done by warping either the template or the image I_1 appropriately.

Thus, to find a match for a template from I_0, in I_1, the Gaussians must be filtered at the appropriate scale and then the image I_1 or the template should be warped appropriately. Now consider the problem of localizing a template T, extracted from I_0, in I_1(see Figure 2). For the purpose of subsequent analysis, assume two corresponding points $(\mathbf{p_0}, \mathbf{q_0})$ of interest in T and I_1 $(\mathbf{p_1}, \mathbf{q_1})$ respectively. To localize the template the following three steps are performed.

1. *Use appropriate Relative Scale:* Filter the template and I_1 with Gaussians whose scale ratio is 2. That is, filter T with a Gaussian of scale 2σ and I_1 with σ.

2. *Account for size change:* Sub-sample T by half. At this point the spatial and intensity relationship between the warped version (filtered and sub-sampled) of template points p_0 and q_0 should be exactly same as the relationships between filtered versions of p_1 and q_1.

3. *Translational Search:* Perform a translational search over I_1 to localize the template.

This three step procedure can be easily extended to match VRs of T and I_1 using Equations 6 and 7. In step(1) generate VRs of T and I_1 using the mentioned filter scale ratios. In step(2) warp the VR of T instead of just the intensity. In step(3) use vector-correlation(Equation 2 at every step of the translational search.

Without loss of generality any arbitrary template T can be localized in any I_1 that contains T scaled by a factor s.

4.1 Matching Queries over Unknown Scale

The aforementioned steps for matching use the assumption that the relative scale between a template and an image is known. However, the relative scale between structures in the database that are similar to a query cannot be determined *a priori*. That is, the query could occur in a database image at some unknown scale. A natural approach would be to search over a range of possible relative scales, the extent and step size being user controlled parameters.

One way of accomplishing this is as follows. First, VRs are generated for each image in the database over a range of scales, say $\frac{1}{4}\sigma, \frac{1}{2\sqrt{2}}\sigma,...,4\sigma$. Then, a VR for the query is generated using Gaussian derivatives of scale σ. The query VR is matched with each of the image VRs, thus traversing a relative scale change of $\frac{1}{4}...4$, in steps of $\sqrt{2}$. For each scale pairing the three step procedure for matching VRs is applied. In the warping step of this procedure either the query or the image is warped depending on the relative scale. If the relative scale between the query and a candidate image is less than 1 the candidate VR is warped and if it is greater than 1 the query VR is warped. After the query is matched with each of the image VRs, the location in the image which has the best correlation score is returned.

It is instructive to note that VR lists over scale are scale-space representations in the sense described by Lindeberg [6]. By smoothing an image with Gaussians at several different scales Lindeberg generates a scale-space representation. While VR lists are scale-space representations, however, they differ from Lindeberg's approach in two fundamental ways. First VRs are generated from derivatives of Gaussians and second, an assumption is made that smoothing is accompanied by changes in size (i.e. the images are scaled versions rather than just smoothed versions of each other). This is the reason warping is required during VR matching across scales. VR lists are proper scale-space representations unlike pyramidal representations [6, 12]

5 Constructing Query Images

The query construction process begins with the user marking salient regions on an object. VRs generated at several scales within these regions are matched with the database in accordance with the description in Section 4. Unselected regions are not used in matching. One way to think about this is to consider a composite template, such as one shown in Figure 1(ii). The unselected regions have been masked out. The composite template preserves inter-region spatial relationships and hence, the structure of the object is preserved. Warping the composite will warp all the components appropriately, preserving relative spatial relationships. That is, both the regions as well as distances between regions are scaled appropriately. Further, there are no constraints imposed on the selection of regions and the regions need not overlap.

Careful design of a query is important. It is interesting to note that marking the entire object does not work very well (see [12] for examples). Marking extremely small regions has also not worked with this database. There are too many coincidental structures that can lead to poor retrieval.

Many of these problems are, however, simplified by having the user interact extensively with the system. Letting the user design queries eliminates the need for detecting the saliency of features on an object. Instead, saliency is specified by the user. In addition, based on the feedback provided by the results of a query, the user can quickly adapt and modify the query to improve performance.

6 Experiments

The choice of images used in the experiments was based on a number of considerations. It is expected that when very dissimilar images are used the system should have little difficulty in ranking the images. For example, if a car query is used with a database containing cars and apes, then it is expected that cars would be ranked ahead of apes. This is borne out by the experiments done to date. Much poorer discrimination is expected if the images are much more 'similar'. For example, man-made vehicles like cars, diesel and steam locomotives should be harder to discriminate. It was, therefore, decided to test the system by primarily using images of cars, diesel and steam locomotives as part of the database.

The database used in this paper has digitized images of cars, steam locomotives, diesel locomotives, apes and a small number of other miscellaneous objects such as houses. Over 300 images were obtained from the internet to construct this database. About 215 of these are of cars, diesel locomotives and steam locomotives. There are about 80 apes and about 12 houses in the database. These photographs, were taken with several different cameras of unknown parameters, and, under varying but uncontrolled lighting and viewing geometry. The objects of interest are embedded in natural scenes such as car shows, railroad stations, country-sides and so on.

Prior to describing the experiments, it is important to clarify what a correct retrieval means. A retrieval system is expected to answer questions such as 'find all cars similar in view and shape to this car' or 'find all steam engines similar in appearance to this steam engine'. To that end one needs to evaluate if a query can be designed such that it captures the appearance of a generic steam engine or perhaps that of a generic car. Also, one needs to evaluate the performance of VR matching under a specified query. In the examples presented here the following method of evaluation is applied. First, the objective of the query is stated and then retrieval instances are gauged against the stated objective. In general, objectives of the form 'extract images similar in appearance to the query' will be posed to the retrieval algorithm.

Several different queries were constructed to retrieve objects of a particular type. It is observed that under reasonable queries at least 60% of m objects underlying the query are retrieved in the top m ranks. Best results indicate retrieval results of up to 85%. This performance compares very well with typical text retrieval systems[2]. To the best of our knowledge other image retrieval systems either need prior segmentation or work on restricted domains. Therefore, accurate comparisons cannot be made.

Several experiments were carried out with the database [12]. The results of the experiments carried out with a car query, a diesel query and a steam query are presented in table 6. The number of retrieved images in intervals of ten is charted in Table 6. The table shows, for example, that there are 16 car images "similar" in view to the car in the query and 14 of these are ranked in the top 20. For the steam query there are 12 "similar" images (as determined by a person), 9 of which are ranked in the top 20. Finally, for the diesel query there are 30 "similar" images, 12 of which are found in the top 20 retrievals. Due to space limitations only the results of the *Car retrieval* are displayed (Figure 3) and analyzed in detail (for the others see [12]).

	No. Retrieved Images				
Query	1-10	11-20	21-30	31-40	41-50
Car	8	6	1	0	1
Steam	7	2	1	0	2
Diesel	7	5	5	6	4

Table 1. Correct retrieval instances for the Car, Steam and Diesel queries in intervals of ten. The number of "similar" images in the database as determined by a human are 16 for the Car query, 12 for the Steam query and 30 for the Diesel query.

The car image used for retrieval is shown in the top left picture of Figure 3. The objective is to 'obtain all similar cars to this picture'. Towards this end a query was marked by the user, highlighting the wheels, side view-mirror and mid section. The results to be read in text book fashion in Figure 3 are the ranks of the retrieved images. The white patches indicate the centroid of the composite template at best match. In the database, there are exactly 16 cars within a close variation in view to the original picture. Fourteen of these cars were retrieved in the top 16, resulting in a 87.5% retrieval. All 16 car pictures were picked up in the top 50. The results also show variability in the shape of the retrieved instances. The mismatches observed in pictures labeled 'car05.tif' and 'car09.tif' occur in VR matching when the relative scale between the query VR and the images is $\frac{1}{4}$.

---

[2] The average retrieval rate for text-based systems is 50%

Fig. 3. Retrieval results for Car.

Wrong instances of retrieval are of two types. The first is where the VR matching performs well but the objective of the query is not satisfied. In this case the query will have to be redesigned. The second reason for incorrect retrieval is mismatches due to the search over scale space. Most of the VR mismatches result from matching at the extreme relative scales.

Overall the queries designed were also able to distinguish steam engines and diesel engines from cars precisely because the regions selected are most similarly found in similar classes of objects. As was pointed out in Section 5 query selection must faithfully represent the intended retrieval, the burden of which is on the user. The retrieval system presented here performs well under it's stated purpose: that is to extract objects of similar shape and view to that of a query.

7 Conclusions and Limitations

This paper demonstrates retrieval of similar objects using vector representations over scale-space. There are several factors that affect retrieval results, including query selection, and the range of scale-space search. The results indicate that this method has sufficient accuracy for image retrieval applications.

One of the limitations of our current approach is the inability to handle large deformations. The filter theorems described in this paper hold under affine deformations and a current step is to incorporate it in to the vector-correlation routine.

While these results execute in a reasonable time they are still far from the high speed performance desired of image retrieval systems. Work is on-going towards building indices of images based on local shape properties and using the indices to reduce the amount of translational search.

Acknowledgments

The authors thank Prof. Bruce Croft and the Center for Intelligent Information Retrieval (CIIR) for continued support of this work. We also thank Jonathan Lim and Robert Heller for systems support. The pictures of trains were obtained from http://www.cs.monash.edu.au/image_lib/trains/. The pictures of cars were obtained from ftp.team.net/ktud/pictures/.

References

1. Myron Flickner, Harpreet Sawhney, Wayne Niblack, Jonathan Ashley, Qian Huang, Byron Dom, Monika Gorkani, Jim Hafner, Denis Lee, Dragutin Petkovix, David Steele, and Peter Yanker: Query By Image and Video Content: The QBIC System. IEEE Computer Magazine, September 1995, pp.23-30.
2. Gösta H. Granlund, and Hans Knutsson: Signal Processing in Computer Vision. Kluwer Academic Publishers, 1995, ISBN 0-7923-9530-1, Dordrecht, The Netherlands.
3. Venkat N. Gudivada, and Vijay V. Raghavan: Content-Based Image Retrieval Systems. IEEE Computer Magazine, September 1995, pp.18-21.
4. P. J. B. Hancock, R. J. Bradley and L. S. Smith: The Principal Components of Natural Images. Network, 1992, 3:61-70.
5. J. J. Koenderink, and A. J. van Doorn: Representation of Local Geometry in the Visual System. Biological Cybernetics, 1987, vol. 55, pp. 367-375.
6. Tony Lindeberg: Scale-Space Theory in Computer Vision. Kluwer Academic Publishers, 1994, ISBN 0-7923-9418-6 , Dordrecht, The Netherlands.
7. R. Manmatha: Measuring Affine Transformations Using Gaussian Filters. Proc. European Conference on Computer Vision, 1994, vol II, pp. 159-164.
8. R. Manmatha and J. Oliensis: Measuring Affine Transform - I, Scale and Rotation. Proc. DARPA IUW, 1993, pp. 449-458, Washington D.C.
9. Rajiv Mehrotra and James E. Gary: Similar-Shape Retrieval In Shape Data Management.IEEE Computer Magazine, September 1995, pp. 57-62.
10. A. Pentland, R. W. Picard, and S. Sclaroff: Photobook: Tools for Content-Based Manipulation of Databases. Proc. Storage and Retrieval for Image and Video Databases II, 1994, Vol.2, 185, SPIE, pp. 34-47, Bellingham, Wash.
11. R. Rao, and D. Ballard: Object Indexing Using an Iconic Sparse Distributed Memory. Proc. International Conference on Computer Vision, 1995, pp. 24-31.
12. S. Ravela, R. Manmatha and E. M. Riseman: Retrieval from Image Databases Using Scale-Space Matching. Technical Report UM-CS-95-104, 1995, Dept. of Computer Science, Amherst, MA 01003.

Optimal Surface Smoothing as Filter Design

Gabriel Taubin[1], Tong Zhang and Gene Golub[2]

[1] IBM T.J.Watson Research Center, P.O.Box 704, Yorktown Heights, NY 10598
[2] Computer Science Department, Stanford University, Stanford, CA 94305.

Abstract. Smooth surfaces are approximated by polyhedral surfaces for a number of computational purposes. An inherent problem of these approximation algorithms is that the resulting polyhedral surfaces appear faceted. Within a recently introduced signal processing approach to solving this problem [7, 8], surface smoothing corresponds to low-pass filtering. In this paper we look at the filter design problem in more detail. We analyze the stability properties of the low-pass filter described in [7, 8], and show how to minimize its running time. We show that most classical techniques used to design finite impulse response (FIR) digital filters can also be used to design significantly faster surface smoothing filters. Finally, we describe an algorithm to estimate the power spectrum of a signal, and use it to evaluate the performance of the different filter design techniques described in the paper.

1 Introduction

The signal processing framework introduced in [7, 8], extends Fourier analysis to *discrete surface signals*, functions defined on the vertices of polyhedral surfaces. As in the method of Fourier Descriptors [9], where a closed curve is smoothed by truncating the Fourier series of its coordinate signals, a very large polyhedral surface of arbitrary topology is smoothed here by low-pass filtering its three surface coordinate signals. And although the formulation was developed mainly for signals defined on surfaces, it is in fact valid for *discrete graph signals*, functions defined on the vertices of directed graphs. Since this general formulation provides a unified treatment of polygonal curves, polyhedral surfaces, and even three-dimensional finite elements meshes, we start this paper by reviewing this formulation in its full generality.

2 Fourier Analysis of Discrete Graph Signals

We represent a *directed graph* on the set $\{1, \ldots, n\}$ of n nodes as a set of *neighborhoods* $\{i^* : i = 1, \ldots, n\}$, where i^* is a subset of nodes which does not contain i. The element of i^* are the *neighbors* of i. A *discrete graph signal* is a vector $x = (x_1, \ldots, x_n)^t$ with one component per node of the graph. A *discrete surface signal* is a discrete graph signal defined on the graph of vertices and edges of a polyhedral surface. We normally use *first order neighborhoods*, were node j is a neighbor of node i if i and j share an edge (or face), but other neighborhood structures can be used to impose certain types of constraints [8].

The Discrete Fourier Transform (DFT) of a signal x defined on a closed polygon of n vertices is obtained by decomposing the signal as a linear combination of the eigenvectors of the Laplacian operator

$$\Delta x_i = \frac{1}{2}(x_{i-1} - x_i) + \frac{1}{2}(x_{i+1} - x_i) .$$ (1)

The DFT of x is the vector $\hat{x}$ of coefficients of the sum. The Laplacian operator must be replaced by another linear operator to define the DFT of a discrete graph signal. This is the same idea behind the method of eigenfunctions of Mathematical Physics [1].

We define the *Laplacian* of a discrete graph signal x by the formula

$$\Delta x_i = \sum_{j \in i^*} w_{ij} (x_j - x_i) ,$$ (2)

where the weights w_{ij} are positive numbers that add up to one for each vertex. These weights can be chosen in many different ways taking into consideration the neighborhoods, but in this paper we will assume that they are not functions of the signal x. Otherwise, the resulting operator is non-linear, and so, beyond the scope of this paper. One particularly simple choice that produces good results is to set w_{ij} equal to the inverse of the number of neighbors $1/|i^*|$ of node i, for each element j of i^*. Other choices of weights are discussed in [7, 8]. Note that the Laplacian of a signal defined on a closed polygon, described in equation (1), is a particular case of these definitions, with $w_{ij} = 1/2$, for $j \in i^* = \{i-1, i+1\}$, for each node i.

If $W = (w_{ij})$ denotes the matrix of weights, with $w_{ij} = 0$ when j is not a neighbor of i, and $K = I - W$, the Laplacian of a discrete signal can be written in matrix form as

$$\Delta x = -Kx .$$ (3)

Although the method applies to general neighborhood structures, in this paper we will restrict our analysis to those cases where the matrix W can be factorized as a product of a symmetric matrix times a positive definite diagonal matrix $W = ED$. In this case the matrix W is a *normal matrix* [3], because the matrix

$$D^{1/2}WD^{-1/2} = D^{1/2}ED^{1/2}$$ (4)

is symmetric. Note that such is the case for the first order neighborhoods of a surface with equal weights $w_{ij} = 1/|i^*|$ in each neighborhood i^*, where E is the *incidence matrix* of the neighborhood structure (a symmetric matrix for first order neighborhoods), the matrix whose ij-th. element is equal to 1 if the nodes i and j are neighbors, and 0 otherwise; and D is the diagonal positive definite matrix whose i-th. diagonal element is $1/|i^*|$. When W is a normal matrix it has all real eigenvalues, and sets of n left and right eigenvectors that form dual bases of n-dimensional space. Furthermore, by construction, W is also a *stochastic matrix*, a matrix with nonnegative elements and rows that add up to one [6]. The eigenvalues of a stochastic matrix are bounded above in magnitude by 1. It follows that the eigenvalues of the matrix K are real, bounded below by 0, and above by 2.

In general, the eigenvectors and eigenvalues of K have no analytic expression, but for filtering operations it is not necessary to compute the eigenvectors explicitly.

If $0 \leq k_1 \leq \cdots \leq k_n \leq 2$ are the eigenvalues of K, $e_1, \ldots, e_n$ a set of corresponding right eigenvectors, and $\delta_1, \ldots, \delta_n$ the associated dual basis of $e_1, \ldots, e_n$, the identity matrix I, and the matrix K can be written as follows

$$I = \sum_{i=1}^{n} e_i \delta_i^t \qquad K = \sum_{i=1}^{n} k_i \, e_i \delta_i^t \, ,$$

and every discrete graph signal x has a unique decomposition as a linear combination of $e_1, \ldots, e_n$

$$x = I \, x = \sum_{i=1}^{n} \hat{x}_i \, e_i \, , \tag{5}$$

where $\hat{x}_i = \delta_i^t x$. We call the vector $\hat{x} = (\hat{x}_1, \ldots, \hat{x}_n)^t$ the Discrete Fourier Transform (DFT) of x.

Note, however, that this definition does not identify a unique object yet. If a different set of right eigenvectors of K is chosen, a different DFT is obtained. To complete the definition, if $W = ED$, with E symmetric and D positive definite diagonal, we impose the right eigenvectors of K to be of unit length with respect to the norm associated with the inner product $\langle x, y \rangle_D = x^t D y$. With this constraint, Parseval's formula is satisfied

$$\|x\|_D^2 = \|\hat{x}\|^2 \, , \tag{6}$$

where the norm on the right hand side is the Euclidean norm. This result will be used in sections 6 and 7.

To filter the signal x is to change its frequency distribution according to a transfer function $f(k)$

$$x' = \sum_{i=1}^{n} f(k_i) \, \hat{x}_i e_i = \left(\sum_{i=1}^{n} f(k_i) \, e_i \delta_i^t \right) x \, . \tag{7}$$

The frequency component of x corresponding the the natural frequency k_i is enhanced or attenuated by a factor $f(k_i)$. For example, the transfer function of an ideal low-pass filter is

$$f_{\text{LP}} = \begin{cases} 1 & \text{for } 0 \leq k \leq k_{\text{PB}} \\ 0 & \text{for } k_{\text{PB}} < k \leq 2 \end{cases} , \tag{8}$$

where k_{PB} is the *pass-band frequency*.

Since there is no efficient numerical method to compute the DFT of a discrete graph signal, the computation can only be performed approximately. To do this the ideal low-pass filter transfer function is replaced by an analytic approximation, usually a polynomial or rational function, for which the computation can be performed in an efficient manner. A wide range of analytic functions of one variable $f(k)$ can be evaluated in a matrix such as K [3]. The result is another matrix $f(K)$ with the same left and right eigenvectors, but with eigenvalues $f(k_1), \ldots, f(k_n)$

$$f(K) = \sum_{i=1}^{n} f(k_i) \, e_i \delta_i^t \, .$$

The main reason why the filtering operation $x' = f(K) \, x$ of equation (7) can be performed efficiently for a polynomial transfer function of low degree, is that when K is sparse, which is the case here, the matrix $f(K)$ is also sparse (but of wider bandwidth), and so, the filtering operation becomes the multiplication of a vector by a sparse matrix.

In Gaussian smoothing the transfer function is the polynomial $f_N(k) = (1 - \lambda k)^N$, with $0 < \lambda < 1$. This transfer function produces shrinkage. The algorithm introduced in [7, 8] is essentially Gaussian smoothing with the difference that the scale factor λ changes from iteration to iteration, alternating between a positive value λ and a negative value μ. This simple modification still produces smoothing, but prevents shrinkage. The transfer function is the polynomial $f_N(k) = ((1 - \lambda k)(1 - \mu k))^{N/2}$, with $0 < \lambda < -\mu$ and N even. The *pass-band frequency* of this filter is defined as the unique value of k in the interval $(0, 2)$ such that $f_N(k) = 1$. Such a value exists when $0 < \lambda < -\mu$, and turns out to be equal to $k_{\text{PB}} = 1/\lambda + 1/\mu$. This polynomial transfer function of degree N results in a linear time and space complexity algorithm. From now on we will refer to this algorithm as the $\lambda - \mu$ algorithm.

3 Fast Smoothing as Filter Design

We are faced with the classical problem of digital filter design in signal processing [5, 4], but with some restrictions. Note that because of the linear complexity constraint discussed above, only polynomial transfer functions (FIR filters) are allowed. We leave the study of rational transfer functions (IIR filters) for the future. And because of space restrictions, of all the traditional FIR filter design methods available in the signal processing literature, we only cover here in some detail the method of windows, which is the simplest one. With this method we can design filters which are significantly faster, or sharper, than those obtain with the $\lambda - \mu$ algorithm for the same degree.

4 Optimizing the $\lambda - \mu$ algorithm

The $\lambda - \mu$ algorithm can be described in a recursive fashion as follows

$$f_N(k) = \begin{cases} 1 & N = 0 \\ (1 - \lambda_N \, k) \, f_{N-1}(k) & N > 0 \end{cases}$$

where $\lambda_N = \lambda$, for N odd, and $\lambda_N = \mu$ for N even. Note that this algorithm requires minimum storage, only one array of dimension n to store the Laplacian of a signal if computed in place, and two arrays of dimension n in general.

To maintain the minimum storage property and the same simple algorithmic structure, one could try to generalize by changing the scale factors λ_N from iteration to iteration in a different way. But if we start with a given pass-band frequency $k_{\text{PB}} = 1/\lambda + 1/\mu$, as it is usually the case when one wants to *design* the filter, there are many values of λ and μ such that $0 < \lambda < -\mu$, that define a filter with the same pass-band frequency. In order for the polynomial $f(k) = (1 - \lambda k)(1 - \mu k)$ to define a low-pass filter in the interval $[0, 2]$ it is necessary that $|f(k)| < 1$ in the stop-band region, so that $f_N(k) = f(k)^N \to 0$ when N grows. Since $f(k_{\text{PB}}) = 1$ and $f(k)$ is strictly decreasing for $k > k_{\text{PB}}$, this condition is equivalent to $f(2) > -1$, which translates into the following constraint on λ

$$\lambda < \frac{-k_{\text{PB}} + \sqrt{(2 - k_{\text{PB}})^2 + 4}}{2(2 - k_{\text{PB}})} . \tag{9}$$

As λ increases, the slope of the filter immediately after the pass-band frequency increases, i.e., the filter becomes sharper, but at the same time instability starts to develop at the other end of the spectrum, close to $k = 2$. If the maximum eigenvalue k_n of the matrix K is significantly less than 2 (which is not usually the case) we only need the filter to be stable in the interval $[0, k_n]$ (i.e., $1 > f(k_n) > -1$), and larger values of λ are acceptable. A good estimate of the maximum eigenvalue of K can be obtained with the Lanczos method [3]. Even if the maximum eigenvalue k_n is not known, the signal x to be smoothed may be band-limited, i.e., the coefficients $\hat{x}_i$ in equation (5) associated with high frequencies are all zero, or very close to zero. This condition may be difficult to determine in practice for a particular signal, but if we apply the algorithm with small λ for a certain number of iterations, the resulting signal becomes in effect band-limited. At this point λ can be increased keeping the pass-band frequency constant, maybe even making the filter unstable, and the algorithm can be applied again with the new values of λ and μ for more iterations. This process of increasing λ keeping the pass-band frequency constant can now be repeated again and again. A moderate speed-up is obtained in this way.

5 Filter Design with Windows

The most straightforward approach to traditional digital filter design is to obtain a trigonometric polynomial approximation of the ideal filter transfer function by truncating its Fourier series. The resulting trigonometric polynomial minimizes the L_2 distance to the ideal filter transfer function among all the trigonometric polynomials of the same degree.

To obtain regular polynomials, not trigonometric ones, we first apply the change of variable $k = 2(1 - \cos(\theta))$. This change of variable is a $1 - 1$ mapping $[0, \pi/2] \to [0, 2]$. Then we extend the resulting function to the interval $[-\pi, \pi]$ as follows

$$h_{\text{LP}}(\theta) = \begin{cases} 0 & \pi/2 \le \theta \le \pi \\ f_{\text{LP}}(2(1 - \cos(\theta))) & 0 \le \theta \le \pi/2 \\ h(-\theta) & -\pi \le \theta \le 0 \,. \end{cases}$$

Note that this function, periodic of period 2π and even, is also an ideal low-pass filter as a function of θ

$$h_{\text{LP}}(\theta) = \begin{cases} 1 & \text{if } |\theta| < \theta_{\text{PB}} \\ 0 & \text{otherwise} \end{cases} ,$$

where θ_{PB} is the unique solution of $k_{\text{PB}} = 2(1 - \cos(\theta_{\text{PB}}))$ in $[0, \pi/2]$. Since $h(\theta)$ is an even function, it has a Fourier series expansion in terms of cosines only

$$h_{\text{LP}}(\theta) = h_0 + 2 \sum_{n=0}^{\infty} h_n \cos(n\theta) \,.$$

Now, it is well known that $\cos(n\theta) = T_n(\cos(\theta))$, where T_n is the n-th. Chebyshev polynomial [2], defined by the three term recursion

$$T_n(w) = \begin{cases} 1 & n = 0 \\ w & n = 1 \\ 2 \, w \, T_{n-1}(w) - T_{n-2}(w) & n > 1 \end{cases}$$

The N-th. polynomial approximation of f_{LP} for $k \in [0, 2]$ is then

$$f_N(k) = \frac{\theta_{\text{PB}}}{\pi} T_0(1 - k/2) + \sum_{n=1}^{N} \frac{2 \sin(n \theta_{\text{PB}})}{n \pi} T_n(1 - k/2) \,. \tag{10}$$

Direct truncation of the series leads to the well-known Gibbs phenomenon, i.e., a fixed percentage overshoot and ripple before and after the discontinuity. As it is shown in section 8, this is one of the problems that makes this technique unsatisfactory. The other problem is that the resulting polynomial approximation does not necessarily satisfy the constraint $f_N(0) = 1$, which is required to preserve the average value of the signal (DC level in classical signal processing, centroid in the case of surfaces). Our experiments show that a desirable surface smoothing filter transfer function should be as close as possible to 1 within the pass-band, and then decrease to zero in the stop-band ($[k_{\text{PB}}, 2]$).

A classical technique to control the convergence of the Fourier series is to use a weighting function to modify the Fourier coefficients. In our case the polynomial approximation of equation (10) is modified as follows

$$f_N(k) = w_0 \frac{\theta_{\text{PB}}}{\pi} T_0(1 - k/2) + w_n \sum_{n=1}^{N} \frac{2 \sin(n \theta_{\text{PB}})}{n \pi} T_n(1 - k/2) \,, \tag{11}$$

where $w_0, w_1, \ldots, w_N$ are the weights that constitute a so called *window*. The polynomial approximation of equation (10) is a particular case of (11), where the weights are all equal to 1. This is called the *Rectangular window*. Other popular windows are, the *Hanning window*, the *Hamming window*, and the *Blackman window.*

$$w_n = \begin{cases} 1.0 & \text{Rectangular} \\ 0.5 + 0.5 \, \cos(n\pi/(N+1)) & \text{Hanning} \\ 0.54 + 0.46 \, \cos(n\pi/(N+1)) & \text{Hamming} \\ 0.42 + 0.5 \, \cos(n\pi/(N+1)) + 0.08 \, \cos(2n\pi/(N+1)) & \text{Blackman} . \end{cases} \tag{12}$$

If the low-pass filter must have a very narrow pass-band region, which is usually the case in the surface smoothing application, then a high degree polynomial is necessary to obtain a reasonable approximation. This is in fact a consequence of the uncertainty principle. The phenomenon can be observed even in the case of the rectangular window. The problem is even worse for the other windows, because they have wider main lobes. To obtain a reasonably good approximation of degree N, the pass-band must be significantly wider than the width of the main lobe of the window. If σ is the width of the main lobe of the window, the resulting filter will be approximately equal to one for $\theta \in [0, \theta_{\text{PB}} - \sigma]$, approximately equal to zero for $\theta \in [\theta_{\text{PB}} + \sigma, \pi]$, and approximately decreasing for $\theta \in [\theta_{\text{PB}} - \sigma, \theta_{\text{PB}} + \sigma]$. Our solution in this case of narrow pass-band frequency, is to design the filter for a small value of N, but with the pass-band frequency increased by σ (no longer the width of the main lobe of the window)

$$f_N(k) = w_0 \frac{(\theta_{\text{PB}} + \sigma)}{\pi} T_0(1 - k/2) + w_n \sum_{n=1}^{N} \frac{2 \sin(n(\theta_{\text{PB}} + \sigma))}{n \pi} T_n(1 - k/2) , \tag{13}$$

and then, eventually iterate this filter $(f(k) = f_N(k)^M)$. The value of σ can be determined numerically by maximizing $f(k_{\text{PB}})$ under the constraints $|f(k)| < 1$ for $k_{\text{PB}} < k \leq 2$. In our implementation, we compute the optimal σ with a local root finding algorithm (a few Newton iterations) so that $f_N(k_{\text{PB}}) = 1$, starting from an interactively chosen initial value. Figure 1 shows some examples of filters designed in this way, compared with filters of the same degree and $\sigma = 0$, and with $\lambda - \mu$ filters of the same degree.

6 How to Choose The Pass-Band Frequency

In this section we are concerned with how to choose the pass-band frequency k_{PB} to prevent shrinkage. As in the classical case, since the DFT $\hat{x}$ of a signal x satisfies Parseval's formula, the value of $\hat{x}_i^2$ can be interpreted as the *energy content* of x in the frequency k_i. Similarly, the sum

$$\sum_{k_i \leq k_{\text{PB}}} \hat{x}_i^2$$

measures the energy content of x in the pass-band. Our criterion is to choose the minimum pass-band frequency such that most of the energy of the signal falls

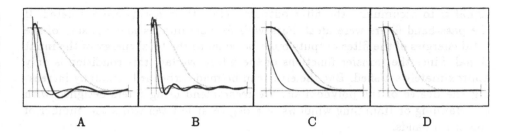

Fig. 1. Filters $f_N(k)$ for $k_{PB} = 0.1$ and $\sigma > 0.0$. (A) Rectangular window, $N = 10$, $\sigma = 0.1353$. (B) Rectangular window, $N = 20$, $\sigma = 0.0637$. (C) Hamming window, $N = 10$, $\sigma = 0.5313$. (D) Hamming window, $N = 20$, $\sigma = 0.2327$. In each of the four cases the thick black line corresponds to the filter described above, the thin black line to the same filter with $\sigma = 0.0$, and the gray line is a $\lambda - \mu$ filter of the same degree and $\lambda = 0.5$.

in the pass-band, i.e., we choose k_{PB} such that

$$\sum_{k_i \leq k_{PB}} \hat{x}_i^2 \geq (1 - \epsilon) \|x\|_D^2 ,$$

where ϵ is a very small number. Of course, since we cannot compute the DFT of x, we cannot minimize this expression exactly. We can only get a rough estimate of the minimizer using the power spectral estimator described in the next section. What value of ϵ to use, and how accurate the estimation should be is application dependent, but in general it should be determined experimentally for a set of typical signals.

7   Power Spectrum Estimation

Ideally, to evaluate the performance of the different low-pass filter algorithms we should measure the DFT of the filter outputs, and check that the high frequency energy content is very small. Since we do not have any practical way of computing the DFT, we estimate the power spectrum of a signal as follows. We partition the interval $[0, 2]$ into a small number of non-overlapping intervals $I^1, \ldots, I^M$, and for each one of this intervals we estimate the energy content of the signal within the interval. We do so by designing a very sharp (high degree) pass-band filter $f^j(k)$ for each interval I^j. The energy content of the signal x within the interval I^j can be estimated by measuring the total energy of the output of corresponding filter applied to the signal

$$\|f^j(K)x\|_D^2 \approx \sum_{k_i \in I^j} \hat{x}_i^2 .$$

By designing all these FIR filters of the same degree, a filter-bank, we can evaluate all of them simultaneously at a greatly reduced computational cost. The only disadvantage is that we need M arrays of the same dimension as the input signal x to accumulate the filter outputs before their norms are evaluated. If the pass-band filters were ideal, Parseval's formula implies that the sum of the total energies of the filter outputs must be equal to the total energy of the input signal. Since the transfer functions of the filters overlap, this condition is only approximately satisfied. But the error can be made arbitrarily small by increasing the degree of the polynomials. We recommend using filters designed with the Hanning or Hamming windows of a degree at lest ten times the number of spectrum bands.

8 Experimental Results

Figure 2 shows the result of applying the filters of figure 1 to the same input surface. The spectrum estimate for the input surface yields the 99.88% of the energy in the band $[0, 0.1]$. This is a typical result for relatively large surfaces, and we have found that a default value $k_{\text{PB}} = 0.1$ produces very good results. But as we pointed out before, the appropriate value for a family of similar signals must be determined experimentally by estimating the spectrum of a typical sample.

The ideal transfer function should be as flat as possible in the pass-band region $(f(k) \approx 1$ for $k \in [0, k_{\text{PB}}])$, and then decrease as fast as possible in the stop-band region $(k \in [k_{\text{PB}}, 2])$. The transfer function of the $\lambda - \mu$ algorithm has this shape, but does not decrease fast enough in the stop-band. The results obtained with rectangular filters are unsatisfactory. The filters designed with the other three windows (Hanning, Hamming, and Blackman), and with increased σ produce transfer functions of similar shape. The Blackman window produces transfer functions that are much flatter in the pass-band, but at the expense of a slower rate of decrease in the stop-band. Hanning and Hamming windows produce similar results, but the Hamming window produces transfer functions with less oscillations. As figure 2 shows, filters designed with the Hamming window produce filters of similar quality as the $\lambda - \mu$ algorithm, but much faster.

References

1. R. Courant and D. Hilbert. *Methods of Mathematical Physics*, volume 1. Interscience, 1953.
2. P.J. Davis. *Interpolation and Approximation*. Dover Publications, Inc., 1975.
3. G. Golub and C.F. Van Loan. *Matrix Computations*. John Hopkins University Press, 2nd. edition, 1989.
4. R.W. Hamming. *Digital Filters*. Prentice Hall, 1989.
5. A.V. Oppenheim and R.W. Schafer. *Digital Signal Processing*. Prentice Hall, Englewood Cliffs, NJ, 1975.
6. E. Seneta. *Non-Negative Matrices, An Introduction to Theory and Applications*. John Wiley & Sons, New York, 1973.
7. G. Taubin. Curve and surface smoothing without shrinkage. In *Proceedings, Fifth International Conference on Computer Vision*, pages 852–857, June 1995.
8. G. Taubin. A signal processing approach to fair surface design. *Computer Graphics*, pages 351–358, August 1995. (Proceedings SIGGRAPH'95).

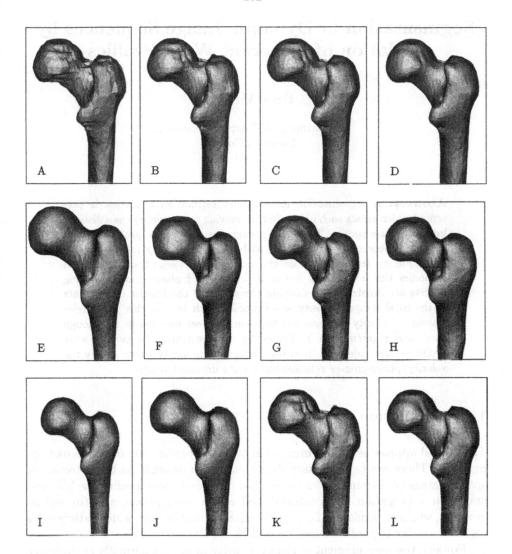

Fig. 2. Filters of Figure 1 applied to the same surface. In all these examples $k_{PB} = 0.1$. (A) Input surface (2565 vertices, 5138 triangles). (B) $\lambda - \mu$ filter $\lambda = 0.5$ $n = 10$. (C) $\lambda - \mu$ filter $\lambda = 0.5$ $n = 20$. (D) $\lambda - \mu$ filter $\lambda = 0.5$ $n = 60$. (E) Rectangular window $\sigma = 0.0$ $n = 10$. (F) Rectangular window $\sigma = 0.0$ $n = 20$. (G) Rectangular window $\sigma = 0.01353$ $n = 10$. (H) Rectangular window $\sigma = 0.06374$ $n = 20$. (I) Hamming window $\sigma = 0.0$ $n = 10$. (J) Hamming window $\sigma = 0.0$ $n = 20$. (K) Hamming window $\sigma = 0.5313$ $n = 10$. (L) Hamming window $\sigma = 0.2327$ $n = 20$.

9. C.T. Zahn and R.Z. Roskies. Fourier descriptors for plane closed curves. *IEEE Transactions on Computers*, 21(3):269–281, March 1972.

Segmentation in Dynamic Image Sequences by Isolation of Coherent Wave Profiles

David Vernon

Department of Computer Science
Maynooth College
Ireland

Abstract. A segmentation and velocity estimation technique is presented which treats each object (either moving or stationary) as a distinct intensity wave profile. The Fourier components of wave profiles — and equally of objects — which move with constant velocity exhibit a regular frequency-dependent phase change. Using a Hough transform which embodies the relationship between velocity and phase change, moving objects are isolated by identifying the subset of the Fourier components of the total image intensity wave profile which exhibit this phase relationship. Velocity is measured by locating local maxima in the Hough space and segmentation is effected by re-constituting the moving wave profile — the object — from the Fourier components which satisfy the velocity/phase-change relationship for the detected velocity.

1 Introduction

Traditional approaches to segmentation typically exploit one of two broad approaches. These are (a) boundary detection, which depends on the detection of spatial intensity discontinuities (using first or second order gradient techniques) and their aggregation into contour-based object descriptions, and (b) region-growing, which depends on the identification of local regions which satisfy some regional similarity predicate (see [1] and [2] for an overview).

Equally, the measurement of object velocity in images normally exploits one of two primary techniques. The first involves the computation of the spatio-temporal gradient, differentiating the (filtered or unfiltered) image sequence with respect to time and subsequently computing the optical flow field (*e.g.* [3]). The second involves the segmentation of the object or feature in question using either region-based gradient (first or second order) filtering and analysis followed either by the computation of the optical flow field or by identification object correspondence, typically by matching contour or region primitives (*e.g.* [4]). Comparisons of the many variations of these approaches and the relationship between them can be found in [5, 6, 7]. A third, lesser-used, approach exploits the regularity in spatiotemporal-frequency representations of the image, such as the spatiotemporal Fourier Transform Domain, resulting from certain types of image motion. Briefly, it can be shown that the spatiotemporal Fourier Transform of an image sequence in which the image content is moving with constant velocity results in a spatiotemporal-frequency representation which is equal to the spatial Fourier Transform of the first image multiplied by a δ-Dirac function in the temporal-frequency domain. This δ-Dirac function is dependent of the image velocity which can be computed if one knows the position of the δ-Dirac function and any spatial frequency [8]. Because this approach is based on image motion, rather than object motion, it normally assumes uniform (zero) background when evaluating object motion. Extensions of the technique have been developed to allow it to cater for situations involving noisy backgrounds [9], several objects [10, 11], and non-uniform cluttered backgrounds [12].

In this paper, we present a segmentation and velocity estimation technique which is based on an alternative formulation of the above spatio-temporal approach. This alternative uses the normal spatial Fourier Transform together with a Hough Transform, rather than the spatiotemporal Fourier Transform. Here, object velocity is computed and segmentation effected by treating each object as a distinct intensity wave profile, with Fourier components, and by identifying the Fourier components which exhibit the magnitude and phase changes which are consistent with detected velocity. This velocity detection is accomplished using an appropriate Hough transform[13] which operates on the phase component of the Fourier Transform of each image in the sequence. Segmentation is effected by re-constituting the moving wave profile — the object — from its identified Fourier components. This approach lends itself to straightforward generalization to types of motion other than the uniform translation in a plane parallel to the image plane, as is normally required. Specifically, the velocity of objects is measured by treating each object (either moving or stationary) as a distinct intensity wave profile, each of which is an additive component of the total image intensity profile, and hence each of which is a solution to the wave equation. The Fourier components of wave profiles — and equally of objects — which move with constant velocity exhibit a regular phase change. The velocity of a moving object is measured by identifying the Fourier components of the total image intensity wave profile which exhibit this phase relationship using an appropriately-designed Hough transform. This Hough transform embodies the relationship between velocity and phase change, and velocity is measured by locating local maxima in the Hough space.

The two major advantages of this technique are that, because the analysis takes place in the Fourier domain, the spatial organization and the visual appearance of the moving object is not significant and, secondly, the formulation presented in this paper lends itself to direct extension for more complex motion. Consequently, objects which are visually or spatially complex and which would be difficult to analyse using either of the traditional spatiotemporal differentiation, feature-based, or region-based approaches can be effectively treated. The objective of this paper is to introduce the alternative formulation and to demonstrate its effectiveness. We will also discuss briefly on-going work in the direct extension of the technique to address situations exhibiting more complex object motion.

2 Overview of the Approach

Consider an image $g(x, y, t)$: a 2-D spatio-temporal representation of the reflectance function of a scene. This image is normally regarded and viewed as a time-varying two-dimensional representation of intensity values. However, the image $g(x, y, t)$ can also be regarded as a time-varying surface. Consider an object O_i to be moving in the image. If we view $g(x, y, t)$ as a time-varying surface, the height of each point on the surface defining the reflectance value at that point, then this object may be viewed as a wave, with a characteristic shape, propagating through the image space with a velocity $v_i(t)$. The velocity function v can, in general, be a function of image coordinates and time: $v(x, y, t)$, that is, it can vary with position and time. For example, consider the motion of an object such as a motor-car travelling toward you on a road. In this paper, however, we will be restricting our attention primarily to the situation where the velocity is constant and parallel to the image plane. This restriction means that the shape of the wave profile does not vary with position and that it propagates with constant velocity: $v_i(t) = v$, a constant. The task then becomes one of isolating the wave (and computing this velocity).

Let us use the general form of the 2-D differential wave equation to model the object O_i or, equivalently, its waveform in image-time space. Thus:

$$\frac{\partial^2 \psi^i(x, y, t)}{\partial x^2} + \frac{\partial^2 \psi^i(x, y, t)}{\partial y^2} = \frac{1}{v_i} \frac{\partial^2 \psi^i(x, y, t)}{\partial t^2}$$

The solution of this wave equation $\psi^i(x, y, t)$ is, in effect, a description of the object as a grey-level wave profile propagating with constant velocity v_i, i.e. $\psi^i(x, y, t) = f^i(x - v_{x_i}t, y - v_{y_i}t)$, where $f^i(x, y)$ is a solution to the wave equation at time $t = 0$ and this solution describes the shape of the wave at time $t = 0$. Our task is to solve this equation for all distinct v_i using only our knowledge of the *total* optical field $\psi = g(x, y, t)$.

Let us assume that the total optical field comprises m objects which we characterize as waves:

$$\psi = g(x, y, t) = \sum_{i=1}^{m} \psi^i(x, y, t) \tag{1}$$

By the principle of linear superposition, the total optical field can also be be decomposed into constituent components:

$$\psi(x, y, t) = \sum_{j=1}^{n} c_j \psi_j(x, y, t)$$

Equally, the wave corresponding to a given object, $\psi^i(x, y, t) = f^i(x - v_{x_i}t, y - v_{y_i}t)$, can be so decomposed:

$$\psi^i(x,y,t) = \sum_{k=1}^{l_i} c_k^i \psi_k^i(x,y,t)$$

where $c_k^i \in \{c_j\}$, $\psi_k^i \in \{\psi_j\}$ and where $\psi_k^i(x,y,t)$ is also a solution to the wave equation.

Recalling that $\psi^i(x,y,t) = f^i(x - v_{x_i}^i t, y - v_{y_i}^i t)$, we have

$$\psi^i(x,y,t) = \sum_{k=1}^{l_i} c_k^i f_k^i(x - v_{x_k}^i t, y - v_{y_k}^i t) \tag{2}$$

such that $v_{x_k} = $ constant, $v_{y_k} = $ constant, $\forall k$: that is, the components of ψ^i all have a constant propagation velocity. Substituting (2) into (1), we have:

$$\psi = g(x,y,t)$$
$$= \sum_{i=1}^{m} \sum_{k=1}^{l_i} c_k^i f_k^i(x - v_{x_k}^i t, y - v_{y_k}^i t)$$

That is, the image $\psi = g(x,y,t)$ is the sum of all the individual wave profiles ψ_i, each wave profile ψ_i comprising components with constant velocity $(v_{x_k}^i, v_{y_k}^i)$. Similarly,

$$\psi(x,y,t) = g(x,y,t)$$
$$= \sum_{j=1}^{n} c_j \psi_j(x,y,t)$$
$$= \sum_{j=1}^{n} c_j f_j(x - v_{x_j} t, y - v_{y_j} t)$$

The task, then, is to decompose the image into constituent components $c_j f_j(x - v_{x_j} t, y - v_{y_j} t)$, each of which satisfies the differential wave equation, and to group them into m sets $c_k^i f_k^i(x - v_{x_k}^i t, y - v_{y_k}^i t)$, $1 \le i \le m$ such that $(v_{x_k}^i, v_{y_k}^i)$ is constant.

We will use the discrete Fourier transform to accomplish the decomposition and the Hough transform to accomplish the grouping.

3 The 2-D Fourier Transform

The Fourier transform $\mathcal{F}(f(x,y)) = F(k_x, k_y)$ of a 2-D function $f(x,y)$ is given by:

$$\mathcal{F}(f(x,y)) = F(k_x, k_y)$$
$$= \int_{-\infty}^{\infty} \int_{-\infty}^{\infty} f(x,y) e^{i(k_x x + k_y y)} dx dy$$

where k_x and k_y are the spatial frequencies in the x and y directions. Note that $F(k_x, k_y)$ is defined on a complex domain with $F(k_x, k_y) = A(k_x, k_y) + iB(k_x, k_y)$ where

$$A(k_x, k_y) = \int_{-\infty}^{\infty} \int_{-\infty}^{\infty} f(x', y') \cos(k_x x' + k_y y') dx' dy'$$

$$B(k_x, k_y) = \int_{-\infty}^{\infty} \int_{-\infty}^{\infty} f(x', y') \sin(k_x x' + k_y y') dx' dy'$$

$F(k_x, k_y)$ may also be expressed in terms of its magnitude and phase:

$$F(k_x, k_y) = |F(k_x, k_y)| \, e^{i\phi(k_x, k_y)}$$

where:

$$|F(k_x, k_y)| = \sqrt{A^2(k_x, k_y) + B^2(k_x, k_y)}$$

$$\phi(k_x, k_y) = \arctan \frac{B^2(k_x, k_y)}{A^2(k_x, k_y)}$$

$|F(k_x, k_y)|$ is the real-valued *amplitude spectrum* and $\phi(k_x, k_y)$ is the real-valued *phase spectrum*. The inverse Fourier transform is given by:

$$f(x, y) = \mathcal{F}^{-1}\left(F(k_x, k_y)\right)$$

$$= \frac{1}{(2\pi)^2} \int_{-\infty}^{\infty} \int_{-\infty}^{\infty} F(k_x, k_y) e^{-i(k_x x + k_y y)} dk_x dk_y$$

$$= \frac{1}{(2\pi)^2} \int_{-\infty}^{\infty} \int_{-\infty}^{\infty} |F(k_x, k_y)| \, e^{i\phi(k_x, k_y)} e^{-i(k_x x + k_y y)} dk_x dk_y$$

In the discrete case, the Fourier transform becomes:

$$\mathcal{F}\left(f(x, y)\right) = F(k_x, k_y)$$

$$= \sum_x \sum_y f(x, y) e^{i(k_x x + k_y y)}$$

and the inverse discrete Fourier transform is:

$$f(x, y) = \mathcal{F}^{-1}\left(F(k_x, k_y)\right)$$

$$= \frac{1}{(2\pi)^2} \sum_{k_x} \sum_{k_y} |F(k_x, k_y)| \, e^{i\phi(k_x, k_y)} e^{-i(k_x x + k_y y)}$$

In effect, $f(x, y)$ can be constructed from a linear combination of elementary functions having the form $e^{-i(k_x x + k_y y)}$, each appropriately weighted in amplitude and phase by a complex factor $F(k_x, k_y)$. However, this construction is valid only for a given time $t = t_0$, say, and, since we are dealing with a wave function $\psi(x, y, t)$ rather than a 2-D spatial image $\psi(x, y)$, we need to develop this formulation of the Fourier and inverse Fourier transforms.

Consider a waveform $f(x, y)$ and an identical waveform shifted to coordinates (x_δ, y_δ), i.e., $f(x - x_\delta, y - y_\delta)$. If the waveform is travelling with constant velocity, then $f(x - x_\delta, y - y_\delta) = f(x - v_x\delta t, y - v_y\delta t)$. The Fourier transform of $f(x - x_\delta, y - y_\delta)$, equivalently $f(x - v_x\delta t, y - v_y\delta t)$, is given by:

$$\mathcal{F}\left(f(x - v_x\delta t, y - v_y\delta t)\right)$$

$$= \mathsf{F}(k_x, k_y)$$

$$= \int_{-\infty}^{\infty}\int_{-\infty}^{\infty} f(x - v_x\delta t, y - v_y\delta t)e^{i(k_x(x - v_x\delta t) + k_y(y - v_y\delta t))}dxdy$$

$$= \int_{-\infty}^{\infty}\int_{-\infty}^{\infty} f(x - v_x\delta t, y - v_y\delta t)e^{i(k_x x + k_y y) - i(k_x v_x\delta t + k_y v_y\delta t)}dxdy$$

$$= \int_{-\infty}^{\infty}\int_{-\infty}^{\infty} f(x - v_x\delta t, y - v_y\delta t)e^{i(k_x x + k_y y)}e^{-i(k_x v_x\delta t + k_y v_y\delta t)}dxdy$$

$$= e^{-i(k_x v_x\delta t + k_y v_y\delta t)}\int_{-\infty}^{\infty}\int_{-\infty}^{\infty} f(x - v_x\delta t, y - v_y\delta t)e^{i(k_x x + k_y y)}dxdy$$

$$= e^{-i(k_x v_x\delta t + k_y v_y\delta t)}\mathsf{F}(k_x, k_y)$$

$$= |\mathsf{F}(k_x, k_y)|\, e^{i\phi(k_x, k_y)}e^{-i(k_x v_x\delta t + k_y v_y\delta t)}$$

Thus, a spatial shift of $(x_\delta, y_\delta) = (v_x\delta t, v_y\delta t)$ of a waveform in the spatial domain, i.e. $f(x, y)$ shifted to $f(x_\delta, y_\delta) = f(v_x\delta t, v_y\delta t)$, only produces a change in the phase of the Fourier components in the frequency domain. This phase change is given by:

$$e^{-i(k_x v_x\delta t + k_y v_y\delta t)}$$

Thus, in order to segment the image into its component waveforms, each of which corresponds to an object moving with constant velocity in the image, we simply need to identify the set of frequency components k_x and k_y which have all been modified by the same phase shift, i.e. $e^{-i(k_x v_x\delta t + k_y v_y\delta t)}$. To accomplish this, we note that the phase spectrum for the shifted wave at time $t + \delta t$ is equal to the phase spectrum of the wave at time t multiplied by the phase change given above:

$$e^{i\phi_{t+\delta t}(k_x, k_y)} = e^{-i(k_x v_x\delta t + k_y v_y\delta t)}e^{i\phi_t(k_x, k_y)}$$

$$= e^{i(\phi_t(k_x, k_y) - (k_x v_x\delta t + k_y v_y\delta t))}$$

Hence:

$$\phi_{t+\delta t}(k_x, k_y) = \phi_t(k_x, k_y) - (k_x v_x\delta t + k_y v_y\delta t)$$

That is, the phase at time $t + \delta t$ is equal to the inital phase at time t minus $(k_x v_x\delta t + k_y v_y\delta t)$. Since we require v_x and v_y, we rearrange as follows:

$$v_y = \frac{1}{k_y\delta t}\left(\phi_t(k_x, k_y) - \phi_{t+\delta t}(k_x, k_y) - k_x v_x\delta t\right)$$

This equation becomes degenerate if $k_y = 0$ in which case we use an alternative re-arrangement as follows:

$$v_x = \frac{(\phi_t(k_x, k_y) - \phi_{t+\delta t}(k_x, k_y))}{k_x \delta t}$$

If we have several images taken at time $t = t_0, t_1, t_2, t_3, ...$, we can compute ϕ_{t_0}, in particular, and $\phi_{t_0+n\delta t}$, in general. Treating the equation above an a Hough transform, with a 2-D Hough transform space defined on v_x, v_y, then we can compute v_y for all possible values of v_x, and for all (known) values of $n, k_x, k_y, \phi_{t_0+n\delta t}(k_x, k_y)$. Local maxima in this v_x, v_y Hough transform space signify Fourier components which comprise waveforms – objects – in the spatial domain which are moving with constant velocity v_x, v_y.

4 Results

Figures 1 through 12 demonstrate the results of applying the technique. Figures 1 and 2 show the first and last images in an eight image sequence featuring a moving dog; figure 3 shows the velocity Hough Transform derived from the phase values of the Fourier Transform of each image in the sequence. The velocity computations are summarized in Table 1. Figures 4 to 6 depict the re-constructed, segmented, image where only frequency components whose phase satisfies the velocity/phase-change relationship embodied in the Hough Transform for the detected velocity are used in the reconstruction; figures 4, 5, and 6 show the results when a minimum phase change threshold of 0.4, 0.3, and 0.2 radians, respectively, is imposed.

Figures 7 through 12 demonstrate the results in a more complex scenario where there is a moving foreground image of a cat superimposed on the background scene depicted in figures 1 and 2. Such a situation arises when, for example, an observer (the cat) views a scene through a window and sees both the external scene and its own reflection.

Image Sequence	Actual Velocity (pixels/frame)		Computed Velocity (pixels/frame)	
	v_x	v_y	v_x	v_y
1	0	3.07	0	3.2
2	0	3.07	0.4	2.8

Table 1. Summary of Measured Velocities

Fig. 1. Image 1 of sequence 1

Fig. 2. Image 8 of sequence 1

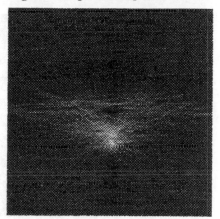

Fig. 3. Hough transform (v_x, v_y) space derived from images 1-8

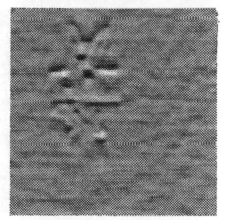

Fig. 4. Re-constructed image (min. phase change of 0.4 radians)

Fig. 5. Re-constructed image (min. phase change of 0.3 radians)

Fig. 6. Re-constructed image (min. phase change of 0.2 radians)

Fig. 7. Image 1 of sequence 2

Fig. 8. Image 8 of sequence 2

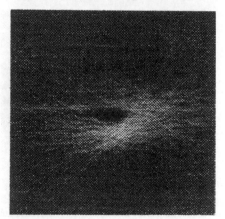

Fig. 9. Hough transform (v_x, v_y) space derived from images 1-8

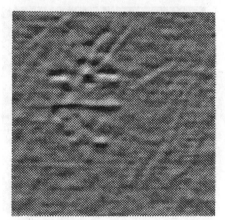

Fig. 10. Re-constructed image (min. phase change of 0.4 radians)

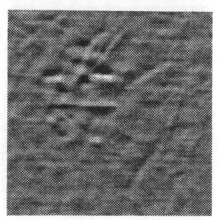

Fig. 11. Re-constructed image (min. phase change of 0.3 radians)

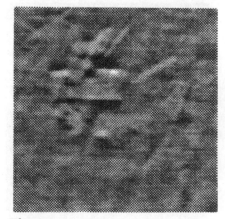

Fig. 12. Re-constructed image (min. phase change of 0.2 radians)

5 Discussion

5.1 Implications of using the Wave Model

The velocity of objects moving parallel to the image plane is measured by treating each object as a distinct intensity wave profile. It is assumed that each wave profile is an additive component of the total image intensity profile and, hence, that each is a solution to the wave equation.

Strictly speaking, this is not a valid assumption since the intensity profile of an object does not add to the intensity profile of the background visual environment; rather, it occludes it. As a result, the occluded part of the background changes as the object moves and this distorts the phase components of the background profile. As it happens, this problem may in fact be useful in circumstances where one is attempting to estimate the motion of transparent or translucent objects where the object and background intensity profile are, to an approximation, additive (such as the situation shown in figures 7 and 8).

5.2 Implications of using the Hough Transform

As noted in the introduction, spatiotemporal-frequency and phase-based approaches to the estimation of object velocity, while not as popular as other approaches, have been successfully used for situations where the objects are translating on a plane parallel to the image plane. The advantage of using the Hough Transform to group the Fourier components rather than a temporal Fourier Transform is that the grouping criterion can be arbitrarily complex (although with consequent increase in computational cost). In this paper we have restricted ourselves to the normal translational motion but the technique can be extended in a very straightforward manner to deal with more complex circumstances as follows.

Object Scaling If the object translates either toward or away from the camera then the perspective lens distortion will result in a scaled image of the object. Such a scaling results in an inverse scaling of the spatial frequencies and this is easily incorporated into the equation defining the Hough Transform. This results in a 3-D Hough Transform defined in terms of v_x, v_y, s: the x and y velocity components and the scaling factor, respectively.

Curvilinear Motion Objects which do not describe translation in a straight line, *i.e.* $v_x = v_x(x, y)$ and $v_y = v_y(x, y)$, can also be estimated if their velocity profiles are continuous. In this case, the object will generate a velocity curve (or crest) in the Hough Transform space rather than a single peak.

Non-Uniform Velocity It is often the case that objects do not move with constant velocity and the velocity is time dependent, *i.e.* $v_x = v_x(t)$ and $v_y = v_y(t)$. In this case, the Hough Transform can be extended by two additional dimensions $v_x(t)$ and $v_y(t)$ to cater for expected velocity profiles. For example, let $v_x = at$ and $v_y = bt$ in which case we have a 4-D transform space defined on v_x, v_y, a, and b.

Rotation about an Axis The situation where an object rotates about a vertical or horizontal axis parallel to the image plane as it translates can be catered for by allowing independent scaling of the object in the horizontal and vertical directions, respectively.

6 Conclusions

A flexible and extendable technique for segmenting and estimating the velocity of objects moving in image sequences has been presented and its efficacy has been demonstrated. It now remains to validate and evaluate the proposed extensions in situations of varying complexity.

References

1. D. Vernon, *Machine Vision* Prentice-Hall International, London (1991).
2. D. Vernon and G. Sandini, *Parallel Computer Vision — The VIS a VIS System*, Ellis Horwood, London (1992).
3. J.H. Duncan and T.-C. Chou, "On the detection and the computation of optical flow", *IEEE Transactions on Pattern Analysis and Machine Intelligence*, **14(3)**, 346-352 (1992).
4. H. Shariat and K.E. Price, "Motion estimation with more than two frames", *IEEE Transactions on Pattern Analysis and Machine Intelligence*, **12(5)**, 417-434 (1990).
5. M. Otte and H.-H. Nagel, "Optical flow estimation: advances and comparisons", *Lecture Notes in Computer Science*, J.O. Eklundh (Ed.), *Computer Vision – ECCV '94*, Springer-Verlag, Berlin, 51-60 (1994).
6. M. Tistarelli, "Multiple constraints for optical flow", *Lecture Notes in Computer Science*, J.O. Eklundh (Ed.), *Computer Vision – ECCV '94*, Springer-Verlag, Berlin, 61-70 (1994).
7. L. Jacobson and H. Wechsler, "Derivation of optical flow using a spatiotemporal-frequency approach", *Computer Vision, Graphics, and Image Processing*, **38**, 29-65 (1987).
8. M.P. Cagigal, L. Vega, P. Prieto, "Object movement characterization from low-light-level images", *Optical Engineering*, **33(8)**, 2810-2812 (1994).
9. M.P. Cagigal, L. Vega, P. Prieto, "Movement characterization with the spatiotemporal Fourier transform of low-light-level images", *Applied Optics*, **34(11)**, 1769-1774 (1995).
10. S. A. Mahmoud, M.S. Afifi, and R. J. Green, "Recognition and velocity computation of large moving objects in images", *IEEE Transactions on Acoustics, Speech, and Signal Processing*, **36(11)**, 1790-1791 (1988).
11. S. A. Mahmoud, "A new technique for velocity estimation of large moving objects", *IEEE Transactions on Signal Processing*, **39(3)**, 741-743 (1991).
12. S.A. Rajala, A. N. Riddle, and W.E. Snyder, "Application of one-dimensional Fourier transform for tracking moving objects in noisy environments", *Computer Vision, Graphics, and Image Processing*, **21**, 280-293 (1983).
13. P.V.C. Hough, 'Method and Means for Recognising Complex Patterrns' U.S. Patent 3,069,654, (1962).

Texture Segmentation Using Local Energy in Wavelet Scale Space

Zhi-Yan Xie and Michael Brady

Robotics Research Group, Department of Engineering Science
University of Oxford, Oxford OX1 3PJ, U.K.

Abstract. Wavelet transforms are attracting increasing interest in computer vision because they provide a mathematical tool for multiscale image analysis. In this paper, we show that i) the subsampled wavelet multiresolution representation is translationally variant; and ii) a wavelet transform of a signal generally confounds the phase component of the analysing wavelet associated with that scale and orientation. The importance of this observation is that commonly used features in texture analysis also depend on this phase component. This not only causes unnecessary spatial variation of features at each scale but also makes it more difficult to match features across scales.
In this paper, we propose a complete 2D decoupled local energy and phase representation of a wavelet transform. As a texture feature, local energy is not only immune to spatial variations caused by the phase component of the analysing wavelet, but facilitates the analysis of similarity of across scales. The success of the approach is demonstrated by experimental results for aerial Infrared Line Scan (IRLS), satellite, and Brodatz images.

1 Introduction

In a multiresolution approach, texture segmentation typically consists of several basic stages: 1) application of a set of wavelet-like filters tuned to different frequency band and orientations; 2) one or more nonlinear operations to the linear filtering outputs; 3) average of the resultant responses to derive "texton" density measure; 4) finally, the segmentation. To choose appropriate filters (wavelets) and adequate nonlinear operations in the first two stages are critical for the subsequent processing.
For a continuous wavelet transform (CWT), it is defined by

$$W_\psi f(a,b) = \frac{1}{\sqrt{a}} \int_{-\infty}^{\infty} f(x)\overline{\psi}(\frac{x-b}{a})\,dx$$

where $a \in \mathbf{R}^+, b \in \mathbf{R}$ are scale and translation parameters and the wavelet $\psi(x)$ can be any bandpass function. If the scale and space parameters of CWT are sampled at $\{a = 2^{-j}; b = 2^{-j}k; j,k \in \mathbf{Z}\}$, and the wavelet $\psi(x)$ is biorthogonal, then the most compact and complete wavelet transform is defined by

$$W_\psi f(j,n) = 2^{\frac{i}{2}} \sum_k f(k)\overline{\psi}(2^j k - n) \overset{\triangle}{=} < f(x), \psi_{j,n}(x) > \qquad (1)$$

where $\psi_{j,n}(x) = 2^{\frac{i}{2}}\psi(2^j k - n)$. Although this representation has been shown useful in image compression because of its completeness and compactness, $W_\psi f(j,n)$ is translationally variant by noticing

$$W_\psi f(j,n) = 2^{\frac{i}{2}} < f(x - x_0), \psi_{j,n}(x) >$$
$$= 2^{\frac{i}{2}} < f(x), \psi_{j,n-2^j x_0}(x) >$$
$$\neq 2^{\frac{i}{2}} < f(x), \psi_{j,n-m}(x) >$$

For an arbitrary translation x_0, $x_0 = 2^{-j}m$; $m \in \mathbf{Z}$ won't in general be satisfied. In other words, if two identical signals were to appear in different positions, their representations $W_\psi f(j,n)$ can be quite different. Such performance may not cause problem for image compression but it badly affects texture segmentation.

The need for non-linear operation after linear filtering has been recognised [5, 1], but there are still no principles determining what kind of operation should be used. Overwhelmingly, the nonlinear operations used to generate features are nonlinear smoothing [3, 5, 7] and energy measure [8, 4]. Although smoothing can remove weak variations, it is also liable to destroy important details. The most commonly used energy measures are full- and half-wave rectification and the square power. However, we show that, in general the features generated by applying these energy measures to the output of the wavelet transform are coupled with the local phase component that depends not only on the analysed image but also on the analysing wavelet at that scale. This is because the output of a wavelet transform oscillate in space depending on the shape of the wavelets. Consequently, their moduli are also affected by the oscillation as illustrated in Figure 1 (top row). Such performance is unacceptable for texture analysis because one wants a uniform feature response in those regions of the image which have uniform texture, while a wavelet is typically an oscillating, wave-like function. Hence, some other nonlinear operation must be found to derive features which can be invariant to *the phase at each scale*.

In this paper, we first define a translationally invariant wavelet transform and propose a definition of 2D local energy and phase in wavelet scale-space. Based on this definition, a complete, decoupled local energy and phase representation of 2D wavelet transforms is presented. As a result of this theory, the local energy is phase independent and used as feature for texture segmentation.

2 Translation Invariant Wavelet Transform

To overcome the translation variant problem of the most compact wavelet transform, we change scale-dependent sampling, i.e. $\{a = 2^{-j}; \ b = 2^{-j}k; \ j, k \in \mathbf{Z}\}$, to a uniform spatial sampling given by $\{a = 2^{-j}; \; b = k; \; j, k \in \mathbf{Z}\}$. Hence, the discrete wavelet transform is redefined by

$$WW_\psi f(j, n) = 2^j \sum_k f(k) \overline{\psi}(2^j(k - n))$$

Clearly, $WW_\psi f(j, n)$ is an oversampled version of $W_\psi f(j, n)$ given in Eqn: 1 and it becomes translationally invariant. Although $WW_\psi f(j, n)$ sacrifices the compactness of the representation compared with $W_\psi f(j, n)$, it provides translational invariance which is essential for texture segmentation, and many other image processing tasks.

Similarly, a translationally invariant 2D separable wavelet transform can be defined by

$$DD_x^j(m, n) = \; < f(x, y), 2^j \phi(2^j(x - m)) 2^j \psi(2^j(y - n)) >$$
$$DD_y^j(m, n) = \; < f(x, y), 2^j \psi(2^j(x - m)) 2^j \phi(2^j(y - n)) >$$
$$DD_d^j(m, n) = \; < f(x, y), 2^j \psi(2^j(x - m)) 2^j \psi(2^j(y - n)) > \quad (2)$$

It can be shown that DD_x, DD_y, DD_d have same orientation emphasis as their subsampled version as given in [6]. More precisely, they give strong response to spatial structures in the horizontal, vertical and diagonal directions[1], and so they are called the horizontal, vertical and diagonal channels, respectively.

3 2D Decoupled Local Energy and Phase

A 1D decoupled local energy and phase representation of a real-valued wavelet transform has been developed using the Hilbert transform [10]. The principal theoretical difficulty in extending the local energy and phase representation of the 1D wavelet transform to 2D is that there does not exist a universal 2D Hilbert transform.

Nevertheless, we propose a definition of the local energy and local phase of a 2D wavelet transform which not only provides a complete representation of a 2D wavelet transform in scale-space, but also facilitates the local energy to be independent from the phase components of the analysing wavelets. Moreover, the relationship between a 2D wavelet transform and its local energy is established both in *scale-space* and in *frequency* space.

3.1 Horizontal and vertical channels

Recall the mother wavelets associated with the horizontal and vertical channels:

$$\Psi^1(x, y) = \phi(x)\psi(y) \qquad \Psi^2(x, y) = \psi(x)\phi(y)$$

---

[1] The orientation of the spatial structure is defined as perpendicular to the direction of maximum gradient.

Since the scaling function $\phi(x)$ is low-pass, $DD_x^j(m,n)$ can be considered as a 1D wavelet transform with respect to $\psi(y)$ for each column (y axis) after first smoothing each row (x axis). Similarly, $DD_y^j(m,n)$ can be considered as a 1D wavelet transform for each row after first smoothing each column. Hence, the local energy and phase can be defined as given in 1D [10]. More precisely,

Definition 1. For a real valued ψ and $f(x,y) \in \mathbf{L}^2(\mathbf{Z}^2)$ the local energy ρ_x, ρ_y and the local phase φ_x, φ_y of $DD_x^j(m,n)$ and $DD_y^j(m,n)$ are given by

$$\rho_x^j(m,n) = \sqrt{[DD_x^j(m,n)]^2 + [H_y\{DD_x^j(m,n)\}]^2}$$

$$\varphi_x^j(m,n) = Atan2\frac{H_y\{DD_x^j(m,n)\}}{DD_x^j(m,n)}$$

$$\rho_y^j(m,n) = \sqrt{[DD_y^j(m,n)]^2 + [H_x\{DD_y^j(m,n)\}]^2}$$

$$\varphi_y^j(m,n) = Atan2\frac{H_x\{DD_y^j(m,n)\}}{DD_y^j(m,n)}$$

where $H_x\{.\}$ ($H_y\{.\}$) denotes the Hilbert transform of the 1D function $DD(m,n)$ when n (m) is fixed.

3.2 Diagonal channel

The diagonal channel of the wavelet transform of a function $f(x,y)$ is given by

$$DD_d^j(m,n) = < f(x,y), 2^{2j}\Psi^3(2^j(x-m), 2^j(y-n)) > \tag{3}$$

The mother wavelet associated with $DD_d^j(m,n)$ is

$$\Psi^3(x,y) = \psi(x)\psi(y) \tag{4}$$

We construct four complex functions as follows

$$G_1(x,y) = [\psi(x) + i\psi_H(x)][\psi(y) + i\psi_H(y)]$$
$$G_2(x,y) = [\psi(x) - i\psi_H(x)][\psi(y) - i\psi_H(y)]$$
$$G_3(x,y) = [\psi(x) + i\psi_H(x)][\psi(y) - i\psi_H(y)]$$
$$G_4(x,y) = [\psi(x) - i\psi_H(x)][\psi(y) + i\psi_H(y)] \tag{5}$$

Noting the following conjugacy relationships,

$$DG_1^j(m,n) = \overline{DG}_2^j(m,n)$$
$$DG_3^j(m,n) = \overline{DG}_4^j(m,n) \tag{6}$$

only one pair $\{DG_k^j(m,n); k=1,3\}$ or $\{DG_k^j(m,n); k=2,4\}$ needs to be considered. In the following discussion, we take the first pair. Substituting $\Psi^3(x,y)$ with $\{G_k(x,y); k=1,3\}$ in Eqn. 3, we generate two complex images by:

$$DG_k^j(m,n) \stackrel{\text{def}}{=} < f(x,y), 2^{2j}G_k(2^j(x-m), 2^j(y-n)) > ; \quad k = 1,3 \tag{7}$$

$$= (f(x,y) * 2^{2j}\overline{G}_k(-2^j x, -2^j y) \tag{8}$$

The properties of functions $\{DG_k^j(m,n); k = 1,3\}$ are essential for deriving the decoupled local energy and local phase representation of the diagonal channel. We present them in the following lemma.

Lemma 2. *For each scale j*

1. *The functions $DG_1^j(m,n)$ can be represented by $DD_d^j(m,n)$ as*

$$DG_1^j(m,n) = DD_d^j(m,n) - H_y\{H_x\{DD_d^j(m,n)\}\} +$$
$$i(H_x\{DD_d^j(m,n)\} + H_y\{DD_d^j(m,n)\}) \qquad (9)$$

and gives a strong response to spatial structures at or close to $\frac{\pi}{4}$.

2. *The functions $DG_3^j(m,n)$ can be represented by $DD_d^j(m,n)$ as*

$$DG_3^j(m,n) = DD_d^j(m,n) + H_y\{H_x\{DD_d^j(m,n)\}\}$$
$$+i(H_x\{DD_d^j(m,n)\} - H_y\{DD_d^j(m,n)\}) \qquad (10)$$

and gives a strong response to spatial structures at, or close to $\frac{3\pi}{4}$.

Proof. See [9].

Now we are in the position to define the local energy and local phase of the diagonal channel:

Definition 3. The function $DG_1^j(m,n)$ is a complex function and can be written as

$$DG_1^j(m,n) = \rho_{x+y}^j(m,n)e^{i\varphi_{x+y}^j(m,n)}$$

where

$$\rho_{x+y}^j(m,n) = \sqrt{[g - H_{xy}\{g\}]^2 + [H_x\{g\} + H_y\{g\}]^2} \qquad (11)$$

$$\varphi_{x+y}^j(m,n) = Atan2[\frac{g - H_{xy}\{g\}}{H_x\{g\} + H_y\{g\}}] \qquad (12)$$

where $g = DD_d^j(m,n)$ and $H_{xy}\{.\}$ denotes the Hilbert transform along x, followed by along y. The $[\rho_{x+y}^j(m,n)]^2$ and $\varphi_{x+y}^j(m,n)$ are called the local energy and the local phase of $DD_d^j(m,n)$, respectively, at or close to $\frac{\pi}{4}$.

Definition 4. The function $DG_3^j(m,n)$ is a complex function and can be written as

$$DG_3^j(m,n) = \rho_{x-y}^j(m,n)e^{i\varphi_{x-y}^j(m,n)}$$

where

$$\rho_{x-y}^j(m,n) = \sqrt{[g + H_{xy}\{g\}]^2 + [H_x\{g\} - H_y\{g\}]^2} \qquad (13)$$

$$\varphi_{x-y}^j(m,n) = Atan2[\frac{g + H_{xy}\{g\}}{H_x\{g\} - H_y\{g\}}] \qquad (14)$$

where $g = DD_d^j(m,n)$ and $[\rho_{x-y}^j(m,n)]^2$ and $\varphi_{x-y}^j(m,n)$ are called the local energy and the local phase of $DD_d^j(m,n)$, respectively, at or close to $\frac{3\pi}{4}$.

Now we have defined four local energy channels for each scale j, denoted by $\rho_x^j(m,n)$, $\rho_y^j(m,n)$, $\rho_{x+y}^j(m,n)$, $\rho_{x-y}^j(m,n)$, which are oriented in the horizontal, vertical and $\frac{\pi}{4}$, $\frac{3\pi}{4}$ directions, respectively. Comparing a patch of local energy and moduli surfaces shown in Figure 1, the phase dependency embedded in the moduli has been removed in the local energy representations. Moreover, the local energy images at different scales become comparable in terms of shape similarity as indicated by the correlation of the images between adjacent scales shown in Figure 1 (bottom row).

3.3 Properties of the local energy and local phase

The following theorem shows that the local energy and local phase defined above provide a complete representation of 2D wavelet transform. The local energy and the wavelet transform are equivalent in frequency domain (conserve energy), but they are very different in scale-space.

Theorem 5. *For a real valued $\psi(x)$ and $f(x,y) \in \mathbf{L}^2(\mathbf{Z}^2)$,*

1. *The wavelet transform $DD_x^j(m,n)$, $DD_y^j(m,n)$ and $DD_d^j(m,n)$ can be represented completely by the local energies and local phases and are given by*

$$DD_x^j(m,n) = \rho_x^j(m,n)\cos\varphi_x^j(m,n)$$
$$DD_y^j(m,n) = \rho_y^j(m,n)\cos\varphi_y^j(m,n)$$
$$DD_d^j(m,n) = \frac{1}{2}(\rho_{x+y}^j(m,n)\cos\varphi_{x+y}^j(m,n) + \rho_{x-y}^j(m,n)\cos\varphi_{x-y}^j(m,n))$$

$$(15)$$

2. *For each scale $j < 0$,*

$$\sum_m \sum_n [DD_x^j(m,n)]^2 = \frac{1}{2}\sum_m \sum_n [\rho_x^j(m,n)]^2$$

$$\sum_m \sum_n [DD_y^j(m,n)]^2 = \frac{1}{2}\sum_m \sum_n [\rho_y^j(m,n)]^2$$

$$\sum_m \sum_n [DD_d^j(m,n)]^2 = \frac{1}{8}\sum_m \sum_n ([\rho_{x+y}^j(m,n)]^2 + [\rho_{x-y}^j(m,n)]^2) \quad (16)$$

Proof. See [9].

The wavelet detail images and their associated local energy images are very different in scale-space: the former confounds the phase component, the latter does not. Further, from Eqns. 15, it is clear that full-, half-wave rectification or squaring of the wavelet transform also confound the phase component.

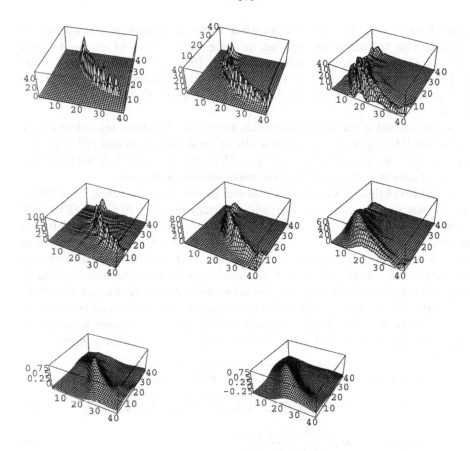

Fig. 1. Comparison between the modulus and the local energy of the wavelet transform. Top row (left to right): the plot of patches of $|DD_d^j|$ of an circular impluse edge, at scale $j = -1, -2, -3$, respectively; middle row: the plot of the local energy ρ_{x-y}^j for the same patch at scale $j = -1, -2, -3$, respectively; bottom row: the plot of the linear correlation coefficients of two local energy images at adjacent scales, $j = -1, -2$ and $j = -2, -3$ respectively.

4 Application to texture segmentation

As shown in the last section, the outputs of real-valued wavelet transform at each scale are coupled with the phase of the wavelet associated with that scale and orientation. As a consequence, squaring, half- and full-wave rectification of the outputs are also phase dependent. Furthermore, it is known that the outputs of wavelet transform at a given location and orientation fail to match across scales according to the local image structures giving rise to the responses. In order to overcome the phase dependency and spatial localisation problems, we propose a four level computation scheme for texture segmentation.

At the **first level**, the 2D **oversampled wavelet transform** is applied to an image. This transform decomposes an image into a stack of images denoted by $DD(\theta, j, x, y)$ at sampled orientation $\theta = \{\theta_1, \cdots, \theta_n\}$ and sampled scale $a = \{2^{-j}; \ j = -1, -2, \cdots, -J\}$. For a 2D separable wavelet transform, an image is decomposed into a pile of images $\{DD_x^j(x, y), DD_y^j(x, y), DD_d^j(x, y); j = -1, -2, \cdots, -J\}$.

The **second level** is a nonlinear operation to remove the phase dependency from each image $DD(\theta, j, x, y)$, to obtain a pile of local energy images $\rho(\theta, j, x, y) = \{\rho_x^j(x, y), \rho_y^j(x, y), \rho_{x+y}^j(x, y), \rho_{x-y}^j(x, y)\}$. This level operates only within a single scale, hence it is also called **intra-scale nonlinear fusion**.

The **third level** derives two texture features in wavelet scale-space, i.e. a multi-scale orientational measure $\alpha(j, x, y)$ and an energy measure $F(j, x, y)$. This level is composed of two sub-processes, namely **inter-scale clustering** and **inter-orientation fusion**. The inter-scale clustering designed to associate the local energy descriptors $\rho(\theta, j, x, y)$ across scales such that, for the resultant new feature image $\rho'(\theta, j, x, y)$, the spatial localisation problem is minimised globally. Unlike the other levels given above, the inter-orientation fusion is not universal. It is specific to each application and to the meaning of different orientation channels. Currently, a simple formula is used to combine four oriented local energy images into quantitative and orientational measures of local energy given by

$$F(j, x, y) = \sqrt{[\rho'(\theta, j, x, y)|_{\theta=0}]^2 + [\rho'(\theta, j, x, y)|_{\theta=\frac{\pi}{2}}]^2}$$

$$+ c * \sqrt{[\rho'(\theta, j, x, y)|_{\theta=\frac{\pi}{4}}]^2 + [\rho'(\theta, j, x, y)]^2|_{\theta=\frac{3\pi}{4}}} \qquad (17)$$

$$\alpha(j, x, y) = \arg\left(\frac{\rho'(\theta, j, x, y)|_{\theta=\frac{\pi}{2}}}{\rho'(\theta, j, x, y)|_{\theta=0}}\right) \qquad (18)$$

The **fourth level** is the segmentation, which is carried out by Gaussian smoothing, clustering and post-processing.

The texture segmentation scheme given above is implemented and has been tested on more than 30 real aerial and satellite images. Typical results are shown in Figure 2. Figure 2 (a), (b) show typical IRLS images taken from a low flying aircraft. The goal (part of a system under development for matching images on successive fly-pasts and matching/constructing a map) is to segment rural and urban areas. The patches in Figure 2 (b) correspond to parks within the surrounding urban area. Figure 2 (c) shows the segmentation of a satellite image taken over Plymouth area, the segmentation result is matched quite well with the map over same area. Finally Figure 2 (d) shows the segmentation of two Brodatz textures [2] (cotton canvas and woolen cloth).

5 Conclusions

In this paper, we developed a complete, decoupled local energy and phase representation of a 2D oversampled wavelet transform. This representation provides

312

(a)   (b)

(c)   (d)

Fig. 2. Examples of texture segmentation results. Texture boundaries are extracted and superimposed on their original images. Top row: Urban regions have been extracted for real IRLS aerial images; bottom row (left to right): Urban regions have been extracted for a satellite image, cotton canvas (in right bottom corner) have been picked up from woolen cloth background for a Brodatz montage image.

- a guide to choose appropriate wavelet by revealing the phase dependency problem associated with widely used real-valued wavelet transforms.
- an approach to construct a complex-valued wavelet from a real-valued wavelet function such that the phase dependency problem can be overcome;
- a method to derive local energy in wavelet scale-space. As a local feature, the local energy is not only immune to spatial variations caused by the phase component of the analysing wavelet, but facilitates the analysis of similarity of across scales;
- a way to formulate 2D Hilbert transform which is still an open problem.

The usefulness of this decoupled local energy and phase representation is demonstrated by its application to segment textures in several classes of natural images.

References

1. A. Blake and A. Zisserman. *Visual Reconstruction*. MIT Press, 1987.
2. Phil Brodatz. *Textures*. Dover Publications, Mineola NY, 1966.
3. A. K. Jain and F. Farrokhnia. Unsupervised texture segmentation using gabor filters. *Pattern Recognition*, 24:1167–1186, 1991.
4. A. Laine and J. Fan. Texture classification by wavelet packet signatures. *IEEE PAMI*, 15(11):1186–1190, 1993.
5. Jitendra Malik and Pietro Perona. Preattentive texture discrimination with early vision mechanisms. *Journal of the Optical Society of America A*, 7(5):923–932, 1990.
6. Stephane G. Mallat. A theory for multiresolution signal decomposition: The wavelet representation. *IEEE PAMI*, 11:671–693, December 1989.
7. T. Randen and J. Husoy. Multiscale filtering for image texture segmentation. *Optical Engineering*, 33:2617–2625, August 1994.
8. A. Sutter, J. Beck, and N. Graham. Contrast and spatial variables in texture segregation:testing a simple spatial-frequency channels model. *Pecept. Psychophys.*, 46:312–332, 1989.
9. Z. Xie. *Multi-scale Analysis and Texture Segmentation*. PhD thesis, University of Oxford, 1994.
10. Zhi-Yan Xie and Michael Brady. A decoupled local energy and phase representation of a wavelet transform. In *VCIP'95 (Visual Communications and Image Processing)*, 1995.

Tracking (1)

Tracking Medical 3D Data with a Deformable Parametric Model

Eric Bardinet, Laurent Cohen, and Nicholas Ayache

^{1} Epidaure project, INRIA,
2004 Route des Lucioles, B.P. 93, 06902 Sophia Antipolis CEDEX, France.
^{2} Ceremade, Université Paris IX - Dauphine,
Place du Maréchal de Lattre de Tassigny, 75775 Paris CEDEX 16, France

Abstract. We present a new approach to surface tracking applied to 3D medical data with a deformable model. It is based on a parametric model composed of a superquadric fit followed by a free form Deformation (FFD), that gives a compact representation of a set of points in a 3D image. We present three different approaches to track a surface in a sequence of 3D cardiac images, from the tracking we infer quantitative parameters which are useful for the physician, like the ejection fraction, the contraction of the heart wall thickness and of the volume during a cardiac cycle or the motion component in the deformation of the ventricle. Experimental results are shown for a tracking and motion analysis of a time sequence of 3D images in Medicine in vivo.

1 Introduction

The analysis of cardiac deformations has received recently a large amount of research in medical image understanding. Indeed, cardiovascular diseases are the first cause of mortality in developed countries. Various imaging techniques make it possible to get dynamic sequences of 3D images (3D + T). The temporal resolution of these techniques is good enough to obtain a sufficient number of images during a complete cardiac cycle (contraction and diastion). These images are therefore wanted to study the behavior of the cardiac system and more they visualize how the heart walls deform. Processing these images opens numerous new applications, like the detection and analysis of pathologies.

The recent developments of imaging, like Nuclear medicine data and Scanner, provide more and more precise resolution in space as well as in time. This means that the data available to the radiologist are larger and larger. To establish a reliable and fast diagnosis, the physician needs models that are defined by a small number of characteristic quantities.

Since it is characteristic of the good health of the heart, the left ventricle motion and deformation has been extensively study studied by medical image processing groups as well as hospitals. Since its creation in 1988, our group has pioneered work in the use of deformable models to extract the left ventricle [2, 5]. Other groups as well have also given various contributions to the understanding of the complex deformation of the ventricle [1, 9, 10, 11].

A parametric model is well suited when dealing with a large amount of data, like for object tracking in a sequence of 3D images. In a previous paper [3], we

Tracking Medical 3D Data with a Deformable Parametric Model

Eric Bardinet[1], Laurent Cohen[2] and Nicholas Ayache[1]

[1] Epidaure project, INRIA
2004 Route des Lucioles, B.P. 93 06902 Sophia Antipolis CEDEX, France.
[2] Ceremade, Université Paris IX - Dauphine
Place du Marechal de Lattre de Tassigny 75775 Paris CEDEX 16, France

Abstract. We present a new approach to surface tracking applied to 3D medical data with a deformable model. It is based on a parametric model composed of a superquadric fit followed by a Free-Form Deformation (FFD), that gives a compact representation of a set of points in a 3D image. We present three different approaches to track surfaces in a sequence of 3D cardiac images. From the tracking, we infer quantitative parameters which are useful for the physician, like the ejection fraction, the variation of the heart wall thickness and of the volume during a cardiac cycle or the torsion component in the deformation of the ventricle. Experimental results are shown for automatic shape tracking and motion analysis of a time sequence of Nuclear Medicine images.

1 Introduction

The analysis of cardiac deformations has given rise to a large amount of research in medical image understanding. Indeed, cardiovascular diseases are the first cause of mortality in developed countries. Various imaging techniques make it possible to get dynamic sequences of 3D images (**3D+T**). The temporal resolution of these techniques is good enough to obtain a sufficient number of images during a complete cardiac cycle (contraction and dilation). These images are perfectly adapted to study the behavior of the cardiac system since they visualize how the heart walls deform. Processing these images opens numerous fields of applications, like the detection and analysis of pathologies.

The recent techniques of imagery, like Nuclear medicine data and Scanner, provide more and more precise resolution in space as well as in time. This means that the data available to the radiologist are larger and larger. To establish a reliable and fast diagnosis, the physician needs models that are defined by a small number of characteristic quantities.

Since it is characteristic of the good health of the heart, the left ventricle motion and deformation has been extensively studied by medical image processing groups as well as hospitals. Since its creation in 1989, our group has pioneered work in the use of deformable models to extract the left ventricle [2, 6, 5]. Other groups as well have also given various contributions to the understanding of the complex deformation of the ventricle [1, 9, 10, 11].

A parametric model is weel-suited when dealing with a huge amount of data like for object tracking in a sequence of 3D images. In a previous paper [3], we introduced a parametric deformable model based on a superquadric fit followed by a Free-Form Deformation (FFD). We show in this paper how we use this parametric model to make an efficient tracking of the LV wall in a sequence of 3D images. The reconstruction and representation of a time sequence of surfaces by a sequence of parametric models will then allow to infer some characteristic parameters which are useful for the physician, like the ejection fraction, the variation of the heart wall thickness and of the volume during a cardiac cycle or the torsion component in the deformation of the ventricle.

2 A parametric model to fit 3D points

In this section, we sketch the deformable model that we use for efficient tracking of the cardiac left ventricle. For more details and references on the complete algorithm, see [3]. In brief, we first fit 3D data with a superellipsoid, and then refine this crude approximation using Free Form Deformations (FFDs).

2.1 Fitting 3D data with superquadrics

The goal of the algorithm is to find a set of parameters such that the superellipsoid best fits the set of data points. Superquadrics form a family of implicit surfaces obtained by extension of conventional quadrics. Superellipsoids are defined by the implicit equation:

$$\left(\left(\left(\frac{x}{a_1} \right)^{\frac{2}{\epsilon_2}} + \left(\frac{y}{a_2} \right)^{\frac{2}{\epsilon_2}} \right)^{\frac{\epsilon_2}{\epsilon_1}} + \left(\frac{z}{a_3} \right)^{\frac{2}{\epsilon_1}} \right)^{\frac{\epsilon_1}{2}} = 1. \tag{1}$$

Suppose that the data we want to fit with the superellipsoid are a set of 3D points $(x_d, y_d, z_d), i = 1, \cdots, N$. Since a point on the surface of the superellipsoid satisfies $F = 1$, where F is the function defined by equation (1), we seek for the minimum of the following energy (see [3] for a geometric interpretation):

$$E(A) = \sum_{i=1}^{N} \left[1 - F(x_d, y_d, z_d, a_1, a_2, a_3, \epsilon_1, \epsilon_2) \right]^2 . \tag{2}$$

2.2 Refinement of the fit with Free Form Deformations (FFDs)

To refine the previous parametric representation, we use a global volumetric deformation called FFD. The main interest of FFDs is that the resulting deformation of the object is just defined by a small number of points. This typical feature allows us to represent voluminous 3D data by models defined by a small number of parameters.

Definition of FFDs FFDs are an application from $\mathbb{R}^3$ to $\mathbb{R}^3$, defined by the tensor product of trivariate Bernstein polynomials. This can be written in a matrix form: $X = BP$, where B is the deformation matrix $ND \times NP$ (ND: number of points on the superellipsoid, NP: number of control points), P is a matrix $NP \times 3$ which contains coordinates of the control points and X is a matrix $ND \times 3$ with coordinates of the model points (see [3] for details).

The inverse problem The superellipsoid fit provides a first parametric approximation of the set of 3D data, which is often a crude one. We use FFDs to refine this approximation. Therefore we need to solve the inverse problem: first compute a displacement field δX between the superellipsoid and the data, and then, after having put the superellipsoid in a 3D box, search the deformation δP of this box which will best minimize the displacement field δX:

$$\min_{\delta P} \| B \delta P - \delta X \|^2 \tag{3}$$

Simultaneous deformation of two surfaces An essential feature of this algorithm is that FFD is a volumetric deformation. This means that several objects can be deformed simultaneously with only one FFD. Using only one model means that the two surfaces are put in a same control point box, and the minimization of equation (3) is done simultaneously on the union of both displacement fields. Moreover, our model gives an interpolation of the 3D deformation everywhere in the volume between the two surfaces. Figure 4 shows the result of the algorithm

	Separate computation	Simultaneous computation
Epicardium	0.007448	0.008236
Endocardium	0.012838	0.014376

Table 1. Least-square errors $\| BP - X \|$ between original data and parametric models. Left column: each model is computed independently. Right column: the two models are computed with one FFD.

for the reconstruction of the epicardium and the endocardium, simultaneously computed with only one FFD. The approximation errors, corresponding to the computation using two FFDs or only one FFD, are presented table 1. One can see that using two FFDs for the two surfaces leads to a better quality of approximation. On the other hand, using only one FFD allows to reduce the number of parameters by half, yielding to a larger compression of the information needed for the description of the parametric model. It also permits to infer from this single FFD, a deformation field over the entire space, due to the volumetric formulation of FFDs. In the particular case of cardiac deformations, it allows to estimate the deformation of any point included in the volume between the epicardium and the endocardium, namely the myocardium.

We show in Figure 1 the effect of the Free Form Deformation applied on the volume between the two superellsoids. To visualize this volume information, we show the image of segments linking these two surfaces. The FFD being computed to obtain simultaneously the epicardium and the endocardium surfaces deforms also the segments that link the surfaces.

3 Dynamic tracking of the left ventricle

In this section, we apply this model to the tracking of the left ventricle in SPECT cardiac images.

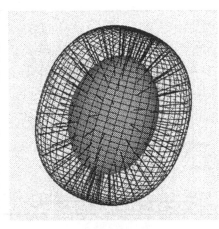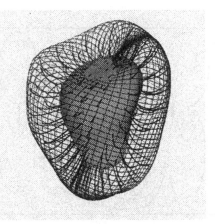

Fig. 1. Volumetric deformation. FFD previously computed simultaneously from the two isosurfaces is applied to rigid links between the superellipsoid models and provides an elastic volumetric deformation of the myocardium.

3.1 Dealing with a time sequence

General deformable models usually need an initialization that is close enough to the solution. This is well suited for tracking in medical images since the deformation between two images is small and the model can start with the solution in the previous image as initialization for the current one.

With our parametric deformable model, the initialization is made automatically through the superquadric fit (see section 2.1), and then refined by the FFD. It is thus possible to make the reconstruction of each data set independently. However, having a previous refined model permits us to get an increasing precision in the reconstruction. This leads to three possible approaches for tracking that are presented in figure 2.

Independent representation This first approach consists of applying to each 3D image the complete model. The advantage is that to define the model at time n, we do not need any previous model information but only the superellipsoid and the control point box for this data.

This approach does not make use of the fact that the results at time n is close to the already computed one at time $n - 1$. This means that there is not a temporal processing but a successive computation of static frames.

Recursive representation This method is a real temporal tracking. The complete model is applied only to the data of the first image, and then for time n, the model is obtained from the one at time $n - 1$. This means that the shape obtained at time $n - 1$ is itself put into a control point box instead of a superellipsoid in section 2.2. It results that the surface at time n is obtained from the

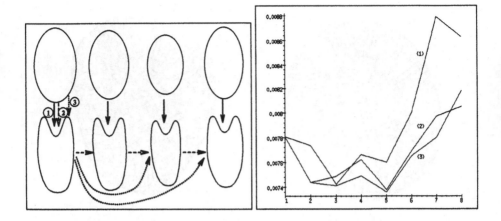

Fig. 2. Left: three different approaches to deal with a temporal sequence. 1 : Data reconstruction at each time step using the superquadric and FFD fit. 2 : Data reconstruction at time step n using the FFD from the model found at time $n-1$. 3 : Data reconstruction at time step n using only the FFD from the model found at time 1. Right: time evolution of the least square error between the data and model for the 8 frames. The three curves correspond to the three approaches. The larger error is obtained with approach 1.

superellipsoid at time 1 iteratively deformed by the sequence of the n first control point boxes. This has the advantage of being more and more precise when time increases since an accumulation of boxes allows the reconstruction of more complex shapes. However, since all previous boxes are needed to reconstruct the data at time n, this may be a difficulty when dealing with a long sequence of images.

Independent representation with a reference deformation The third approach is a trade-off between the two previous ones. The complete model is applied only to the data of the first image, and then for time n, the model is obtained from the one at time 1. This means that the first reconstruction at time 1 is considered as a reference deformation of the superellipsoid. At time n, this reference shape is put into a control point box like in section 2.2. It results that the surface at time n is obtained from the superellipsoid at time 1, followed by two deformations defined by the reference control point box and the current box. This has the advantage of both previous approaches. The approximation is more precise, being the iteration of two boxes and each data set can be retrieved from only one box and the first box and superellipsoid parameters. This is thus independent of the length of the time sequence.

Since in practical applications. this method is as precise as the second one, as shown in figure 2, this is the one we have chosen for the result presented in the next section.

3.2 Application to the left ventricle tracking in spatio-temporal data (3D+T)

We present in this section applications of the tracking algorithm on 3D+T cardiac images.

The models were computed on the following time sequence: Nuclear medicine data (SPECT sequence), with 8 successive time frames during one cardiac cycle. Each image is a volume of $64 \times 64 \times 64$ voxels. The original 3D images are visualized as a series of 2D cross-sections (according to the Z axis) in figure 3.

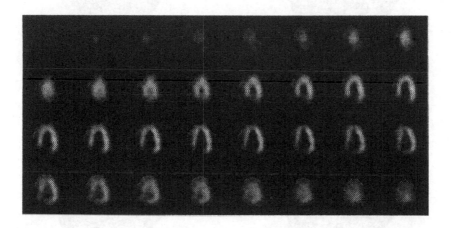

Fig. 3. 3D image of the left ventricle (SPECT data). Order of sections: from left to right and from top to bottom.

Morphological segmentation and representation of the data In order to get time sequences of 3D points which correspond to the anatomical structure that we want to track (epicardium and endocardium of the cardiac left ventricle), and therefore fit our model on these sets of points, we have to segment the original data. As one can visually remark on Figures 3, this is not an easy task because the SPECT images are quite noisy.

To obtain an accurate and robust segmentation, we must combine thresholding with mathematical morphology and connected components analysis (as in Hoehne [8]). We first choose a threshold which grossly separates the ventricle (high values) from the rest of the image. The same value is chosen for the whole sequence of images. Then we extract the largest connected component for each of the resulting 3D binary images, and perform a equal number of erosions and dilations (morphological closings). This last operation is necessary to bridge little gaps and smooth the overall segmentation. Finally, the extraction of the sequence of isosurfaces from that last sequence of images provides the sequence of sets of 3D points that we need as input for the complete reconstruction and tracking algorithm.

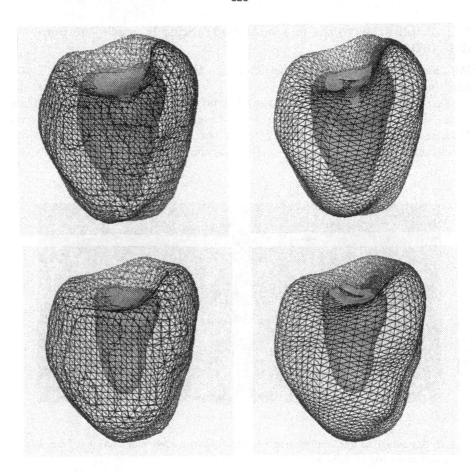

Fig. 4. Time sequence of the epicardium and the endocardium. Left: isosurfaces obtained by data segmentation (4500 + 1500 points). Right: representation by two parametric models (2 × 130 parameters).

To understand the complex behavior of the cardiac muscle, we have to recover the deformations of both the internal and external walls of the myocardium, namely the endocardium and the epicardium. It is possible to represent the complete myocardium by only one deformable superellipsoid-based model (as in Figure 2). However, by recovering the large concavity which corresponds to the ventricular cavity, which means a strong displacement constraint, it follows that all the other constraints are negliged, thus involving a smoothing effect on the surface model. On the other hand, the shape of the myocardium looks very much like two deformed concentric ellipsoids, and it is thus natural to use two models to recover the two cardiac walls.

Figure 4 represents the dynamic sequence of the segmented and reconstructed surfaces, using one model as explained in Section 2.2.

3.3 Quantitative analysis

The reconstruction and representation of a time sequence of surfaces by a sequence of parametric models permits to extract some characteristic parameters and give interpretation or diagnosis on the patient. In the major domain of cardio-vascular diseases, and especially to study the cardiac muscle, the useful parameters for diagnosis are mainly the ejection fraction, the variation of the heart wall thickness and of the volume during a cardiac cycle or the torsion component in the deformation of the ventricle. Similar parameters are also obtained in the work of [10] to quantify the left ventricle deformation.

With this goal in mind, we use our sequence of models to compute the volume of the ventricular cavity, and also to extract the time trajectory of each point of the surface during a cardiac cycle. The assumption made is that the deformation of a point in the parameterization of the surface corresponds to the deformation of the material point of the tissue. Of course, some other constraints could be added to get a better correspondance between the two physical surfaces. In [2, 5], local geometrical properties based on curvature are used to improve the matching between two curves, surfaces or images in a context of registration. However since the deformation is nonrigid, the differential constraints are not always very significant.

Volume evolution To evaluate the ejection fraction, we need a way to compute the evolution in the time sequence of the ventricle cavity volume. We use the discrete form of the Gauss integral theorem to calculate the volume of a region bounded by a grid of points. More details on this formula can be found in [4]. We applied this calculation of endocardium volumes to the sequences of both data points and parametric models obtained in the previous sections. Once we have the values of the volume along a cardiac cycle, we can easily obtain the ejection fraction (calculated precisely as : $\dfrac{Vd - Vc}{Vd}$, with Vd volume at dilation (end diastole), Vc volume at contraction (end systole), see for example [7]. The results presented in figure 5 show that:

- The evolution of the volume has the expected typical shape found in medical litterature [7]. Moreover, estimation of the ejection fraction on our example gives a value of 68%, that is in the range of expected values from medical knowledge [7].
- The volumes found for data and models are almost identical as seen from the error curve. The relative average error along the cycle is 0.42%. This proves that our model is robust with respect to the volume estimation. Of course, the ejection fraction is also obtained with a very small relative error (0.19%).
- The volume evolution found for initial superellipsoid models before FFD, have also a very similar shape. However, there is a size ratio due to the overestimation of the volume before the FFD. This ratio is almost constant in time, which makes possible to get a good estimate of the ejection fraction directly from the initial model. This proves that the superellipsoid model

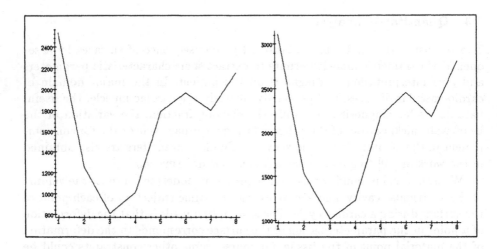

Fig. 5. Endocardium volume during the cardiac cycle. Left: volume of the data. Right: volume of the superellipsoid model.

provides a good global estimate of the shape. Also, the volume of the superellipsoid can be obtained analytically from its set of parameters without the previous discrete approximation (for details, see [4]).

Trajectories Listing the successive positions of a parametric point of the deformed surface model along the time sequence, we obtain the trajectory followed by this point. Figure 6 shows the trajectories of the node points between the end diastole and the end systole. One can see that the model catches the characteristic twist component of the motion. This torsion has been quantified by the decomposition of the displacement vectors in cylindrical coordinates:

$$\begin{pmatrix} x \\ y \\ z \end{pmatrix} \rightarrow \begin{pmatrix} \rho = \sqrt{x^2 + y^2} \\ \theta = Arcos(\dfrac{x}{\sqrt{x^2 + y^2}}) \\ z = z \end{pmatrix} \qquad (4)$$

The z-axis for the cylindrical representation correspond to the z-axis of inertia of the superellipsoid model. To measure the torsion, we compute the difference of the θ parameters for the two points that represent the same parametric point during the contraction, Figure 7 represents the mean values of $\theta' - \theta$ (in radians) along the different latitudes. One can see that the torsion is in the range 10 - 12 degrees which is the expected range.

The pointwise tracking of the deformation permits to give an evaluation of the velocity field during the sequence. The visualization of these displacements by different colors, according to their range, on the surface shows up clearly areas on the ventricle where the deformation is weak (see Figure 6). This visualization could be used by the physician to help localize pathologies like infarcted regions.

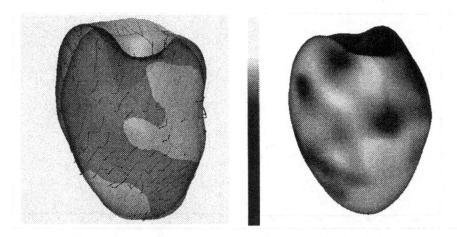

Fig. 6. Left: trajectories of the model's points during a cardiac cycle. The two surfaces represent the the models at end of diastole (dilation) and systole (contraction). Right: range of the displacements of the model's points during a cardiac cycle.

Wall thickness Another important feature that is useful for the diagnosis is the evolution of the wall thickness during the cardiac contraction. Computing only one model to recover the deformation of both the epicardium and the endocardium permits to calculate easily this parameter. Figure 7 shows the evolution of the wall thickness over time for a given point. This thickness has been computed as the difference of the ρ parameters for two parametric points on the epicardium and the endocardium. Figure 8 represents the volumetric deformation of a volume element inside the myocardium muscle. This element is defined by two corresponding rectangle elements on each of the two parameterized surfaces (epicardium and endocardium). The nodes of these rectangles are linked by curvilinear segments that show the volumetric effect of the FFD (see Section 2.2).

4 Conclusion

We presented a new approach to surface tracking applied to 3-D medical data with a deformable model. It is based on a parametric model that gives a compact representation of a set of points in a 3-D image. Three approaches were presented to use this model in order to track efficiently the left ventricle walls in a sequence of 3D images during a cardiac cycle. The model is able to track simultaneously the endocardium and the epicardium. Experimental results have been shown for automatic shape tracking in time sequences.

The reconstruction and representation of a time sequence of surfaces by a sequence of parametric models has then permitted to infer some characteristic parameters which are useful for the physician.

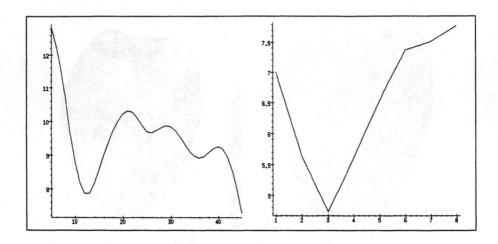

Fig. 7. Left: mean torsion during the cardiac cycle along the z-axis (0 and 50 represent the two poles of the parameterization). Right: evolution of the wall thickness during the cardiac cycle for one point of the volumetric model.

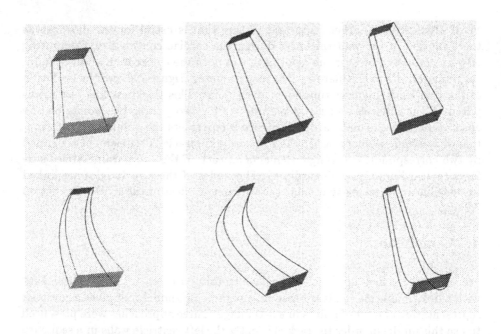

Fig. 8. Volumetric deformation of a volume element inside the myocardium during the cardiac cycle (3 time steps). Top: volume element between the superellipsoid models. Bottom: volume element between the final models, after the volumetric deformation.

We currently consider adding to our work hard displacement constraints like the ones available from the recent imaging technique denoted tagged MRI.

References

1. A. Amini, R. Owen, P. Anandan, and J. Duncan. Non-rigid motion models for tracking the left ventricular wall. In *Information processing in medical images*, Lecture notes in computer science, pages 343–357, 1991. Springer-Verlag.
2. N. Ayache, I. Cohen, and I. Herlin. *Medical Image Tracking*, chapter 20. MIT Press, 1992.
3. E. Bardinet, L. Cohen, and N. Ayache. Fitting of iso-surfaces using superquadrics and free-form deformations. In *Proceedings IEEE Workshop on Biomedical Image Analysis (WBIA)*, Seattle, Washington, June 1994.
4. E. Bardinet, L.D. Cohen, and N. Ayache. Analyzing the deformation of the left ventricle of the heart with a parametric deformable model. Research report, IN-RIA, Sophia-Antipolis, February 1996. (to appear).
5. S. Benayoun, C. Nastar, and N. Ayache. Dense non-rigid motion estimation in sequences of 3D images using differential constraints. In *Proceedings Conference on Computer Vision, Virtual Reality and Robotics in Medecine (CVRMed)*, pages 309–318, Nice, France, April 1995.
6. L.D. Cohen and I. Cohen. Finite element methods for active contour models and balloons for 2-D and 3-D images. *IEEE Transactions on Pattern Analysis and Machine Intelligence*, 15(11), November 1993.
7. M. Davis, B. Rezaie, and F. Weiland. Assessment of left ventricular ejection fraction from technetium-99m-methoxy isobutyl isonitrile multiple-gated radionuclide angiocardoigraphy. *IEEE Transactions on Medical Imaging*, 12(2):189–199, June 1993.
8. K. Höhne and W. Hanson. Interactive 3D segmentation of MRI and CT volumes using morphological operations. *Journal of Computer Assisted Tomography*, 16(2):285–294, March 1992.
9. D. Metaxas and D. Terzopoulos. Shape and nonrigid motion estimation through physics-based synthesis. *IEEE Transactions on Pattern Analysis and Machine Intelligence*, pages 580–591, June 1993.
10. J. Park, D. Metaxas, and A. Young. Deformable models with parameter functions : application to heart-wall modeling. In *Proceedings IEEE Computer Society Computer Vision and Pattern Recognition (CVPR)*, pages 437–442, June 1994.
11. P. Shi, A. Amini, G. Robinson, A. Sinusas, C. Constable, and J. Duncan. Shape-based 4D left ventricular myocardial function analysis. In *Proceedings IEEE Workshop on Biomedical Image Analysis (WBIA)*, pages 88–97, Seattle, June 1994.

EigenTracking: Robust Matching and Tracking of Articulated Objects Using a View-Based Representation

Michael J. Black[1] and Allan D. Jepson[2]

[1] Xerox Palo Alto Research Center, 3333 Coyote Hill Road, Palo Alto, CA 94304
black@parc.xerox.com

[2] Department of Computer Science, University of Toronto, Ontario M5S 3H5
and Canadian Institute for Advanced Research
jepson@vis.toronto.edu

Abstract. This paper describes a new approach for tracking rigid and articulated objects using a view-based representation. The approach builds on and extends work on eigenspace representations, robust estimation techniques, and parameterized optical flow estimation. First, we note that the least-squares image reconstruction of standard eigenspace techniques has a number of problems and we reformulate the reconstruction problem as one of robust estimation. Second we define a "subspace constancy assumption" that allows us to exploit techniques for parameterized optical flow estimation to simultaneously solve for the view of an object and the affine transformation between the eigenspace and the image. To account for large affine transformations between the eigenspace and the image we define an EigenPyramid representation and a coarse-to-fine matching strategy. Finally, we use these techniques to track objects over long image sequences in which the objects simultaneously undergo both affine image motions and changes of view. In particular we use this "EigenTracking" technique to track and recognize the gestures of a moving hand.

1 Introduction

View-based object representations have found a number of expressions in the computer vision literature, in particular in the work on eigenspace representations [10, 13]. Eigenspace representations provide a compact approximate encoding of a large set of training images in terms of a small number of orthogonal basis images. These basis images span a subspace of the training set called the eigenspace and a linear combination of these images can be used to approximately reconstruct any of the training images. Previous work on eigenspace representations has focused on the problem of object recognition and has only peripherally addressed the problem of tracking objects over time. Additionally, these eigenspace reconstruction methods are not invariant to image transformations such as translation, scaling, and rotation. Previous approaches have typically assumed that the object of interest can be located in the scene, segmented, and transformed into a canonical form for matching with the eigenspace. In this paper we will present a robust statistical framework for reconstruction using the eigenspace that will generalize and extend the previous work in the area to ameliorate some of these problems. The work combines lines of research from object recognition using eigenspaces, parameterized optical flow models, and robust estimation techniques into a novel method for tracking objects using a view-based representation.

There are two primary observations underlying this work. First, standard eigenspace techniques rely on a least-squares fit between an image and the eigenspace [10] and this can lead to poor results when there is structured noise in the input image. We reformulate the eigenspace matching problem as one of robust estimation and show how it overcomes the problems of the least-squares approach. Second, we observe that rather than try to represent all possible views of an object from all possible viewing positions, it is more practical to represent a smaller set of canonical views and allow a parameterized transformation (eg. affine) between an input image and the eigenspace. This allows a *multiple-views plus transformation* [12] model of object recognition. What this implies is that matching using an eigenspace representation involves both estimating the view as well as the transformation that takes this view into the image. We formulate this problem in a robust estimation framework and simultaneously solve for the view and the transformation. For a particular view of an object we define a *subspace constancy assumption* between the eigenspace and the image. This is analogous to the "brightness constancy assumption" used in optical flow estimation and it allows us to exploit parameterized optical flow techniques to recover the transformation between the eigenspace and the image. Recovering the view and transformation requires solving a non-linear optimization problem which we minimize using gradient descent with a continuation method. To account for large transformations between model and image we define an EigenPyramid representation and a coarse-to-fine matching scheme. This method enables the tracking of previously viewed objects undergoing general motion with respect to the camera. This approach, which we call *EigenTracking*, can be applied to both rigid and articulated objects and can be used for object and gesture recognition in video sequences.

2   Related Work

While eigenspaces are one promising candidate for a view-based object representation, there are still a number of technical problems that need to be solved before these techniques can be widely applied. First, the object must be located in the image. It is either assumed that the object can be detected by a simple process [9, 10] or through global search [9, 13]. Second, the object must be segmented from the background so that the reconstruction and recognition is based on the object and not the appearance of the background. Third, the input image must be be transformed (for example by translation, rotation, and scaling) into some canonical form for matching. The robust formulation and continuous optimization framework presented here provide a local search method that is robust to background variation and simultaneously matches the eigenspace and image while solving for translation, rotation, and scale.

To recognize objects in novel views, traditional eigenspace methods build an eigenspace from a dense sampling of views [6, 7, 10]. The eigenspace coefficients of these views are used to define a surface in the space of coefficients which interpolates between views. The coefficients of novel views will hopefully lie on this surface. In our approach we represent views from only a few orientations and recognize objects in other orientations by recovering a parameterized transformation (or warp) between the image and the eigenspace. This is consistent with a model of human object recognition that suggests that objects are represented by a set of views corresponding to familiar orientations and that new views are transformed to one of these stored views for recognition [12].

To track objects over time, current approaches assume that simple motion detection and tracking approaches can be used to locate objects and then the eigenspace matching verifies the object identity [10, 13]. What these previous approaches have failed to exploit is that the eigenspace itself provides a representation (i.e. an image) of the object that can be used for tracking. We exploit our robust parameterized matching scheme to perform tracking of objects undergoing affine image distortions and changes of view.

This differs from traditional image-based motion and tracking techniques which typically fail in situations in which the viewpoint of the object changes over time. It also differs from tracking schemes using 3D models which work well for tracking simple rigid objects. The EigenTracking approach encodes the appearance of the object from multiple views rather than its structure.

Image-based tracking schemes that emphasize learning of views or motion have focused on region contours [1, 5]. In particular, Baumberg and Hogg [1] track articulated objects by fitting a spline to the silhouette of an object. They learn a view-based representation of people walking by computing an eigenspace representation of the knot points of the spline over many training images. Our work differs in that we use the brightness values within an image region rather than the region outline and we allow parameterized transformations of the input data in place of the standard preprocessing normalization.

3 Eigenspace Approaches

Given a set of images, eigenspace approaches construct a small set of basis images that characterize the majority of the variation in the training set and can be used to approximate any of the training images. For each $n \times m$ image in a training set of p images we construct a 1D column vector by scanning the image in the standard lexicographic order. Each of these 1D vectors becomes a column in a $nm \times p$ matrix A. We assume that the number of training images, p, is less than the number of pixels, nm and we use Singular Value Decomposition (SVD)[3] to decompose the matrix A as

$$A = U \Sigma V^T. \tag{1}$$

U is an orthogonal matrix of the same size as A representing the principle component directions in the training set. Σ is a diagonal matrix with singular values $\sigma_1, \sigma_2, \ldots, \sigma_p$ sorted in decreasing order along the diagonal. The $p \times p$ orthogonal matrix V^T encodes the coefficients to be used in expanding each column of A in terms of the principle component directions.

If the singular values σ_k, for $k \geq t$ for some t, are small then, since the columns of U are orthonormal, we can approximate some new column $\mathbf{e}$ as

$$\mathbf{e}^* = \sum_{i=1}^{t} c_i U_i. \tag{2}$$

---

[3] Other approaches have been described in the literature (cf. [10]).

Sample images from the training set:

First few principle components:

Fig. 1. Example that will be used to illustrate ideas throughout the paper.

where the c_i are scalar values that can be computed by taking the dot product of $\mathbf{e}$ and the column U_i. This amounts to a projection of the input image, e, onto the subspace defined by the t basis vectors.

For illustration we constructed an eigenspace representation for soda cans. Figure 1 (top row) shows some example soda can images in the training set which contained 200 images of Coke and 7UP cans viewed from the side. The eigenspace was constructed as described above and the first few principle components are shown in the bottom row of Figure 1.[4] For the experiments in the remainder of the paper, 50 principle components were used for reconstruction. While fewer components could be used for recognition, EigenTracking will require a more accurate reconstruction.

4 Robust Matching

The approximation of an image region by a linear combination of basis vectors can be thought of as "matching" between the eigenspace and the image. This section describes how this matching process can be made robust.

Let $\mathbf{e}$ be an input image region, written as a $nm \times 1$ vector, that we wish to match to the eigenspace. For the standard approximation $\mathbf{e}^*$ of $\mathbf{e}$ in Equation (2), the coefficients c_i are computed by taking the dot product of $\mathbf{e}$ with the U_i. This approximation corresponds to the least-squares estimate of the c_i [10]. In other words, the c_i are those that give a reconstructed image that minimizes the squared error $E(\mathbf{c})$ between $\mathbf{e}$ and $\mathbf{e}^*$ summed

---

[4] In this example we did not subtract the mean image from the training images before computing the eigenspace. The mean image corresponds to the first principle component resulting in one extra eigenimage.

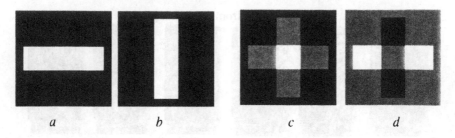

Fig. 2. A simple example. *(a, b):* Training images. *(c, d):* Eigenspace basis images.

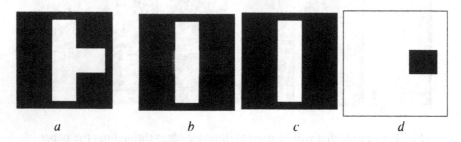

Fig. 3. Reconstruction. *(a):* New test image. *(b):* Least-squares reconstruction. *(c):* Robust reconstruction. *(d):* Outliers (shown as black pixels).

over the entire image:

$$E(\mathbf{c}) = \sum_{j=1}^{n \times m} (\mathbf{e}_j - \mathbf{e}_j^*)^2 = \sum_{j=1}^{n \times m} \left(\mathbf{e}_j - \left(\sum_{i=1}^{t} c_i U_{i,j} \right) \right)^2. \tag{3}$$

This least-squares approximation works well when the input images have clearly segmented objects that look roughly like those used to build the eigenspace. But it is commonly known that least-squares is sensitive to gross errors, or "outliers" [8], and it is easy to construct situations in which the standard eigenspace reconstruction is a poor approximation to the input data. In particular, if the input image contains structured noise (eg. from the background) that can be represented by the eigenspace then there may be multiple possible matches between the image and the eigenspace and the least-squares solution will return some combination of these views.

For example consider the very simple training set in Figure 2 (*a* and *b*). The basis vectors in the eigenspace are shown in Figure 2 (*c*, *d*).[5] Now, consider the test image in Figure 3*a* which does not look the same as either of the training images. The least-squares reconstruction shown in Figure 3*b* attempts to account for all the data but this cannot be done using a linear combination of the basis images. The robust formulation described below recovers the dominant feature which is the vertical bar (Figure 3*c*) and to do so, treats the data to the right as outliers (black region in Figure 3*d*).

---

[5] We subtracted the mean from each of Figure 2 *a* and *b* and included the constant image in the expansion basis.

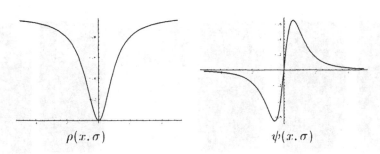

$$\rho(x,\sigma) \qquad\qquad\qquad \psi(x,\sigma)$$

Fig. 4. Robust error norm (ρ) and its derivative (ψ).

To robustly estimate the coefficients $\mathbf{c}$ we replace the quadratic error norm in Equation (3) with a robust error norm, ρ, and minimize

$$E(\mathbf{c}) = \sum_{j=1}^{n \times m} \rho\left(\left(\mathbf{e}_j - \left(\sum_{i=1}^{t} c_i U_{i,j}\right)\right), \sigma\right). \qquad (4)$$

where σ is a scale parameter. For the experiments in this paper we take ρ to be

$$\rho(x,\sigma) = \frac{x^2}{\sigma + x^2}, \qquad \frac{\partial}{\partial x}\rho(x,\sigma) = \psi(x,\sigma) = \frac{2x\sigma^2}{(\sigma^2 + x^2)^2}. \qquad (5)$$

which is a robust error norm that has been used extensively for optical flow estimation [3, 4]. The shape of the function, as shown in Figure 4, is such that it "rejects", or downweights, large residual errors. The function $\psi(x,\sigma)$, also shown in Figure 4, is the derivative of ρ and characterizes the influence of the residuals. As the magnitudes of residuals ($\mathbf{e}_j - \mathbf{e}_j^*$) grow beyond a point their influence on the solution begins to decrease and the value of $\rho(\cdot)$ approaches a constant.

The value σ is a scale parameter that affects the point at which the influence of outliers begins to decrease. By examining the ψ-function we see that this "outlier rejection" begins where the second derivative of ρ is zero. For the error norm used here, this means that those residuals where $|(\mathbf{e}_j - \mathbf{e}_j^*)| > \sigma/\sqrt{3}$ can be viewed as outliers.

The computation of the coefficients $\mathbf{c}$ involves the minimization of the non-linear function in Equation (4). We perform this minimization using a simple gradient descent scheme with a continuation method that begins with a high value for σ and lowers it during the minimization (see [2, 3, 4] for details). The effect of this procedure is that initially no data are rejected as outliers then gradually the influence of outliers is reduced. In our experiments we have observed that the robust estimates can tolerate roughly 35 − 45% of the data being outliers.

4.1 Outliers and Multiple Matches

As we saw in Figure 3 it is possible for an input image to contain a brightness pattern that is not well represented by any single "view". Given a robust match that recovers the

| Image | Least Squares | Robust 1 | Outliers 1 | Robust 2 | Outliers 2 |

Fig. 5. Robust matching with structured noise.

"dominant" structure in the input image, we can detect those points that were treated as outliers. We define an outlier vector, or "mask", **m** as

$$m_j = \begin{cases} 0 & |(\mathbf{e}_j - \mathbf{e}_j^-)| \leq \sigma/\sqrt{3} \\ 1 & \text{otherwise.} \end{cases}$$

If a robust match results in a significant number of outliers, then additional matches can be found by minimizing

$$E(\mathbf{c}) = \sum_{j=1}^{n \times m} m_j \, \rho \left(\left(\mathbf{e}_j - \left(\sum_{i=1}^{t} c_i U_{i,j} \right) \right) . \sigma \right). \tag{6}$$

4.2 Robust Matching Examples

An example will help illustrate the problems with the least-squares solution and the effect of robust estimation. Figure 5 shows an artificial image constructed from two images that were present in the training data (the image is 2/3 Coke can and 1/3 7UP can). It is impossible to reconstruct the entire input image accurately with the eigenspace despite the fact that both parts of the image can be represented independently. The least-squares solution recovers a single view that contains elements of both possible views. The robust estimation of the linear coefficients results in a much more accurate reconstruction of the dominant view (Figure 5, Robust 1). Moreover, we can detect those points in the image that did not match the reconstruction very well and were treated as outliers (black points in Figure 5, Outliers 1) Equation (6) can be used to recover the view corresponding to the outliers (Figure 5, Robust 2) and even with very little data, the reconstructed image reasonably approximates the view of the 7UP can.

5   EigenSpaces and Parametric Transformations

The previous section showed how robust estimation can improve the reconstruction of an image that is already aligned with the eigenspace. In this section we consider how to achieve this alignment in the first place. It is impractical to represent all possible views of an object at all possible scales and all orientations. One must be able to recognize a familiar object in a previously unseen pose and hence we would like to represent a small set of views and recover a transformation that maps an image into the eigenspace. In the previous section we formulated the matching problem as an explicit non-linear parameter estimation problem. In this section we will simply extend this problem formulation with the addition of a few more parameters representing the transformation between the image and the eigenspace.

To extend eigenspace methods to allow matching under some parametric transformation we to formalize a notion of "brightness constancy" between the eigenspace and the image. This is a generalization of the notion of brightness constancy used in optical flow which states that the brightness of a pixel remains constant between frames but that its location may change. For eigenspaces we wish to say that there is a view of the object, as represented by some linear combination of the basis vectors, U_i, such that pixels in the reconstruction are the same brightness as pixels in the image given the appropriate transformation. We call this the *subspace constancy assumption*.

Let $U = [U_1, U_2, \ldots, U_t]$, $\mathbf{c} = [c_1, c_2, \ldots, c_t]^T$, and

$$U\mathbf{c} = \sum_{i=1}^{t} c_i U_i. \tag{7}$$

where $U\mathbf{c}$ is the approximated image for a particular set of coefficients. $\mathbf{c}$. While $U\mathbf{c}$ is a $nm \times 1$ vector we can index into it as though it were an $n \times m$ image. We define $(U\mathbf{c})(\mathbf{x})$ to be the value of $U\mathbf{c}$ at the position associated with pixel location $\mathbf{x} = (x, y)$.

Then the robust matching problem from the previous section can be written as

$$E(\mathbf{c}) = \sum_{\mathbf{x}} \rho(I(\mathbf{x}) - (U\mathbf{c})(\mathbf{x}), \sigma). \tag{8}$$

where I is an $n \times m$ sub-image of some larger image. Pentland *et al.* [11] call the residual error $I - U\mathbf{c}$ the distance-from-feature-space (DFFS) and note that this error could be used for localization and detection by performing a global search in an image for the best matching sub-image. Moghaddam and Pentland extend this to search over scale by constructing multiple input images at various scales and searching over all of them simultaneously [9]. We take a different approach in the spirit of parameterized optical flow estimation. First we define the subspace constancy assumption by parameterizing the input image as follows

$$I(\mathbf{x} + \mathbf{u}(\mathbf{x}, \mathbf{a})) = (U\mathbf{c})(\mathbf{x}), \quad \forall \mathbf{x}. \tag{9}$$

where $\mathbf{u}(\mathbf{x}, \mathbf{a}) = (u(\mathbf{x}, \mathbf{a}), v(\mathbf{x}, \mathbf{a}))$ represents an image transformation (or motion), u and v represent the horizontal and vertical displacements at a pixel, and the parameters $\mathbf{a}$ are to be estimated. For example we may take $\mathbf{u}$ to be the affine transformation

$$u(\mathbf{x}, \mathbf{a}) = a_0 + a_1 x + a_2 y$$
$$v(\mathbf{x}, \mathbf{a}) = a_3 + a_4 x + a_5 y$$

where x and y are defined with respect to the image center. Equation (9) states that there should be some transformation, $\mathbf{u}(\mathbf{x}, \mathbf{a})$, that, when applied to image region I, makes I

look like some image reconstructed using the eigenspace. This transformation can be thought of as *warping* the input image into the coordinate frame of the training data.

Our goal is then to simultaneously find the **c** and **a** that minimize

$$E(\mathbf{c}, \mathbf{a}) = \sum_{\mathbf{x}} \rho(I(\mathbf{x} + \mathbf{u}(\mathbf{x}, \mathbf{a})) - (U\mathbf{c})(\mathbf{x}), \sigma). \tag{10}$$

As opposed to the exhaustive search techniques used by previous approaches [9, 13], we derive and solve a continuous optimization problem.

First we rewrite the left hand side of Equation (9) using a first order Taylor series expansion

$$I(\mathbf{x}) + I_x(\mathbf{x})u(\mathbf{x}, \mathbf{a}) + I_y(\mathbf{x})v(\mathbf{x}, \mathbf{a}) = (U\mathbf{c})(\mathbf{x})$$

where I_x and I_y are partial derivatives of the image in the x and y directions respectively. Reorganizing terms gives

$$I_x(\mathbf{x})u(\mathbf{x}, \mathbf{a}) + I_y(\mathbf{x})v(\mathbf{x}, \mathbf{a}) + (I(\mathbf{x}) - (U\mathbf{c})(\mathbf{x})) = 0. \tag{11}$$

This is very similar to the standard optical flow constraint equation where the $U\mathbf{c}$ has replaced $I(\mathbf{x}, t - 1)$ and $(I - U\mathbf{c})$ takes the place of the "temporal derivative".

To recover the coefficients of the reconstruction as well as the transformation we combine the constraints over the entire image region and minimize

$$E(\mathbf{c}, \mathbf{a}) = \sum_{\mathbf{x}} \rho(I_x(\mathbf{x})u(\mathbf{x}, \mathbf{a}) + I_y(\mathbf{x})v(\mathbf{x}, \mathbf{a}) + (I(\mathbf{x}) - (U\mathbf{c})(\mathbf{x})), \sigma) \tag{12}$$

with respect to **c** and **a**. As in the previous section, this minimization is performed using a simple gradient descent scheme with a continuation method that gradually lowers σ. As better estimates of **a** are available, the input image is warped by the transformation $\mathbf{u}(\mathbf{x}, \mathbf{a})$ and this warped image is used in the optimization. As this warping registers the image and the eigenspace, the approximation $U\mathbf{c}$ gets better and better. This minimization and warping continues until convergence. The entire non-linear optimization scheme is described in greater detail in [2].

Note that this optimization scheme will not perform a global search to "find" the image region that matches the stored representation. Rather, given an initial guess, it will refine the pose and reconstruction. While the initial guess can be fairly coarse as described below, the approach described here does not obviate the need for global search techniques but rather compliments them. In particular, the method will be useful for tracking an object where a reasonable initial guess is typically available.

EigenPyramids. As in the case of optical flow, the constraint equation (11) is only valid for small transformations. The recovery of transformations that result in large pixel differences necessitates a coarse-to-fine strategy. For every image in the training set we construct a pyramid of images by spatial filtering and sub-sampling (Figure 6). The images at each level in the pyramid form distinct training sets and at each level SVD is used to construct an eigenspace description of that level.

The input image is similarly smoothed and subsampled. The coarse-level input image is then matched against the coarse-level eigenspace and the values of **c** and **a** are

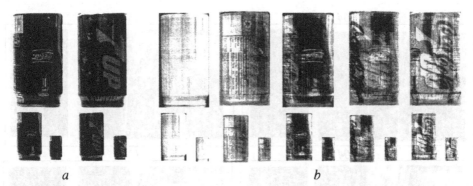

a   *b*

Fig. 6. Example of EigenPyramids. *a:* Sample images from the training set. *b:* First few principle components in the EigenPyramid.

estimated at this level. The new values of **a** are then projected to the next level (in the case of the affine transformation the values of a_0 and a_3 are multiplied by 2). This **a** is then used to warp the input image towards the eigenspace and the value of **c** is estimated and the a_i are refined. The process continues to the finest level.

6 EigenTracking

The robust parameterized matching scheme described in the previous section can be used to track objects undergoing changes in viewpoint or changes in structure. As an object moves and the view of the object changes, we recover both the current view of the object and the transformation between the current view and the image. It is important to note that no "image motion" is being used to "track" the objects in this section. The tracking is achieved entirely by the parameterized matching between the eigenspace and the image. We call this *EigenTracking* to emphasize that a view-based representation is being used to track an object over time.

For the experiments here a three-level pyramid was used and the value of σ started at $65\sqrt{3}$ and was lowered to a minimum of $15\sqrt{3}$ by a factor of 0.85 at each of 15 stages. The values of **c** and **a** were updated using 15 iterations of the descent scheme at each stage, and each pyramid level. The minimization was terminated if a convergence criterion was met. The algorithm was given a rough initial guess of the transformation between the first image and the eigenspace. From then on the algorithm automatically tracked the object by estimating **c** and **a** for each frame. No prediction scheme was used and the motion ranged from 0 to about 4 pixels per frame. For these experiments we restricted the transformation to translation, rotation, and scale.

6.1 Pickup Sequence

First we consider a simple example in which a hand picks up a soda can. The can undergoes translation and rotation in the image plane (Figure 7). The region corresponding to the eigenspace is displayed as white box in the image. This box is generated by projecting the region corresponding to the eigenspace onto the image using the inverse of the

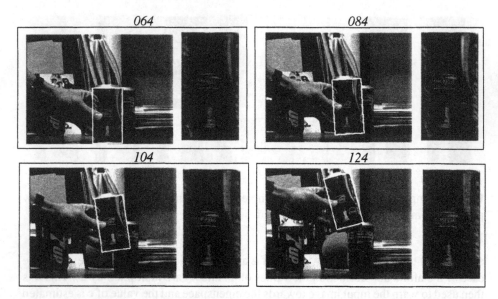

Fig. 7. Pickup Sequence. EigenTracking with translation and rotation in the image plane. Every 20 frames in the 75 frame sequence are shown.

estimated transformation between the image and the eigenspace. This projection serves to illustrate the accuracy of the recovered transformation. Beside each image is shown the robust reconstruction of the image region within the box.

6.2 Tracking a Rotating Object

Figure 8 shows the tracking of a soda can that translates left and right while moving in depth over 200 frames. While the can is changing position relative to the camera it is also undergoing rotations about its major axis. What this means is that the traditional brightness constancy assumption of optical flow will not track the "can" but rather the "texture" on the can. The subspace constancy assumption, on the other hand, means that we will recover the transformation between our eigenspace representation of the can and the image. Hence, it is the "can" that is tracked rather than "texture".

More details are provided to the right of the images. On the left of each box is the "stabilized" image which shows how the original image is "warped" into the coordinate frame of the eigenspace. Notice that the background differs over time as does the view of the can, but that the can itself is in the same position and at the same scale. The middle image in each box is the robust reconstruction of the image region being tracked. On the right of each box (in black) are the "outliers" where the observed image and the reconstruction differed by more than $\sigma/\sqrt{3}$.

6.3 Articulated Motion and Gesture Recognition

A final example considers the problem of recognizing hand gestures in video sequences in which the hand is moving. We define a simple set of four hand gestures illustrated in

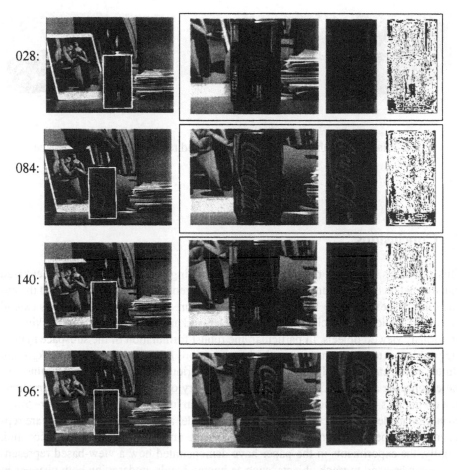

028:

084:

140:

196:

Fig. 8. EigenTracking with translation and divergence over 200 frames. The soda can rotates about its major axis while moving relative to the camera.

Figure 9. A 100 image training set was collected by fixing the wrist position and recording a hand alternating between these four gestures. The eigenspace was constructed and 25 basis vectors were used for reconstruction. In our preliminary experiments we have found brightness images to provide sufficient information for both recognition and tracking of hand gestures (cf. [9]).

Figure 10 shows the tracking algorithm applied to a 100 image test sequence in which a moving hand executed the four gestures. The motion in this sequence was large (as much as 15 pixels per frame) and the hand moved while changing gestures. The figure shows the backprojected box corresponding to the eigenspace model and, to the right, on top, the reconstructed image. Below the reconstructed image is the "closest" image in the original training set (taken to be the smallest Euclidean distance in the space of coefficients). While more work must be done, this example illustrates how eigenspace approaches might provide a view-based representation of articulated objects. By allowing parameterized transformations we can use this representation to track and recognize human gestures.

Fig. 9. Examples of the four hand gestures used to construct the eigenspace.

7 Conclusions

This paper has described robust eigenspace matching, the recovery of parameterized transformations between an image region and an eigenspace representation, and the application of these ideas to EigenTracking and gesture recognition. These ideas extend the useful applications of eigenspace approaches and provide a new form of tracking for previously viewed objects. In particular, the robust formulation of the subspace matching problem extends eigenspace methods to situations involving occlusion, background clutter, noise, etc. Currently these problems pose serious limitations to the usefulness of the eigenspace approach. Furthermore, the recovery of parameterized transformations in a continuous optimization framework provides an implementation of a *views+transformation* model for object recognition. In this model a small number of views are represented and the transformation between the image and the nearest view is recovered. Finally, the experiments in the paper have demonstrated how a view-based representation can be used to track objects, such as human hands, undergoing both changes in viewpoint and and changes in pose.

References

1. A. Baumberg and D. Hogg. Learning flexible models from image sequences. In J. Eklundh, editor, *ECCV-94*, vol. 800 of *LNCS-Series*, pp. 299–308, Stockholm, 1994.
2. M. J. Black and A. D. Jepson. EigenTracking: Robust matching and tracking of articulated objects using a view-based representation. Tech. Report T95-00515, Xerox PARC, Dec. 1995.
3. M. Black and P. Anandan. The robust estimation of multiple motions: Affine and piecewise-smooth flow fields. *Computer Vision and Image Understanding*, in press. Also Tech. Report P93-00104, Xerox PARC, Dec.1993.
4. M. J. Black and P. Anandan. A framework for the robust estimation of optical flow. In *ICCV-93*, pp. 231–236, Berlin, May 1993.
5. A. Blake, M. Isard, and D. Reynard. Learning to track curves in motion. In *Proceedings of the IEEE Conf. Decision Theory and Control*, pp. 3788–3793, 1994.
6. A. F. Bobick and A. D. Wilson. A state-based technique for the summarization and recognition of gesture. In *ICCV-95* , pp. 382–388, Boston, June 1995.

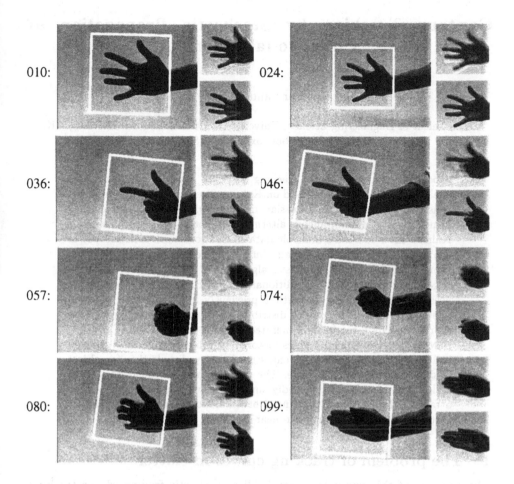

Fig. 10. Tracking and recognizing hand gestures in video.

7. C. Bregler and S. M. Omohundro. Surface learning with applications to lip reading. *Advances in Neural Information Processing Systems 6*, pp. 43–50, San Francisco, 1994.

8. F. R. Hampel, E. M. Ronchetti, P. J. Rousseeuw, and W. A. Stahel. *Robust Statistics: The Approach Based on Influence Functions*. John Wiley and Sons, New York, NY, 1986.

9. B. Moghaddam and A. Pentland. Probabilistic visual learning for object detection. In *ICCV-95*, pp. 786–793, Boston., June 1995.

10. H. Murase and S. Nayar. Visual learning and recognition of 3-D objects from appearance. *International Journal of Computer Vision*, 14:5–24, 1995.

11. A. Pentland, B. Moghaddam, and T. Starner. View-based and modular eigenspaces for face recognition. In *CVPR-94*, pp. 84–91, Seattle, June 1994.

12. M. J. Tarr and S. Pinker. Mental rotation and orientation-dependence in shape recognition. *Cognitive Psychology*, 21:233–282, 1989.

13. M. Turk and A. Pentland. Face recognition using eigenfaces. In *CVPR-91*, pp. 586–591, Maui, June 1991.

Contour Tracking by Stochastic Propagation of Conditional Density

Michael Isard and Andrew Blake

Department of Engineering Science, University of Oxford, Oxford OX1 3PJ, UK.
{misard, ab}@robots.ox.ac.uk, 01865 273919

Abstract. The problem of tracking curves in dense visual clutter is a challenging one. Trackers based on Kalman filters are of limited use; because they are based on Gaussian densities which are unimodal, they cannot represent simultaneous alternative hypotheses. Extensions to the Kalman filter to handle multiple data associations work satisfactorily in the simple case of point targets, but do not extend naturally to continuous curves. A new, stochastic algorithm is proposed here, the CONDENSATION algorithm — Conditional Density Propagation over time. It uses 'factored sampling', a method previously applied to interpretation of static images, in which the distribution of possible interpretations is represented by a randomly generated set of representatives. The CONDENSATION algorithm combines factored sampling with learned dynamical models to propagate an entire probability distribution for object position and shape, over time. The result is highly robust tracking of agile motion in clutter, markedly superior to what has previously been attainable from Kalman filtering. Notwithstanding the use of stochastic methods, the algorithm runs in near real-time.

1 The problem of tracking curves in clutter

The purpose of this paper is to establish a stochastic framework for tracking curves in visual clutter, and to propose a powerful new technique — the CONDENSATION algorithm. The new approach is rooted in strands from statistics, control theory and computer vision. The problem is to track outlines and features of foreground objects, modelled as curves, as they move in *substantial* clutter, and to do it at, or close to, video frame-rate. This is challenging because elements in the background clutter may mimic parts of foreground features. In the most severe case, the background may consist of objects similar to the foreground object, for instance when a person is moving past a crowd. Our framework aims to dissolve the resulting ambiguity by applying probabilistic models of object shape and motion to analyse the video-stream. The degree of generality of these models must be pitched carefully: sufficiently specific for effective disambiguation but sufficiently general to be broadly applicable over entire classes of foreground objects.

1.1 Modelling shape and motion

Effective methods have arisen in computer vision for modelling shape and motion. When suitable geometric models of a moving object are available, they can be matched effectively to image data, though usually at considerable computational cost [17, 26, 18]. Once an object has been located approximately, tracking it in subsequent images becomes more efficient computationally [20], especially if motion is modelled as well as shape [12, 16]. One important facility is the modelling of curve segments which interact with images [29] or image sequences [19]. This is more general than modelling entire objects but more clutter-resistant than applying signal-processing to low-level corners or edges. The methods to be discussed here have been applied at this level, to segments of parametric B-spline curves [3] tracking over image sequences [8]. The B-spline curves could, in theory, be parameterised by their control points. In practice this allows too many degrees of freedom for stable tracking and it is necessary to restrict the curve to a low-dimensional parameter x, for example over an affine space [28, 5], or more generally allowing a linear space of non-rigid motion [9].

Finally, probability densities $p(x)$ can be defined over the class of curves [9], and also over their motions [27, 5], and this constitutes a powerful facility for tracking. Reasonable default functions can be chosen for those densities. However, it is obviously more satisfactory to measure the actual densities or estimate them from data-sequences $(x_1, x_2, \ldots)$. Algorithms to do this assuming Gaussian densities are known in the control-theory literature [13] and have been applied in computer vision [6, 7, 4].

1.2 Sampling methods

A standard problem in statistical pattern recognition is to find an object parameterised as x with prior $p(x)$, using data z from a single image. (This is a simplified, static form of the image sequence problem addressed in this paper.) In order to estimate x from z, some information is needed about the conditional distribution $p(z|x)$ which measures the *likelihood* that a hypothetical object configuration x should give rise to the image data z that has just been observed. The data z could either be an entire grey-level array or a set of sparse features such as corners or, as in this paper, curve fragments obtained by edge detection. The posterior density $p(x|z)$ represents all the knowledge about x that is deducible from the data. It can be evaluated in principle by applying Bayes' rule to obtain

$$p(x|z) = kp(z|x)p(x) \qquad (1)$$

where k is a normalisation constant that is independent of x. In the general case that $p(z|x)$ is multi-modal $p(x|z)$ cannot be evaluated simply in closed form: instead iterative sampling techniques can be used.

The first use of such an iterative solution was proposed by Geman and Geman [11] for restoration of an image represented by mixed variables, both continuous (pixels) and discrete (the 'line process'). Sampling methods for recovery of a parametric curve x by sampling [24, 14, 25] have generally used spatial Markov processes as the underlying probabilistic model $p(x)$. The basic method is *factored sampling* [14]. It is useful when the conditional observation probability $p(z|x)$ can be evaluated pointwise and sampling it is not feasible and when, conversely, the prior $p(x)$ can be sampled but not evaluated. The algorithm estimates means of properties $f(x)$ (e.g. moments) of the posterior $p(x|z)$ by first generating randomly a sample $(s_1, s_2, \ldots)$ from the density $p(x)$ and then weighting with $p(z|x)$:

$$E[f(x)|z] \approx \frac{\sum_{n=1}^{N} f(s_n) p(z|s_n)}{\sum_{n=1}^{N} p(z|s_n)} \qquad (2)$$

where this is asymptotically ($N \to \infty$) an unbiased estimate. For example, the mean can be estimated using $f(x) = x$ and the variance using $f(x) = xx^T$. If $p(x)$ is a spatial Gauss-Markov process, then Gibbs sampling from $p(x)$ is used to generate the random variates $(s_1, s_2, \ldots)$. Otherwise, for low-dimensional parameterisations as in this paper, standard, direct methods can be used for Gaussians[1] — we use rejection sampling [21]. Note that, in the case that the density $p(z|x)$ is normal, the mean obtained by factored sampling would be consistent with an estimate obtained more conventionally, and efficiently, from linear least squares estimation. For multi-modal distributions which cannot be approximated as normal, so that linear estimators are unusable, estimates of mean x by factored sampling continue to apply.

Sampling methods have proved remarkably effective for recovering static objects, notably hands [14] and galaxies [24], in clutter. The challenge addressed here is to do this over time, estimating $x(t)$ from time-varying images $z(t)$.

1.3 Kalman filters and data-association

Spatio-temporal estimation, the tracking of shape and position over time, has been dealt with thoroughly by Kalman filtering, in the relatively clutter-free case in which $p(z|x)$ can satisfactorily be modelled as Gaussian [16, 12, 23] and can be applied to curves [27, 5]. These solutions work relatively poorly in clutter which easily 'distracts' the spatio-temporal estimate $\hat{x}(t)$. With simple, discrete features such as points or corners combinatorial data-association methods can be effective, including the 'JPDAF' [2, 22] and the 'RANSAC' algorithm [10]. They allow several hypotheses about which data-elements 'belong' to the tracked object to be held simultaneously, and less plausible hypotheses to be progressively pruned. Data association methods do not, however, apply to moving curves where the features are continuous objects, and a more general methodology is demanded.

1.4 Temporal propagation of conditional densities

The Kalman filter as a recursive linear estimator is a very special case, applying only to Gaussian densities, of a more general probability density propagation process. In continuous time this can be described in terms of diffusion [15], governed by a 'Fokker-Planck' equation [1], in which the density for $x(t)$ drifts and spreads under the action of a stochastic model of its dynamics. The random component of the dynamical model leads to spreading — increasing uncertainty

---

[1] Note: the presence of clutter causes $p(z|x)$ to be non-Gaussian, but the prior $p(x)$ may still happily be Gaussian, and that is what will be assumed in our experiments.

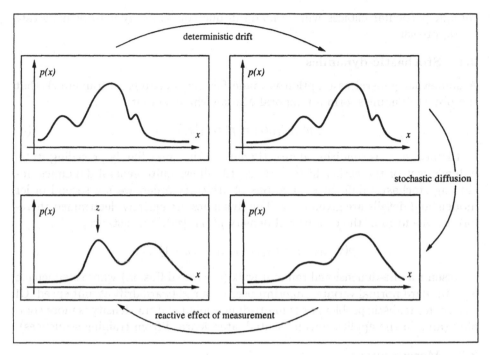

Fig. 1. *Probability density propagation. Propagation is depicted here as it occurs over a discrete time-step. There are three phases: drift due to the deterministic component of object dynamics; diffusion due to the random component; reactive reinforcement due to measurements.*

— while the deterministic component causes a drift of the mass of the density function, as shown in figure 1. The effect of external measurements $z(t)$ is to superimpose a reactive effect on the diffusion in which the density tends to peak in the vicinity of measurements.

In the simple Gaussian case, the diffusion is purely linear and the density function evolves as a Gaussian pulse that translates, spreads and is reinforced, remaining Gaussian throughout. The Kalman filter describes analytically exactly this process. In clutter, however, when measurements have a non-Gaussian, multi-modal conditional distribution, the evolving density requires a more general representation. This leads to a powerful new approach to tracking, developed below, in which a sparse representation of the density for $x(t)$ is carried forward in time. No mean position or variance is computed explicitly, though they and other properties can be computed at any time if desired.

2 Discrete-time propagation of state density

For computational purposes, the propagation process must be set out in terms of discrete time t. The state of the modelled object at time t is denoted x_t and its history is $\mathbf{x}_t = (x_1, x_2, \ldots, x_t)$. Similarly the set of image features at time t is z_t with history $\mathbf{z}_t = (z_1, \ldots, z_t)$. Note that no functional assumptions (linearity, Gaussianity, unimodality) are made about densities, in the general treatment, though particular choices will be made in due course in order to demonstrate the approach.

2.1 Stochastic dynamics

A somewhat general assumption is made for the probabilistic framework that the object dynamics form a temporal Markov chain so that:

$$p(x_{t+1}|\mathbf{x}_t) = p(x_{t+1}|x_t) \tag{3}$$

— the new state is conditioned directly only on the immediately preceding state, independent of the earlier history. This still allows quite general dynamics, including stochastic difference equations of arbitrary order; we use second order models and details are given later. The dynamics are entirely determined therefore by the form of the conditional density $p(x_{t+1}|x_t)$. For instance,

$$p(x_{t+1}|x_t) = \exp{-(x_{t+1} - x_t - 1)^2/2},$$

represents a one-dimensional random walk (discrete diffusion) whose step length is a standard normal variate, superimposed on a rightward drift at unit speed. Of course, for realistic problems x is multi-dimensional and the density is more complex (and, in the applications presented later, learned from training sequences).

2.2 Measurement

Observations z_t are assumed to be independent, both mutually and with respect to the dynamical process, and this is expressed probabilistically as follows:

$$p(\mathbf{z}_t, x_{t+1}|\mathbf{x}_t) = p(x_{t+1}|\mathbf{x}_t) \prod_{i=1}^{t} p(z_i|x_i). \tag{4}$$

Note that integrating over x_{t+1} implies the mutual conditional independence of observations:

$$p(\mathbf{z}_t|\mathbf{x}_t) = \prod_{i=1}^{t} p(z_i|x_i). \tag{5}$$

The observation process is therefore defined by specifying the conditional density $p(z_t|x_t)$ at each time t, and later, in computational examples, we take this to be a time-independent function $p(z|x)$. Details of the shape of this function, for applications in image-stream analysis, are given in section 4.

2.3 Propagation

Given a continuous-valued Markov chain with independent observations, the rule for propagation of conditional density $p(x_t|\mathbf{z}_t)$ over time is:

$$p(x_{t+1}|\mathbf{z}_{t+1}) = k_{t+1} \, p(z_{t+1}|x_{t+1})p(x_{t+1}|\mathbf{z}_t) \tag{6}$$

where

$$p(x_{t+1}|\mathbf{z}_t) = \int_{x_t} p(x_{t+1}|x_t)p(x_t|\mathbf{z}_t) \tag{7}$$

and k_{t+1} is a normalisation constant that does not depend on x_{t+1}.

The propagation rule (6) should be interpreted simply as the equivalent of the Bayes rule (1) for inferring posterior state density from data, for the time-varying case. The effective prior $p(x_{t+1}|\mathbf{z}_t)$ is actually a prediction taken from the posterior $p(x_t|\mathbf{z}_t)$ from the previous time-step, onto which is superimposed one time-step from the dynamical model (Fokker-Planck drift plus diffusion as in figure 1), and this is expressed in (7). Multiplication in (6) by the conditional measurement density $p(z_{t+1}|x_{t+1})$ in the Bayesian manner then applies the reactive effect expected from measurements (figure 1).

3 The CONDENSATION algorithm

In contrast to the static case in which the prior $p(x)$ may be Gaussian, the effective prior $p(x_{t+1}|\mathbf{z}_t)$ in the dynamic case is not Gaussian when clutter is present. It has no particular known form and therefore cannot apparently be represented exactly in the algorithm. The CONDENSATION algorithm solves this problem by doing altogether without any explicit representation of the density function itself. Instead, it proceeds by generating sets of N samples from $p(x_t|\mathbf{z}_t)$ at each time-step. Each sample s_t is considered as an (s_t, π_t) pair, in which s_t is a value of x_t and π_t is a corresponding sampling probability. Suppose a particular s_t is drawn randomly from $p(x_t|\mathbf{z}_t)$ by choosing it, with probability π_t, from the set of N samples at time t. Next draw s_{t+1} randomly from $p(x_{t+1}|x_t = s_t)$, one time-step of the dynamical model, starting from $x_t = s_t$, a Gaussian density to which standard sampling methods apply. A value s_{t+1} chosen in this way is a fair sample from $p(x_{t+1}|\mathbf{z}_t)$. It can then be retained as a pair (s_{t+1}, π_{t+1}) for the N-set at time $t + 1$, where $\pi_{t+1} = p(z_{t+1}|x_{t+1} = s_{t+1})$. This sampling scheme is the basis of the CONDENSATION algorithm and details are given in figure 2. In practice, random variates can be generated efficiently, using binary search, if, rather than storing probabilities π_t, we store cumulative probabilities c_t as shown in the figure. At any time t, expected values $E[f(x_t)|\mathbf{z}_t]$ of properties of the state density $p(x_t|\mathbf{z}_t)$ can be evaluated by applying the rule (2) from the factored sampling algorithm.

4 Probabilistic parameters for curve tracking

In order to apply the CONDENSATION algorithm, which is general, to the tracking of curves in image-streams, specific probability densities must be established both for the dynamics of the object and for the measurement process. As mentioned earlier, the parameters x denote a linear transformation of a B-spline curve, either an affine deformation, or some non-rigid motion. The dynamical model and learning algorithm follow established methods [6, 7]. The model is a stochastic differential equation which, in discrete time, is

$$x_{t+1} = Ax_t + B\omega_t \qquad (8)$$

where A defines the deterministic component of the model and ω_t is a vector of independent standard normal random variables scaled by B so that BB^T is the

---

Iterate

At time-step $t + 1$, construct the nth of N samples as follows:

1. Generate a random number $r \in [0, 1]$, uniformly distributed.
2. Find, by binary subdivision on m, the smallest m for which $c_t^{(m)} \leq r$.
3. Draw a random variate $s_{t+1}^{(n)}$ from the density $p(x_{t+1} | x_t = s_t^{(m)})$, assumed Gaussian so direct sampling is possible.

Store samples $n = 1, .., N$ as $(s_{t+1}^{(n)}, \pi_{t+1}^{(n)}, c_{t+1}^{(n)})$ where

$$c_{t+1}^{(0)} = 0$$
$$\pi_{t+1}^{(n)} = p(z_{t+1} | x_{t+1} = s_{t+1}^{(n)})$$
$$c_{t+1}^{(n)} = c_{t+1}^{(n-1)} + \pi_{t+1}^{(n)}$$

and then normalise by dividing all cumulative probabilities $c_{t+1}^{(n)}$ by $c_{t+1}^{(N)}$, i.e. so that $c_{t+1}^{(N)} = 1$.

If required, mean properties can be estimated at any time t as

$$E[f(x) | \mathbf{z}_t] \approx \sum_{n=1}^{N} \pi_t^{(n)} f(s_t^{(n)}).$$

For example, if the mean configuration $\hat{x}$ is required for graphical display, the above rule is used with $f(x) = x$.

Fig. 2. *The* CONDENSATION *algorithm.*

process noise covariance. The model can clearly be re-expressed as a temporal Markov chain as follows:

$$p(x_{t+1} | x_t) = \exp -\frac{1}{2} \| B^{-1}(x_{t+1} - Ax_t)) \|^2. \tag{9}$$

In practice, we use second order models, where x_t, A and B are replaced by

$$\begin{pmatrix} x_t \\ x_{t+1} \end{pmatrix}, \quad \begin{pmatrix} 0 & I \\ A_0 & A_1 \end{pmatrix} \quad \text{and} \quad \begin{pmatrix} 0 & 0 \\ 0 & B \end{pmatrix}$$

respectively. Coefficients are learned from sequences of images. An untrained tracker is used to follow training motions against a relatively clutter-free background. The tracked sequence in the form $(x_1, x_2, ...)$ is then analysed [6, 7] by Maximum Likelihood Estimation to generate estimates of A_0, A_1 and B, thus defining the model for use by the CONDENSATION algorithm. A set of sample values for time-step $t = 0$ must be supplied to initialise the algorithm. If the prior density $p(x_0)$ is Gaussian, direct sampling may be used for initialisation, otherwise it is possible simply to allow the density to settle to a steady state $p(x_\infty)$ in the absence of object measurements.

4.1 Observations

The measurement process defined by $p(z_t|x_t)$ is assumed here to be stationary in time (though the CONDENSATION algorithm does not require this) so a static function $p(z|x)$ is to be specified. As yet we have no capability to estimate it from data, though that would be ideal, so some reasonable assumptions must be made.

Measurements z arising from a curve x are image-edge fragments obtained by edge-detection along curve normals. We assume that noise and distortions in imaging z are local, so in order to determine $p(z|x)$ it is necessary only to examine image pixels near the image curve which we denote (with mild abuse of notation) $x(s), 0 \leq s \leq 1$. The corresponding measurement sequence is then denoted $z(s)$, where $z(s)$ for each s is the detected edge on the normal at $x(s)$ that lies closest to the curve x. To allow for measurement failures and clutter, the measurement density is modelled as a robust statistic, a truncated Gaussian:

$$p(z|x) = \exp\left\{ -\frac{1}{2\sigma^2} \int_0^1 \phi(s)ds \right\} \tag{10}$$

where

$$\phi(s) = \begin{cases} |x(s) - z(s)|^2 & \text{if } |x(s) - z(s)| < \delta \\ \rho & \text{otherwise} \end{cases} \tag{11}$$

and ρ is a penalty constant, related to the probability of failing to find a feature, either on the curve or the background. Note that ϕ is constant at distances greater than δ from the curve, so δ acts as a maximum scale beyond which it is unnecessary to search for features. In practice, of course, the integral is approximated as a sum over discrete sample intervals of s.

5 Applying the CONDENSATION algorithm to video-streams

5.1 Tracking a multi-modal distribution

In order to test the CONDENSATION algorithm's ability to represent a multi-modal distribution, we collected a 70 frame (2.8 second) sequence showing a cluttered room with three people in it, facing the camera. The person initially on the right of the image moves to the left, in front of the other two. A template was drawn, using an interactive package, to fit around the head and shoulders of a person, and we constructed an affine space of deformations of that template. A motion model was learned by tracking a single person walking around the room; background subtraction was necessary to ensure accurate tracking past the clutter. Results of running the CONDENSATION algorithm are shown in figure 3. Since the feature of interest is primarily x translation, only the distribution of the parameter corresponding to x coordinate has been plotted, however it is clear that the people are of slightly different sizes and heights, and this is modelled in the full distribution. No background subtraction or other preprocessing is used; the input is the raw video stream. Initialisation is performed simply by iterating the stochastic model in the absence of measurements, and it can be seen that this corresponds to a roughly Gaussian distribution on x coordinate at the first time-step. The distribution rapidly collapses onto the three peaks present in the image, and tracks them correctly, despite temporary difficulties while the people occlude each other. The time-step used for tracking is frame rate (40 ms) since the motion is fairly slow; in the figure, distributions are plotted only every 80 ms for clarity. The stationary person on the left has the highest peak in the distribution; this is to be expected since he is standing against a clutter-free background, and so his outline is consistently detectable. The experiment was run using a distribution of $N = 1000$ samples.

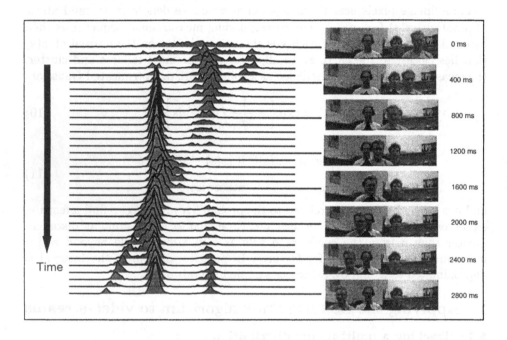

Fig. 3. *Tracking a multi-modal distribution. A histogram of the horizontal translation component of the distribution is plotted against time. The initial distribution is roughly Gaussian, but the three peaks are rapidly detected and tracked as one person walks in front of the other two.*

5.2 Tracking rapid motions through clutter

Next we collected a 500 field (10 second) sequence showing a girl dancing vigorously to a Scottish reel against a highly cluttered background, in order to test the CONDENSATION algorithm's agility when presented with rapid motions. We drew a head-shaped template and constructed an affine space to represent its allowable deformations. We also collected a training sequence of dancing against a mostly uncluttered background, from which we trained a motion model for use when the CONDENSATION tracker was applied to test data including clutter.

Figure 4 shows some stills from the clutter sequence, with tracked head positions from preceding fields overlaid to indicate motion. The contours are plotted

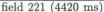

field 221 (4420 ms)   field 265 (5300 ms)

Fig. 4. *Maintaining tracker agility in clutter. A sequence of 500 fields (10 seconds) was captured showing a dancer executing rapid motions against a cluttered background. The dancer's head was then tracked through the sequence. Representative fields are shown, with preceding tracked head positions to give an indication of the motion. The tracked positions are shown at 40 ms intervals. The distribution consists of $N = 100$ samples.*

at 40 ms intervals. The model parameters are estimated by the mean of the distribution at each time-step. The distribution consists of $N = 100$ samples. The distribution was initialised by hand near the dancer's position in the first field, as 100 samples do not sweep out enough of the prior to locate the initial peak reliably. It would be equally feasible to begin with a larger number of samples in the first field, and reduce the size of the distribution when the dancer had been found (this technique was used in section 5.3).

Figure 5 shows the centroid of the head position estimate as tracked by both the CONDENSATION algorithm and a Kalman filter. The CONDENSATION tracker correctly estimated the head position throughout the sequence, but after about 40 fields (0.80 s), the Kalman filter was distracted by clutter, never to recover.

Although it is expected that the posterior distribution will be largely unimodal throughout the sequence, since there is only one dancer, figure 6 illustrates the point that it is still important for robustness that the tracker is able to represent distributions with several peaks. After 920 ms there are two distinct peaks, one caused by clutter, and one corresponding to the dancer's head. At this point the clutter peak has higher posterior probability, and a unimodal tracker like the Kalman filter would discard the information in the second peak, rendering it unable to recover; however the CONDENSATION algorithm does recover, and the dancer's true position is again localised after 960 ms.

5.3 Tracking complex jointed objects

The preceding sequences show motion taking place in a model space of at most 4 dimensions, so in order to investigate tracking performance in higher dimensions, we collected a 500 field (10 second) sequence of a hand translating, rotating,

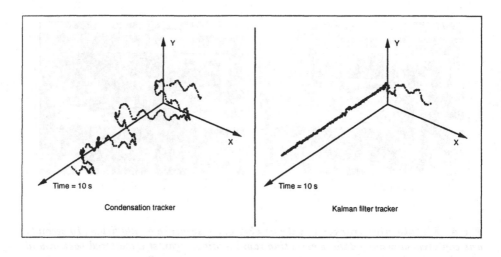

Fig. 5. *The Condensation tracker succeeds where a Kalman filter fails. The centroid of the state estimate for the sequence shown in figure 4 is plotted against time for the entire 500 field sequence, as tracked by first the* CONDENSATION *tracker, then a Kalman filter tracker. The* CONDENSATION *algorithm correctly estimates the head position throughout the sequence. The Kalman filter initially tracks correctly, but is rapidly distracted by a clutter feature and never recovers.*

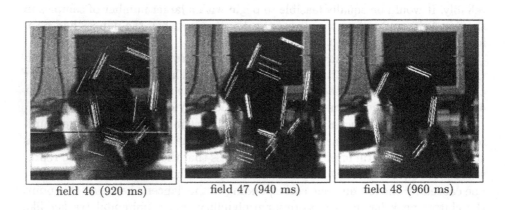

Fig. 6. *Recovering from tracking failure. Detail from 3 fields of the sequence illustrated in figure 4. Each sample from the distribution is plotted on the image, with intensity scaled to indicate its posterior probability. Most of the samples, from a distribution of $N = 100$, have too low a probability to be visible. In field 46 the distribution has split into two distinct peaks, the larger attracted to background clutter. The distribution converges on the dancer in field 48.*

and flexing its fingers independently, over a highly cluttered desk scene. We constructed a twelve degree of freedom shape variation model and an accompanying motion model with the help of a Kalman filter tracking in real time against a plain white background, using signed edges to help to disambiguate the finger boundaries.

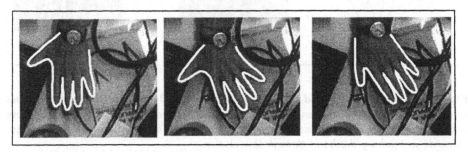

Fig. 7. *Tracking a flexing hand across a cluttered desk. Representative stills from a 500 field (10 second) sequence of a hand moving over a highly cluttered desk scene. The fingers and thumb flex independently, and the hand translates and rotates. The distribution consists of $N = 500$ samples except for the first 4 fields, when it decreases from 1500 samples to aid initialisation. The distribution is initialised automatically by iterating on the motion model in the absence of measurements.*

Figure 7 shows detail of a series of images from the tracked 500 field sequence. The distribution is initialised automatically by iterating the motion model in the absence of measurements. The initialisation is performed using $N = 1500$ samples, but N is dropped gradually to 500 over the first 4 fields, and the rest of the sequence is tracked using $N = 500$. Occasionally one section of the contour locks onto a shadow or a finger becomes slightly misaligned, but the system always recovers. Figure 8 shows just how severe the clutter problem is — the hand is immersed in a dense field of edges. The CONDENSATION algorithm succeeds in tracking the hand despite the confusion of input data.

6 Conclusions

Tracking in clutter is hard because of the essential multi-modality of the conditional measurement density $p(z|x)$. In the case of curve tracking, multiple-hypothesis tracking is inapplicable and a new approach is needed. The CONDENSATION algorithm is a fusion of the statistical factored sampling algorithm for static, non-Gaussian problems with a stochastic differential equation model for object motion. The result is an algorithm for tracking rigid and non-rigid motion which has been demonstrated to be far more effective in clutter than comparable Kalman filters. Performance of the CONDENSATION algorithm improves as the sample size parameter N increases, but computational complexity is $O(N \log N)$. Impressive results have been demonstrated for models with 4 to 12 degrees of freedom, even when $N = 100$. Performance in several cases was

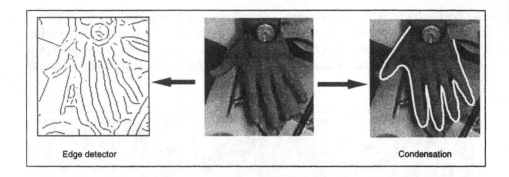

Fig. 8. *Localising the hand in a dense edge map. Detail of a field from the hand sequence. The result of running a directional Gaussian edge detector shows that there are many clutter edges present to distract the system. The* CONDENSATION *algorithm succeeds in tracking the hand through this clutter.*

improved still further with increased $N = 1000$. The system currently runs with $N = 50$ in real-time (25Hz) on a desk-top graphics workstation (Indy R4400SC, 200 MHz).

The authors would like to acknowledge the support of the EPSRC. They are also grateful for discussions with Roger Brockett, Brian Ripley, David Reynard, Simon Rowe and Andrew Wildenberg, and for experimental assistance from Sarah Blake.

References

1. K. J. Astrom. *Introduction to stochastic control theory.* Academic Press, 1970.
2. Y. Bar-Shalom and T.E. Fortmann. *Tracking and Data Association.* Academic Press, 1988.
3. R.H. Bartels, J.C. Beatty, and B.A. Barsky. *An Introduction to Splines for use in Computer Graphics and Geometric Modeling.* Morgan Kaufmann, 1987.
4. A. Baumberg and D. Hogg. Generating spatiotemporal models from examples. In *Proc. BMVC*, 413–422, 1995.
5. A. Blake, R. Curwen, and A. Zisserman. A framework for spatio-temporal control in the tracking of visual contours. *Int. Journal of Computer Vision*, 11(2):127–145, 1993.
6. A. Blake and M.A. Isard. 3D position, attitude and shape input using video tracking of hands and lips. In *Proc. Siggraph*, 185–192. ACM, 1994.
7. A. Blake, M.A. Isard, and D. Reynard. Learning to track the visual motion of contours. *Artificial Intelligence*, 78:101–134, 1995.
8. R. Cipolla and A. Blake. The dynamic analysis of apparent contours. In *Proc. 3rd Int. Conf. on Computer Vision*, 616–625, 1990.
9. T.F. Cootes, C.J. Taylor, A. Lanitis, D.H. Cooper, and J. Graham. Building and using flexible models incorporating grey-level information. In *Proc. 4th Int. Conf. on Computer Vision*, 242–246, 1993.

10. M.A. Fischler and R.C. Bolles. Random sample consensus: a paradigm for model fitting with application to image analysis and automated cartography. *Commun. Assoc. Comp. Mach.*, 24:381–95, 1981.

11. Stuart Geman and Donald Geman. Stochastic Relaxation, Gibbs Distributions, and the Bayesian Restoration of Images. *IEEE Trans. Pattern Analysis and Machine Intelligence*, 6(6):721–741, 1984.

12. D.B. Gennery. Visual tracking of known three-dimensional objects. *Int. Journal of Computer Vision*, 7:3:243–270, 1992.

13. C.G. Goodwin and K.S. Sin. *Adaptive filtering prediction and control*. Prentice-Hall, 1984.

14. U. Grenander, Y. Chow, and D. M. Keenan. *HANDS. A Pattern Theoretical Study of Biological Shapes*. Springer-Verlag. New York, 1991.

15. U. Grenander and M.I. Miller. Representations of knowledge in complex systems (with discussion). *J. Roy. Stat. Soc. B.*, 56:549–603, 1993.

16. C. Harris. Tracking with rigid models. In A. Blake and A. Yuille, editors, *Active Vision*, 59–74. MIT, 1992.

17. D. Hogg. Model-based vision: a program to see a walking person. *Image and Vision Computing*, 1(1):5–20, 1983.

18. D.P. Huttenlocher, J.J. Noh, and W.J. Rucklidge. Tracking non-rigid objects in complex scenes. In *Proc. 4th Int. Conf. on Computer Vision*, 93–101, 1993.

19. M. Kass, A. Witkin, and D. Terzopoulos. Snakes: Active contour models. In *Proc. 1st Int. Conf. on Computer Vision*, 259–268, 1987.

20. D.G. Lowe. Robust model-based motion tracking through the integration of search and estimation. *Int. Journal of Computer Vision*, 8(2):113–122, 1992.

21. W.H. Press, S.A. Teukolsky, W.T. Vetterling, and B.P. Flannery. *Numerical Recipes in C*. Cambridge University Press, 1988.

22. B. Rao. Data association methods for tracking systems. In A. Blake and A. Yuille, editors, *Active Vision*, 91–105. MIT, 1992.

23. J.M. Rehg and T. Kanade. Visual tracking of high dof articulated structures: an application to human hand tracking. In J-O. Eklundh, editor, *Proc. 3rd European Conference on Computer Vision*, 35–46. Springer-Verlag, 1994.

24. B.D. Ripley and A.L. Sutherland. Finding spiral structures in images of galaxies. *Phil. Trans. R. Soc. Lond. A.*, 332(1627):477–485, 1990.

25. G. Storvik. A Bayesian approach to dynamic contours through stochastic sampling and simulated annealing. *IEEE Trans. Pattern Analysis and Machine Intelligence*, 16(10):976–986, 1994.

26. G.D. Sullivan. Visual interpretation of known objects in constrained scenes. *Phil. Trans. R. Soc. Lond. B.*, 337:361–370, 1992.

27. D. Terzopoulos and D. Metaxas. Dynamic 3D models with local and global deformations: deformable superquadrics. *IEEE Trans. Pattern Analysis and Machine Intelligence*, 13(7), 1991.

28. S. Ullman and R. Basri. Recognition by linear combinations of models. *IEEE Trans. Pattern Analysis and Machine Intelligence*, 13(10):992–1006, 1991.

29. A. Yuille and P. Hallinan. Deformable templates. In A. Blake and A. Yuille, editors, *Active Vision*, 20–38. MIT, 1992.

Learning Dynamics of Complex Motions from Image Sequences

David Reynard[1], Andrew Wildenberg[1], Andrew Blake[1] and John Marchant[2]

[1] Department of Engineering Science, University of Oxford, Oxford OX1 3PJ, UK
[2] Silsoe Research Institute, Bedford, UK.

Abstract. The performance of Active Contours in tracking is highly dependent on the availability of an appropriate model of shape and motion, to use as a predictor. Models can be hand-built, but it is far more effective and less time-consuming to learn them from a training set. Techniques to do this exist both for shape, and for shape and motion jointly. This paper extends the range of shape and motion models in two significant ways. The first is to model jointly the random variations in shape arising within an object-class and those occuring during object motion. The resulting algorithm is applied to tracking of plants captured by a video camera mounted on an agricultural robot. The second addresses the tracking of coupled objects such as head and lips. In both cases, new algorithms are shown to make important contributions to tracking performance.

1 Introduction

The use of Kalman filters [1] to track the motion of objects in real time is now a standard weapon in the arsenal of Computer Vision [14, 11, 8, 10, 16, 3]. A crucial consideration for effective real-time performance is that some form of dynamical model be identified [13, 5, 2] and used as a predictor. In many cases, available models are deterministic — based on ordinary differential equations. However, to be usable in a Kalman filtering framework it is crucial that the model contain both deterministic and stochastic components — stochastic differential equations. Such models can be learned effectively from training data [9, 5].

In this paper we develop two significant elaborations for stochastic dynamical models. The first concerns modelling object classes for objects in motion. The second addresses the efficient modelling of couplings between tracked objects.

1.1 Shape and Motion Variability

The first problem addressed by this paper concerns learning dynamical models which represent both *class* and *dynamical* variability. Class variability arises from the differences between objects in a given class which can be learned effectively by Principal Components Analysis applied to a linear curve parameterisation [6]. With moving objects, dynamical variability must also be considered, modelling changes that occur during motion, due to projective effects and actual physical disturbances.

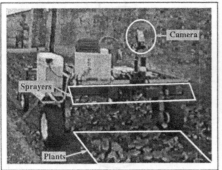

Fig. 1. (a) *An autonomous tractor incorporates a downward pointing camera to monitor plants passing under an array of spray nozzles. Sprayed chemicals can then be directed onto or away from plants, as appropriate.* (b) **Plant tracking.** *A mobile vehicle sees plants (cauliflowers in this instance) passing through the field of view of a downward pointing camera. The plants' motion and shape are captured by a dynamic contour, shown in white.*

Previous systems for learning dynamics have not addressed the important distinction between these two sources of variability [5, 2]. If the two are lumped together into one dynamical model, there will be insufficient constraint on motion variability. A tracker incorporating such a model as a predictor will allow temporal motion/shape changes to range over the *total* modelled variability, both class and temporal. This is inappropriate. Rather, modelled class variability should apply only as a tracker is re-initialised on a new object. Once tracking is underway, the identity of the object does not change, so class variability should be suspended. Instead, the dynamical variability model should take over, relatively tightly constrained as it allows only the variability normally associated with motion. Maintaining the distinction between the two sources of variability is of considerable practical importance.

In a robotic application, a mobile tractor-robot monitoring the motion of plants in a downward pointing camera controls the application of sprayed chemicals from a spray-nozzle array (figure 1 (a)). Moving plants in the video stream are tracked by a dynamic contour tracker (figure 1 (b)). A dynamical model was learnt from video sequences of 36 plants. Inter-class variability proves to be considerably greater than motion variability in this kind of data (figure 2). Modelling them independently ought therefore to enhance temporal stability during tracking. This will indeed prove to be so.

1.2 Coupled Motions

The second problem addressed here is that of learning the dynamics of a coupled pair of systems. The point of attempting to model the coupling is that stability of tracking may be enhanced by the implied constraints. For instance, Moses *et al* [12] built a system to track lip motion and deformation, but it proved difficult, given that lips are long and thin, to stabilise them to horizontal translation of the head. This phenomenon is illustrated in figure 3. Here we show that this

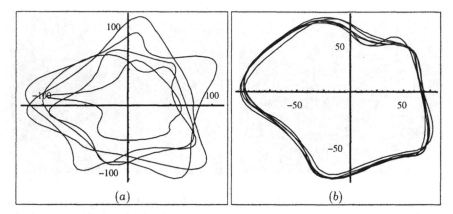

Fig. 2. Class-variability swamps motion-variability. *(a) Measured class-variation over five plants is considerable. In contrast (b) variation due to the motion of one individual plant, as it passes through the camera's field of view, is relatively small.*

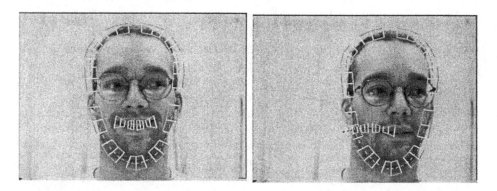

Fig. 3. Uncoupled head and lip tracking is unsuccessful. *The two images show the results of simultaneous head and lip tracking when the dynamics of the two systems are independent. In the first image, the lip tracker follows the movement of the lips well, but as soon as the head begins to translate the position of the lips is lost. This shows that some sort of coupling is necessary.*

defect can be largely removed by using a prior model for motion in which one object, in this case the head, is regarded as primary and the other (the lips) as secondary, driven by the primary.

Modelling this coupling fully and installing it, without approximation, in a Kalman filter tracker — "strong coupling" — incurs a severe increase in computational cost. We therefore propose an approximate "weak coupling" in which state covariance is constrained during tracking (see [16] for a related idea), leading to greatly reduced computational cost. Weak coupling is tested experimentally and found, in practice, to give results that differ remarkably little from the strongly coupled system.

2 Modelling Class and Motion Variability

The distinction between class and motion variability is important. In this section multidimensional, second order dynamical models will be extended to model these two sources of variability independently. The resulting, augmented model, once installed as the predictor for a tracker, performs much more robustly than when all variability is modelled jointly.

The joint dynamical model is a second order stochastic differential equation (SDE) which, in the discrete-time form, is:

$$\mathbf{X}_{n+2} - \overline{\mathbf{X}} = A_1(\mathbf{X}_{n+1} - \overline{\mathbf{X}}) + A_0(\mathbf{X}_n - \overline{\mathbf{X}}) + B\mathbf{w}_n \qquad (1)$$

where $\mathbf{X}$ is an N-dimensional state vector parameterising the configuration (position and shape) of the modelled contour. The mean configuration $\overline{\mathbf{X}}$, is a constant in this simple model. Coefficients A_0 and A_1 fix the deterministic components of the model [13]. The stochastic component of the model is a noise source $\mathbf{w}_n$, a vector of N independent unit normal variables, coupled into the system via the $N \times N$ matrix B. An algorithm for learning systems of this kind is known [5, 4] but an improved algorithm is needed to model the two sources of variability independently.

2.1 Extended Dynamical Model

The improved model extends the state vector by regarding the mean configuration $\overline{\mathbf{X}}$ as a variable, rather than a constant as above. Now the dynamical model, most conveniently written in block-matrix form, is:

$$\begin{pmatrix} \mathbf{X}_{n+1} \\ \mathbf{X}_{n+2} \\ \overline{\mathbf{X}}_{n+1} \end{pmatrix} = \begin{pmatrix} 0 & I & 0 \\ A_0 & A_1 & I - A_0 - A_1 \\ 0 & 0 & I \end{pmatrix} \begin{pmatrix} \mathbf{X}_n \\ \mathbf{X}_{n+1} \\ \overline{\mathbf{X}}_n \end{pmatrix} + \begin{pmatrix} 0 \\ B\mathbf{w}_n \\ 0 \end{pmatrix} \qquad (2)$$

This simply augments the original model (1) with an additional equation stating $\overline{\mathbf{X}}$ is constant over time. It might appear that this is identical to the old model. The crucial difference is that $\overline{\mathbf{X}}$ is no longer known *a priori*, but is estimated on-line. Once this dynamical system is installed as the predictor for a Kalman filter tracker, $\overline{\mathbf{X}}$ is initialised with its *estimated* mean value $\overline{\mathbf{X}}_0$ and associated variances and covariances which are obtained by statistical estimation from training data. The value of $\overline{\mathbf{X}}_n$ converges rapidly in the Kalman filter and remains fixed, reinitialised only when a new object is to be tracked.

2.2 Learning Algorithm

One of the main theoretical results of this paper is the learning algorithm for the extended system dynamics. Given objects labelled $\gamma = 1, \ldots, \Gamma$, each observed in motion for N_γ timesteps as data sequences $\mathbf{X}_n^\gamma$, the problem is to estimate global parameters for the dynamics A_0, A_1, B and mean configurations for each object $\overline{\mathbf{X}}^\gamma$ by jointly maximising log likelihood over all the parameters:

$$L(\{\mathbf{X}_n^\gamma\}|A_0, A_1, B, \overline{\mathbf{X}}^1, \ldots, \overline{\mathbf{X}}^\Gamma) = -\frac{1}{2} \sum_{\gamma=1}^{\Gamma} \sum_{n=1}^{N_\gamma-2} |B^{-1}R_n^\gamma|^2 \tag{3}$$

$$- \left(\sum_\gamma (N_\gamma - 2)\right) \log \det B,$$

where R_n^γ is an instantaneous measure of error of fit:

$$R_n^\gamma = (\mathbf{X}_{n+2}^\gamma - \overline{\mathbf{X}}^\gamma) - A_0(\mathbf{X}_n^\gamma - \overline{\mathbf{X}}^\gamma) - A_1(\mathbf{X}_{n+1}^\gamma - \overline{\mathbf{X}}^\gamma) \tag{4}$$

The nonlinearity introduced by the product terms $A_0\overline{\mathbf{X}}$ and $A_1\overline{\mathbf{X}}$ of R_n^γ can be removed by defining a parameter

$$D^\gamma = (I - A_0 - A_1)\overline{\mathbf{X}}^\gamma \tag{5}$$

so that the instantaneous error becomes $R_n^\gamma = \mathbf{X}_{n+2} - A_0\mathbf{X}_n - A_1\mathbf{X}_{n+1} - D^\gamma$. Object means $\overline{\mathbf{X}}^\gamma$ can always be obtained explicitly if desired, provided $(I - A_0 - A_1)$ is non-singular, by solving for them in equation (5). (Singularity arises in certain special cases, for instance the simple harmonic oscillator, when estimated dynamical parameters are non-unique.) The mean and covariance of the set of object-means $\overline{\mathbf{X}}^\gamma$, $\gamma = 1, \ldots, \Gamma$ are computed using the normal definitions to use as initial mean and covariance for the new state variable $\overline{\mathbf{X}}$ (mean-shape) in the tracking procedure.

Solution

The solution to the maximum likelihood problem is given here but, for space considerations, the proof is omitted. The following set of $\Gamma + 2$ equations is solved simultaneously for the estimated values of parameters A_0, A_1 and D^γ:

$$S_2^\gamma - A_0 S_0^\gamma - A_1 S_1^\gamma - D^\gamma(N_\gamma - 2) = 0, \quad \gamma = 1, \ldots, \Gamma, \tag{6}$$
$$S_{20} - A_0 S_{00} - A_1 S_{10} = 0$$
$$S_{21} - A_0 S_{01} - A_1 S_{11} = 0$$

where the S_i^γ, S_{ij} are certain moments of the data-set, defined below. This leaves only the random process coupling parameter B to be determined. In fact B cannot be determined uniquely, but any solution of $BB^T = C$, where

$$C = \frac{1}{\sum_\gamma (N_\gamma - 2)} \sum_{\gamma=1}^{\Gamma} \sum_{n=1}^{N_\gamma-2} R_n^\gamma (R_n^\gamma)^T \tag{7}$$

is acceptable. It remains to define the moments of the data-set[3]:

$$S_{ij} = \sum_{\gamma=1}^{\Gamma} \sum_{n=1}^{N_\gamma - 2} \mathbf{X}_{n+i}^{\gamma} (\mathbf{X}_{n+j}^{\gamma})^T - \Gamma \sum_{\gamma=1}^{\Gamma} \frac{S_i^{\gamma} (S_j^{\gamma})^T}{N_\gamma - 2}, \quad i, j = 0, 1, 2 \qquad (8)$$

$$S_i^{\gamma} = \sum_{n=1}^{N_\gamma - 2} \mathbf{X}_{n+i}^{\gamma}, \quad i = 0, 1, 2. \qquad (9)$$

2.3 Results

The extended system with independently modelled class and motion variability has been applied to the agricultural robotics problem described earlier. The outlines of plants are described by B-spline curves parameterised here over a 6-dimensional affine space. The curves are estimated over time in a standard Kalman filter tracker [3]. In figure 4, three different trackers are demonstrated. The first uses a "reasonable" default model, not learnt, but predicting constant velocity and uniform driving noise. The second uses a "joint variability" model in which class and motion variability are lumped together. Because, class variability swamps motion variability for this data (see figure 2), the resulting tracker has unnecessarily weak constraints on temporal shape change. Lastly, we demonstrate a tracker based on the new "independent variability" model which exploits the prior knowledge on shape and motion as thoroughly as possible.

3 Tracking Coupled Objects

Here we explore mechanisms for allowing two coupled objects, one primary and one secondary, to be tracked simultaneously and cooperatively. The aim is to devise an efficient mechanism that applies the coupling approximately but avoids the high computational cost associated with exact coupling.

3.1 Coupled Systems

The configurations of the objects in the coupled system are represented by vectors $\mathbf{X}^{(1)}, \mathbf{X}^{(2)}$. General, coupled, linear, stochastic dynamics are represented by:

$$\begin{pmatrix} \mathbf{X}^{(1)} \\ \mathbf{X}^{(2)} \end{pmatrix}_{k+2} = \mathcal{A}_0 \begin{pmatrix} \mathbf{X}^{(1)} \\ \mathbf{X}^{(2)} \end{pmatrix}_k + \mathcal{A}_1 \begin{pmatrix} \mathbf{X}^{(1)} \\ \mathbf{X}^{(2)} \end{pmatrix}_{k+1} + B\mathbf{w}_k \qquad (10)$$

where $\mathbf{w}$ is a vector of unit normal noise terms coupled into the system via the matrix B and $BB^T = C$, as before. The matrices $\mathcal{A}_0, \mathcal{A}_1$ are damping and elastic coefficients respectively, each composed of submatrices:

$$\mathcal{A}_i = \begin{pmatrix} A_i^{1,1} & A_i^{1,2} \\ A_i^{2,1} & A_i^{2,2} \end{pmatrix}, \quad i = 0, 1. \qquad (11)$$

---

[3] Moments defined here are rather different from the moments used in [5, 2] despite the similarity of the S_{ij} notation.

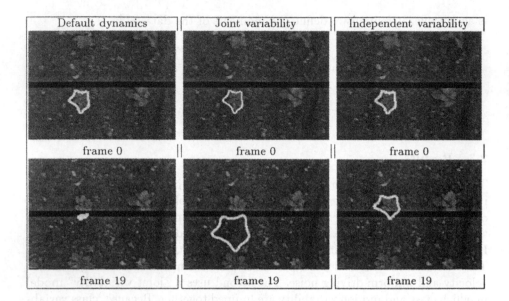

Default dynamics	Joint variability	Independent variability
frame 0	frame 0	frame 0
frame 19	frame 19	frame 19

Fig. 4. Joint variability modelling is too weak. *Results are shown here for the three trackers. In this case the tracked plant is rather smaller than average and there is an occluding feature (a simulated shadow). The "joint variability" tracker reverts, after a short time, towards the grand mean of the training set and hence loses lock on the plant. The "independent variability" tracker however adjusts rapidly, via its additional mean-configuration variable $\overline{\mathbf{X}}$, to the small size of this plant and continues to track successfully.*

Each submatrix contains coefficients either for one of the objects or for the cross-couplings between objects.

Frequently, as with head and lip motion, one object $\mathbf{X}^{(1)}$ is primary and drives a secondary object $\mathbf{X}^{(2)}$. The motion of the primary object is assumed independent of the secondary, and is modelled stand-alone so that

$$A_i^{1,2} = 0, \ i = 0, 1. \tag{12}$$

However, the motion of the secondary system $\mathbf{X}^{(2)}$ has two components, one caused by the primary system, $\mathbf{X}^{(1)}$, and one caused by its own independent motion, $\mathbf{Y}$, with dynamics given by

$$\mathbf{Y}_{k+2} = A_0^{2,2}\mathbf{Y}_k + A_1^{2,2}\mathbf{Y}_{k+1} + B^{(2)}\mathbf{w}_k^{(2)} \tag{13}$$

and a direct coupling is added to represent the influence of the primary process:

$$\mathbf{X}^{(2)} = \mathbf{Y} + \mu\mathbf{X}^{(1)}, \tag{14}$$

where μ is a constant matrix of the appropriate dimension. Combining (13) and (14) and comparing with (10) leads to constraints on the dynamical parameters:

$$A_i^{2,1} = \mu(A_i^{1,1} - A_i^{2,2}) \ i = 0, 1. \tag{15}$$

3.2 Learning Coupled Dynamics

With the constraints (12) and (15), the maximum likelihood learning algorithm turns out to be nonlinear, with a consequent need for iteration and the possibility of multiple solutions. To avoid this we propose that (15) be relaxed, retaining only (12). (This can be interpreted physically as allowing a velocity coupling in addition to the positional coupling (14).) The resulting learning algorithm is now linear.

Given its isolation, coefficients for the motion of the primary object are learned exactly as for a single object, as earlier and in [5]. Dynamics of the secondary object and its coupling to the primary are obtained (proof omitted here) by solving the following simultaneous equations:

$$S_{20}^{(2,2)} - A_0^{2,2} S_{00}^{(2,2)} - A_1^{2,2} S_{10}^{(2,2)} - A_0^{2,1} S_{00}^{(2,1)} - A_1^{2,1} S_{01}^{(2,1)} = 0$$
$$S_{21}^{(2,2)} - A_0^{2,2} S_{01}^{(2,2)} - A_1^{2,2} S_{11}^{(2,2)} - A_0^{2,1} S_{10}^{(2,1)} - A_1^{2,1} S_{11}^{(2,1)} = 0$$
$$S_{20}^{(2,1)} - A_0^{2,2} S_{00}^{(2,1)} - A_1^{2,2} S_{10}^{(2,1)} - A_0^{2,1} S_{00}^{(2,2)} - A_1^{2,1} S_{01}^{(2,2)} = 0.$$

where moments are defined:

$$S_{ij}^{(p,q)} = \sum_{n=1}^{m-2} \mathbf{X}_{n+i}^{(p)} \mathbf{X}_{n+j}^{(q)}, \quad i,j = 0,1,2, \quad p,q = 1,2.$$

(The algorithm given here is for the case of known mean configuration $\overline{\mathbf{X}}^{(1)} = 0, \overline{\mathbf{X}}^{(2)} = 0$.) The covariance C is calculated from instantaneous error R_n as earlier and [5].

3.3 Computational Complexity for Filtering

In the visual tracking application, the above dynamical system is used as a predictor in a Kalman filter [7, 1] for curve tracking — a dynamic contour tracker [15, 3]. In the filter, the prediction step covariance $\mathcal{P}$ evolves, in the conventional manner, as:

$$\mathcal{P}_{k+1|k} = \begin{pmatrix} \mathcal{A}_{1,1} & 0 \\ \mathcal{A}_{2,1} & \mathcal{A}_{2,2} \end{pmatrix} \begin{pmatrix} P_{1,1} & P_{1,2} \\ P_{1,2}^T & P_{2,2} \end{pmatrix}_{k|k} \begin{pmatrix} \mathcal{A}_{1,1}^T & \mathcal{A}_{2,1}^T \\ 0 & \mathcal{A}_{2,2}^T \end{pmatrix} + C,$$

a critical step from the point of view of computational load. Even though $\mathcal{A}_{1,2} = 0$ and $\mathcal{A}_{2,1}$ is sparse, the covariance $\mathcal{P}_{k+1|k}$ remains dense, leaving unabated the full computational load of $6\alpha n^3$ operations, where α is a constant and n is the filter's dimension. This compares with αn^3 operations to track an isolated object, or $2\alpha n^3$ for two uncoupled objects. So the price of representing the coupling faithfully is a 3-fold increase in computation.

The measurement process suffers from similar costs and as before, the amount of computation to run coupled systems is $6\gamma n^3$ where γn^3 is the amount of computation required for an isolated object. Again, coupling incurs a 3-fold cost increase.

3.4 Approximating the Coupling

We refer to the full solution of the Kalman filtering problem as "strongly coupled". A "weakly coupled" approximation to the full algorithm is constructed by treating the state of the primary system as if it were known exactly, for the purpose of determining its effect on the secondary system. In other words, equation (14) is replaced by

$$\mathbf{X}^{(2)} = \mathbf{Y} + \mu \hat{\mathbf{X}}^{(1)}$$

where $\hat{\mathbf{X}}^{(1)}$ is the *estimate* of the primary system's state. The effect of the primary system on the secondary becomes entirely deterministic. Covariance terms $P_{1,1}$ and $P_{2,2}$ become mutually independent and, moreover, $P_{1,2} = P_{2,1}^T = 0$. Covariances must be computed continuously to maintain the Kalman gain matrices, so this is where the weak coupling allows substantial savings in computation. Of course, the restricted P matrices computed in this way cannot be claimed to approximate the true covariances, and since they are used to define Kalman gains, those gains are substantially altered by the approximation. However, these are open loop gains and it is well known large changes in open loop gain often cause only small perturbations in closed loop response. This explains somewhat the accuracy of results with the weak coupling reported below.

Weak coupling can be generalised to larger numbers of objects tracking simultaneously, with or without object hierarchy. In general, if a single filter has a computational burden of Γ, then, in the strongly coupled case, with m objects, a burden of $\Gamma(m^3 + m^2)/2$ is incurred. In the weakly coupled case it is reduced to Γm.

3.5 Head and Lip Tracking

Previously Moses *et al* [12] designed a contour tracker to follow an intensity valley between the lips. The valley proved to be a robust feature under a wide range of lighting conditions but because of the extreme aspect ratio of the mouth, the tracker tended to be unstable to horizontal translation (figure 3). The instability problem is tackled here by a weak coupling between the lip as secondary object to the head. A five dimensional space was used to model the head outline, comprising translation in the x and y direction, uniform scaling, and rotation about the vertical axis (using two degrees of freedom not one, to allow for silhouette effects).

Figure 5 shows the effect of tracking when coupling is enabled. In this case the strongly coupled algorithm was run. The coupling is such that, appropriately enough, the horizontal motion of the head and its rotation affect the horizontal location of the lips. The size of the head also affects the size of the lips, allowing for zoom-coupling. The figure clearly demonstrates that coupling is effective. Figure 6 shows that weak coupling produces results similar to strong coupling, and that without coupling, the tracker fails at an early stage. Figure 7 displays lip deformations as a function of time for strong and weak coupling. Two affine components of motion are plotted: horizontal and vertical scaling. The most obvious feature is that, as before, the uncoupled tracker loses lock after about

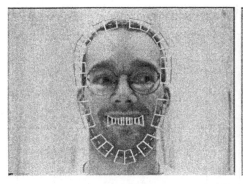
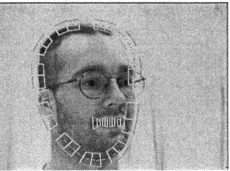

Fig. 5. Strongly-coupled algorithm is successful. *This set of images shows that the poor tracking performance obtained from an isolated mouth tracker (figure 3) is substantially redeemed by appropriate coupling to head motion.*

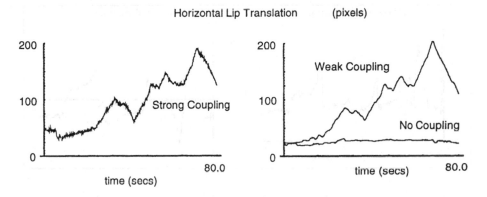

Fig. 6. Lip tracking is successful only when coupled to the head tracker. *The first graph displays lip translation as a function of time, in the case of strong coupling. An independent experiment has confirmed that this result is broadly accurate. The second graph shows that without coupling the tracker fails at an early stage but that with weak coupling accurate tracking is obtained.*

20 seconds. It is also clear that weak coupling produces smoother estimates of the lip shape than strong coupling.

4 Conclusions

Work presented here reinforces the growing acceptance that careful modelling of object shape, motion and environment is crucial to effective performance of visual processes in general and visual tracking in particular. Stochastic differential equations are particularly useful as models for this purpose because they are the basis of prediction in the Kalman filter. Moreover, procedures exist for learning such models from examples.

More specifically, stochastic modelling for tracking has been advanced in two particular regards in this paper. The first is in clearly separating object vari-

Uncoupled Tracking

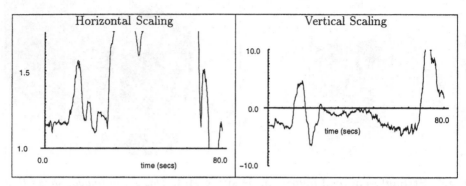

Strongly Coupled Tracking

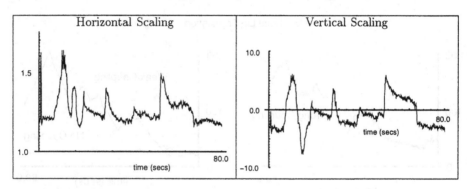

Weakly Coupled Tracking

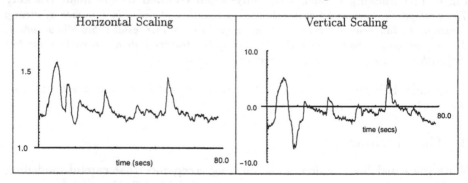

Fig. 7. Comparative performance of tracking algorithms. *This figure shows two affine components of the lip motion as a function of time. The top pair of graphs show how lock is lost very early in the sequence (after about 20.0 sec) in the absence of any coupling. The final pair shows that when the coupling is weak, performance is close to that in the strongly coupled case.*

ations arising within a class of objects from those that occur temporally, during motion. The second is in building coupled motion models to track objects whose motions are not independent, for instance head and lips. In both case, appropriate modelling leads to substantial enhancements of tracking performance.

References

1. Y. Bar-Shalom and T.E. Fortmann. *Tracking and Data Association.* Academic Press, 1988.
2. A. Baumberg and D. Hogg. Generating spatiotemporal models from examples. In *Proc. BMVC*, 413–422, 1995.
3. A. Blake, R. Curwen, and A. Zisserman. A framework for spatio-temporal control in the tracking of visual contours. *Int. Journal of Computer Vision*, 11(2):127–145, 1993.
4. A. Blake, M. Isard, and D. Reynard. Learning to track the visual motion of contours. *Artificial Intelligence*, 78:101–133, 1995.
5. A. Blake and M.A. Isard. 3D position, attitude and shape input using video tracking of hands and lips. In *Proc. Siggraph*, 185–192. ACM, 1994.
6. T.F. Cootes, C.J. Taylor, A. Lanitis, D.H. Cooper, and J. Graham. Buiding and using flexible models incorporating grey-level information. In *Proc. 4th Int. Conf. on Computer Vision*, 242–246, 1993.
7. A. Gelb, editor. *Applied Optimal Estimation.* MIT Press, Cambridge, MA, 1974.
8. D.B. Gennery. Visual tracking of known three-dimensional objects. *Int. Journal of Computer Vision*, 7:3:243–270, 1992.
9. C.G. Goodwin and K.S. Sin. *Adaptive filtering prediction and control.* Prentice-Hall, 1984.
10. C. Harris. Tracking with rigid models. In A. Blake and A. Yuille, editors, *Active Vision*, 59–74. MIT, 1992.
11. D.G. Lowe. Robust model-based motion tracking through the integration of search and estimation. *Int. Journal of Computer Vision*, 8(2):113–122, 1992.
12. Y. Moses, D. Reynard, and A. Blake. Determining facial expressions in real time. In *Fifth International Conference on Computer Vision*, 296–301, Massachusetts, USA, 1995.
13. A. Pentland and B. Horowitz. Recovery of nonrigid motion and structure. *IEEE Trans. Pattern Analysis and Machine Intelligence*, 13:730–742, 1991.
14. R.S. Stephens. Real-time 3D object tracking. *Image and Vision Computing*, 8(1):91–96, 1990.
15. R. Szeliski and D. Terzopoulos. Physically-based and probabilistic modeling for computer vision. In B. C. Vemuri, editor, *Proc. SPIE 1570, Geometric Methods in Computer Vision*, 140–152, San Diego, CA, July 1991. Society of Photo-Optical Instrumentation Engineers.
16. D. Terzopoulos and R. Szeliski. Tracking with Kalman snakes. In A. Blake and A. Yuille, editors, *Active Vision*, 3–20. MIT, 1992.

Acknowledgement: We are grateful for the support of the BBSRC and the Rhodes Trust.

Grouping and
Segmentation

Quantitative Analysis of Grouping Processes

Amnon Amir and Michael Lindenbaum

Computer Science Department, Technion-Haifa 32000, ISRAEL

amir,mic@cs.technion.ac.il

Abstract. This paper presents a quantitative approach to grouping. A generic grouping method, which may be applied to many domains, is given, and an analysis of the expected grouping quality is one. The grouping method is divided into two parts: constructing a graph representation of the geometric relations in the data set, and then finding the 'good' partition of the graph into groups. Both stages are implemented using known statistical tools such as Ward's SPRT algorithm and the Maximum Likelihood test. The resulting grouping quantitative analysis allows some, not all, of these properties to be calculated, and gives some insight into the computational efforts to the expected grouping quality. To our best knowledge, such an analysis is not, reasoning process given here for the first time. The quantitative grouping also some-what demonstrated for some different grouping tasks and domains. Experimental results show the ability of the generic algorithm to provide a successful algorithm in a specific domain.

Keywords: Grouping Analysis, Perceptual Grouping, Performance Prediction, Quantic Grouping Algorithm, Graph Clustering, Maximum Likelihood, Ward's, SPRT.

1 Introduction

This paper presents a quantitative approach to grouping that contains a generic grouping method, and focuses on analyzing the relation between the information available to the grouping process and the corresponding grouping result.

The proposed method separates between two components of the grouping process. In grouping, cues that are used and the associated mechanism that combines them into a partition of the data set. Our grouping process is based on a special graph representation, in which the vertices are the observed data elements (edges, pixels, etc.), and the arcs contain the grouping information and are estimated by cues. (Others e.g. [1, 4, 11], have used graphs for grouping algorithms, but we use it differently here.) The hyper-graph based grouping is a partition of the graph which maximizes a functional over all the possible partitions.

In contrast to most other grouping methods, which depend on the domain in which the grouping is done, this grouping mechanism is domain independent.

Good cues are essential for successful grouping, but finding them is not our aim here. Instead we consider that cues are given, model them as random variables, and quantify their reliability using the properties of the corresponding distribution. Moreover, we suggest a general method, called the cue enhancement

Quantitative Analysis of Grouping Processes

Arnon Amir and Michael Lindenbaum

Computer Science Department, Technion, Haifa 32000, ISRAEL
arnon,mic@cs.technion.ac.il

Abstract. This paper presents a quantitative approach to grouping. A generic grouping method, which may be applied to many domains, is given, and an analysis of its expected grouping quality is done. The grouping method is divided into two parts: Constructing a graph representation of the geometric relations in the data set, and then finding the "best" partition of the graph into groups. Both stages are implemented using known statistical tools such as Wald's SPRT algorithm and the *Maximum Likelihood* criterion. The accompanying quantitative analysis shows some relations between the data quality, the reliability of the grouping cues and the computational efforts, to the expected grouping quality. To our best knowledge, such an analysis of a grouping process is given here for the first time. The synthesis of specific grouping algorithms is demonstrated for three different grouping tasks and domains. Experimental results show the ability of this generic approach to provide successful algorithm in specific domains.

Keywords : Grouping Analysis, Perceptual Grouping, Performance Prediction, Generic Grouping Algorithm, Graph Clustering, Maximum Likelihood, Wald's SPRT.

1 Introduction

This paper presents a quantitative approach to grouping, which contains a generic grouping method, and focuses on analyzing the relation between the information available to the grouping process and the corresponding grouping quality.

The proposed method separates between two components of the grouping method: the *grouping cues* that are used and the *grouping mechanism* that combines them into a partition of the data set. Our grouping process is based on a special graph representation, in which the vertices are the observed data elements (edges,pixels, etc.) and the arcs contain the grouping information and are estimated by cues. (Others, e.g. [8, 4, 11], have used graphs for grouping algorithms, but we use it differently here.) The hypothesized grouping is a partition of the graph which maximize a functional over all the possible partitions. In contrast to most other grouping methods, which depend on the domain in which the grouping is done, this grouping mechanism is domain independent.

Good cues are essential for successful grouping, but finding them is not our aim here. Instead we consider the cues as given, model them as random variables, and quantify their reliability using the properties of the corresponding distribution. Moreover, we suggest a general method, called the *cue enhancement procedure*, for improving the reliability of grouping cues, and show an interesting tradeoff between the computational efforts and the achievable reliability of the enhanced cue.

Unlike other grouping methods, the proposed method provides, for the first time, some relations between the quality of the available data, the computational effort invested, and the grouping performance, quantified by several measures.

2 The Grouping Task and its Graph Representation

The *grouping task* is a partitioning problem. Let $S = \{v_1, v_2, \ldots, v_N\}$ be the set of data elements, which may consist, for example, of the boundary points in an image. S is naturally divided into several *groups* (disjoint subsets) so that all data elements in the same group belong to the same object, lie on the same smooth curve, or associated with each other in some other manner. $S = S_0 \cup S_1 \cup S_2 \cup \ldots \cup S_M$. In the context of the grouping task the data set is given but its partition is unknown and should be inferred from indirect information given in the form of grouping cues [1].

Grouping cues are the building blocks of the grouping process and shall be treated as the only source of information available for this task. The grouping cues are domain-dependent and may be regarded as scalar functions $C(A)$ defined over subsets $A \subset S$ of the data feature set. Such cue functions should be discriminative, and should also be invariant to change of the viewing transformation and robust to noise [6]. At this stage we consider only bi-feature cues, defined over data subsets including two elements ($|A| = 2$). Bi-feature cues may be either the cues used by most common grouping processes, or the result of the *cue enhancement procedure*, which accumulates statistical information by using multi-feature cues, and is described in Sec 5. From now on we shall use the notation $C(e)$ for the bi-feature cues, where $e = (u, v)$, $u, v \in S$ is also the arc connects the nodes u, v in the following graph representation.

A *reliability measure* for grouping cues, which is domain-independent, is specified as follows: Consider the cue function to be a random variable, the distribution of which depends on whether the two data features belong to the same group or not. For binary cues, which provides only negative or positive answers, this dependency is simply quantified by two error probabilities: ϵ_{miss} is the probability that the cue $C(A)$ indicates a wrong negative answer, and ϵ_{fa} is the probability that the cue indicates a wrong positive answer (false alarm). If both $\epsilon_{miss} = 0$ and $\epsilon_{fa} = 0$, then $C(A)$ is an ideal cue. This characterization can sometimes be calculated using analytical models (e.g. [6]), and can always be approximated using Monte-Carlo experimentations.

Both the unknown partition into groups, and the data available from the cues, are represented using graphs. The nodes of all the graphs are the observed data elements, $V = S$, but the arcs may take different meanings, as explained

---

[1] We should also mention, that according to another grouping concept the hypothesized groups are not necessarily disjoint. We believe that at least some of the tools developed here are useful for the other approaches.

in Figure 1. The unknown partition, which is to be determined, is represented by the *target graph*, $G_t = (V, E_t)$, composed of several disconnected complete subgraphs (cliques). Every clique represents a different object (or group). A graph with this characterization is called a *clique graph* and the class of such graphs is denoted $\mathcal{G}_c$. We shall denote by $E_c(V_j)$ the arcs of a complete subgraph (clique) $V_j \subseteq V$. Knowing that $G_t \in \mathcal{G}_c$, the grouping algorithm should provide a *hypothesis graph*, $G_h = (V, E_h) \in \mathcal{G}_c$, which should be as close as possible to G_t.

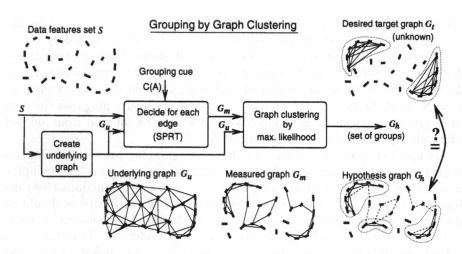

Fig. 1. The proposed grouping process: The image is a set of data features (edgels in this illustration) every one of which is represented by a node of a graph. The designer should decide about a cue and about the set of feature-pairs to be evaluated using this cue. This set of feature-pairs is specified by the arcs of the *underlying graph* G_u. The first step of the algorithm is to use *grouping cues* to decide, for every feature pair in $G_u = (V, E_u)$, if both data features belong to the same group. These decisions are represented by the a *measured graph* $G_m = (V, E_m)$: every arc corresponds to a positive decision (hence $E_m \subseteq E_u$). The known reliability of these decisions is used in the second and last step to find a maximum likelihood partitioning of the graph, which is represented by the hypothesized (clique) graph G_h. A main issue considered in this paper is the relation between this hypothesis G_h and the ground truth *target graph*, G_t, which is unknown.

3 The Generic Grouping Algorithm

The algorithm consists of two main stages: cue evaluation for (many) feature pairs and maximum likelihood graph partitioning. Before these stages, two decisions should be made by the designer. The first is to choose an appropriate grouping cue. The second one is to choose the set of feature-pairs to be evaluated using this cue. This set of feature-pairs is specified by the arcs of the *underlying graph* G_u. In principle, all feature pairs, corresponding to a *complete underlying graph*, $G_u = (V, E_c(V))$, should be evaluated using the cue, in order to extract the maximal information. Some cues are meaningful, however, only for near or adjacent data elements (e.g. while looking for smooth curves). In such cases, a locally connected underlying graph should be used. Another consideration which affects the choice of the underlying graph is the reliability of the grouping process and the computational effort invested in it. At this stage we assume that both the cue and the associated adequate "topology" are either given or chosen intuitively.

In the first stage of the grouping process, every feature pair, $e = (u, v)$, corresponds to an arc in $G_u = (V, E_u)$ is considered, and the cue function is used to decide whether the two data features belong to the same group, or not. The positive decisions are represented as the arcs of the *measured graph* $G_m = (V, E_m)$. G_m carries the information accumulated in the first stage to the second one.

Recall that every arc-decision made in the first stage is modeled as a binary random variable, the statistics of which depends on whether the two data features belong to the same group or whether they not. In the context of this paper, the cue decisions assumed to be independent and identically distributed, and are characterized by two error probabilities [2]:

$$\epsilon_{miss}(e) = Prob(e \in (E_t \cap E_u) \backslash E_m) \qquad \epsilon_{fa}(e) = Prob(e \in E_m \backslash (E_t \cap E_u)) \quad (1)$$

The likelihood of the measurement graph, G_m, for every candidate hypothesis $G = (V, E) \in \mathcal{G}_c$, is then given by

$$L\{G_m | G\} = \prod_{e \in E_u} L\{e|E\} \quad \text{where} \quad L\{e|E\} = \begin{cases} \epsilon_{miss} & \text{if } e \in E \backslash E_m \\ \epsilon_{fa} & \text{if } e \in E_m \backslash E \\ 1 - \epsilon_{miss} & \text{if } e \in E \cap E_m \\ 1 - \epsilon_{fa} & \text{if } e \notin E \cup E_m . \end{cases} \quad (2)$$

We propose now to use the maximum likelihood principle, and to hypothesize the most likely clique graph

$$G_h = \arg \max_{G \in \mathcal{G}_c} L\{G_m | G\}. \quad (3)$$

The maximum likelihood criterion specifies the (not necessarily unique) grouping result, G_h, but is not a constructive algorithm. We therefore address the theoretical aspect and the practical side separately.

From the theoretical point of view, we shall now assume that the hypothesis which maximizes the likelihood may be found, and address our main question: **"what is the relation between the result G_h, and the unknown target graph G_t?"** This question is interesting because it is concerned with predicting the grouping performance. If we can show that these two graphs are close in some sense, then it means that algorithms which use the maximum likelihood principle

---

[2] This definition of $(\epsilon_{miss}, \epsilon_{fa})$, in terms of the graph notation, is identical to the previous one, which refers to the cue reliability.

have predictable expected behavior and that even if we can't know G_t, the grouping hypothesis G_h they produces is close enough to the true partitioning. This question is considered in the next section.

From the practical point of view, one should ask if this optimization problem can be solved in a reasonable time. Some people use simulated annealing to solve similar problems [4]. Others use heuristic algorithms [9]. We developed a heuristic algorithm which is based on finding seeds of the groups, which are (almost) cliques in G_m. Then it makes iterative modifications, using a greedy policy, until a (local) maximum of the likelihood function is obtained. In our experiments this algorithm performs nicely. More details can be found in [2].

4 Analysis of The Grouping Quality

This section quantifies some aspects of the similarity between the unknown scene grouping (represented by G_t), and the hypothesized maximum-likelihood grouping (represented by G_h). We provide a fundamental claim, and two of its results. The fundamental claim provides a necessary condition, satisfied by any partition selected according to the maximum likelihood principle. Consider two nodes-disjoint subsets V_i, V_j of the graph $G = (V, E)$, and denote their cut by $J(V_i, V_j) = \{e = (u, v) | u \in V_i, v \in V_j\}$. Let $l_u(V_i, V_j) = |J(V_i, V_j) \cap E_u|$ denote the cut width relative to the underlying graph. Similarly, let $l_m(V_i, V_j) = |J(V_i, V_j) \cap E_m|$ denote the cut width relative to the measurement graph ($l_m \leq l_u$). Then,

Claim 1. necessary condition: *Let $G_h = (V, E_h)$, $V = \{V_1 \cup V_2 \cup \ldots\}$, $E_h = \{E_c(V_1) \cup E_c(V_2) \cup \ldots\}$ be the maximum likelihood hypothesis (satisfying eq. (3)), and let $\alpha = \left(1 + \frac{\log(\epsilon_{fa}/(1 - \epsilon_{miss}))}{\log(\epsilon_{miss}/(1 - \epsilon_{fa}))}\right)^{-1}$ Then,*
1. For any bisection of any group $V_i = V_i' \cup V_i''$ ($V_i' \cap V_i'' = \emptyset$),

$$l_m(V_i', V_i'') \geq \alpha l_u(V_i', V_i'').$$

2. For any two groups V_i, V_j, $i \neq j$,

$$l_m(V_i, V_j) \leq \alpha l_u(V_i, V_j).$$

Proof. For proving the first part, consider the likelihood ratio between two hypotheses: One is G_h and the other, denoted $\tilde{G}_h$, is constructed from G_h by separating V_i into two different groups, V_i' and V_i''. Denote $l_m = l_m(V_i', V_i'')$, $l_u = l_u(V_i', V_i'')$. Then

$$\frac{L\{G_m | G_h\}}{L\{G_m | \tilde{G}_h\}} = \prod_{e \in J(V_i', V_i'') \cap E_u} \frac{Pr\{e | E_h\}}{Pr\{e | \tilde{E}_h\}} = \left(\frac{1 - \epsilon_{miss}}{\epsilon_{fa}}\right)^{l_m} \left(\frac{\epsilon_{miss}}{1 - \epsilon_{fa}}\right)^{l_u - l_m}.$$

(arcs of $E_u \setminus J(V_i', V_i'')$ do not affect that ratio, and therefore are not counted). This likelihood ratio is an increasing function of l_m and is larger than 1, for $l_m \geq \alpha l_u$. Therefore, if the claim is not satisfied, then $\tilde{G}_h$ is more likely then G_h which contradicts the assumption that (3) holds. The second part of the claim is proved in a similar manner.

Qualitatively, the claim shows that a maximum likelihood grouping must satisfy local conditions between many pairs of feature subsets. It further implies that a grouping error, either in the form of adding an alien data feature to a group (denoted *addition error*) or deleting its member (denoted *deletion error*), requires a substantial number of

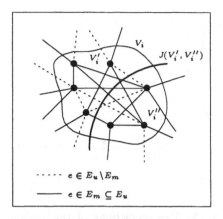

Fig. 2. The cut involved in splitting a group into two (proof 1).

false alarms, or misses, respectively. The parameter α, specifying the fraction of cut edges required to merge two subsets reflects the expected error types ; if $\epsilon_{fa} = \epsilon_{miss}$, then $\alpha = 0.5$, while if $\epsilon_{fa} > \epsilon_{miss}$ then $\alpha > 0.5$. This claim implies that choosing a sufficiently dense underlying graph can significantly improve the grouping performance, and compensate for unreliable cues.

Two cases were considered in [2]: A complete underlying graph, and a locally connected underlying graph. A complete underlying graph provides the maximal information. Therefore, it may lead to excellent grouping accuracy. The next claim, given here as an example, bounds the probability of getting k' addition errors or more:

Claim 2. *Let G_u be a complete graph. Let S_i and V^* denote a true group and a maximum likelihood hypothesized group containing at least k nodes of S_i. Then, the probability that V^* contains k' nodes or more which are alien to S_i, is at most*

$$p_{k'-aliens} \leq \sum_{j=k'}^{N-k} \binom{N-k}{j} \sum_{i=k_{min}(j)}^{kj} \binom{kj}{i} \epsilon_{fa}^i (1 - \epsilon_{fa})^{kj-i} \quad (k_{min}(j) = \lceil \alpha k j \rceil)$$

$$(4)$$

(for proof see [2]). This upper bound is plotted in Figure 3(Right). Other results in [2] provides the bound for the probability of k-deletion errors, the expected number of such errors, and more. These results simply state that if the original group S_i is big enough and $\epsilon_{miss}, \epsilon_{fa}$ are small enough, it is very likely that the maximum likelihood partition will include one group for each object, containing most of S_i, and very few aliens. Experimental results for these two grouping error types are given in Figure 6 (d,e) and discussed in Section 6.

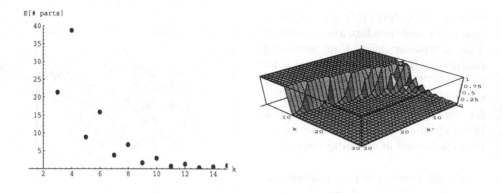

Fig. 3. Two predictions of the analysis: Left: A k-connected curve-like group (e.g. smooth curve) is likely to break into a number of sub-groups. The graph shows an upper bound on the expected number of sub-groups versus the minimal cut size in the group, k. Here the group size (length) is 400 elements, $\epsilon_{miss} = 0.14$ and $\epsilon_{fa} = 0.1$ (typical values for images like Figure 8 (a)). It shows how increasing connectivity quickly reduces the false division of this type of groups. Right: Upper bound on the probability for adding any k' alien data features to a group of size k, using a complete underlying graph (claim 2). This probability is negligible for $k > 15$ (Here $\epsilon_{miss} = \epsilon_{fa} = 0.2$).

Another prediction, given in Figure 3 (Left), corresponds to a second case, where the underlying graph is not a complete graph, but a locally-connected one where every node is connected to the closest k data features. This is useful, for example, for the common curve-like groups, in which all data features are ordered along some curve (e.g the smooth curve experiment). The grouping process may divide a long curve into a number of fragments. An upper bound on the expected number of these parts is shown in Figure 3(Left).

5   The Cue Enhancement Procedure (CEP)

The performance of the grouping algorithm depends very much on the reliability of the cues available to it. This section shows how the reliability of a grouping cue can be significantly improved by using statistical evidence accumulation techniques. This method is not restricted only to our grouping algorithm, and can be used also in other grouping algorithms. Two of the three given grouping examples (co-linearity and smoothness) use this procedure.

The CEP considers one pair of data features, $e = (u, v)$, $e \in E_u$, at a time, and uses some of the other data features in order to decide whether or not this pair is *consistent* (belongs to the same true group). Its result serves as a very reliable *binary-cue*, $C(e)$, as defined in Section 2. The key to CEP is the use of *Multi-feature cues*, $C(A)$, associated with three data features or more. The important observation here is that if the pair, $e = (u, v)$, is not consistent, then any larger subset A, $u, v \in A$ is not consistent as well. Therefore, the (multi-feature) cue, $C(A)$, which test for the consistency of A, carries some statistical information on the consistency of e. To evaluate $C(e)$, the CEP draws several random data subsets, $A_1, A_2, \ldots$, of size $k > 2$, which contain the pair e, and find $c_j = C(A_j)$, $j = 1, 2, \ldots$. $C(A)$ is a deterministic function of A, but c_j may be considered as an instance of a random variable, c, as some part of A is randomly selected.

The statistics of c depends on the data pair e, and in particular, on its consistency. A conclusive reliable decision on the consistency of e is determined adaptively and efficiently by a well-known method for statistical evidence integration: Wald's *Sequential Probability Ratio Test (SPRT)* algorithm [10].

The distribution of c depends on an unknown binary parameter (the consistency of e), which takes the value of ω_0 (false) or ω_1 (true). The SPRT quantifies the evidence obtained from each trial by the log-likelihood ratio function of its result $h(c) = \ln \frac{p_1(c)}{p_0(c)}$ where $p_i = Pr\{c|\omega_i\}$ $i = 0, 1$ are the probability functions of the two different populations and c is the value assigned to the random variable in this trial (see Figure 5). When several trials are taken, the log-likelihood function of the composite event $\bar{c} = (c_1, c_2, \ldots, c_n)$ should be considered. If, however, the trials are independent then this composite log-likelihood function becomes $\sigma_n = \sum_{j=1}^{n} h(c_j)$. The sum σ_n serves as the statistics by which the decision is made. The SPRT-based cue enhancement procedure is summarized in Figure 5. The upper and lower limits, $a > 0 > b$, depend only on the required

---

For every feature pair $e = (u, v)$ in the underlying graph:

1. **Set the evidence accumulator, σ, and the trials counter, n, to 0.**
2. **Randomly choose $k - 2$ data features $x_3, \ldots, x_k \in S \backslash \{u, v\}$**
3. **Calculate $c = C(\{u, v, x_3, \ldots, x_k\})$.**
4. **Update the evidence accumulator $\sigma = \sigma + \log \frac{P_1(c)}{P_0(c)}$.**
5. **if $\sigma \geq a$ or if $n \geq n_0$ and $\sigma > 0$, output: (u, v) is consistent.**
   if $\sigma \leq b$ or if $n \geq n_0$ and $\sigma < 0$, output: (u, v) is inconsistent.
   else, repeat (2)-(5)

Fig. 4. The SPRT-based Cue Enhancement Procedure (CEP)

cue reliability $\epsilon_{miss}, \epsilon_{fa}$ (defined in eq. 1), which are specified by the user, and do not depend on the distribution of the random variable c. We calculate a, b using a practical approximation, proposed by Wald [10], which is very accurate when $\epsilon_{miss}, \epsilon_{fa}$ are small: $a = \log(\frac{1-\epsilon_{miss}}{\epsilon_{fa}})$ $b = \log(\epsilon_{miss}(1 - \epsilon_{fa}))$. The derivation of $P_1(c), P_0(c)$, which depends on $\epsilon_{miss}, \epsilon_{fa}$ defined before, and on some combinatorial considerations, is given in [2].

The basic SPRT algorithm terminates with probability one and is optimal in the sense that it requires a minimal expected number of tests to obtain the required decision error [10]. This expected number of tests is given by:

$$E\{n|\omega_0\} = [a\epsilon_{fa}+b(1-\epsilon_{fa})]/\eta_0 \qquad\qquad E\{n|\omega_1\} = [a(1-\epsilon_{miss})+b\epsilon_{miss}]/\eta_1$$

$$(5)$$

where η_0, η_1 are the conditional expected amounts of evidence from a single test: $\eta_i = E\{h(c)|\omega_i\}$ $i = 0, 1$. The maximal allowed trials number, n_0, is set to be few times larger than $E\{n\}$.

Claim 3. *Given that*

(a) *The statistics of the cue values evaluated over all data subsets containing a consistent (inconsistent) arc is approximately the same, and*
(b) *The cues extracted from two random subsets including the same feature pair are independent identically distributed random variables,*

then the CEP can identify the consistency of the feature pair e within any specified error tolerance, ϵ_{miss}, ϵ_{fa}, irrespective of the reliability of the basic cue, $C(A)$.

Arbitrarily high performance is practically impossible because it requires a large number of trials leading to a contradiction of the independence assumption. Therefore, the reliability of the basic cue, $P_0(c), P_1(c)$, is important to achieve a lower expected-number of trials, $E\{n\}$. Indeed, our experiments show that the SPRT significantly improves the cue reliability but that the achievable error rate is not arbitrarily small (see Section 6). In the co-linearity experiments, $C(A)$ is a binary cue that depends on a threshold. A threshold cause a tradeoff between the miss to the false alarm ratios of the cue. For any given required reliability of the CEP , $(\epsilon_{miss}, \epsilon_{fa})$, we use eq. (5) to find the optimal threshold level, which minimize $E\{n\}$ (See Figure 5).

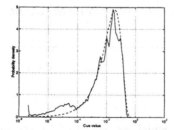 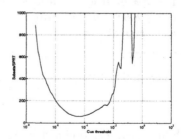

Fig. 5. Left: The two distributions of the co-linearity multi-feature cue, $p_0(A)$ (dashed), and $p_1(A)$ (solid). Although these two are very similar, their populations can be distinguished with less than 5% error (see Figure 6(d,e,f)). Right: The expected number of trials needed for the cue enhancement procedure as a function of the selected cue threshold. The optimal cue threshold correspond to the minima of this curve, as shown in Figure 5 (Right).

For a constant specified reliability $(\epsilon_{miss}, \epsilon_{fa})$, the expected run-time of the cue enhancement procedure is constant. The expected total run-time for evaluating all the arcs of the *underlying graph*, G_u, is, therefore, linear in the number of arcs.

6   Simulation and experimentation

This section presents three different grouping applications, implemented in three different domains, as instances of the generic grouping algorithm described above (see Table 1). The aim of these examples is to show that useful grouping algorithms may be obtained as instances of the generic approach and to examine the performance predictions against experimental results. (For all the technical details, more results and other examples please refer to [2].) The first example

Table 1. The three instances of the generic grouping algorithm

	The 1st example (Co-linear points)	The 2nd example (Smooth Curves)	The 3rd example (Motion Segm)
data elements	points in R^2	edgels	patches of Affine optical flow
grouping cues	co-linearity	co-circularity and proximity	consistency with Affine motion
Cue's extent	global	local	global
Enhanced cue	subsets of 3 points	subsets of 3 edgels	–
underlying graph	complete graph	locally connected graph	a complete graph
grouping mechanism:	maximum likelihood graph clustering (same program)		

is of grouping points by co-linearity cues. Given a set of points in R^2 (or in R^n), the algorithm should partition the data into co-linear groups (and one background set). To remove any doubt, we do not intend to propose this example as an efficient (or even reasonable) method for detecting co-linear clusters (Hough transform or RANSAC, for example, are better methods). We have chosen this example because it is a characteristic example of grouping tasks associated with globally valid cues (and complete underlying graphs). Moreover, it provides a convenient way for measuring grouping performance, the quantification and prediction of which is our main interest here.

We consider synthetic random images containing randomly drawn points (e.g Figure 6 (a)). A typical grouping result is shown in Figure 6 (b,c). Few of the quantitative results show the effect of the cue reliability on the overall grouping quality (Figure 6(d,e)) and on the CEP computational time (Figure 6(f)). Regardless the choice of $(\epsilon_{miss}, \epsilon_{fa})$, all the 5 lines were always detected as the 5 largest groups in our experiments. The selection of $(\epsilon_{miss}, \epsilon_{fa})$ does affects, however, the overall grouping quality. This is measured by counting the addition errors and the deletion errors, as shown in Figures 6 (d), and (e), respectively.

The second example is the grouping of edgels by smoothness and proximity. Starting from an image of edgels, (data feature = edge location + gradient direction), the algorithm should group edgels which lie on the same smooth curve. This is a very useful grouping task, considered by many researchers (see, e.g [3, 12, 4, 7]). We test this procedure both on synthetic and real images, and the results are very good in both cases (see Figure 8 and Figure 9).

The third grouping algorithm is based on common motion. The data features are pixel blocks, which should be grouped together if their motion obeys the same rule, that is if the given optical flow over them is consistent with one Affine motion model [5, 1]. Technically, every pixel block is represented by its location and six parameters of the local Affine motion model (calculated using Least Squares). No cue enhancement is used here, and the cue is not very reliable: typical error probabilities are $\epsilon_{miss} = 0.35$ and $\epsilon_{fa} = 0.2$. Still, the results are comparable to those obtained by a domain specific algorithm [1]. The final clustering result is shown in Figure 7 (Right).

7 Discussion

The goal of this work is to provide a theoretical framework and a generic algorithm that may be applied to various domains and that have predictable performance. The proposed algorithm relies on established statistical techniques such as sequential testing and maximum likelihood, which are well known. However, this paper is distinctive from previous approaches because it provides, for the first time, an analysis of the use of these principles, which relates the expected grouping quality to the cue reliability, the connectivity used, and in some cases the computational effort invested. We did not limit ourselves to the theoretical study: three grouping applications, in different domains, are implemented as instances of the generic grouping algorithm. Although we made an argument against visually judging the merits of vision algorithm, we would like to indicate here that our results are similar to those obtained by domain specific methods (e.g. [7, 4] for smoothness based grouping). Note that G_m may also be used to create a saliency map, by specifying the saliency of a feature as the degree of the corresponding node in G_m (e.g Figure 8(d),9(d)). This is also comparable with other's results (e.g. [3]). Its suitability for figure-ground discrimination is now under study.

From our analysis and experimentation it is apparent that higher connectivity of the objects in G_u can enhance the grouping quality. Therefore, the selection of cues should consider, in addition to their reliability, also their spatial extent. Another consideration is the use of multi-feature cues, and the cue enhancement possibility (CEP).

Our analysis of the computational complexity is not complete. We still do not have complexity results for the second stage, of finding the maximum likelihood partition. This task is known to be difficult, and for now we use a heuristic algorithm, which gave good results in our experiments. Another research direction is to use our methodology in the context of a different grouping notion, different than partitioning, by which the hypothesized groups are not necessarily disjoint.

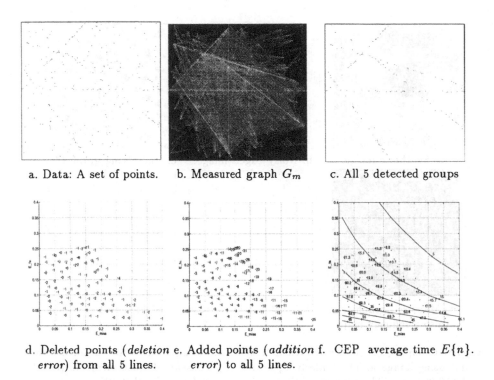

a. Data: A set of points.   b. Measured graph G_m c. All 5 detected groups

d. Deleted points (*deletion* e. Added points (*addition* f. CEP average time $E\{n\}$.
error) from all 5 lines.   error) to all 5 lines.

Fig. 6. Grouping of co-linear points, and its analysis. The data, (a), is associated with five lines, contains 30 points in the vicinity of each of them, and 150 "noise" points. The grouping result is near-optimal, and is close to the predictions for a complete underlying graph. Quantitative results show how the resulting grouping quality depends on the cue reliability (d)(e). Every point represents a complete grouping process and is labeled by the total addition/deletion errors. The average number of trials needed to achieve this enhanced cue reliability by the CEP, $E\{n\}$, is given in (f) near every point, and is compared to the predicted value, given by the labeled curves.

a. One image of a sequence.   b. Final image segmentation.

Fig. 7. Image segmentation into regions consistent with the same Affine motion parameters. Grouping is done on the optical-flow image (see Sec. 6). A post-processing stage use the obtained grouping to calculate an Affine motion model for every group, and to classify each pixel to the nearest model (The same post-processing used in [1]. Black pixels were not classified). The underlying graph is a complete graph of about 600 nodes (180,000 arcs), and the runtime is about 5 minutes.

383

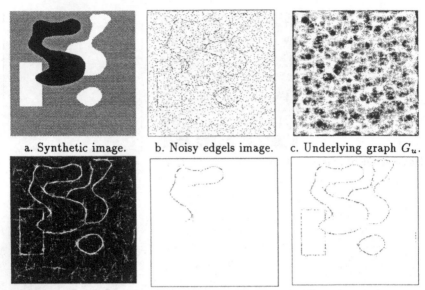

a. Synthetic image.   b. Noisy edgels image.   c. Underlying graph G_u.

d. Measured graph G_m. e. Largest detected group. f. All 14 detected groups.
(also a saliency map)

Fig. 8. Grouping of smooth curves in a synthetic image. Edges and gradient where found using image a. The underlying graph, G_u, consists of 5,000 elements (edgels), and 110,000 arcs. The processing time is 3 min on a Super-Spark CPU.

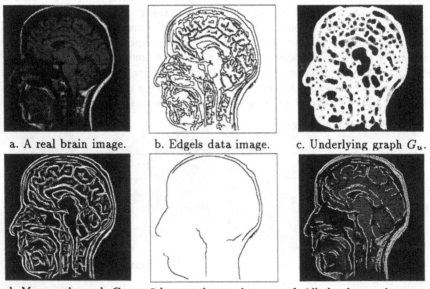

a. A real brain image.   b. Edgels data image.   c. Underlying graph G_u.

d. Measured graph G_m e. 5 largest detected groups. f. All the detected groups,
(also a saliency map)   superimposed on image a.

Fig. 9. Grouping of smooth curves in a brain image. The underlying graph, G_u, consists of 10,400 edgels and 230,000 arcs. The processing time is about 10 minutes on a Super-Spark CPU.

Acknowledgments:

We would like to thank John Wang for providing us the Flowers garden optical flow data, and the anonymous referees for their helpful comments. This research was supported by the fund for the promotion of research at the Technion.

References

1. ADELSON, E. H., AND WANG, J. Y. A. Representing moving images with layers. Tech. Rep. 279, M.I.T, May 1994.
2. AMIR, A., AND LINDENBAUM, M. The construction and analysis of a generic grouping algorithm. Tech. Rep. CIS-9418, Technion, Israel, Nov. 1994.
3. GUY, G., AND MEDIONI, G. Perceptual grouping using global saliency-enhancing operators. In *ICPR-92* (1992), vol. I, pp. 99–103.
4. HERAULT, L., AND HORAUD, R. Figure-ground discrimination: A combinatorial optimization approach. *PAMI 15*, 9 (Sep 1993), 899–914.
5. JACOBS, D. W., AND CHENNUBHOTLA, C. Finding structurally consistent motion correspondences. In *ICPR-94, Jerusalem* (1994), pp. 650–653.
6. LOWE, D. G. *Perceptual Organization and Visual Recognition.* Kluwer Academic Pub., 1985.
7. SHA'ASHUA, A., AND ULLMAN, S. Grouping contours by iterated pairing network. *Neural Information Processing Systems (NIPS) 3* (1990).
8. SHAPIRO, L. G., AND HARALICK, R. M. Decomposition of two-dimensional shapes by graph-theoretic clustering. *PAMI 1*, 1 (Jan. 1979), 10–20.
9. VOSSELMAN, G. *Relational Matching*, third ed. Lect. Notes in CS. Springer, 1992.
10. WALD, A. *Sequencial Analysis*, third ed. Wiley Publications in Statistics. 1947(1952).
11. WU, Z., AND LEAHY, R. An optimal graph theoretic approach to data clustering: Theory and its application to image segmentation. *PAMI 15*, 11 (Nov 1993), 1001–1113.
12. ZISSERMAN, A., MUNDY, J., FORSYTH, D., AND LIU, J. Class-based grouping in perspective images. In *ICCV-95, MIT* (1995), pp. 183–188.

Geometric Saliency of Curve Correspondences and Grouping of Symmetric Contours

Tat-Jen Cham and Roberto Cipolla

Department of Engineering
University of Cambridge
Cambridge CB2 1PZ, England

Abstract. Dependence on landmark points or high-order derivatives when establishing correspondences between geometrical image curves under various subclasses of projective transformation remains a shortcoming of present methods. In the proposed framework, geometric transformations are treated as smooth functions involving the parameters of the curves on which the transformation basis points lie. By allowing the basis points to vary along the curves, hypothesised correspondences are *freed* from the restriction to fixed point sets. An optimisation approach to localising *neighbourhood-validated* transformation bases is described which uses the deviation between projected and actual curve neighbourhood to iteratively improve correspondence estimates along the curves. However as transformation bases are inherently localisable to different degrees, the concept of *geometric saliency* is proposed in order to quantise this localisability. This measures the sensitivity of the deviation between projected and actual curve neighbourhood to perturbation of the basis points along the curves. Statistical analysis is applied to cope with image noise, and leads to the formulation of a *normalised basis likelihood*. Geometrically salient, neighbourhood-validated transformation bases represent hypotheses for the transformations relating image curves, and are further refined through curve support recovery and geometrically-coupled active contours. In the thorough application of this theory to the problem of detecting and grouping *affine* symmetric contours, good preliminary results are obtained which demonstrate the independence of this approach to landmark points.

1 Introduction

Establishing correspondences between points in images or 3D structure models in databases is often an integral part of solving key problems in computer vision. The choice of correspondences is largely determined by the localisability of the points, and methods such as cornerness detectors [7] have been proposed to extract these interest points from models and images. In cases such as stereo vision and motion, correspondences between interest points are hypothesised through the use of intensity correlation techniques [1] with the assumption that point features in the different images do not differ much.

Methods based solely on the geometry of the contours are certainly more applicable in problems such as object recognition and pose identification. Various strategies for identifying salient points and regions on contours have been suggested [10, 5] but they are seldom invariant beyond similarity transformations. More appropriately, *landmark* points such as corners and inflection points are used, eg. to compute algebraic invariants for comparison purposes [15]. Although less computationally expensive than Hough methods [1] and more stable than using high-ordered differential invariants [8], the shortfall of this approach lies in the dependence on the availability and accurate localisation of landmark points.

Similar difficulties are also encountered in the detection of image contours which are geometrically related by properties such as parallel or skewed symmetry [11, 9, 6]. The recovery of the geometric transformations in these situations will facilitate the grouping of geometrically related contours which may be fragmented due to occlusion and background clutter. For general classes of objects, the grouping provided by geometry is an enhancement over traditional methods of perceptual grouping. Results by West and Rosin [13], Zerroug and Nevatia [14] and Zisserman *et al.* [16] for the recovery of the contours of surfaces of revolution and canal surfaces are promising, although the correspondences are found only if specific image features are located.

In this paper, a general theory of geometric curve-based correspondence is proposed. We initially discuss general approaches to establishing curve correspondences, followed by introducing transformation bases as functions of curve parameters. We further outline a process to localise transformation bases in order to satisfy the curve neighbourhood at the basis points. The concept of *geometric saliency* is proposed to evaluate the discriminatory power of different transformation bases, and statistical analysis is carried out to identify salient bases in the presence of noise, resulting in the derivation of a *normalised basis likelihood*. Finally, the theory is applied to the problem of detecting and grouping affine symmetric curves.

2 Theory

2.1 Formation and Parameterisation of Transformation Bases

The unique point-to-point correspondence between two sets of image curves related by up to a *2D projective* transformation may be obtained once an initial N point correspondences in general position have been established together with P non-invariant contour derivatives at these points, where $2N + P \geq d$ and $d \leq 8$ is the number of degrees of freedom available in the transformation class. The collections of minimum number of pairs of corresponding points and associated derivatives required to fix the free parameters (ie. the equality $2N + P = d$ holds) form the **bases** of the transformation (Fig. 1). See [12] for an in-depth discussion.

Associated with every transformation basis are matrices which relate corresponding points in the basis. Since in our framework all points lie on curves, we designate these pairs of corresponding points as $(\boldsymbol{x}_{1j}(t_{1j}), \boldsymbol{x}_{2j}(t_{2j})), j = 1, .., N$, where t_{ij} is the parameterisation for the curve on which $\boldsymbol{x}_{ij}$ lie. Then

$$\boldsymbol{A}_{12} : \boldsymbol{x}_{1j}(t_{1j}) \longmapsto \boldsymbol{x}_{2j}(t_{2j}), \qquad j = 1, .., N \qquad (1)$$

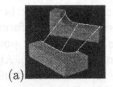 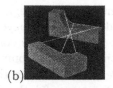 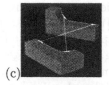

(a)   (b)   (c)

Fig. 1. The affine transformation bases in (a) and (b) are formed from 3 point correspondences, whereas the basis in (c) is formed from 2 point correspondences and the associated tangents. However only (b) and (c) correctly defines the relation between the two image curves.

$$A_{21} : x_{2j}(t_{2j}) \longmapsto x_{1j}(t_{1j}), \qquad j = 1, .., N \qquad (2)$$

where A_{12} is the matrix mapping points $x_{1j}(t_{1j})$ to $x_{2j}(t_{2j})$ and A_{21} is the matrix for the inverse mapping. However, *the matrices which transform x_{1j} to x_{2j} and vice-versa are derived via these points*:

Proposition 1. (Transformation as a function of curve parameters)
A transformation matrix A_{12} which is derived from a transformation basis comprising of N correspondences between two sets of curves and P contour derivatives may be represented by a tensor function M operating on the $2N$-dimensional curve parameter space, ie.

$$M : \mathbb{R}^{2N} \longrightarrow \mathbb{R}^{2N+P}$$
$$(t_1, t_2) \longmapsto A_{12}(t_1, t_2) \qquad (3)$$

where $t_1 = [t_{11}\ t_{12}\ \ldots\ t_{1N}]^T$ is a vector of curve parameters for the first set of curves, and $t_2 = [t_{21}\ t_{22}\ \ldots\ t_{2N}]^T$ the vector of curve parameters for the second set.

The $2N$-dimensional space of curve parameters may be considered a generalisation of Van Gool *et al.* 's [12] 2D *Arc-Length Space* (used for comparing pairs of points on curves) to higher dimensions for establishing multiple correspondence pairs.

Bases for Affine Symmetry. As we previously described in [4], there are three degrees of freedom in an affine symmetry transformation, more intuitively expressed as the location of the symmetry axis (2 DOF), and the angle of skew (1 DOF). We explained how *a single point correspondence together with the tangents at these points form an affine symmetric basis.*

A pair of corresponding points $x_i(t_i) = [x_i(t_i)\ y_i(t_i)]^T$, $i = 1, 2$ is related by:

$$x_2\, Z_2 = x_1\, Z_1 \qquad (4)$$

where Z_1 and Z_2 are matrices containing the symmetry parameters ϕ (the orientation of the axis of symmetry), C (the perpendicular distance of the origin to the axis), and α (the angle of skew). *These parameters are functions of the basis correspondences $x_1(t_1)$ and $x_2(t_2)$ and hence are functions of the arc-length curve parameters t_1 and t_2.* Further details may be found in [3].

2.2 Correspondences Formed with Neighbourhood of Curves

The use of landmark points in establishing transformation bases for curve matching is disadvantageous in that the transformations may be sensitive to errors in localisation of these points. We propose instead an approach based on matching *regions* of curves, in which (i) hypothesised corresponding points (which need not be restricted to landmark points) are allowed to vary for limited distances along the contours; (ii) the neighbourhood of curves at these points should not only be used to verify the validity of the hypothesised transformation (via the *edge-continuity assumption*), but also be *integrated* into an optimisation process to *localise* the correspondences; and (iii) the rest of the curves are gradually incorporated to improve the accuracy should the neighbourhood of the correspondences be well matched after the initial optimisation process, *and* appear *geometrically salient* (defined later in Sect. 2.3).

Localisation of Curve Neighbourhood Correspondences. In order to have a measure of the differences between two sets of curves, we define the *deviation* between curves C_1 and C_2:

Definition 2. (Deviation)
*The **Deviation** D_{12} from curve C_1 to curve C_2 is defined as the integral of squared distances from points on C_1 perpendicular to C_2, ie.*

$$D_{12} = \int_{C_1} \|x_1(\tau) - C_2\|^2 \, d\tau \tag{5}$$

where τ is the arc-length parameterisation for curve C_1, and $x_1(\tau)$ is the position vector for the point with parameter τ on curve C_1. See Fig. 2(a).

Note that this is not invariant to the interchange of curves, ie. $D_{12} \neq D_{21}$, but it is suitable to our analysis as well as for practical purposes.

(a)　　　　　　　　　　　　　　　(b)

Fig. 2. (a) The deviation of C_1 to C_2 is the integral of squared distances from points on C_1 perpendicular to C_2, as defined by definition 2. (b) A neighbourhood-validated basis comprising of 2 correspondences is shown.

From (3), it is observed that as basis points slide along the associated curves, the transformation basis and matrix change accordingly. If the basis defines a correct hypothesis for the transformation relating the two sets of curves, the projection of one set of curves via the transformation will match the second set of curves exactly. Since not all parts of the curves are necessarily available in the images, bases will have to be classified according to the *curve neighbourhood support* available at the basis points:

Definition 3. (Neighbourhood-Validated Basis)
*A neighbourhood-validated basis, or **NV-basis**, is a transformation basis which satisfies the requirement that the projection of the curve neighbourhood at all basis points match exactly the curve neighbourhood at their respective corresponding basis points. This is a necessary but insufficient condition for the basis to represent the correct transformation between the two sets of curves. See Fig. 2(b).*

The goal of *localising* the correct set of correspondences to define the transformation between two sets of curves can therefore be formulated as an optimisation problem on the $2N$ dimensional domain of curve parameters. However this set of of correspondences is not unique since any pair of actual corresponding points between the two curves may be used. We instead reformulate the optimisation problem:

Proposition 4. (Correspondence localisation as a minimisation)
Suppose a transformation basis comprising N correspondences is needed to establish the transformation relating two sets of curves. Given N points with the curve parameters t_1 on the one set of curves, an NV-basis may be found by recovering the corresponding points with curve parameters t_2 on the opposite set of curves through solving the minimisation problem

$$\min_{t_2} \left\{ \sum_{j=1}^{N} \int_{N\{t_{1j}\}} \|A_{12}(t_1, t_2)\, x_{1j}(\tau) - C_{2j}\|^2 \, d\tau \right\} \tag{6}$$

*where A_{12} is as defined in (3), C_{ij} is the curve on which x_{ij} lies, and $N\{t_{ij}\}$ is the curve neighbourhood of arbitrary size about t_{ij}. The N points on the first set of curves are the fixed basis points (or **pivot points**), whereas the N corresponding points on the second set of curves are the **free** basis points.*

Optimisation Process for the Localisation of Correspondences. We outline this *correspondence localisation* process in terms of a *prediction-verification* mechanism as follows:

1. Initially extract from both sets of curves a number of interest points (eg. points with large cornerness [7]).

2. Hypothesise a transformation basis by assuming correspondences between N pairs of points on the two sets of curves and using P contour derivatives at these points,

3. Project the curve neighbourhood at the pivot points via the hypothesised transformation and compute the total deviation between the projected curve neighbourhood and the second set curves;

4. Predict the best locations on the second set of curves for the free basis points by minimising the deviation with an optimisation algorithm operating on the N dimensional space of curve parameters;

5. Cycle through steps 2 and 4 until a minima in deviation is reached, or if the estimation diverges.

6. Determine the *geometric saliency* (Sect. 2.3) and *normalised basis likelihood* (Sect. 2.4) of the hypothesised transformation basis to decide if the basis should be admitted.

Figure 3 demonstrates this optimisation process for an affine symmetry basis.

(a)   (b)   (c)

Fig. 3. (a) An affine symmetric transformation basis representing the initial hypothesis for the transformation relating the pair of image curves; (b) the free basis point is localised by an optimisation process; (c) the final transformation basis is obtained which correctly represents the affine symmetry between the two contours.

2.3   Geometric Saliency of Transformation Bases

In [2], Brady discussed the need in vision problems to work from locations offering the tightest constraints ('seeds') to locations of decreasing constraint. For example in point registration, corners are considered first followed by edges. In the case of geometric curves, curvature is often cited as a measure of its constraint potential [5]. We show that this is *not necessarily so*, since the localisability of correspondences is *both* dependent on shape as well as the transformation class and defining parameters, especially in cases when curvature does not undergo a monotonic mapping.

It may be shown that the localisation of correspondence can be poor even when there is insignificant deviation of projected and actual curve neighbourhood. For an example, consider the case of bilateral symmetry for which a transformation basis comprises of a point-to-point correspondence and no derivatives. In an attempt to establish the transformation relating the two sets of symmetric curves in Fig. 4(a), a wide range of incorrect bases which have strong neighbourhood support may be formed from a pivot point with a free corresponding point – it is impossible to localise the correct basis accurately. A more discriminating transformation basis is shown in Fig. 4(b). Note that the curvatures at the basis points in Fig. 4(b) are smaller than the curvatures in Fig. 4(a), contrary to expectations that points with larger curvatures are more salient.

In order to select the transformation bases with greater discriminatory power, we derive a quantitative definition for the *geometric saliency* of a correct transformation basis.

Derivation of Geometric Saliency. Consider a NV-basis subjected to perturbation of its free basis points. Since the pivot points are fixed, the transformation matrix A relating pivot points to free points in (1) will be rewritten as a function

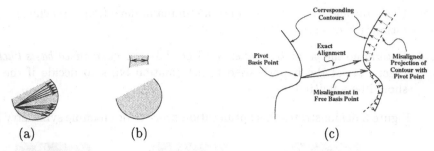

Fig. 4. In (a), it is impossible to localise the correct bilateral symmetric NV-basis. (b) shows a highly localisable NV-basis which correctly represents the bilateral symmetry. In (c), when a free basis point in an NV-basis is perturbed by a small error, the resultant deviation between the projected and actual curve neighbourhood may be expressed in the form of equation 7.

of only the curve parameters associated with the free points, ie. $A(t_2)$. Additionally, the deviation D_{12} from (5) is defined as a function of the perturbation, and written $D(\epsilon)$, where $\epsilon = [\ \epsilon_1\ \epsilon_2\ \dots\ \epsilon_N\ ]^T$ is a vector of the perturbation in t_2. Finally, all curve parameters are assumed to be expressed in terms of arc-length.

For *small* ϵ, the deviation between the projected curve neighbourhood of the pivot points and the curve neighbourhood of the free basis points for the NV-basis may be written as

$$D(\epsilon) = \sum_{j=1}^{N} \int_{a_j}^{b_j} r_j^2\ d\tau \tag{7}$$

$$r_j = x_{1j}^T(\tau)\ (A(t_2) - A(t_2 + \epsilon))^T\ k_j(t_2, \tau) \tag{8}$$

where b_j and a_j demarcate the parameter boundaries for the curve neighbourhood of pivot point j, and k_j are the unit normals to the *projected* curve neighbourhood of pivot point j. The curve neighbourhood of the free basis points do not appear in this equation since for a NV-basis they are obtained simply by projecting the curve neighbourhood of the pivot points. See Fig. 4(c).

Taking derivatives,

$$\nabla_\epsilon D(\epsilon) = -2 \sum_{j=1}^{N} \int_{a_j}^{b_j} r_j\ \nabla_\epsilon \left(x_{1j}^T(\tau) A(t_2 + \epsilon)^T k_j(t_2, \tau) \right)\ d\tau \tag{9}$$

Since $\epsilon = 0$ represents the least-squares solution for (7), $r_j = 0$ and hence $\nabla_\epsilon D(0) = 0$. The Hessian matrix $H(\epsilon)$ for D at $\epsilon = 0$ is non-zero but has a simplified, *positive semi-definite* form:

$$H(0) = 2 \sum_{j=1}^{N} \int_{a_j}^{b_j} G_j\ G_j^T\ d\tau \tag{10}$$

$$G_j = \nabla_\epsilon \left(x_{1j}^T(\tau) A(t_2)^T k_j(t_2, \tau) \right) \tag{11}$$

where G represents the Jacobian of $r_j(\epsilon, t_2, \tau)$ at $\epsilon = 0$. Furthermore since $H(0)$

is *positive semi-definite* from (10), all its eigenvalues are real. We therefore define our *geometric saliency* as follows:

Definition 5. (Geometric Saliency)
*Given a NV-basis for which $H(\epsilon)$ is the Hessian function of the deviation $D(\epsilon)$ as defined in (7), the **Geometric Saliency** S of this basis is defined as the smallest eigenvalue of $H(0)$, ie.*

$$S = \min(\lambda) \tag{12}$$

$$\text{subject to} \qquad H(0)\,\epsilon = \lambda\,\epsilon, \qquad \lambda \geq 0 \tag{13}$$

The associated normalised eigenvector of S, denoted by v_S, represents the normalised ϵ which causes the minimum increase in the deviation $D(\epsilon)$.

We noted that the form of the first derivative matrix 'M' used in the Plessey corner finder [7] is similar to $H(0)$ except that it is specifically two-dimensional and applied to the image domain.

The $D(\epsilon)$ energy functional surfaces in 2D curve parameter space is graphically illustrated for various bases representing *similarity* (scaled Euclidean) transformations in Fig. 5. These bases are formed with two point correspondences and no contour derivatives (four degrees of freedom represented by a 2D translation, a rotation and an isotropic scaling). The eigenvectors associated with the geometric saliencies are in the directions of the most gradual ascent at the minimum point. It is seen that the transformation basis in Fig. 5(a) is preferred to the basis in Fig. 5(b) by virtue of its larger geometric saliency and hence greater localisability. This is in spite of the fact that they both represent the correct similarity transformation relating the two curves.

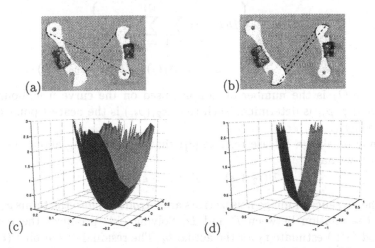

Fig. 5. The similarity (scaled Euclidean) NV-bases shown in (a) and (b) are formed with two point correspondences, and the deviation functionals are shown respectively in (c) and (d). The geometric saliencies are related to the directions of the most gradual ascent.

Geometric Saliency for Affine Symmetry. The Hessian derived in (10) reduces to a single value since only one pair of corresponding points is involved, and is therefore equal to the geometric saliency. Complete details of the derivation may be found in [3]. Figure 6 shows some transformation bases for affine symmetry. The geometric saliency of the transformation basis in Fig. 6(b) is near zero because the ellipse formed by the base of the key is affinely equivalent to a circle, which has infinite symmetry axes.

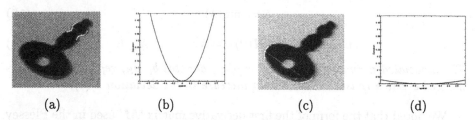

    (a)   (b)   (c)   (d)

Fig. 6. The affine symmetry bases in (a) and (c) are formed with one pair of point correspondences. Basis (c) has a near zero geometric saliency since all points on an ellipse are affine-symmetric to each other.

2.4 Statistical Analysis and Normalised Basis Likelihood

When image noise is present, even NV-bases are expected to contain residual deviation in the curve neighbourhood. In this section, we therefore consider the essential task of judging the localisation errors involved when a basis hypothesis is presented.

In a discrete implementation, the minimisation problem given in (6) may be expressed as

$$\min_{t_2} \left\{ \mathcal{D}(t_2) = \sum_{j=1}^{N} \sum_{i=1}^{Q_j} s_{ji}^T s_{ji} \right\} \tag{14}$$

where
$$s_{ji} = A(t_2) \, x_{1j}(\tau_{ji}) - x_{2j}(\rho_{ji}) \tag{15}$$

and where Q_j is the number of samples used on the curve neighbourhood of basis point j. ρ_{ji} is determined such that $x_{2j}(\rho_{ji})$ is the nearest point on curve C_{2j} to $x_{1j}(\tau_{ji})$.

It may be shown (please refer to [3]) that by linearising about the solution $\mathcal{D}(t_2) = 0$ we have

$$\mathcal{D}(t_2 + \epsilon) = D(\epsilon) = \epsilon^T H(0) \, \epsilon \tag{16}$$

In the presence of image noise, errors are introduced into the terms x_{ij}, which distort the relation between $\mathcal{D}$ and D. Solving (14) produces the *maximum-likelihood* (ML) estimate $\hat{t}_2$ for the actual t_2. The residual deviation $\mathcal{D}(\hat{t}_2)$ arises from a combination of errors in the measurement of s_{ij}'s and the error in the estimation of t_2. From the relation $\hat{t}_2 = t_2 + \delta t_2$, δt_2 represents the *misalignment* error. The error δt_2 is dominant along v_S, the eigenvector related to the geometric saliency.

Since the distribution of noise is not known, it is useful to gauge the accuracy of $\hat{t}_2$ by comparing $\mathcal{D}(\hat{t}_2)$ with the deviation $D(\delta t)$ which will arise solely through a misalignment error. Intuitively, we would expect that NV-bases with larger geometric saliencies to be better localised despite having a greater amount of residual deviation. This is easily understood from Figs. 7(a,b), where the deviation functionals shown in Figs. 6(b,d) are subjected to independent errors in the measurement of s_{ij}.

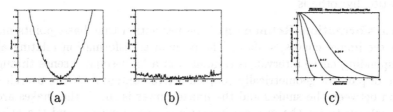

(a)	(b)	(c)

Fig. 7. The deviation functionals in (a) and (b) are possible scenarios in which the deviation functionals from Fig. 6(b,d) are distorted by errors in the measurement of the image contours. Figure (c) shows the $\sqrt{\text{Deviation}}$ – Normalised-Basis-Likelihood plots for geometric saliencies of 1, 5 and 20.

We make the assumption that $\epsilon = \delta t_2$ is Gaussian-distributed. Furthermore, instead of relating ϵ to the probability density distribution of $D(\epsilon)$ which is highly non-trivial, we relate the deviation $D(q\,v_S)$ to the error q along v_S – the dominant direction of ϵ, ie. assume that $\epsilon = q\,v_S$.

Let

$$z = q\sqrt{v_S^T H(0)\, v_S} \qquad (17)$$

Then
$$\begin{aligned} E\{D(q\,v_S)\} = Var\{z\} &= E\{z^2\} \\ &= Var\{q\}\, v_S^T H(0) v_S = Var\{q\}\, S \end{aligned} \qquad (18)$$

Since q is assumed to be Gaussian, so is z. Hence

$$p(z) = \frac{1}{\sigma_z\sqrt{2\pi}}\exp\left(-\frac{z^2}{2\sigma_z^2}\right) = \frac{1}{\sigma_q\sqrt{2\pi S}}\exp\left(-\frac{z^2}{2S\sigma_q^2}\right) \qquad (19)$$

where $\sigma_z = Var(z)$ and $\sigma_q = Var(q)$. We can therefore define a *normalised basis likelihood* value based on (19):

Definition 6. (Normalised Basis Likelihood)
Given a hypothesis for an NV-basis with geometric saliency S and residual deviation $\mathcal{D}$, the normalised basis likelihood L is defined from (19) by assigning σ_q to be unity and assuming likelihoods to be equal when $\mathcal{D} = 0$:

$$L = \exp\left(-\frac{\mathcal{D}}{2S}\right) \qquad (20)$$

L lies between 0 and 1, with $L = 1$ when $\mathcal{D} = 0$. See Fig. 7(c).

From Fig. 7(c) it is apparent that with increasing geometric saliency a greater amount of residual deviation is allowed without a decrease in the likelihood.

2.5 Constraint Propagation and the Refinement of Correspondences

Further refinement of hypotheses may be achieved through a cyclic constraint propagation process:

1. **Curve Support Recovery.** Starting from pivot points, validation checks are propagated along the pivot curves. The goal is to recover the maximum interval of curve correspondences which are in consensus with the transformation hypothesis.

2. **Transformation Refinement.** The restriction that basis points must lie on the image curves is lifted. The optimal transformation relating all corresponding curve intervals is estimated in a least-squares sense through the use of pairs of geometrically-coupled active contours[1]. If the overall deviation between the snakes and the image curves is small, the snakes are used to substitute for the original intervals of image curves, and the algorithm returns to the support recovery step.

Figure 8 shows this process in action.

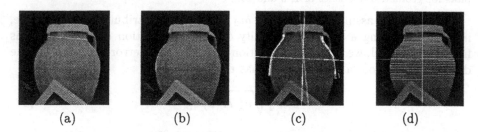

| (a) | (b) | (c) | (d) |

Fig. 8. In (a), an NV-basis is hypothesised. The curve support recovery proceeds as shown in (b). (c) shows the refinement of the transformation estimate via geometrically-coupled affine-symmetric active contours, with the eventual result obtained in (d). In practice, the process is repeated a number of times.

3 Implementation of a Grouping Process for Affine Symmetric Contours

The grouping of skew-symmetric contours are carried out in the following way:

1. Edgel chains are initially extracted via the Canny edge-detector, to which B-splines are fitted.

2. A number of interest points on curves are extracted and basis hypotheses are made between these points and the *regions* around them, and localised via the process described in Sect. 2.2.

---

[1] A pair of geometrically-coupled active contours is treated as a single active contour operating in a canonical frame together with a set of transformation parameters mapping the canonical frame to the two image frames. This work will be published in due course.

396

3. Basis hypotheses are selected if empirical thresholds for geometric saliency and normalised basis likelihood are exceeded.

4. The maximum amount of connected curve support is gathered and the transformation parameters iteratively refined as described in Sect. 2.5.

Note that this is a preliminary implementation. No additional grouping of sets of curves with similar transformation parameters is performed as yet. This is expected to be implemented in due course.

4 Preliminary Results

The affine-symmetric contour-grouping algorithm is applied to the images shown in Fig. 9. B-splines have been fitted to the contour chains obtained from the Canny edge detector, and a number of interest points are selected on these splines.

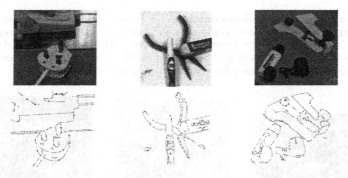

Fig. 9. A number of test images are shown in the top row, and splines fitted to the detected edges are shown in the bottom row. The extracted interest points are marked on the contours with crosses.

Figures 10(a,b,c) show NV-bases which have high normalised basis likelihoods and large geometric saliencies. Further trimming of these 'seeds' is possible by considering the amount of curve support available outside the basis curve neighbourhood. The initial extent of curve support obtained is shown in Figs. 10(d,e,f).

After the application of curve-support / transformation-refinement cycle, we obtain the final results shown in Fig. 11.

5 Conclusions and Future Work

The main aim of this research is to avoid the use of landmark points or high-order differential invariants when establishing correspondences. In the framework of geometrical curves, transformations defined by transformation bases are functions of the curve parameters. Point correspondences are not restricted to matches between isolated point sets, but formed as initial broad matches between intervals of curves. Transformation bases represent hypotheses for the transformation

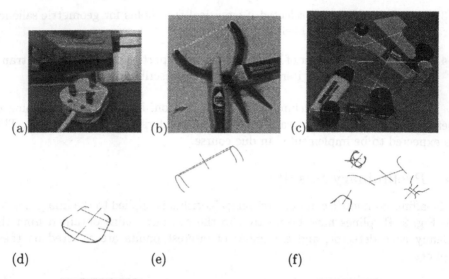

(d)   (e)   (f)

Fig. 10. (a,b,c) show the affine-symmetric NV-bases found from the image contours with have large normalised basis likelihoods and geometric saliencies. Further trimming is achieved by considering the amount of curve support obtained – The strong bases and curve support are shown in (d,e,f) together with the symmetry axes and angles of skew. Different valid hypotheses for the same contours exists in (f).

Fig. 11. The final results obtained after iterating through curve-support recovery and transformation refinement cycles. Corresponding points are marked as shown.

relating curves and may be localised through an optimisation process, terminating in either hypothesis rejection or recovery of neighbourhood-validated bases. We have shown that NV-bases are localisable to different degrees, giving rise to the notion of geometric saliency. In the presence of noise, the accuracy of the estimated correspondences is dependent on the geometric saliency since a highly salient basis is more robust to misalignment errors – the normalised basis likelihood formalises this accuracy. A method is proposed to further validate the hypotheses by recovering additional curve support, and to refine transformation estimates through the use of geometrically-coupled active contours. Preliminary results obtained from applying this theory to the grouping of affine symmetric contours are shown. In the immediate future, further grouping of geometrically-related sets of curves will be implemented.

Acknowledgements. Our appreciation goes to Dr. Andrew Zisserman for providing much valuable feedback. We would also like to thank Jonathan Lawn, Jun Sato and Gabriel Hamid for help and useful discussions. Tat-Jen Cham is supported on an Overseas Research Scholarship and a Cambridge Commonwealth Trust Bursary.

References

1. D.H. Ballard and C.M. Brown. *Computer Vision.* Prentice– Hall, New Jersey, 1982.
2. J.M. Brady. Seeds of perception. In *Proc. 3rd Alvey Vision Conf.*, pages 259–265, Cambridge, Sep 1987.
3. T. J. Cham and R. Cipolla. Geometric saliency of curve correspondences and grouping of symmetric contours. Technical Report CUED/F-INFENG/TR 235, University of Cambridge, Oct 1995.
4. T. J. Cham and R. Cipolla. Symmetry detection through local skewed symmetries. *Image and Vision Computing*, 13(5):439–450, June 1995.
5. M. A. Fischler and H. C. Wolf. Locating perceptually salient points on planar curves. *IEEE Trans. Pattern Analysis and Machine Intell.*, 16(2):113–129, Feb 1994.
6. R. Glachet, J.T. Lapreste, and M. Dhome. Locating and modelling a flat symmetrical object from a single perspective image. *CVGIP: Image Understanding*, 57(2):219–226, 1993.
7. C. Harris. Geometry from visual motion. In A. Blake and A. Yuille, editors, *Active Vision*. MIT Press, 1992.
8. J. L. Mundy and A. Zisserman. *Geometric Invariance in Computer Vision.* Artificial Intelligence. MIT Press, 1992.
9. P. Saint-Marc, H. Rom, and G. Medioni. B-spline contour representation and symmetry detection. *IEEE Trans. Pattern Analysis and Machine Intell.*, 15(11):1191–1197, Nov 1993.
10. J. L. Turney, T. N. Mudge, and R. A. Volz. Recognizing partially occluded parts. *IEEE Trans. Pattern Analysis and Machine Intell.*, 7(4):410–421, July 1985.
11. F. Ulupinar and R. Nevatia. Perception of 3-D surfaces from 2-D contours. *IEEE Trans. Pattern Analysis and Machine Intell.*, 15(1):3–18, 1993.
12. L. J. Van Gool, T. Moons, E. Pauwels, and A. Oosterlinck. Semi-differential invariants. In J. L. Mundy and A. Zisserman, editors, *Geometric Invariance in Computer Vision*, pages 157–192. MIT Press, 1992.
13. G. A. W. West and P. L. Rosin. Using symmetry, ellipses and perceptual groups for detecting generic surfaces of revolution in 2D images. In *Applications of Artificial Intelligence*, volume 1964, pages 369–379. SPIE, 1993.
14. M. Zerroug and R. Nevatia. Using invariance and quasi-invariance for the segmentation and recovery of curved objects. In J.L. Mundy and A. Zisserman, editors, *Proc. 2nd European-US Workshop on Invariance*, pages 391–410, 1993.
15. A. Zisserman, D. A. Forsyth, J. L. Mundy, and C. A. Rothwell. Recognizing general curved objects efficiently. In J. L. Mundy and A. Zisserman, editors, *Geometric Invariance in Computer Vision*, pages 228–251. MIT Press, 1992.
16. A. Zisserman, J. Mundy, D. Forsyth, J. Liu, N. Pillow, C. Rothwell, and S. Utcke. Class-based grouping in perspective images. In *Proc. 4th Int. Conf. on Computer Vision*, pages 183–188, 1995.

Computing Contour Closure

James H. Elder[1] and Steven W. Zucker[2]

[1] NEC Research Institute, Princeton, NJ, U.S.A.
[2] Centre for Intelligent Machines, McGill University, Montréal, Canada.

Abstract. Existing methods for grouping edges on the basis of local smoothness measures fail to compute complete contours in natural images: it appears that a stronger global constraint is required. Motivated by growing evidence that the human visual system exploits contour closure for the purposes of perceptual grouping [6, 7, 14, 15, 25], we present an algorithm for computing highly closed bounding contours from images. Unlike previous algorithms [11, 18, 26], no restrictions are placed on the type of structure bounded or its shape. Contours are represented locally by tangent vectors, augmented by image intensity estimates. A Bayesian model is developed for the likelihood that two tangent vectors form contiguous components of the same contour. Based on this model, a sparsely-connected graph is constructed, and the problem of computing closed contours is posed as the computation of shortest-path cycles in this graph. We show that simple tangent cycles can be efficiently computed in natural images containing many local ambiguities, and that these cycles generally correspond to bounding contours in the image. These closure computations can potentially complement region-grouping methods by extending the class of structures segmented to include heterogeneous structures.

1 Introduction

We address the problem of computing closed bounding contours in real images. The problem is of interest for a number of reasons. A basic task of early vision is to group together parts of an image that project from the same structure in a scene. Studies of perceptual organization have demonstrated that the human visual system exploits a range of regularities in image structure to solve this task [12, 14, 28]. Inspired in part by these studies, algorithms have been developed to apply continuity, cocircularity and smoothness constraints to organize local edges into extended contours [2, 10, 17, 19, 22, 24]. However, despite persuasive psychophysical demonstrations of the role of contour closure in perceptual organization [6, 7, 14, 15, 25], closure constraints for computer vision algorithms have been largely ignored (although see [1, 11]). This is surprising, since existing algorithms, while capable of producing extended chains of edges, are seldom successful in grouping complete contours. Closure is a potentially powerful cue that could complement smoothness cues to allow more complete contours to be computed.

In natural images, contours are fragmented due to occlusions and other effects, making local grouping cues weak and unreliable. While these local cues may be summed or otherwise integrated along a contour to form global measures of "smoothness", "likelihood" or "salience", such a global measure is as weak as its weakest local constituent. This is illustrated in Fig. 1(a): the most plausible continuation of a contour viewed locally may be clearly incorrect when viewed in global context. This error is not revealed in a summation of local grouping cues over the curve, since both the correct and the incorrect continuations lead to similar measures. A global feature is needed which is far more sensitive to such local errors. Closure is a potentially powerful feature because a single local error will almost certainly lead to a low measure of closure.

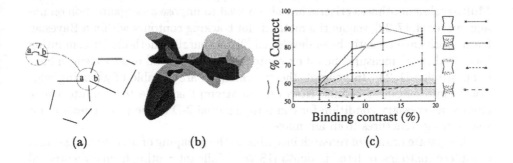

(a)   (b)   (c)

Fig. 1. (a) Locally, the most plausible continuation of fragment *a* is through fragment *b*. Given global context, fragments instead group to form simple cycles with high measures of closure. **(b)** A region grouping algorithm would segment this image into 12 disjoint regions, yet human observers see two overlapping objects. Regularities of the object boundaries must be exploited. **(c)** Psychophysical data for shape identification task. Subjects must discriminate between a fragmented concave shape (shown) and a 1-D equivalent convex shape (not shown), at very low contrasts. Results show that contour closure cues greatly enhance performance. No effect of texture cues is observed. From [8].

The computation of closed contours is also potentially useful for grouping together image *regions* which project from common structures in the scene: objects, parts of objects, shadows and specularities. Existing techniques for region grouping apply homogeneity or smoothness constraints on luminance, colour or texture measures over regions of the image (e.g. 16, 20, 21). These techniques have inherent limitations. While a region-grouping algorithm would segment the image of Fig. 1(b) into 12 disjoint components, human observers perceive two irregularly painted, overlapping objects. Since these are nonsense objects, our inference cannot be based on familiarity. We must be using the geometry of the boundaries to group the objects despite their heterogeneity.

This situation is not artificial. Objects are often highly irregular in their surface reflectance functions, and may be dappled in irregular ways by shadows and specularities. While surface markings, shadows and specularities fragment image regions into multiple components, geometric regularities of bounding contour persist. Contour closure is thus potentially important for segmentation because it broadens the class of structures that may be segmented to include such heterogeneous structures. Interestingly, recent psychophysical experiments [8] suggest that contour grouping cues such as closure may be more important than regional texture cues for the perceptual organization of 2-D form (Fig. 1(c)).

2   Previous Work

The problem of contour grouping has been approached in many different ways. Multi-scale smoothness criteria have been used to impose an organization on image curves [4,17,22], sequential methods for tracking contours within a Bayesian framework have recently been developed [2] and parallel methods for computing local "saliency" measures based on contour smoothness and total arclength have been studied [1, 24]. In general, these techniques are capable of grouping edge points into extended chains. However, no attempt is made to compute *closed* chains; a necessary condition for computing global 2-D shape properties and for segmenting structures from an image.

A separate branch of research investigates the grouping of *occlusion* edges into complete contours, ordered in depth [18,26]. While interesting from a theoretical point of view, a large fraction of the edges in real images are not occlusion edges, and a recent study [5] suggests that it is not possible to locally distinguish occlusion edges from other types of edges. It is our view that algorithms for grouping contours must work for all types of structure in an image (e.g. objects, shadows, surface markings).

Jacobs [11] has studied the problem of inferring highly-closed convex cycles of line segments from an image, to be used as input for a part-based object recognition strategy. Given the generality of boundary shape, it is clearly of great interest to determine whether bounding contours can be recovered without such restrictive shape constraints. Most similar to our work is a very recent study by Alter [1] on the application of shortest-path algorithms to the computation of closed image contours. While similar in concept, these two independent studies differ substantially in their implementation.

3   Overview of the Algorithm

Our goal here is to recover cycles of edge points which bound two-dimensional structures in an image. The algorithm is to be fully automatic and no restrictions are placed on the type of structure bounded or its shape. Since no constraint of disjointness is imposed, in principle the bounding contours of an entire object, its parts, markings, and shadows can be recovered.

Image contours are represented locally as a set of tangent vectors, augmented by image intensity estimates. A Bayesian model is developed to estimate the likelihoods that tangent pairs form contiguous components of the same image contour. Applying this model to each tangent in turn allows the possible continuant tangents for each tangent to be sorted in order of their likelihood. By selecting for each tangent the 6 most likely continuant tangents, a sparse (6-connected), weighted graph is constructed, where the weights are the computed pairwise likelihoods.

By assuming independence between tangent pair likelihoods, we show that determining the most probable tangent cycle passing through each tangent can be posed as a shortest path computation over this sparse graph. We can therefore use standard algorithms (e.g. Dijkstra's algorithm [23]) to solve this problem in low-order polynomial time.

4 Extended Tangents

Edges are detected by a multi-scale method which automatically adapts estimation scale to the local signal strength and provides reliable estimates of edge position and tangent orientation [9]. In addition to the geometric properties of position and orientation, we make use of local image intensity estimates provided by our edge detector.

Due to uncertainty induced by discretization and sensor noise, contours generate noisy, laterally-displaced local edges (Fig. 2(a)). Tracing a contour through these local tangents generates a curve corrupted by wiggles due to sensing artifacts. Also, due to blurring of the luminance function at the imaging and estimation stages, edge estimates near corners and junctions are corrupted (Fig. 2(b)).

(a)   (b)

Fig. 2. (a) The set of raw tangent estimates for a contour. Imposing an ordering on these local tangent estimates generates a contour distorted by sampling artifacts. (b) The smoothing of the image at the sensing and estimation stages corrupts tangent estimates near contour junctions and corners.

Achieving a more reliable local representation requires more global constraints. Here, we introduce a method for refining local edge information based on an *extended tangent representation*, which represents a curve as a sequence of disjoint line segments. Each local edge in the image generates a tangent line passing through the edge pixel in the estimated tangent direction. The subset of tangent estimates which are 8-connected to the local edge and which lie within an ϵ-neighbourhood of the local tangent line are identified with the extended tangent model. The algorithm for selecting extended tangents to approximate a contour is illustrated in Fig. 3. Given a connected set of local edges, the longest line segment which faithfully models a subset of these is determined. This subset is then subtracted from the original set. This process is repeated for the connected subsets thus created until all local edges have been modeled.

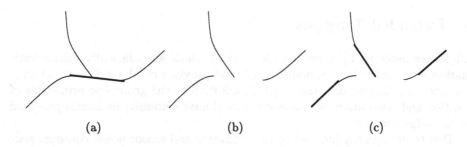

(a)   (b)   (c)

Fig. 3. Computing the extended tangent representation. (a) For each connected set of edge pixels, the subset of pixels underlying the longest extended tangent is selected. (b) The edge points thus modeled are subtracted. (c) The process is repeated for each connected set of edge pixels thus spawned.

Since the extended tangents selected must be consistent with the global geometry of the curves, they provide more accurate estimates of contrast and tangent orientation than do the corrupted local edges near the junction. The extended tangent algorithm is conceptually simpler than most methods for computing polygonal approximations [20], and does not require preprocessing of the edge map to link local edges into ordered lists, as is required for most other methods (e.g. 11, 17, 20).

5 A Bayesian Model for Tangent Grouping

The extended tangent representation leads naturally to a representation for global contours as tangent sequences:

Definition 1. A *tangent sequence* $t_1 \rightarrow ... \rightarrow t_n$ is an injective mapping from a finite set of integers to a set of extended tangents.

The injective property restricts our definition to sequences which do not pass through the same extended tangent twice. The identification of extended tangents with integers imposes an *ordering* on the tangents which distinguishes a tangent sequence from an arbitrary clutter of tangents. If a contour bounds a 2-D structure in the image, this sequence will come back on itself. Thus bounding contours are represented as *cycles* of extended tangents, $t_1 \rightarrow ... \rightarrow t_n \rightarrow t_1 \rightarrow ...$

By this definition, any ordered set of tangents in an image can form a tangent sequence. In order to compute bounding contours, some measure of the likelihood of a tangent sequence must be established. For this purpose, we develop a Bayesian model for estimating the posterior probability of a tangent sequence given data on the geometric and photometric relations between adjacent tangent tuples of the sequence.

We will begin by assuming that tangent links are independent: i.e.

$$p(t_1 \rightarrow ... \rightarrow t_n) = p(t_1 \rightarrow t_2)p(t_2 \rightarrow t_3)...p(t_{n-1} \rightarrow t_n)$$

This approximation will greatly simplify the computation of tangent sequence likelihoods, reducing likelihood estimation for a tangent sequence to the problem of estimating the likelihoods of its constituent links. The likelihood that two tangents project from the same contour is modeled as the probability of their rectilinear completion (Fig. 4), so that the probability of a link depends on the following observables (see Sections 6 and 7 for details of the model):

1. The lengths l_1 and l_2 of the extended tangents.
2. The length r of the straight-line interpolant.
3. The 2 orientation changes θ_a and θ_b induced by the interpolation.
4. The differences in estimated image intensity Δi_h, Δi_l on the bright side and the dark side of the tangents, respectively.

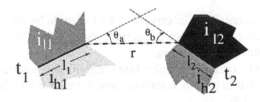

Fig. 4. Rectilinear interpolation model.

Setting $o = \{l_1, l_2, r, \theta_a, \theta_b, \Delta i_h, \Delta i_l\}$, Bayes' theorem can be used to express the posterior probability of a link from tangent t_1 to t_2 (called the "link hypothesis") in terms of the likelihoods of the observables:

$$p(t_1 \rightarrow t_2 | o) = \frac{p(o | t_1 \rightarrow t_2)p(t_1 \rightarrow t_2)}{p(o)}$$

Letting $t_1 \nrightarrow t_2$ represent the hypothesis that t_2 is *not* the continuant of t_1 (the "no-link hypothesis"), the evidence $p(o)$ can be expanded as

$$p(o) = p(o|t_1 \rightarrow t_2)p(t_1 \rightarrow t_2) + p(o|t_1 \nrightarrow t_2)p(t_1 \nrightarrow t_2).$$

It is convenient to rewrite the posterior probability as

$$p(t_1 \rightarrow t_2|o) = \frac{1}{(1 + LP)}$$

where

$$L = \frac{p(o|t_1 \nrightarrow t_2)}{p(o|t_1 \rightarrow t_2)} \qquad P = \frac{p(t_1 \nrightarrow t_2)}{p(t_1 \rightarrow t_2)}$$

The *prior ratio* P represents the ratio of the probability that a curve ends at t_1, to the probability that the curve continues. For most images, curves are expected to continue over many tangents. It is therefore appropriate to choose a large value for the prior ratio: in our experiments we use $P = 50$.

The *likelihood ratio* L represents the ratio of the likelihood of the observables given that t_2 *is not* a continuant of t_1 to their likelihood given that t_2 *is* a continuant of t_1. Models for these likelihoods are developed in the next two sections.

6   Link Hypothesis Likelihoods

In order to model the link hypothesis likelihoods $p(o|t_1 \rightarrow t_2)$ we must consider the distinct events that can split the image curve into two separate extended tangents t_1 and t_2. The three possible hypotheses for a *tangent split* considered are termed respectively the *curvature*, *interruption* and *corner* hypotheses:

Curvature The contour is curving smoothly: two tangents are needed to model the local edges to ϵ accuracy. Relatively small values for r, θ_a and θ_b are expected.

Interruption The contour is interrupted, for example by an occlusion, shadow, or loss of contrast. We expect potentially large values for r, but again relatively small values for θ_a and θ_b.

Corner The contour corners sharply: two tangents are generated on either side of the corner. We expect a relatively small value for r, but possibly large values for θ_a and θ_b.

Since each of these hypotheses generates different expectations for the observables, the corresponding link hypothesis likelihoods are decomposed into likelihoods for the 3 disjoint events:

$$\begin{aligned}
p(o|t_1 \rightarrow t_2) = {} & p(o|t_1 \rightarrow t_2, \text{ curvature})p(\text{curvature}) \\
& + p(o|t_1 \rightarrow t_2, \text{interruption})p(\text{interruption}) \\
& + p(o|t_1 \rightarrow t_2, \text{ corner})p(\text{corner})
\end{aligned}$$

In a natural world of piecewise-smooth objects, the curvature hypothesis is the most likely. For our experiments we assign

$$p(\text{curvature}) = 0.9 \text{ and } p(\text{interruption}) = p(\text{corner}) = 0.05.$$

Combining the observables l_1, l_2 and r into a normalized gap length r',

$$r' = \frac{r}{\min\{l_1, l_2\}}$$

we write a summarized set of observables as $o' = \{r', \theta_a, \theta_b, \Delta b, \Delta d\}$. Approximating these as conditionally independent on the 3 tangent split hypotheses, we use half-Gaussian functions to model the link hypothesis likelihoods for each observables o_i:

$$p(o_i | t_1 \rightarrow t_2) = \frac{\sqrt{2}}{\sqrt{\pi}\sigma_{o_i}} e^{-\frac{o_i^2}{2\sigma_{o_i}^2}} \quad , \quad o_i > 0.$$

The scale constants σ_{o_i} used in this paper are shown in Table 1.

	σ_r (pixels)	$\sigma_{r'}$ (pixels)	$\sigma_{\theta_a} = \sigma_{\theta_b}$ (deg)	$\sigma_{\Delta b} = \sigma_{\Delta d}$ (grey levels)
curvature	2	-	10	20
interruption	-	0.5	10	20
corner	2	-	90	20

Table 1. Scale constants for link hypothesis likelihood functions

7 No-Link Hypothesis Likelihoods

Modelling the position of a tangent as a uniform distribution over the image domain, for an $L \times L$ image, and $r << L$ we can approximate the no-link likelihood for r as $p(r) \approx \frac{4r}{L^2}$ [5]. No-link likelihood functions for θ_a and θ_b follow immediately from the assumption of isotropic tangent directions:

$$p(\theta_a | t_1 \not\rightarrow t_2) = p(\theta_b | t_1 \not\rightarrow t_2) = \frac{1}{\pi} \quad , \quad 0 \leq \theta_a, \theta_b \leq \pi.$$

Modelling image intensity i as a uniform distribution, $0 \leq i \leq 255$, no-link likelihood functions for Δi_h and Δi_l, can be derived [5]:

$$p(\Delta i_h | t_1 \rightarrow t_2) = \frac{2}{255}(1 - \Delta i_h / 255) \quad p(\Delta i_l | t_1 \rightarrow t_2) = \frac{2}{255}(1 - \Delta i_l / 255)$$

8 Constructing a Sparse Graph

Since tangent tuple observations provide only a weak means for discriminating probable from improbable links (Fig. 1(a)), it is essential that the data structure from which tangent cyles are computed represent multiple potential continuations for each tangent. This also allows for the co-occurrence of overlapping cycles, which, as we shall see, occur frequently in natural images.

We construct an appropriate data structure by computing the likelihoods for all tangent pairs, and then selecting for each tangent the m most likely continuant tangents. In this way, we represent the image as a sparse, weighted graph in which each vertex represents a tangent, directly connected to m other tangents. Tangent links now become edges in this graph, weighted by the computed posterior probabilities. We set $m = 6$ for the experiments presented here. [3]

The complexity of building the sparse graph is $O(n^2)$, where n is the number of tangents in the image. In practice, this procedure takes approximately 1 minute of computation on a Sparc 10 for a moderately complex 300×400 image.

9 Maximum Likelihood Cycles

We set as our goal the computation of the maximum likelihood cycle for each tangent in this tangent graph. By the identification between the vertices and edges of this graph and the extended tangents and tangent links of the contour model, these cycles correspond to highly closed contours in the image. Note that since the graph is sparse, not all tangents lie on a cycle. Also, the same cycle will often be computed by more than one tangent. These duplicate cycles are easily detected and ignored. Finally, observe that any given tangent may lie on many cycles, but each cycle must form the best closure for at least one tangent in the image.

Since tangent links are assumed to be independent, the likelihood of a tangent sequence is simply the product of the probabilities of its constituent links:

$$p(t_1 \to ... \to t_n) = p(t_1 \to t_2)p(t_2 \to t_3)...p(t_{n-1} \to t_n)$$

Taking the natural logarithm, we have

$$\log p(t_1 \to ... \to t_n) = \log p(t_1 \to t_2) + \log p(t_2 \to t_3) + ... + \log p(t_{n-1} \to t_n)$$

Since all of the terms on the right-hand side are negative, maximizing this sum corresponds to minimizing the sum of the absolute values of the terms. The problem of computing a maximum likelihood sequence can therefore be expressed as a minimization of the absolute sum of the log likelihoods for each link. Thus computing the maximum likelihood cycle for each tangent is a shortest-path problem, and can be solved using standard techniques, such as Dijkstra's algorithm [23], allowing the maximum likelihood cycles for all tangents to be computed in $O(n^2 \log n)$ operations, where n is the number of tangents in the image.

---

[3] Since a tangent may link to another tangent in 2 possible ways (contrast reversals are allowed), the size of the graph is effectively doubled.

10 Topological Constraints

Up to this point we have ignored the topology of bounding contours: self-intersecting contours have not been ruled out. Demanding that each extended tangent appear at most once in any tangent sequence eliminates many intersecting cycles (Fig. 5(a)), but there remain those cycles which intersect but which do not overlap the same tangent twice (Fig. 5(b-c)). [4]

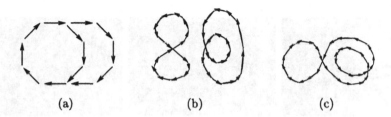

(a)   (b)   (c)

Fig. 5. Topological constraints on tangent cycles. (**a**) Each tangent may appear at most once in a tangent sequence. (**b**) By constraining cycles to have unit rotation index, "figure eights" and "double loops" are eliminated. (**c**) More complex intersecting contours with unit rotation index are not restricted.

A subclass of these self-intersecting contours can be eliminated by constraining the rotation index [3] of the underlying contour to be

$$\frac{1}{2\pi} \sum_{1}^{n} (\theta_{ai} + \theta_{bi}) = \pm 1.$$

This eliminates the more common types of erroneous cycles, such as "figure eights" (rotation index 0) and "double loops" (rotation index ± 2) (Fig. 5(b)). However, more complex self-intersecting contours with rotation index $= \pm 1$ are still possible (Fig. 5(c)), and we do not detect these.

These topological constraints cannot be embodied in the weights of the tangent graph, since the rotation index cannot be computed until a complete cycle has been determined. However, by assuming an upper bound on the number of different rotation indices that may be generated from each tangent, Dijkstra's algorithm can be generalized to incorporate these constraints. This modification does not change the complexity of the algorithm, which requires on the order of 1-2 minutes to complete in our experiments.

---

[4] The rim of a smooth solid *can* generate a self-intersecting contour in the image [26]. However, the closure algorithm developed here is not restricted to the recovery of occlusion boundaries. By imposing the constraint of non-intersection, we narrow the class of boundaries which may be recovered. The recovery of more complex boundaries will likely require information about the *type* of structure bounded.

11 Results

Figs. 6 and 8 show the bounding contours computed for two different images. For clarity, only cycles for extended tangents over 10 pixels in length are shown, and cycles which share one or more extended tangents are shown in separate images. Note that many overlapping cycles are computed, supporting our earlier observation that a constraint of disjointness is too restrictive for natural images.

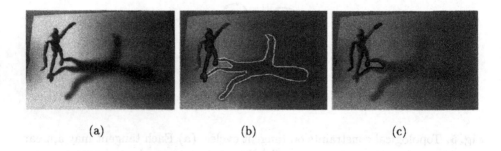

(a)   (b)   (c)

Fig. 6. (a) Image of mannequin and shadow. **(b-c)** Tangent cycles detected in mannequin/shadow image.

Fig. 6 shows the bounding contours computed for the image of the mannequin casting a shadow. The bounding contour of the cast shadow is recovered nearly perfectly. The boundary of the mannequin itself is generally recovered, although the right arm and left forearm are pinched off. The left forearm is recovered as a separate bounding contour in Fig. 6(b), as is the attached shadow on the right arm. Finally, the contour of the "hole" formed by the legs of the mannequin and its shadow is recovered as a bounding contour.

The image of a totem pole shown in Fig. 7(a) (courtesy of David Lowe) poses a greater challenge. Fig. 7(b) shows the edge groupings computed by the multiple hypothesis tracking method of Cox *et al.*, and Fig. 7(c) shows the groupings computed by Lowe's smoothness criteria [17]. While both methods group edges into extended contours, neither method recovers complete contours which could bound structural units in the image.

The tangent cycles selected as bounding contours by our closure computation are shown in Fig. 8. Major structure boundaries are identified, including the teeth, mouth, eyeball and left shoulder of the human figure and the tongue, lips, eye and eyebrow of the wolf figure, as well as various shadows and decorative markings. An error can be seen in Fig. 8(d), where the eyebrow of the human figure and the lips of the wolf figure have been grouped as a single structure.

The results of these experiments show that bounding contours can be computed as cycles of tangents in the image. While some errors are made, on the

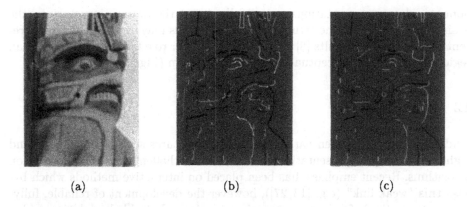

(a)   (b)   (c)

Fig. 7. (a) Noisy image of a totem pole (courtesy of David Lowe). **(b)** Edges grouped by multiple-hypothesis tracking [2]. **(c)** Edges grouped by Lowe's smoothness criteria [17].

whole these tangent cycles correspond to the boundaries of two-dimensional structures in the scene: objects, object parts, surface markings and shadows.

Note that closure computations select the tangent cycle bounding the mouth of the human figure in the totem pole (shown in white in Fig. 8(a)), even though the region thus enclosed is highly heterogeneous, encompassing a second closed contour (shown in black). While the many edges within the region of the mouth would cause region-grouping schemes to carve the structure up into smaller com-

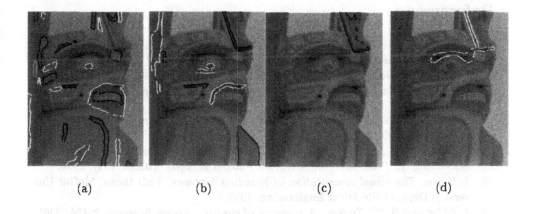

(a)   (b)   (c)   (d)

Fig. 8. Closed contours of totem pole image. The bounding contour of the mouth of the human figure (shown in white in **(a)**) is recovered despite the heterogeneity of the region it bounds.

ponents, closure computations successfully detect the unitary structure on the basis of regularities in the structure boundary. This may explain in part the recent psychophysical results [8] suggesting a stronger role for boundary cues than regional cues in the perceptual organization of form (Fig. 1(c)).

12 Conclusion

Bridging the gap between early visual data structures such as edge maps and higher-level shape representations is a significant challenge for computer vision algorithms. Recent emphasis has been placed on interactive methods which bypass this "weak link" (e.g. [13,27]), however the development of reliable, fully-automatic methods for contour grouping remains a desirable but elusive goal.

In this paper, we have proposed that the strong global constraint of contour closure may significantly aid in achieving this goal. To support our argument, we have developed an algorithm for computing closed bounding contours as topologically simple cycles of contour tangents. Unlike previous algorithms [11], this closure algorithm does not impose hard constraints on the shape of the image structures to be recovered. Since no constraint of disjointness is imposed, overlapping contours and abutting structures may be computed. Experiments indicate that this closure algorithm generally succeeds in segmenting two-dimensional structures from a variety of real images. While these closure computations do not produce a complete description of the image, they may serve to complement region-grouping methods by extending the class of segmented structures to include heterogeneous structures.

References

1. T.D. Alter. *The Role of Saliency and Error Propagation in Visual Object Recognition*. PhD thesis, MIT, 1995.
2. I.J. Cox, J.M. Rehg, and S. Hingorani. A Bayesian multiple-hypothesis approach to edge grouping and contour segmentation. *Int. J. Comp. Vision*, 11(1):5–24, 1993.
3. M.P. do Carmo. *Differential Geometry of Curves and Surfaces*. Prentice-Hall, Englewood Cliffs, NJ, 1976.
4. G. Dudek and J.K. Tsotsos. Recognizing planar curves using curvature-tuned smoothing. In *Proc. 10th Int. Conf. on Pattern Recogn.*, Atlantic City, 1990.
5. J. Elder. *The visual computation of bounding contours*. PhD thesis, McGill University, Dept. of Electrical Engineering, 1995.
6. J. Elder and S. W. Zucker. A measure of closure. *Vision Research*, 34(24):3361–3370, 1994.
7. J. Elder and S.W. Zucker. The effect of contour closure on the rapid discrimination of two-dimensional shapes. *Vision Research*, 33(7):981–991, 1993.
8. J. Elder and S.W. Zucker. Boundaries, textures and the perceptual binding of fragmented figures. In *European Conf. on Visual Perception*. Tubingen, Germany, 1995.

9. J. Elder and S.W. Zucker. Local scale control for edge detection and blur estimation. In *Lecture Notes in Computer Science*, New York, 1996. Proc. 4th European Conf. on Computer Vision, Springer Verlag.

10. W.T. Freeman. *Steerable Filters and Local Analysis of Image Structure*. PhD thesis, MIT Media Lab, 1992.

11. D.W. Jacobs. Finding salient convex groups. In I.J. Cox, P. Hansen, and B. Julesz, editors, *Partitioning Data Sets*, volume 19 of *DIMACS (Series in Discrete Mathematics and Theoretical Computer Science)*. 1995.

12. G. Kanizsa. *Organization in Vision*. Praeger, New York, 1979.

13. M. Kass, A. Witkin, and D. Terzopoulos. Snakes: Active contour models. *Proc. 1st Int. Conf. Comp. Vision*, pages 259–268, 1987.

14. K. Koffka. *Principles of Gestalt Psychology*. Harcourt, Brace & World, New York, 1935.

15. I. Kovacs and B. Julesz. A closed curve is much more than an incomplete one: Effect of closure in figure-ground discrimination. *Proc. Natl. Acad. Sci. USA*, 90:7495–7497, 1993.

16. Y.G. Leclerc. Constructing simple stable descriptions for image partitioning. *Int. J. Computer Vision*, 3:73–102, 1989.

17. D.G. Lowe. Organization of smooth image curves at multiple scales. *Int. J. Comp. Vision*, 3:119–130, 1989.

18. M. Nitzberg, D. Mumford, and T. Shiota. Filtering, segmentation and depth. *Lecture Notes in Computer Science*, 662, 1993.

19. P. Parent and S.W. Zucker. Trace inference, curvature consistency, and curve detection. *IEEE Trans. Pattern Anal. Machine Intell.*, 11:823–839, 1989.

20. T. Pavlidis. *Structural Pattern Recognition*. Springer-Verlag, Berlin, 1977.

21. E.M. Riseman and M.A. Arbib. Computational techniques in the visual segmentation of static scenes. *Comp. Graph. Image Proc.*, 6:221–276, 1977.

22. E. Saund. Symbolic construction of a 2-d scale-space image. *IEEE Trans. Pattern Anal. Machine Intell.*, 12(8):817–830, 1990.

23. R. Sedgewick. *Algorithms in C*. Addison-Wesley, Reading, Mass., 1990.

24. A. Sha'ashua and S. Ullman. Structural saliency: The detection of globally salient structures using a locally connected network. In *Proc. 2nd Int. Conf. on Computer Vision*, pages 321–327, Tampa, Florida, 1988. IEEE Computer Soc. Press.

25. A. Treisman and S. Gormican. Feature analysis in early vision: Evidence from search asymmetries. *Psychol. Rev.*, 95:15–48, 1988.

26. L.R. Williams. *Perceptual completion of occluded surfaces*. PhD thesis, University of Massachusetts, Amherst, Mass., February 1994.

27. A. Yuille and P. Hallinan. Deformable templates. In A. Blake and A. Yuille, editors, *Active Vision*. MIT Press, Cambridge, Mass., 1992.

28. S.W. Zucker. The diversity of perceptual grouping. In M. Arbib and A. Hanson, editors, *Vision, Brain and Cooperative Computation*, pages 231–261. MIT, Cambridge, Mass., 1986.

Visual Organization of Illusory Surfaces

Davi Geiger[1] and Krishnan Kumaran[2]

[1] Courant Institute, New York University, New York NY 10012, USA
[2] Rutgers University, Piscataway NJ 08854, USA

Abstract. A common factor in all illusory contour figures is the perception of a surface occluding part of a background. These surfaces are not constrained to be at constant depth and they can cross other surfaces. We address the problem of how the image organizations that yield illusory contours arise. Our approach is to iteratively find the most salient surface by (i) detecting occlusions; (ii) assigning salient-surface-states, a set of hypothesis of the local salient surface configuration; (iii) applying a Bayesian model to diffuse these salient-surface-states; and (iv) efficiently selecting the best image organization (set of hypothesis) based on the resulting diffused surface.

We note that the illusory contours arise from the surface boundaries and the amodal completions emerge at the overlapping surfaces. The model reproduces various qualitative and quantitative aspects of illusory contour perception.

1 Introduction

The first experiments with stimuli in which contours are perceived without intensity gradients (see Figure 3) are described by Schumann [26]. Kanizsa [17] has described striking visual experiments to relate the perception of illusory contours to various perceptual phenomena such as occlusions, transparency, depth sensations, brightness contrast and object recognition. These contours reveal visual grouping capacity and visual cortical activity [5] [28][23] that have provoked scientific interest for a century. Two questions natural to ask are (a) how these contours are perceived, i.e., how the shape of these illusory contours arise ? and (b) why these contours are seen and not others among all possible ones, i.e., how a visual organization is selected ?

1.1 Previous approaches

Many computational models have been proposed to describe the formation of illusory contours. The following ones take the primary view of extending the intensity edges, Ullman[27], Shashua and Ullman [25], Heitger & von der Heydt [12], Grossberg & Mingolla[10] and Grossberg [11], Guy & Medioni[9], Kellman & Shipley[18] , Mumford [20], Williams & Jacobs [30]. The approach of regions is presented by Nitzberg & Mumford [22] [21] and Brady & Grimson[3]. All the above work have very little discussion on how a particular illusory contour is selected, but rather focus on the shape of the contour. Williams and Hanson

[29], argue that the topological validity of the completed surfaces can be ensured by enforcing the labeling scheme of Huffman[14]. Topological validity, however, is too weak to constrain the solution space.

Our surface reconstruction approach is most closely related to Nitzberg and Mumford work, though our energy function is substantially different. Also, unlike previous authors, we address the problem of selecting a visual organization with a model that can be approximately computed.

1.2 Our approach

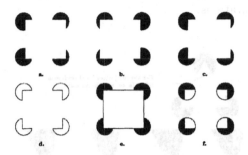

Fig. 1. Kanizsa stereo pair. Different possible interpretations of the Kanizsa square figure. Cross-fusing (b.) and (c.) leads to percept (e.) while cross-fusing (a.) and (b.) leads to percept (f.). Percept (d.) is obtained from monocular images when interpreted "literally", i.e. without the illusory contours.

When the Kanizsa square image is shown (see figure 1 a) various possible visual organizations, salient surfaces, are plausible. One can organize the cues to perceive a white square in front of four dark circles (see figure 1 c), or alternatively, various black shapes distributed over the white background (see figure 1 d). When stereo pairs are available (see figure 1 a,b,c), other organizations emerge, such as the "amodal completion", which gives the percept of four dark holes with a white square being occluded by the white region (see figure 1 f).

We address the problems of why and how the salient surface arises. Our approach is to (i) detect occlusions; (ii) assign salient-surface-states at these locations that reflect a particular surface configuration; (iii) apply a Bayesian model to diffuse this local surface information; (iv) define an energy measure for each diffusion to select the salient surface (the image organization). We note that the illusory contours arise from the surface boundaries. Thus we do not propagate/extend intensity edges.

2 A salient diffused surface

In this section the analysis is confined to select and diffuse a single salient surface, and in the next section we generalize it to multiple surfaces.

415

2.1 Junctions, occlusions and data assignment

It is well known that when occlusions are detected, an immediate sensation of depth change occurs. This can occur in monocular images due to features such as line endings and junctions. In stereo vision, occlusions along an epipolar line in one eye correspond to depth discontinuities in the other[7]. Occlusions are typically perceived where junctions or line stops occur. Thus, we start by identifying the local occlusion cues such as T-junctions, Y-junctions, corners and line endings. Each of these cues could suggest various local occlusion scenarios, e.g. corners have multiple occlusion interpretations (see figure 2). Initial information is then provided in these regions in the form of salient-surface-states representing local surface configurations.

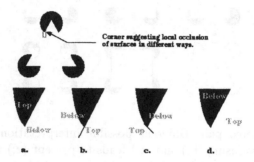

Fig. 2. Each corner has five (5) local salient surface interpretations. (a.) Convex shape. Salient black (on top). (b.,c.) Two different T-junction interpretations. (Extension shown to denote depth edge) (d.) Concave shape. Salient white (on top). The last one is the null hypothesis, i.e., no salient surface.

Salient-surface-state:

We define a binary field $\sigma_{i,j} \in [0,1]$ representing the salient-surface-state at pixel (i,j) on the image. More precisely, $\sigma_{ij} = 1$ indicates that the pixel (i,j) is salient, and $\sigma_{ij} = 0$ indicates that it is not. Initially the information is only availble at occlusion. We define $\sigma^q_{0_{ij}}$ as the initial data, where q indicates various possible hypothesis (to be defined), and a binary parameter $\lambda_{ij} = 0,1$, which indicates where a junction is present or not.

Corners (as T-junctions): At a corner, five (5) different possible configurations (hypothesis) for $\sigma^q_{0_{ij}} = 1,0$ must be analyzed (see figure 2), i.e., $q = 1,2,3,4,5$. Different from previous work, we are conjecturing a mechanism where even at corners a T-junction is hypothesized. This mechanism will be of key importance to obtain the formation of illusory surfaces.

T- and Y- junctions: These 3-junctions admit any of the three surface patches on top, with the angles biasing the probability. A null hypothesis must be always added.

End-Stopping: At an end stop of a bar, a surface discontinuity is suggested, the bar is being occluded at that location. Thus, just one interpretation is provided and the null hypothesis, i.e., $q = 1, 2$.

It is plausible that each configuration $\sigma_{0_{ij}}^q$ has a different probability to occur and a statistical analysis of scenes would be necessary to estimate $P(I|\{\sigma_{0_{ij}}^q, \lambda_{ij}^s\})$. For simplification, we will set all of them equal, i.e., we will consider the image to be conditionally independent on the parameters $\{\sigma_{0_{ij}}^q, \lambda_{ij}\}$.

2.2 Diffusion process from sparse data

In order to diffuse the salient-surface-state values from the sparse junction locations, we fit piecewise smooth surfaces using the prior distribution

$$P(\{\sigma_{ij}\}) = C_1 \prod_{ij} e^{-[\mu_{ij}^v(\sigma_{ij} - \sigma_{i,j+1})^2 + \mu_{ij}^h(\sigma_{ij} - \sigma_{i+1,j})^2]},$$

where C_1 is a normalization constant, and the diffusion coefficients μ_{ij} depend upon intensity edges as we will discuss. Assuming a Gaussian error we derive a posterior Gibbs distribution associated to an energy functional $E(\sigma)$ given by

$$E^{Diffusion}(\sigma) = \sum_{ij} [\lambda_{ij}(\sigma_{ij} - \sigma_{0_{ij}}^q)^2 + \mu_{ij}^v(\sigma_{ij} - \sigma_{i,j+1})^2 + \mu_{ij}^h(\sigma_{ij} - \sigma_{i+1,j})^2], \quad (1)$$

The first term in the energy functional imposes a fit to the sparse input data ($\lambda_{ij} = 0, 1$) while the other terms perform a controlled diffusion of this data to nearby locations if there are no discontinuities, i.e. $\mu's \neq 0$. Note that each set of hypothesis $\{\sigma_{0_{ij}}^q\}$ yields a different diffusion result.

Smoothing coefficients:
In order to prevent smoothing along intensity boundaries, we define the smoothing coefficients as

$$\mu_{ij}^h = \mu(1 - e_{ij}^h) \quad \text{and} \quad \mu_{ij}^v = \mu(1 - e_{ij}^v), \quad (2)$$

where $e^{h(v)} = 1$ indicates the presence of a horizontal edge (or vertical edge). The intensity edge map efficiently allow for neighbor pixels with the same "color" to interact and to block the ones of different "color".

Illusory contours and a description of the algorithm:
We propose the following method to estimate the locations of surface discontinuities, which provide the illusory contours.

(i) Compute μ_{ij} from the edge map;
(ii) Estimate $\{\sigma_{ij}\}$ by minimizing $E^{Diffusion}(\sigma)$;
(iii) Apply a threshold function on $\{\sigma_{ij}\}$ such that if $\sigma_{ij} > 0.5$ it is mapped to 1, otherwise to zero.

This last step identifies likely discontinuity locations as $\sigma_{ij} = 0.5$ iso-contours, i.e. the 0.5 transition points of the salient-surface-state values. Step (ii) we use two methods to minimize for $E(\sigma)$. An exact, but computationally expensive, Cholevsky decomposition and an approximate one, very fast and parallel, based on renormalization group (averaging process). The description can be found in [19] where we also show the solution σ_{ij} is bounded by the initial data $\sigma_{0_{ij}}$. Thus, $\sigma_{ij} \in [0,1]$ if $\sigma_{0_{ij}} \in [0,1]$ and the σ_{ij} can be interpreted as the degree of saliency (salient-surface-states).

Curvature Remark: Note that no curvature term has been considered in this model (just nearest neighbor interactions) and so, the desired smoothness is captured by allowing continuous solutions for σ_{ij} and the (simple) non-linearity of the model (thresholding).

2.3 Salient Organization

We now define a measure to evaluate the saliency of the surface obtained from the occlusion cues. Each image organization is defined by a distinct set of hypothesis $\{\sigma_{0_{ij}}^q, \lambda_{ij}\}$, and the reconstruction of $\{\sigma_{ij}\}$ proceeds by minimizing the energy given by equation (1). We then propose that the favored organizations (saliency) are the ones with salient-surface-states $\{\sigma_{ij}\}$ with lowest energy value of

$$S(\sigma) = E^{Saliency}(\sigma(\{\sigma_{0_{ij}}^q, \lambda_{ij}\})) = \sum_{ij} [\sigma_{ij} (1 - \sigma_{ij})] / N^2, \qquad (3)$$

where the image size is assumed to be $N \times N$. Notice that $\sigma = \sigma(\{\sigma_{0_{ij}}^q, \lambda_{ij}\})$ is evaluated by minimizing (1). This criterion basically favors organizations where regions have been more clearly segregated by the diffusion process, i.e., where the final states are closer to 1's and 0's (the minimum energy is obtained when $\sigma_{ij} = 0, 1$), and penalizes for ambiguities $\sigma_{ij} = 0.5$.

Complexity: Let us say we have P junctions, each one having Q possible configuration states. The total number of hypothesis is Q^P. Thus, the computational complexity is exponential with the number of junctions. One may reduce the number of hypothesis by only allowing salient hypothesis that are "color coherent", like only assigning saliency $\sigma_{0_{ij}} = 1$ to white regions. This reduces Q, the possible hypothesis for each junction, but still leads to an exponential growth on the total number of hypothesis (as long as $Q \neq 1$). We now offer an approach to this organization problem where various methods are known to approximately compute the best organization in polynomial (very fast) time.

Matrix notation: In matrix notation the formula (1) can be written as

$$E(\sigma) = (\sigma - (\mathbf{A} - \mathbf{S})^{-1} \Sigma \sigma_0^q)^T (\mathbf{A} - \mathbf{S})(\sigma - (\mathbf{A} - \mathbf{S})^{-1} \Sigma \sigma_0^q) \qquad (4)$$

where we have neglected an additive constant . Here, $\sigma_0^{\mathbf{P}}$ and σ are vectors containing the values of $\sigma_{0_{ij}}^q$ and σ_{ij} respectively, cast into a single column vector, i.e.,

$$\sigma_k = \sigma_{i+Nj} = \sigma_{ij} \quad \text{or} \quad \sigma_k = \sigma_{int(k/N),mod(k,N)}, \cdot$$

The superscript T represents transposes and $\mathbf{I}$ is the identity matrix. We have introduced the variable $s_{ij} = 1 - \lambda_{ij}$. $\mathbf{S}$ is a diagonal matrix with diagonal entries s_{ij} and $\Sigma = (\mathbf{I} - \mathbf{S})$. The matrix $(\mathbf{A} - \mathbf{S})$ is symmetric, block tri-diagonal, with the following structure.

$$(\mathbf{A} - \mathbf{S}) = \begin{pmatrix} \mathbf{A}_1 - \mathbf{S}_1 & \mathbf{D}_1^T & \mathbf{0} & \cdots \\ \mathbf{D}_1 & \mathbf{A}_2 - \mathbf{S}_2 & \mathbf{D}_2^T & \ddots \\ \mathbf{0} & \mathbf{D}_2 & \mathbf{A}_3 - \mathbf{S}_3 & \ddots \\ \vdots & & \ddots & \ddots & \ddots \end{pmatrix}$$

where the matrices $\mathbf{D}_i$ and $\mathbf{A}_i - \mathbf{S}_i$ are given by

$$\mathbf{D}_i = \begin{pmatrix} -\mu_{i-1,1}^v & 0 & \cdots \\ 0 & -\mu_{i-1,2}^v & 0 \\ \vdots & & 0 & \ddots \end{pmatrix}, \quad \mathbf{A}_i - \mathbf{S}_i = \begin{pmatrix} \chi_{i1} & -\mu_{i1}^h & 0 \\ -\mu_{i1}^h & \chi_{i2} & \ddots \\ 0 & \ddots & \ddots \end{pmatrix}$$

where $\chi_{ij} = \lambda_{ij} + \mu_{i-1,j}^h + \mu_{ij}^h + \mu_{i,j-1}^v + \mu_{ij}^v$ and i is a row index.

Since the energy is quadratic and positive definite, the minimum of the energy is 0 and the minimum energy configuration is given by

$$\sigma = (\mathbf{A} - \mathbf{S})^{-1} \Sigma \sigma_0^q . \tag{5}$$

The organization selection process:

We now procede to elaborate, within an efficient representation, the problem of visual organization selection. We draw this formulation from previous work on optimization by Karmarkar [16].

Let us assume we have $p = 1, ..., P$ junctions and that for each junction we have $q = 1, ..., Q$ hypothesis (salient states). Let us define the binary field $\phi_{p,q} = 0, 1$ that is 1 when junction p is at state q. For each junction only one hypothesis is accepted (to select one salient surface), i.e., $\sum_{q=1}^{Q} \phi_{pq} = 1$. Let us represent each hypothesis as $\sigma_{0_p}^q$ and thus,

$$\sigma_0^q = \sum_{p=1}^{P} \sigma_{0_p}^q .$$

Using (5) we can write the energy (3) as

$$S(\{\phi_{pq}\}) = \frac{1}{N^2} \sum_{p=1}^{P} \sum_{q=1}^{Q} \phi_{pq}((\mathbf{A}-\mathbf{S})^{-1}\Sigma\sigma_{0_p}^q)^T(\mathbf{1} - \sum_{r=1}^{P}\sum_{s=1}^{Q}\phi_{rs}(\mathbf{A}-\mathbf{S})^{-1}\Sigma\sigma_{0_r}^s)$$

$$= \frac{1}{N^2} \sum_{p,q}^{P,Q} \phi_{pq}(\sigma_{0_p}^q)^T\mathbf{M}^{-T}\mathbf{1} - \sum_{p,q,r,s}^{P,Q,P,Q} \phi_{pq}\phi_{rs}(\sigma_{0_p}^q)^T\mathbf{M}^{-T}\mathbf{M}^{-1}\sigma_{0_s}^r, \quad (6)$$

where $\mathbf{M}^{-1} = (\mathbf{A}-\mathbf{S})^{-1}\Sigma$, and under the constraint $\sum_{q=1}^{Q}\phi_{pq} = 1$. Vast literature exists on solving this problem. The interior point method proposed by Karmarkar, e.g. [16]. A related mechanical statistical methods, where ϕ_{pq} is averaged out, was proposed in [6] [8]. Also a related method is mean field annealing where early work [2] already introduces. These methods lead to a solution of the form

$$\bar{\phi}_{pq} = \frac{e^{-\beta(h_{p,q} - \sum_{r,s}\bar{\phi}_{rs}k_{p,q,r,s})}}{\sum_{q=1}^{Q} e^{-\beta(h_{p,q} - \sum_{r,s}\bar{\phi}_{rs}k_{p,q,r,s})}},$$

where $h_{pq} = (\sigma_{0_p}^q)^T\mathbf{M}^{-T}\mathbf{1}$ and $k_{p,q,r,s} = (\sigma_{0_p}^q)^T\mathbf{M}^{-T}\mathbf{M}^{-1}\sigma_{0_s}^r$, and β is an annealing parameter. For $\beta = 0$ the solution is $\bar{\phi}_{pq} = 1/Q$ and as $\beta \to \infty$ a solution is obtained.

Alternatively, in this paper, we simply inspect various natural organizations to examine the associated energy $S(\sigma)$ and select the best one.

We note that analogous models could be constructed for various vision modalities. For example in texture discrimination, following the "Texton Theory" proposed by Julesz [15]. The features instead of junctions would be "textons", their interpretations would build a set of "states" respectively, while the interactions between textons would be dictated by a discontinuity preserving diffusion process. Much work remains to be done.

3 Experiments and Results

The model provides a way to test qualitative and quantitative aspects of illusory surface perception. We first present the results of the reconstruction for "classical" illusory surfaces (see figure 3). Then we investigate various other geometrical parameters[23][24] of illusory surfaces, taking the kanizsa square as the starting point. All results presented in this article have been obtained using $\mu = 0.01$ and a threshold of 0.5. !dvi

4 Multiple surfaces and amodal completion

We have so far described a model for reconstructing the salient surface and applicable to modal completion, i.e, when the perceived illusory surface is seen in front. We now extend the model to account for multiple surfaces and amodal

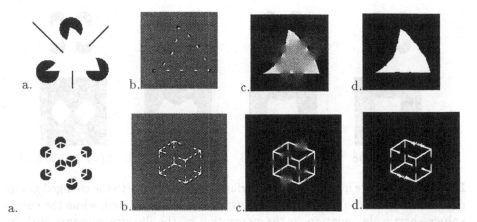

Fig. 3. Two examples of illusory surface perception, stressing the surface aspect of the illusory figures. Each of the above figures depict a. The original figure. b. Input data - white=1; black=0; gray=0.5. c. The reconstruction. d. Thresholding.

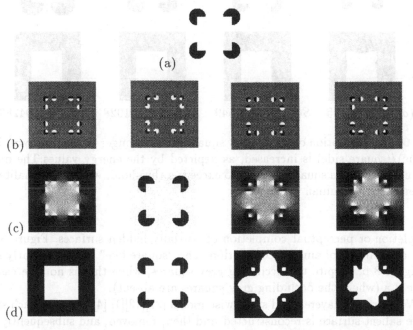

(e) $S(\sigma) = 0.389704$ $S(\sigma) = 0.000000$ $S(\sigma) = 0.552932$ $S(\sigma) = 0.527047$

Fig. 4. Different ways of interpreting each corner lead to different surface shapes. The 2 lowest $S(\sigma)$ values give "natural" perceptions. (a) Original figure; (b) Input data (white - $\sigma_{0_{ij}} = 1, \lambda_{ij} = 1$; black-$\sigma_{0_{ij}} = 0, \lambda_{ij} = 1$; gray-$\sigma_{0_{ij}} = 0.5, \lambda_{ij} = 0$); (c) Reconstruction σ_{ij}; (d) Thresholded values from (c); (e) Energy $S(\sigma)$.

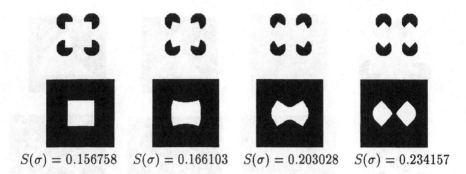

$$S(\sigma) = 0.156758 \quad S(\sigma) = 0.166103 \quad S(\sigma) = 0.203028 \quad S(\sigma) = 0.234157$$

Fig. 5. The change in the perceived surface leads gradually to the changed grouping of features as the vertices of the illusory square are rotated, while the energy values show a degradation in the perception of the illusory contours with increasing curvature.

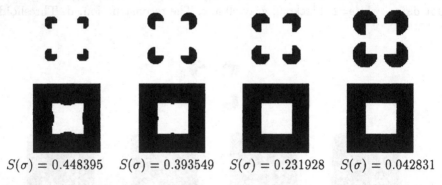

$$S(\sigma) = 0.448395 \quad S(\sigma) = 0.393549 \quad S(\sigma) = 0.231928 \quad S(\sigma) = 0.042831$$

Fig. 6. The perception of the illusory square grows stronger as the ratio γ=(circle radius)/(square side) is increased, as depicted by the energy values.The model only converges to a square if the ratio exceeds a threshold, which is in qualitative agreement with human perception.

completion or perceptual completion of partially hidden surfaces. Figure 8 depicts an example of amodal completion. The "square like" shape is clearly seen in Figure 8 b, despite the occluding grey squares, while that is not the case in figure 8 a (where the occluding grey squares are absent).

We propose a layered and stage wise model [22][13][1] [4] where at each stage s, the salient-surface is reconstructed and then, removed, and subsequently the same process continues to the next stage until all surfaces have been reconstructed. Multiple surfaces can be assigned to each image location due to the layered reconstruction. Then, the overlapping regions correspond to amodal completions of partially occluded surfaces.

Updating μ_{ij}^s: Once the salient surface is reconstructed, at stage s, it is removed so that concealed surfaces can be reconstructed. T The intensity edges associated to the first surface need also to be removed since the surface is no longer there to block the interactions. The removal process updates the intensity edges of the previously reconstructed surface to zero.

Updating $\sigma^s_{0_{ij}}$**:** Once the salient surface is reconstructed, at stage s, the occlusion information must be reconsidered. The removed surface is no longer there and therfore the cues for saliency are no longer valid. One must choose the salienct-surface-states accordingly.

The experiments are shown in figures 7 and 8.

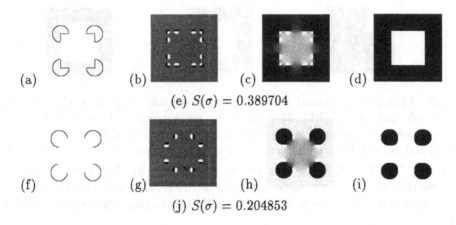

(a)   (b)   (c)   (d)

(e) $S(\sigma) = 0.389704$

(f)   (g)   (h)   (i)

(j) $S(\sigma) = 0.204853$

Fig. 7. Multi layer reconstruction for the Kanizsa square figure. First layer: (a) Initial values of $\mu^{s=top}_{ij}$; (b) Sparsity map - $\lambda_{ij} = 1$ at non-gray pixels and $\lambda_{ij} = 0$ in the gray region. (c) Reconstruction σ_{ij}; (d) values from (c) thresholded at 0.5; (e) Energy values. Removing the Kanizsa square, and reconstructing the Second Layer (Amodal completion): (f), (g), (h), (i), (j) are the same quantities for the second layer.

Acknowledgements

The authors are grateful to valuable discussions and insights offered by Bela Julesz, Ilona Kovacs and Barton L. Anderson of the Laboratory of Vision Research at Rutgers and Bob Shapley of the Dept. of Psychology at NYU. Part of the work was done while Davi Geiger was visiting the Isaac Newton Institute for Mathematics, Cambridge, UK.

References

1. E. Adelson and P. Anandan. Ordinal characteristics of transparency. In Proc. AAAI Workshop on Qualitative Vision, pages 552–573, Los Angeles, CA, 1990.

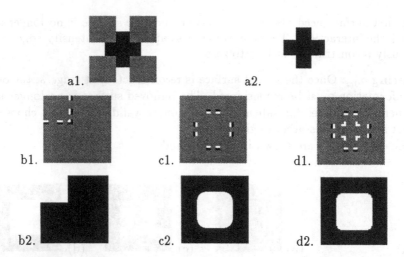

Fig. 8. Example of amodal completion. The square shape is easily seen in a1., while not in a2. (b1....d2.) First, we reconstruct each one of the gray squares, which is perceived as the salient. Reconstruction of the surfaces of a1. (b1,c1,d1) Initial data: white $-\sigma_{0_{ij}}^{s=1,2} = 1, \lambda_{ij}^s = 1$; black$-\sigma_{0_{ij}}^{s=1,2} = 0, \lambda_{ij}^s = 1$; grey$-\sigma_{0_{ij}}^{s=1,2} = 0.5, \lambda_{ij}^s = 0$. Next we reconstruct the black square. (b2,c2,d2) Reconstructions σ_{ij}^s of successive completions. Here, the model provides an interesting result. The corners, which are hidden by the gray squares, are rounded off

2. D. J. Amit, H. Gutfreund, and H. Sompolinsky. Spin-glass models of neural networks. *Physical Review A*, 32:1007–1018, 1985.
3. M. Brady and W. E. L. Grimson. The perception of subjective surfaces. A.I. Memo No. 666, AI Lab., MIT, Nov. 1982.
4. T. Darrell and A. Pentland. Cooperative robust estimation using layers of support. *IEEE Trans. PAMI*, PAMI-17(5):474–487, 1995.
5. R. von der Heydt. E. Peterhans and G. Baumgartner. Neuronal responses to illusory contour stimuli reveal stages of visual cortical processing. In *Visual Neuroscience*, pages 343–351. Cambridge U.P, 1986.
6. D. Geiger and F. Girosi. Parallel and deterministic algorithms for mrfs: surface reconstruction. *IEEE Trans. PAMI*, PAMI-13(5):401–412, May 1991.
7. D. Geiger, B. Ladendorf, and A. Yuille. Binocular stereo with occlusion. In *2nd ECCV*, Santa Marguerita, Italy, May 1992. Springer-Verlag.
8. D. Geiger and A. Yuille. A common framework for image segmentation. *International Journal of Computer Vision*, 6:3:227–243, August 1991.
9. G.Guy and G. Medioni. Inferring global perceptual contours from local features. In *Proc. Image Understanding Workshop DARPA*, September 1992.
10. S. Grossberg and E. Mingolla. Neural dynamics of perceptual grouping:textures, boundaries and emergent segmentations. *Perception & Psychophysics*, 38(2):141–170, 1985.
11. Stephen Grossberg. 3-d vision and figure-ground separation by visual cortex. *Perception & Psychophysics*, 55(1):48–120, 1994.

424

12. F. Heitger and R. von der Heydt. A computational model of neural contour processing: Figure-ground segregation and illusory contours. *Proceedings of the IEEE*, pages 32–40, 1993.
13. S. Hsu, P. Anandan, and S. Peleg. Accurate computation of optical flow by using layered motion representation. In *ICPR*, pages 743–746, Oct 1994.
14. D.A. Huffman. A duality concept for the analysis of polyhedral scenes. In *Machine Intelligence*, volume 6. Edinb. Univ. Press, Edinb., U.K., 1971.
15. Bela Julesz. Textons: the elements of texture perception, and their interactions. *Nature*, 290:91–97, March 1981.
16. A. Kamath and N. Karmarkar. A continuous approach to compute upper bounds in quadratic minimization problems with integer constraints. In C. A. Floudas and P. M. Pardalos, ed., *Recent Adv. in Global Opt.*, volume 2, 229–241, 1991.
17. G. Kanizsa. *Organization in Vision*. Praeger, New York, 1979.
18. P.J. Kellman and T.F. Shipley. A theory of visual interpolation in object perception. *Cognitive Psychology*, 23:141–221, 1995.
19. K. Kumaran, D. Geiger, and L. Gurvits. Illusory surfaces and visual organization. *Network:Computation in Neural Systems*, 7(1), February 1996.
20. D. Mumford. Elastica and computer vision. In C. L. Bajaj, editor, *Algebraic Geometry and Its Applications*. Springer-Verlag, New York, 1993.
21. Nitzberg, Mumford, and Shiota. *Filtering, Segmentation, and Depth*. Springer-Verlag, New York, 1993.
22. M. Nitzberg and D. Mumford. The 2.1-d sketch. In *Proceedings of the International Conference on Computer Vision*, pages 138–144. IEEE, DC, 1990.
23. D. L. Ringach and R. Shapley. The dynamics of illusory contour integration. *Investigative Ophthalmology and Visual Science*, 35,#2:4196–, 1994.
24. D. L. Ringach and R. Shapley. Similar mechanisms for illusory contour and amodal completion. *Investigative Ophthalmology and Visual Science*, 35,#2:1089–, 1994.
25. A. Shashua and S. Ullman. Structural saliency: The detection of globally salient structures using a locally connected network. In *Proceedings of the International Conference on Computer Vision*, pages 321–327, 1988.
26. F. Shumann. Einige beobachtungen uber die zusammenfassung von gesichtseindrucken zu einheiten. *Physiologische Studien*, 1:1–32, 1904.
27. S. Ullman. Filling in the gaps: The shape of subjective contours and a model for their generation. *Biological Cybernetics*, 25:1–6, 1976.
28. R. von der Heydt, E. Peterhans, and G. Baumgartner. Illusory contours and cortical neuron responses. *Science Washington*, 224:1260–1262, 1984.
29. L. R. Williams and A.R. Hanson. Perceptual completion of occluded surfaces. *Proc. of IEEE Computer Vision and Pattern Recognition*, 1994.
30. L. R. Williams and D.W. Jacobs. Stochastic completion fields: A neural model of illusory contour shape and salience. *Proc. of 5th Intl. Conf. on Comp. Vision*, 1995.

Stereo

Uncalibrated Relief Reconstruction and Model Alignment from Binocular Disparities

Jonas Gårding[a], John Porrill[b], John P. Frisby[b], John E. W. Mayhew[b]

[a] Computational Vision and Active Perception Laboratory (CVAP)
Dept. of Numerical Analysis and Computing Science
KTH (Royal Institute of Technology), S-100 44 Stockholm, Sweden
Email: Jonas.Garding@bion.kth.se

[b] AI Vision Research Unit, Sheffield University,
P.O. Box 603, Psychology Building,
Sheffield S10 2UR, United Kingdom.
Email: J.P.Frisby@sheffield.ac.uk

Abstract. We propose a computational scheme for uncalibrated reconstruction of scene structure up to a relief transformation from binocular disparities. This scheme, which we call retinal disparity correction
(RDC), is motivated both by computational considerations and by psychophysical observations regarding human stereoscopic depth perception.
We describe an implementation of RDC and demonstrate its performance experimentally. As an example of applications of RDC, we show
how it can be used to align a range-image and object model with an
uncalibrated disparity field.

1 Introduction

The process of stereoscopic depth perception in humans and machine vision comprises two major computational steps. First, the correspondence between the
left and right images of points in three-dimensional space must be established,
resulting in a disparity map which may be sparse or dense. Then, the disparity
map must somehow be interpreted in terms of the depth structure of the scene.
In this paper we consider the second of these two steps, i.e. the problem of
disparity interpretation.

The classical approach to interpretation of stereo images is to perform a careful camera calibration, which is then used to reconstruct the three-dimensional
structure of the scene by intersecting the left and right visual rays at each point.
This technique was originally developed and refined by photogrammetrists, and
it continues to be an important and powerful tool in circumstances that allow
an accurate calibration to be performed; see e.g. [18].

However, in many situations a careful calibration step is impractical, and
there has recently been a great interest in approaches that require no calibration
as all [3, 11], and there now exists a rich literature on the subject. Typically, these
methods produce structure up to an arbitrary projective or affine transformation.

Uncalibrated Relief Reconstruction and Model Alignment from Binocular Disparities

Jonas Gårding[1], John Porrill[2], John P Frisby[2], John E W Mayhew[2]

[1] Computational Vision and Active Perception Laboratory (CVAP)
Dept. of Numerical Analysis and Computing Science
KTH (Royal Institute of Technology), S-100 44 Stockholm, Sweden
Email: Jonas.Garding@bion.kth.se

[2] AI Vision Research Unit, Sheffield University
P.O. Box 603, Psychology Building
Sheffield S10 2UR, United Kingdom
Email: J.P.Frisby@sheffield.ac.uk

Abstract. We propose a computational scheme for uncalibrated reconstruction of scene structure up to a relief transformation from binocular disparities. This scheme, which we call *regional disparity correction* (RDC), is motivated both by computational considerations and by psychophysical observations regarding human stereoscopic depth perception. We describe an implementation of RDC, and demonstrate its performance experimentally. As an example of applications of RDC, we show how it can be used to align a three-dimensional object model with an uncalibrated disparity field.

1 Introduction

The process of stereoscopic depth perception in human and machine vision comprises two major computational steps. First, the correspondence between the left and right images of points in three-dimensional space must be established, resulting in a disparity map which may be sparse or dense. Then, the disparity map must somehow be interpreted in terms of the depth structure of the scene. In this paper we consider the second of these two steps, i.e., the problem of disparity interpretation.

The classical approach to interpretation of stereo images is to perform a careful camera calibration, which is then used to reconstruct the three-dimensional structure of the scene by intersecting the left and right visual rays of each point. This technique was originally developed and refined by photogrammetrists, and it continues to be an important and powerful tool in circumstances that allow an accurate calibration to be performed; see e.g. [18].

However, in many situations a separate calibration stage is impractical, and there has recently been a great interest in approaches that require no calibration at all [6, 11], and there now exists a rich literature on the subject. Typically, these methods produce structure up to an arbitrary projective or affine transformation.

Another useful approach is possible in applications in which it is known that there exists a ground plane (or other surface) in the scene. By computing disparities relative to this surface, many computational simplifications are achieved. Early applications of the approach were intended for obstacle detection [4], but more recently this idea has been put in a more general framework [3, 17].

It has also been pointed out that disparities can be useful for segmentation or control of attention [10, 23] without attempting to reconstruct anything at all. These methods use raw disparity measurements to filter out regions for which the depth differs significantly from that of the region of interest, and for this purpose uncalibrated horizontal disparities are typically good enough.

In this paper we propose an alternative approach to disparity interpretation, which we call *regional disparity correction* (RDC). This approach originates from our work on modelling certain aspects of human stereoscopic depth perception, and it is a generalization of the model proposed in [9]. Like the raw disparity approach, RDC avoids explicit estimation of physical viewing parameters such as fixation distance. In contrast, however, RDC allows recovery of both general projective structure and of the metric depth ordering of the scene, and it is therefore potentially useful as a basis e.g. for object recognition and other tasks that can be aided by a characterization of the three-dimensional shape of objects in the scene. A more detailed account of the present work is given in [8].

2 Regional Disparity Correction (RDC)

In short, RDC comprises the following three steps:

1. Approximate the *vertical* component of disparity by a quadratic polynomial $\hat{v}(x, y)$.

2. Compute a "correction polynomial" $g(x, y)$ by a certain reshuffling of the coefficients of $\hat{v}(x, y)$.

3. Compute *affine nearness* $\rho(x, y) = h(x, y) + g(x, y)$.

It will be shown that $\rho(x, y)$ is itself a projective reconstruction of the scene, but more importantly, it allows the structure of the scene to be recovered up to a *relief transformation*, which preserves the depth ordering of the scene. Before describing the method in detail, we shall briefly discuss the psychophysical observations and computational considerations that motivate the approach.

2.1 Human Stereoscopic Depth Perception

Two aspects of human stereoscopic depth perception are particularly relevant for the RDC model; the role of vertical disparities, and the geometric nature of the depth percept.

The fact that the vertical component of disparity plays a significant role in human stereoscopic depth perception was first pointed out by Helmholtz [12].

He found that an array of vertical threads arranged in the fronto-parallel plane appears significantly curved, but that the perceived curvature can be eliminated by attaching beads to the threads. Presumably the beads allow the visual system to estimate vertical disparities, which allows the horizontal disparities of the threads to be interpreted correctly. Another well-known demonstration of the influence of vertical disparities is due to Ogle [21], who showed that unilateral vertical magnification (achieved by inserting a horizontal cylindrical lens in front of one eye) induces an apparent slant of the fronto-parallel plane. More recently, a number of researchers have investigated the effect of vertical disparity manipulations in a variety of circumstances, and there now exists a substantial body of empirical data; see e.g. [7] for a review.

Concerning the geometric aspects of stereoscopic depth perception, several studies (e.g. [14, 27]) have found that human performance in tasks involving estimation of metric structure from binocular disparities is remarkably poor, even in the presence of a richly structured disparity field. We interpret this as an indication that metric reconstruction is not the primary purpose of human stereopsis.

2.2 Computational Motivation

The RDC model takes an intermediate position between fully calibrated metric recovery and weakly calibrated recovery of projective structure.

We believe that it is beneficial to use as much knowledge about the viewing geometry as is available, without relying on unrealistic assumptions or computationally demanding calibration procedures. In this sense, the calibrated approach uses too much information, whereas the weakly calibrated approach uses too little. The RDC model is based on fixating binocular vision, typical of both human vision and anthropomorphic robot vision systems [23, 24]. In such systems the viewing geometry is constrained in a number of ways. The extrinsic geometry has essentially only three degrees of freedom if Donder's law is assumed, and it seems reasonable to assume that most of the intrinsic parameters are relatively stable or change only slowly over time (perhaps with the exception of the focal length). Nevertheless, it would be unrealistic to assume perfect fixation with zero cyclovergence etc., so we explicitly incorporate small-angle deviations from the idealized geometry in the RDC model.

Concerning the end result of the disparity interpretation process, there are good computational reasons for considering alternatives to metric reconstruction. It has been shown [22, 10] that the quality of metric reconstruction depends crucially on an accurate estimate of the fixation distance d (or equivalently the vergence angle between the two optical axes). RDC avoids this difficulty by not attempting to recover metric depth, and hence not having to estimate d. Instead, RDC recovers depth up to a relief transformation by computations performed only in the disparity domain, i.e., by adding small corrections to the horizontal disparities. This means that errors in the estimated disparity vectors are never magnified e.g. by division by small scale factors or by non-linear operations, in contrast to the case of recovery of metric structure.

3 Description and Analysis of RDC

The RDC method described here extends the model proposed in [9] in several ways, notably by allowing unknown fixation errors, unknown cyclovergence and unknown focal length.

A schematic representation of the binocular viewing geometry is shown in Figure 1. We represent visual space with respect to a virtual cyclopean eye, constructed such that the cyclopean visual axis (the Z axis) bisects the left and right visual axes. We explicitly model deviations from this idealized model by including (small) rotation angles (ω_x, ω_z), where ω_x is a relative rotation around the X axis and hence represents a vertical fixation error, and ω_z is a relative rotation around the Z axis, representing cyclovergence.

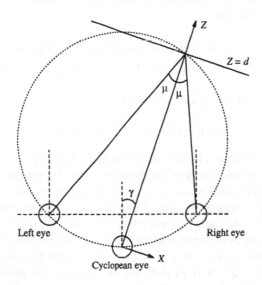

Fig. 1. Idealized representation of the viewing geometry. The plane of the drawing is referred to as the fixation plane. The Vieth-Müller circle (dotted) through the fixation point and the eyes indicates a part of the point horopter, i.e., the locus of points that yield zero disparity. μ is the vergence half-angle, and γ is the angle of asymmetric gaze.

We use a standard pinhole projection model, i.e., for each camera

$$x = f\frac{X}{Z}, \qquad y = f\frac{Y}{X}, \tag{1}$$

where (x, y) are image coordinates, f is the focal length[3] and (X, Y, Z) are (left-handed) camera coordinates with the Z axis pointing toward the scene.

We use a flow approximation of disparity. This is equivalent to assuming that the angles ω_x and ω_z as well as μ (vergence) are small enough to be represented

---

[3] We assume the f is unknown but equal in the left and right images.

by a first-order approximation. As shown in [9], the error in this approximation is negligible for realistic viewing distances.

To simplify the notation, we define the inverse distance function $\lambda(x,y) = 1/Z(x,y)$. We relate the translation and rotation parameters $(t_x, t_y, t_z; \omega_x, \omega_y, \omega_z)$ to the viewing geometry by

$$\omega_y = t_x/d = t_x\lambda_0.$$
$$(t_x, t_y, t_z) = -I(\cos\gamma, 0, \sin\gamma).$$

This generalized definition of the fixation distance $d = 1/\lambda_0$ in terms of t_x and ω_y makes sense even if the optical axes do not cross. The only assumption made here is thus that $t_y = 0$, i.e., that there is no vertical displacement between the cameras. Applying these definitions to the standard image flow equations [19], we obtain the disparity as

$$h = x_r - x_l \approx \dot{x} = \rho + (I\sin\gamma)\lambda\, x + (I\cos\gamma)\lambda_0\, x^2/f + \omega_z\, y + \omega_x\, xy/f,$$
$$v = y_r - y_l \approx \dot{y} = (I\sin\gamma)\lambda\, y + (I\cos\gamma)\lambda_0\, xy/f - \omega_z\, x + \omega_x\, (f + y^2/f),$$

where we have defined the *affine nearness*

$$\rho(x,y) = fI\cos\gamma\,(\lambda_0 - \lambda(x,y)). \tag{2}$$

For reasons which will be elaborated in Section 4, the purpose of RDC is to estimate $\rho(x,y)$. The key to achieving this aim is the observation that the vertical disparity depends very weakly on $\lambda(x,y)$, i.e., the depth structure of the scene. In fact, for symmetric vergence this small dependency vanishes completely, so that $v(x,y)$ encodes only the camera geometry. The idea is therefore to use $v(x,y)$ to "calibrate" $h(x,y)$ in order to estimate $\rho(x,y)$. We do this by approximating $v(x,y)$ in some region by the vertical disparity field corresponding to the average plane $\pi : Z = PX + QY + R$ in that region. Due to the weak dependency on depth, the error in this approximation can be expected to be significantly smaller than what would be the case if we had instead attempted to approximate $h(x,y)$. In this way RDC is different from methods based on subtracting the disparity of a reference surface [3, 17].

In terms of the image coordinates, we have

$$\lambda_\pi(x,y) = \frac{1}{Z(x,y)} = \frac{f - Px - Qy}{fR}. \tag{3}$$

The corresponding disparity field can be expressed as

$$h_\pi(x,y) = fI\cos\gamma\,(\lambda_0 - \lambda_\pi(x,y)) - g_\pi(x,y), \tag{4}$$
$$v_\pi(x,y) = A + Bx + Cy + Exy + Fy^2, \tag{5}$$

where

$$g_\pi = -Cx + By - Ex^2 - Fxy,$$

and (A, B, C, E, F) are constants depending on the viewing geometry and the parameters (P, Q, R) of the plane. They are estimated directly from $v(x, y)$ by linear least squares. Affine nearness can now be estimated by computing

$$\hat{\rho}(x, y) = h(x, y) + g_\pi(x, y).$$

By substituting (4), we can express this estimate as

$$\hat{\rho}(x, y) = \rho(x, y) + \epsilon(x, y), \tag{6}$$

where

$$\epsilon(x, y) = I \sin \gamma \, (\lambda(x, y) - \lambda_\pi(x, y))x \tag{7}$$

is the estimation error. This error is the product of three factors which are generally small: the deviation from symmetric vergence, the deviation from the average plane, and the horizontal image eccentricity. Consequently, $\hat{\rho}(x, y)$ is typically a good estimate of $\rho(x, y)$.

4 Affine Nearness and Relief Transformations

The affine nearness $\rho(x, y)$ contains important and useful information about the structure of the world. From (2) and (1) we can solve for the relation between $(x, y, \rho(x, y))$ and 3D-points (X, Y, Z). In homogeneous coordinates, we obtain the invertible relation

$$\begin{pmatrix} X \\ Y \\ Z \\ 1 \end{pmatrix} \cong \begin{pmatrix} 1 & 0 & 0 & 0 \\ 0 & 1 & 0 & 0 \\ 0 & 0 & 0 & f \\ 0 & 0 & -1/L \, f\lambda_0 \end{pmatrix} \begin{pmatrix} x \\ y \\ \rho \\ 1 \end{pmatrix}, \tag{8}$$

where $L = I \cos \gamma$, and the symbol $\cong$ denotes equality up to an arbitrary scale factor. This relation demonstrates that the surface in $\mathbb{P}^3$ with homogeneous coordinates $(x, y, \rho(x, y), 1)$ is in fact a projective reconstruction of the scene. Hence, if for example $\rho(x, y)$ is flat, then so is the corresponding physical surface in the scene.

However, $\rho(x, y)$ is distinctively different from the physical surface in other ways, and there is no choice of parameters (λ_0, L, f) for which they become identical. It is therefore more useful to consider the equivalence class of surfaces $(X(x, y), Y(x, y), Z(x, y))$ that can be reconstructed from $\rho(x, y)$ by "guessing" the parameters (λ_0, L, f). From (8) we see that two such surfaces, corresponding to (λ_0, L, f) and (λ'_0, L', f') respectively, are related by the transformation

$$\begin{pmatrix} X' \\ Y' \\ Z' \\ 1 \end{pmatrix} \cong \begin{pmatrix} 1 & 0 & 0 & 0 \\ 0 & 1 & 0 & 0 \\ 0 & 0 & c & 0 \\ 0 & 0 & b & a \end{pmatrix} \begin{pmatrix} X \\ Y \\ Z \\ 1 \end{pmatrix}, \tag{9}$$

where

$$a = L/L' > 0, \qquad b = \frac{f'\lambda_0' L' - f\lambda_0 L}{fL'}, \qquad c = f/f' > 0.$$

In the following, the matrix in (9) will be denoted by $T_{a,b,c}$. It is easily verified that $T_{a,b,c}$ defines a *transformation group*, which in turn defines an *equivalence class* of 3-D shapes compatible with a given affine nearness $\rho(x, y)$.

We shall refer to $T_{a,b,c}$ as a generalized[4] *relief transformation*. Transformations of this type have a long history in vision, and have been considered e.g. in [12, 15, 16]. Perhaps the most important property of a relief transformation is that it preserves the *depth ordering* of the scene.[5] Hence, knowing the structure of the scene up to $T_{a,b,c}$ entails knowing what is in front of what, in addition to the general projective properties such as coplanarity and collinearity.

5 Experimental Evaluation

In this section we shall give some quantitative examples of the performance of RDC, using the following procedure:

1. Choose the binocular camera geometry $(f, d, I, \gamma, \omega_x, \omega_z)$

2. Generate a random cloud of points in space around the fixation point

3. For each 3-D point, generate its left and right projection using the pinhole camera model (i.e., *not* the flow approximation)

4. Perform RDC, i.e., fit a quadratic function to the vertical disparity field, and use this function to compute the estimate $\hat{\rho}(x, y)$ of affine nearness

5. Reconstruct the 3-D points from $\hat{\rho}(x, y)$ by resolving the relief ambiguity, i.e., by applying (8) using the true values of (d, L, f)

6. Compute the difference between the reconstructed points and the true 3-D points.

The reason for performing the metric 3-D reconstruction is to obtain a quantitative and geometrically intuitive measure of the quality of the estimate $\hat{\rho}(x, y)$.

As shown in Table 1, in the noise-free case the reconstruction errors from RDC, which originate from the flow approximation of disparity and the error term $\epsilon(x, y)$, are on the order of a millimeter or less with a viewing geometry representative of human vision. It is perhaps more interesting to consider what happens under the more realistic assumption that the disparity vectors are noisy. The columns labeled $\sigma = 1$ show the corresponding data obtained after adding Gaussian noise to each disparity vector. It is evident that RDC handles this noise level relatively gracefully.

---

[4] The definition of a relief transformation given in [9] corresponds to $T_{a,b,1}$, i.e., the case of known focal length.

[5] Strictly speaking, the depth order is preserved on either side of the singularity $a + bZ = 0$, but not across it. However, any physically valid relief reconstruction (X', Y', Z') must satisfy $Z' > 0$ for all points, which implies $a + bZ > 0$.

Viewing geometry	n	RDC error		Raw disparity error	
		$\sigma = 0$	$\sigma = 1$	$\sigma = 0$	$\sigma = 1$
Symmetric vergence;	5	0.037	2.681	2.437	2.412
$\gamma = \omega_x = \omega_z = 0$	10	0.041	1.002	2.416	2.726
	100	0.043	0.929	2.812	2.848
Asymmetric vergence	5	0.385	1.682	5.096	5.240
with cyclovergence;	10	0.400	1.249	9.033	9.796
$\gamma = 25°$, $\omega_z = 5°$	100	0.464	1.257	10.879	10.875

Table 1. Average distances (in cm) between reconstructed and true 3-D points, for two different viewing geometries and varying number n of points. The fixation distance was 50cm, the interocular baseline was 6cm, and the 3-D points were randomly distributed in a box 40cm wide and 20cm deep centered around the fixation point. The columns labelled *RDC error* indicate the result when depth was reconstructed from estimated affine nearness $\hat{\rho}(x, y)$. As a comparison, the columns labelled *Raw disparity error* show the result of neglecting the vertical disparities and using $\hat{\rho}(x, y) \approx h(x, y)$. σ indicates the standard deviation (in pixels) of Gaussian noise, which was added independently to the horizontal and vertical components of each disparity vector (in a 512×512 image).

6 Model Alignment

In applications such as object recognition, the need often arises to align an object model with image or 3-D data using hypothesized correspondences between model and data features [13, 28]. When only a monocular image is available, the alignment is by necessity done from 3-D model features to 2-D image features. A binocular image pair, however, opens the possibility of performing the alignment in 3-D [20], if the binocular system can be calibrated with high enough accuracy.

In this section we propose an intermediate approach which combines the strengths of both 2-D and 3-D methods, namely model alignment based on affine nearness. This approach provides more information and hence better disambiguation than monocular data, while avoiding the need for accurate camera calibration.

We assume that the model consists of a list of 3-D points, and that there exists a hypothesized correspondence between each model point (X_i, Y_i, Z_i) and a point on the affine nearness surface $(\hat{x}, \hat{y}, \hat{\rho})$ which has been estimated by RDC. In order to align the object we need to estimate (i) the pose (R, t) of the object relative to the cyclopean coordinate system, and (ii) the parameters (d, L, f) needed for resolving the relief ambiguity (Section 4).

Simple equation counting tells us that three point correspondences may suffice, but we shall generally use more points and perform a least squares estimation. More precisely, we wish to minimize the goal function

$$\delta^2(R, \mathbf{t}, d, L, f)) = \sum \left(\frac{1}{\sigma_x^2}(\hat{x}_i - x_i)^2 + \frac{1}{\sigma_y^2}(\hat{y}_i - y_i)^2 + \frac{1}{\sigma_\rho^2}(\hat{\rho}_i - \rho_i)^2 \right),$$
(10)

where $(\hat{x}_i, \hat{y}_i, \hat{\rho}_i)$ are the estimated data features, $(\sigma_x, \sigma_y, \sigma_\rho)$ their associated covariances, and

$$x_i = f\frac{X_i}{Z_i + d}, \qquad y_i = f\frac{Y_i}{Z_i + d}, \qquad \rho_i = \frac{fL}{d}\frac{Z_i}{Z_i + d}.$$

(X_i, Y_i, Z_i) are the coordinates of a model point rotated to be parallel to the cyclopean system and translated such that the model point $(0, 0, 0)$ coincides with the fixation point. These coordinates are related to the given model points by a rotation R and a translation $\mathbf{t}$.

To simplify this minimization problem, we use an iterative procedure in which we in each iteration first minimize δ^2 with respect to the viewing parameters (d, L, f) only, and then use these estimated parameters to reconstruct the 3D points and to perform a 3D-to-3D pose estimation. To obtain an initial pose estimate, we use a simple monocular method based on a scaled orthography approximation [5].

The computations performed in each iteration are relatively simple. For the pose estimation the optimal estimate can be computed by singular value decomposition of a 3×3 matrix [2]. For the viewing parameter estimation it is easily shown (by differentiating δ^2 with respect to L and f) that for a given estimate of d, the optimal estimates of f and L are given by

$$f = \frac{\frac{1}{\sigma_x^2}\sum\frac{\hat{x}_i X_i}{Z_i + d} + \frac{1}{\sigma_y^2}\sum\frac{\hat{y}_i Y_i}{Z_i + d}}{\frac{1}{\sigma_x^2}\sum\frac{X_i^2}{(Z_i + d)^2} + \frac{1}{\sigma_y^2}\sum\frac{Y_i^2}{(Z_i + d)^2}},$$
(11)

and

$$L = \frac{d}{f}\frac{\sum\frac{\hat{\rho}_i Z_i}{Z_i + d}}{\sum\frac{Z_i^2}{(Z_i + d)^2}}.$$
(12)

Hence, the problem of estimating the viewing parameters for a given pose estimate can be reduced to a 1-D minimization problem in d, which can be solved numerically with a very limited computational effort.

We have found that the scheme typically converges in 2–5 iterations, after which the estimates change very slowly. Numerical values obtained with the data set corresponding to the second viewing geometry and $n = 10$, $\sigma = 1$ in Table 1 are shown in Table 2.

Since this alignment method produces explicit estimates of the parameters (d, L, f) needed to resolve the relief ambiguity, it is of course possible to apply

Iteration	d	L	f	δ^2
Initial	100	10.59	2.21	$3.8 \cdot 10^{-3}$
1	55.9	6.02	1.19	$3.0 \cdot 10^{-4}$
2	51.8	5.65	1.06	$1.2 \cdot 10^{-4}$
3	49.4	5.45	1.00	$8.4 \cdot 10^{-5}$
5	47.5	5.36	0.95	$7.0 \cdot 10^{-5}$
10	47.1	5.43	0.94	$6.4 \cdot 10^{-5}$
50	47.2	5.48	0.93	$6.2 \cdot 10^{-5}$
True	49.8	5.44	1.00	

Table 2. Estimation of viewing parameters by first applying RDC to a noisy disparity field and then aligning five model points. The cyclovergence is 5°, and the remaining true parameter values are shown in the bottom row. The other rows show the estimates after each iteration.

the resulting transformation to the entire disparity field. Figure 2 shows a stereo image pair of a calibration cube, in which the intersections of the white rulings were matched to produce a sparse disparity field. A model containing the seven visible points of the top corner of the cube was used in the procedure described above, and the estimated parameters were then used to reconstruct the entire cube. The result is shown in Figure 3.

Fig. 2. Stereo pair showing a calibration cube (arranged for cross-eyed fusion).

7 Future Work

Future work includes integration of RDC with the stereo matching stage; the matching and the construction of the parametric representation $\hat{v}(x,y)$ of the vertical disparity field should be performed in parallel, so that early estimates of $\hat{v}(x,y)$ can be used to constrain the epipolar geometry which reduces the matching ambiguity and hence allows more matches to be found, which in turn can be used to improve the estimate $\hat{v}(x,y)$, and so on.

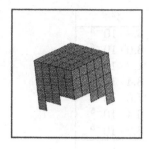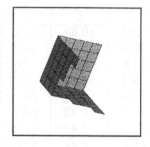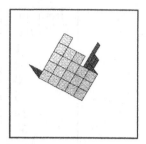

Fig. 3. Three different views of a cube reconstructed from affine nearness by aligning seven model points around the central corner of the cube.

Moreover, in this paper we have not explored the "R" in RDC, i.e., we have always used the vertical disparity field in the entire image as a reference. There are, however, indications [26, 25, 1] that the human visual system may use more localized representations, which would also make sense from a computational point of view.

References

1. W. Adams, J.P. Frisby, D. Buckley, H. Grant, J. Gårding, S.D. Hippisley-Cox, and J. Porrill, "Pooling of vertical disparities by the human visual system", *Perception*, 1996. (To appear).
2. K.S. Arun, T.S. Huang, and S.D. Blostein, "Least-squares fitting of two 3-D point sets", *IEEE Trans. Pattern Anal. and Machine Intell.*, vol. 9, pp. 698–700, 1987.
3. J.R. Bergen, P. Anandan, K.J. Hanna, and R. Hingorani, "Hierarchical model-based motion estimation", in *Proc. 2nd European Conf. on Computer Vision*, vol. 508 of *Lecture Notes in Computer Science*, pp. 237–252, Springer Verlag, Berlin, 1992.
4. S. Carlsson and J.-O. Eklundh, "Object detection using model based prediction and motion parallax", in *Proc. 1st European Conf. on Computer Vision* (O. Faugeras, ed.), vol. 427 of *Lecture Notes in Computer Science*, pp. 297–306, Springer Verlag, Berlin, Apr. 1990. (Antibes, France).
5. D.F. DeMenthon and L.S. Davis, "Model-based object pose in 25 lines of code", in *Proc. 2nd European Conf. on Computer Vision* (G. Sandini, ed.), vol. 588 of *Lecture Notes in Computer Science*, pp. 335–343, Springer-Verlag, May 1992.
6. O. Faugeras, "What can be seen in three dimensions with an uncalibrated stereo rig?", in *Proc. 2nd European Conf. on Computer Vision* (G. Sandini, ed.), vol. 588 of *Lecture Notes in Computer Science*, pp. 563–578, Springer-Verlag, May 1992.
7. J.P. Frisby, H. Grant, D. Buckley, J. Gårding, J.M. Horsman, S.D. Hippisley-Cox, and J. Porrill, "The effects of scaling vertical disparities on the perceived amplitudes of three-dimensional ridges", May 1995. (Submitted for publication).
8. J. Gårding, J. Porrill, J.P. Frisby, and J.E.W. Mayhew, "Uncalibrated relief reconstruction and model alignment from binocular disparities", Tech. Rep. ISRN

438

KTH/NA/P--96/02--SE, Dept. of Numerical Analysis and Computing Science, KTH (Royal Institute of Technology), Jan. 1996.

9. J. Gårding, J. Porrill, J.E.W. Mayhew, and J.P. Frisby, "Stereopsis, vertical disparity and relief transformations", *Vision Research*, vol. 35, pp. 703–722, Mar. 1995.

10. W.E.L. Grimson, A.L. Ratan, P.A. O'Donnell, and G. Klanderman, "An active visual attention system to play "Where's Waldo?"", in *Proc. Workshop on Visual Behaviors*, (Seattle, Washington), pp. 85–90, IEEE Computer Society Press, June 1994.

11. R. Hartley, R. Gupta, and T. Chang, "Stereo from uncalibrated cameras", in *Proc. IEEE Comp. Soc. Conf. on Computer Vision and Pattern Recognition*, (Champaign, Illinois), pp. 761–764, June 1992.

12. H.L.F. von Helmholtz, *Treatise on Physiological Optics*, vol. 3. (trans. J.P.C Southall, Dover, New York 1962), 1910.

13. D.P. Huttenlocher and S. Ullman, "Recognizing solid objects by alignment with an image", *Int. J. of Computer Vision*, vol. 5, pp. 195–212, 1990.

14. E.B. Johnston, "Systematic distortions of shape from stereopsis", *Vision Research*, vol. 31, pp. 1351–1360, 1991.

15. J.J. Koenderink and A.J. van Doorn, "Geometry of binocular vision and a model for stereopsis", *Biological Cybernetics*, vol. 21, pp. 29–35, 1976.

16. J.J. Koenderink and A.J. van Doorn, "Affine structure from motion", *J. of the Optical Society of America A*, vol. 8, pp. 377–385, 1991.

17. R. Kumar, P. Anandan, and K. Hanna, "Shape recovery from multiple views: a parallax based approach", in *Proc. Image Understanding Workshop*, (Monterey, CA), 1994.

18. M. Li and D. Betsis, "Head-eye calibration", in *Proc. 5th International Conference on Computer Vision*, (Cambridge, MA), pp. 40–45, June 1995.

19. H.C. Longuet-Higgins and K. Prazdny, "The interpretation of a moving retinal image", *Proc. Royal Society London B*, vol. 208, pp. 385–397, 1980.

20. J.E.W. Mayhew and J.P. Frisby, eds., *3D Model Recognition from Stereoscopic Cues*. MIT Press, 1991.

21. K.N. Ogle, *Researches in Binocular Vision*. Saunders, Philadelphia, 1950.

22. T.J. Olson, "Stereopsis for verging systems", in *Proc. IEEE Comp. Soc. Conf. on Computer Vision and Pattern Recognition*, (New York), pp. 55–66, 1993.

23. T.J. Olson and D.J. Coombs, "Real-time vergence control for binocular robots", *Int. J. of Computer Vision*, vol. 7, no. 1, pp. 67–89, 1991.

24. K. Pahlavan, T. Uhlin, and J.-O. Eklundh, "Dynamic fixation", in *Proc. 4th Int. Conf. on Computer Vision*, (Berlin, Germany), pp. 412–419, May 1993.

25. B.J. Rogers and J.J. Koenderink, "Monocular aniseikona: a motion parallax analogue of the disparity-induced effect", *Nature*, vol. 322, pp. 62–63, 1986.

26. S.P. Stenton, J.P. Frisby, and J.E.W. Mayhew, "Vertical disparity pooling and the induced effect", *Nature*, vol. 309, pp. 622–623, June 1984.

27. J.S. Tittle, J.T. Todd, V.J. Perotti, and J.F. Norman, "Systematic distortion of perceived three-dimensional structure from motion and binocular stereopsis", *J. of Experimental Psychology: Human Perception and Performance*, vol. 21, no. 3, pp. 663–678, 1995.

28. S. Ullman and R. Basri, "Recognition by linear combinations of models", *IEEE Trans. Pattern Anal. and Machine Intell.*, vol. 13, pp. 992–1006, 1991.

Dense Depth Map Reconstruction : A Minimization and Regularization Approach which Preserves Discontinuities

Luc ROBERT and Rachid DERICHE

INRIA. B.P. 93. 06902 Sophia-Antipolis. FRANCE.
Phone : (+33) 93 65 78 32 - Fax: (+33) 93 65 78 45
E-mail: {lucr,der}@sophia.inria.fr

Abstract. We present a variational approach to dense stereo reconstruction which combines powerful tools such as regularization and multi-scale processing to estimate directly depth from a number of stereo images, while preserving depth discontinuities. The problem is set as a regularization and minimization of a nonquadratic functional. The *Tikhonov* quadratic regularization term usually used to recover smooth solution is replaced by a function of the gradient depth specifically derived to allow depth discontinuities formation in the solution. Conditions to be fulfilled by this specific regularizing term to preserve discontinuities are also presented. To solve this problem in the discrete case, a PDE-based explicit scheme for moving iteratively towards the solution has been developed. This approach presents the additional advantages of not introducing any intermediate representation such as disparity or rectified images: depth is computed directly from the grey-level images and we can also deal with any number (greater than two) of cameras. Promising experimental results illustrate the capabilities of this approach.

1 Introduction

Over the years numerous algorithms for passive stereo have been proposed, which use different strategies:

Feature-based: Those algorithms establish correspondences between features extracted from the images, like edge pixels [20, 23, 18], line segments [17, 4] or curves [8, 25] for instance. Their main advantage is to yield accurate information and to manipulate reasonably small amounts of data, thus gaining in time and space complexity. Their main drawback is the sparseness of the recovered depth information.

Area-based: In these approaches, dense depth maps are provided by correlating the grey levels of image patches in the views being considered, assuming that they present some similarity [19, 13]. These methods are well adapted for relatively textured areas; however, they generally assume that the observed scene is locally fronto-parallel, which causes problems for slanted surfaces and in particular near the occluding contours of the objects. Lastly, the matching process does not take into account the edge information.

Energy-based: A third kind of approach which does not suffer any of the inconvenients presented above, consists of expressing the correspondence problem as a minimization and regularization one [6, 30]. An iterative solution of the discrete version of the associated Euler-Lagrange equation is then used in order to estimate depth.

The method which we present in this paper follows the third strategy with the important following issues:

- It computes depth directly from the grey-level images intensities. No intermediate process such as rectification [5] or disparity estimation is used. The system of cameras is supposed to be calibrated, and the depth information is directly issued as a depth function $m \rightarrow Z(m)$ of the image point.

- The method addresses the problem of accurately determining depth near discontinuities. It is well known that using the classical *Tikhonov* regularization approach [28] by considering a quadratic regularizing term in the energy function, leads to smoothing the depth image across the discontinuities, yielding a destruction of these important characteristics in the resulting depth image. We address this important problem by replacing the energy quadratic regularizing term by a function specially designed in order to allow the minimization process to preserve the original discontinuities in the depth map. It is shown that in this case, the minimization process involves an isotropic smoothing step in the homogeneous regions (i.e with small depth gradient), and an anisotropic smoothing step in the inhomogeneous regions (i.e high depth gradient).

- In order to speed up convergence and avoid possible local minima, a multi-scale approach is also used.

2 Formalism of the matching process

2.1 Notations for one camera

We assume that the imaging system follows the pinhole model. The projection matrix $\tilde{\mathbf{P}}$ of a camera with respect to a reference frame $\mathcal{R}_w$ is computed during a calibration phase. It allows finding $m = [u, v]^T$, projection onto the retina of the point $M = [X, Y, Z]^T$ expressed in world coordinates (s is a scale factor):

$$[su, sv, s]^T = \tilde{\mathbf{P}}[X, Y, Z, 1]^T$$

The 3×4 matrix $\tilde{\mathbf{P}}$ can be decomposed as follows:

$$\tilde{\mathbf{P}} = [\mathbf{J} \ \mathbf{0_3}] \begin{bmatrix} \mathbf{R} & \mathbf{t} \\ \mathbf{0_3^T} & 1 \end{bmatrix} = [\mathbf{J} \ \mathbf{0_3}] \mathbf{D}$$

The 4×4 matrix $\mathbf{D}$ involving $\mathbf{R}$ and $\mathbf{t}$ changes world coordinates into camera coordinates (see figure 1). It represents the *extrinsic* information about the camera. $\mathbf{0_3}$ is the null 3-vector. The other matrix contains the *intrinsic* information [29]: the 3×3 matrix **J** changes camera coordinates into pixel units. It is well known that in the generic case,

$$\mathbf{J} = \begin{bmatrix} \alpha_u & -\alpha_u \cot\theta & u_0 \\ 0 & \alpha_v/\sin\theta & v_0 \\ 0 & 0 & 1 \end{bmatrix}$$

where θ is the angle between pixel rows and columns.

Thus, the camera coordinates of a point M in the 3D space can be derived from its projection m and its depth:

$$[X, Y, Z]^T = Z\mathbf{J}^{-1}[u, v, 1]^T$$

All these matrices can be computed with good accuracy with the method described in [24].

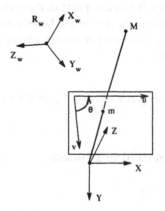

Fig. 1. The pinhole model and the different frames

2.2 Correspondence

In the remainder of this article, we consider a calibrated stereo rig. All the quantities introduced in the previous paragraph and related to one camera are assigned the index of the considered camera.

If m_i, m_j are the two-dimensional projections of a three-dimensional point M on retinae i, j, we can easily express m_j as a function $f_{i \to j}$ of (m_i, Z_i). Indeed,

$$
\begin{bmatrix} su_j \\ sv_j \\ s \end{bmatrix} = Z_i \left[\mathbf{J}_j \; \mathbf{0}_3 \right] \mathbf{D}_j \mathbf{D}_i^{-1} \begin{bmatrix} \mathbf{J}_i^{-1} \begin{bmatrix} u_i \\ v_i \\ 1 \end{bmatrix} \\ 1 \end{bmatrix}
$$

Thus, finding a correspondent for a point of image i is completely equivalent to finding depth Z_i at that point.

3 The energy function

Let us assume for a while that the world is made of Lambertian objects, i.e. of objects that look equally bright from all viewing directions. Finding a correspondent in image 2 for each point of image 1 turns into finding a Z function that minimizes:

$$
M_{12}(Z) = \iint \left\| \mathbf{I}_1(m_1) - \mathbf{I}_2(f_1(m_1, Z(m_1))) \right\|^2 dm_1
$$

($\mathbf{I}_i$ is the intensity in image i). This is of course not sufficient, since it may first lead to chaotic solutions, in which each point of image 1 finds its correspondent in image 2 independently from its neighbors. We need to add a constraint on the shape of the depth function. As a consequence, the matching problem is expressed as a constrained minimization one, where the minimized functional is the following:

$$
E(Z) = M_{12}(Z) + \lambda S(Z)
$$

The first term holds for similarity of the image intensities at corresponding points. In practice, the use of additional attributes like intensity gradient or rgb grey-level images helps establishing the right correspondences by reducing ambiguity. More generally, we have:

$$
M_{12}(Z) = \iint \sum_k \left\| \mathbf{F}_1^k(m_1) - \mathbf{F}_2^k(f_1(m_1, Z(m_1))) \right\|^2 dm_1
$$

where $\mathbf{F}_i^k$ is a field of attributes, scalar – e.g. intensity – or vectorial – e.g. intensity gradient – extracted from image i.

If we have more than two views, we can very easily take into account the information of all the images, by defining the similarity term of the functional as

$$
M(Z) = \sum_j M_{1j}(Z)
$$

where j varies from 2 to the number of images.

Please note that the simplicity of this expression emerges from the fact that we are searching for a depth function Z defined with respect to one reference image, which is matched with any other image. With classical techniques based on a disparity representation, things would not be as simple. Indeed, to define disparity, one needs two images.

In the last section of the article, we will show the importance of using more than two views in image-based surface reconstruction.

The second term stands for the constraint to be applied on the depth map. A classical constraint is the smoothness assumption on the resulting depth map. That is the case for the well-known *Tikhonov* regularization term [28]:

$$S(Z) = \int\int |\nabla_{m_1} Z|^2 dm_1$$

This term leads to a solution where discontinuities of the depth function are smoothed. This is not really desirable if one wants to preserve these discontinuities and recover the original scene as accurately as possible. The next section is devoted to presenting an original approach which tackles in an efficient way this important problem.

4 Regularizing the solution and preserving discontinuities

In order to preserve the discontinuities while regularizing the solution, a natural way to proceed is to forbid regularizing and smoothing across such discontinuities. One way of taking into account these technical remarks is by looking for a function $\Phi(.)$ such that the following regularization term:

$$S(Z) = \int\int \Phi(|\nabla_{m_1} Z|) dm_1$$

preserves those discontinuities (For instance, a quadratic function as the one used in the *Tikhonov* case [28], does clearly not correspond to such type of functions).

In this section, we summarize our variational approach to 3D recovery from stereo images. This approach is inspired from the approaches developed for image restoration purpose in [26, 21, 31, 7, 9, 10, 3]. A detailed review of all these approaches can be found in [12].

The key idea to deal with such a problem is first to consider the functional to be minimized, written as follows:

$$E(Z) = \int\int F(u, v, Z, \Phi(|\nabla_m Z|)) du dv$$

A necessary condition for it to be extremal is the derived Euler-Lagrange differential equation:

$$F_Z - \frac{\partial}{\partial u} F_{Z_u} - \frac{\partial}{\partial v} F_{Z_v} = 0$$

This yields the following equation:

$$\sum_k \left(\mathbf{F}_1^k(m) - \mathbf{F}_2^k(f_1(m, Z(m))) \right) . \frac{\partial \mathbf{F}_2}{\partial Z} - \frac{\lambda}{2} \left(div(\frac{\phi'(|\nabla Z|)}{|\nabla Z|} \nabla Z) \right) = 0 \quad (1)$$

Where the term *div* denotes the divergence operator. Developing and simplifying, the term on the left can be rewritten as

$$\sum_k \left(\mathbf{F}_1^k(m) - \mathbf{F}_2^k(f_1(m, Z(m))) \right) . \frac{\partial \mathbf{F}_2}{\partial Z} - \frac{\lambda}{2} \left(\frac{\Phi'(|\nabla_m Z|)}{|\nabla_m Z|} Z_{\xi\xi} + \Phi''(|\nabla_m Z|) Z_{\eta\eta} \right)$$

$$(2)$$

where Φ' and Φ'' represent respectively the first and second derivatives of $\Phi(s)$ with respect to the parameter s. $Z_{\eta\eta}$ represents the second order directional derivatives of $Z(m)$ in the direction of the gradient $\eta = \frac{\nabla Z}{|\nabla Z|}$, and $Z_{\xi\xi}$ is the second order directional derivatives of $Z(m)$ in the direction ξ orthogonal to the gradient. Boundary conditions have also to be considered in order to solve this equation.

In order to regularize the solution and preserve discontinuities, one would like to smooth isotropically the solution inside homogeneous regions and preserve the discontinuities in the inhomogeneous regions. Assuming that the function $\Phi''(.)$ exists, the condition on smoothing in an isotropic way inside homogeneous regions can be achieved by imposing the following conditions on the $\Phi(.)$ function:

$$\lim_{|\nabla_m Z| \to 0} \frac{\Phi'(|\nabla_m Z|)}{|\nabla_m Z|} = \lim_{|\nabla_m Z| \to 0} \Phi''(|\nabla_m Z|) = \Phi''(0) > 0 \qquad (3)$$

Therefore, at the points where the depth gradient is small, Z is solution of:

$$\sum_k \left(\mathbf{F}_1^k(m) - \mathbf{F}_2^k(f_1(m, Z(m))) \right) \cdot \frac{\partial \mathbf{F}_2}{\partial Z} - \frac{\lambda}{2} \Phi''(0)(Z_{\xi\xi} + Z_{\eta\eta}) = 0$$

This process corresponds to the case where the function $\Phi(s)$ is quadratic [28]. Note that the coefficients are required to be positive, otherwise the regularization part will act as an inverse heat equation notably known as an instable process.

In order to preserve the discontinuities near inhomogeneous regions presenting a strong depth gradient, one would like to smooth along the isophote and not across them. This leads to stopping the diffusion in the gradient direction, i.e setting the weight $\Phi''(|\nabla_m Z|)$ to zero, while keeping a stable diffusion along the tangential direction to the isophote, i.e setting the weight $\frac{\Phi'(|\nabla_m Z|)}{|\nabla_m Z|}$ to some positive constant:

$$\lim_{|\nabla_m Z| \to \infty} \Phi''(|\nabla_m Z|) = 0 \; ; \quad \lim_{|\nabla_m Z| \to \infty} \frac{\Phi'(|\nabla_m Z|)}{|\nabla_m Z|} = \beta > 0 \qquad (4)$$

Therefore, at the points where the depth gradient is strong, Z will be the solution of the following equation:

$$\sum_k \left(\mathbf{F}_1^k(m) - \mathbf{F}_2^k(f_1(m, Z(m))) \right) \cdot \frac{\partial \mathbf{F}_2}{\partial Z} - \frac{\lambda}{2} \beta Z_{\xi\xi} = 0$$

which yields Z as a regularized solution in the ξ direction. Note that the positiveness of the β coefficient is also required to generate a stable smoothing process in the ξ direction.

Unfortunately, the two conditions of (4) cannot be satisfied simultaneously by a function $\Phi(|\nabla_m Z|)$. However, the following conditions can be imposed in order to decrease the effects of the diffusion along the gradient more rapidly than those associated with the diffusion along the isophotes:

$$\lim_{|\nabla_m Z| \to \infty} \Phi''(|\nabla_m Z|) = \lim_{|\nabla_m Z| \to \infty} \frac{\Phi'(|\nabla_m Z|)}{|\nabla_m Z|} = \lim_{|\nabla_m Z| \to \infty} \frac{\Phi''(|\nabla_m Z|)}{\frac{\Phi'(|\nabla_m Z|)}{|\nabla_m Z|}} = 0$$

$$(5)$$

The conditions given by Equations (3) and (5) are those one would like to impose in order to deal with a regularization process while preserving the existing discontinuities. As it has been shown very recently in [3], these conditions are also sufficient to prove that the model is well-posed mathematically, and the existence and uniqueness of a solution is also guaranteed by these conditions.

A certain number of functions have already been proposed in the literature in order to address the problem of discontinuities. Table 4 illustrates the most commonly used functions. One can see easily that only the last three functions fulfill all the conditions mentioned above. The *Tikhonov* function and the *Aubert* function will be used in our experimental section.

Author	$\Phi(s)$	$\Phi'(s)/s$	$\Phi''(s)$
Perona-Malik [22]	$\frac{-k^2}{2}(e^{-(s/k)^2} - 1)$	$e^{-(s/k)^2}$	$(1 - 2(\frac{s}{k})^2)e^{-(\frac{s}{k})^2}$
Perona-Malik [22]	$\frac{k^2}{2}log(1 + (s/k)^2)$	$\frac{1}{1+(\frac{s}{k})^2}$	$\frac{k^2(k^2-s^2)}{(k^2+s^2)^2}$
Geman et Reynolds [14]	$\frac{(s/k)^2}{1+(s/k)^2}$	$\frac{2k^2}{(k^2+s^2)^2}$	$-\frac{2k^2(-k^2+3s^2)}{(k^2+s^2)^3}$
Alvarez [1]	...	$g(s)$	$(1 - h(s))g(s)$
Tikhonov [28]	$s^2/2$	1	1
Green [15]	$logcosh(s/k)$	$\frac{tanh(s/k)}{ks}$	$k^{-2}\left(cosh(\frac{s}{k})\right)^{-2}$
Rudin [26]	s	$\frac{1}{s}$	0
Aubert [10]	$\sqrt{1 + (s/k)^2} - 1$	$\frac{1}{\sqrt{\frac{k^2+s^2}{k^2}}}k^{-2}$	$\frac{|k|}{(k^2+s^2)^{3/2}}$

5 Minimization of the Energy Function

This section presents the numerical scheme developed to solve the Euler-Lagrange equation (1,2) associated to the energy function. A time-dependent approach has been developed to solve this non-linear PDE. We consider the associated evolution equation, or equivalently the gradient descent method. This leads us to consider the following equation:

$$Z_t = \sum_k \left(\mathbf{F}_1^k(m) - \mathbf{F}_2^k(f_1(m, Z(m)))\right)\cdot\frac{\partial \mathbf{F}_2}{\partial Z} - \frac{\lambda}{2}\left(\frac{\Phi'(|\nabla_m Z|)}{|\nabla_m Z|}Z_{\xi\xi} + \Phi''(|\nabla_m Z|)Z_{\eta\eta}\right)$$

The corresponding explicit numerical scheme is then implemented:

$$\begin{cases} Z_{i,j}^{n+1} = Z_{i,j}^n + \Delta t \left(\sum_k(\mathbf{F}_1^k(m) - \mathbf{F}_2^k(f_1(m, Z(m))))\cdot\frac{\partial \mathbf{F}_2}{\partial Z} - \frac{\lambda}{2}(\frac{\Phi'(|\nabla_m Z|)}{|\nabla_m Z|}Z_{\xi\xi} + \Phi''(|\nabla_m Z|)Z_{\eta\eta})\right)_{i,j}^n \\ \oplus \text{ Boundary conditions on the depth} \end{cases} \quad (6)$$

Finally, we apply a Gauss-Seidel relaxation method for moving iteratively towards the solution of this problem [27].

5.1 Discretization Scheme

In the following we present the consistent way to discretize the divergence term that appears in the equation (1).

Denoting by θ the angle that the unit gradient $\nabla Z/(|\nabla Z|)$ makes with the x axis, we have the well known expressions:

$$Z_{\xi\xi} = sin(\theta)^2 Z_{xx} - 2sin(\theta)cos(\theta)Z_{xy} + cos(\theta)^2 Z_{yy}$$
$$Z_{\eta\eta} = cos(\theta)^2 Z_{xx} + 2sin(\theta)cos(\theta)Z_{xy} + sin(\theta)^2 Z_{yy} \tag{7}$$

In order to deal with a consistent discrete approximation of these second directional derivatives, we proceed as follows:

We look for some constant $(\lambda_i)_{i=0..4}$ and $(\lambda'_i)_{i=0..4}$ such that:

$$Z_{\xi\xi i,j} \simeq \begin{bmatrix} \lambda_4 & \lambda_2 & \lambda_3 \\ \lambda_1 & -4\lambda_0 & \lambda_1 \\ \lambda_3 & \lambda_2 & \lambda_4 \end{bmatrix} * Z \text{ and } Z_{\eta\eta i,j} \simeq \begin{bmatrix} \lambda'_4 & \lambda'_2 & \lambda'_3 \\ \lambda'_1 & -4\lambda'_0 & \lambda'_1 \\ \lambda'_3 & \lambda'_2 & \lambda'_4 \end{bmatrix} * Z \tag{8}$$

where the sign $*$ denotes the convolution operation. Using The Taylor expansion of (8) up to the second order and identifying the coefficients with (7) yields the values of the following λ_i and λ'_i parameters we are looking for:

$$\begin{cases} \lambda_1 = 2\lambda_0 - \frac{\delta^x Z}{|\nabla Z|}^2 \\ \lambda_2 = 2\lambda_0 - \frac{\delta^y Z}{|\nabla Z|}^2 \\ \lambda_3 = \frac{1}{2}(-\frac{\delta^y Z}{|\nabla Z|}\frac{\delta^x Z}{|\nabla Z|} + 1 - 2\lambda_0) \\ \lambda_4 = \frac{1}{2}(1 - 2\lambda_0 + \frac{\delta^y Z}{|\nabla Z|}\frac{\delta^x Z}{|\nabla Z|}) \end{cases} \qquad \begin{cases} \lambda'_1 = 2\lambda'_0 - \frac{\delta^y Z}{|\nabla Z|}^2 \\ \lambda'_2 = 2\lambda'_0 - \frac{\delta^x Z}{|\nabla Z|}^2 \\ \lambda'_3 = \frac{1}{2}(\frac{\delta^y Z}{|\nabla Z|}\frac{\delta^x Z}{|\nabla Z|} + 1 - 2\lambda'_0) \\ \lambda'_4 = \frac{1}{2}(1 - 2\lambda'_0 - \frac{\delta^y Z}{|\nabla Z|}\frac{\delta^x Z}{|\nabla Z|}) \end{cases}$$

Then denoting:

$$c_\eta = \Phi''(|\nabla Z|) \quad c_\xi = \frac{\Phi'(|\nabla Z|)}{|\nabla Z|} \tag{9}$$

We obtain the following coefficients $(\Sigma_{Zi})_{i=0..4}$ for the divergence term:

$$\left(div(\frac{\phi'(|\nabla Z|)}{|\nabla Z|}\nabla Z)\right) = (c_\eta Z_{\eta\eta} + c_\xi Z_{\xi\xi})_{i,j} \simeq \begin{bmatrix} \Sigma_{Z4} & \Sigma_{Z2} & \Sigma_{Z3} \\ \Sigma_{Z1} & \Sigma_{Z0} & \Sigma_{Z1} \\ \Sigma_{Z3} & \Sigma_{Z2} & \Sigma_{Z4} \end{bmatrix}$$

Where:

$$\Sigma_{Z0} = -4(c_\xi \lambda_0 + c_\eta \lambda'_0)$$

$$\Sigma_{Z1} = 2(c_\xi \lambda_0 + c_\eta \lambda'_0) - (c_\xi \frac{\delta^x Z}{|\nabla Z|}^2 + c_\eta \frac{\delta^y Z}{|\nabla Z|}^2)$$

$$\Sigma_{Z2} = 2(c_\xi \lambda_0 + c_\eta \lambda'_0) - (c_\xi \frac{\delta^y Z}{|\nabla Z|}^2 + c_\eta \frac{\delta^x Z}{|\nabla Z|}^2) \qquad (10)$$

$$\Sigma_{Z3} = -(c_\xi \lambda_0 + c_\eta \lambda'_0) + \frac{c_\xi}{2}(1 - \frac{\delta^y Z}{|\nabla Z|}\frac{\delta^x Z}{|\nabla Z|}) + \frac{c_\eta}{2}(1 + \frac{\delta^y Z}{|\nabla Z|}\frac{\delta^x Z}{|\nabla Z|})$$

$$\Sigma_{Z4} = -(c_\xi \lambda_0 + c_\eta \lambda'_0) + \frac{c_\xi}{2}(1 + \frac{\delta^y Z}{|\nabla Z|}\frac{\delta^x Z}{|\nabla Z|}) + \frac{c_\eta}{2}(1 - \frac{\delta^y Z}{|\nabla Z|}\frac{\delta^x Z}{|\nabla Z|})$$

Therefore, the divergence term $c_\eta Z_{\eta\eta} + c_\xi Z_{\xi\xi}$ can be implemented as a convolution of the image Z by a 3*3 mask. It is important to note that the coefficients of this convolution mask are not constant. The values of the Σ_{Zi} depend on the gradient $|\nabla Z|$ through the function Φ. Hence, in the regions where the depth is homogeneous, $|\nabla Z|$ is small and the two coefficients c_η and c_ξ are equal to a constant c (this is due to the constraints imposed on the function $\Phi(.)$) and if one note $\lambda = \lambda_0 + \lambda'_0$, the coefficients for the divergence term simplify to:

$$\begin{cases} \Sigma_{w0} = -4c\lambda \qquad \Sigma_{w1} = c(2\lambda - 1) \quad \Sigma_{w2} = c(2\lambda - 1) \\ \\ \Sigma_{w3} = c(1 - \lambda) \quad \Sigma_{w4} = c(1 - \lambda) \end{cases}$$

As expected, these coefficients depend on the λ parameter. Here are some configurations for the divergence filter: These filters look like the classical Laplacian operator.

	$\lambda = \frac{1}{4}$	$\lambda = \frac{1}{2}$	$\lambda = \frac{3}{4}$	$\lambda = 1$
filter	$\begin{matrix} \frac{3c}{4} & -\frac{c}{2} & \frac{3c}{4} \\ -\frac{c}{2} & -c & -\frac{c}{2} \\ \frac{3c}{4} & -\frac{c}{2} & \frac{3c}{4} \end{matrix}$	$\begin{matrix} \frac{c}{2} & 0 & \frac{c}{2} \\ 0 & -2c & 0 \\ \frac{c}{2} & 0 & \frac{c}{2} \end{matrix}$	$\begin{matrix} \frac{c}{4} & \frac{c}{2} & \frac{c}{4} \\ \frac{c}{2} & -3c & \frac{c}{2} \\ \frac{c}{2} & \frac{c}{2} & \frac{c}{4} \end{matrix}$	$\begin{matrix} 0 & c & 0 \\ c & -4c & c \\ 0 & c & 0 \end{matrix}$

Table 1. Filters obtained for different values of λ

This is what we expected in the design of the regularization scheme: In homogeneous depth regions, the regularization must allow an isotropic smoothing. In inhomogeneous regions the configuration of the mask will not look like a Laplacian operator but as a directional second derivative in the direction normal to the gradient. This will perform an anisotropic diffusion i.e a diffusion only along the ISO-depth and not across the depth discontinuities.

Another directional numerical scheme has also been developed. It uses only 6 neighbors and not all the 8 neighbors as the one presented here and inspired from the discretization scheme proposed by L.Alvarez in [2] for multi-scale image analysis purpose. For more details about these two numerical schemes, one can refer to [11] and [16] where the two numerical schemes are presented in details.

6 Further Implementation Details

The following two additional points have to be mentioned regarding the implementation part:

First, it appears that the only use of the intensity field is not sufficient for dealing with real images, where the Lambertian assumption is almost always violated. The use of intensity gradient information generally helps a great deal in getting close to the correct solution. We have studied the evolution of the minimization process using different additional fields of attributes such as: intensity gradient along the epipolar line, intensity gradient vector, Laplacian of the intensity. It turns out that the intensity gradient vector field gives the best results.

Secondly, a multi scale approach speeds up convergence and helps avoiding local minima. The initial depth function at the coarsest scale is assigned a constant value. Then, the solution at each scale is used as initial value for the next, finer scale. In practice, we start with 32×32 images and double the image size at each level of the pyramid until we reach the final resolution.

7 Experimental Results

This section illustrates the experimental results that we have obtained on triplet of synthetic views of a pyramid. The baseline of cameras 1 and 2 (resp. 1 and 3) is aligned with the X (resp. Y) axis of the pyramid, so the epipolar lines have the directions represented in Figure 2 (top). Image intensity is very different from one stage of the pyramid to the next, and at each stage it is almost constant. The small variations due to lighting simulation cannot be perceived when looking at the images.

The three bottom lines of Figure 2 represent depth maps $((u,v,Z(u,v))$ (left column) and dense 3D reconstructions (center and right columns) obtained, from top to bottom,

1. with cameras 1 and 2, and smoothing using the *Tikhonov* regurarization term (see how strong the smoothing is across the discontinuities)

2. with cameras 1, 2 and 3, and smoothing using the *Tikhonov* regurarization term (see how the use of a third camera improves the results but we still have the smoothing across the discontinuities)

3. with cameras 1, 2 and 3, and smoothing using the *Aubert* regurarization term (see how the use of this function improves the preservation of the discontinuities.)

The top two lines show the importance of using more than two images. Indeed, in the binocular case, there is almost no matching information along the epipolar lines, except close to the vertical edges. Thus, the shape is recovered only at these places. Elsewhere, since the image information is very weak, the depth-recovery process almost completely relies on the regularization term. In the trinocular case, depth is recovered everywhere on the pyramid. However, the smoothing term is too important for the small intensity variations within each pyramid level to be of any use to the algorithm. In the third case, we see how the algorithm can recover the shape of the pyramid when it also takes into account depth discontinuities.

8 Conclusion

A variational approach to dense stereo reconstruction has been presented. It allows to estimate directly the depth from a number of stereo images, while preserving depth discontinuities. The problem has been set as a regularization and minimization of a non quadratic functional. The *Tikhonov* regularization term usually used to recover smooth solution has been replaced by a function of the gradient depth specifically derived to allow depth discontinuities formation in the solution. The conditions that must be be fulfilled by this specific regularizing term to preserve discontinuities have been presented and a PDE based explicit scheme for moving iteratively towards the solution has been developed. Promising experimental results have been obtained. These results clearly illustrated the capabilities of this promising approach: direct depth recovery, depth discontinuities preservation and possibility to use more than 2 cameras.

References

1. L. Alvarez, P.L. Lions, and J.M.. Morel. Image selective smoothing and edge detection by nonlinear diffusion (ii). *SIAM J, Numer Anal*, 29(3):845–866, 1992.
2. L. Alvarez and F. Morales. Affine multiscale analysis of corners. *Prepublicacion, Ref: 9402, Universidad de Las Palmas de Gran Canaria*, 1994.
3. G. Aubert and L. Lazaroaia. A variational method in image recovery. TR 423, URA au CNRS no 168,laboratoire J-A Dieudonne, Université de Nice Sophia-Antipolis-CNRS, June 1995.
4. N. Ayache and B. Faverjon. Efficient Registration Of Stereo Images by Matching Graph Descriptions of Edge Segments. *The Int'l Journal of Comp. Vision*, 1(2):107–131, April 1987.
5. N. Ayache and C. Hansen. Rectification of Images for Binocular and Trinocular Stereovision. In *Proc. Int'l Conf. on Pattern Recognition*, pages –, October 1988. 9th, Rome, Italy.
6. S.T. Barnard. Stochastic Stereo Matching Over Scale. *The Int'l Journal of Comp. Vision*, 3(1):17–32, May 1989.
7. L. Blanc-Ferraud, P. Charbonnier, G. Aubert, and M. Barlaud. Nonlinear image processing : Modeling and fast algorithm for regularization with edge detection. TR 1, Laboratoire I3S, URA 1376 du CNRS, Université de Nice Sophia-Antipolis-CNRS, Janvier 1995.
8. A.T. Brint and M. Brady. Stereo Matching of Curves. In *IARP – 1st Int'l Workshop on Multi-sensor Fusion and Environment Modelling*, October 1989.
9. P. Charbonnier. *Reconstruction d'image : Régularisation avec prise en compte des discontinuités*. Thèse, Université of Nice Sophia-Antipolis, September 1994.
10. P. Charbonnier, G. Aubert, M. Blanc-Ferraud, and M. Barlaud. Two-Deterministic Half-Quadratic Regularizatiob Algorithms for Computed Imaging. In *Proc. IEEE Int. Conf. on Image Proc. ICIP*, volume II of III, pages 168–172, Austin(Texas), November 13-16 1994.
11. P. Delacourt. Applications des EDP en Analyse Multi-Echelle et en Traitement d'Images. Rapport de stage de DEA. proposé et encadré par R. Deriche, INRIA, Université Nice-Sophia Antipolis, France, Juin 1995.
12. R. Deriche and O. Faugeras. Les EDP en traitement des images et vision par ordinateur. Technical Report 2697, INRIA, November 1995.
13. P. Fua. Combining Stereo and Monocular Information to Compute Dense Depth Maps that Preserve Depth Discontinuities. In *Proc. Int'l Joint Conf. on Artificial Intelligence*, pages –, Sydney, Australia, August 1991.

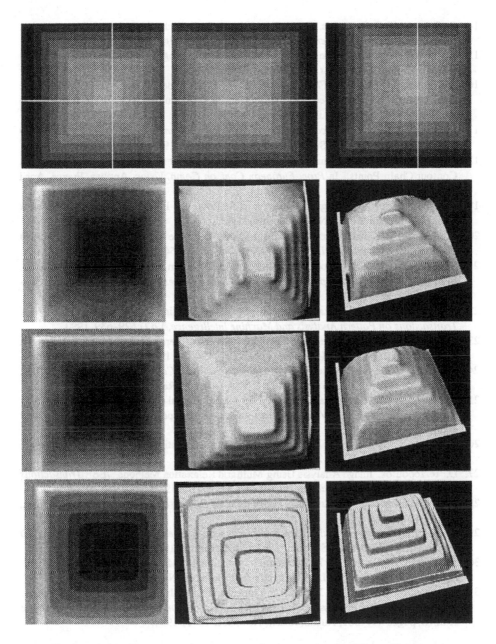

Fig. 2. Triple of images, depth maps and reconstructed surfaces (see text).

14. S. Geman and G. Reynolds. Constrained Restauration and the Recovery of discontinuities. *IEEE Transactions on Pattern Analysis and Machine Intelligence*, 14, 1992.
15. P.J. Green. Bayesian Reconstruction From Emission Tomography Data Using A Modified EM Algorithm. *IEEE Trans. Med. Imaging*, MI-9, 1990.
16. P Kornprobst. Calcul de Flot Optique avec Préservation des Discontinuités. Rapport de stage de DEA. Proposé et encadré par R. Deriche et G. Aubert, INRIA, Université Nice-Sophia Antipolis, France, Juin 1995.
17. F. Lustman. *Vision Stéréoscopique Et Perception Du Mouvement En Vision Artificielle*. PhD thesis, Université de Paris-Sud Centre d'Orsay, December 1988.
18. A. Meygret, M. Thonnat, and M. Berthod. A Pyramidal Stereovision Algorithm Based on Contour Chain Points. In *Proc. European Conf. on Comp. Vision*, Antibes, France, April 1990. Springer Verlag.
19. H.K. Nishihara and T. Poggio. Stereo Vision for Robotics. In *Int'l Symposium on Robotics Research*, Bretton Woods, New Hampshire, 1983.
20. Y. Ohta and T. Kanade. Stereo by Intra- and Inter-Scanline Search. *IEEE Transactions on Pattern Analysis and Machine Intelligence*, 7, No 2:139–154, 1985.
21. S. Osher and L. Rudin. Shocks and Other Nonlinear Filtering Applied to Image Processing. In *Proc SPIE Vol 1567 : Applications of Digital Image Processing XIV*, pages 414–431, San-Diego, 1991.
22. P. Perona and J. Malik. Scale-space and edge detection using anisotropic diffusion. *IEEE Transactions on Pattern Analysis and Machine Intelligence*, 12(7):629–639, 1990.
23. S.B. Pollard, J.E.W. Mayhew, and J.P. Frisby. PMF: a Stereo Correspondance Algorithm using a Disparity Gradient Limit. *Perception*, 14:449–470, 1985.
24. L Robert. Camera calibration without feature extraction. *Computer Vision, Graphics, and Image Processing*, 1995. to appear, also INRIA Technical Report 2204.
25. L. Robert and O.D. Faugeras. Curve-Based Stereo: Figural Continuity And Curvature. In *Proc. Int'l Conf. on Comp. Vision and Pattern Recognition*, pages 57–62, Maui, Hawai, June 1991. IEEE.
26. L. Rudin, S. Osher, and E. Fatemi. Nonlinear total variation based noise removal algorithms. *Physica D*, 60:259–268, 1992.
27. D. Terzopoulos. Image Analysis Using Multigrid Relaxation Methods. *IEEE Transactions on Pattern Analysis and Machine Intelligence*, 8(2):129–139, March 1986.
28. A.N. Tikhonov and V.Y. Arsenin. In *Solutions of Ill-Posed Problems*. Wiley, New-York, 1977.
29. G. Toscani. *Systèmes de Calibration et Perception du Mouvement en Vision Artificielle*. PhD thesis, Université de Paris-Sud Centre d'Orsay, December 1987.
30. N. Yokoya. Stereo Surface Reconstruction by Multiscale-Multistage Regularization. Technical Report TR-90-45, ETL Electrotechnical Laboratory, November 1990.
31. Y.L. You, M. Kaveh, W.Y. Xu, and A. Tannenbaum. Analysis and Design of Anisotropic Diffusion for Image Processing. In *Proc. IEEE Int. Conf. on Image Proc. ICIP-94*, volume II of III, pages 497–501, Austin(Texas), November 13-16 1994.

Stereo Without Search

Carlo Tomasi and Roberto Manduchi

Computer Science Department
Stanford University, Stanford, CA 94305

Abstract. In its traditional formulation, stereo correspondence involves both searching and selecting. Given a feature in one scanline, the corresponding scanline in the other image is searched for the positions of similar features. Often more than one candidate is found, and the correct one must be selected. The problem of selection is unavoidable because different features look similar to each other. Search, on the other hand, is not inherent in the correspondence problem. We propose a representation of scanlines, called *intrinsic curves*, that avoids search over different disparities. The idea is to represent scanlines by means of local descriptor vectors, without regard for where in the image a descriptor is computed, but without losing information about the contiguity of image points. In fact, intrinsic curves are the paths that the descriptor vector traverses as an image scanline is traversed from left to right. Because the path in the space of descriptors ignores image position, intrinsic curves are invariant with respect to disparity under ideal circumstances. Establishing stereo correspondences is then reduced to the selection of one among few match candidates, a task simplified by the contiguity information carried by intrinsic curves. We analyze intrinsic curves both theoretically and for real images in the presence of noise, brightness bias, contrast fluctuations, and moderate geometric distortion. We report preliminary experiments.

1 Introduction

The computation of stereo correspondences has traditionally been associated with a search over all possible disparities: for every point in the left scanline the corresponding right scanline is searched for a similar point. In this paper we show that search over disparities is not inherent in the correspondence problem. The way out of search is associative memory, and essentially inverts the way images are represented. Rather than storing image intensities by their position in the scanline, the usual array $I(x)$, we can store scanline positions by their appearance: in a sense, $x(I)$. Then, image points that look similar are stored in the same place. If both scanlines are stored in the same memory, correspondences are trivially established, because corresponding points share the same memory locations. Occlusions are also easily found as points that live alone in some location. There are two problems with this scheme: ambiguity and disguise.

---

[0] This research was supported by the National Science Foundation under contract IRI-9496205.

Ambiguity means that different image points can look the same, so memory locations can be crowded, and one match must be selected among many. Disguise occurs when corresponding points in the two scanlines look different because of the viewpoint change or because of image noise. In this case, points that should go in the same memory location do not. We deal with disguise by analyzing possible changes between scanlines. This analysis tells us where to look next if a memory location is missing a point. Ambiguity is addressed by a twofold strategy. On the one hand, it is reduced by encoding image appearance with descriptors that are richer than the mere image intensity I: each image location is described by a whole vector of parameters. On the other hand, the resolution of the remaining ambiguity is made easier by preserving contiguity information with the descriptors. Consider traversing a scanline in one of the two images. The vector of descriptors traces a curve in some space, and points that are nearby in the scanline are also nearby in the representation. Contiguity then helps selecting among similar match candidates: when two points look similar, we look around them, and attempt matching entire curve segments at once, rather than isolated points.

To illustrate the approach, here is one simple version of intrinsic curve for, say, the left scanline. A lowpass filtered version of the image intensity $l(x)$ and its derivative $l'(x)$ are computed everywhere (solid lines in figure 1 (b) and (c)) and are plotted against each other (solid lines in figures 2 (a) and (b)). When plotting l' versus l we lose track of space, that is, of the coordinate x which merely parameterizes the curve $l'(l)$. This parameter is stored for later use, but it plays no role in the shape of the curve. If $l(x)$ is replaced by a shifted replica $r(x) = l(x + d)$, the curve of figure 2 (b) remains the same. Because of this invariance to displacements, we call the curve of figure 2 (b) an *intrinsic curve*. More general geometric transformations $r(x) = l(\alpha(x))$ between l and r can deform an intrinsic curve, but the deformations can be predicted. The dashed curves in figures 1 and 2 show the construction of the intrinsic curve for the scanline $r(x)$ taken from a different viewing position.

Ambiguities cause intrinsic curves to self-intersect or overlap, and cannot be avoided. The selection process just mentioned is therefore unavoidable. On the other hand, the richer the description is, that is, the higher the dimensionality in which an intrinsic curve lives, the less likely self-intersections are.

In the next section, we present a theory of intrinsic curves. Section 3 show how real images differ from the ideal case, and section 4 shows preliminary experiments.

2   Intrinsic Curves: Theory

An efficient procedure for matching two signals is to consider a vectorial description of the local intensity variation at every point. Then two points from the two images are match candidates if the local descriptions are "close" to each other. A similar idea is at the basis of the stereo algorithms of Kass [14], Jones and Malik [11], and Weng, Ahuja and Huang [30]. In this section we define intrinsic

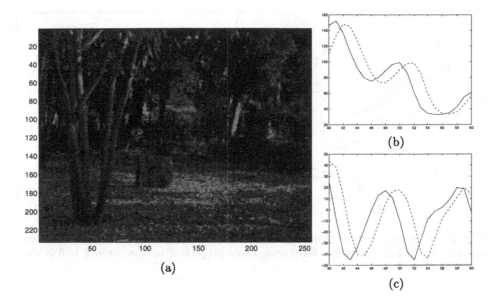

Fig. 1. (a) Test image "Trees" from SRI - frame 1. (b) Lowpass-filtered scanline 68, pixels 40–60 (solid line: frame 1, dashed line: frame 2), and (c) its derivative.

curves more generally. We also identify the geometric mappings

$$r(x) = l(\alpha(x)) \tag{1}$$

between the two images that are *compatible* with any particular way of building intrinsic curves, in the sense that they leave the curves unaltered. In other words, intrinsic curves are invariant with respect to compatible mappings. Finally, we investigate geometrical and topological properties of intrinsic curves.

Definition of an *intrinsic curve*. Suppose that the N operators $P_1, \ldots, P_N$ are applied to the intensity signal $l(x)$ to produce the new signals $p_n(x) = [P_n l](x)$ for $n = 1, \ldots, N$. The vector

$$\mathbf{p}(x) = (p_1(x), \ldots, p_N(x)) \tag{2}$$

describes a curve C in R^N parameterized by the real variable x:

$$C = \{\mathbf{p}(x), \, x \in R\} . \tag{3}$$

C is called the *intrinsic curve* generated by $l(x)$ through the operators $P_1, \ldots, P_N$ (see figures 1, 2).

It is crucial to notice that the curve C lives in R^N, not R^{N+1}: the image coordinate x is not a component of the curve. Spatial information is lost when going from image l to curve C, and it is exactly this loss of information that makes C invariant to a suitable class of geometric transformations, as discussed in the following subsection.

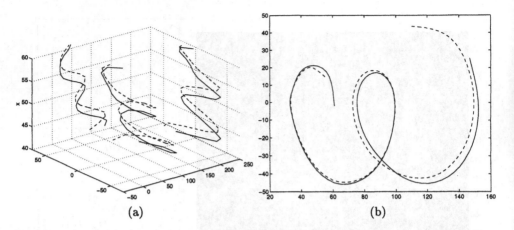

(a)   (b)

Fig. 2. (a) Intrinsic curve formation: the signals of figure 1 (b) and (c) are plotted against each other, forming a 3-D curve whose projection on the plane $x = 0$ is the intrinsic curve (b).

2.1 Compatible Mappings

While any reparametrization of C leaves C unchanged, reparametrizing the generator signal $l(x)$ to $l(\alpha(x))$ can in general modify C. For instance, if $p_1(x) = l(x)$ and $p_2(x) = l'(x)$, where the prime denotes differentiation, the new components of C after the change $x \to \alpha(x)$ become $\tilde{p}_1(x) = l(\alpha(x))$ and $\tilde{p}_2(x) = \alpha'(x)l'(\alpha(x))$ so that $\tilde{\mathbf{p}}(x)$ traces a new curve that is modulated by $\alpha'(x)$ in its second component. It therefore makes sense to ask what reparametrizations $x \to \alpha(x)$ leave C unaltered.

Definition . A mapping $x \to \alpha(x)$ is said to be *compatible* with the operators $P_1, \ldots, P_N$ if for any signal $l(x)$ the intrinsic curve generated by $l(x)$ is equal to the intrinsic curve generated by $l(\alpha(x))$.

Examples

Constant Displacement. Let the operators P_n in (2) be shift-invariant: $l(x) \to l(x + d) \Rightarrow p_n(x) \to p_n(x + d)$. The constant displacements $\alpha(x) = x + d$ are compatible with shift-invariant operators.

Affine Mapping. The affine mappings of the form $\alpha(x) = ax + d$ are compatible with the operators $p_0(x) = l(x)$ and

$$p_n(x) = [P_n l](x) = \frac{\left(\frac{\mathrm{d}^n}{\mathrm{d}x^n} l(x)\right)^{(n+1)/n}}{\frac{\mathrm{d}^{n+1}}{\mathrm{d}x^{n+1}} l(x)}$$

for $n > 0$, defined wherever $\frac{d^{n+1}}{dx^{n+1}} l(x) \neq 0$. This is proved immediately by noting that

$$\frac{d^n}{dx^n}[l(ax + d)] = a^n [\frac{d^n l}{dx^n}](ax + d) .$$

Semi-Commutative Mapping. If the mapping $x \rightarrow \alpha(x)$ is regarded as an operator A applied to $l(x)$, that is, $[Al](x) = l(\alpha(x))$, then $\alpha(x)$ is compatible with operators $P_1, \ldots, P_N$ if there is a change-of-variable operator $[Dq](x) = q(\delta(x))$ such that for every n we have $P_n A = D P_n$ where δ is a diffeomorphism independent of n. In fact, in this case, the mapping $x \rightarrow \alpha(x)$ simply reparameterizes the intrinsic curve. Both previous examples are special cases of semi-commutative mappings.

Thus, intrinsic curves can be regarded as invariants with respect to the set of compatible mappings, and provide a more general description than "classical" invariants such as function moments [5],[26],[19],[24]. Affine mappings are a popular model for the transformation between the two images of a stereo pair [15],[13],[6], and shift-invariant filters are often used for image descriptors [14],[11],[12],[18],[17]. The fact that in general affine mappings are not compatible with shift-invariant operators is therefore important. This was pointed out in [14], where a clever analysis of the effect of filtering a signal undergoing affine geometrical distortion is carried out. From the results of [14], we can assume that the intrinsic curves are approximately invariant with diffeomorphisms $x \rightarrow \alpha(x)$, so long as the supports of the filters' kernels are narrow and $\alpha(x)$ is close to the identity function. In the remainder of this section we assume that the mapping $\alpha(x)$ is a diffeomorphism (which, in particular, implies that it is monotone and continuous). In addition, we assume throughout this paper that both the input signals $l(x)$, $r(x)$ and the operators P_n are continuous, so that the intrinsic curves are connected.

If the transformation between left and right image were just a mapping $\alpha(x)$ compatible with the operators $P_1, \ldots, P_N$, stereo matching would be nearly trivial. In fact, to determine $\alpha(x)$ from the observation of $l(x)$ and of $r(x) = l(\alpha(x))$, the intrinsic curves are first computed from the two signals. For each signal, the parametrization (3) is stored, so that every point on either curve can be traced back to its image coordinate x via table lookup. Because of compatibility, the two intrinsic curves coincide. For every point $\mathbf{p}$ that belongs to both of them, the corresponding image coordinates are a match, with the sole exception of points where the intrinsic curves self-intersect.

2.2 Geometrical and Topological Properties of Intrinsic Curves

Our definition of intrinsic curves is quite general. Their properties depend on the characteristics of the operators $\{P_n\}$ in (2). In this section, we concentrate on the case $N = 2$ with the following choice for these operators:

$$p_1(x) = [P_1 l](x) = l(x) \quad \text{and} \quad p_2(x) = [P_2 l](x) = l'(x) . \tag{4}$$

Vector $\mathbf{p}(x)$ is thus composed by the first two terms of the Taylor expansion of $l(x)$ around x, and each point on the intrinsic curve generated by $l(x)$ represents a description of the local behavior of $l(x)$. With this choice, intrinsic curves are defined on a plane, reminiscent of the *phase space* of systems theory [2]. We can define an orientation at $\mathbf{p}$ by computing the unit-norm tangent $\mathbf{t}(\mathbf{p})$ to the curve:

$$\mathbf{t}(\mathbf{p}) = \frac{(l'(x), l''(x))}{\sqrt{(l'(x))^2 + (l''(x))^2}} \, . \tag{5}$$

The values of $\mathbf{t}(\mathbf{p})$ depend on the position of $\mathbf{p}$ as follows:

- If $\mathbf{p}$ lies in the upper open half-plane, where $l'(x) > 0$, $\mathbf{t}(\mathbf{p})$ assumes values in the right open half-circle $\{(p_1, p_2); p_1^2 + p_2^2 = 1; p_1 > 0\}$. When $\mathbf{p}$ lies on the lower open half-plane, $\mathbf{t}(\mathbf{p})$ is in the left open half-circle.
- If $\mathbf{p}$ lies on the axis of the abscissas, where $l'(x) = 0$, then $\mathbf{t}(\mathbf{p}) = (0, \pm 1)$. In other words, when crossing the axis of the abscissas, the curve forms an angle of $\pm \pi/2$ with it.

Note that if $l'(x) = l''(x) = 0$ (e.g., in a segment where the signal is constant), $\mathbf{p}$ is singular with respect to x [25]. In such a case, the tangent can be defined by continuity. On the other hand, an intrinsic curve can be singular only on the axis of the abscissas. From these rules it follows that *intrinsic curves are naturally oriented clockwise*: they are traversed left-to-right in the upper half-plane and right-to-left in the lower. Furthermore, *any loop must intersect the axis of the abscissas*.

In general, we may consider intrinsic curves $\mathcal{C}$ in R^M of the form $\mathbf{p}(x) = (l(x), l'(x), l''(x), \ldots, l^{(M)}(x))$. Each point of the curve represents a local description of the function $l(x)$ by an M-term Taylor expansion. The topological properties described above apply to the projection of $\mathcal{C}$ onto each plane $(l^{(n)}, l^{(n+1)})$, *i.e.* to the curve generated by $\mathbf{P}[l](x) = (l^{(n)}(x), l^{(n+1)}(x))$.

2.3 Intrinsic Curve Reparametrization

Intrinsic curves are continuous curves on the plane or in a space of higher dimension. For computation, on the other hand, intrinsic curves must have a discrete representation. To this end, we now introduce the *arc length parametrization*, which leads to a variable-rate image sampling that emphasizes "busy" parts of the image. We assume hereafter that $l(x)$ has support in the segment $[x_0, x_1]$.

The length of the arc $\mathcal{C}(\mathbf{p}_a, \mathbf{p}_b)$ from $\mathbf{p}_a = \mathbf{p}(x_a)$ to $\mathbf{p}_b = \mathbf{p}(x_b)$ is equal to

$$\text{arc length } \mathcal{C}(\mathbf{p}_a, \mathbf{p}_b) = \int_{x_a}^{x_b} \sqrt{(l'(x))^2 + (l''(x))^2} \, dx \, . \tag{6}$$

The arc length parametrization is then $s(x) = $ arc length $\mathcal{C}(\mathbf{p}_0, \mathbf{p}(x))$, where $\mathbf{p}_0 = \mathbf{p}(x_0)$. It is instructive to study the relation between $s(x)$ and $l(x)$. We have that

$$\frac{d}{dx} s(x) = \sqrt{(l'(x))^2 + (l''(x))^2}. \tag{7}$$

Hence, we may expect that a variation Δs of the new parameter will correspond to a large variation Δx if $l'(x)$ and $l''(x)$ are small (*i.e.*, in parts of the curve that lie close to the horizontal axis), and to a small Δx when $l'(x)$ and $l''(x)$ are large (*i.e.*, when the curve is far from the horizontal axis). This observation suggests a sort of "adaptive" sampling paradigm for $l(x)$. Assume to sample the curve $\mathcal{C}$ at constant-width intervals, that is, by keeping the arc length of the segments $\mathcal{C}(\mathbf{p}_i, \mathbf{p}_{i+1})$ constant. This procedure corresponds to sampling signal $l(x)$ on a nonuniform grid: the grid will be less dense in areas characterized by small values of $l'(x)$ and $l''(x)$ (where the signal is "flat"), and denser if $l'(x)$ and $l''(x)$ are larger (where the signal "busyness" is higher). This looks like a useful sampling strategy for signal matching. In fact, it is well known (see e.g. [9]) that a match is expected to be less robust (with respect, for example, to noise and to quantization errors) in regions where the signal is "flat". The adaptive sampling procedure leads to concentrating estimates in reliable areas(see figure 3).

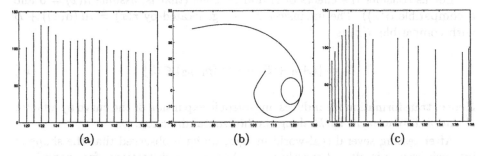

$$(a) \qquad\qquad (b) \qquad\qquad (c)$$

Fig. 3. A signal sampled on a uniform grid (a) and on an nonuniform grid (c) induced by the uniform arc length sampling of the intrinsic curve (b).

3   Deviations from the Ideal Case

Intrinsic curves of corresponding scanlines $l(x)$ and $r(x)$ related by a compatible mapping

$$r(x) = l(\alpha(x)) \tag{8}$$

are identical. In reality, however, $l(x)$ and $r(x)$ can differ for the following reasons.

No mapping. In certain cases, $\alpha(x)$ may not even exist, such as when regions of $l(x)$ or $r(x)$ are occluded.

Incompatible mapping. The mapping $x \to \alpha(x)$ that relates $l(x)$ and $r(x)$ as in (8) is not compatible with respect to the operators $P_1, \ldots, P_N$ that are used to build the intrinsic curves. For instance, affine transformations $x \to ax + d$ are not compatible with the operator $\mathbf{P}[l](x) = (l(x), l'(x))$. In fact, if $\mathcal{C}_l$ is the intrinsic curve generated by $l(x)$, the intrinsic curve $\mathcal{C}_r$ generated by $r(x) = $

$l(ax + d)$ is $C_r = \{(p_1, ap_2) : (p_1, p_2) \in C_l\}$ which is a vertically expanded $(a > 1)$ or compressed $(a < 1)$ version of C_l.

Photometric distortion and noise. The constant-brightness hypothesis implied by relation (8) is not satisfied. A convenient model that accounts for both geometric and photometric distortion is the following (see [9] for a general discussion of related issues):

$$r(x) = Al(ax + d) + B + n(x) . \tag{9}$$

In this model, A and B represent the difference in contrast and brightness between the two images, and are either constant or varying slowly with respect to the dynamic of the signal. The term $n(x)$ represents "noise", that is, any discrepancy independent of the signals. The terms a and d represent geometric distortion and, in particular, d is the inter-frame disparity we are after.

Let us consider the effects of A and B alone (that is, assume $n(x) = 0$ and a compatible $\alpha(x)$). The intrinsic curve C_r generated by $r(x) = Al(\alpha(x)) + B$ with compatible $\alpha(x)$ is

$$C_r = \{(Ap_1 + B, Ap_2) : (p_1, p_2) \in C_l\} . \tag{10}$$

Hence, transformation (9) induces an isotropic expansion of the curve by a factor A and a displacement by B along the horizontal direction.

After testing several real-world images, we have observed that the shape of intrinsic curves is altered mostly after photometric distortions, for example as a consequence of the different viewing position of the two cameras, optical attenuation and sensitivity of the image sensors [29], [14], [30]. Large geometric distortions that give raise to vertical dilation or shrinking of the intrinsic curve are less likely to happen than photometric distortions [1]. Consequently, we believe that the terms A and B in our model are dominant over the geometric distortion related to a.

The effects of both brightness bias B and noise $n(x)$ can be neutralized by preprocessing both signals with a zero-mean filter with an otherwise lowpass frequency response. The contrast difference term A is then the dominant remaining term, and the point $\mathbf{p}_r$ on C_r corresponding to a given point $\mathbf{p}_l$ on C_l is collinear with $\mathbf{p}_l$ and with the origin. Hence, candidates for $\mathbf{p}_r$ are among the points $\{\hat{\mathbf{p}}_r\}$ of C_r lying on the "radial line" passing through both the origin and $\mathbf{p}_l$. We then select the "right" correspondence within $\{\hat{\mathbf{p}}_r\}$ according to a number of criteria which may be local (proximity in the phase space) or global (ordering and coherence principles). This is described in [28] where we also present a data structure for the efficient access to points that lie along a given radial line.

Our procedure leaves image dilation or shrinking (modeled by the term a in equation (9)) and a possible leftover intensity bias as the only terms of our model (9) that have not been accounted for. An analysis of the inaccuracy due to neglecting such distortion terms may be found in [28].

4 A Possible Matching Algorithm

In the ideal case, intrinsic curves from different images coincide, except for occlusions. In reality, because of the phenomena discussed in section 3, intrinsic
curves are only close to each other, and matching points can be found along the
radial line. A "pathological" case is when the operators P_n are shift-invariant,
and $l(x)$ is periodic. In such a case, the intrinsic curve is closed, and infinite
instances for $\alpha(x)$ are available. This fact reflects the inherent ambiguity in the
match of periodic signals. For non-periodic signals, ambiguity can be somewhat
reduced by enriching the description of signals, as noticed also in [14] and [11].
The intrinsic curve representation makes such a notion apparent from a topological standpoint; for example, using only two operators, the intrinsic curves lie
in a plane, and self-intersections are to be expected. With three operators, the
curves live in a 3-D space, where a path is less likely to cross itself.

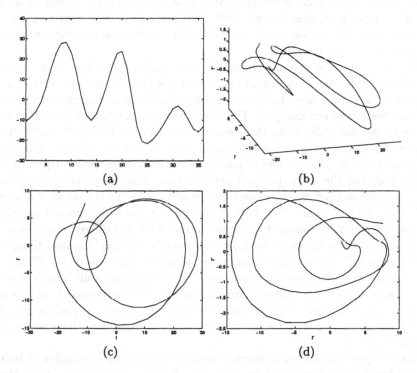

Fig. 4. A signal $l(x)$ (a), the intrinsic curve generated by $l(x)$ in the phase space
(l, l', l'') (b), and its projections onto the (l, l') plane (c) and onto the (l', l'') plane (d).

Figure 4 shows the intrinsic curve relative to a signal $l(x)$ in the 3-D phase
space (l, l', l''), together with its projections on planes (l, l') (our usual intrinsic
curve in the plane) and (l', l''). It is interesting to note (from (5)) that the phase
of any point of the curve in the plane (l', l'') coincides with the phase of the

tangent to the curve in (l, l') at the corresponding point.

In our algorithm, for each point $\mathbf{p}_l(x)$ in C_l we select the candidates $\{\hat{\mathbf{p}}_r = \mathbf{p}_r(\hat{x})\}$ of C_r lying on the radial line passing through $\mathbf{p}_l$ in the plane (l, l'). Then, we "rate" each candidate with its distance d to $\mathbf{p}_l(x)$ in the 3-D space (l, l', l''):

$$d(\mathbf{p}_l, \hat{\mathbf{p}}_r) = \sqrt{(l(x) - r(\hat{x}))^2 + (l'(x) - r'(\hat{x}))^2 + (l''(x) - r''(\hat{x}))^2}. \qquad (11)$$

However, the closeness in the phase space alone is not sufficient to guarantee a robust pointwise match of the curves. In fact, approximate ambiguity may occur even in the 3-D phase space when the curve loops close to itself (see figure 4). Hence, to pick the right candidate among $\{\hat{\mathbf{p}}_r\}$, i.e., to resolve the ambiguity, we rely on the available contextual information, as shown in the following.

In the literature, ambiguity is typically resolved by imposing constraints on the disparity field, such as *uniqueness, ordering* (or *monotonicity* [7]), and *smoothness* [20],[8], [23],[3],[22],[21]. Note that also other algorithms that make use of vectorial local descriptions ([14], [11]) need to impose constraints on the disparity field: the notion of "closeness" in the representation space is not itself sufficient for a reliable match.

The main novelty of our approach is that *disparity values never enter our procedure to solve the ambiguity*. In fact, we work only on intrinsic curves, which have lost track of space: the inverse mapping $\mathbf{p} \to x$ is determined only after the matches have been assigned. The new constraint we impose to resolve ambiguities comes naturally from the consideration that under ideal conditions (no photometric distortion, compatible geometric distortion) curves C_l and C_r are identical. Let s be the arc length parameter on C_l (see section 2.3), and let Δs be the length of the arc $C(\mathbf{p}_{l_1}, \mathbf{p}_{l_2})$ between two sampling points $\mathbf{p}_{l_1}, \mathbf{p}_{l_2}$ on C_l. Then, barring occlusions, we expect the length of the (oriented) arc $C(\mathbf{p}_{r_1}, \mathbf{p}_{r_2})$ on C_r to be "close" to Δs. Note that we still rely on the constraints of uniqueness and monotonicity, as parameter s is monotone with x, but we do not need any other quality of the disparity field. In other words, we simply expect that, while $\mathbf{p}_l$ moves along C_l, the corresponding point $\mathbf{p}_r$ moves similarly on C_r. We will call such a constraint the *coherence principle*. The important point here is that two corresponding points $\mathbf{p}_l$ and $\mathbf{p}_r$ differ not because of the disparity, but because of noise and distortions. Similarly the arc lengths of $C(\mathbf{p}_{l_1}, \mathbf{p}_{l_2})$ and $C(\mathbf{p}_{r_1}, \mathbf{p}_{r_2})$ reflect changes of appearance, not image distances. Details of the algorithm can be found in [28].

For our preliminary experiments, we have chosen three couples of images, namely two frames from the test sequence "Trees" from SRI, two from the sequence "Library", from the movie "Wings of Desire" directed by Wim Wenders, and two from the "Castle" sequence from CMU (figure 5). In all cases, the camera was moving roughly horizontally (*i.e.*, parallel to the scanlines). In the images of figure 5, the part of image above the white line belongs to the first frame, the one below belongs to the second frame.

The left part of sequence "Trees" exhibits a very articulated disparity field, induced by the sharp depth discontinuities along the boundaries of the branches

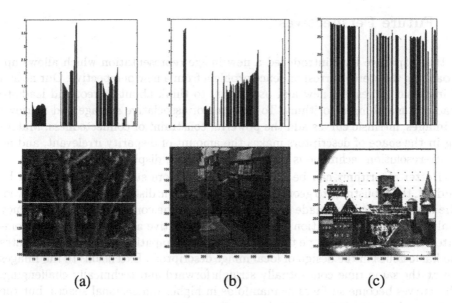

Fig. 5. Sequences (a) "Trees", (b) "Library" and (c) "Castle" with the computed disparity fields relative to one scanline (below white line: frame 1, above white line: frame 2).

of the trees. No post–processing (e.g., filtering) has been applied to the results. Note that in some parts we have produced a dense disparity field while in other ones no measurement was available. This is due to the following reasons: (i) the intrinsic curves are sampled with uniform arc length period, which induces the nonuniform sampling period of the measurement field, and (ii) the disparity is computed only where reliable segments of points were found. As pointed out in section 2.3, dense measurements are characteristic of high signal busyness regions, where disparity estimates are more reliable. Figure 5 (a) shows that the computed disparity field follows the depth discontinuities of the scene very tightly. Sequence "Library" (figure 5 (b)) is characterized by a wide disparity range (from less than 2 pixels corresponding to the back of the room, to approximately 10 pixels at the edge of the bookshelves). Both the disparity jump corresponding to the standing person's head (pixels 260–300) and the ramp corresponding to the books on the shelf are detected by our system. The distinguishing feature of sequence "Castle" (figure 5 (c)) is the very large disparity overall: more than 25 pixels. In systems that search over disparities, this large displacement is usually handled by multi–resolution techniques. Our approach, which matches scanlines in the space of descriptors, shows that multi–resolution is not a conceptual necessity.

In these experiments, the variance of the estimates is substantial. However, the measurements are very dense, and a simple post–processing (e.g., median filtering [11]) would "clean" the computed disparity field effectively.

5 Future Perspectives

In this paper we have introduced a new image representation which allows approaching the stereo correspondence problem from a new perspective. Our notion of intrinsic curves is a new and useful way to think about stereo, and leads to practical matching algorithms. To the idea of associative storage and retrieval of images, intrinsic curves add the powerful constraint of connectedness. Matching in the space of descriptors makes the amount of disparity irrelevant, and no multi-resolution technique is needed even for large displacements.

Better algorithms can be devised, richer or more stable descriptors can be studied, the robustness to geometric and photometric distortion can be improved. The descriptors can be made even richer through the concepts of local frequency analysis and multi-resolution descriptions, both active areas of research in computer vision today. We hope that the concept of compatible mappings elucidates the basic issues in the design of local image descriptors. Extensions to full images are at the same time conceptually straightforward and technically challenging. The curves become surfaces or manifolds in higher dimensional spaces, but the basis for the matching remains the same.

Finally, we would like to outline an intriguing direction of research that we are starting to investigate for the detection of occlusions. Any stereo algorithm must cope with occlusions. A number of researchers have dealt with the problem of occlusions in stereo [27], [16], [11], [4], [7], [10]. The robust and accurate detection of occlusions, however, seems still an open problem.

With intrinsic curves, an occlusion manifests itself as an arc of one curve that is not matched in the other. Just before and just after the unmatched arc the curves are expected to coincide. This situation appears clearly in figures 6(b) and 7(b), where the intrinsic curves of the signals of figures 6(a) and 7(a) are depicted: occlusions stand out as "anomalous" loops in one of the intrinsic curves.

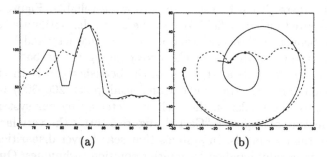

(a)   (b)

Fig. 6. Scanline 95 from the image of figure 1 (a), pixels # 74–94 (solid line: $l(x)$, dashed line: $r(x)$). (a) Intensity. The part of $l(x)$ form pixel 79 to pixel 81 is not matched by $r(x)$. (b) Intrinsic curves. The arc of C_l between the two circled points is not matched in C_r.

464

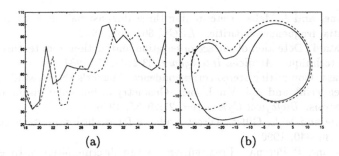

(a)   (b)

Fig. 7. Scanline 92 from the image of figure 1 (a), pixels 18–38 (solid line: $l(x)$, dashed line: $r(x)$). (a) Intensity. The part of $r(x)$ form pixel 23 to pixel 25 is not matched by $l(x)$. (b) Intrinsic curves. The arc of C_r between the two circled points is not matched in C_l.

In general, the presence of an unmatched loop is not by itself sufficient evidence of occlusion. Loops may be produced sometimes by noise, and we must look for a more robust topological characterization. However, it is clear that an occlusion manifests itself as a "perturbation" of only one of the two intrinsic curves in a limited region. It seems therefore that the detection and analysis of occlusions should be easier in this setting, rather than observing the profiles of the two signals in their "natural" spatial domain. In other words, the phase space is the appropriate place to look whether two signals match – or they don't.

References

1. R.D. Arnold and T.O. Binford. Geometric constraints in stereo vision. *SPIE* 238:281–292, 1978.
2. V.I. Arnold. *Ordinary Differential Equations.* MIT Press, 1990.
3. H.H. Baker and T.O. Binford. Depth from edge and intensity based stereo. *IJCAI*, 631–636, 1981.
4. P.N. Belhumeur and D. Mumford. A Bayesian treatment of the stereo correspondence problem using half–occluded regions. *CVPR*, 506–512, 1992.
5. A. Blake and C. Marinos. Shape from texture: Estimation, isotropy and moments. *Artificial Intelligence*, 45:323–380, 1990.
6. M. Campani and A. Verri. Motion analysis from first–order properties of optical flow. *CVGIP: Image Understanding*, 56(1):90–107, 1992.
7. D. Geiger, B. Ladendorf, and A. Yuille. Occlusions and binocular stereo. *EECV*, 425–433, 1992.
8. W.E.L. Grimson. Computational experiments with a feature based stereo algorithm. *PAMI*, 7(1):17–34, 1985.
9. W. Förstner. Image Matching. In R.M. Haralick and L.G. Shapiro, *Computer and Robot Vision.* Addison–Wesley, 1992.
10. S.S. Intille and A.F. Bobick. Disparity–space images and large occlusion stereo. *ECCV*, 179–186, 1994.
11. D.G. Jones and J. Malik. A computational framework for determining stereo correspondence from a set of linear spatial filters. *EECV*, 395–410, 1992.

12. D.G. Jones and J. Malik. Determining three–dimensional shape from orientation and spatial frequency disparities. *EECV*, 661–669, 1992.
13. K. Kanatani. Detection of surface orientation and motion from texture by a stereological technique. *Artificial Intelligence*, 23:213–237, 1984.
14. M.H. Kass. Computing stereo correspondence. Master's thesis, M.I.T., 1984.
15. J.J. Koenderink and A.J. Van Doorn. Geometry of binocular vision and a model for stereopsis. *Biological Cybernetics*, 21:29–35, 1976.
16. J.J. Little and W.E. Gillet. Direct evidence for occlusion in stereo and motion. *ECCV*, 336–340, 1990.
17. J. Malik and P. Perona. Preattentive texture discrimination with early vision mechanisms. *JOSA – A*, 7(5):923–932, 1990.
18. S.G. Mallat. A theory for multiresolution signal decomposition: The wavelet representation. *PAMI*, 11(7):674–693, 1989.
19. R. Manmatha. A framework for recovering affine transforms using points, lines or image brightness. *CVPR*, 141–146, 1994.
20. D. Marr and T. Poggio. Cooperative computation of stereo disparity. *Science*, 194:283–287, 1976.
21. G. Medioni and R. Nevatia. Segment–based stereo matching. *CVGIP*, 31:2–18, 1985.
22. Y. Ohta and T. Kanade. Stereo by intra– and inter–scanline search using dynamic programming. *PAMI*, 7(2):139–154, 1985.
23. S.B. Pollard, J.E. Mayhew, and G.P. Frisby. PMF: A stereo correspondence algorithm using a disparity gradient limit. *Perception*, 14:449–470, 1985.
24. J. Sato and R. Cipolla. Extracting the affine transformation from texture moments. *ECCV*, 165–172, 1994.
25. D.J. Struik. *Lectures on Classical Differential Geometry*. Dover, 1988.
26. B.J. Super and A.C. Bovik. Shape–from–texture by wavelet–based measurement of local spectral moments. *CVPR*, 296–301, 1992.
27. P.S. Toh and A.K. Forrest. Occlusion detection in early vision. *ICCV*, 126–132, 1990.
28. C. Tomasi and R. Manduchi. Stereo without search. Tech. Rep. STAN-CS-TR-95-1543, Stanford, 1995.
29. A. Verri and T. Poggio. Against quantitative optical flow. *ICCV*, 171–180, 1987.
30. J. Weng, N. Ahuja, and T.S. Huang. Matching two perspective views. *PAMI*, 14(8):806–825, 1992.

Recognition/
Matching/
Segmentation

Informative Views and Sequential Recognition

Tal Arbel and Frank P. Ferrie

Centre for Intelligent Machines, McGill University, Montreal, Quebec,
CANADA H3A 2A7

Abstract. In this paper we introduce a method for distinguishing between informative and uninformative viewpoints as they pertain to an active observer whose aim is to identify an object in a known environment. The method is based on a probabilistic inverse theory using a probabilistic framework where assertions are represented by conditional probability density functions. Experimental results are presented showing how the resulting algorithm can be used to distinguish between informative and uninformative viewpoints, rank a sequence of images on the basis of their information (e.g. to generate a set of characteristic views), and sequentially identify an unknown object.

1 Introduction

Consider an active agent charged with the task of roaming the environment in search of some particular object. It has an idea of what it is looking for, at least at some general level, but resources are limited so it must act purposein in when carrying out its task[1]. In particular, the agent needs to assess what it sees and quickly determine whether or not the information is useful so that it can evoke alternate strategies (the next place to look for example). Key to this requirement is the ability to make and qualify assertions while taking into account prior expectations about the environment. In this paper we show how the problem is cast in probabilistic terms from the point of view of inverse theory [2]. Assertions are presented by conditional probability density functions, which we refer to as belief distributions, that relate the likelihood of a particular hypothesis given a set of measurements. What is particularly important about the methodology is that it yields a prescription for generating these distributions, taking into account the different sources of uncertainty that enter into the process. Based on this result we show how the resulting distributions can be used to (i) assess the quality of a viewpoint on the basis of the assertions it generates, and (ii) successfully recognize an unknown object by accumulating evidence as the probabilities level.

Specifically, we show how uncertainty, conditions, prior expectations such that the shape of the resulting belief distribution can vary greatly. Recently, very delta-like is the interpretation tends towards certainty. In contrast, ambiguous or poor interpretations consistently tend towards very broad or flat distributions [3]. We exploit this characteristic to define the notion of an informative viewpoint, i.e. a view which gives rise to assertions that have a high probability according to their associated belief distribution. There are at least two applications for this result. First, in the case of an active observer, viewpoints can be chosen so as to maximize the distribution associated with an object of interest. This does not specify how to choose an informative viewpoint, but can be used as a figure of

_____
1 Strategies for gaze planning are operationally defined [4, 5].

Informative Views and Sequential Recognition

Tal Arbel and Frank P. Ferrie

Centre for Intelligent Machines, McGill University, Montréal, Québec,
CANADA H3A 2A7

Abstract. In this paper we introduce a method for distinguishing between informative and uninformative viewpoints as they pertain to an active observer seeking to identify an object in a known environment. The method is based on a generalized inverse theory using a probabilistic framework where assertions are represented by conditional probability density functions. Experimental results are presented showing how the resulting algorithms can be used to distinguish between informative and uninformative viewpoints, rank a sequence of images on the basis of their information (e.g. to generate a set of characteristic views), and sequentially identify an unknown object.

1 Introduction

Consider an active agent charged with the task of roaming the environment in search of some particular object. It has an idea of what it is looking for, at least at some generic level, but resources are limited so it must act purposefully when carrying out its task [1]. In particular, the agent needs to assess what it sees and quickly determine whether or not the information is useful so that it can evolve alternate strategies (the next place to look for example). Key to this requirement is the ability to make and quantify assertions while taking into account prior expectations about the environment. In this paper we show how the problem be cast in probabilistic terms from the point of view of inverse theory [2]. Assertions are represented by conditional probability density functions, which we refer to as *belief distributions*, that relate the likelihood of a particular hypothesis given a set of measurements. What is particularly important about the methodology is that it yields a precise recipe for generating these distributions, taking into account the different sources of uncertainty that enter into the process. Based on this result we show how the resulting distributions can be used to (i) assess the quality of a viewpoint on the basis of the assertions it generates and (ii) sequentially recognize an unknown object by accumulating evidence at the probabilistic level.

Specifically, we show how uncertainty conditions prior expectations such that the shape of the resulting belief distribution can vary greatly, becoming very delta-like as the interpretation tends towards certainty. In contrast, ambiguous or poor interpretations consistently tend towards very broad or flat distributions [3]. We exploit this characteristic to define the notion of an *informative viewpoint*, i.e. a view which gives rise to assertions that have a high probability according to their associated belief distribution. There are at least two applications for this result. First, in the case of an active observer, viewpoints can be chosen so as to maximize the distribution associated with an object of interest. This does not specify *how* to choose an informative viewpoint[1], but can be used as a figure of

---

[1] Strategies for gaze planning are operationally defined [4, 5].

merit for a particular choice. Second, in the case of an off-line planner, it is often advantageous to be able to pre-compute a set of characteristic views to aid in recognition [6]. A good strategy here would be to select the n best views of an object ranked according to its belief distribution.

Although viewpoints can be labelled as either informative or uninformative, ambiguous cases where there is "reasonable" belief in more than one interpretation still exist. It becomes apparent that evidence from more than one viewpoint is needed. This leads to a sequential recognition strategy that seeks to improve uncertain interpretations by accumulating evidence over several views. We show that such evidence can be accumulated by histogramming votes from each viewpoint and picking the hypothesis with the highest score. This strategy is appropriate provided that clear "winner" hypotheses prevail in a largely view-invariant manner.

This brings us to the problem of obtaining the belief distributions. Here we consider the recognition problem itself, focusing on a model-based approach. Specifically, model-based recognition focuses on matching an unknown model, which is computed on-line from sensory data, with a predetermined model computed off-line and residing in a database of known objects [7]. What differentiates approaches is largely a matter of the kinds of models used to represent objects in the scene and how models are matched. Our interest is in three-dimensional object recognition in which objects are represented by parametric shape descriptors (i.e. models) such as superellipsoids [8, 9, 10, 11], deformable solids [12], and algebraic surfaces [13]. In our context, models are constructed through a process of *autonomous exploration* [4, 5] in which a part-oriented, articulated description of an object is inferred through successive probes with a laser range-finding system. The set-up used to perform experiments consist of a two-axis laser range-finder mounted on the end-effector of an inverted PUMA-560 ma-

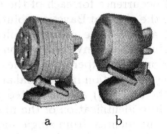

a   b

Fig. 1. (a) Laser range finder image of a pencil sharpener rendered as a shaded image. (b) An articulated, part-oriented model of the sharpener using superellipsoid primitives; 8 superellipsoids are used, one corresponding to each of the parts of the object.

nipulator. For any particular viewpoint, such as the one shown in Figure 1a, a process of bottom-up shape analysis leads to an articulated model of the object's shape (Figure 1b) in which each part is represented by a superellipsoid primitive [11]. Associated with each primitive is a covariance matrix $\mathbf{C}$ which embeds the uncertainty of this representation and which can be used to plan subsequent gaze positions where additional data can be acquired to reduce this uncertainty further [4]. A system which automatically builds object models based on this principle is reported in [5, 14].

Many approaches have been advocated for the problem of *matching* models. The majority of these employ various metrics to measure the distance between models in the appropriate parameter spaces, e.g., Mahalanobis distance [15], dot product [12] to mention but a few. These strategies rarely include both the uncertainties in the parameters of the measured models and the ambiguities of the representations in the database. However, when fitting a model to data that are noisy, there is an inherent lack of uniqueness in the parameters that describe the model. In these cases it is impossible to make a definitive statement as to which model fits the data best [4]. For this reason, rather than choose external constraints that would force the choice of one model over another, it would be more instructive to embed the uncertainty in the chosen description into the representation. This is precisely the approach that we have taken in computing the belief distribution.

Our methodology is based on a probabilistic inverse theory first introduced by Tarantola in [2]. Earlier work has shown how this theory can be used to methodically synthesize belief distributions corresponding to each model hypothesis, $\mathcal{H}_)$, given the parameters corresponding to the unknown model, $\mathcal{M}$, computed from the current measurement D_j, i.e. $P(\mathcal{H}_)|\mathcal{M}_{\mathcal{D}_|})$ [3]. This procedure explicitly accounts for uncertainties arising from the estimation of the unknown model parameters, database model parameters, and prior expectations on the frequency of occurrence for each of the database entries. In this case, the solution reduces to the classical Bayesian solution, similar to the result obtained by Subrahmonia et al. [13] - the primary difference being in the techniques used to obtain the solution. The inverse solution forces all sources of knowledge to be made explicit prior to the experiment. The method provides a more general recipe for combining information in a formal and structured fashion. In addition, they (and many others [7, 16]) are interested in constructing a discriminant that makes an absolute identification of the measured object. We argue that making assessments about identity from single measurements can be erroneous. We are more interested in assessing the *quality* of the identification from a particular viewpoint and to communicate this belief to other processes to determine whether further sampling is required.

The sequential recognition strategy therefore seeks to combine information at the level of the belief distribution. That is, given two data sets D_j and D_{j+1} corresponding to different viewpoints we seek a conjunction of $P(\mathcal{H}_)|\mathcal{M}_{\mathcal{D}_|})$ and $P(\mathcal{H}_)|\mathcal{M}_{\mathcal{D}_{|+\infty}})$ that is equivalent to $P(\mathcal{H}_)|\mathcal{M}_{\mathcal{D}_|+\mathcal{D}_{|+\infty}})$. Although the theory formally defines conjunction, such an operation requires knowing how a change in viewpoint conditions the respective belief distributions. Later on, we will show that if the maximum likelihood hypothesis [2] is largely invariant over a sequence of trials, then a robust interpretation can be made by tabulating the votes for each one and picking the hypothesis with the highest score. We also show that this invariance can be maximized by using the structure of the belief distribution to filter out uninformative hypotheses.

The remainder of the paper is organized as follows. We begin in Section 2 by describing how to distinguish between informative and uninformative viewpoints. We then introduce the general inverse theory in Section 2.1 and explain how to apply the theory to the problem of recognizing parametric models. We then indicate how the theory can be used to label viewpoints as informative or uninformative in Section 2.2. In Section 3 these results are combined to form an incremental recognition scheme, and in Section 4 we show how it can be applied to the recognition of complex, multi-part objects. Finally, we conclude in Section 5 with a summary of the results and a pointer to future applications.

---

[2] This refers to the hypothesis that the correct answer is the one with the highest belief.

2 Determining Which Viewpoints are Informative

In order to be able to determine whether a viewpoint is informative or not, the recognition engine should quantify the identification by producing a degree of confidence in the hypotheses, rather than establish an absolute identity for the unknown object. In this fashion, views with stronger hypotheses in terms of a significantly higher degree of confidence in one model than the others, can be considered informative. Viewpoints associated with low confidence levels in the hypotheses are considered uninformative. In the next sections, we will illustrate how the inverse theory can be used to generate confidence in various hypotheses, and illustrate how it can be used to distinguish between informative and uninformative viewpoints within the context of model-based object recognition.

2.1 The Inverse Problem Theory

In lieu of a single maximum likelihood solution, we seek a method that generates a measure of confidence in various hypotheses within the context of an object recognition problem. Like all inverse problems, the recognition problem is ill posed in that, i) several models can give rise to identical measurements and, ii) experimental uncertainty gives rise to uncertain measurements. As a result it is not possible to identify the unknown object uniquely. There are various ways of conditioning ill posed problems, but these all require strong, and often implicit, a priori assumptions about the nature of the world. As a result a method may work well only in specific cases and because of the hidden implicit nature of the conditioning assumptions, cannot be easily modified to work elsewhere.

For this reason we have adopted the very general inverse problem theory of Tarantola [2]. In it the sources of knowledge used to obtain inverse solutions are made explicit, so if conditioning is required, the necessary assumptions about that knowledge are apparent and can be examined to see if they are realistic. The theory uses probability density functions to represent the following sources of knowledge:

1. Knowledge given by a theory which describes the physical interaction between models $\mathbf{m}$ and measurements $\mathbf{d}$, denoted $\theta(\mathbf{d}, \mathbf{m})$,
2. Knowledge about the model from measurements, denoted $\rho_D(\mathbf{d})$.
3. Information from unspecified sources about the kinds of models which exist in the world (namely that there are a discrete number of them). We denote this knowledge $\rho_M(\mathbf{m})$. Knowledge like this is a powerful constraint and can be used to eliminate many of the unconstrained solutions.

The Inverse Solution The theory postulates that our knowledge about a set of parameters is described by a probability density function over the parameter space. This requires us to devise appropriate density functions in order to represent what we know about the world. The solution to the inverse problem then becomes a simple matter of combining the sources of information. Tarantola defines the logical *conjunction* of states of information such that the solution to the inverse problem is given by the theory AND the measurements AND any a priori information about the models. With this definition we can therefore combine the information from the joint prior probability density function $\rho(\mathbf{d}, \mathbf{m})$ and the theoretical probability density function $\theta(\mathbf{d}, \mathbf{m})$ to get the a posteriori state of information

$$\sigma(\mathbf{d}, \mathbf{m}) = \frac{\rho(\mathbf{d}, \mathbf{m})\, \theta(\mathbf{d}, \mathbf{m})}{\mu(\mathbf{d}, \mathbf{m})} \tag{1}$$

where $\theta(\mathbf{d}, \mathbf{m}) = \theta(\mathbf{d}|\mathbf{m})\, \mu_M(\mathbf{m})$ and $\rho(\mathbf{d}, \mathbf{m}) = \rho_D(\mathbf{d})\, \rho_M(\mathbf{m})$ over the joint space $M \times D$, where M refers to the *model space* and D, the *data space*. The so called non-informative probability density $\mu(\mathbf{d}, \mathbf{m}) = \mu_D(\mathbf{d})\mu_M(\mathbf{m})$ represents the reference state of information. For our purposes we will assume that all the non-informative densities are uniform over their respective spaces.

Accordingly, (1) is more general that the equations obtained through traditional approaches, but degenerates to the classical Bayesian solution under the aforementioned conditions. The a posteriori information about the model parameters is given by the marginal probability density function:

$$\sigma(\mathbf{m}) = \rho_M(\mathbf{m}) \int_D \frac{\rho_D(\mathbf{d})\, \theta(\mathbf{d}|\mathbf{m})}{\mu_D(\mathbf{d})}\, d\mathbf{d}. \tag{2}$$

The Part Recognition Problem In the system we have constructed, range measurements are taken, surfaces are reconstructed, segmented into parts, and individual models are fit to each part. We will treat *the whole system as a measuring instrument.* Given some model $\mathbf{m}$ in the scene, range measurements are taken and from these an *estimate* of the model $\mathbf{d}$ is obtained, which we call a *measurement of the model* in the scene.

1. **Information Obtained from Physical Theories**
   We first formulate an appropriate distribution to represent what is known about the physical theory that predicts estimates of the model parameters given a model in the scene. Such a theory is too difficult to formulate mathematically given the complications of our system. We therefore build an empirical theory through a process called the *training* or learning stage. Here, Monte Carlo experiments are run on measures of a known model exactly as in traditional statistical pattern classification methods. The conditional probability density function $\theta(\mathbf{d}|\mathbf{m})$ is calculated for each model $\mathbf{m}$ by assuming a multivariate normal distribution. Therefore, the equation for $\theta(\mathbf{d}|\mathbf{m})$ is:

$$\theta(\mathbf{d}|\mathbf{m}) = N(\mathbf{d} - \mathbf{m}, \mathbf{C}_T) \tag{3}$$

   where N is the multivariate normal distribution, with a covariance matrix, $\mathbf{C}_T$, describing estimated modelling errors for a model $\mathbf{m}$.

2. **Information Obtained from Measurements**
   Much of the knowledge we have about a problem comes in the form of experimental measurements. In our system [5], we obtain an estimate of the observed model parameters $\mathbf{d}_{obs}$, and also an estimate of their uncertainty in the covariance operator $\mathbf{C}_d$. The assumption we make is that the multivariate normal distribution $N(\mathbf{d} - \mathbf{d}_{obs}, \mathbf{C}_d)$ represents our knowledge of the measurements. The probability density function representing this information is the conditional probability density function $\nu(\mathbf{d}_{obs}|\mathbf{d})$, such that:

$$\nu(\mathbf{d}_{obs}|\mathbf{d}) = \rho_D(\mathbf{d})/\mu_D(\mathbf{d}) = N(\mathbf{d} - \mathbf{d}_{obs}, \mathbf{C}_d) \tag{4}$$

3. **A Priori Information on Model Parameters**
   In the current context, there are a discrete number of reference models, $\mathbf{m}_i, i = 1 \ldots M$. The probability density function used to convey this knowledge is

$$\rho_M(\mathbf{m}) = \sum_i P(\mathbf{m}_i)\, \delta(\mathbf{m} - \mathbf{m}_i), \tag{5}$$

   where $P(\mathbf{m}_i)$ is the a priori probability that the ith model occurs.

4. Solution to the Inverse Problem

Substituting the probability density functions (3), (4), and (5) into (2) gives us the final equation for the a posteriori probability density function

$$\sigma(\mathbf{m}) = \sum_i P(\mathbf{m}_i) N(\mathbf{d}_{obs} - \mathbf{m}_i, \mathbf{C}_D)\, \delta(\mathbf{m} - \mathbf{m}_i). \tag{6}$$

where $\mathbf{C}_D = \mathbf{C}_d + \mathbf{C}_T$. This density function is comprised of one delta function for each model in the database. Each delta function is weighted by the *belief* $P(\mathbf{m}_i) N(\mathbf{d}_{obs} - \mathbf{m}_i, \mathbf{C}_D)$ in the model $\mathbf{m}_i$. The final distribution represents the "state of knowledge" of the parameters of $\mathbf{m}_i$. The beliefs in each of the reference models are computed by convolving the normal distributions in (3) and (4). The advantage of the method is that rather than establish a final decision as to the exact identity of the unidentified object, it communicates the degree of confidence in assigning the object to each of the model classes. It is then up to the interpreter to decide what may be inferred from the resulting distribution.

The methodology introduced applies to the recognition of any parametric primitive. For our purposes, superellipsoid models were chosen because of the range of shapes they can represent as well as their computational simplicity. However, representations based on superquadrics pose a number of problems due to degeneracies in shape and orientation.

2.2 Determining Which Viewpoints are Informative using the Inverse Theory

Figure 2 shows by example how the resulting belief distribution can be used to differentiate between informative and uninformative viewpoints. In this case, one can see that the system is able to distinguish the cylinder from a block with great ease, if the cylinder is measured from an informative viewpoint. However, if measured from an uninformative viewpoint, there is little confidence in either model. In this case, the beliefs are in fact below the numerical precision of the system, and therefore become zeros.

The problem of distinguishing between the two kinds of states becomes one of determining the threshold, below which one can safely state that the beliefs are in fact insignificant. It is obvious that cases where the beliefs in all the models are zero are uninformative. However, this threshold depends on the numerical precision of the system. In this sense, it is chosen externally. We therefore feel justified in raising this threshold to one that excludes other low confidence states. The expectation is that this will eliminate false positive states, as they are generally occur with low belief. One can determine this cutoff point empirically, by observing the belief distributions from different viewpoints, and noting if there is a clear division between the clear winner states and the low confidence states. A bi-modal distribution would indicate that an application of a predefined threshold can easily distinguish between these states.

Figure 3 illustrates the logarithm of the beliefs resulting from recognizing 36 different single-view samples of each of six models in a database: a Big Sphere, a Block, a Cylinder, a Lemon, a Small Sphere and a Round Block. The results indicate the bi-modality of the belief distribution.

3 Sequential Recognition

Provided that the low belief states have been identified, we wish to make a statement about the remaining beliefs. Even though the majority of the cases

The Database

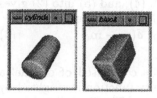

Measured Model	View 1	View 2	View 3	View 4
Belief in cylinder	2.237	0.009181	0.0	0.0
Belief in block	0.0	0.0	0.0	0.0
	a) Informative		b) Uninformative	

At the left of this figure are the two reference models in the data base: the cylinder and the square block. To their right are measured models of the *cylinder* obtained after scanning its surface from 4 different viewing positions. Below each model one can find the unnormalized belief distributions obtained when attempting to recognize each of the measured models.

Fig. 2. (a) Informative and (b) Uninformative Views of a Cylinder.

can be clearly divided into informative and uninformative states, there are still ambiguous cases where a "significant" belief in more than one model exists. Because of these situations, it becomes apparent that evidence from more than one viewpoint is needed. The question becomes: how do we accumulate evidence from different views, when the evidence is in the form of a conditional probability density function? The immediate response is given by the theory (Section 2.1) which formally defines the operation of conjunction of information, i.e. the belief distributions. To state this more formally, we denote belief distributions corresponding to each model hypothesis, $\mathcal{H}_{\rangle}$, given the parameters of the unknown model, $\mathcal{M}$, computed from the measurement, D_j, by $P(\mathcal{H}_{\rangle}|\mathcal{M}_{\mathcal{D}_|})$. Then, given two data sets D_j and D_{j+1} corresponding to different viewpoints we seek a conjunction of $P(\mathcal{H}_{\rangle}|\mathcal{M}_{\mathcal{D}_|})$ and $P(\mathcal{H}_{\rangle}|\mathcal{M}_{\mathcal{D}_{|+\infty}})$ that is equivalent to $P(\mathcal{H}_{\rangle}|\mathcal{M}_{\mathcal{D}_|+\mathcal{D}_{|+\infty}})$. An active agent would then gather sufficient evidence in this fashion until the composite belief distribution associated with a particular hypothesis exceeds a predefined level of acceptability.

Although the theory formally defines conjunction, such an operation requires knowing how a change in viewpoint conditions the respective belief distributions, as they are normalized with respect to a global frame of reference. As a result, relative values between the views are meaningless. The normalizing factor is some unknown function of viewpoint, and is difficult to obtain analytically.(See [17].) As a result, the beliefs are not normalized, making it difficult to compare the values from different viewpoints in a sensible fashion. As well, in situations where the beliefs are "close" in value, it becomes impossible to establish a clear winner.

For this reason, we have chosen not to select a "winner" in ambiguous situations, and state that all beliefs above a threshold indicate equally likely hypotheses. We illustrate this philosophy by binarizing the conditional probability

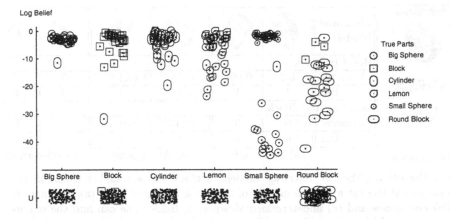

Above are the results from attempting to recognize 36 different single-view samples of each of the models in the database. The beliefs in the different models are represented by different symbols, each symbol indicating the true model used during that trial. The level of numerical underflow of the system is represented by a "U" on the $y - axis$. Because so many trials fall into this category they are marked with a simple point, *except* when the belief is for the true model used in the trial.

Fig. 3. Log of beliefs in the Big Sphere, Block, Cylinder, Lemon, Small Sphere, and Round Block.

density function values at each view, such that all beliefs above the threshold become ones. In this fashion, we have divided the possible results to include:

1. *Informative states*: states with one clear winner (a single positive value).
2. *Uninformative states*: states without a clear winner. These include:
    a) *Ambiguous states*: states with more than one possible winner (more than one single positive value).
    b) *Undetermined states*: states with no winners (all zero values).

It is important to note that ambiguous states are, in fact, undetermined states that lie above the chosen threshold. In theory, careful choice of cutoff level should eliminate these states as well (without eliminating a large number of informative states). Figure 4 illustrates these different states in the case of a square block. Here, the system is asked to identify a square block from different views, and correctly distinguish it from a similar rounder one. This example indicates that the results match human intuition. The clear winners, or informative states, in Figure 4a indicate that the system is able to identify the block despite wide variations in its three dimensions. The ambiguous cases (Figure 4b) occur when the resulting models are rounder in shape. Here, the system has trouble differentiating between the models. In fact, these models resemble the rounded block more than the square one. In the third case (Figure 4c), the system does not have significant belief in any of the models. Intuitively, one can see that these models are not similar to either reference model.

In order to communicate the validity of all hypotheses above a particular threshold, the beliefs are binarized at a threshold value. By normalizing our confidence values in this manner, combining them from different viewpoints becomes straightforward. Should the maximum likelihood hypothesis prevail in a

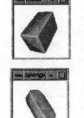

Ref. Models

Measured Model						
Belief in Block						
Unnorm.	0.2	0.007	2.0×10^{-13}	3.4×10^{-13}	0	0
Bin.	1	1	1	1	0	0
Belief in Rnd Block						
Unnorm.	0	0	5.8×10^{-6}	0.002	0	0
Bin.	0	0	1	1	0	0
	a) Informative		b) Ambiguous		c) Undetermined	

On the left are the two reference models: a block and a rounded block. In the first row of the table are the models of the block measured from (a) informative, (b) ambiguous and (c) undetermined viewpoints. Below, one can find the unnormalized, and binarized belief distributions (a threshold of 10^{-13}) obtained when attempting to recognize each of the measured models.

Fig. 4. Informative, Ambiguous, and Undetermined States for the Block.

largely view-invariant manner, then after a sequence of trials, a robust interpretation can be made by tabulating the votes for each one, represented by the binarized beliefs, and picking the hypothesis with the highest score. In this fashion, a clear winner should emerge. In addition, the confidence in the incorrect models should become insignificant.

Figure 5 illustrates an attempt at sequentially recognizing the square block at 40° increments. As in the previous example, the square and round blocks are used as reference models. The raw beliefs are binarized by imposing a threshold of 10^{-13}. Notice that the ambiguous case quickly becomes insignificant with the increase of evidence in the correct model. After only 9 iterations, the clear winner emerges, casting all doubt aside.

View Angle	0°	40°	80°	120°	160°	200°	240°	280°	320°	Final Score
Model								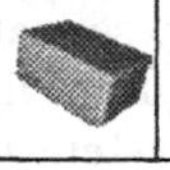		
Belief in Block										
Unnorm	2.0×10^{-13}	0	0.2	0.03	0	0.1	0	0.03	0.001	
Bin	1	0	1	1	0	1	0	1	1	6
Belief in Rnd Block										
Unnorm	5.8×10^{-6}	0	0	0	0	0	0	0	0	
Bin.	1	0	0	0	0	0	0	0	0	1

Displayed above are the 9 models resulting from sequentially measuring the square block at 40° increments. From top to bottom, one can see the viewing angle, the measured model, the unnormalized and binarized (threshold of 10^{-13}) belief distribution resulting from attempting to recognize each of the measured models. The final distribution is the histogram of the binarized distributions.

Fig. 5. Incremental Recognition of a Block.

4 Application to Complex Objects

In practical recognition scenarios objects rarely correspond to the simple shapes depicted in Figure 5. A more realistic object model, such as the one shown earlier in Figure 1, accounts for the fact that objects are often comprised of multiple parts that can be articulated in different ways. This suggests a recognition by parts approach using the sequential recognition strategy developed in the preceding sections. However the task becomes much more difficult because of the complex self-occlusions by different parts of the object. Still, the results suggest that the sequential recognition strategy outlined is sufficiently robust to cope with this added complexity.

In these experiments, two articulated models were used: a potato-head toy consisting of two ears, two eyes, a nose and a head, and an alarm clock with two bells, two legs, a cylindrical face and a back. In addition, six single-part "distractors" were placed in the database in order to render the recognition task more difficult. These objects consisted of: two spheres (rad $= 20mm$, rad $= 25mm$), a block, a cylinder, a lemon, and a block with rounded edges. The objects were chosen for the experiments because they consisted of parts that generally conformed well to non-deformable superellipsoids, with the exception of the toy head whose shape was tapered. The parts varied in size and shape, so as not to be clustered together too tightly in five-dimensional feature space. However, their distributions overlapped sufficiently enough in several dimensions so that the recognition procedure was challenged in its discrimination task.

Training automatically produced object class representatives, by measuring the object numerous times. Each individual model was created by scanning the object from several views in an exploration sequence [4, 5]. Here, each object was scanned using a laser range-finder, segmented into its constituent parts, a superellipsoid model was fit to each part, and the resulting parameters stored. In order to create the representatives in the database, 24 samples of each single-part object, 10 samples of the potato-head, and 7 samples of the alarm clock were used. Figure 6 illustrates the actual potato-head and alarm used in recognition experiments, and the representative models of each object that result from training.

The first result (Figure 7) shows that the system is able to successfully recognize instances of articulated parts of a complex object with only partial information available, even with the added effects of self-occlusion. It also indicates that, for most models, an external threshold retained most of the correct states, confirming that the system had high confidence in the correct identifications. In addition, the majority of the false-positive assertions were eliminated. This confirms the hypothesis that, because the beliefs are bi-modal in nature, the application of an external threshold can be used to successfully distinguish between informative and uninformative viewpoints. An active observer can then assess these results from a particular viewpoint and determine if further sampling is necessary.

Table 1a shows the results of now accumulating this evidence over the sequence of views. A similar result is obtained for the alarm clock object (binarized belief distribution not shown) and shown in Table 1b. The results show that the majority of the evidence in the incorrect models was removed after application of an external threshold. The exceptions to this rule are the potato-head's head and the face of the alarm clock, where the majority of the evidence in the correct model was eliminated as well. This indicates the possibility that

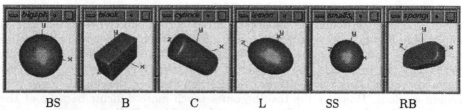

BS   B   C   L   SS   RB

Displayed above are reference objects that result from training acting as "distractors" for the recognition procedure: a big sphere (BS), a block (B), a cylinder (C), a lemon (L), a smaller sphere (SS), and a rounded block (RB).

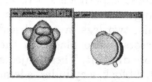

(a) Real potato-head and alarm clock used in experiment.

(b) Reference potato-head and alarm clock models created by training.

Fig. 6. The reference parts resulting from training.

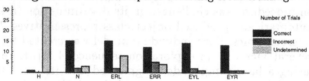

Threshold = 0.00001

Displayed above is the belief distribution of the potato-head measured from single view-points. The parts of the potato-head are: a head (H), a nose (N), a left ear (ERL), a right ear (ERR), a left eye (EYL), and a right eye (EYR). Here, labelling one eye as the other, or one ear as the other was considered to be a correct identification. Zero values are defined application of a threshold of 0.00001.

Fig. 7. Matching samples of the potato-head taken from single viewpoints.

the choice of threshold was not appropriate for these parts However, even with a uniform threshold, the results indicate that the correct assertion is obtained with the combination of a threshold to remove false assertions and the accumulation of information from a series of views to remove the ambiguous cases. In fact, if one were to choose a winner based on a maximum likelihood scheme of the accumulated evidence, the results would be correct for all models[3].

5 Discussion and Conclusions

In this paper, we have introduced a method for distinguishing between informative and uninformative viewpoints and for assessing the beliefs associated with a particular set of assertions based on this data. The importance of this result

---

[3] We have treated the left and right eyes of the potato-head as being instances of the same class. A similar rule was applied to the left and right ears, as well as the left and right bells and the left and right legs of the alarm clock.

	H	N	EAR	EYE	BS	B	C	L	SS	RB
H	1	0	0	0	0	0	0	0	0	0
N	0	16	2	12	0	0	0	0	0	0
ERL	0	1	15	1	0	0	0	0	0	0
ERR	0	1	13	1	2	0	0	0	3	0
EYL	0	3	0	16	0	0	0	0	0	0
EYR	0	3	0	14	0	0	0	0	0	0

a) Accumulation of evidence in potato-head, threshold = 0.00001

	F	BA	BELL	LEG	H	N	EAR	EYE	BS	B	C	L	SS	RB
F	1	0	0	0	0	0	0	0	0	0	0	0	0	0
BA	0	6	0	0	0	0	0	0	0	0	0	0	0	0
RBL	0	0	4	0	0	1	2	1	0	0	1	0	3	0
LBL	0	0	6	0	0	3	0	1	0	0	1	0	2	0
RL	0	0	0	13	0	0	0	0	0	0	0	0	0	0
LL	0	0	0	15	0	0	0	0	0	0	0	0	0	0

b) Accumulation of evidence in alarm clock, threshold = 0.0001

Displayed above are the tables describing the accumulation of evidence from 32 single-view recognition experiments. Each row describes the histogram of the binarized belief distributions for a particular measured model. In a), the measured models include the parts of the potato-head (see Figure 7 for notation.) In b), the measured models include the parts of the alarm clock: the face (F), the back (BA), the right bell (RBL), the left bell (LBL) the right leg (RL), and the left leg (LL). The columns refer to the reference models, including the alarm clock parts: the face (F), the back (BA), the legs (LEG), and the bells (BELL), the potato-head parts and the single-part objects. Zero values are defined by a) b) a threshold of 0.0001.

Table 1. Histogram of binarized belief distributions for the potato-head and the alarm clock after 32 single-view iterations.

is that it provides a basis by which an external agent can assess the quality of the information from a particular viewpoint, and make informed decisions as to what action to take using the data at hand. Our approach was based on a generalized inverse theory [2] using a probabilistic framework where assertions are represented by conditional probability density functions (belief distributions). The importance of the method is that it provides a formal recipe for representing and combining all prior knowledge in order to obtain these distributions. We have illustrated how to apply the theory to solve a 3-D model-based recognition problem and have shown how the resulting belief distributions can be used to assess the quality of the interpretation. An important characteristic of the resulting belief distributions is that they are bi-modal, simplifying the problem of determining how to distinguish between informative and uninformative viewpoints.

A major strength of the method is its potential for a wide variety of applications. For example, an active recognition agent can choose viewpoints that will maximize the belief distribution associated with an object of interest. We have not specified *how* to choose this viewpoint, but the method can be used to determine if the particular choice leads to a sufficient level of information. Another important application of the methodology is a strategy for off-line computation of a pre-computed set of characteristic views. One can rank these views according to the belief distributions, and then store the *n* best views. Predefining these views speeds up on-line computations by directing the active agent's attention to informative views, thereby reducing the search space of viable hypotheses. These and other topics are currently under investigation in our laboratory.

References

1. Ed. Aloimonos, Y., "Purposive, qualitative, active vision", *CVGIP: Image Understanding*, vol. 56, no. 1, pp. 3–129, 1992, special issue.
2. Albert Tarantola, *Inverse Problem Theory: Methods for Data Fitting and Model Parameter Estimation*, Elsevier Science Publishing Company Inc., 52, Vanderbuilt Avenue, NewYork, NY 10017, U.S.A., 1987.
3. Tal Arbel, Peter Whaite, and Frank P. Ferrie, "Recognizing volumetric objects in the presence of uncertainty", in *Proceedings 12th International Conference on Pattern Recognition*, Jerusalem, Israel, Oct 9-13 1994, pp. 470–476, IEEE Computer Society Press.
4. Peter Whaite and Frank P. Ferrie, "From uncertainty to visual exploration", *IEEE Transactions on Pattern Analysis and Machine Intelligence*, vol. 13, no. 10, pp. 1038–1049, Oct. 1991.
5. Peter Whaite and Frank P. Ferrie, "Autonomous exploration: Driven by uncertainty", in *Proceedings, Conference on Computer Vision and Pattern Recognition*, Seattle, Washington, June 21-23 1994, Computer Society of the IEEE, pp. 339–346, IEEE Computer Society Press.
6. K. Bowyer and C. Dyer, "Aspect graphs: An introduction and survey of recent results", in *Close Range Photogrammetry Meets Machine Vision, Proc. of SPIE*, 1990, vol. 1395, pp. 200–208.
7. Farshid Arman and J.K. Aggarwal, "Model-based object recognition in dense-range images - a review", *ACM Computing Surveys*, vol. 25, no. 1, pp. 5–43, apr 1993.
8. A. H. Barr, "Superquadrics and angle preserving transformations", *IEEE Computer Graphics and Applications*, vol. 1, no. 1, pp. 11–23, Jan. 1981.
9. R. Bajcsy and F. Solina, "Three dimensional object recognition revisited", in *Proceedings, 1ST International Conference on Computer Vision*, London,U.K., June 1987, Computer Society of the IEEE, IEEE Computer Society Press.
10. Narayan S. Raja and Anil K. Jain, "Recognizing geons from superquadrics fitted to range data", *Image and Vision Computing*, April 1992.
11. Frank P. Ferrie, Jean Lagarde, and Peter Whaite, "Darboux frames, snakes, and super-quadrics: Geometry from the bottom up", *IEEE Transactions on Pattern Analysis and Machine Intelligence*, vol. 15, no. 8, pp. 771–784, Aug. 1993.
12. A. Pentland and S. Sclaroff, "Closed form solutions for physically based shape modelling and recognition", in *IEEE Transactions on Pattern Analysis and Machine Intelligence: Special Issue on Physical Modeling in Computer Vision*, T. Kanade and K. Ikeuchi, Eds., July 1991, vol. 13(7), pp. 715–729.
13. Jayashree Subrahmonia, David B. Cooper, and Daniel Keren, "Practical reliable bayesian recognition of 2D and 3D objects using implicit polynomials and algebraic invariants", LEMS 107, Brown University LEMS, Laboratory fo Engineering Man/Machine systems, Division of Engineering, Brown University, Providence, RI 02912, USA, 1992.
14. A. Lejeune and F.P. Ferrie, "Partitioning range images using curvature and scale", in *PROC. IEEE Computer Society Conference on Computer Vision and Pattern Recognition*, New York City, New York, June 15-17 1993, pp. 800–801.
15. Daniel Keren, David Cooper, and Jayashree Subrahmonia, "Describing complicated objects by implicit polynomials", Tech. Rep. 102, Brown University LEMS, Laboratory for Engineering Man/Macine Systems, Division of Engineering, Brown University, Providence RI 021912 USA, 1992.
16. Roland T. Chin and Charles R. Dyer, "Model-based recognition in robot vision", *Computing Surveys*, vol. 18, no. 1, pp. 67–108, mar 1986.
17. Tal Arbel and Frank P. Ferrie, "Parametric shape recognition using a probabilistic inverse theory", *Pattern Recognition Letters*, p. TBA, 1996.

Unsupervised Texture Segmentation using Selectionist Relaxation

Philippe Andrey and Philippe Tarroux

Groupe de BioInformatique, Département de Biologie
Ecole Normale Supérieure, 46 rue d'Ulm, F-75230 Paris Cedex 05

Abstract. We introduced an unsupervised texture segmentation method, the selectionist relaxation, relying on a Markov Random Field (MRF) texture description and a genetic algorithm based relaxation scheme. It has been shown elsewhere that this method is convenient for achieving a parallel and reliable estimation of MRF parameters and consequently a correct image segmentation. Nevertheless, these results have been obtained with an order 2 model on artificial textures. The purpose of the present work is to extend the use of this technique to higher orders and to show that it is suitable for the segmentation of natural textures, which require orders higher than 2 to be accurately described. The results reported here have been obtained using the generalized Ising model but the method can be easily transposed to other models.

1 Introduction

Texture segmentation is an important issue for computer vision. Since texture is often related to the nature and composition of natural objects, it offers a reliable and efficient way to identify them. Thus, segmentation of the external scene on the basis of textural information leads to a partition of the world in more insightful classes than those based on other criteria. However, the question of how to achieve a fast and reliable segmentation of a natural scene on the basis of its textural information is far from being solved.

The various existing approaches to texture segmentation [16] can be distinguished on the basis of their underlying texture descriptions, which can be either structural, feature-based, or stochastic model-based. Among the last category, the Markov Random Field (MRF) approach to texture modeling [5, 6, 7] offers local and parsimonious texture descriptions. These properties combined with the equivalence between MRFs and Gibbs random fields [3] has allowed a widespread use of MRFs for image processing [7].

However, methods relying on MRF texture modeling require an estimation of MRF parameters for each texture present in the image. In the unsupervised segmentation mode, which is the more realistic, a circular problem arises: the parameter estimates required to segment the image can only be computed from an already segmented image. A first solution to this problem is to assume that the image can be partitioned into homogeneously textured blocks large enough to provide reliable parameter estimates, which are subsequently used for segmentation through the optimization of a criterion such as the likelihood [4, 15], the a posteriori probability [14, 12, 15], or the classification error [14]. This approach may suffer from the complexity of real images, which usually contain unequally sized and irregularly shaped textured regions. A second solution consists in iterating an estimation/segmentation cycle [17], during which segmentation and estimates are alternatively refined. The cycle is repeated with different number of textures, and the optimal number is determined using a model fitting criterion. This renders the unknown number of textures critical in this approach.

Hence, the combined problem of estimation and segmentation represents a large combinatorial problem, that is usually solved using computationally expensive relaxation methods. Among relaxation methods for solving large optimization problems, those based on genetic algorithms (GAs) [10, 11] seem to be well adapted to image processing. GAs are parallel search methods inspired by natural evolution. A population of candidate solutions is updated iteratively by genetic operators, the effect of which is to spread out good solutions within the population while permanently generating possibly better ones. GAs not only achieve an explicit parallel exploration of the search space by exploiting a population of solutions, hence avoiding being stuck into local optima, but also an implicit parallel testing of each characteristic of a given solution, which has been demonstrated as one of the main reason of their efficiency [11].

Furthermore, within the context of image segmentation, the parallel exploration of the search space combines with the parallelism inherent to the structure of the image. Each solution in the population provides a local answer to the question of how to assign a label to a given texture. One solution is thus associated with each point of the image and the parallel evaluation of the solutions goes together with the search for the right label. Unlike in the standard GA, each individual compete and/or recombine almost exclusively with its neighbors. This population structure is here suggested by the fact that, except for boundary sites, close sites are likely to share common textural characteristics.

The method we use here, termed selectionist relaxation (SR), is an unsupervised segmentation method based on a fine-grained distributed GA [13, 1, 2]. Each pixel of the image is associated with a solution in the form of a unit linking a set of MRF parameters to a label. In the present work, textures are modeled using the generalized Ising model [8, 6, 9]. SR has been shown to achieve correct segmentations of order 2 synthetic textures [2]. However, natural textures cannot accurately be described by order 2 models. The purpose of the present work is to extend the SR approach to higher order texture descriptions, which is a necessary prerequisite for natural texture segmentation.

2 Background on MRF Texture Modeling

A $N_R \times N_C$ texture sample is considered as a realization of a stochastic process $\boldsymbol{X} = \{X_s, s \in S\}$, wherein $S = \{s = (i, j) : 0 \leq i < N_R, 0 \leq j < N_C\}$ is the set of sites in the image, and each X_s is a random variable with values in the gray level set $G = \{0, \cdots, g - 1\}$. Each site is assigned a symmetrical set of neighboring sites, the number of which depends on the order of the model

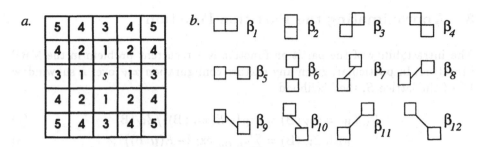

Fig. 1. (a) Neighboring sites of site s. At order n, the neighborhood contains all the sites that are assigned a digit less than or equal to n. (b) Clique families. The 12 pair clique families of the fifth-order model are shown with their associated parameters.

(Fig. 1a). A set of reciprocally neighboring sites is called a clique. Because the generalized Ising model is a pairwise interaction model, only pair cliques will be of interest herein. The set of pairwise cliques is $C = \cup_{i=1}^{p} C_i$, p being the number of clique families. At order 5, for example, there are 12 such families (Fig. 1b).

In the Gibbs/MRF approach to texture modeling, each possible outcome x of X is assigned a probability given by

$$P(X = x) = \exp\{-E(x)\}/Z ,\qquad (1)$$

wherein the energy $E(x)$ is the sum of clique potentials and the normalizing constant Z is the partition function

$$E(x) = \sum_{c \in C} V_c(x) ,\qquad (2)$$
$$\text{and } Z = \sum_{y \in \Omega} \exp\{-E(y)\} ,\qquad (3)$$

$\Omega = G^{|S|}$ being the state space of X.

For the generalized Ising model, the potential of a clique $c = \{s,t\} \in C_i$ is defined as [8, 6, 9]

$$V_c(x) = (1 - 2\delta_{x_s,x_t})\beta_i ,\qquad (4)$$

wherein β_i is the parameter associated to the ith clique family (Fig. 1b), and δ is the Kronecker symbol. The attractive or repulsive nature of the interactions between neighboring pixels, as well as the strength of these interactions, are governed by the signs and the absolute values of clique parameters.

Letting $B = (\beta_1, \cdots, \beta_p)$ denote a vector of model parameters, and defining the vector $K(x) = (\kappa_1(x), \cdots, \kappa_p(x))$ by

$$\kappa_i(x) = \sum_{\{s,t\} \in C_i} 1 - 2\delta_{x_s,x_t} ,\qquad (5)$$

the energy of any configuration x, given B, can be conveniently rewritten as

$$E(x; B) = B \cdot K(x)^t .\qquad (6)$$

3 Approximating the Partition Function

The intractability of the partition function is a recurrent problem in the MRF literature. In particular, given any texture configuration x_W over a subwindow W of the lattice S, the likelihood

$$P(X_W = x_W; B) = \exp\{-E(x_W; B)\}/Z_W(B) , \tag{7}$$
$$\text{with } Z_W(B) = \sum_{y \in \Omega_W} \exp\{-E(y; B)\} , \tag{8}$$

cannot be determined. Indeed, $Z_W(B)$ has no simple analytical form as a function of the β_is, and it cannot be numerically computed either because of the huge number of configurations over W ($|\Omega_W| = g^{|W|}$).

Texture window likelihood evaluation is essential to SR. To overcome the awkwardness of the partition function, we propose to use the approximation $Z_W(B) \approx \tilde{Z}_W(B)$ introduced in [2], with

$$\tilde{Z}_W(B) = g^{n-2} \left\{ g^2 - ng(g-2) \sum_{i=1}^{p} \beta_i + 2n(g-1) \sum_{i=1}^{p} \beta_i^2 + \frac{n^2}{2}(g-2)^2 \left[\sum_{i=1}^{p} \beta_i \right]^2 \right\} \tag{9}$$

wherein g is the number of gray levels and n is the size of the window W, which is assumed to be toroidal.

This expression is derived by approximating the generic term in $Z_W(B)$ by its second-order expansion

$$\exp\{-E(y; B)\} \approx 1 - E(y; B) + 1/2\, E(y; B)^2 , \tag{10}$$

and by then summing up all the approximated terms.

Clearly, from (6), the error due to (10) vanishes as the β_is $\to 0$, and so does the summed resulting error. Accordingly, (9) can be interpreted as a high-temperature approximation of the partition function. The temperature T that usually characterizes the Gibbs distribution (1) can indeed be considered as implicitly incorporated within the parameters. Thus, for any parameter vector B, one can define

$$B^* = BT . \tag{11}$$

It then appears that, because all β_is $\to 0$ when $T \to \infty$, the error due to (10) vanishes at high temperatures. This is illustrated in Fig. 2, which plots the relative error defined by

$$\text{err}(T) = \left| Z(B) - \tilde{Z}(B) \right| / Z(B) = \left| Z(B^*/T) - \tilde{Z}(B^*/T) \right| / Z(B^*/T) . \tag{12}$$

In the remainder of the paper, we return to the use of B (instead of B^* and T), considering T as an implicit scaling factor: the condition $\beta_i \ll 1$, necessary to keep the approximation error reasonable, will be interpreted as an absorption of T within the parameters themselves, according to (11).

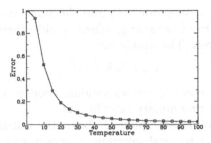

Fig. 2. Plot of the relative approximation error as a function of temperature ($g = 8$ gray levels, $n = 3 \times 3$ pixels, model at order 5 with $\beta_i = 1(i \leq 4)$, and $\beta_i = -1(i > 4)$).

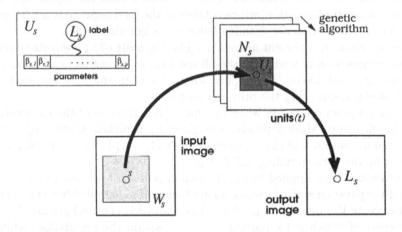

Fig. 3. Overview of selectionist relaxation. The labelization of each site of the input image is achieved by a unit (detailed in the inset). The population of units is iteratively updated through a fine-grained GA.

4 Selectionist Relaxation

Selectionist relaxation is an unsupervised segmentation method whereby the transformation of an input image into an output image is computed by a population of units iteratively evolving through a distributed GA (Fig. 3).

A unit $U_s = (B_s, L_s)$ is assigned to each site s of the image. L_s is a candidate texture label for site s, while B_s is a candidate vector of model parameters. Unit U_s is assigned a fitness value $f(U_s)$ that measures the adequacy of B_s to the local texture data in the $w \times w$ window W_s surrounding site s. $f(U_s)$ is given by the (approximated) log-likelihood

$$f(U_s) = -E(x_{W_s}; B_s) - \log(\tilde{Z}_{W_s}(B_s)) . \qquad (13)$$

At the beginning of the segmentation process, each label L_s, $s \in S$, is set to the raster scan index of site s, and each model parameter vector B_s is chosen at random in the set $[-\delta; \delta]^p$. Because unit fitness relies on the approximation of the partition function, unit parameters must be small enough to prevent large approximation errors from occurring, which would otherwise gives unreliable log-likelihood evaluations. The simple rule

$$\delta = \alpha/w^2 \ (\alpha < 1) \tag{14}$$

has been used with success. α values yielding good segmentation results are coarsely determined in preliminary experiments.

Once it has been build as explained above, the initial population of units $U(0)$ is then submitted to local selection, crossover and mutation, leading to population $U(1)$. Since some units are copied while others are discarded during selection, labels are not the same in $U(0)$ and in $U(1)$: the segmented image has been modified. This relaxation cycle is iterated until the segmented image reaches a stable configuration, which is taken as the final segmentation. Selection, crossover, and mutation are detailed below for a generic site s.

During selection, the eight nearest neighboring units of U_s compete to replace it. This competition is a winner-take-all mechanism known as local tournament selection: the unit having the highest fitness value in the neighborhood N_s is copied at site s, replacing the previous unit U_s.

Crossover is an operation whereby unit U_s inherits one of the parameters of a neighboring unit. More precisely, a neighbor U_r is picked in the neighborhood of U_s, and a position k is chosen at random in $\{1, \cdots, p\}$. The value $\beta_{s,k}$ is then replaced by the corresponding value $\beta_{r,k}$.

Mutation is then applied to U_s. A random position l is chosen on the vector B_s, and the parameter $\beta_{s,l}$ is added a small amount m sampled from the uniform distribution in the interval $[-\mu; \mu]$. In GAs, mutation is a background operator, the purpose of which is to maintain diversity within the population, while not having large disruptive effects. To prevent such effects from occurring, we define

$$\mu = \epsilon\delta \ (\epsilon < 1) \ , \tag{15}$$

As is common practice with GAs, the importance given to mutation is chosen on an empirical basis. A small value such as $\epsilon = 0.02$ gave satisfactory results.

Each of the three operators (selection, crossover, and mutation) synchronously affects all the units on the grid S. Each time step, however, the units having at least one neighbor with a label differing from their own label are not processed as described above. These EDGE units correspond to the units located at the borders of the homogeneous label blobs that are growing during the relaxation process as a result of local selection from one site to the next.

To prevent crossover between units that correspond to different textures, EDGE units do not undergo crossover. They are not affected by mutation either, which would otherwise results in large instability of label boundaries because of sudden fitness changes. Unlike selection at a NON-EDGE unit site, which only involves the nearest neighboring units, selection at an EDGE unit site additionally involves a remote unit picked anywhere in the population. Consequently, labels are not only spreading over the image from one site to the next, but they are also allowed to jump to remote sites, so that distinct regions with the same texture can be assigned a same label even though they are not connected.

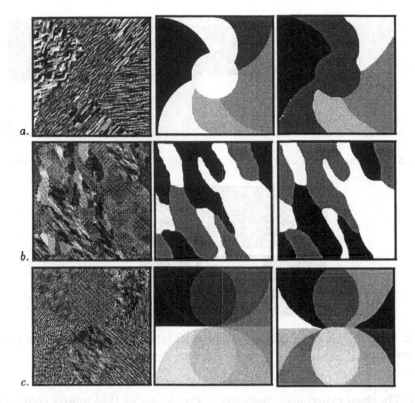

Fig. 4. Segmentation results on three 256×256 synthetic images containing, respectively, 5 (a), 3 (b), and 10 (c) textures with 8 gray levels. Samples of the generalized Ising model at order 5 were synthesized using the Gibbs Sampler [8] over 100 steps. From the left to the right: input images, ideal segmentations, and segmentation results.

5 Segmentation Results

The ability of SR to segment images containing MRF generated textures is illustrated on three synthetic texture patches (Fig. 4, Left). In each case, the algorithm was run for 500 time steps. As explained in the previous section, α (for initialization) and ϵ (for mutation) have been determined in preliminary experiments. The values $\alpha = 0.001$ and $\epsilon = 0.02$ are kept constant throughout all experiments. The third and last SR parameter is the size w of the windows W_s that are used to evaluate unit fitnesses. The same value $w = 15$ has been used in each experiment. The number of textures in the images and their associated parameters are determined automatically during the relaxation process.

Figure 4a illustrates the segmentation of an image that contain 5 textures with the same global diagonal orientation. In this image, each texture forms a unique, totally connected region. Figure 4b reports the segmentation of an image with 3 textures, each of which forms several, unconnected regions. Note how, thanks to the long-distance unit interactions occurring during selection at EDGE unit sites, all the regions having the same texture are assigned the same

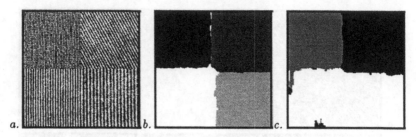

Fig. 5. Segmentation at order 5 (b) and at order 2 (c) of four natural textures (a).

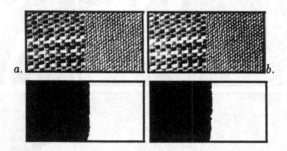

Fig. 6. Segmentation of Brodatz textures without (a) and with (b) an artificial gradient of illumination.

label. The third experiment, reported in Fig. 4c, shows that images containing a larger number of textures can be dealt with as well: in this particular case, the image to be segmented is made of 10 textures.

SR has also been tested on natural textures. The input image (Fig. 5a) contains 4 distinct textures of textile fibers. Histogram quantization has been done before segmentation to reduce the number of gray levels from 256 to 4. In this experiment, the size of the texture windows has been increased to $w = 31$.

As can be assessed from Fig. 5b, the segmentation is satisfactory, in the sense that the exact number of regions is found as well as their locations. Note that boundary errors are greater than in the preceding experiments, which is expected given the more than four times as big texture windows in the present case. For comparison, Fig. 5c shows the segmentation result obtained with the second-order model. In this case, each site has only 8 neighbors (Fig. 1a), and there are 4 distinct clique families (1–4 in Fig. 1b). Even though some textural features are captured, such as orientation, the second-order model cannot deal with other properties such as the different spatial frequencies of the two vertically oriented textures in the lower part of the image.

The segmentation at order 5 is all the more acceptable that it has overcome the inherent difficulty offered by this image, which is due to the non stationarity of the textures. In particular, the result is not affected by the variations in illumination. To further investigate the robustness of the algorithm with regards to such non stationarities, SR has been tested on an image whereupon an artificial gradient in illumination has been added. As can be assessed from Fig. 6, the segmentation is robust with regards to this gradient.

Execution times increase with the size of the image since the number of units in the population is equal to the number of pixels. On a sequential computer, 100 relaxation steps are achieved within about 5 mn for a 256×256 image.

6   Discussion - Conclusion

In a recent work [2], we have used SR and an approximation of the partition function to segment texture samples of the second-order generalized Ising model in unsupervised mode. The selectionist scheme is efficient because to achieve a correct segmentation, it is sufficient to spread out, over each texture, the unit with the greatest likelihood among a population of units, even when this unit does not bear the absolute optimal parameter set for that texture. In the present paper, these results have been extended with the following two contributions:

First, it has been shown that the previously obtained results generalize to higher order instances of the generalized Ising model. Segmentation experiments covering various texture region numbers, sizes, and geometries, have been reported. Unlike previously proposed solutions to the unsupervised texture segmentation problem, SR does not make any assumption concerning the textural contents of the image. Besides the two parameters α and ϵ, a coarse empirical tuning of which was performed off-line once for all, the only external parameter of the algorithm is the size w of the texture windows whereupon the evaluation of the units is based. This parameter is easily tuned because it is naturally dependent on the coarseness of the textures.

Second, the ability of the method to deal with high order instances of the generalized Ising model allowed us to tackle the problem of segmenting natural textures. Up to now, the use of this model has been hampered by the computational awkwardness of the partition function. The approximation introduced in [2] and used here allows to overcome this difficulty. Using this approximation at high orders, it is shown here that the generalized Ising model exhibits interesting properties for texture segmentation, in particular by being insensitive to the variations in illumination. This property is of course inherited from the definition of the model clique potentials (4). The autobinomial model [5], a more popular model for texture segmentation based on MRF texture modeling [12], does not exhibit such a fundamental property.

However, the all-or-none clique potentials of the generalized Ising model strongly limit its range of application. In particular, the necessity to quantify input images upon a small number of gray levels may result in an important loss of textural information. It is noted that in SR, the performance results from two components. The first one is the GA-based relaxation scheme. The second one is the texture description used. One attractive feature of this method is that, should the global performance be unsatisfactory because of an inadequate texture model, the extension to a more appropriate representation is straightforward. This property, inherent to any GA, results from the fact that the genetic operators are problem independent, the fitness function being the problem specific component of such algorithms. Further research with SR will be concerned by such an extension to MRF models that may be more appropriate than the generalized Ising model to capture the complexity of natural textures.

References

1. P. Andrey and P. Tarroux. Unsupervised image segmentation using a distributed genetic algorithm. *Patt. Recogn.*, 27(5):659–673, 1994.
2. P. Andrey and P. Tarroux. Unsupervised segmentation of Markov random field modeled textured images using selectionist relaxation. Technical report BioInfo-95-03, Groupe de BioInformatique, Ecole Normale Supérieure, Paris, August 1995.
3. J. Besag. Spatial interaction and the statistical analysis of lattice systems. *J. Roy. Stat. Soc., Ser. B*, 36:192–236, 1974.
4. F. S. Cohen and Z. Fan. Maximum likelihood unsupervised textured image segmentation. *CVGIP: Graph. Models Image Process.*, 54(3):239–251, 1992.
5. G. R. Cross and A. K. Jain. Markov random field texture models. *IEEE Trans. Patt. Anal. Mach. Intell.*, 5(1):25–39, 1983.
6. H. Derin and H. Elliott. Modeling and segmentation of noisy and textured images using Gibbs random fields. *IEEE Trans. Patt. Anal. Mach. Intell.*, 9(1):39–55, 1987.
7. R. C. Dubes and A. K. Jain. Random field models in image analysis. *J. Appl. Stat.*, 16(2):131–164, 1989.
8. S. Geman and D. Geman. Stochastic relaxation, Gibbs distributions, and the Bayesian restoration of images. *IEEE Trans. Patt. Anal. Mach. Intell.*, 6(6):721–741, 1984.
9. G. L. Gimel'farb and A. V. Zalesny. Probabilistic models of digital region maps based on Markov random fields with short- and long-range interaction. *Patt. Recogn. Lett.*, 14:789–797, 1993.
10. D. E. Goldberg. *Genetic Algorithms in Search, Optimization and Machine Learning.* Addison Wesley, Reading, MA, 1989.
11. J. H. Holland. *Adaptation in Natural and Artificial Systems.* University of Michigan Press, Ann Arbor, MI, 1975.
12. R. Hu and M. M. Fahmy. Texture segmentation based on a hierarchical Markov random field model. *Signal Process.*, 26:285–305, 1992.
13. B. Manderick and P. Spiessens. Fine-grained parallel genetic algorithms. In J. D. Schaffer, editor, *Proc. Third Intl. Conf. on Genetic Algorithms*, pages 428–433. Morgan Kaufmann, San Mateo, CA, 1989.
14. B. S. Manjunath and R. Chellappa. Unsupervised texture segmentation using Markov random field models. *IEEE Trans. Patt. Anal. Mach. Intell.*, 13(5):478–482, 1991.
15. H. H. Nguyen and P. Cohen. Gibbs random fields, fuzzy clustering, and the unsupervised segmentation of textured images. *CVGIP: Graph. Models Image Process.*, 55(1):1–19, 1993.
16. T. R. Reed and H. J. M. du Buf. A review of recent texture segmentation and feature extraction techniques. *CVGIP: Image Underst.*, 57(3):359–372, 1993.
17. C. S. Won and H. Derin. Unsupervised segmentation of noisy and textured images using Markov random fields. *CVGIP: Graph. Models Image Process.*, 54(4):308–328, 1992.

Robust Affine Structure Matching for 3D Object Recognition

Todd A. Cass

Xerox Palo Alto Research Center
Palo Alto, CA 94304 USA, cass@parc.xerox.com

Abstract. This paper considers a model-based approach to identifying and locating known 3D objects from 2D images. The method is based on geometric feature matching of the model and image data, where both are represented in terms of *local geometric features*. This paper extends and refines previous work on feature matching using transformation constraint methods by detailing the case of full 3D objects represented as point features and developing geometric algorithms based on *conservative approximations* to the previously presented general algorithm which are much more computationally feasible.

Keywords: *Object Recognition, Geometric Feature Matching, Pose Recovery, Pose Refinement, Affine Structure, Model Indexing, Computational Geometry.*

1 Introduction

This paper considers the task of identifying and locating known 3D objects in 2D images of a scene containing them by matching geometric representations of a model of the object to geometric representations of the image data—a task often called *model-based object recognition*. We take a *model-based* approach to *interpreting* subsets of sensory data corresponding to *instances* of objects in the scene in terms of *models* of the objects known *a priori*. Both the model and the image data are represented in terms of sets of local geometric features[1] consisting of geometric elements like points, line segments, and curve normals. An approach based on geometric feature matching is used to interpret the image by finding both the model to data feature correspondences and a 3D pose of the model which could produce the observed instance.

Geometric feature matching is difficult because the correspondences between the model features and the corresponding subset of data features are unknown, there is clutter from data features with no model correspondence, there are model features which have no corresponding data feature due to occlusion and other processes, and there is geometric uncertainty in the measurement of the data features arising from the imaging process, the feature representation process, and scene events such as occlusion.

---

[1] Throughout this paper we will follow the convention that terms being effectively defined will appear in sans-serif bold, while items being emphasized for clarity will appear in *italics*.

This paper considers the case of 3D objects and 3D transformations, greatly extending and refining previously reported techniques for geometric matching based on discrete constraints and methods from computational geometry first presented for the case of 2D objects and planar transformations and 2D and 3D objects and 3D transformations[5, 6, 8].

We present a technique for feature matching based on analyzing constraints on feasible feature matchings in transformation space. It applies to affine feature matching of 2D and 3D point sets and 2D and 3D affine motion. The main contribution is a method for carefully projecting the constraints in the high-dimensional transformation space, an 8-dimensional space for 3D affine motion, down to a two dimensional transformation space. This approximation is conservative in that it will never exclude feasible matches. This approximate projection then allows the use of planar geometric algorithms (which is very important due to the great difficulty of geometric computations in higher dimensions) for analyzing these transformation constraints to find feasible matchings. Efficient algorithms for this planar geometric analysis are presented, and experimental results of their application to matching are presented.

1.1 The Case of 3D Models and 2D Data

This section will first give a brief outline of some of the notation used and the key ideas of this geometric feature matching method; a more detailed explanation can be found in other papers[6, 7, 9]. The local geometric features used here consist of points. Denote a model feature $\mathbf{m}$ by a vector representing the feature's position and denote the position of a data feature by $\mathbf{d}$. Denote the bounded uncertainty[3, 13] region U^i for the data feature position measurement, for feature $\tilde{\mathbf{d}}_i$. We let $\tilde{\mathbf{d}}_i$ denote the measured value of a model feature and $\mathbf{d}_i$ denote the value without error. The true position of $\mathbf{d}_i$ falls in the set U^i, which contains $\tilde{\mathbf{d}}_i$.

A single transformed model feature is aligned with a data feature when $T[\mathbf{m}_i] \in U^j$, so the set of transformations T geometrically feasible for a particular model and data feature match $(\mathbf{m}_i, \tilde{\mathbf{d}}_j)$ is the set of transformations $\mathcal{F}^{ij} = \{T \in \Omega | T[\mathbf{m}_i] \in U^j\}$. Define a match as a pairing of a model feature and a data feature, and a match set, denoted $\mathcal{M}$, as an arbitrary set of matches—some subset of all possible matches. A match set is called geometrically consistent if and only if $\bigcap_{(i,j)} \mathcal{F}^{ij} \neq \emptyset$ for (i,j) such that $(\mathbf{m}_i, \mathbf{d}_j) \in \mathcal{M}$.

With each transformation T we associate the function $\varphi(T) = \{(\mathbf{m}_i, \tilde{\mathbf{d}}_j) | T[\mathbf{m}_i] \in U^j\}$ defined to be the set of feature matches aligned by the transformation T. We call the match set $\varphi(T)$ a maximal geometrically-consistent match set. A match set $\mathcal{M}$ is a maximal geometrically-consistent match set (or, a maximal match set) if it is the largest geometrically consistent match set at some transformation T. Maximal match sets are the only sets of feature correspondences we need to consider, and in the cases of interest here there are only a polynomial number of them.

Finally, we define the notion of a geometric consistency measure, $\Phi(T)$, characterizing the number of matches feasible for a given transformation T. The simplest measure is a count of the number of matches feasible at a transformation $\Phi^{\mathcal{M}}(T) = |\varphi(T)|$; we use a more sophisticated measure which is the minimum of the number of distinct model features and distinct data features appearing in the feasible match set[17] given by

$$\Phi_T(T) = \min\left(\left|\{\mathbf{m}_i; (\mathbf{m}_i, \tilde{\mathbf{d}}_k) \in \varphi(T)\}\right|, \left|\{\mathbf{d}_j; (\mathbf{m}_l, \tilde{\mathbf{d}}_j) \in \varphi(T)\}\right|\right) \qquad (1)$$

which approximates the size of a one-to-one feature correspondence between the model and a data subset.

This paper will focus on using linear bounds on geometric uncertainty. This formulation has been applied to 2D models and transformations and 2D data[3, 4, 7] and 2D models and data and 3D transformations[6, 14]. Here we consider 3D models and transformations and 2D data. The 3D rigid transformation and scaled orthographic projection mapping a model point $\mathbf{m}_i' \in \mathbb{R}^3$ onto a data point $\mathbf{d}_j' \in \mathbb{R}^2$ can be expressed by the transformation $\mathbf{Sm}_i' + \mathbf{t} = \mathbf{d}_j'$ where

$$\mathbf{S} = \sigma\mathbf{PR} = \begin{bmatrix} s_{11} & s_{12} & s_{13} \\ s_{21} & s_{22} & s_{23} \end{bmatrix}$$ with $0 < \sigma \in \mathbb{R}$ a positive scale factor, $\mathbf{P}$ a 2×3 projection matrix simply eliminating the third row of $\mathbf{R}$, where $\mathbf{R} \in SO_3$, and $\mathbf{t}, \mathbf{d}_j' \in \mathbb{R}^2$. From $\mathbf{S} = \sigma\mathbf{PR}$ and the fact the $\mathbf{R}$ is an orthonormal matrix we have the quadratic constraint that

$$\mathbf{SS}^t = \sigma^2(\mathbf{PR})(\mathbf{R}^t\mathbf{P}^t) = \begin{bmatrix} \sigma^2 & 0 \\ 0 & \sigma^2 \end{bmatrix}. \qquad (2)$$

For convenience we will refer to $\mathbf{S}$ as the linear component of the transformation, and $\mathbf{t}$ the translation component.

In essence the approach is to define a model for geometric uncertainty in the data features which in turn imposes constraints on the geometrically feasible transformations aligning model and data feature sets. This frames geometric feature matching as a problem in computational geometry, specifically constructing and exploring arrangements of constraints on geometrically feasible (i.e. structurally consistent) matchings of model features to data features.

For each data feature formulate a convex polygonal uncertainty region bounding the measurement uncertainty consisting of the set of points $\mathbf{x}'$ characterized by a set of linear inequalities[3] $(\mathbf{x}' - \tilde{\mathbf{d}}_j')^t\hat{\mathbf{n}}_l \geq -\epsilon^l$. For feature match $(\mathbf{m}_i', \tilde{\mathbf{d}}_j')$ this imposes constraints on feasible transformations of the form:

$$(\mathbf{Sm}_i' + \mathbf{t} - \tilde{\mathbf{d}}_j')^t\hat{\mathbf{n}}_l \geq -\epsilon^l \qquad (3)$$

for $l = 1, ..., k$ where $\hat{\mathbf{n}}_l$ is the unit normal vector and ϵ_l the scalar distance describing each linear constraint for $l = 1, ..., k$. This describes a polytope of feasible transformation parameters, $\mathcal{F}^{ij}$, mapping $\mathbf{m}_i'$ to within the uncertainty bounds of $\tilde{\mathbf{d}}_j'$. Each possible feature match $(\mathbf{m}_i', \tilde{\mathbf{d}}_j')$ yields such a constraint. Feature matching can be accomplished by searching an arrangement of the boundaries of the polytopes $\mathcal{F}^{ij}$ for all i and j to find sets of transformation parameters with a high geometric consistency measure, i.e. satisfying many of the match constraints. Conceptually the constraints forming the boundaries of the polytopes $\mathcal{F}^{ij}$ partition the parameter space into cells. Exploring the arrangement then involves finding the cells with high geometric consistency.

In the case of 3D affine transformation and orthographic projection (where we relax the quadratic constraints) there are 8 transformation parameters. If there are M model features and N data features then the arrangement of the bounding constraints for each feasible polytopes $\mathcal{F}^{ij}$ has size $O(M^8 N^8)$[11]; constructing and exploring it seems a formidable task. Nonetheless, an important point is that finding *all* geometrically feasible matchings of model and data features within the uncertainty bounds is a polynomial-time problem, in contrast to worst-case exponential-time approaches such as popular feature correspondence-space searches.

2 Approximating the Constraint Arrangement

2.1 The Image of 3D Points Under Scaled-Orthography

Unfortunately constructing and exploring arrangements of hyperplanes in high dimensional spaces is extremely difficult, and computationally infeasible in this case. There is, however, considerable structure to the feature matching problem which we can exploit to make the computations more feasible and to allow the introduction of careful approximations to greatly reduce the computation required to find large geometrically consistent match sets. One approach to this is to work directly with the constraints in the s-parameter space spanned by the coordinates s_{ij}. This approach was outlined for the 3D case previously[6]. This section introduces an elegant formulation which through convenient choice of representations and careful approximation reduces this from a constraint problem in an 8D parameter space to a constraint problem in a 2D parameters space which is much more computationally feasible.

Let $\{\mathbf{m}'_i\}_{i=0}^{M-1}$ be a set of 3D model points, and $\{\mathbf{d}'_i\}_{i=0}^{M-1}$ be the set of points forming their image without error from an arbitrary viewpoint under scaled-orthography such that $\mathbf{S}\mathbf{m}'_i + \mathbf{t} = \mathbf{d}'_i$. Distinguish three model points $\mathbf{m}'_{\pi_0}$, $\mathbf{m}'_{\pi_1}$, and $\mathbf{m}'_{\pi_2}$ and their corresponding image points $\mathbf{d}'_{\pi_0}$, $\mathbf{d}'_{\pi_1}$, and $\mathbf{d}'_{\pi_2}$ which we will call primary model features and primary data features, respectively. Define $\mathbf{m}'_i \neq \mathbf{m}'_{\pi_0}, \mathbf{m}'_{\pi_1}, \mathbf{m}'_{\pi_2}$ and $\mathbf{d}'_i \neq \mathbf{d}'_{\pi_0}, \mathbf{d}'_{\pi_1}, \mathbf{d}'_{\pi_2}$ secondary model features and secondary data features, respectively. Finally, define $\mathbf{m}_i = (\mathbf{m}'_i - \mathbf{m}'_{\pi_0})$ and $\mathbf{d}_i = (\mathbf{d}'_i - \mathbf{d}'_{\pi_0})$ for all i which defines new coordinate frames for the model and the image data with $\mathbf{m}'_{\pi_0}$ and $\mathbf{d}'_{\pi_0}$ as origins, respectively. The new primary features are thus $\mathbf{m}_{\pi_1}$, $\mathbf{m}_{\pi_2}$, $\mathbf{d}_{\pi_1}$, and $\mathbf{d}_{\pi_2}$ and we now have $\mathbf{S}\mathbf{m}_i = \mathbf{d}_i$ for all i.

Next, express the model point $\mathbf{m}_i$ using $\mathbf{m}_{\pi_1}$, $\mathbf{m}_{\pi_2}$, and $\hat{\mathbf{n}} = \frac{\mathbf{m}_{\pi_1} \times \mathbf{m}_{\pi_2}}{|\mathbf{m}_{\pi_1} \times \mathbf{m}_{\pi_2}|}$ as basis vectors[2]. Assuming $\mathbf{m}_{\pi_1}$ and $\mathbf{m}_{\pi_2}$ are linearly independent there exist three constants a_i, b_i, and c_i such that

$$\mathbf{m}_i = a_i \mathbf{m}_{\pi_1} + b_i \mathbf{m}_{\pi_2} + c_i \hat{\mathbf{n}} = \begin{bmatrix} \mathbf{m}_{\pi_1} & \mathbf{m}_{\pi_2} \end{bmatrix} \begin{bmatrix} a_i \\ b_i \end{bmatrix} + c_i \hat{\mathbf{n}}.$$ When the three primary

model and data features are exactly aligned by some linear transformation we have $S [m_{\pi_1} \ \ m_{\pi_2}] = [d_{\pi_1} \ \ d_{\pi_2}]$. In this case it can be shown[9] that the image of any secondary model feature m_i under the imaging transformation S which aligns the primary features (see also Jacobs[18]) can be expressed:

$$Sm_i = c_i \begin{bmatrix} \xi_1 \\ \xi_2 \end{bmatrix} + [d_{\pi_1} \ \ d_{\pi_2}] \begin{bmatrix} a_i \\ b_i \end{bmatrix} \qquad (4)$$

where ξ_1 and ξ_2 are parameters completely determining the viewpoint of the model. The image of m_i under the matrix S thus has two components. The component due to the planar component of m_i, given by $[d_{\pi_1} \ \ d_{\pi_2}] [a_i \ b_i]^t$, and the component due to the non-planar component of m_i, given by $c_i [\xi_1 \ \ \xi_2]^t$.

Let's call equation (4) the reprojection equation for a secondary model feature m_i with respect to the primary feature alignments. Note that by choosing m'_{π_0} and d'_{π_0} as origins we have eliminated the translation component of the transformation, for now.

When the transformation given by S and t corresponds to 3D affine transformation and orthographic projection, the vector $[\xi_1 \ \ \xi_2]^t$ can take on any value corresponding to different viewpoints and affine transformations. When the correspondence between the primary features is used to compute a *rigid* alignment transformation then the components $[\xi_1 \ \ \xi_2]^t$ must satisfy two quadratic constraints, having at most two solutions[16, 2] yielding rigid 3D motion and scaled orthographic projection. This follows directly[10] from equations (2) and (4) and is closely related to similar quadratic constraints in other formulations[16, 2].

2.2 Incorporating the Effects Of Uncertainty

Central to the reprojection equation and the computation of the rigid three-point alignment transformation are the three primary model and data feature matches. The effect of geometric uncertainty in these primary features is significant and must be handled carefully, in addition to uncertainty in the secondary model features. We will follow an approach representing the measured data feature and the error as explicit terms[12]. When there is geometric uncertainty in the measured position of the data features the transformation mapping model feature m'_i to the measured data feature $\tilde{d}'_j$ must satisfy $Sm'_i + t = \tilde{d}'_j = d'_j + e'_j$, where again d'_j represents the (unknown and unmeasurable) true position of the data feature, e'_j the measurement error, and $d'_j + e'_j = \tilde{d}'_j$ represents the *observed* location of the data feature. As before we match m'_{π_0} and d'_{π_0} as corresponding origins in the model and data coordinate frames, respectively, and define $m_i = (m'_i - m'_{\pi_0})$ for all i and $\tilde{d}_j = (\tilde{d}'_j - d'_{\pi_0})$ for all j. We then have $Sm_i = \tilde{d}_j = d_j + e_j$ where we define $e_j = e'_j - e'_{\pi_0}$.

When the reprojection or alignment transformation is computed from the *measured* geometry of the primary data features $\tilde{d}_{\pi_1}$ and $\tilde{d}_{\pi_2}$ the reprojection of a secondary model feature is given by $Sm_i = c_i \begin{bmatrix} \tilde{\xi}_1 \\ \tilde{\xi}_2 \end{bmatrix} + [\tilde{d}_{\pi_1} \ \ \tilde{d}_{\pi_2}] \begin{bmatrix} a_i \\ b_i \end{bmatrix}$

where in the case of rigid alignment $\tilde{\xi}_1$ and $\tilde{\xi}_2$ are computed from the noisy primary data features $\tilde{d}_{\pi_1}$ and $\tilde{d}_{\pi_2}$. In the affine case $\tilde{\xi}_1 = \xi_1$ and $\tilde{\xi}_2 = \xi_2$

can be anything. The planar component of the reprojection has a simple linear form[12] given by $[\tilde{\mathbf{d}}_{\pi_1} \quad \tilde{\mathbf{d}}_{\pi_2}]\begin{bmatrix} a_i \\ b_i \end{bmatrix} = [\mathbf{d}_{\pi_1} \quad \mathbf{d}_{\pi_2}]\begin{bmatrix} a_i \\ b_i \end{bmatrix} + [\mathbf{e}_{\pi_1} \quad \mathbf{e}_{\pi_2}]\begin{bmatrix} a_i \\ b_i \end{bmatrix}$. The non-planar component of the reprojection has a complicated non-linear dependence on the uncertainty in the primary data features[10] which is a problem when computing $\tilde{\xi}_1$ and $\tilde{\xi}_2$ for a rigid transformation. We will only be concerned with 3D affine transformation, and will constrain ξ_1 and ξ_2 without explicitly computing them.

3 Geometric Feature Matching

We now consider a sequence of formulations which make successively more approximations to reduce the dimensionality of the geometric constraints. We emphasize that these approximations are *conservative* in that they will never exclude a feasible transformation. They are also careful approximations in that they do not include *too many* infeasible transformations. These formulations illustrate that the geometric feature matching problem has both a purely combinatorial component as well as a geometric component and the different balance between these components.

An 8D Constraint Problem: Starting with the equation for an aligned feature match incorporating error $\mathbf{Sm}_i' + \mathbf{t} - \tilde{\mathbf{d}}_j' = -\mathbf{e}_j'$, from the definition of the uncertainty bounds we have $\hat{\mathbf{n}}_i^t \mathbf{e}_p' \leq \epsilon^l$, for $l = 1, ..., k$ and for all p, which directly leads to the inequalities (3). These inequalities describe an 8D constraint arrangement of size $O(M^8 N^8)$ which can be explored in $O(M^8 N^8)$ time[11].

A 6D Constraint Problem: Next hypothesize the correspondence between one model feature $\mathbf{m}_{\pi_0}'$ and one data feature $\tilde{\mathbf{d}}_{\rho_0}'$ and let these points be the origin for the model and image coordinate frames, respectively. We subscript the primary data feature $\tilde{\mathbf{d}}_{\rho_0}$ differently than the model feature $\mathbf{m}_{\pi_0}$ to emphasize that their correspondence to one another is hypothesized, not known. As before we define $\mathbf{m}_i = (\mathbf{m}_i' - \mathbf{m}_{\pi_0}')$ and $\tilde{\mathbf{d}}_i = (\tilde{\mathbf{d}}_i' - \tilde{\mathbf{d}}_{\rho_0}')$ for all i, and the transformation aligning $\mathbf{m}_i$ and $\tilde{\mathbf{d}}_j$ must satisfy $\mathbf{Sm}_i - \mathbf{d}_j = 0$ or equivalently $\mathbf{Sm}_i - \tilde{\mathbf{d}}_j = -\mathbf{e}_j$ where we define $\mathbf{e}_j = \mathbf{e}_j' - \mathbf{e}_{\rho_0}'$. Using again the uncertainty bound $\hat{\mathbf{n}}_i^t \mathbf{e}_p' \leq \epsilon^l$, for $l = 1, ..., k$ and for all p we then have constraints for a single match on the 6 s-parameters which comprise the linear component of the transformation[6]: $(\mathbf{Sm}_i - \tilde{\mathbf{d}}_j)^t \hat{\mathbf{n}}_l \geq -2\epsilon^l$ for $l = 1, ..., k$. This constraint arrangement is of size $O(M^6 N^6)$, and we see that we have shifted around where the computation must be done. For each choice of π_0 and ρ_0 we analyze the arrangement of s-parameter constraints, of size $O(M^6 N^6)$, but must perform a combinatorial search through the MN possible different primary matches $(\mathbf{m}_{\pi_0}, \tilde{\mathbf{d}}_{\rho_0})$ for $\pi_0 = 1, ..., M - 1$, $\rho_0 = 1, ..., N - 1$, and for each match do the s-parameter constraint analysis described above. This is also an approximation because each uncertainty region shares the error vector $\mathbf{e}_{\rho_0}$ which can only take on one global value. Thus we are computing approximate geometric consistency, but conservatively in the sense that the bounds are conservative thus we will never underestimate true global consistency in the arrangement.

A 2D Constraint Problem: Next assume the correspondence between three primary model and data features, $(\mathbf{m}'_{\pi_0}, \tilde{\mathbf{d}}_{\rho_0})$, $(\mathbf{m}'_{\pi_1}, \tilde{\mathbf{d}}_{\rho_1})$, and $(\mathbf{m}'_{\pi_2}, \tilde{\mathbf{d}}_{\rho_2})$, and represent the model features in terms of the natural affine coordinate frame as discussed in section 2.1. In order for a feature match $(\mathbf{m}_i, \mathbf{d}_j)$ to be feasible we must have that $c_i \begin{bmatrix} \xi_1 \\ \xi_2 \end{bmatrix} + [\,\mathbf{d}_{\pi_1} \quad \mathbf{d}_{\pi_2}\,] \begin{bmatrix} a_i \\ b_i \end{bmatrix} = \mathbf{d}_j$ for some value of the viewpoint parameters ξ_1 and ξ_2. Using the actual observed data features and accounting for uncertainty we must have $c_i \begin{bmatrix} \xi_1 \\ \xi_2 \end{bmatrix} + [\,\tilde{\mathbf{d}}_{\pi_1} \quad \tilde{\mathbf{d}}_{\pi_2}\,] \begin{bmatrix} a_i \\ b_i \end{bmatrix} - \tilde{\mathbf{d}}_j = [\,\mathbf{e}_{\pi_1} \quad \mathbf{e}_{\pi_2}\,] \begin{bmatrix} a_i \\ b_i \end{bmatrix} - \mathbf{e}_j$ and expanding the individual error terms[12] we get $c_i \begin{bmatrix} \xi_1 \\ \xi_2 \end{bmatrix} + [\,\tilde{\mathbf{d}}_{\pi_1} \quad \tilde{\mathbf{d}}_{\pi_2}\,] \begin{bmatrix} a_i \\ b_i \end{bmatrix} -$ $\tilde{\mathbf{d}}_j = -\mathbf{e}'_j + \mathbf{e}'_{\pi_0} + [\,\mathbf{e}'_{\pi_1} \quad \mathbf{e}_{\pi_2}\,'] \begin{bmatrix} a_i \\ b_i \end{bmatrix} - [\,\mathbf{e}'_{\pi_0} \quad \mathbf{e}_{\pi_0}\,'] \begin{bmatrix} a_i \\ b_i \end{bmatrix}$. A bound on the uncertainty region for the sum of the errors on the right hand side of this equation is given by the Minkowsky sum of the individual uncertainty regions for $\mathbf{e}'_{\rho_0}$, $\mathbf{e}'_{\rho_1}$, $\mathbf{e}'_{\rho_2}$, and $\mathbf{e}'_j$, which leads to constraints for each match on the two viewpoint parameters, ξ_1 and ξ_2.[2]

$$(c_i \begin{bmatrix} \xi_1 \\ \xi_2 \end{bmatrix} + [\,\tilde{\mathbf{d}}_{\rho_1} \quad \tilde{\mathbf{d}}_{\rho_2}\,] \begin{bmatrix} a_i \\ b_i \end{bmatrix} - \tilde{\mathbf{d}}_j)^t \hat{\mathbf{n}}_l \geq -\epsilon'(|a_i + b_i - 1| + |a_i| + |b_i| + 1) \quad (5)$$

The inequality (5) represent conservative constraints on the parameters ξ_1 and ξ_2 feasible for feature match $(\mathbf{m}_i, \tilde{\mathbf{d}}_j)$ when the primary model and data features arc approximately aligned; these constraints are based directly on bounds on measurement error for $\mathbf{d}'_{\rho_0}$, $\mathbf{d}'_{\rho_1}$, $\mathbf{d}'_{\rho_2}$, and $\mathbf{d}'_j$.

After this change to affine bases we simply analyze a 2D constraint arrangement for the two components ξ_1 and ξ_2 which is now an arrangement of convex constraint polygons, one for each secondary feature match. Each polygon's shape is a scaled copy of the basic data uncertainty region and its size is roughly proportional to $(|a_i| + |b_i| + 1)/c_i$. This arrangement of MN such polygons is now only of size $O(M^2 N^2)$, however we increased the combinatorial component to time $O(M^3 N^3)$ because we must search over the possible triples of primary model and data features which form the affine bases. These primary matches must be correctly matched to find geometrically consistent secondary matches. This formulation for deriving constraints on ξ_1 and ξ_2 for each match is closely related to an approach working directly in the 6D space of s-parameters s_{ij} using the constraints of section 3 and making a careful projection of the constraints onto a lower dimensional space[8].

In the special case where the model features are coplanar we have that $c_i = 0$ for all i, and 3D rigid motion corresponds to a 2D affine transformation. If we use the vector $\mathbf{m}_{\pi_1}$ and its perpendicular $\mathbf{m}_{\pi_1}^\perp$ as a basis to represent the coplanar model features, the image of the model over all viewpoints is described by

---

[2] The bound on the feasible viewpoint parameters ξ_1 and ξ_2 is essentially the same as the bounds on the reprojection of a secondary model feature by an alignment transformation computed from three primary matches in the case of *co-planar* model features[12].

$\mathbf{Sm}_i = [\,\mathbf{d}_{\rho_1} \quad \boldsymbol{\eta}\,][\,a_i \quad b_i\,]^t$ where the two parameters $\boldsymbol{\eta} = [\,\eta_1 \quad \eta_2\,]^t$ completely characterized the viewpoint. This leads to constraints on the feasible parameters $\boldsymbol{\eta}$ for a given secondary match $(\mathbf{m}_i, \tilde{\mathbf{d}}_j)$:

$([\,\tilde{\mathbf{d}}_{\rho_1} \boldsymbol{\eta}\,]\begin{bmatrix} a_i \\ b_i \end{bmatrix} - \tilde{\mathbf{d}}_j)^t \hat{\mathbf{n}}_l \geq -\epsilon^l(|a_i - 1| + |a_i| + 1)$. We then have $O(M^2 N^2)$ combinatorial component, and a planar constraint analysis in the η_1-η_2 plane as before. See Cass[9] for more detail, and Cass[6, 8] and Hopcroft[14] for related approaches. Similar constraints can be developed for the case of scaled planar motion[7].

4 An Algorithm for Square Uncertainty Bounds

We can now pull together the ideas from the previous sections to develop an algorithm for geometric matching of 3D models to 2D data. A basic algorithm for geometric feature matching consists of performing a combinatorial search through triples of feature matches each defining affine bases for the model and data points. For each such primary match triple, compute for each secondary feature match the constraint polygon for the set of feasible transformation parameters ξ_1 and ξ_2 and analyze the 2D constraint arrangement to find match sets with a high conservative estimate of geometric consistency.

Consider the constraint formulation using axial square uncertainty bounds for the position of the data features characterized by $|\hat{\mathbf{x}}^t \mathbf{e}'_j| \leq \epsilon$ and $|\hat{\mathbf{y}}^t \mathbf{e}'_j| \leq \epsilon$ for all j. Assume we have selected a triple of primary feature matches yielding $(\mathbf{m}_{\pi_1}, \tilde{\mathbf{d}}_{\rho_1})$ and $(\mathbf{m}_{\pi_2}, \tilde{\mathbf{d}}_{\rho_2})$ as primary matches. In this case equation (5) yields

$|(c_i \begin{bmatrix} \xi_1 \\ \xi_2 \end{bmatrix} + [\,\tilde{\mathbf{d}}_{\rho_1} \quad \tilde{\mathbf{d}}_{\rho_2}\,]\begin{bmatrix} a_i \\ b_i \end{bmatrix} - \tilde{\mathbf{d}}_j)^t \hat{\mathbf{x}}| \leq \epsilon(|a_i + b_i - 1| + |a_i| + |b_i| + 1)$ and

$|(c_i \begin{bmatrix} \xi_1 \\ \xi_2 \end{bmatrix} + [\,\tilde{\mathbf{d}}_{\rho_1} \quad \tilde{\mathbf{d}}_{\rho_2}\,]\begin{bmatrix} a_i \\ b_i \end{bmatrix} - \tilde{\mathbf{d}}_j)^t \hat{\mathbf{y}}| \leq \epsilon(|a_i + b_i - 1| + |a_i| + |b_i| + 1)$ as constraints

on the *unknown* vector $[\,\xi_1 \quad \xi_2\,]^t$ feasible with respect to the secondary feature match $(\mathbf{m}_i, \tilde{\mathbf{d}}_j)$. The set of feasible viewpoint parameters $[\,\xi_1 \quad \xi_2\,]^t$ for the feature match $(\mathbf{m}_i, \tilde{\mathbf{d}}_j)$ is bounded by a square of side $\frac{2\epsilon}{c_i}(|a_i + b_i - 1| + |a_i| + |b_i| + 1)$ centered at the point $\frac{1}{c_i}(\tilde{\mathbf{d}}_j - [\,\tilde{\mathbf{d}}_{\rho_1} \quad \tilde{\mathbf{d}}_{\rho_2}\,][\,a_i \quad b_i\,]^t)$ in the ξ_1-ξ_2 plane.

For every secondary match $(\mathbf{m}_i, \tilde{\mathbf{d}}_j)$ we have such a square constraint on the feasible non-linear component $[\,\xi_1 \quad \xi_2\,]^t$. There are $(M - 3)(N - 3)$ of them forming an arrangement with $O(M^2 N^2)$ cells in the ξ_1-ξ_2 plane. Each cell of this arrangement corresponds to a set of values for the parameters $[\,\xi_1 \quad \xi_2\,]^t$ for which a set of feature matches are consistent with respect to these approximate bounds. Finding cells consistent for many feature matches yield match sets and transformations which are very good candidates for large, fully geometrically consistent match sets. An important point, however, is that although there are $O(M^2 N^2)$ cells, finding the cells of maximal coverage in this arrangement, e.g. computing $\Phi^{\mathcal{M}}(T) = |\varphi(T)|$ defined in section 1.1 can be done in $O(MN \lg MN)$ time as discussed in section 4.1. It is more difficult to compute the measure $\Phi(T)$, with $T = [\,\xi_1 \quad \xi_2\,]^t$, described in equation (1). We do not know of an

$O(MN \lg MN)$ time algorithm for this, although we use a heuristic algorithm with very good performance.

4.1 Analyzing the Planar Constraint Arrangement

Define the set $\mathcal{F}_\xi^{ij}$ to be the set of parameters (ξ_1, ξ_2) satisfying equation (5) for feature match $(\mathbf{m}_i, \tilde{\mathbf{d}}_j)$; $\mathcal{F}_\xi^{ij}$ is an axial square in the ξ_1-ξ_2 plane. Also define $\varphi_\xi(\xi_1, \xi_2) = \{(\mathbf{m}_i, \tilde{\mathbf{d}}_j) | (\xi_1, \xi_2) \in \mathcal{F}_\xi^{ij}\}$. Then analogous to equation (1) we have $\Phi_\xi(\xi_1, \xi_2)$ and similarly $\Phi_\xi^M = |\varphi_\xi(\xi_1, \xi_2)|$. Computing the match count Φ_ξ^M is straightforward. Compute the MN axial squares for the matches as described in section 4. We seek a point in the ξ_1-ξ_2 plane of maximal coverage by the squares. This can be computed, for MN axial rectangles in time $O(MN \lg MN)$ using a segment tree[19]. To compute $\Phi_\xi(\xi_1, \xi_2)$ we use a heuristic extention to the same $O(MN \lg MN)$ time algorithm. It is heuristic in that it can only be guaranteed to produce a lower bound to the maximal value of $\Phi_\xi(\xi_1, \xi_2)$, but performs very well empirically.

4.2 A Tradeoff: Geometric Consistency vs. Algorithmic Complexity

There is an important tradeoff we made in the previous section between finding fully geometrically consistent match sets, and the nature of the algorithm required to do so. In effect we moved part of the computation away from the complex computational geometric problem of analyzing constraint arrangements in higher dimensional spaces, to a simple combinatorial search through triples of primary matches, leaving the much smaller and simpler computational geometric problem of analyzing planar arrangements. Analyzing arrangements is hard algorithmically, combinatorial searches are easy; so this is a good trade. This also allows us to incorporate other methods for reducing the computation required like grouping of primary data triples. For this reduction we traded somewhat relaxed geometric consistency constraints. It is, however, a careful approximation and a *conservative estimate* of full geometric consistency; we will always overestimate the consistency.

5 Testing the Method

The method was tested on both synthetic data and real image data, both for the cases of 2D object undergoing 3D motion using the η-plane variation, and 3D objects undergoing 3D motion using the ξ-plane variation. Here we present a representative sample of tests for 3D objects in 3D. The size of the secondary feature match constraint squares in the ξ-plane was computed using a Gaussian approximation method[3] instead of the absolute bound as discussed in section 4 because the performance was better.

---

[3] Instead of the uniform distribution of e'_{ρ_0}, e'_{ρ_1}, and e'_{ρ_2} assume they are indentically distributed Gaussian random variables with variance $\sigma_e^2 = \epsilon^2/4$. Our bounded regions enclose approximately distances within 2σ. The vector ξ is then a linear function of these random variables and is thus also a Gaussian random variable. The variance of ξ is then $\sigma_{\xi_i}^2 = \frac{4\epsilon^2}{c_i^2}[(a_i + b_i - 1)^2 + a_i^2 + b_i^2 + 1)]$.

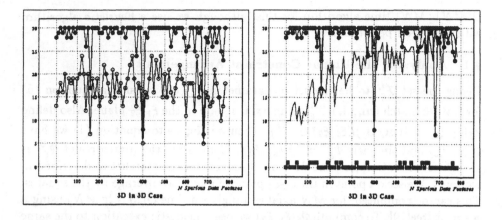

Fig. 1. Results from synthetic tests. In the left graph, the curve with solid dots shows the number correctly matched features aligned in the image within 2ϵ, the curve with open dots shows the number correctly matched features aligned in the image within ϵ. The value plotted is the maximum, over three different correctly chosen primary match groups. In the right graph, the open dots show correct matches aligned within 2ϵ, the filled dots show the consistency measure for matches aligned within 2ϵ (the two curves are almost coincident) in both cases for correct primary matches. The curve labelled with squares (at the bottom) shows the number of correct matches aligned within 2ϵ, and the plain curve shows the consistency measure for matches aligned within 2ϵ, in both these latter graphs for the case of incorrect primary matches. Here $M = 30$, $\epsilon = 5$, the image size was 1024^2, and $N = 1,800$.

The synthetic model objects consisted of points generated uniformly in in a sphere in the 3D in 3D case, of about the size of the image dimensions. The models were imaged synthetically by applying a random transformation and projection to the image plane, then adding independent error to each point uniformly distributed over a square of side 2ϵ. Spurious data features were then generated uniformly in a circle the size of the image. The experimental results show that the method works well for finding transformations with high geometric consistency, as shown in figure 1.

Tests of the method have also been performed on data derived from real images of both 2D and 3D objects. For a 3D object, two black metal bookends where fastened together to construct an interesting non-planar object. To speed up the experiments, a rough grouping was done by hand so as to avoid the long combinatorial search through primary matches, though no specific primary matches were specified. See figure 2.

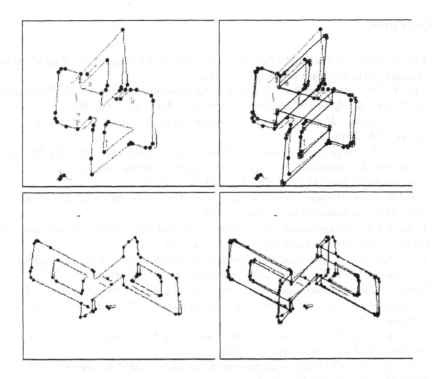

Fig. 2. Experiments with real images. On the left are the image features, connected by line segments. On the right is the best model transformation, shown by overlaying the model. Note that no verification has been performed—shown are the raw hypotheses computed assuming 3D affine transformations.

6   Conclusion

We briefly note some related work. The most similar geometric matching method to ours is the alignment method[16] in it's more careful form accounting for the effects of data feature error [1]. Our approach does not enforce the quadratic constraints on ξ while alignment does; however our approach requires all secondary matches to use the same ξ, while the alignment method does not. For 3D objects, the ξ-plane method, for $M \leq N$, takes $O(M^4 N^4 \lg N)$ time, while the alignment method takes $O(M^4 N^3 \lg N)$. For 2D objects the η-plane method takes time $O(M^3 N^3 \lg N)$ and alignment again takes $O(M^4 N^3 \lg N)$, yet by randomly choosing primary matches, they both take $O(N^3 M \lg N)$ expected time.

We note there are other approaches based on ideas from computational geometry and grounded in real recognition systems focused on 2D objects and affine transformations[15, 14, 20].

The author thanks Tao Alter, David Jacobs, Dan Huttenlocher, and William Ruckledge for helpful comments and discussions.

References

1. Tao D. Alter. Robust and efficient 3d recognition by alignment. Master's thesis, Massachusetts Institute of Technology, 1992.
2. Tao D. Alter. 3-d pose from 3 points using weak-perspective. *IEEE Transactions on Pattern Analysis and Machine Intelligence*, 16(8):802–808, 1994.
3. Henry S. Baird. *Model-Based Image Matching Using Location*. MIT Press, Cambridge, MA, 1985.
4. Thomas M. Breuel. An efficient correspondence based algorithm for 2d and 3d model based recognition. A.I. Memo No. 1259, Artificial Intelligence Laboratory, Massachusetts Institute of Technology, October 1990.
5. Todd A. Cass. Parallel computation in model-based recognition. Master's thesis, Massachusetts Institute of Technology, 1988.
6. Todd A. Cass. *Polynomial-Time Geometric Matching for Object Recognition*. PhD thesis, Massachusetts Institute of Technology, 1992.
7. Todd A. Cass. Polynomial-time geometric matching for object recognition. In *Proceedings of the European Conference on Computer Vision*, 1992.
8. Todd A. Cass. Robust geometric matching for 3d object recognition. In *Proceedings Int. Conf. on Pattern Recognition*, pages A477–A482, Jerusalem, Israel, October 1994.
9. Todd A. Cass. Robust affine structure matching for 3d object recognition. *Submitted for publication review*, Oct 1995.
10. Todd A. Cass. The three-point pose solution under scaled-orthography. *Submitted for publication review*, Oct 1995.
11. Herbert Edelsbrunner. *Algorithms in Combinatorial Geometry*. Springer-Verlag, 1987.
12. W. Eric L. Grimson, Daniel P. Huttenlocher, and David W. Jacobs. Affine matching with bounded sensor error: A study of geometric hashing & alignment. Technical Report A.I. Memo 1250, Artificial Intelligence Laboratory, Massachusetts Institute of Technology, 1990.
13. W. Eric L. Grimson and Tomas Lozano-Perez. Localizing overlapping parts by searching the interpretation tree. *IEEE Transactions on Pattern Analysis and Machine Intelligence*, 9(4):469–482, July 1987.
14. Michael J. Hopcroft. *A Geometrical Approach to Model-Based Vision*. PhD thesis, Cornell University, 1995.
15. D. Huttenlocher, G. Klanderman, and W. Rucklidge. Comparing images using the hausdorff distance. *IEEE Transactions on Pattern Analysis and Machine Intelligence*, 15(9):850–863, 1993.
16. D. Huttenlocher and S. Ullman. Recognizing solid objects by alignment with an image. *International Journal of Computer Vision*, 5(7):195–212, 1990.
17. Daniel P. Huttenlocher and Todd A. Cass. Measuring the quality of hypotheses in model-based recognition. In *Proceedings of the European Conference on Computer Vision*, Genova, Italy, May 1992.
18. David W. Jacobs. 2d images of 3d oriented points. In *Proceedings IEEE Conf. on Computer Vision and Pattern Recognition*, New York, NY, June 1993.
19. E. P. Preparata and M. Shamos. *Computational Geometry–An Introduction*. Springer-Verlag, New York, 1985.
20. William Rucklidge. Locating objects using the hausdorff distance. In *Proceedings of the International Conference on Computer Vision*, Cambridge, MA,, June 1995.

Automatic Face Recognition: What Representation?

Nicholas Costen[1], Ian Craw[2] *, Graham Robertson[2] and Shigeru Akamatsu[1]

[1] ATR Human Information Processing Research Laboratories,
2-2 Hikaridai, Seika-cho, Soraku-gun, Kyoto 619-02, Japan.
[2] Department of Mathematical Sciences, University of Aberdeen,
Aberdeen AB9 2TY, Scotland.

Abstract. A testbed for automatic face recognition shows an eigenface coding of shape-free texture, with manually coded landmarks, was more effective than correctly shaped faces, being dependent upon high-quality representation of the facial variation by a shape-free ensemble. Configuration also allowed recognition, these measures combine to improve performance and allowed automatic measurement of the face-shape. Caricaturing further increased performance. Correlation of contours of shape-free images also increased recognition, suggesting extra information was available. A natural model considers faces as in a manifold, linearly approximated by the two factors, with a separate system for local features.

1 Aims

In machine based face recognition, a *gallery* of faces is first enrolled in the system and coded for subsequent searching. A *probe* face is then obtained and compared with each face in the gallery; recognition is noted when a suitable match occurs. The challenge of such a system is to perform recognition of the face despite transformations, such as changes in angle of presentation and lighting, common problems of machine vision, and changes also of expression and age which are more special. The need is thus to find appropriate codings for a face which can be derived from (one or more) images of it, and to determine in what way, and how well two such codings match, before the faces are declared the same.

A number of face recognition systems have become available recently which propose solutions to these problems, and a natural concern has been the systems's overall performance [13, 7, 10, 3, 12, 11]. Although the choice of coding and matching strategies differ significantly, the greatest source of variability is probably the selection of the faces to test, and the choice of transformation between target and probe over which the system performs recognition. The FERRET database, a potential standard, is currently only available within the USA.

In this paper we seek to avoid some of these difficulties by fixing a matching strategy and testing regime, and concentrating on the first of these problems; to find effective codes for recognition. Our concern is then no longer how well we

---

* This work was in part supported by EPSRC (GR/H75923 and GR/J04951 to IC).

can recognise; indeed for our purposes, a testing regime with a low recognition rate is of most interest: our interest is in *comparing* different coding strategies.

2 Coding via Principal Component Analysis

We contrast simple image-based codings with *eigenface* codings, derived from Principal Component Analysis. Eigenface codings were used to demonstrate pattern completion in a net based context [9, Page 124], to represent faces economically [8], and explicitly for recognition [13]. Much subsequent work has been based on eigenfaces, either directly, or after preprocessing [5, 12, 11].

While undoubtedly successful in some circumstances, the theoretical foundation for the use of eigenfaces is less clear. Formally, Principal Component Analysis assumes that face images, usually normalised in some way, such as co-locating eyes, are usefully considered as (raster) vectors. A given set of such faces (an *ensemble*), is then analysed to find the "best" ordered basis for their span. Some psychological theories of face recognition start from such a norm-based coding; an appropriate model may be a "face manifold" [5], and the usual normalisation is then seen as a local linear approximation, or chart, for this manifold. Since a chart is a local diffeomorphism, and has its range in a linear space, the average of two sufficiently close normalised faces should also be a face.

Clearly existing normalisation techniques approximate this property, but a more elaborate one [14] has recently become prominent as the way to perform a "morph" between two faces. Landmarks give a description of each face's shape; there is a natural way to average landmark positions, and then to map the face texture onto the resulting shape. We describe this as a decomposition into shape *configuration* and shape-free *texture* vectors. The main aim of our paper is to show that this coding produces significantly better recognition results.

Our methodology starts with face images on which a collection of landmarks have been located. Our first tests use manual location; automatic location, and the corresponding results are discussed in Section 4. Eigenfaces are computed from an ensemble of faces which have no further rôle; the gallery and probe faces are coded in these terms. Each probe face has one other image in the gallery, the *target*; our interest is in when the target best matches the probe.

Fourteen images of each of 27 people were acquired under fairly standardised conditions, referred to as Conditions 1 to 14. An initial set of 10 images was acquired on a single occasion: those in Conditions 1 to 4 were lit with good flat controlled lighting; later conditions have increasingly severe lighting and pose variations. Four images were acquired between one and eight weeks later: the first, Condition 11, in lighting conditions similar to Condition 1; subsequent ones with increasing differences. Condition 14 is the only image lit with a significant amount of natural, uncontrolled, light. Fig. 1 shows some of the variability.

The 27 images in Condition 1 provide our fixed gallery. The remaining 13 conditions provide 351 probes. Using all faces as probes avoids that differences in ease of recognition. An additional 50 faces were collected only in Condition 1 and are used as ensemble images, from which the eigenfaces are generated.

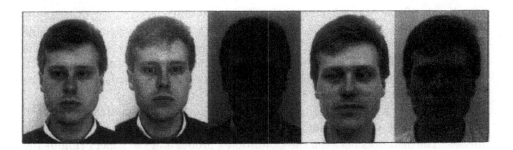

Fig. 1. Conditions 1,4,8,12 and 14.

All the images are processed in the same way. Thirty-four landmarks are found manually, giving a triangulation, or face *model*, part of which can be seen in Fig. 2. A (uniformly) scaled Euclidean transformation is applied, minimising the error between these and corresponding points on a reference face, giving *normalised* images. The background is removed and the histogram of the remaining pixel values flattened. When the face image includes the hair, featural information in the hair can allow good short term recognition. To avoid this, the face model is used to extract an image, containing "inner features" only, as in Fig. 2. Essentially all the results reported are for such images of 2557 pixels.

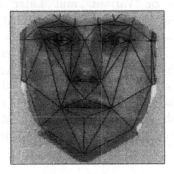

Fig. 2. Inner face showing the facial locations. The mask is enlarged to show the points.

A Principal Component Analysis is performed on the resulting ensemble, obtaining eigenvalues and unit eigenvectors (or *eigenfaces*) of the image cross-correlation matrix. The orthonormality of the eigenfaces allows the computation of the weight of any (normalised) face on each eigenface, giving an n-tuple or code. A coded probe image and the gallery codes are then compared. One method uses nearest neighbour matching in the ensemble span, and a natural metric is the Euclidean distance, leading to template matching within the span. Another natural choice is the Mahalanobis distance, where $d(\mathbf{x}, \mathbf{y})^2 = \sum \lambda_i^{-1} (x_i - y_i)^2$, where $\{\lambda_i\}$ is the sequence of eigenvalues. This treats variations on each axis as equally significant, arguably better for discrimination.

A more robust scheme uses match strength to reduce false acceptances. One such has a sequence of match scores $\{c_j\}$ between the gallery images and the probe[10]. The best match gives the lowest score, c_0; the next c_1. The mean μ and standard deviation σ of the sequence excluding c_0 are calculated and define two inequalities, $c_0 < c_1 - t_1\sigma$ and $c_0 < \mu - t_1\sigma$, for fixed thresholds t_1 and t_2, which must both be met to accept a match. A correct match is reported as a *clear* hit if this criterion is met, and *just* a hit otherwise, similarly for misses. The distances between the Condition 2 probes and the gallery (minus the target) set t_1 and t_2 which were calculated for each probe and the largest values independently chosen. This ensured that in the best, base, condition there were no "clear misses"; although conservative, there are cases where these do occur.

3 Results

We group Conditions 2, 3 and 4, describing this as "Immediate" recognition. Conditions 5, 6 and 7 form a similar set, called "Variant", with small changes in lighting and position. More fundamental lighting changes distinguish the "Lighting" group, Conditions 8, 9 and 10. Finally the four conditions with delayed image acquisition, are called "Later". A weighted average gives the "Overall" value; since the latter conditions are more important, the "Lighting" group has twice the weight of "Immediate" or "Variant", and "Later" four times the weight.

Our main interest is the comparison between scaled Euclidean normalisation, and the more intrusive shape-free form; and the contrast between these and a pure correlation approach. Initial testing used the ensemble of 50 faces described above. However, using the approximate vertical symmetry in individual faces by creating 50 "mirror" faces, reflected about the vertical facial mid line [8] gave a noticeable improvement in recognition, and all results use this "doubled" ensemble. Table 1, Method 'Mah' gives results against which subsequent performance is compared, obtained using all 99 eigenfaces from this ensemble.

	Hit				Miss						
	Clear			Just			Just			Clear		
Method:	Mah	Euc	Cor	Mah	Euc	Cor	Mah	Euc	Cor	Mah	Euc	Cor
Immediate:	90.1	82.7	31.3	9.9	14.8	7.4	0.0	2.5	1.2	0.0	0.0	0.0
Variant:	67.9	34.6	55.6	22.2	45.7	35.8	9.9	19.5	8.6	0.0	0.0	0.0
Lighting:	17.3	3.7	11.1	48.1	29.6	42.0	34.6	66.7	46.9	0.0	0.0	0.0
Later:	34.3	16.7	23.1	32.4	28.7	40.7	33.3	54.9	36.1	0.0	0.0	0.0
Overall:	40.2	22.9	31.3	32.3	29.2	36.9	27.5	47.9	31.7	0.0	0.0	0.0

Table 1. Match percentages from 351 trials per method. Scaled Euclidean normalised in all cases . Method 'Mah': matching with Mahalanobis distance. Method 'Euc': matching with Euclidean distance. Method 'Cor': matching by correlation of the images. Hair has been *excluded* from the match.

Our first comparisons are between the Mahalanobis distance and identical tests using Euclidean distance or thirdly a correlation of the whole of the (masked) face images, all shown in Table 1. The Mahalanobis distance is clearly most effective, confirming that the eigenface formulation, with its variance properties, is worthwhile here. The advantage is smallest in the "Immediate" group, where simple template matching is expected to perform well; but even here, the effect on the "Clear Hits" is noticeable. The alternative baseline uses the whole of the relevant image information, including that lost when the images are projected onto the span of the ensemble. It is clear that this projection looses significant information. Matching using the Mahalanobis distance more than makes up for this loss and we thus adopt this as our baseline.

The theoretical considerations in Section 2 suggest that the distortion of a face into a shape-free or texture vector may provide more effective coding. The normalisation texture-maps each face to a standard shape, here the average of the ensemble images. We used linear interpolation based on the model in Fig. 2; although simpler than Bookstein's thin plate spline warps [11], the procedure was more effective. The results given in Table 2, Method 'T', are directly comparable to Table 1 and suggest that shape-free normalisation is slightly *better* than the scaled Euclidean version, despite deliberately ignoring the shape information.

| | Hit | | | | | | Miss | | | | | |
| | Clear | | | Just | | | Just | | | Clear | | |
Method:	T	S	S–T	T	S	S–T	T	S	S–T	T	S	S–T
Immediate:	95.1	39.5	90.1	4.9	46.9	9.9	0.0	13.6	0.0	0.0	0.0	0.0
Variant:	64.2	23.5	71.6	29.6	58.0	25.9	6.2	17.3	2.5	0.0	1.2	0.0
Lighting:	18.0	27.2	40.7	51.9	54.3	50.6	29.6	18.5	8.6	0.0	0.0	0.0
Later:	28.7	19.4	42.6	46.5	59.3	46.3	25.0	21.3	11.1	0.0	0.0	0.0
Overall:	28.7	23.7	50.4	41.3	56.7	41.1	21.3	19.4	8.5	0.0	0.1	0.0

Table 2. Match percentages from 351 trials per method. Matching with Mahalanobis distance in all cases . Method 'T': using shape-free texture. Method 'S': using shape or configuration (20 most variable eigenshapes). Method 'S–T': combining shape and texture. Hair has been *excluded* from the match.

The shape-free advantage may reflect superior matching of the distorted images, rather than superior coding. A shape-free normalisation was used on the ensemble and a scaled Euclidean normalisation on the gallery and probes, and *vice versa*. The results in Fig. 3 show that the determining factor is the ensemble standardisation method, suggesting the advantage for shape-free-faces reflects superior representation of the faces, not just better matching.

The data discarded by shape-free normalisation can also be used for recognition, performing Principal Component Analysis on the landmark locations. This was done as already described, applying a scaled Euclidean transformation to remove position effects, and then, if necessary, removing the points relating to the hair. The shapes of the ensemble images then provided suitable principal components (*eigenshapes*). The data are highly correlated; after the first 15 or 20 eigenshapes the eigenvalues become small. The number of principal components used to code the shape was varied and the hit rates are shown in Fig. 4. Both with and without hair, recognition peaks when 20 components are included; the peak results for the configuration without hair are given in Table 2, Method 'S'.

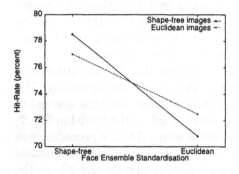

Fig. 3. Recognition using different normalisations for ensemble and test images: hit rates for Euclidean and shape–free.

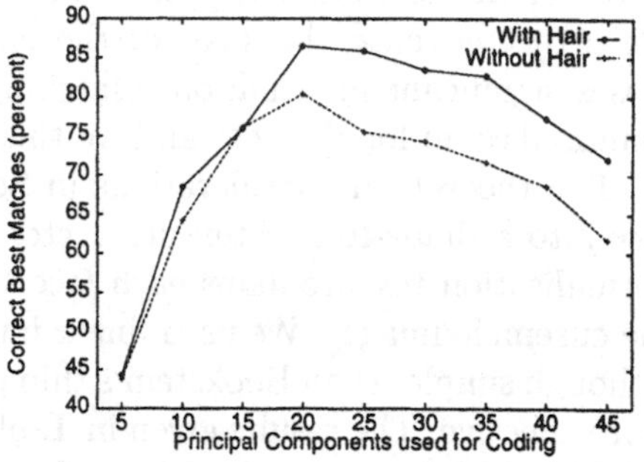

Fig. 4. Recognition using shape or configuration: hit rates for variable numbers of initial principal components.

These tests show a real advantage in representing faces by shape-free texture. This may extend to matches between the images themselves. Because the normalisation uses a relatively small number of points, the match may be underestimated; to compensate, the correlation between a probe and each gallery image was optimized by varying a scaled Euclidean transform of the normalised probe. There was a very noticeable advantage for preprocessing using a laplacian transformation, a 3×3 matrix often thought of as a sharpening operator. The results for the shape-free laplacian images are given in Table 3, showing very good and constant recognition. However, this is very slow even with the small gallery here; optimizing the match required comparing each image with each gallery member 50 times. These results again show the advantages of a shape-free representation; it ensures that all sections of the laplacian-processed images can be aligned at once. In contrast, when shape is still in the images, different sections of the probe face compete to match sections of the gallery images.

Coding using either texture or face shape gives reasonable recognition. If these measures are relatively independent, a combination may be effective. Principal Component Analysis was performed separately on the shape and shape-free images. Independence was assessed by rank correlations of distances between each probe and the *other* gallery images (reducing outlier effects). The average Spearman rank correlations are positive but modest with a maximum value (for the "Immediate" images) of 0.267. This suggests that shape and texture describe dis-similar properties; the positive correlation may reflect landmark location errors. The shape and texture distances for each probe were combined using a root

Method:	Hit Clear Cor	Car	Com	Hit Just Cor	Car	Com	Miss Just Cor	Car	Com	Miss Clear Cor	Car	Com
Immediate:	85.1	95.1	97.5	14.8	4.9	2.5	0.0	0.0	0.0	0.0	0.0	0.0
Variant:	67.9	80.2	88.9	30.9	18.5	11.1	1.2	1.2	0.0	0.0	0.0	0.0
Lighting:	38.3	48.1	67.9	50.6	46.9	29.6	11.1	4.9	2.5	0.0	0.0	0.0
Later:	28.7	58.3	66.7	62.0	34.3	32.4	9.3	7.4	0.9	0.0	0.0	0.0
Overall:	41.0	62.4	72.6	51.2	32.1	26.3	7.8	5.4	1.1	0.0	0.0	0.0

Table 3. Match percentages from 351 trials per method. Method 'Cor': shape-free normalised, matching by full correlation of laplacian images. Method 'Car': shape and texture, matching with Mahalanobis distance with the images 156 % caricatures. Method 'Com': shape and texture, matching with Mahalanobis distance combined with full correlation of laplacian images. Hair has been *excluded* from the match.

mean square, after rescaling the individual distances so the sum of each set was unity. The results in Table 2, Method 'S–T', are thus comparable with Table 1, but combine locally linearised shape with (shape-free) texture information.

The distinct shape and texture components of the face allow it to be *caricatured*. Face shape is coded as a set of position vectors, each the displacement of a landmark from its position in the average face. Scaling the displacements by an amount k gives a caricatured shape, with $k = 100\%$ representing the veridical; the face image is then texture-mapped to this shape. In humans, familiar faces are recognised better with modest caricatures (about 110 %) [1]. Image texture can also be caricatured by displacing the grey levels in a shape-free face away from the mean for each pixel; an example is shown in Fig. 5. Similar modest caricatures are extracted by a Radial Basis Function network using feature-distances [2], as RBFs extract distinctive sets of features. However, a Principal Components Analysis technique, with veridical coding, has greater freedom as it allows investigation of the coding giving the most effective caricatures.

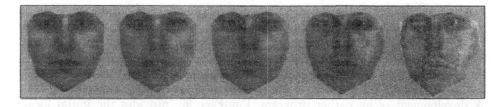

Fig. 5. An image caricatured on shape and texture at 41, 64, 100, 156 and 244 percent.

The faces were caricatured on shape and texture before recognition using the inner face and deriving t_1 and t_2 from the veridical images. The tests show the effects of recognizing the images with a scaled Euclidean normalised Principal Components Analysis. This yields the notably small caricature effect shown by

Fig. 6, while independent shape and shape-free Principal Components Analysis, as shown in Fig. 7, give a strong effect with peak recognition at about 150 %. The peak recognition rates are shown in Table 3. This difference in the caricature effect is only seen if the images are caricatured against independent shape and texture averages. Caricaturing images against the average of the Principal Components Analysis, regardless of the type of normalisation used, gives approximately equal effects. The advantage of the shape-free manipulation is that it allows equivalent transformations in both image-space (as evidenced by the human data) and also in the Principal Components Analysis linearisation.

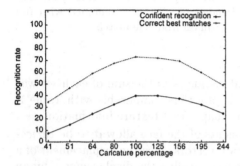 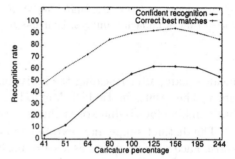

Fig. 6. Confident and total hit rates for shape-and-texture caricatured faces, recognised as Euclidean-normalised. Hair has been *excluded* from the match.

Fig. 7. Confident and total hit rates for shape-and-texture caricatured faces, recognised as separate shape and texture. Hair has been *excluded* from the match.

In a final result, all three matching methods; shape, texture and shape-free correlation, are combined using a root mean square. The results, in Table 3, Method 'Com', suggest there remains relevant information which has not been coded using caricature techniques; but we again emphasize that the optimized correlation takes impractically long, and does not scale well for larger galleries.

4   Facial shape-finding

If our coding process is to operate automatically the landmarks must be located. Given the location of enough landmarks to provide a scaled Euclidean normalisation of a new face, we sequentially generate refined shape estimates. A development of an earlier program, FindFace [6], provides the initial locations, and initialises a bootstrapping procedure to locate the remainder given the ensemble. Each set of landmark locations on a face defines a corresponding shape-free face; we choose those locations on our new face for which the shape-free version has the highest correlation with the average shape-free face.

To optimize efficiently needs the Principal Components Analysis orthogonal decomposition of face shape. Fitting these components successively gives an effective means of navigating in shape space. Starting with the average model, new models were built by varying the shape on the first Principal Component over a range of up to two standard deviations, so applying an active shape model [4]. The resulting model was used to distort the probe to shape-free form; this was then correlated with the shape-free average texture to measure the appropriateness of the model. A simple hill-climbing algorithm sequentially derived the 20 most variable component. The fitting was performed upon the whole, masked, face, including the hair; this gave the most accurate and consistent point-definitions.

When the points so found were used as the input to the complete system, including a caricature of 156 %, it gave the values shown in Table 4. There was a significant caricature effect, suggesting location consistency on the same face; the recognition rate for veridical images was 65.7 %, with 34.7 % clear hits.

	Hit		Miss	
	Clear	Just	Just	Clear
Immediate:	86.4	12.3	1.2	0.0
Variant:	67.9	24.7	7.4	0.0
Lighting:	37.0	34.6	24.2	1.2
Later:	30.6	37.0	32.4	0.0
Overall:	41.9	32.5	25.3	0.3

Table 4. Match percentages from 351 trials. Automatic shape and texture, matching with Mahalanobis distance on 156 % caricatures. Hair has been *excluded* from the match

5 Conclusions

We have attempted to show that a greater consideration of the nature of Principal Component Analysis yields advantages in recognition. Doing so, moving from scaled Euclidean normalised images to the combined configuration and texture images reduces misses three-fold without adding extra information. Caricaturing the images can improve this, by distorting them to emphasize their already atypical aspects. This may not change the ordering of matches, but does increase the separation. This advantage for shape-free Principal Components remains even if the probe is not itself shape-free, again suggesting that this is a representational advance. This decomposition of the face into configuration and texture, and then into Principal Components also allows the efficient location of facial features.

The clear advantage for Mahalanobis over Euclidean distance provides evidence that Principal Component Analysis is a more appropriate coding of faces than raw images; and that something more sophisticated than simple template matching is occurring. Since the Mahalanobis distance pays equal attention to all components, no particular band of eigenfaces should best code the images; once variability is accounted for, the eigenfaces should be equally important. Within limits, this was found; thus we used all the eigenfaces in the tests described here.

Overall we believe we have shown that Principal Component Analysis, implemented under the influence of a manifold model of "face space", separating configural and textural information, has proved of value in coding for recognition; this could be of relevance when constructing psychological models of face recognition. We do not advocate it as a universal code; the observations of very high levels of recognition with shape-free contour matching and when this is combined with the shape-and-texture output show that not all the facial information has been captured. This suggests that psychological implications of this work are late in the processing chain, when the face is being considered as a whole. One model selects a small group of possible matches with local chart-based shape and texture, and uses contour correlation for the final decision.

References

1. Benson, P. and Perrett, D.: 1994, Visual processing of facial distinctiveness, *Perception* **23**, 75–93.
2. Brunelli, R. and Poggio, T.: 1993a, Caricatural effects in automated face perception, *Biological Cybernetics* **69**, 235–241.
3. Brunelli, R. and Poggio, T.: 1993b, Face Recognition: Features versus Templates, *IEEE: Transactions on Pattern Analysis and Machine Intelligence* **15**, 1042–1052.
4. Cootes, T., Taylor, C., Cooper, D. and Graham, J.: 1995, Active shape models – their training and application, *Comp. Vis. and Image Understanding* **61**, 38–59.
5. Craw, I.: 1995, A manifold model of face and object recognition, *in* T. Valentine (ed.), *Cognitive and Computational Aspects of Face Recognition*, Routledge, London, chapter 9, pp. 183–203.
6. Craw, I., Tock, D. and Bennett, A.: 1992, Finding face features, *Proceedings of ECCV-92*, pp. 92–96.
7. Edelman, S., Reisfield, D. and Yeshurun, Y.: 1992, Learning to recognise faces from examples, *Proceedings of ECCV-92*, pp. 787–791.
8. Kirby, M. and Sirovich, L.: 1990, Application of the Karhunen-Loève procedure for the characterisation of human faces, *IEEE: Transactions on Pattern Analysis and Machine Intelligence* **12**, 103–108.
9. Kohonen, T., Oja, E. and Lehtiö, P.: 1981, Storage and processing of information in distributed associative memory systems, *in* G. Hinton and J. Anderson (eds), *Parallel models of associative memory*, Erlbaum, Hillsdale N.J., chapter 4.
10. Lades, M., Vorbrüggen, J., Buchmann, J., Lange, J., v. d. Malsburg, C., Würtz, R. and Konen, W.: 1993, Distortion invariant object recognition in the dynamic link architecture, *IEEE Transactions on Computers* **42**, 300–311.
11. Lanitis, A., Taylor, C. and Cootes, T.: 1994, An automatic face identification system using flexible appearance models, *BMVC 1994*, pp. 65–74.
12. Pentland, A., Moghaddam, B. and Starner, T.: 1994, View-based and modular eigenspace for face recognition, *Proceedings of IEEE Computer Society Conference on Computer Vision and Pattern Recognition*, pp. 84–91.
13. Turk, M. and Pentland, A.: 1991, Eigenfaces for recognition, *Journal of Cognitive Neuroscience* **3**, 71–86.
14. Ullman, S.: 1989, Aligning pictorial descriptions: An approach to object recognition, *Cognition* **32**, 193–254.

Genetic Search for Structural Matching

Andrew D.J. Cross, Richard C. Wilson and Edwin R. Hancock
Department of Computer Science, University of York
York, Y01 5DD, UK

Abstract. This paper describes a novel framework for performing relational graph matching using genetic search. The fitness measure is the recently reported global consistency measure of Wilson and Hancock. The basic measure of relational distance underpinning the technique is Hamming distance. Our standpoint is that genetic search provides a more attractive means of performing stochastic discrete optimisation on the global consistency measure than alternatives such as simulated annealing. Moreover, the action of the optimisation process is easily understood in terms of its action in the Hamming distance domain. We provide some experimental evaluation of the method in the matching of aerial stereograms.

1 Introduction

Discrete optimisation problems [4, 5, 6, 10, 11, 12, 19] are of pivotal importance in high and intermediate level vision where symbolic interpretations must be assigned to relational descriptions of image entities [2, 13, 14, 15, 16]. Although the theory of continuous optimisation is mature [4], discrete or configurational optimisation is still in its infancy [5, 10, 11, 12]. The main difficulties stem from the fact that objective functions defined over a set of discontinuous states are prone to develop local optima [6]. It is for this reason that techniques such as simulated annealing [10, 6, 7], mean, field annealing [19] and most recently genetic search [5, 11, 12] have been developed to overcome some of the local convergence problems.

Broadly speaking discrete or configurational optimisation problems can be approached according to two distinct methodologies. The first of these effectively corresponds to transforming the discrete optimisation problem into a continuous one [4, 19]. In other words, the discrete symbolic assignments which are the goal of computation are replaced by a continuous representation. Algorithms falling into this category include mean-field annealing [19] and probabilistic relaxation [4]. The second class of algorithm retains the discrete assignment representation, but avoid local optima by incorporating a stochastic element into the update process [1, 10, 6, 5, 11, 12]. Perhaps the most popular algorithm falling into this category is the simulated annealing idea of Kirkpatrick [10], which as been exploited with seminal impact by Geman and Geman in the context of low-level vision [6]. Gidas has addressed the problem of slow convergence by developing a multi-resolution Markov model that efficiently tracks the optima of the cost function from coarse to fine detail in a resolution pyramid [7]. A more recent addition to the family of stochastic optimisation methods is genetic search [5, 11, 12]. Rather than being motivated by the heat-bath analogy of simulated annealing [10, 1, 6], genetic search appeals to ideas concerning chromosomal evolution [5].

Although genetic search is a new and imperfectly understood optimisation method, it offers certain attractive computational features. Basic to genetic search is the idea of maintaining a population of alternative global solutions to the discrete optimisation problem in-hand. The initial population may be generated in a number of different ways, but should in some sense uniformly sample the feasible solution space. Associated with each of the different solutions is a cost function which in keeping with the evolutionary analogy is termed the "fitness" [5]. Genetic updates involve three distinct stages. Crossover involves randomly selecting pairs of solutions from the current population and interchanging the symbols at corresponding configuration sites with a uniform probability [12]. Mutation aims to introduce new information into the population by randomly updating the component symbols for individual solutions [12] and takes place with a uniform probability. The net effect of modifying the population in this way is to randomly sample the landscape of the fitness function. Configurations generated by crossover and mutation are subjected to a stochastic selection process in order to avoid convergence to a local optimum [11]. The probability that a modified configuration enters the population is computed on the basis of the fitness measure.

In many ways genetic search provides an interesting compromise between the continuous transformation of the discrete optimisation problem [4, 19] and its realisation by simulated annealing [1, 10, 6]. In the first instance, the selection process is analogous to the Metropolis [1] algorithm employed in the sampling of Gibbs distributions in simulated annealing. Moreover, maintaining a population of solutions each with an associated fitness measure, naturally bridges the gap between the idea of having continuous optimisation variables and discrete ones. One of the unique features of genetic search is the possibility of crossover. This effectively provides a means of mixing existing solutions to provide a diversity of new ones. In this way locally consistent sub-solutions may be combined to generate a globally consistent solution. If effectively controlled, this feature can provide convergence advantages over simulated annealing [3].

It is for these reasons that we would like to exploit genetic search in this paper. Our principal interest is in the matching of symbolic relational descriptions [13, 14, 16, 18]. Here we aim for find a discrete matching configuration that optimises a measure of relational consistency. Symbolic approaches to the relational matching problem have proved of perennial popularity since Barrow and Popplestone's pioneering work which located consistent matches by searching for cliques of the association graph [2]. Difficulties associated with matching inexact relational structures representing imperfectly segmented or cluttered scenes soon became evident [13, 14, 18]. One way of circumventing these difficulties is to pose the matching process as one of minimising a relational distance measure [13, 14, 18]. This measure should be capable of gauging both matching inconsistencies and structural errors. The idea of quantifying relational inexactness in this way has been pursued by Shapiro and Haralick who attempt to minimise the number of inconsistently matched cliques [13]. Structural errors are accommodated by inserting dummy nodes into the relational graphs without penalty. Sanfeliu and Fu's structural edit operations are more complex [14]. Separate heuristic costs are associated with the operations of node relabelling, node deletion and node reinsertions; there are additional costs for the analogous operations on edges. In each of these methods consistent matches are located by deterministic search.

Viewed from the perspective of relational matching, our genetic search process also has a number of novel features. We gauge consistency using a matching probability defined over connected subgraphs of the relational structures. The development of this consistency measure commences from an objective Bayesian model of matching errors and has been extensively reported elsewhere by Wilson and Hancock [15, 16]. Compared with this earlier work, the novel aspect reported in this paper resides in the use of the relational matching probability as a fitness measure in genetic search. By maintaining a population of matches, we have a natural mechanism for simultaneously enumerating different ambiguous solutions. Crossover allows us to mix these solutions in such a way as to combine consistent subgraphs to form a more globally consistent solution. This can offer accelerated convergence, since standard discrete relaxation algorithms only propagate constraints over a distance of one neighbourhood with each iteration. Selection operations stochastically refine the population on the basis of probabilities derived from the Wilson and Hancock [15, 16] consistency measure. Structural errors are removed by graph edit operations [14, 17] aimed at increasing the fitness of match during the selection phase. Finally, it is worth mentioning that since probabilistic selection is a critical stage in genetic optimisation, the availability of an objective Bayesian measure greatly simplifies the search procedure. Moreover, the population of solutions can be regarded as sampling the probability distribution for consistent relational matches.

The outline of this paper is as follows. In Section 2 we review the basic ingredients of the Wilson and Hancock relational consistency measure [15, 16] and describe how the optimisation of this measure may be mapped onto a genetic search procedure. Section 3 provides some performance evaluation on synthetic data. Section 4 shows the utility of the genetic matching procedure in the registration of aerial stereograms. Finally, Section 5 offers some conclusions.

2 Genetic Search

Genetic search [5, 11, 12] provides a very natural way of locating globally consistent relational matches. In essence the approach relies on generating a population of random global matching configurations. These undergo cross-over, mutation and selection to locate the match that optimises a fitness measure. Mutation operations ensure that the fitness landscape is uniformly sampled. Crossover, introduces diversity by mixing partially consistent solutions; if effectively controlled this can accelerate the merging of consistent subgraphs. Selection stochastically selects from the population so as to locate the solution of optimum fitness in a manner analogous to the Metropolis algorithm [1]. However, since the algorithm commences from a set of random matches, accurate initialisation is not an issue of critical importance. One of the novel features of our genetic search process is the incorporation of a deterministic hill-climbing stage. This additional step is applied to the fitness measure once mutations have occurred and is used to accelerate convergence to the nearest optimum of the average consistency measure. In this way sub-optimal solutions may be rapidly rejected by the selection process.

2.1 Fitness

Our adopted fitness measure is the recently reported Bayesian relational consistency measure of Wilson and Hancock [16]. The basic idea underlying this measure is to compare the symbolic matches residing on the neighbourhoods of a data graph with their counterparts in a model graph. Suppose that the data graph $G_1 = (V_1, E_1)$ has node-set V_1 and edge-set E_2. The basic unit of comparison is the neighbourhood which consists of the nodes connected to a centre-object j by data-graph edges, i.e. $C_j = j \cup \{i|(i,j) \in E_1\}$. If the model graph is denoted by $G_2 = (V_2, E_2)$, then the state of match between the two graphs is represented by the function $f : V_1 \to V_2$. The matched realisation of the neighbourhood C_j is represented by the configuration of symbols $\Gamma_j = \cup_{i \in C_j} f(i)$. Wilson and Hancock's basic idea was to invoke the concept of a label-error process to facilitate the comparison of the matched neighbourhoods in the data graph with their counterparts in a model graph. The consequence of this model is that the consistency between the matched data graph neighbourhood Γ_j and the model graph neighbourhood S_k is gauged by the Hamming distance $H(\Gamma_j, S_k) = \sum_{l \in S_k} (1 - \delta_{f(l),l})$. These Hamming distances are used to compute a global probability of match using the following formula

$$P_G = \frac{1}{|V_1| \times |V_2|} \sum_{j \in V_1} (1-p)^{|C_j|} \sum_{k \in V_2} \exp[-\beta H(\Gamma_j, S_k)] \qquad (1)$$

The exponential constant appearing in the above expression is related to the uniform probability of matching errors, i.e. $\beta = \ln \frac{1-p}{p}$.

2.2 Initial Population Generation

Key to genetic search is the idea of maintaining a population of alternative solutions, each with a computed fitness value. The initial choice of the trial solutions which undergo genetic refinement may be made in a number of ways. Here we choose the initial matching configurations so as to uniformly sample the feasible search space. From the perspective of structural matching, this has a number of advantages. Firstly, it means that we obviate the need for accurate initialisation, which has proved to be a perennial problem in the application of iterative labelling schemes. Secondly, it means that our matching scheme is purely symbolic. An alternative to selecting a uniformly distributed initial population is to adopt a bias towards the matches suggested by unary measurements.

The choice of population size determines the rate of convergence for genetic search [11, 12]. There is a tradeoff between the sampling the fitness landscape with a fine granularity and the computational overheads associated with maintaining a large population of solutions. Since Hamming distance is the basic ingredient of our consistency model, we have appealed to a simple pattern space-model [9] to select the population size. We have demanded that the spacing of the initial solutions is less than the average Hamming distance between random graph pairs. For a uniformly distributed population of initial configurations, the distribution of inter-pattern Hamming distance is binomial with mean $|V_1|(1 - \frac{1}{|V_2|})$. This means that the population size is approximately equal to the number of nodes in the data graph.

2.3 Crossover

Crossover is the process which mixes the pool of solutions to produce new ones. If effectively controlled, the process can be used to combine pairs of suboptimal or partially consistent matches to produce one of improved global consistency. Typically, deterministic updating of the match will propagate constraints only over the distance of one neighbourhood with each iteration. Crossover can accelerate this process by combining disconnected yet internally consistent subgraphs from the individual solutions in the pool.

The standard crossover procedure involves selecting at random pairs of global matching configurations from the current population. Random matches at corresponding sites in the match are then interchanged with uniform probability $\frac{1}{2}$; we term this probabilistic crossover. However, this crossover mechanism will not necessarily facilitate the merging of locally consistent subgraphs. Moreover, the process also ignores the underlying structure of the graphs. A better strategy is to combine the solutions by physically dividing the graphs into two disjoint subgraphs. In this way internally consistent portions of the individual solutions may be exchanged. at the structural level.

Initially, when the genetic population consists of random configurations, crossover results in a distribution of step-size which is binomial with a mean Hamming distance of $|V_1|(1 - \frac{1}{|V_2|})$. At later epochs, when the genetic population is dominated by a few ambiguous solutions, the principal mode of the step-size distribution will be at zero Hamming distance with a number of submodes corresponding to the distance between different solutions.

2.4 Mutation

A further randomisation stage is applied to the individual matches to introduce new information into the population of global matches through a process of mutation. This is effected by randomly swapping the matches on individual sites with a uniform probability. In order to sample the probability distribution for relational matches in equation (1), we perform mutations with probability p, i.e. the prevailing value of the label-error probability. The effect of mutation operations can again be understood by reference to our simplistic pattern-space model. The distribution of Hamming distance associated with uniform mutation will again be binomial. The mean Hamming distance step-size is $|V_1|p$ while the variance is $|V_1|p(1-p)$. In other words, $\frac{1}{p}$ is the average number of mutation steps required to transform the different initial solutions into one-another.

2.5 Population Refinement

The crossover and mutation stages of genetic search take place without reference to the value of the fitness measure; they simple maintain a diverse population of matches for further refinement. In our matching process the refinement process is effected using both hill-climbing and selection operations. Both stages of the algorithm are aimed at optimising the global configurational probability measure P_G.

Hill Climbing The aim in performing hill climbing operations is to restore consistency to graphs modified by the crossover and mutation operations. Although this can be effected by stochastic means, this is time consuming. The hill climbing stage involves iteratively reconfiguring the graphs modified by crossover or mutation to maximise the value of P_G. This process has the effect of locating the nearest local optima of the global consistency measure. It therefore redistributes the population of solutions to reside at the modes of this fitness measure. Suboptimal modes become increasingly unlikely as they are removed from the population by the stochastic selection operations. This process not only accelerates convergence, it also diminishes the requirements for a large population of graphs.

Unmatchable nodes One of the critical ingredients in effective relational matching is the way in which unmatchable entities or clutter are accommodated. Conventionally, there are two principal ways in which the effect of clutter can be neutralised. The first of these is to retain clutter nodes as an integral part of the graphs, but to explicitly label them as null-matched [15]. The alternative is to follow a graph-edit philosophy and to remove the clutter nodes, recomputing the edge-set of the graph if necessary [14, 17]. As recently demonstrated by Wilson and Hancock, the main advantage of graph edit operations is that if effectively controlled, they can overcome relational fragmentation due to severe levels of clutter [17]. Although effective when subgraph matching is being attempted, the null-labelling technique has a greater sensitivity to noise.

We therefore choose to control clutter using a graph-edit process which allows nodes to be deleted and reinstated [17]. This process is incorporated into the hill-climbing stage in the following way. Each node in turn is deleted from the graph and the edge-set recomputed. For our experimental evaluation of the method, we have chosen to use Delaunay graphs representing Voronoi tessellations of the image plane. Here the node deletion process corresponds to removing a particular Voronoi cell and growing adjacent cells to fill the vacated space. This process effectively modifies the edge-set of the associated Delaunay graph. Following Wilson and Hancock [17], by adopting the Delaunay representation we simplify the graph-editing process by lifting the requirement for an explicit set of edge-edit operations of the sort employed by Sanfeliu and Fu [14]. Our decision concerning node deletion or re-insertion is based on the value of P_G. If the value of P_G increases due to the deletion process, then the node is edited from the graph. If, on the other hand, the value of P_G increases as a result of node re-insertion at a later stage, then it is reinstated.

Selection The hill-climbing and node deletion operations are purely deterministic processes which effectively bring about local improvements in matching consistency. These operations would otherwise prove time consuming if pursued by stochastic means. The final stochastic element of genetic search is the selection process. The aim here is to randomly admit the configurations refined by the hill climbing process to the population on the basis of their fitness measure.

The probability distribution defined in equation (1) lends itself naturally to the definition of a population membership probability. Suppose that $P_G^{(i)}$ denotes the global configurational probability for the ith member of the pool (population) of graphs. By normalising the sum of clique configuration probabilities over the population of matches, we arrive at the following probability for randomly admitting the ith solution to the pool of graphs $\mathcal{P}$ with probability

$$P_s = \frac{P_G^{(i)}}{\sum_{i \in \mathcal{P}} P_G^{(i)}} \qquad (2)$$

The final optimal match is located by selecting the graph for which $P_G^{(i)}$ is maximum. The idea of maintaining a population of alternative weighted matching configurations effectively bridges the conceptual gap between classical discrete relaxation methods [8] and continuous labelling algorithms such as probabilistic relaxation [4] or mean-field annealing [19].

3 Convergence Behaviour

Our aim in this Section is to evaluate the degree to which our method is sensitive to the various optimisation parameters. There are two parameters that control the genetic search procedure. The first of these is the population size, while the second is the mutation probability. Our aim is to investigate the sensitivity of the convergence-rate to systematic variation of these two parameters. Figure 1a shows the number of iterations to convergence as a function of the population size for a graph of 40 nodes. The main feature of this graph is the existence of a critical population size. Once the population size exceeds 20 graphs, convergence has taken place within 5 iterations. These convergence results are many orders of magnitude faster than those obtained if we omit the hill-climbing stage or adopt a probabilistic rather than geometric crossover strategy.

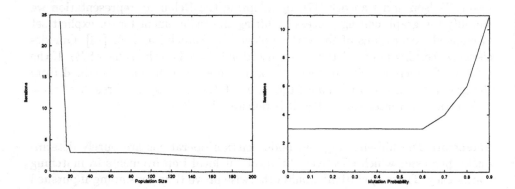

Fig. 1. a) Number of iterations to converge for a given population size. b) Iterations to converge for a given mutation probability.

The second parameter of the optimisation scheme is the mutation probability. Normally, this is taken to be the label-error probability p used in the computation of global fitness P_G. Figure 1b shows the number of iterations to convergence when this constraint on the mutation probability is relaxed. Provided that $p <$ 0.6, then the algorithm appears to be completely insensitive to the mutation probability. In fact, we find that mutation is a relatively insignificant component of the algorithm. Most of the effectiveness of the algorithm is derived from the crossover, hill-climbing and selection stages. Finally, we have performed some experiments to measure the covergence-rate as a function of graph size. For graphs of up to 50 nodes we have found empirically the that number of Hamming distance computations required to locate a graph isomorphism rises as $|V_1|^{2.13}$.

It is also interesting to study how the distribution of the fitness value over the pool of graphs evolves with iteration number. This is illustrated in Figure 2a. Initially P_G is concentrated close to the origin. As the genetic search proceeds, the mode of the histogram slowly moves to larger value of P_G. However, of greater importance to the genetic search procedure, the largest fitness value increases at a more rapid rate. In fact, the rightmost column of Figure 2a illustrates how the frequency of the fittest match increases with iteration number. The ground truth solution is in fact first encountered in the third iteration and has saturated the genetic population by the tenth iteration. Finally, it is interesting to note that the mode of the histogram corresponds in a statistical sense to the typical results achievable by of deterministic search. Figure 2a therefore underlines some of the convergence advantages of genetic search over its deterministic counterpart.

To underline some of the points made in the previous section, we have plotted the distribution of Hamming distance between the ground-truth solution and the individual members of the population of graphs. Figure 2b shows the Hamming distance distribution as a function of iteration number for the 40 node problem studied in Figure 1a. Initially, the minimum Hamming distance is 36, while the mode occurs at 39. As the genetic search process iterates the modal Hamming distance moves towards zero as the correct solution saturates the population.

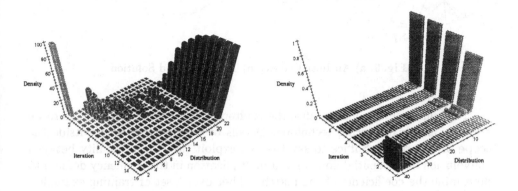

Fig. 2. a) Cost Function Distribution as a function of iteration number. b) Hamming Distance Distribution as a function of iteration number.

Figure 3 illustrates some typical matching results on synthetic graphs. Figure 3a shows a representative solution from the initial population. The left-hand graph is the model while the right-hand graph is the data; lines between the two graphs indicate matches. The data graph has been obtained by adding random clutter to the model and perturbing the nodes with Gaussian position errors. The original model graph contains 20 nodes while the corrupted data graph contains 40 nodes. Figure 3b shows the fittest match from the genetic population after 3 iterations. There are two features worth noting. Firstly, the overall consistency of match has improved. The lines connecting the nodes in the data and model graphs are no longer randomly distributed. Secondly, the added clutter nodes have all been correctly identified and deleted from the data graph; they appear as disjoint points in the right-hand graph of Figure 3b. The overall accuracy of match in this example is 100%.

4 Matching Experiments with Aerial Stereograms

Our experimental vehicle is provided by the matching of aerial stereograms of suburban areas. A typical image pair is shown in Figure 4. The image registration problem is posed as one of matching rooftops. In this section we briefly review the image processing operations required so that we can abstract the registration process in terms of genetic graph matching, before proceeding to detail algorithm performance.

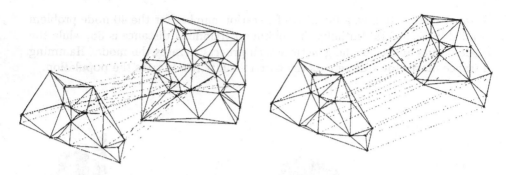

Fig. 3. a) An Initial Guess. b) The Recovered Solution

To commence we must localise the rooftop features to be matched. Here we adopt a matched filtering technique. Details of the algorithm are outside the scope of this paper. Suffice to say that we exploit the Fourier duality between convolution in the spatial domain and multiplication in the frequency domain to determine the coefficients of the matched filter over a set of training examples. The training examples are chosen on the basis of their angle, size and shape. Once the matched filter is to hand, we can attempt to identify typical rooftop structures by locating maxima of the image convolution. We localise significant rooftop responses by applying a simple thresholding technique to the convolution output and locating the centroids of the associated connected components. A neighbourhood structure is established by computing the Voronoi tessellation of the centroids. The Delaunay triangulation of the Voronoi regions provides us with a relational structure for matching. The Voronoi centroids are marked as crosses in Figure 4.

The lines between the two graphs in Figure 5 represent the final matches between the two graphs. This example represents a less demanding test than the synthetic matching experiment in Figure 4a There is only one spurious node, which appears in the left-hand image, due to thresholding errors in the centroid location process. This node is correctly identified as clutter and is deleted. The remaining nodes are all correctly matched. Although the segmental input to this matching problem is good, the regularity of the graphs poses potential problems of ambiguity. The genetic search process is effective in overcoming these and locates a solution of maximum global consistency.

5 Conclusions

To conclude, we have shown how the optimisation of the relational consistency measure of Wilson and Hancock [15, 16] naturally maps onto genetic search. Moreover, the physical variable underpinning this consistency measure, namely Hamming distance, allows us to picture the action of the various stages of the search process in an intuitive way. Based on an extensive simulation study, we

Fig. 4. Aerial stereograms with detected houses marked with crosses.

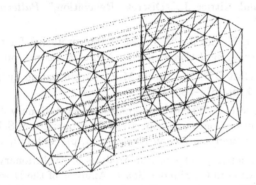

Fig. 5. Matching results (Popsize=100)

evaluate the performance of the resulting optimisation process. This yields a number of interesting results. Empirical results suggest that the number of computations required to locate a consistent solution rises in a polynomial manner with graph size. Despite being stochastic in nature a recent comparative study has shown that the convergence is more rapid than simulated annealing [3]. It is also interesting to note that the method requires a population size which is approximately equal to the number of nodes in the graphs under match. In addition to the population size the only parameter of the method is the mutation probability. Again, convergence is not critically dependant upon the choice of parameter.

The genetic search procedure therefore represents a relatively effective stochastic optimisation process that is not sensitive to the choice of parameters.

References

1. Aarts E. and Korst J., "Simulated Annealing and Boltzmann Machines", *John Wiley and Sons, New York*, 1989.

2. Barrow H.G. and Poplestone R.J., "Relational Descriptions in Picture Processing", *Machine Intelligence*, **6**, 1971.

3. Cross A.D.J. and Hancock E.R., "Relational Matching with Stochastic Optimisation" *IEEE International Symposium on Computer Vision*, pp. 31–36, 1995.

4. Faugeras O.D. and Berthod M., "Improving Consistency and Resolving Ambiguity in Relaxation Labelling", *IEEE PAMI*, **3**, pp. 412–424, 1981.

5. Fogel D.B., "An Introduction to Simulated Evolutionary Optimisation", *IEEE Transactions on Neural Networks*, **5**, pp. 3–14, 1994.

6. Geman S. and Geman D., "Stochastic relaxation, Gibbs distributions and Bayesian restoration of images," *IEEE PAMI*, **PAMI-6** , pp.721–741, 1984.

7. Gidas B., "A Re-normalisation-Group Approach to Image Processing Problems", *IEEE PAMI*, **11**, pp. 164–180, 1989.

8. Hancock, E.R. and Kittler J., "Discrete Relaxation," *Pattern Recognition*, **23**, pp.711–733, 1990.

9. Hancock, E.R. and Kittler J., "A Bayesian Interpretation for the Hop field Network," *IEEE International Conference on Neural Networks*, pp. 341–346, 1993.

10. Kirkpatrick S., Gelatt C.D. and Vecchi M.P., "Optimisation by Simulated Annealing", *Science*, **220**, pp. 671–680, 1983.

11. Qi, X.F. and Palmieri F., "Theoretical Analysis of Evolutionary Algorithms with an Infinite Population in Continuous Space: Basic Properties of Selection and Mutation" *IEEE Transactions on Neural Networks*, **5**, pp. 102–119, 1994.

12. Qi, X.F. and Palmieri F., "Theoretical Analysis of Evolutionary Algorithms with an Infinite Population in Continuous Space: Analysis of the Diversification Role of Crossover" *IEEE Transactions on Neural Networks*, **5**, pp. 120–129, 1994.

13. Shapiro L.G. and Haralick R.M., "A Metric For Comparing Relational Descriptions", *IEEE PAMI*, **7**, pp 90-94, 1985.

14. Sanfeliu A. and Fu K.S., "A Distance Measure Between Attributed Relational Graphs for Pattern Recognition", *IEEE SMC*, **13**, pp 353–362, 1983.

15. Wilson R.C and Hancock E.R, "Graph Matching by Discrete Relaxation", *Pattern Recognition in Practice IV, edited by E Gelsema and L Kanal*, pp 165–176, 1994.

16. Wilson R.C., Evans A.N. and Hancock E.R., "Relational Matching by Discrete Relaxation", *Image and Vision Computing*, **13**, pp. 411–421, 1995.

17. Wilson R.C. and Hancock E.R., "Relational Matching with Active Graph Structures", *Proceedings of the Fifth International Conference on Computer Vision*, pp. 450-456, 1995.

18. Wong A.K.C. and You M., "Entropy and Distance of Random Graphs with Application to Structural Pattern Recognition", *IEEE PAMI*, **7**, pp 599–609, 1985.

19. Yuille A., "Generalised Deformable Models, Statistical Physics and Matching Problems", *Neural Computation*, **2**, pp. 1-24, 1990.

Automatic Selection of Reference Views for Image-Based Scene Representations*

Václav Hlaváč[1], Aleš Leonardis[2], Tomáš Werner[1]

[1] Czech Technical University, Faculty of Electrical Engineering
Department of Control Engineering
12135 Prague 2, Karlovo náměstí 13, Czech Republic
{hlavac,werner}@vision.felk.cvut.cz
[2] Technical University Vienna
Department of Pattern Recognition and Image Processing
A-1040 Wien, Treitlstrasse 3, Austria
(also at the University of Ljubljana, Slovenia)
ales@prip.tuwien.ac.at

Abstract. The problem addressed in this paper is related to displaying a real 3-D scene from any viewpoint. To display a scene, a relatively sparse set of 2-D *reference views* is stored. The images that are in-between the reference views are obtained by *interpolation* of coordinates and brightness (colour). This approach is able to generate the scene representation and render images automatically and efficiently even for complex scenes of 3-D objects. This is possible since the processing time does not depend on the complexity of the scene as there is no attempt to understand the semantics of images.

In this paper we present *a novel approach* to automatically determine a minimal set of views from which the complete scene can be rendered. The method consists of two procedures: view-interval growing and selection. The first procedure independently searches for the intervals from which large portions of the scene can be rendered. These intervals are then passed to the selection procedure, which selects the minimal set of necessary views. The selection procedure is posed as an optimization problem that minimizes the number of reference views and the error due to the interpolation.

* We are primarily indebted to Prof. Roger Hersch from the EPFL Lausanne in Switzerland who initiated the whole project and mediated T. Werner's support from the Swiss National Fund, grant 83H-036863. We were supported by the Grant Agency of the Czech Republic, grants 102/93/0954, 102/95/1378, and the EU grant Copernicus No. 1068. A. Leonardis' work was supported partly by the Austrian National Fonds zur Förderung der wissenschaftlichen Forschung under grant S7002MAT, the Ministry of Science and Technology of Republic of Slovenia (Project J2-6187), and by U.S.–Slovene Joint Board (Project #95-158). We are grateful to Tomáš Pajdla, Walter Kropatsch, and Radim Šára for discussions.

1 Task Formulation

Traditional approaches for displaying a 3-D scene from any viewpoint use a full 3-D model of the scene. The drawback is that the methods for automatic 3-D model reconstruction can only cope with simple man-made objects.

Recently, alternative methods have appeared using *image-based scene representations*. In this case, the scene is not represented by a 3-D model but rather by a collection of 2-D views. The research activities focus either on the geometry of the problem [UB91, Sha94, LF94], or on the image interpolation problem [CW93, SD95, WHH95]. However, none of the works address the problem of how to determine an optimal set of views from which the scene can be rendered to a sufficient degree of accuracy.

This paper proposes a *novel paradigm* of how to automatically select the set of reference views from the set of all views of a scene for displaying purposes. In further text, the term *primary views* denotes captured 2-D images that constitute the input information about the scene. By *reference views* we understand the smallest subset of the original set of primary views that is sufficient to accurately interpolate all other primary views.

Let us illustrate the above idea on an example. Imagine, that a 3-D rigid object is fixed on a table that can turn and slant. A camera looks at the object on the table from a fixed point. We can capture and store 2-D views covering the whole hemisphere of possible views. Assume for the moment that we already have a set of reference views. The reference views should encompass the crucial information needed for displaying. Other views can be computed by interpolation from the reference views, where the interpolation combines both position (coordinates) and brightness or colour. These images are called *interpolated views*.

There are four issues related to the formulated task [WHH95]: (1) How to predict the position and the intensity of a point in the new view if the positions and the intensities of corresponding points in the reference views are known? (2) How to determine the visibility of points in the new view? (3) How to find the optimal set of necessary reference views? (4) How to find the correspondence between reference views? In this paper we primarily concentrate on *issue 3*.

A "brute force" method would represent a scene by generating a densely sampled set of primary views. Our intention is to do better and replace many primary views by a much sparser set of reference views. Practically, we can think of several possible setups related to the capturing of primary views: (1) A 3-D object can be placed on a turn-and-slant table and captured by a camera from a fixed viewpoint. (2) Several dozens of cameras can be placed around the captured object. This setup was proposed for the so called virtualized reality [KNR95]. (3) The source of primary views can also be a video sequence. An uncalibrated camera moving along an unknown trajectory is the most general case.

Note, that the practical experiments reported in this paper fall in the first category, however, our paradigm applies to all three categories.

2 New Paradigm for Selection of Reference Views

The *basic idea* is to group close and similar primary views and represent them by the reference views. Let us first assume the simple case where the object makes a 1-DOF movement with respect to the camera. This corresponds to an object placed on a turn-table and observed by a camera from a fixed distance. Moreover, let all possible primary views lie on a view circle. Let us assume that a densely sampled set of primary views is captured. Similar and neighbouring views on the view circle can be used to create the *viewing interval* of the view angle. Limits of the interval correspond to two 2-D images. The images within the interval can be interpolated from the extremal 2-D images. If there is a close similarity between the interpolated and the primary views, we can represent the whole viewing interval by its two extremal views. They will be called reference views. The similarity between interpolated and primary views is based on an approximately identical visual appearance. The *visual appearance dissimilarity* (VAD) measure is used for comparing interpolated and primary views. The VAD is described in more detail in Section 2.1. The VAD should be below a predefined precision for all views in the interval.

Since we do not want to be affected by the starting point during the growth of the viewing interval, we start the process of creating the interval from each primary view independently. The result is a set of candidate intervals that can potentially overlap. To select the best set that covers the whole range of views is the goal of an optimization task.

In this paper, the stress is put on the method of computing reference views. We worked out the simplest case where primary views lie on the view circle. The generalization to the view sphere case seems possible. A few remarks on the generalization will be discussed in Section 4.

2.1 The Visual Appearance Dissimilarity

The *visual appearance dissimilarity* (VAD) assesses the closeness of the match between a primary view and its interpolation created from some reference views. In other words, it expresses the error due to replacing the real image by the interpolated one. The word *dissimilarity* can hardly be defined in absolute terms. It is related to the purpose the interpolated images will serve to and must be designed with this purpose in mind. For instance, a human observer can tolerate slight geometrical distortion of the whole image, but is very disturbed by some spatially small artifacts in the images. For the demonstration of our approach, we chose the following simple VAD function.

Let $v(x,y)$ denote the primary view and $i(x,y)$ the interpolated view. Let $\mathcal{O}_i$ denote the set of points that could not be interpolated in $i(x,y)$ because of occlusion. A complementary set $\mathcal{O}_i^C$ denotes the points that are not occluded. We can define a VAD as a scalar function that is basically independent on image interpretation:

$$\text{VAD}(v,i) = \alpha \, |\mathcal{O}_i| + \beta \sum_{(x,y)\in\mathcal{O}_i^C} |v(x,y) - i(x,y)| \, , \tag{1}$$

where α and β weight the influence of occlusion and of the error due to incorrect matching or interpolation, respectively. $|.|$ is the number of elements of a set. Another possibility is to use the sum of squared pixelwise differences or other measures [Cha89].

2.2 Constructing the Set of Plausible Viewing Intervals

Let us consider the simple 1-DOF case where primary views lie on the view circle. Then the *plausible viewing interval* (PVI) is a viewing interval such that all primary views inside it can be interpolated from the two primary views corresponding to PVIs endpoints with the VAD smaller than some predefined value ϵ.

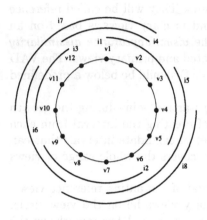

Fig. 1. Example of primary views and PVIs.

Let us focus now on the construction of the set of PVIs. The input is a set of all primary views, $v_1, \ldots, v_N$. The neighbour of the view v_N is the view v_1. The desired output is the set of the PVIs. In Fig. 1, the dots on the circle represent primary views, and the circular arcs illustrate constructed PVIs, $i_1, \ldots, i_8$.

We use a technique similar to region growing. Each PVI has a starting and an ending primary view. Initially, we have a degenerate case where each PVI consists of only one primary view. Then we can try to extend the PVI by adding the left neighbouring primary view to the existing PVI. If the expanded interval is to be the PVI, we should check if all interpolated intermediate views and the corresponding primary views satisfy the condition[3] that the image VAD $< \epsilon$. If the left neighbour fails, we try to merge the right neighbour. If the right neighbour also fails, then the final PVI is obtained.

The interval merging process results in one PVI for each primary view. Now, the selection procedure, described in section 2.3, has the task to find those PVIs that best cover the whole view circle.

All the intervals are first grown to their full extent and then passed to the selection module. As a consequence of the selection process, eventually very few of the intervals emerge as sufficient to cover the view-space. We call this strategy *Build(intervals)-then-Select* because it grows all the intervals fully and independently, and then discards the redundant ones. However, the computational cost can be reduced. Instead of growing all the intervals completely, it is possible to discard some of the redundant intervals even before they are fully grown [Leo93]. This suggests incorporating the selection procedure into the procedure of growing intervals. We call this approach *Build(intervals)-and-Select*.

---

[3] In current experiments, we check the VAD for all intermediate images. However, repeating the VAD check for all primary views within the PVI every time the interval is extended is time consuming. If the current PVI was interpolable from extremal images, it is likely that, after merging one new primary view to it, we might not check every primary view for the $VAD < \epsilon$ again.

2.3 Selection of the Optimal Set of Plausible Viewing Intervals

The selection is a discrete optimization problem and its formulation is similar to the one presented in [Leo93].

The set of PVIs which has been generated may be highly redundant. Thus the selection procedure selects a subset of best PVIs and rejects the superfluous ones. The optimization of an objective function including the information about the competing PVIs is carried out. The objective function is proposed in the following form:

$$F(\mathbf{i}) = \mathbf{i}^T \mathbf{C} \mathbf{i} \qquad (2)$$

The vector $\mathbf{i}^T = [i_1, i_2, \ldots, i_R]$ denotes a set of PVIs, where i_i is a *presence-variable* having the value 1 for the presence and 0 for the absence of the i-th

Fig. 2. The experimental setup for capturing the sequence.

PVI in the resulting set of PVIs. The diagonal terms of the matrix $\mathbf{C}$ express the cost-benefit value for a particular PVI

$$c_{ii} = K_1 s_i - K_2 \|\epsilon_i\| - K_3 V_i \; , \qquad (3)$$

where V_i is the number of reference views plus the data volume needed for correspondence per PVI, s_i is the number of primary views that are covered by the interval, and $\|\epsilon_i\|$ is the error measure calculated for the interval. The coefficients K_1, K_2, and K_3 adjust the contribution of the three terms.

The off-diagonal terms handle the interaction between the overlapping PVIs.

$$c_{ij} = \frac{-K_1 |D_i \cap D_j| + K_2 \|\epsilon_{ij}\| + K_3 \theta_{ij}}{2} \; , \quad \|\epsilon_{ij}\| = \max(\sum_{D_i \cap D_j} \epsilon_i, \sum_{D_i \cap D_j} \epsilon_j)$$

$$(4)$$

D_i denotes the domain of the i-th PVI, i.e., the primary views that are covered by the i-th PVI. $\theta_{ij} = 1$ if the intervals i and j are adjacent, i.e., if the end view of the first interval is equal to the start view of the other interval or vice versa. Otherwise, $\theta_{ij} = 0$.

The objective function takes into account the overlap between the different PVIs which may be completely or partially overlapped. However, we consider only the pairwise overlaps in the final solution. The matrix $\mathbf{C}$ is symmetric, and depending on the overlap, it can be sparse or banded. These properties of matrix $\mathbf{C}$ can be used to reduce the computations needed to calculate the value of $F(\mathbf{i})$.

We have now formulated the selection problem in such a way that its solution corresponds to the global extreme of the objective function. Maximization of the objective function $F(\mathbf{i})$ belongs to the class of combinatorial optimization problems (quadratic Boolean problem). Since the number of possible

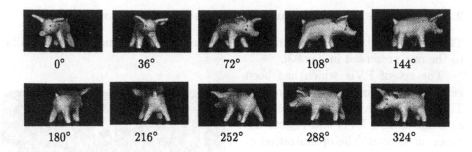

Fig. 3. Ten views from the PIG sequence 36 degrees apart.

solutions increases exponentially with the size of the problem, the exhaustive search is usually not tractable. Various methods have been proposed for finding a global "optimum" of a class of nonlinear objective functions. Among these methods are winner-takes-all strategy, simulated annealing, microcanonical annealing, mean field annealing, Hopfield networks, continuation methods, and genetic algorithms. We are currently using two different optimization methods. One is a simple greedy algorithm that is computationally very efficient, the other one is tabu search [GL93, SL95]. Tabu search is computationally a little more demanding but it provides consistently better results than the greedy algorithm.

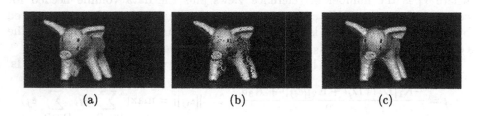

Fig. 4. A pair of reference views and the interpolated view. (a, c) are the reference views ($8°$ and $16°$ in the captured sequence), (b) is the interpolated view (corresponding to $12°$ in the sequence).

3 Experiments

In this section, we show the practical feasibility of our approach on an experiment which was carried out on real data and for one degree of freedom of the object's motion. The *Build(intervals)-then-Select* control strategy was used to select the optimal set of reference views from a large set of primary views.

The setup which was used to capture primary views is shown in Fig. 2. The object to be captured, a small clay sculpture of a pig about 6 cm long, was placed on a turn-table. The object was viewed by the camera from the distance 2 metres (the focal length of the camera lens was 100 mm, CCD chip had the side 12.7 mm). That means that the projection was close to orthographic.

We captured 180 primary views of the object. Each neighbouring pair of views was 2 degrees apart. Ten primary views of the sequence are shown in Fig. 3. The size of the images was 150 × 80 pixels.

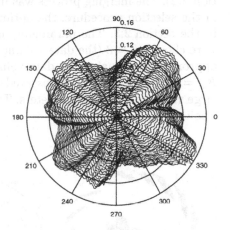

Assuming for a moment that a set of reference views of these 180 primary views have already been selected, the remaining views can be constructed by view interpolation as described, e.g., in [LF94, WHH95]. The example triplet of two reference and one interpolated views is shown in Fig. 4. In the interpolated view, the pixels detected by the stereo matcher as occluded can be seen as black pixels inside the boundary of the object. In these pixels the view interpolation fails.

Fig. 5. The plot of VAD defined by Eq. (5) for the PIG sequence.

The view interpolation requires that the correspondence is available for each pair of neighbouring reference views. For that, we used the binocular stereo matcher [CHMR92]. This algorithm provides correspondence as well as occlusion maps. The algorithm requires that the epipolar lines be parallel to scanlines in the matched image pair, which is ensured by our experimental setup (the projection is close to orthographic) without additional rectification.

For the VAD, we used only the first term of Eq. (1). The reason for this simplification is that the interpolated views have always slight geometrical distortion in our current implementation, which degrades the result of the direct pixelwise image comparison (as in Eq. (1)). Thus we used

$$\text{VAD}(v, i) = \frac{|\mathcal{O}_i|}{|\mathcal{N}_i|}, \tag{5}$$

where $\mathcal{O}_i$ is the set of pixels in the interpolated image i that cannot be computed due to occlusion, and $\mathcal{N}_i$ is the set of all pixels in the interpolated image i. This function is very simple, e.g., the primary view v does not occur explicitly in it. In most cases, the function is almost constant while i moves from the first reference view to the second one. The VAD (more accurately, its mean value over the interval between the pair of reference views) for our object is plotted in Fig. 5. Here, the interpolated views were constructed from a pair of reference views using correspondence obtained by the stereo matcher [CHMR92]. The angle in the plot in polar coordinates represents the angle of the first view of the reference view pair. Each curve was obtained for different angular distance between the two reference views (the most inner curve corresponds to the distance 1°, the most outer one to 30°, the step is 1°). The radius represents the mean value of the VAD over the interpolated interval.

PVIs were built, starting independently from each primary view (see section 2.2). The merging process was limited by the condition VAD < 0.05. Then, in the selection procedure, the optimal set of PVIs was found as it is described in the section 2.3. The parameters in the objective function (Eq. (3) and (4)) were chosen $V_i = 3$ (the data volume needed for storing the correspondence was approximately equal to the data volume needed for storing one image), $K_1 = 1$, $K_2 = 1$, and $K_3 = 1$. We observed that the results remain the same even for large variations of the parameters. To find a solution of Eq. (2), the tabu search was used.

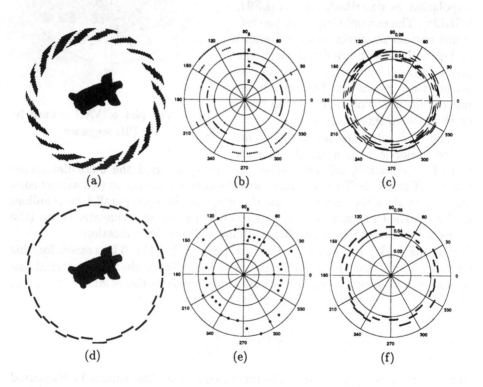

Fig. 6. The results of building PVIs (a, b, c) and selecting the optimal set of the built PVIs (d, e, f).

The result of the process is shown in Fig. 6. Subfigures (a, b, c) show the result of the growing process, subfigures (d, e, f) show the results of the selection process. The intervals are depicted as arcs in (a, d). The radius coordinate has no other meaning here than to help visualizing the overlapping intervals. The top view of the object is superimposed. The length of the PVIs is plotted in (b, e). The intervals are depicted as dots. The radius represents the numbers of views the PVI covers. The VAD for each PVI is plotted in (c, f).

4 Generalization to the Case of the Whole View Sphere

The automatic selection of reference views for image-based scene representations has been demonstrated for the 1-DOF viewing directions (view circle). There is no principal obstacle to generalize the paradigm to more difficult 2-DOF case, where possible viewpoints lie on a view sphere. The selection procedure remains the same. The only difference is in creating plausible view patches (analogous to current PVIs). The plausible view patches can be created from primary views independently using similar approaches like in region-growing based segmentation.

The growing process becomes very computationally intensive. Therefore, more attention should be paid to utilizing *Build(intervals)-and-Select* control strategy, where the selection procedure is started already before the patches are fully grown. This rules out apparently redundant regions before they are fully grown.

5 Conclusions and Future Work

The contribution of this paper is the paradigm formulation that allows to automatically obtain the optimal set of 2-D reference views. This is relevant to currently frequently studied image-based scene representations.

Experiments on real data have demonstrated the result of using our algorithm for a set of 180 primary views. Yet how should the quality of the results be evaluated? The problem is that this evaluation is closely related to VAD. Although we are aware that our current VAD is probably inadequate for real requirements and thus opened for further research, the main idea of this paper can be separated from the specific dissimilarity measure.

In the future, we would like to dedicate more attention to VAD. The difficulty is the discrepancy between the expected visual appearance of the interpolated views (complicated and not well understood process studied in cognitive sciences) and extremely simplified dissimilarity function used for comparing interpolated and primary views. If we had better VAD, we would just incorporate it in the current framework. More extensive quantitative comparisons will be necessary as well as further improvements of geometrical validity of interpolated views. In fact, the latter is another separate track of our current research.

The extension to 2-DOF (view hemisphere) is the step we plan to do next. Using the *Build(intervals)-and-Select* control strategy, we would like to lower the computational demands so that the 2-DOF case would become practical.

References

[CW93]   S.E. Chen and L. Williams. View interpolation for image synthesis. In *Proceedings of the SIGRAPH'93*, pages 279–288, 1993.

[GL93]   F. Glover and M. Laguna. Tabu search. In C. R. Reeves, editor, *Modern heuristic techniques for combinatorial problems*, pages 70–150. Blackwell Scientific Publications, 1993.

[Cha89]   Shi-Kuo Chang. *Principles of Pictoral Information Systems Design*. Prentice-Hall, Englewood Cliffs, New Jersey, USA, 1989.

[CHMR92]   Ingemar J. Cox, Sunita Hingorani, Bruce M. Maggs, and Satish B. Rao. Stereo without disparity gradient smoothing: a Bayesian sensor fusion solution. In *British Machine Vision Conference*, pages 337–346, Berlin, 1992. Springer-Verlag.

[KNR95]   T. Kanade, P.J. Narayanan, and P.W. Rander. Virtualized reality: Concepts and early results. In *Proceedings of the Visual Scene Representation Workshop, Boston, MA., USA, June 24*, page 8. available via http://www.cis.upenn.edu/ēero/VisSceneRep95.html, 1995.

[Leo93]   A. Leonardis. *Image Analysis Using Parametric Models*. PhD thesis, University of Ljubljana, Slovenia, March 1993.

[LF94]   Stéphane Laveau and Olivier Faugeras. 3-D scene representation as a collection of images. In *Proc. of 12th International Conf. on Pattern Recognition, Jerusalem, Israel*, pages 689–691, October 9–13 1994. Also the technical report, available on ftp.inria.fr.

[SD95]   S.M. Seitz and C.R. Dyer. Physically-valid view synthesis by image interpolation. In *Proceedings of the Visual Scene Representation Workshop, Boston, MA., USA, June 24*, page 8. available via http://www.cis.upenn.edu/ēero/VisSceneRep95.html, 1995.

[Sha94]   A. Shashua. Trilinearity in visual recognition by alignment. In *Proceedings of the 3rd European Conference on Computer Vision, Stockholm, Sweden*, volume 1, pages 479–484. SV, 1994.

[SL95]   M. Stricker and A. Leonardis. ExSel++: A general framework to extract parametric models. In V. Hlaváč and R. Šára, editors, *6th CAIP'95*, number 970 in Lecture Notes in Computer Science, pages 90–97, Prague, Czech Republic, September 1995. Springer.

[UB91]   S. Ullman and R. Basri. Recognition by linear combination of models. *IEEE Transactions of Pattern Analysis and Machine Intelligence*, 13(10):992–1005, October 1991.

[WHH95]   T. Werner, R.D. Hersch, and V. Hlaváč. Rendering real-world objects using view interpolation. In *ICCV95*, pages 957–962, Boston, USA, June 1995. IEEE Press.

Object Recognition Using Subspace Methods

Daniel P. Huttenlocher, Ryan H. Lilien and Clark F. Olson

Department of Computer Science, Cornell University, Ithaca, NY 14853, USA

Abstract. In this paper we describe a new recognition method that uses a subspace representation to approximate the comparison of binary images (e.g. intensity edges) using the Hausdorff fraction. The technique is robust to outliers and occlusion, and thus can be used for recognizing objects that are partly hidden from view and occur in cluttered backgrounds. We report some simple recognition experiments in which novel views of objects are classified using both a standard SSD-based eigenspace method and our Hausdorff-based method. These experiments illustrate how our method performs better when the background is unknown or the object is partially occluded. We then consider incorporating the method into an image search engine, for locating instances of objects under translation in an image. Results indicate that all but a small percentage of image locations can be ruled out using the eigenspace, without eliminating correct matches. This enables an image to be searched efficiently for any of the objects in an image database.

1 Introduction

Appearance based recognition using subspace methods has proven successful in a number of visual matching and recognition systems (e.g. [2, 6, 4, 3]). The central idea underlying these methods is to represent images in terms of their projection into a relatively low-dimensional space which captures the important characteristics of the objects to be recognized. The most effective applications of these methods have been to problems in which objects are fully visible, against a uniform background, and are nearly correctly registered with each other. For example, a particularly successful application is the recognition of faces from mugshots, where the head is generally about the same size and location in the image, and the background is a fixed color [4]. The main reason for these limitations is that when extraneous information from the background of an unknown image is projected into the subspace, it tends to cause incorrect recognition results. This is a common problem in any window-based matching technique, where background pixels included in a matching window can significantly alter the match.

In this paper we describe a new subspace recognition method that is designed to handle objects which appear in cluttered images and may be partly hidden from view, without prior segmentation of the objects from the background or registration of the image. This method is based on using subspace techniques to approximate the generalized Hausdorff measure [1], which measures the degree of resemblance between binary images (bitmaps). We present some simple experiments which demonstrate that the method performs well when the background is unknown or when the object is partially occluded, including in cases where methods based on the SSD break down. More importantly, we can detect when the approximation to the generalized Hausdorff measure is likely to select an incorrect match. In addition, we need not assume that the location of the unknown object in the image is known. Instead we can incorporate our eigenspace matching methods into an image search engine, enabling the vast majority of image locations to be ruled out for all of the models in a large database.

2 Subspace approximation of SSD

Let I denote a two-dimensional image with N pixels, and let x be its representation as a (column) vector in scan line order. Given a set of training or model images, I_m, $1 \leq m \leq M$, define the matrix $X = [x_1 - c, \ldots, x_M - c]$, where x_m denotes the vector representation of I_m, and c is the average of the x_m's. The average image is subtracted from each x_m so that the predominant eigenvectors of XX^T will capture the maximal variation of the original set of images. In many applications of subspace methods, the x_m's are normalized in some fashion prior to forming X, such as making $\|x_m\| = 1$, to prevent the overall brightness of the image from affecting the results.

The eigenvectors of XX^T are an orthogonal basis in terms of which the x_m's can be rewritten (and other, unknown, images as well). Let λ_i, $1 \leq i \leq N$, denote the ordered (from largest to smallest) eigenvalues of XX^T and let e_i denote each corresponding eigenvector. Define E to be the matrix $[e_1, \ldots, e_N]$. Then $g_m = E^T(x_m - c)$ is the rewriting of $x_m - c$ in terms of the orthogonal basis defined by the eigenvectors of XX^T (the original x_m is just the weighted sum of the eigenvectors). It is straightforward to show that $\|x_m - x_n\|^2 = \|g_m - g_n\|^2$ [3], because distances are preserved under an orthonormal change of basis. That is, the sum of squared differences (SSD) of two images can be computed using the distance between the eigenspace representations of the two images.

The central idea underlying the use of subspace methods is to approximate x_m using just those eigenvectors corresponding to the few largest eigenvalues, rather than all N eigenvectors,

$$x_m \approx \sum_{i=1}^{k} g_{m_i} e_i + c$$

for $k << N$ (where g_{m_i} denotes the i-th element of the vector g_m). This low-dimensional representation is intended to capture the important characteristics of the set of training images. Let $f_m = (g_{m_1}, \ldots, g_{m_k}, 0, \ldots, 0)$ and $r_m = (0, \ldots, 0, g_{m_{k+1}}, \ldots, g_{m_N})$, so that $g_m = f_m + r_m$. That is, f_m is the vector of coefficients corresponding to the first k terms in the sum, and r_m is the vector of remaining coefficients. The SSD, $\|x_m - x_n\|^2$, is then approximated as $\|f_m - f_n\|^2$. As this representation uses just the k predominant eigenvectors, it is not necessary to compute all N eigenvalues and eigenvectors of XX^T (which would be quite impractical as N is usually many thousands).

3 Approximating the Hausdorff Fraction

In this section we describe a subspace method for approximating the generalized Hausdorff measure. Note that we are now restricting the discussion to binary images, which can be thought of as representing sets of feature points on a grid (i.e., a binary image is 1 for points in the set and 0 otherwise). First we review the generalized Hausdorff measure. Given two point sets $\mathcal{P}$ and $\mathcal{Q}$, with m and n points respectively, and a fraction, $0 \leq f \leq 1$, the generalized Hausdorff measure is defined in [1, 5] as

$$h_f(\mathcal{P}, \mathcal{Q}) = f^{\text{th}}_{p \in \mathcal{P}} \min_{q \in \mathcal{Q}} \|p - q\|, \tag{1}$$

where $f^{\text{th}}_{p \in \mathcal{P}} g(p)$ denotes the f-th quantile value of $g(p)$ over the set $\mathcal{P}$. For example, the 1-th quantile value is the maximum (the largest element), and the $\frac{1}{2}$-th quantile value is the median. Equation (1) generalizes the classical Hausdorff distance, which *maximizes* over $p \in \mathcal{P}$. In other words, the classical distance uses the maximum element rather than some chosen rank.

The generalized Hausdorff measure is asymmetric (as is the classical distance). Given a fraction, f, and two point sets, $\mathcal{P}$ and $\mathcal{Q}$, $h_f(\mathcal{P}, \mathcal{Q})$ and $h_f(\mathcal{Q}, \mathcal{P})$ can attain very different values. For example, there may be points of $\mathcal{P}$ that are not near any points of $\mathcal{Q}$, or vice versa. We can also use a bidirectional form of this measure, $h_{fg}(\mathcal{P}, \mathcal{Q}) = \max(h_f(\mathcal{P}, \mathcal{Q}), h_g(\mathcal{Q}, \mathcal{P}))$. The bidirectional form is not robust to large amounts of image clutter, but it is useful in uncluttered images and for verification of hypotheses.

The generalized Hausdorff measure has been used for a number of matching and recognition problems. One means of using the measure is to specify a fixed distance, d, and then determine the resulting fraction of points that are within that distance. In other words, to find the largest f such that $h_f(\mathcal{P}, \mathcal{Q}) \leq d$. Intuitively, this measures what portion of $\mathcal{P}$ is near $\mathcal{Q}$, for some fixed neighborhood size, d. This has been termed "finding the fraction for a given distance." It measures how well two sets match, with larger fractions being better matches.

The subspace method that we present in this paper is based on finding the fraction for a given distance. Assume that the points of $\mathcal{P}$ and $\mathcal{Q}$ have integral coordinates and let P be a binary image denoting the set $\mathcal{P}$, with a 1 in P corresponding to a point that is in $\mathcal{P}$ and a zero otherwise. Similarly for Q and $\mathcal{Q}$. We are interested in what fraction of the 1's in P are near (within d of) 1's in Q. Let Q^d be the dilation of Q by a disk of radius d (i.e., each 1 in Q is replaced by a "disk" of 1's of radius d). The fraction for a given distance d is then

$$\Phi_d(P, Q) = \frac{\#(P \wedge Q^d)}{\#(P)} \tag{2}$$

where $\#(S)$ denotes the number of 1's in a binary image S, and $\wedge$ denotes the logical and (or the product) of two bitmaps. This measure has also been termed the Hausdorff fraction. It is the fraction of points in P that lie within distance d of points in Q.

Given two binary images, I_m and I_n, if we let x_m be the representation of I_m as a column vector and x_n be the representation of I_n^d (the dilated I_n) then $\Phi_d(I_m, I_n)$ can be computed as follows,

$$\Phi_d(I_m, I_n) = \frac{x_m^T x_n}{||x_m||^2}$$

The Hausdorff fraction, Φ_d, can be approximated using the subspace approximation to the correlation. First we look at the relation between the correlation of two images and their representations in eigenspace, where, as above, g_m and g_n are the rewriting of x_m and x_n in the new coordinate system defined by the eigenvectors E of XX^T.

$$
\begin{aligned}
x_m^T x_n &= (x_m - c + c)^T (x_n - c + c) \\
&= (x_m - c)^T (x_n - c) + (x_m - c)^T c + (x_n - c)^T c + ||c||^2 \\
&= g_m^T g_n + x_m^T c + x_n^T c - ||c||^2
\end{aligned}
$$

The last step follows from $g_m^T g_n = (x_m - c)^T E E^T (x_n - c) = (x_m - c)^T (x_n - c)$ (i.e., dot products are preserved under an orthogonal change of basis).

As in the SSD-based eigenspace methods, we approximate g_m and g_n using just the first k coefficients, which we denote by f_m and f_n. Thus we note that $g_m^T g_n = (f_m + r_m)^T (f_n + r_n) = f_m^T f_n + r_m^T r_n$, because all the cross terms are zero. Hence the error in using $f_m^T f_n$ as an approximation for $g_m^T g_n$ is $r_m^T r_n$. While we cannot compute this error term efficiently we can bound its magnitude by $||r_m|| \cdot ||r_n||$ which can be computed efficiently. Therefore the true correlation $x_m^T x_n$ lies in the range $f_m^T f_n + x_m^T c + x_n^T c - ||c||^2 \pm ||r_m|| \cdot ||r_n||$.

To construct the Hausdorff eigenspace for a set of *binary* "model" images, $x_1, \ldots, x_M$, form the matrix $X = [x_1 - c, \ldots, x_M - c]$, where c is the centroid of the x_m's. Compute and save the first k eigenvectors of XX^T (i.e., those corresponding to the k largest eigenvalues). For each of the x_m's, compute $f_m = (g_{m_1}, \ldots, g_{m_k})$, where $g_{m_i} = e_i^T (x_m - c)$. Then compute $x_m^T c$ and $||x_m||^2$. Save this vector and two scalars for each x_m. This is all the information needed to match the set of models to each unknown image.

Once the above information has been computed and saved for each model image, an unknown image is processed by dilating it by d, forming the vector x_n from this dilated image, and computing f_n and $x_n^T c$. An explicit search of all of the models can be performed by computing the approximation to the Hausdorff fraction for each x_m and the (dilated) unknown x_n,

$$\hat{F}_m = \frac{f_m^T f_n + x_m^T c + x_n^T c - ||c||^2}{||x_m||^2} \tag{3}$$

Note that each of the terms in this expression was computed and stored in forming the eigenspace, except for $f_m^T f_n$. Thus the matching only requires a dot product of two k length vectors (just as in the traditional eigenspace matching techniques), plus a division and a few additions.

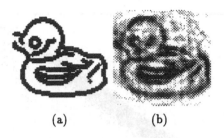

(a)　　　　　　(b)

Fig. 1. Error introduced by the subspace approximation. (a) The dilated edges of an unknown image. (b) The edges after they have been projected into the eigenspace and then reconstructed using only the first 76 eigenvectors.

3.1 Error in the Approximation

The amount of error in $\hat{F}_m$ as an approximation to Φ_d can be bounded by $\varepsilon_m = (\|r_m\| \cdot \|r_n\|)/(\|x_m\|^2)$, so Φ_d lies in the interval $[\hat{F}_m - \varepsilon_m, \hat{F}_m + \varepsilon_m]$ (of course the true fraction can never be greater than 1).

One issue with approximating the Hausdorff fraction is that the unknown image is not necessarily well approximated by the eigenspace, because all of the model images are undilated and the unknown image is dilated. For "thin" features like intensity edges, the dilated images are quite different in appearance and thus are not necessarily well represented by the eigenspace. Empirically we have determined that there is a smaller residual error in the reconstructed images when the model images are dilated and the unknown image is not dilated than when the reverse is the case. Thus, we approximate the Hausdorff fraction $\Phi_d(P, Q) = \#(P \wedge Q^d)/\#(P)$ by $\#(P^d \wedge Q)/\#(P)$. This approximation is quite good for small d, which is generally the case as we use $d = 1$ in order to allow for uncertainty in the edge pixel locations.

In practice, the errors in the estimated fraction are considerably smaller than the error bound given above would predict. This is because the error bound is the worst case, which occurs when the two vectors are pointing in exactly the same direction and all of the errors multiply together. For cases where the true Hausdorff fraction is not large, the estimated fraction is typically very close to the true fraction (within $\pm.05$). In order to examine the errors in the subspace approximation to the Hausdorff fraction, we ran an experiment using a subset of the image set from [3]. This set of images consists of 20 different three-dimensional objects. 60 views of each object were created by placing each object on a turntable and capturing an image at regularly spaced rotations of the turntable. We downsampled these images to 64 × 64 pixels and used the even numbered views as the model image set and the odd numbered views as the unknown image set. In these experiments we used the 76 most significant eigenvectors to approximate the set of training images. Fig. 1 shows an example of a dilated image from this data set and the reconstruction of this image after projecting it into the eigenspace. Fig. 2 shows a plot of the approximate Haus-

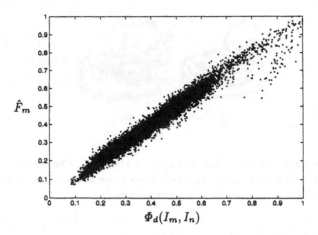

Fig. 2. Plot of the correct fraction versus the estimated fraction in the image subspace for an experiment with 100 model images and 100 unknown images.

dorff fraction versus the true Hausdorff fraction for 10,000 pairs of model images with unknown images (that were not part of the training set).

Note that as the true fraction, $\Phi_d(I_m, I_n)$, becomes large, the approximate fraction, $\hat{F}_m$, sometimes underestimates the correct value. The reason for this is that, in closely correlated images, r_m and r_n will have similar directions, which results in $\hat{F}_m$ being less than $\Phi_d(I_m, I_n)$. In the extreme case, if the dilated model image was exactly the same as the unknown image, then $\Phi_d(I_m, I_n)$ would be underestimated by $\frac{\|r_m\|^2}{\|x_m\|^2}$ since r_m and r_n would be the same. Of course, we will never reach this extreme since the model images are dilated and the unknown images are not.

4 Matching experiments

We now consider some experiments to evaluate the discrimination ability of these matching techniques. We are particularly interested in comparing the performance of these techniques with previous techniques using grey-level images (e.g., [3]) when the background is unknown or when the object is partially occluded. These experiments used the image set from [3], with 30 evenly spaced views of each of 20 objects as the model set and 30 other evenly spaced views of each of the same objects as the set of unknown images. The backgrounds in all the images are dark black.

Each of the 600 unknown views (not used in constructing the eigenspace) was classified as one of the 20 objects by finding the closest matching model view in the eigenspace. That is, a trial was considered successful if the best match was from the same object as the unknown, regardless of the viewpoint of the unknown image and the best matching model image. For the grey-level matching both the model images and unknown images were normalized such that each has a magnitude of one. We selected as the best match for an unknown image, the

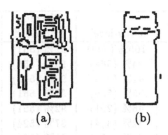

(a)   (b)

Fig. 3. An example where the directed Hausdorff fraction yields an incorrect match, but the bidirectional measure does not. (a) The unknown image. (b) The incorrect match.

model image with the minimum approximate SSD computed using the method described in Section 2 For the binary matching we computed edge maps for each image and selected the model image with the largest approximate Hausdorff fraction $\hat{F}_m$ as the best match for each unknown image.

First it should be noted that using the actual Hausdorff fraction, Φ_d, to select the best matching view did not exhibit perfect performance in selecting the correct object. It was successful in 96% of the trials (575 of 600). The reason that the true Hausdorff fraction was unsuccessful was typically due to unknown images that had dense edges, such that the fraction of model pixels that were near image pixels was very high. This is because of the asymmetry of the Hausdorff distance, which only measures how well the model is accounted for by the image, and not vice versa. Fig. 3 shows an example. In this case, the sparse edges of the incorrect match were well matched by the unknown image, but reverse was not true. The bidirectional Hausdorff measure yields better results for this case (99% correct), since the images are uncluttered. This is analogous to the SSD performing better in uncluttered images; both the SSD and the bidirectional Hausdorff measure take advantage of the excess clutter to rule out possible matches, which results in neither being robust to significant image clutter.

Using the unperturbed images the grey-level matching techniques have perfect performance, while the Hausdorff subspace matching techniques are successful in 551 of 600 trials. Of the 49 unsuccessful trials, 23 were also unsuccessful when the true Hausdorff fraction was used to find the best match. One model accounted for 28 of the unsuccessful trials, with 8 other models accounting for the remaining unsuccessful trials. It is important to note that we can detect when the approximate Hausdorff match is likely to be incorrect. For the successful trials, the difference between the largest $\hat{F}_m$ for a view of the correct object and the largest $\hat{F}_m$ for a view of any other object was .234 on average. In contrast, for the unsuccessful trials this difference was .015 on average, with a maximum value of .090. We should thus consider not only the match with the largest approximate Hausdorff fraction, but also any matches with approximate Hausdorff fractions that are nearly as large. The subspace version of the bidirectional measure was

Image change	Grey-level	Directed Hausdorff	Bidirectional Hausdorff
Unperturbed	100% (600)	92% (551)	98% (589)
Background=50	94% (564)	93% (556)	97% (580)
Background=100	41% (248)	90% (542)	91% (546)
Shift by 50	95% (568)	92% (551)	98% (589)
Shift by 100	48% (291)	92% (551)	98% (589)
25% occlusion	52% (314)	87% (524)	94% (565)
50% occlusion	49% (291)	83% (501)	85% (507)

Table 1. Summary of results for the subspace image matching experiments using the normalized correlation of grey-level images and the Hausdorff fraction of edge images. The results show the percentage (number) of trials out of 600 that were successful.

successful in 589 of 600 trials.

We next considered unknown images where the background had been changed to a uniform non-zero value. The overall image was still normalized to be a vector of unit length. When the background of the unknown images was changed changed to 50, the grey-level techniques were successful in 564 of 600 trials. When the background value was changed to 100, the grey-level techniques were successful in only 248 of 600 trials. These changes yielded little difference for the Hausdorff techniques, yielding 556 and 542 successful trials, respectively. When the grey-levels in the entire image were shifted up by 50 and 100 values, the grey-level techniques were successful in 568 and 291 trials, respectively. Such a shift had no effect on the the Hausdorff matching techniques.

Finally we returned to images with a uniform background, but in which the object had been occluded. We occluded 25% of the object by setting the upper, left quarter of the image to the background color in the grey-level images and by erasing the edge pixels in this region for the edge images. In this experiment, the grey-level techniques were successful in 314 trials, while the Hausdorff techniques were successful in 524 trials. When the entire left half of the image was occluded, the grey-level techniques yielded 291 successful trials and the Hausdorff techniques yielded 501 successful trials.

Table 1 summarizes the subspace results for the grey-level matching techniques and for both the directed and bidirectional Hausdorff matching techniques. The Hausdorff matching techniques have uniformly good performance, whereas the grey-level techniques break down when the background or the total brightness is changed and when the object is partially occluded.

5 Image search

In many applications the positions of the object(s) that may be present in an image are not known. Moreover, current segmentation methods cannot reliably

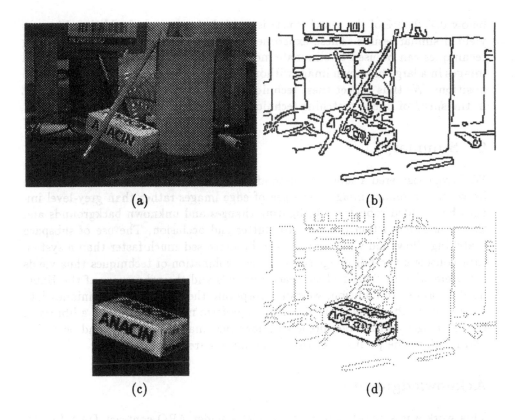

(a)　　　　　　　　　　(b)

(c)　　　　　　　　　　(d)

Fig. 4. A cluttered image with some occlusion that was used to test the image search. (a) The original image. (b) The edges detected in the image. (c) The best matching view of the Anacin box. (d) The edges of the Anacin box overlaid on the full edge image at the location of the best match.

determine the regions of an image that correspond to separate objects, except in very simple cases. In this section we consider the simple experiment of using the eigenspace approximations to rule out those locations (translations) in an unknown image that are a poor match in the subspace. As long as the vast majority of the locations and models are eliminated, without eliminating the correct matches, we can use standard techniques to check the remaining hypotheses. We depend on the fact that the approximate Hausdorff fraction is nearly always close to (within ±0.05 of) the true fraction to avoid ruling out correct matches (see Fig. 2).

Fig. 4 shows an example of the kind of image that was searched in these experiments. The instance present in this image is the Anacin box, which is partially occluded. In this case the best match shown in the figure yielded a true Hausdorff fraction of 0.702 and the subspace methods yield an estimated fraction of 0.727. When we eliminate all translations that yield a best estimated fraction below 0.7, 99.3% of the search space is pruned. A number of additional images yielded similar results. These experiments indicate that the subspace matching techniques can be used to eliminate most of the possible positions of the model images in a large unknown image without performing the full correlation at these positions. We thus expect these techniques to yield a considerable improvement in the speed of image matching techniques using the Hausdorff fraction.

6 Summary

We have considered a subspace method of approximating the Hausdorff fraction between two binary images. The use of edge images rather than grey-level images has yielded robustness to lighting changes and unknown backgrounds and the Hausdorff fraction is robust to clutter and occlusion. The use of subspace matching allows individual matches to be processed much faster than a system that considers the full image space. This combination of techniques thus yields a system with the speed of subspace methods and the robustness of the Hausdorff measure. In addition, we can incorporate these matching techniques into an image search engine. This allows us to perform matching between a library of model images and a large unknown image that may have clutter and occlusion and in which the positions of the model images are unknown.

Acknowledgments

This work was supported in part by ARPA under ARO contract DAAH04-93-C-0052 and by National Science Foundation PYI grant IRI-9057928.

References

1. D. P. Huttenlocher, G. A. Klanderman, and W. J. Rucklidge. Comparing images using the Hausdorff distance. *IEEE Transactions on Pattern Analysis and Machine Intelligence*, 15(9):850–863, September 1993.
2. M. Kirby and L. Sirovich. Application of the Karhunen-Loeve procedure for the characterization of human faces. *IEEE Transactions on Pattern Analysis and Machine Intelligence*, 12(1):103–108, January 1990.
3. H. Murase and S. K. Nayar. Visual learning and recognition of 3-d objects from appearance. *International Journal of Computer Vision*, 14:5–24, 1995.
4. A. Pentland, B. Moghaddam, and T. Starner. View-based and modular eigenspaces for face recognition. In *Proceedings of the IEEE Conference on Computer Vision and Pattern Recognition*, pages 84–91, 1994.
5. W. J. Rucklidge. Locating objects using the Hausdorff distance. In *Proceedings of the International Conference on Computer Vision*, pages 457–464, 1995.
6. M. A. Turk and A. P. Pentland. Face recognition using eigenfaces. In *Proceedings of the IEEE Conference on Computer Vision and Pattern Recognition*, pages 586–591, 1991.

Detecting, localizing and grouping repeated scene elements from an image

Thomas Leung and Jitendra Malik

Department of Electrical Engineering and Computer Sciences
University of California at Berkeley, Berkeley, CA 94720
email: leungt@cs.berkeley.edu, malik@cs.berkeley.edu

Abstract. This paper presents an algorithm for detecting, localizing and grouping instances of repeated scene elements. The grouping is represented by a graph where nodes correspond to individual elements and arcs join spatially neighboring elements. Associated with each arc is an affine map that best transforms the image patch at one location to the other. The approach we propose consists of 4 steps: (1) detecting "interesting" elements in the image; (2) matching elements with their neighbors and estimating the affine transform between them; (3) growing the element to form a more distinctive unit; and (4) grouping the elements. The idea is analogous to tracking in dynamic imagery. In our context, we "track" an element to spatially neighboring locations in one image, while in temporal tracking, one would perform the search in neighboring image frames.

1 Introduction

This paper presents an algorithm for detecting, localizing and grouping instances of repeated scene elements. The notion of scene elements is intended to be quite general and includes as special cases:

- texels on a surface as shown in Figure 1(a). Other examples might be spots on a leopard's skin or the pattern of brickwork on a pavement.
- discrete structures such as the eyes on a person's face.

The goal of our approach is to compute a representation where the individual elements are detected, localized and grouped. The grouping is represented by a graph where nodes correspond to individual elements and arcs join spatially neighboring elements. With each arc r_{ij} is associated an affine map A_{ij} that best transforms the image patch $I(x_i)$ to $I(x_j)$. This affine transform implicitly defines a correspondence between points on the image patches at x_i and x_j.

What is this representation good for? The two short term objectives are:

1. Grouping of these repeated elements is a useful goal in itself. Certainly this provides an operationalization of one of the Gestalt laws of grouping, that based on similarity.
2. Knowledge of the affine transforms between corresponding patches, or equivalently knowledge of the point correspondences between fiducial points on the image patches, can be used to recover 3D scene structure. The mathematics for shape recovery can either follow analysis of shape from texture [4, 10, 11] or structure from motion, or variations of the ideas therein.

In the long term, our primary driving application is recognition. For that the vital importance of grouping has long been noted, and recovery of 3D structural relationships can be useful as well. A more novel aspect is that given a knowledge of the 3D scene structure, we can also hope to recover what one of the scene elements looks like from a canonical view (say fronto-parallel). The appearance from such a canonical view is distinctive and invariant (up to a scale factor) to camera pose and could be used in an iconic matching scheme to provide another visual cue for recognition.

2   Relevant previous work

The most relevant previous work is that on texture processing and on reconstruction from repeated structure.

2.1   Texture processing

Texture has often been defined to be some repeating pattern, so it is natural to think of the problem of finding repeated elements as being a problem of finding regions with coherent spatial texture. This is simplistic–we will discuss below the similarities and differences between texture processing and our work.

First, we talk about the similarities. As pointed out by Malik and Rosenholtz [10, 11], shape from texture is a cue to 3D shape very similar to binocular stereopsis and structure from motion. All of these cues are based on the information available in multiple perspective views of the same surface in the scene. In the context of shape from texture, two nearby patches appear slightly differently because of the slightly different geometrical relationships that they have with respect to the observer's eye or camera. We thus get multiple views in a *single* image. This naturally leads to a two-stage framework (1) computing the "affine texture distortion" from the image, and (2) interpreting the "texture distortion" to infer the 3D orientation and shape of the scene surface. When we deal with repeated scene elements, clearly the idea of multiple views of the "same" element being available in one image carries over.

The differences are enumerated as follows:

1. The notion of texture includes very irregular arrangements where one may not meaningfully speak of a basic unit, a texel, that is repeated across the image. Some of the techniques in texture analysis, such as looking at the response of multi-scale, multi-orientation filter outputs are of course applicable. However we exclude such textures from our consideration of arrangments of repeated scene elements.

2. We want to deal with the cases where there are only a small number of repeated elements (e.g. two eyes). In this case, the notion of texture is usually not appropriate and coherence of filter outputs is too weak a grouping cue.

3. In the case of texture analysis, one usually ignores the precise positional relationships. A crude summary of the psychophysics results of Julesz, Beck, Treisman and others is that in preattentive vision, the positional relationships between the textons do not matter, only the densities need to be considered. This insight has carried over to the quasilinear filtering models that now dominate the field—spatial averages of filter outputs are considered [9].

In our context of finding repeated scene elements, *the precise positional relationships between the features on a single element do matter: we can talk about precise point to point correspondences between features on one element and another.*

In summary, we have a problem formulation that excludes a certain class of textures—stochastic textures with ill defined texture elements—but considering only repeated elements enables a notion of point to point correspondence that is now *exactly* like that in motion or stereopsis.

2.2 Reconstruction from repeated structure

It has been pointed out [1, 6, 7, 12] that structures that repeat in a single image of a scene are equivalent to multiple views of a single instance of a structure. If there are many identical objects available, then one can even calibrate the camera and recover structure up to a similarity ambiguity. Though arrived at independently, fundamentally this represents a variation of the same idea as expressed by Malik and Rosenholtz in the texture context [10, 11]. But, here, the notion of point to point correspondence is exactly the same as in the motion/stereopsis context.

However, in previous work on this topic, the early vision operations of finding image correspondences have not been addressed. The work in this paper makes this process feasible.

3 The Algorithm

An intuitive understanding of the algorithm can be obtained rather simply by an analogy to tracking in dynamic imagery. Suppose, one has identified one of the elements somehow. To identify others like it, we "track" it by searching for it in neighboring locations in the image plane, just as one would in the temporal tracking context where we search in neighboring image frames. A temporally tracked object would be expected to change in size and shape because of its differing pose with respect to the camera. That change, if small enough, can be conveniently modeled as an affine transform. For finding the best match, variations on the theme of maximizing cross-correlation or minimizing SSD (sum of squared differences) have been demonstrated [1] to be simple and robust. Once the object has been located in the next time frame, one can then use a new template based on its new appearance to search for it in the subsequent frame and so on. Reliable, robust temporal tracking has now become a core technology in dynamic computer vision.

The success of temporal tracking suggested that a multi-directional spatial tracking approach to finding repeated scene elements should work well. The difficult part is that of finding a good initialization—it is not easy as in the temporal tracking context. Our approach was to look for "interesting" windows–ones with enough image intensity variation. The algorithm consists of the following steps:

1. Detection of interesting windows in the image. These windows constitute the possible candidates for the repeating elements. The criterion is that they are distinct enough so that matching can be done easily and unambiguously.

---

[1] Weaknesses of correlation/SSD approaches for very large changes in foreshortening as in wide-baseline stereo are not relevant in this setting.

2. For each window, we look at the neighboring regions to search for similar structures. This search allows for small affine deformations.
3. The size of the unit is allowed to change to maximize the distinctiveness of the element.
4. The whole repeating structure is grouped together with the individual elements marked.

3.1 Detection of distinctive elements

The first step of our algorithm is to look for distinctive elements in the image. We break up the whole image into overlapping windows. At each window, we compute the second moment matrix:

$$\Sigma_W = \sum_W \nabla I(x) \nabla I^T(x) \tag{1}$$

where ∇ is the gradient operator.

The second moment matrix tells us how much spatial intensity variation is present in the window. It has been used before in early vision processing as in [3, 5]. The eigenvalues, k_1 and k_2 (assume $k_1 > k_2$), of Σ_W represent the amount of "energy" in the two principal directions. For 2D structures, k_1 and k_2 will be large and comparable in magnitude; k_1 will be large and k_2 will be close to zero for 1D structures; both k_1 and k_2 will be small if little variation is present.

Using the two eigenvalues, we can give further characterization for each window. For a 2-D pattern, k_1 and k_2 will be similar:

$$\frac{k_1}{k_2} < r_1 \tag{2}$$

Here, r_1 is a constant. It describes the ratio of the energy in the two directions of the spatial structure. This tells apart a 2D from a 1D pattern.

In general, there are two types of 1D structures. They can be an edge or a pattern in one direction (a flow), e.g. the vertical stripes on a shirt. However, they can be distinguished by looking at the sign of the gradient vectors. Consider an edge versus a bar. The gradient vectors in the neighborhood of an edge will all point in the same direction, while the gradient vectors point in opposite directions on the two sides of a bar. Making use of this observation, we can tell a flow from an edge. Let v denote the dominant orientation in the window. This is the direction of the eigenvector corresponding to the large eigenvalue of Σ_W. Let x_1 be the pixel locations where the gradient makes an acute angle with the dominant direction, i.e. $v^T \nabla I(x_1) > 0$. Similarly, let x_2 be the locations where the gradient makes an obtuse angle with v, i.e. $v^T \nabla I(x_2) < 0$. Define

$$\Sigma_1 = \sum_{x_1} \nabla I \nabla I^T \quad \text{and} \quad \Sigma_2 = \sum_{x_2} \nabla I \nabla I^T \tag{3}$$

For a flow, Σ_1 and Σ_2 will have similar energy, while for an edge, most of the energy will be concentrated in one of them. Therefore, an 1-D pattern has to satisfy the following requirements:

$$\frac{k_1}{k_2} > r_1 \quad \text{and} \quad \frac{\max(\sum \sigma(\Sigma_1), \sum \sigma(\Sigma_2))}{\min(\sum \sigma(\Sigma_1), \sum \sigma(\Sigma_2))} < r_2 \tag{4}$$

Here, $\sigma(A)$ refers to the spectrum of the matrix A. r_1 is the same as in Equation 2. It tells us that we are dealing with a 1D pattern. r_2 is a constant describing the ratio of energy in Σ_1 and Σ_2. It distinguishes a 1D flow pattern from an edge.

In all our experiments, r_1 and r_2 are fixed. We set $r_1 = 4$, meaning that a pattern is defined to be 2D if the energy in the dominant direction is no more than 4 times larger than the energy in the orthogonal direction. The value of r_2 is fixed to be 2, meaning that when the ratio of the energy in Σ_1 and Σ_2 is no more than 2, the pattern is declared to be 1D flow, as opposed to an edge.

3.2 Finding Matches

After identifying possible candidates, we examine each of them to determine if they belong to a repeating structure. This is done by searching for similar patterns in the neighborhood. A small amount of affine transform is allowed to take into account surface shapes [2]. A rough registration is obtained by finding the neighboring patches where the SSDs to our candidate are small. We then use a differential approach [3, 8] to estimate the affine transform which will bring the two patches in better correspondence:

$$Err = \sum(I(\boldsymbol{x}) - \tilde{I}(A\boldsymbol{x} + d))^2 \tag{5}$$

$$\{A, d\} = \arg \min_{\{A, d\}} Err \tag{6}$$

$\tilde{I}$ is the intensity at the neighboring location. Linearizing,

$$A \approx I + \Delta A \quad \text{and} \quad d \approx 0 + \delta d \tag{7}$$

$$Err \approx \sum(I(\boldsymbol{x}) - \tilde{I}(\boldsymbol{x}) - \nabla\tilde{I}(\boldsymbol{x})^T \Delta A\boldsymbol{x} + \nabla\tilde{I}(\boldsymbol{x})^T \delta d)^2 \tag{8}$$

ΔA and δd can be estimated by solving a system of linear equations. See [8] for details. We then iterate by warping the image and refining the affine transform. Similarity of the two patches are measured in a normalized way. Define

$$D(\boldsymbol{x}) = I(\boldsymbol{x}) - \tilde{I}(A\boldsymbol{x} + d) \tag{9}$$

$$\sigma_I = \sqrt{\sum(I(\boldsymbol{x}) - \overline{I}(\boldsymbol{x}))^2} \tag{10}$$

$$\sigma_e = \sqrt{\sum(D(\boldsymbol{x}) - \overline{D}(\boldsymbol{x}))^2} \tag{11}$$

Two patches are declared to be similar if

$$E_{sim} = \frac{\sigma_e}{\sigma_I} < \tau_e \tag{12}$$

---

[2] The repeated elements are related by a projective transformation, but an affine transform is a very good approximation for neighboring elements.

σ_I can be interpreted as the contrast of the image [3] and σ_e the variation of the error. $\frac{\sigma_e}{\sigma_I}$ can then be thought of as the inverse of the signal-to-noise ratio. τ_e is a constant threshold taken to be 0.2 in all our experiments. Notice that this error measure is invariant to both constant scaling of image brightness and an illumination offset.

The output of this procedure is a list of elements which form units for repeating structures in the image. Associated with each element are the neighboring patches which match well with the element, together with the affine transform relating them.

3.3 Growing the Pattern

The initial size of the window is chosen quite arbitrarily [4], and thus each element may not correspond to the most distinctive unit. Therefore, we allow the elements to change in size. The criterion for change is that it will decrease the matching error (Eq. 12). This is measured against the neighboring patches that we obtained in the previous step. With the affine transform, we can warp each individual neighboring patch to one which registers with the element. We iteratively increase the height and width of the window and measure the total error. If the total error decreases, we enlarge the window and continue the process until the error stops decreasing or a maximum allowed size is reached. The outcome would be a set of elements which are more representative to the underlying repeating structure.

3.4 Grouping Elements

The final step is to group the elements together. We start at each basic unit, and look inside its 8 neighboring windows for similar elements. In each of the 8 windows, we find the patch which matches the element best. The affine transform between them is also computed. If the error between the neighboring patch and the element is smaller than τ_e (Eq. 12), we group the two patches together.

When we propagate the growing procedure outward, the size and shape of the basic element in the image will change. This is very well illustrated in Figure 1. Owing to the slanting of the surface, the glass windows look smaller when we go towards the far end of the building. The affine transform tells us exactly how the size and shape change. So, if we initially have a rectangular window, the neighboring elements are best described by a parallelogram in the image plane. We approximate the parallelogram by a rectangle, defined by the mid-points of the 4 edges of the parallelogram. This notion of changing the window size is very important because we are grouping *scene* elements, not image elements. Because of the surface shape, the scene elements will "look" different in the image.

4 Results

The results of applying the algorithm on a number of images are presented in this section. In all our experiments, we set $r_1 = 4$ (Eq. 2), $r_2 = 2$ (Eq. 4) and $\tau_e = 0.2$ (Eq. 12).

---

[3] In psychophysics, contrast is usually defined as $\sigma_I / \bar{I}$. This has no effect on our similarity measure, because σ_e will be scaled by the same factor and the ratio will not change.

[4] A multi-scale analysis may help in choosing the optimal window size.

Figure 1 shows an image of a building. Architectural objects usually have a very large number of repeating elements, like windows or doors. The rectangle marked in (a) indicates where the initial element is. This unit is found automatically by the "interesting-window" search described in Section 3.1 and the size automatically adjusted to give a more distinctive element. The crosses are the locations of the elements obtained by the grouping process. Notice that the spacing of the crosses are smaller as the units get thinner when we propagate to the far end of the building. This is where the importance of changing the window size and shape by the affine transform shows up. In (b), the grouped region is displayed. This provides us with a segmentation of the surface patch.

The algorithm also works well on curved surfaces, where affine distortion between neighboring elements is significant. Figures 2 and 3 are some textile images. In Figure 2, the repeating unit chosen by the algorithm was 2 units of the actual pattern; in Figure 3, it was one unit. Note that there is an inherent ambiguity here because the 2-unit element chosen in Figure 2 is also a repeating pattern. In Figure 3, note that the front of the shirt is isolated from the arms. This is to be expected because the grouping process assumes small changes from one element to its neighbor. This offers us a useful segmentation technique where the shirt front, left arm and right arm would emerge as separate groups. Work on human figure recognition [2] have shown the usefulness of such segmentation into parts.

Figure 4 shows another type of grouping. Here, we have a simple repeating pattern. Traditionally, this would not be regarded as texture as it does not have significant 2D spatial extent. Instead, it is a distinctive element in the scene. The grouping works well too and the repeating units are found.

Figure 5 is an example of a yet different kind of image. Here the pattern is that of an oriented flow field. Note that repeating units are found though obviously the breakup into individual elements is somewhat artificial. However, the knowledge that these have predominantly 1D structure could be used to perform a more suitable grouping operation in a post-processing stage.

5 Conclusion

We have demonstrated a technique to detect, localize and group instances of repeated scene elements in an image. The result is a graph where nodes correspond to individual elements and arcs join spatially neighboring elements. Associated with each arc is an affine map that best transforms the image patch at one location to the other. Results on a variety of images are shown. They include 2D structures on planar and curved surfaces and 1D flows on curved surfaces.

A number of extensions of this work follow immediately. A multi-scale analysis would be useful to group elements of different scales. At present, the initial partition into windows is arbitrary. In principle, this should be chosen based on the scale of the features we are looking for.

Another direction of work that will be done is to use the affine transforms and the point correspondences in the patches to infer surface shape. Techniques from shape from texture [4, 10, 11] or structure from motion will be useful.

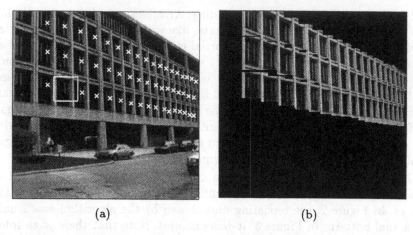

(a)   (b)

Fig. 1. Architectural image. The rectangle in (a) is the unit we found and where the growing process starts. The crosses are the locations of the units grouped together. The segmented region is shown in (b). Notice that the spacings between the crosses become smaller as we propagate to the far end of the building. This is capturing well the change in size and shape of the units in the scene.

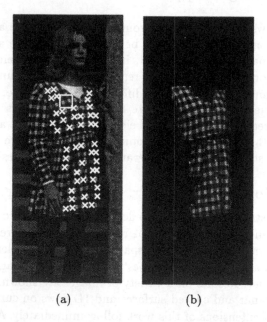

(a)   (b)

Fig. 2. A textile image on a curved surface. The rectangle in (a) is the initial unit we found. The crosses are the locations of the units grouped together. In (b), the segmented region is shown. Notice that the rectangle includes two units in the actual pattern. This is due to the inherent ambiguity in defining a repeating unit, namely, 2 units together is still a repeating element.

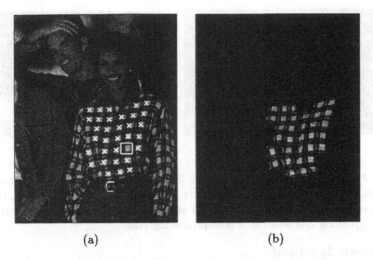

Fig. 3. Another example of a textile image on a curved surface.

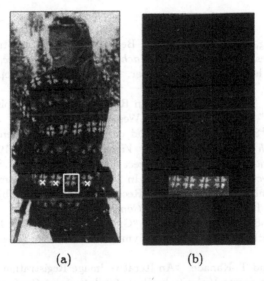

Fig. 4. A different type of grouping is shown in this figure. The rectangule in (a) shows a distinctive element in the image. Traditionally, this would not be considered as texture because there is no significant 2D spatial extent. Our algorithm works well in this situation too. The crosses in (a) mark the instances of the unit.

As we have pointed out, the ultimate goal of our work is object recognition. This algorithm can provide grouping and 3D structural information about the scene. Given the 3D structural relationships, we can recover what one of the scene elements looks like from a canonical view (say fronto-parallel). The appearance from such a canonical view is distinctive and invariant (up to a scale factor) to camera pose and could be used in an iconic matching scheme to provide another visual cue for recognition.

(a)   (b)

Fig. 5. An oriented 1D flow field. Notice that repeating units are found though obviously the breakup into individual elements is somewhat artificial. However, the knowledge that these have predominantly 1D structure could be used to perform a more suitable grouping operation in a post-processing stage.

Acknowledgement

This work was supported by the Digital Library Grant IRI-9411334 and a Berkeley Fellowship.

References

1. M. Armstrong, A. Zisserman, and P. Beardsley. "Euclidean Structure from Uncalibrated Images". In *Proc. British Machine Vision Conference*, 1994.
2. M. Fleck, D. Forsyth, and C. Bregler. "Finding Naked People". *To appear in ECCV*, 1996.
3. W. Förstner. "Image Matching". In R. Haralic and L. Shapiro, *"Computer and Robot Vision"*, chapter 16. Addison-Wesley, 1993.
4. J. Gårding. "Surface orientation and curvature from differential texture distortion". In *Fifth Intl. Conf. Computer Vision*, pages 733–739, 1995.
5. J. Gårding and T. Lindeberg. "Direct Estimation of Local Surface Shape in a Fixating Binocular Vision System". In *Proc. Euro. Conf. Computer Vision*, 1994.
6. J. Liu, J. Mundy, and E. Walker. "Recognizing Arbitrary Objects from Multiple Projections". In *Proc. Asian Conf. Computer Vision*, 1993.
7. J. Liu, J. Mundy, and A. Zisserman. "Grouping and Structure Recovery for Images of Objects with Finite Rotational Symmetry". In *Proc. Asian Conf. Computer Vision*, 1995.
8. B.D. Lucas and T. Kanade. "An Iterative Image Registration Technique with an Application to Stereo Vision". In *Proc. 7th Intl. Joint Conf. on Art. Intell.*, 1981.
9. J. Malik and P. Perona. "Preattentive Texture Discrimination with Early Vision Mechanisms". *J. Opt. Soc. Am. A*, 7(5):923–932, 1990.
10. J. Malik and R. Rosenholtz. "Recovering Surface Curvature and Orientation from Texture Distortion: a Least Squares Algorithm and Sensitivity Analysis". *Proc. Third Euro. Conf. Computer Vision*, pages 353–364, Sweden, 1994.
11. J. Malik and R. Rosenholtz. "Computing Local Surface Orientation and Shape from Texture for Curved Surfaces". *To appear in the International Journal of Computer Vision*, 1996.
12. J.L. Mundy and A. Zisserman. "Repeated structures: image correspondence constraints and 3D structure recovery.". In J.L. Mundy, A. Zisserman, and D.A. Forsyth, editors, *Applications of invariance in computer vision*. 1994.

Image Recognition with Occlusions

Tyng-Luh Liu[1]   Mike Donahue[2]   Davi Geiger[1]   Robert Hummel[1]

[1] Courant Institute, New York University, New York NY 10012, USA
[2] IMA, University of Minnesota, Minneapolis MN 55455, USA

Abstract. We study the problem of how to detect "interesting objects" appeared in a given image, I. Our approach is to treat it as a function approximation problem based on an over-redundant basis. Since the basis (a library of image templates) is over-redundant, there are infinitely many ways to decompose I. To select the "best" decomposition we first propose a global optimization procedure that considers a concave cost function derived from a "weighted L^p norm" with $0 < p \leq 1$. This concave cost function selects as few coefficients as possible producing a sparse representation of the image and handle occlusions. However, it contains multiple local minima. We identify all local minima so that a global optimization is possible by visiting all of them. Secondly, because the number of local minima grows exponentially with the number of templates, we investigate a greedy "L^p Matching Pursuit" strategy.

1   Introduction

In the field of signal processing and computer vision an input signal or image is a function f over some subset of $\mathbb{R}$ or $\mathbb{R}^2$. To manipulate and analyze f, it is useful to introduce a linear decomposition into basis elements f_j, i.e., $f = \sum_j c_j f_j$. An example of a well known and useful decomposition of this type is the Fourier series expansion.

We study the object recognition problem via a robust template decomposition approach. Let the image to be recognized be I and the template library be $\mathcal{L}$. The task of image recognition is reduced to a function approximation problem of the form

$$I(x) = \sum_j \sum_i c_{ij} A_i(\tau_j)(x) = \sum_{i,j} c_{ij} T_{ij}(x) \qquad (1)$$

where $\tau_j \in \mathcal{L}$, $T_{ij} = A_i(\tau_j)$ denotes an affine transformation applied to the template τ_j, and c_{ij} is the choice of coefficients that "best" decompose the image. Typically the library $\mathcal{L}$ is large, in order to accommodate many possible situations and also consider the possible (affine) transformations. Thus, we have an over-redundant basis leading to infinite many solutions, c_{ij}, to this problem. That is not the case for the Fourier decomposition.

Let us illustrate the problem of function decomposition with over-redundant library. Say our basis consists of sinusoids and functions of the form $1/(k + x)$ ($k \in \mathbb{N}$). Assume that $f(x) = \sin 2x + \frac{4}{(3+x)}$ is our target function (our image). It is clear that only two terms from the prototype library are required to represent $f(x)$. However, one could write $f(x)$ using either sinusoids alone or as combinations of $1/(k + x)$ alone, but either representation would require many terms. The problem is to formulate a coefficient selection criterion and a method to compute the coefficients that yields compact representations.

1.1 Coefficient selection, concave cost function, and optimization

Our approach [7, 13] is to construct an objective function $F(\mathbf{c})$ that when minimized selects a best representation, $\mathbf{c}^*$, from among all solutions $\mathbf{c}$ that satisfy the constraint $I(x) = \sum_j \sum_i c_{ij} A_i(\tau_j)(x)$. We require

1. **Sparse Representation:** represent (decompose) an image using as few templates as possible in order to have an economical (minimal) representation. Field [9] also argued for sparse representations in the brain.
2. **Occlusions:** allow for partial occlusions, i.e., the cost of fitting a template must take into account that portions of the template may have a "bad match".
3. **Noise:** model noise via "noise templates" accounting for the difference between the template fit and the image. This leads us to search for cost functions that escalate with the magnitude of c_{ij}, but should not dominate the first condition, i.e., the rate of increase in cost as a function of $|c_{ij}|$ should decrease.

The above consideration leads us naturally to adopt concave objective functions. In particular, we will primarily study the objective function

$$F_p(\mathbf{c}) = \sum_{j=1}^{M} \sum_{i=1}^{N} \omega_{ij} |c_{ij}|^p , \tag{2}$$

where N is the number of possible (affine) transformations and M is the size of the template library. The scalars ω_{ij}'s are positive, e.g., they may be set to 1 or to the inverse of the template and image variances.

The sparsity of templates suggests $\mathbf{p} = 0$ to count the number of templates (weighted by ω_{ij}). Noise templates should be paid according to how large the "repair" is, i.e., how large the error c_{ij} is. The balance between both processes, sparsity of the templates and noise modeling leads to values of $0 < \mathbf{p} \le 1$.

The objective function is non-convex, and in fact the optimization problem will generally have multiple local minima, making the optimization more difficult. We will show that it is possible to characterize all local minima and obtain the global one by visiting them. Since the number of local minima grows exponentially with the size of the template library we consider an alternative greedy algorithm. Recently, Chen and Donoho [3, 4] studied the overcomplete signal representation problems with L^1 norm optimization. Their method is based on linear programming, which is efficient, but only applies to the $\mathbf{p} = 1$ case and still leads to a slow algorithm. Coifman and Wickerhauser [5], modeled an entropy like function, $\sum_{i,j} |c_{ij}|^2 log|c_{ij}|^2$ with more constraints on the the coefficients c_{ij} square-sum to 1.

Comparison with principal component analysis/Eigenfaces: Our approach is fundamentally different from the "eigenfaces" approach (PCA approach) [16]. In our case the basis functions are fixed and the adaptation of the method is on choosing the appropriate coefficients (from a redundant basis), a non-linear process. In the PCA approach the choice of basis functions, a linear process, is where the adaptation occurs. PCA works well only when the task function is a simple linear superposition of the basis functions.

1.2 Matching pursuit

Inspired by Mallat and Zhang's work [14] we consider a matching pursuit strategy where, at each stage, the criterion of best selection is based on minimizing an image residue. In regression statistics, this decomposition method is known as *Projection Pursuit Regression* , a non-parametric method that is concerned with "interesting" projections of high dimensional data (see Friedman and Stuetzle [10], Huber [11]). Recently, Bergeaud and Mallat [2] used the (L^2) matching pursuit with a redundant family of Gabor oriented wavelets to approximate images and produce compact decompositions for the main features of images.

The original matching pursuit is based on the standard L^2 (Hilbert space) method. We propose an L^p matching pursuit with $0 < p \leq 1$, to improve the robustness. With $0 < p \leq 1$, we lose the structure of inner product but the notion of a template "closest" to the image is recaptured via the cost function.

2 Template Library and Image Coordinates

We must first establish a well-defined over-redundant library of templates containing many non-canonical templates as well as one canonical template. A canonical template is a trivial template with zero gray-level value pixels everywhere except one pixel at the extreme left and top corner that its gray-level value is 1. Moreover, we will assume we can apply a set of affine transformations to each template, indeed we will restrict ourselves to translations. Clearly, this single canonical template plus a set of all translations form a basis for the image space.

Coordinate transformations: Suppose we have now created a template library $\mathcal{L} = \{\tau_j : j = 1...M\}$ for some application, where we will use $\epsilon_1 \equiv \tau_1$ to represent the canonical template. Let the image to be recognized be I of dimension N and each template τ_j be of dimension N_T (we assume that both N and N_T are perfect square numbers). Furthermore, let $P = \{p_1, p_2, \cdots, p_N\}$ and $Q = \{q_1, q_2, \cdots, q_{N_T}\}$ be the pixel sets of I and any τ_j, respectively. (We order the pixels from top to bottom and left to right.) Let the translation $A_i(\tau_j)$ indicate that the first template pixel q_1 is positioned at the i-th pixel $p_i \in P$. The mapping formula for A_i is such that $q_r \mapsto p_k = p_{k(r,i)}$ where [3] $k = i + (\lfloor \frac{r-1}{\sqrt{N_T}} \rfloor \times N) + (r - 1 - \lfloor \frac{r-1}{\sqrt{N_T}} \rfloor \times \sqrt{N_T})$. Denote $T_{ij} = A_i(\tau_j)$ and $e_{i1} = T_{i1} = A_i(\epsilon_1)$ [4], then we

---

[3] The expression $\lfloor x \rfloor$ denotes the greatest integer less than or equal to x.

[4] Note that $e_{i1}(p_j) = e_i(p_j) = \delta_{ij}$, where $\delta_{ij} = 1$ for $i = j$ and $\delta_{ij} = 0$ otherwise.

have $T_{ij}(p_k) = \tau_j(q_r)$. Using these notations, one can write the decomposition equation (1) as

$$I(p_k) = \sum_{i=1}^{N} c_{i1} e_{i1}(p_k) + \sum_{j=2}^{M} \sum_{i=1}^{N} c_{ij} T_{ij}(p_k) = \sum_{\lambda=1}^{N} c_\lambda e_\lambda(p_k) + \sum_{\lambda=N+1}^{M.N} c_\lambda T_\lambda(p_k) \quad (3)$$

where $\lambda = \lambda(i,j) = (j-1) \times N + i$. We may write $I[k]$, $e_\lambda[k]$ and $T_\lambda[k]$ instead of $I(p_k)$, $e_\lambda(p_k)$ and $T_\lambda(p_k)$, respectively, for simplification.

3 Optimization Problem and Solution

Equation (3) can be written in matrix notation as $Tc = I$ where

$$T = \begin{pmatrix} e_1[1] & \cdots & e_N[1] & T_{N+1}[1] & \cdots & T_{MN}[1] \\ e_1[2] & \cdots & e_N[2] & T_{N+1}[2] & \cdots & T_{MN}[2] \\ \vdots & \ddots & \vdots & \vdots & \ddots & \vdots \\ e_1[N] & \cdots & e_N[N] & T_{N+1}[N] & \cdots & T_{MN}[N] \end{pmatrix},$$

$$c = (c_1, c_2, \ldots, c_{MN})^t \quad \text{and} \quad I = (I[1], I[2], \ldots, I[N])^t.$$

Note that if the prototype library forms a basis (linearly independent), then $M = 1$, and there is no freedom in choosing the coefficients (c_λ); the coefficients are uniquely determined by the constraint. If there are linear dependencies in the prototype library, then $M > 1$, the prototype library over-spans, and the set of all solutions (c_λ) to the constraint forms an $(M-1)N$ dimensional affine subspace in the $M.N$-dimensional coefficient space. Let S denote this solution space, i.e., $\dim(S) = (M-1)N$. Using the above matrix notations, our optimization problem can be formulated as:

$$\underset{c}{Min}\, F_p(c) = \underset{c}{Min} \sum_{\lambda=1}^{MN} \omega_\lambda |c_\lambda|^p \quad \text{subject to the constraint } Tc = I \quad (4)$$

where $T \in \mathbb{R}^{N \times M.N}$, $c \in \mathbb{R}^{M.N}$, $I \in \mathbb{R}^N$, $M > 1$. The next result is shown in [7, 13], or previously stated in [8].

Proposition 1 *All the local minima of L^p-cost function in (4) occur at the vertices of a polytope. This polytope is constructed from the intersection of the affine subspace S and a cube defined by the origin and bounded in each axis by d_λ. d_λ can be as large as $(F_p(c_0)/\omega_\lambda)^{1/p}$, where c_0 is any solution to $Tc = I$.*

4 One Template Matching and Simulations

If we want to find a specific face in an image, then it suffices to use only one face-template. In these cases the non-canonical template represents a key feature and the canonical templates e_λ represents non-interest elements , e.g., noise. Let us assume that this particular template be τ_2 of size N_T ($\tau_1 \equiv e_1$) and A_i be the translation, that is, $A_i(\tau_2) = T_{i2} = T_{N+i}$. This says that we look for a decomposition of the form:

$$I(x) = c_{N+i}T_{N+i}(x) + \sum_{\lambda=1}^{N} c_\lambda e_\lambda(x). \tag{5}$$

It is clear that $c_\lambda = I[\lambda]$ if pixel p_λ is not covered by T_{N+i}. So, the equation (1) can be restricted to the region where T_{N+i} is located.

$$\begin{pmatrix} e_{A_i(1)}[A_i(1)] & \cdots & e_{A_i(N_T)}[A_i(1)] & T_{N+i}[A_i(1)] \\ e_{A_i(1)}[A_i(2)] & \cdots & e_{A_i(N_T)}[A_i(2)] & T_{N+i}[A_i(2)] \\ \vdots & \ddots & \vdots & \vdots \\ e_{A_i(1)}[A_i(N_T)] & \cdots & e_{A_i(N_T)}[A_i(N_T)] & T_{N+i}[A_i(N_T)] \end{pmatrix} \begin{pmatrix} c_{A_i(1)} \\ \vdots \\ c_{A_i(N_T)} \\ c_{N+i} \end{pmatrix} = \begin{pmatrix} I[A_i(1)] \\ \vdots \\ I[A_i(N_T)] \end{pmatrix}$$

where $e_i[j] = e_i(p_j) = \delta_{ij}$. Recall that $T_{N+i}[A_i(r)] = \tau_2[r]$ (and $A_i(1) = i$). We can also assume that $\tau_2[r] \neq 0$ for $r = 1, \ldots, N_T$, since otherwise we can redefine either τ_2 or the pixel ordering to get a smaller value for N_T.

It follows from Proposition 1 that the local minima of $F_p(c)$ can be found by setting c_{N+i}, $c_{A_i(1)}$, $\ldots$, $c_{A_i(N_T)}$ to zero one at a time. If we set $c_{N+i} = 0$ then we get $c_\lambda = I[\lambda]$ for all λ. This is the "pure noise" solution. The first nontrivial (template using) solution sets $c_{A_i(1)} = 0$. This forces the template coefficient $c_{N+i} = I[A_i(1)]/\tau_2[1]$, from which it follows that $c_{A_i(r)} = I[A_i(r)] - c_{N+i}\tau_2[r]$, for $r = 2, \ldots, N_T$. The solution determined by setting $c_{A_i(r)} = 0$ ($2 \leq r \leq N_T$) can be calculated in an analogous fashion.

The optimal cost of the match of the template in the (translation) position i is the smallest of the values of $F_p(c)$ across all $N_T + 1$ solutions (c). One performs a similar analysis for all template translations, and finds the position which generated the smallest match cost. Note that in the case of one template matching, the L^p-norm decomposition problem is actually the same as p-norm minimization.

4.1 Simulations

We have designed a sequence of experiments focused on the effects of noise and occlusions to demonstrate both the weighted and unweighted (all ω_λ's are set to 1) L^p decomposition methods are superior to the conventional correlation techniques. The weights used in the weighted scheme are defined as $\omega_{\lambda(i,j)} = 1/([\sum_{k=1}^{N_T} |\tau_j[k]|^p][\sum_{k=1}^{N_T} |I[A_i(k)]|^p])$, for $0 < \mathbf{p} \leq 1$.

The experiments consist of numerous trials on random images with fixed occlusion size and fixed noise variance. The latter determines the signal-to-noise ratio (SNR) for the experiment, defined here as the ratio of the standard deviation of the image to the standard deviation of the noise.

Each trial has four components: an image, a template, an occlusion, and noise. The image is 64 pixels wide by 64 pixels high, randomly generated using an uncorrelated uniform distribution across the range $(-256, 256)$. The template is a 4 pixel by 4 pixel subimage of the image. After selecting the template, a portion of the image from which the template is drawn is "occluded" by redrawing from the same distribution that formed the image, i.e., from an uncorrelated uniform distribution with range $(-256, 256)$. (Occlusion sizes range from 0–14 pixels, from a total subimage size of 16 pixels.) Finally, noise is added to the (occluded) image, drawn from an uncorrelated Gaussian mean-zero random variable.

Translates of the template are compared against the noisy, occluded image, using both weighted and unweighted L^p-norm decomposition method. (Because both the template and the image are drawn from zero-mean random variables, there is little difference between 2-norm error minimization and standard correlation.) For each method the translation position yielding the best score is compared with the position of the original subimage from which the template was formed. If the two agree then the match is considered successful, otherwise the match fails for the trial in question.

5 Multiple Templates and Matching Pursuit

In this section, we proceed to elucidate the matching pursuit method for the case of multiple templates. The basic idea is to devise a greedy iterative method where at each stage only one template is selected and thus, we can rely on the previous section result. In this section we will also consider, for comparison, a cost function based on the LTS (Least Trimmed Squares, Rousseuw 1983, 1984, [15]).

5.1 Review

We briefly review the (L^2) matching pursuit below. Suppose it is given a signal f, and a library of functions $D = \{g_\gamma\}_{\gamma \in \Gamma}$ where Γ is a set of index tuples and D represents a large, over-redundant family of functions. A "best" matching library element to the residual signal structures at each stage is decided by successive approximations of the residual signal with orthogonal projections on elements in the library. That is, say at stage n, for any element $g_\gamma \in D$, we consider

$$R^{n-1}f = < R^{n-1}f, g_\gamma > g_\gamma + R^n f \qquad (6)$$

where $R^n f$ is the n-th residue after approximating $R^{n-1}f$ in the direction of g_γ (assume that the initial residue is the function f, i.e. $R^0 f = f$). The matching pursuit strategy is to find g_{γ^*} that minimizes $\|R^n f\|$ (or the g_{γ^*} closest to $R^{n-1}f$), i.e. $\|R^{n-1}f - < R^{n-1}f, g_{\gamma^*} > g_{\gamma^*}\|_{L^2} = \underset{\gamma \in \Gamma}{Min} \|R^{n-1}f - < R^{n-1}f, g_\gamma > g_\gamma\|_{L^2}$.

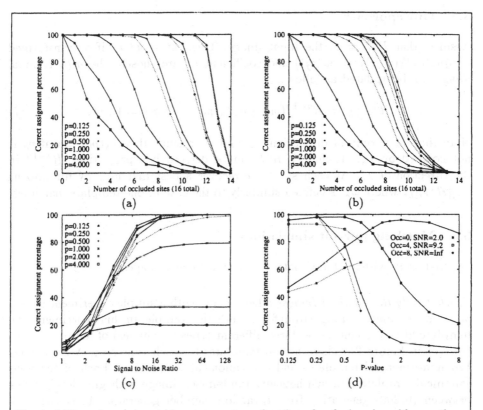

Fig. 1. (a)Experimental matching accuracy as a function of occlusion size with no noise. The solid curves correspond to our proposed weighted L^p decomposition, dashed to the unweighted L^p decomposition. Note that smaller values of p outperform larger values providing nearly 100% correct results with p = 0.125 for occlusions as large as 11 (out of 16) pixels. Also given a p value the weights (normalization) can help improve the results. (b) Experimental matching accuracy as a function of occlusion size at a Signal to noise Ratio (SNR) of 37. The solid curves correspond to our proposed method, dashed to the unweighted L^p method. Here we note that p = 0.125 still performs very well, although good results can not be obtained if the occlusion is larger than half the template size. Notice that the results using larger values of p are less affected by noise, especially those with p > 1. (c) Experimental matching accuracy as a function of noise level at a fixed occlusion size of 5 (out of 16) pixels. Note again that larger values of p produce results which are less sensitive to noise. For example, the results for p = 0.125, which are best for large SNR, are poorest for SNR of less than about 3. The solid (dashed) curves correspond to weighted (not weighted) L^p method. (d) Experimental matching accuracy as a function of p, at various noise levels and occlusion sizes. For an occlusion size of 4/16 and a SNR of 9.2, the best p value for our weighted L^p method is somewhere between 0.25 and 0.5. The solid (dashed) curves correspond to weighted (not weighted) L^p method.

5.2 Our approach

Assume that $R^0 I = I$, the input image. Then, at stage n, if a transformed template $T_\lambda (= T_{ij} = A_i(\tau_j))$ and coefficient c_λ are chosen, the n-th residual image can be updated as follows:

$$R^n I(p_k) = R^{n-1} I(p_k) - c_\lambda T_\lambda(p_k) \qquad \text{for } k = 1...N. \qquad (7)$$

Note that T_λ is only of dimension N_T and we assume that $T_\lambda(p_k) = 0$ if p_k is not covered by T_λ. From (7), $R^n I$ can be derived by "projecting" $R^{n-1} I$ in the direction of T_λ. At each stage, we recover a best matching by minimizing $\omega_\lambda \| R^n I \|_{L^p}$ where ω_λ is defined similarly to the case of one-template matching.

5.3 Matching pursuit simulations

We first work with synthetic data and then with real images.

Synthetically Randomized Images : Let's begin with a simple experiment to test our template matching algorithm for a synthetic example. In this experiment, the template library $\mathcal{L}$ consists of three different types (or shapes) of templates ((a), (b), (c) in Figure 2). There are 40 templates for each type so that $\mathcal{L}$ includes 120 non-canonical templates and one canonical template ϵ_1. Each of the non-canonical template is a synthetically randomized image with gray-level values between $(0, 200)$ generating from a random number generator. To construct a test image I_1 (as in Figure 2-(d)), we first select one non-canonical template randomly from each template type in $\mathcal{L}$ to form the base (exact) image then add noise and an occluded square derived from uniform distribution in $(0, 10)$ and $(245, 255)$, respectively. The threshold values used in simulation vary with respect to the value of $\mathbf{p}$ for L^p matching pursuit and α for LTS matching pursuit. We see that both methods can handle occlusions (e.g. see Figure 2-(e) R_1). Our experiment results suggest for $\mathbf{p} \in (0.25, 0.75)$ and $\alpha \in (0, 51, 0.75)$, both the L^p and LTS methods are rather robust. But, as shown in Figure 2-(f) R_2, both methods failed to recognize the occluded object for $\mathbf{p} \geq 0.75$ and for $\alpha \geq 0.75$.

Face Recognition : A small library of face templates has been established (see Figure 3 (a)-(f)). The dimension of all the six templates is 64×64. Numerous experiments have been carried out to test our algorithm. To illustrate, consider the three real images, I_1 - I_3, in Figure 4 (a)-(c). We obtained decomposition results R_1, R_2 and R_3 shown in Figure 4, for $\mathbf{p} = 0.25$. (Similar results are derived for $\mathbf{p} = 0.50$ and 0.75.) When $\mathbf{p} = 2$, it is indeed the L^2 matching pursuit method and the recognition results are R_4, R_5 and R_6. Our proposed L^p matching pursuit has the robustness advantage over the L^2 one. In case that an image contains objects with large occlusions (like I_3), the LTS may fail to recognize them as shown in 4-(l). In addition, the L^p is more efficient than LTS regarding to the computation complexity.

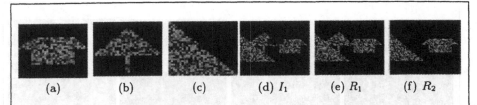

(a)   (b)   (c)   (d) I_1 (e) R_1 (f) R_2

Fig. 2. (a), (b), (c) are synthetic template type 1, type 2 and type 3, respectively. (d) Test image I_1 with noise added and occlusion (e) Result of the decomposition for the L^p with $p = 0.25$, and also for the LTS with $\alpha = 0.51$. (α is the robust constant of LTS.) (f) Results once the breakdown limits are reached, and occluded templates are not recognized. For example, L^p, with $p = 0.75$ and LTS with $\alpha = 0.75$.

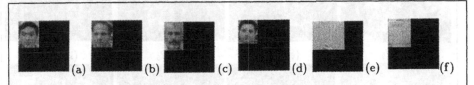

(a)   (b)   (c)   (d)   (e)   (f)

Fig. 3. (a) - (f) are the face and book templates used in the face recognition simulation.

References

1. J. Ben-Arie and K. R. Rao, *"On the Recognition of Occluded Shapes and Generic Faces Using Multiple-Template Expansion Matching"*, Proceedings IEEE International Conference on Pattern Recognition, New York City, 1993.
2. F. Bergeaud and S. Mallat, *"Matching Pursuit of Images"*, SPIE, Orlando, 1995.
3. S. Chen and D. Donoho, *"Atomic Decomposition by Basis Pursuit"*, Technical Report, Stanford University, May, 1995.
4. S. Chen and D. Donoho, *"Basis Pursuit"*, TR, Stanford University, Nov. 1994.
5. R. Coifman and V. Wickerhauser, *"Entropy-based Algorithms for Best Basis Selection"*, IEEE Transactions on Information Theory, vol. 38, no. 2, 1992.
6. T. H. Cormen, C. E. Leiserson and R. L. Rivest, *Introduction to Algorithms*, McGraw-Hill, 1990.
7. M. J. Donahue and D. Geiger, *"Template Matching and Function Decomposition Using Non-Minimal Spanning Sets"*, Technical Report, Siemens, 1993.
8. H. Ekblom, *"L_p-methods for Robust Regression"*, BIT 14, p.22-32, 1973.
9. D. Field, *"What Is the Goal of Sensory Coding"*, Neural Comp. 6, p.559-601, 1994.
10. J. H. Friedman and W. Stuetzle, *"Projection Pursuit Regression"*, Journal of the American Statistical Association, vol. 76, p.817-823, 1981.
11. P. J. Huber, *"Projection Pursuit"*, The Ann. of Stat., vol. 13, No.2, p.435-475, 1985.
12. P. J. Huber, *Robust Statistics*, John Wiley & Sons, New York, 1981.
13. T. Liu, M. Donahue, D. Geiger and R. Hummel, *"Sparse Representations for Image Decomposition"*, Technical Report, CS, Courant Institute, NYU, 1996.

565

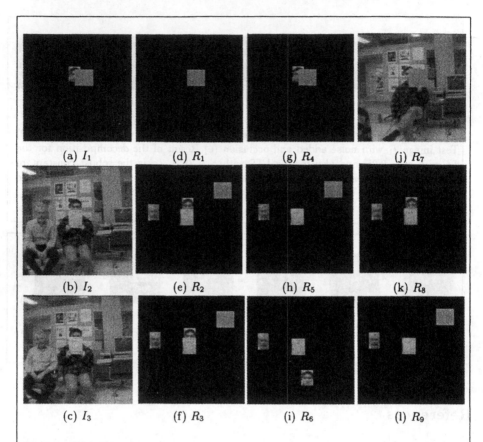

(a) I_1 (d) R_1 (g) R_4 (j) R_7

(b) I_2 (e) R_2 (h) R_5 (k) R_8

(c) I_3 (f) R_3 (i) R_6 (l) R_9

Fig. 4. (a)-(c) The test images, where some templates are present with small distortions (scale and viewing angle), noise and occlusions. (d)-(f) Image decomposition for L^p matching pursuit with $\mathbf{p} = 0.25$ (similar results are obtained for p up to 0.75). (g) - (i) Image decomposition for $\mathbf{p} = 2.0$ and recognition is destroyed (this is equivalent to use correlations methods, like in the L^2 matching pursuit). Note that false recognition occurs in (i). (j)-(l) Image decomposition with LTS.

14. S. Mallat and Z. Zhang, *"Matching Pursuit with Time-Frequency Dictionaries"*, IEEE Trans. on Signal Processing, Dec. 1993.
15. P. J. Rousseeuw and A. Leroy, *Robust Regression and Outlier Detection*, John Wiley, New York, 1987.
16. M. Turk and A. Pentland, *"Eigenfaces for Recognition"*, J. of Cognitive Neuroscience, vol. 3, p.71-86, 1991.
17. B. Uhrin, *"An Elementary Constructive Approach to Discrete Linear l_p-approximation, $0 < p \leq +\infty$"*, Colloquia Mathematica Societatis János Bolyai, 58. Approximation Theory, Kecskemét, 1990.

Silhouette-Based Object Recognition with Occlusion through Curvature Scale Space

Farzin Mokhtarian

Department of Electronic and Electrical Engineering
University of Surrey, Guildford, England GU2 5XH
E-mail: *F.Mokhtarian@ee.surrey.ac.uk*

Abstract. A complete and practical system for object recognition with occlusion has been developed which is very robust with respect to noise and local deformations of shape as well as scale changes and rigid motions of the objects. The system has been tested on a wide variety of 3-D objects with different shapes and surface properties. No restrictive assumptions have been made about the shapes of admissible objects. An industrial application with a controlled environment is envisaged. The *Curvature Scale Space* technique [4, 5] is used to obtain a novel *multi-scale segmentation* of the image contour and the model contours using curvature zero-crossing points. Multi-scale segmentation renders the system substantially more robust with respect to noise and local shape differences. *Object indexing* [9] is used to narrow down the search-space and avoid an exhaustive investigation of all model segments. A local matching algorithm applies *candidate generation, selection, merging, extension* and *grouping* to select the best matching models.

1 Introduction

Object representation and recognition is one of the central problems in computer vision. Normally, a reliable, working vision system must be able to **a)** effectively segment the image and **b)** recognize objects in the image using their representations. A complete and robust isolated object recognition system was described in [6, 3]. This paper describes a complete, working vision system which segments the image effectively using a light-box setup and recognizes occluded objects in the image reliably using their curvature scale space (or *CSS*) representations [4, 5]. The CSS representation is based on the *scale space image* concept proposed in [10]. It is an organization of curvature zero-crossing points on a contour at multiple scales.

It is assumed that the recognition system developed here may be used for recognition of occluded 3-D objects. In particular, it is assumed that a number of objects are placed on a light-box directly in front of a camera and that the task is to recognize each object. We believe that this particular task is interesting for the following reasons:

– Despite the constraints placed on the environment, *no* constraints have been placed on object shapes or types. Furthermore, environment constraints are not difficult to satisfy in many recognition tasks (such as industrial settings).

- Every 3-D object resting on a flat surface and viewed by a fixed camera, has a few stable positions, each of which can be modeled by a 2-D contour.
- Recognition can still become challenging due to arbitrary shapes of objects, noise, and local deformations of shape which can be caused by perspective projection, segmentation errors and non-rigid material.

The existing literature on shape representation and recognition is quite large. A survey of some recent work can be found in [9]. It should be noted that projective invariants [8] have received attention recently as tools for object recognition.

In general, a shortcoming of some object representation techniques is that the features extracted from the objects and used for matching are too local and therefore the resulting system is not robust with respect to noise and local deformations of shape. In those systems in which less local features are used, the utilized features are not necessarily inherent features of the object and therefore have weak discriminative power.

Sections 2 through 8 of this paper explain various aspects of the object recognition system developed. Section 2 briefly explains how the segmentation of an image using a light-box system is accomplished. Section 3 reviews the CSS representation as a multi-scale organization of the inherent features of a planar curve. Section 4 shows how multi-scale segmentation of a 2-D contour using curvature zero-crossing points may be accomplished. Section 5 describes a fast, local matching algorithm. Section 6 proposes a procedure for estimating the transformation parameters. Section 7 shows how the image-model curve distance can be computed. Section 8 describes a procedure for efficient optimization of the transformation parameters. Section 9 presents the results and an evaluation of the system. Section 10 contains the concluding remarks.

2 Image Segmentation

The use of a light-box setup makes the segmentation of the image reasonably straightforward. The same threshold value T was used to effectively segment all input images. A salt-and-pepper noise removal procedure was applied to the resulting binary image in order to remove isolated noise. The next step is to apply a process of region growing followed by shrinking to the image in order to fill in cracks and small holes. The resulting binary image always had only one connected region of 1-pixels which corresponded to the objects. Next, boundary pixels belonging to the region of 1-pixels are detected. The final step is to recover the image coordinates of the boundary points.

3 The Curvature Scale Space Representation

A CSS representation is a multi-scale organization of the invariant geometric features (curvature zero-crossing points and/or extrema) of a planar curve (here, only curvature zero-crossings were used). The CSS representation of a planar curve represents that curve uniquely modulo scaling and a rigid motion [2]. To compute it, the curve Γ is first parametrized by the arc length parameter u:

$$\Gamma(u) = (x(u), y(u)).$$

It is assumed that the input curve is initially represented by a polygon with possibly many vertices. Therefore only the coordinates of the vertices of the polygon need be given. If the distances between adjacent vertices of the polygon are all equal, then an arc length parametrization of the curve is already available. Otherwise, that polygon is sampled to obtain a new list of points such that the distances between points adjacent on the list are all equal *on the original polygon*. An *evolved version* Γ_σ of Γ can then be computed. Γ_σ is defined by [4]:

$$\Gamma_\sigma = (X(u, \sigma), Y(u, \sigma))$$

where

$$X(u, \sigma) = x(u) \otimes g(u, \sigma) \qquad Y(u, \sigma) = y(u) \otimes g(u, \sigma)$$

where $\otimes$ is the convolution operator and $g(u, \sigma)$ denotes a Gaussian of width σ. The process of generating evolved versions of Γ as σ increases from 0 to ∞ is referred to as the *evolution* of Γ. This technique is suitable for removing noise from a planar curve. Evolving contours can be considered an early form of active contours (snakes) [1] since they are similar in behaviour to snakes without any external constraints.

The CSS representation contains curvature zero-crossings or extrema extracted from evolved versions of the input curve. In order to find such points, we need to compute curvature accurately and directly on an evolved version Γ_σ of a planar curve. It can be shown that curvature κ on Γ_σ is given by [5]:

$$\kappa(u, \sigma) = \frac{X_u(u, \sigma)Y_{uu}(u, \sigma) - X_{uu}(u, \sigma)Y_u(u, \sigma)}{(X_u(u, \sigma)^2 + Y_u(u, \sigma)^2)^{1.5}}$$

where

$$X_u(u, \sigma) = \frac{\partial}{\partial u}(x(u) \otimes g(u, \sigma)) = x(u) \otimes g_u(u, \sigma)$$

$$X_{uu}(u, \sigma) = \frac{\partial^2}{\partial u^2}(x(u) \otimes g(u, \sigma)) = x(u) \otimes g_{uu}(u, \sigma)$$

and

$$Y_u(u, \sigma) = y(u) \otimes g_u(u, \sigma) \qquad Y_{uu}(u, \sigma) = y(u) \otimes g_{uu}(u, \sigma).$$

The function defined implicitly by $\kappa(u, \sigma) = 0$ is the CSS image of Γ. Note that:

- The CSS image is stored as a binary image in which each row corresponds to a specific value of σ and each column to a specific value of u.

- A brute force computation of a CSS image can be inefficient. The method usually used is to *track* the zero-crossings in the CSS image: at each scale curvature is computed only in a small neighborhood of each location where a zero-crossing was detected at the previous scale. This is possible since for a small change in σ, the change in location of any curvature zero-crossing on the curve is also small.
- For all values of σ larger than a σ_c, evolved curves Γ_σ will be simple and convex. This suggests that the computation can stop as soon as σ_c is reached or as soon as no more curvature zero-crossings are detected on Γ_σ [7].

For examples of CSS images, see [5, 6].

4 Multi-Scale Segmentation of 2-D Contours

The basic idea behind the segmentation scheme is to divide the input contour into primitive segments to be used by a local matching algorithm (described in section 5). Curvature zero-crossing points are the natural feature points to divide the contour since their locations are invariant with respect to rotation, scaling and translation of the contour. The main issue, therefore, is the issue of scale: *which scale should be chosen for the detection of curvature zero-crossing points on the input contour?* If the scale chosen is too small, the segmentation may be affected by noise and local distortions of shape, and if it is too large, important structure on the contour may be lost.

The solution used here was motivated by the main underlying concept of the curvature scale space representation: *utilize information from multiple scales rather than prefer a single scale.* Therefore the segmentation of the input contour is also carried out at multiple scales. The procedure is as follows:

- Start the segmentation at the lowest scale of the CSS image and end at a medium scale since the segments discovered at high scales are not useful.
- Segment the contour using the curvature zero-crossing points detected at the lowest scale and add all segments (defined by their left and right endpoints expressed in arc-length values) to a segment-list. As each higher scale is considered, again detect all curvature zero-crossing points at that scale *but* add a new segment if it does not already exist in the segment-list.

Care must be taken to account for the movement of curvature zero-crossing points in the CSS image. Therefore an auxiliary segment-list is also used which always records the updated values (across scales) of the left and right endpoints of each segment in the original segment-list. To check for existence, the auxiliary segment-list is searched. When extracted segments are written to the output file, the original segment-list is used since the segments in the auxiliary segment-list become very small at the maximum of the corresponding CSS zero-crossing contour. This multi-scale segmentation scheme is substantially more robust with respect to noise and local shape differences.

5 Local Matching through Curvature Scale Space

Due to occlusion, the matching algorithm employed is a local one and consists of several stages. This section describes those stages in the sequence in which they are carried out.

5.1 Rescaling

Model contours and the image contour are rescaled so that they just fit in a unit square. The model contours are further rescaled so that they reflect the relative sizes of model objects when viewed at the same distance. Model contour rescaling is carried out off-line. As a result of image and model contour rescaling, the possible scale changes from model contours to the image contour become predictable which helps to define an *admissible space* for the scale-factors (In principle, since the distance from the camera to the light-box is known, the scale-factors are also known, but that information was not used here: the system is therefore allowed to recover the correct scale-factors as a result of the recognition process). The multi-scale segmentation procedure described in section 4 is then used to segment the model contours and the image contour.

5.2 Candidate Generation and Filtering

Due to occlusion, all possible local matches must be considered (note however, that very small segments on either the image contour or the model contours are discarded). In order to avoid an exhaustive search of all model contour segments, *object indexing* [9] is employed to render the initial search more efficient. After segmentation, each model contour segment is rescaled so that each has the same length $\mathcal{L}$ (subject to constraints imposed by the admissible space defined in the previous subsection). Average curvature is then computed for each of those segments and used to create an index-table for all the model contour segments. All the computation is carried out off-line.

Once the segmentation of the image contour is completed, each image contour segment is also rescaled so that each has the same length $\mathcal{L}$ (again subject to constraints imposed by the admissible space). Average curvature is also computed for each of those segments. The average curvatures now serve as indices into the model contour segment index-table to recover a more likely (and smaller) set of potentially matching model contour segments. A candidate is generated for the possible match of each image contour segment and the corresponding model contour segments recovered from the index-table.

Transformation parameter optimization is then applied (as described in section 8) to the generated candidates in order to refine the initial estimate of those parameters. *This step is crucial since the accuracy of segment distance calculation depends greatly on the accuracy of the transformation parameters.* For each candidate, *segment-dist* is defined as the average point distance between the image-model contour segments (see section 7) and used as a measure of the goodness of fit between the two segments. A number of candidates with low *segment-dist* values are then selected for further processing.

5.3 Candidate Merging

Initial candidates correspond to simple segments delimited by neighboring curvature zero-crossing points. Nevertheless, it is possible for the visible boundary of an object in the input image to be divided into several neighboring or even overlapping segments. It is therefore necessary to merge those initial candidates which satisfy several criteria intended to measure candidate compatibility. It follows that two candidates c_1 and c_2 will be merged if they satisfy the following criteria:

- c_1 and c_2 must be valid (not previously merged) and different candidates.
- c_1 and c_2 must correspond to the same model.
- The transformation parameters of c_1 and c_2 should be *roughly* the same.
- The corresponding segments of c_1 and c_2 must be neighboring or overlapping.
- The scale factor associated with the new candidate must be admissible.
- The new candidate must have a low *segment-dist* value.

When two candidates are merged, the corresponding segments will be the union of the old segments. The old candidates are invalidated. Candidate merging will continue until no two candidates can be found which satisfy the merging criteria.

5.4 Candidate Extension

In general, the intersection point of two object boundaries in the input image does not coincide with an endpoint of a curvature zero-crossing segment. Therefore in order to find the exact location of such intersection points, it is necessary to gradually extend the contour segments associated with the merged candidates as long as a good fit between the image and model segments can be observed. Extension is first carried out at the right endpoint until mismatch error is too large and then carried out at the left endpoint. It is assumed that in general, object intersection points are a subset of the curvature maxima on the image contour (this is true except in hypothetical situations). First, all curvature extrema are located on a slightly smoothed version of the input image contour. Then, the following procedure is applied at each endpoint of each candidate:

- Extend the image contour segment to the next curvature maximum.
- The corresponding model contour segment is extended accordingly.
- Determine new transformation parameters and the new value of *segment-dist* for the candidate being extended.
- Determine the number of points k in a small neighborhood of the endpoint which are far from the image contour.

Extension stops if either the new candidate no longer has a low *segment-dist* value, or the new value of *segment-dist* rises sharply compared to previous value, or k rises above an acceptable limit. When extension stops, tests are carried out to detect a borderline case (k is just above the acceptable limit or value of *segment-dist* is just above the cut-off threshold). If so, the current endpoint becomes the final endpoint. Otherwise, the previous endpoint becomes the final endpoint.

5.5 Candidate Grouping

The next step in matching is to group compatible but disjoint candidates. The tests applied to determine compatibility are the same as the tests in section 5.3 except that the fourth test is not applied. It is certainly possible that, due to occlusion, an object in the scene may appear as two or more disjoint components in the input image. The goal of this step is to identify such situations to aid in the process of recognition.

5.6 Candidate Selection

What remains is to select the *best* candidates using an appropriate criterion. As stated earlier the value of *segment-dist* for each candidate is the average point distance between the contour segments associated with that candidate. This is a suitable measure of how well the shapes of those contour segments match. Another measure of the significance of a candidate is its *support*. Candidate support is defined as the length of the image contour segment associated with the candidate (note that if two disjoint candidates are found in section 5.5 to be compatible, the support of each candidate is increased by the length of the image contour segment associated with the other candidate). Define:

$$candidate - cost = \frac{segment - dist}{candidate - support}.$$

Note that a candidate with a lower cost is a *better* candidate. The following procedure is then used to select the best candidates:

- Determine the cost of each candidate.
- Select the valid candidate with the lowest cost.
- Disqualify all candidates whose corresponding image contour segment overlaps with the image contour segment of the chosen candidate or the image contour segment of any candidate compatible (see section 5.5) with the chosen candidate.
- Find any image contour segments delimited by negative curvature minima which do not overlap with the image contour segments associated with any chosen candidates or candidates compatible with them, and which fit well with the model associated with the chosen candidate. Examples are straight line segments which do not occur in valid candidates.
- Disqualify all candidates whose corresponding image contour segment overlaps with any of the image contour segments discovered in the previous step.
- Determine the final fit of the model associated with the chosen candidates using all relevant image contour segments and map the model to the image space.

- Disqualify the chosen candidate and all candidates compatible with it.
- If any valid candidates remain, go to the second step above, otherwise STOP.

Note that this procedure is independent of number of objects in input image.

6 Solving for the Transformation Parameters

When mapping a model curve segment to an image curve segment, it is possible to obtain many pairs of points on those segments in order to compute an initial approximation for the transformation parameters since the correspondence between arc length values on the curve segments is known. It is assumed that the transformation to be solved for consists of uniform scaling, rotation and translation in x and y. Let $\mathcal{X} = (x_j, y_j)$ be a set of η points on the image curve and let $\Xi = (\xi_j, \psi_j)$ be the set of corresponding points on the model curve. The parameters of the following transformation:

$$x_j = a\xi_j + b\psi_j + c \qquad y_j = -b\xi_j + a\psi_j + d \qquad (1)$$

must be solved for. A *Least-Squares Estimation* method is used to estimate values of a, b, c and d. Let the *dissimilarity measure* Ω which measures the difference between the model curve segment and the image curve segment be defined by:

$$\Omega = \sum_{j=1}^{\eta} \left(x_j^t - x_j^c\right)^2 + \left(y_j^t - y_j^c\right)^2$$

where (x_j^c, y_j^c) is the closest point on the image curve to transformed model curve point (x_j^t, y_j^t). Using equation (1) to eliminate x_j^t and y_j^t yields:

$$\Omega = \sum_{j=1}^{\eta} \left(a\xi_j + b\psi_j + c - x_j^c\right)^2 + \left(-b\xi_j + a\psi_j + d - y_j^c\right)^2.$$

Let $\mathcal{P} = (a, b, c, d)$ be the vector defined by the transformation parameters. The solution of $\frac{\partial \Omega}{\partial \mathcal{P}} = 0$ is the least-squares estimate of those parameters. To compute that estimate, determine the partial derivatives of Ω with respect to each of a, b, c and d and set those partial derivatives to zero. The result is a linear system of four equations in four unknowns which is solved to obtain estimates for a, b, c and d:

$$a = \frac{\sum \xi_j x_j^c + \sum \psi_j y_j^c - \frac{1}{\eta}\sum x_j^c \sum \xi_j - \frac{1}{\eta}\sum y_j^c \sum \psi_j}{\sum \xi_j^2 + \sum \psi_j^2 - \frac{1}{\eta}\sum \xi_j \sum \xi_j - \frac{1}{\eta}\sum \psi_j \sum \psi_j}$$

$$b = \frac{\sum \psi_j x_j^c - \sum \xi_j y_j^c + \frac{1}{\eta}\sum y_j^c \sum \xi_j - \frac{1}{\eta}\sum x_j^c \sum \psi_j}{\sum \xi_j^2 + \sum \psi_j^2 - \frac{1}{\eta}\sum \xi_j \sum \xi_j - \frac{1}{\eta}\sum \psi_j \sum \psi_j}$$

$$c = \frac{\sum x_j^c - a\sum \xi_j - b\sum \psi_j}{\eta}$$

$$d = \frac{\sum y_j^c + b\sum \xi_j - a\sum \psi_j}{\eta}.$$

7 Measuring Image-Model Curve Distances

Once an estimate of the transformation parameters is available, it is possible to map the model curve to the space of the image curve. It is then useful to measure the image-model curve segment distance for two reasons:

- Different model curves are mapped to the image curve in order to determine which model curve is locally closest to the image curve. This is accomplished by measuring image-model curve segment distances.
- The computation of the image-model curve segment distance is essential to transformation parameter optimization as described in section 8.

The image-model curve segment distance is computed by determining the closest point on the image curve segment (not necessarily a vertex) to each vertex of the model curve segment, and averaging the corresponding distances.

8 Optimizing the Transformation Parameters

The least-squares estimate of the transformation parameters computed in section 6 is, in general, *not* the optimal estimate. This is because the image-model point correspondences are not precise due to noise and local shape distortions. Nevertheless, it is possible to optimize those parameters as following:

- Let $D_p = \infty$.
- Compute the least-squares estimate of the parameters using the technique described in section 6 and use it to map the model curve to the image curve.
- Determine a new set of corresponding points on the image curve as described in section 7 and compute the new image-model curve distance D_n.
- If $D_p - D_n < \varepsilon$, then STOP.
- Let $D_p = D_n$ and go to the second step above.

In this system, it was possible to compute the optimal parameters with less than 1% error using at most 10 iterations of the procedure described above.

9 Results and Discussion

A total of 15 model objects and seven input images were used to evaluate the object recognition system described in this paper. Three of those images and the system's corresponding output are shown in this section. The model objects are as following: *bottle, clip, fork, key, monkey wrench* (two model contours were used), *panda,* two *connector cases, screw-driver, scissors, spoon, vase, wire-cutter* and two regular *wrenches* (two model contours were used for each). Therefore a total of 18 model contours were used. Each model contour was acquired off-line by either manually reading and entering the coordinates of points on the contour or obtaining an image of the isolated model object, segmenting the image and recovering the contour. Each model contour was represented by 200 points. The segmentation of each model contour was also computed off-line. The exact same starting scale and final scale were used to compute the segmentation of each model contour. About 10-20 segments were extracted from each model contour.

Due to the light-box setup used, the images obtained had high contrast. As a result, thresholding followed by preprocessing was successful in properly segmenting each of the input images after which the bounding contours were recovered. Figure 9.1 shows three input images and the contours recovered from those. The left column shows the original input images, and the right column shows only the outermost contours recovered from those images after thresholding, processing and boundary detection (see section 2). Note that only the outermost contours were used by the system to arrive at recognition results even though the inner contours are visually significant to human viewers. This was done to demonstrate the recognition power of the system. Each image contour was represented by 300 points. The exact same starting scale and final scale were used to compute the segmentation of each image contour. About 20-40 segments were extracted from each image contour.

The input images depicted scenes of varying complexity. The scene depicted in the top row of figure 9.1 contains 4 objects and can be considered to be of medium complexity. The scenes depicted in the middle and bottom rows of figure 9.1 contain 6 and 8 objects respectively and can be considered difficult. The system was tested on each of the three inputs. Each of the objects in each input image was recognized correctly by the system which also determined the correct scale, location and pose of each object. *Note that none of the internal parameters of the program were modified from one run to the next: the exact same system produced the correct result for each input image.*

The system was implemented in *C* and ran on a *SiliconGraphics Crimson* workstation. Execution times of 3.9 seconds, 12.0 seconds, and 13.1 seconds were obtained for the top, middle and bottom scenes of figure 9.1 respectively. These execution times indicate that the system is very fast given the complexity of the tasks it must perform. The left column in figure 9.2 shows the recognition results reached by the system for the 3 input images. Note that in each sub-figure the model contours are shown using a thin line and the image contour is shown using a thick line. The system was very robust in each case despite the presence of noise and local deformations of shape due to segmentation errors, non-rigid material in some objects, and perspective projection. Note that the right column in figure 9.2 also shows the segmentation points discovered during recognition. In almost all cases, the system was able to determine exact locations of those points.

10 Conclusions

This paper presented a complete and practical system for object recognition with occlusion which is very robust with respect to noise and local deformations of shape (due to perspective projection, segmentation errors and material of a non-

576

Figure 9.1. Input images and recovered contours

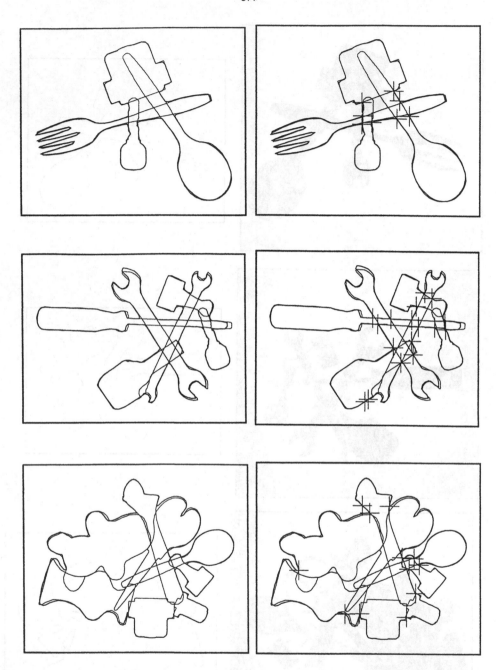

Figure 9.2. Recognition results for input scenes

rigid nature) as well as scale, position and orientation changes of the objects. The system was tested on a wide variety of 3-D objects with different shapes and surface properties. A light-box setup was used to obtain silhouette images which are segmented to obtain object boundaries. The *Curvature Scale Space* technique was then used to obtain a multi-scale segmentation of the image contour and the model contours using curvature zero-crossing points. This method made the system robust with respect to noise and local shape differences. A local matching algorithm applied *candidate generation, selection, merging, extension* and *grouping* to select the best matching models. Efficient transformation parameter optimization is used to map candidate models to the image space and directly measure the model-data quality of match. It is also used to compute the optimal pose for each selected model.

Acknowledgments

This work was supported by the NTT Basic Research Laboratories, Tokyo, the Laboratory for Computational Intelligence at the University of British Columbia, Vancouver, and the Vision, Speech, and Signal Processing Laboratory at the University of Surrey, England.

References

1. M. Kass, A. Witkin, and D. Terzopoulos. Snakes: active contour models. In *Proc International Conference on Computer Vision*, pages 259–268, 1987.
2. F. Mokhtarian. Fingerprint theorems for curvature and torsion zero-crossings. In *Proc IEEE Conference on Computer Vision and Pattern Recognition*, pages 269–275, San Diego, CA, 1989.
3. F. Mokhtarian. Silhouette-based isolated object recognition through curvature scale space. *IEEE Trans Pattern Analysis and Machine Intelligence*, 17(5), 1995.
4. F. Mokhtarian and A. K. Mackworth. Scale-based description and recognition of planar curves and two-dimensional shapes. *IEEE Trans Pattern Analysis and Machine Intelligence*, 8(1):34–43, 1986.
5. F. Mokhtarian and A. K. Mackworth. A theory of multi-scale, curvature-based shape representation for planar curves. *IEEE Trans Pattern Analysis and Machine Intelligence*, 14(8):789–805, 1992.
6. F. Mokhtarian and H. Murase. Silhouette-based object recognition through curvature scale space. In *Proc International Conference on Computer Vision*, pages 269–274, Berlin, 1993.
7. F. Mokhtarian and S. Naito. Scale properties of curvature and torsion zero-crossings. In *Proc Asian Conference on Computer Vision*, pages 303–308, Osaka, Japan, 1993.
8. C. A. Rothwell, D. A. Forsyth, A. Zisserman, and J. L. Mundy. Extracting projective structure from single perspective views of 3d point sets. In *Proc International Conference on Computer Vision*, Berlin, 1993.
9. F. Stein and G. Medioni. Structural indexing: Efficient 3-d object recognition. *IEEE Trans Pattern Analysis and Machine Intelligence*, 14:125–145, 1992.
10. A. P. Witkin. Scale space filtering. In *Proc International Joint Conference on Artificial Intelligence*, pages 1019–1023, Karlsruhe, Germany, 1983.

A Focused Target Segmentation Paradigm *

Dinesh Nair and J. K. Aggarwal

Computer and Vision Research Center
The University of Texas at Austin,
Austin, Texas 78712-1084, USA.
email: aggarwaljk@mail.utexas.edu

Abstract. In this paper we present new algorithms for target detection/segmentation in second generation Forward Looking Infra-Red (FLIR) images. An initial detection algorithm that models the background using Weibull functions, is used to identify candidate target locations in the image. A two-stage focused analysis of each candidate target location is then performed to get an accurate representation of the target boundary. A region-growing procedure is used to get an initial estimate of the target region, which is then combined with salient edge information in the image to arrive at a more accurate representation of the target boundary. The region and edge integration is done using a novel method that uses a Bayes' minimum risk classification approach. Finally, to reduce the false alarm rate, a higher level interpretation module is used to classify the detected areas as man-made or natural objects using geometric and FLIR-intensity based features extracted from the target.

1 Introduction

A central problem in computer (machine) vision, and automatic target recognition (ATR) in particular, is obtaining robust descriptions of the *objects of interest* (OOI) in an image. Vital to this problem is the need for good segmentation techniques that identify the regions in an image that are occupied by the objects. In ATR applications, it is important to get an accurate and precise representation of the boundary of the tactical targets [1]. Since targets are usually characterized by their shape and the gray scale representation of the segmented target, segmentation results directly affect the performance of the system.

In this paper, we present a segmentation method which uses low level target segmentation techniques along with higher level interpretation techniques for robust target detection and segmentation. The strategy adopted here is the following: (1) An initial detection algorithm is used to identify regions in the image that are candidate location of objects of interest (targets) by modeling the background using Weibull functions. At this stage all possible object locations in the image are identified, at the expense of a high false alarm rate. Concurrently, a *salient edge image*, which contains all the salient edge information in

---

* This work was supported by the Army Research Office Contracts DAAH-94-G-0417 and DAAH 049510494.

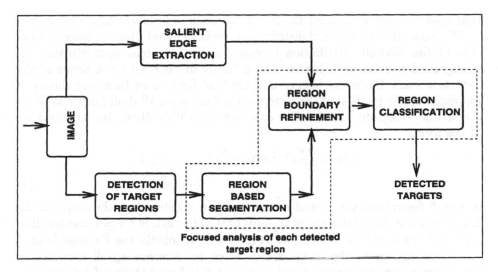

Fig. 1. The general flow of the target segmentation paradigm.

the image, is obtained from the original image. Saliency of edge segments in an image is defined by its length (L), average contrast (C) and smoothness (Δk), as described in [3]. Salient edge segments are those which are more significant than the others in the image. By considering only salient edges, the effect of clutter in an image is reduced.

(2) The initial detection stage is followed by a *focused* analysis of the candidate target areas. The objective at this stage is to get a good representation of the image attributes of the target region under consideration, such as boundary information, size, compactness, etc. The focused analysis consists of a two-stage process. A region-growing procedure is used to get an initial estimate of the target region, which is then combined with the edge information in the corresponding region of the *salient edge image* to get an accurate estimate of the target boundary. (3) In the final stage a higher level interpretation technique is used to identify the OOI from the candidates in the image. A schematic of the overall approach is shown in figure 1.

The rest of the paper is organized as follows: Section 2 describes the algorithm developed to detect the initial candidate locations of the targets in the image. In section 3, the focused segmentation of each candidate target location is presented. Section 4 describes the higher interpretation module and the types of features extracted from the image to discriminate between man-made objects and natural background. Finally, in section 5, a summary of this study is presented.

2 Initial Detection Of Target Locations

In this section, we present a method for better representation of the intensity distribution of the background in second generation FLIR images and a scheme that uses this information for the initial detection of probable target locations.

To model the background distribution, we have found that the density function of the Weibull distribution (predominantly Rayleigh) approximates the background clusters well (better than a Gaussian distribution in terms of the detection rate). Since the background tends to form more than one cluster, a piecewise approximation of the histogram is done using Weibull functions to arrive at the background model. The approximating Weibull density functions are of the form:

$$f(x) = \frac{A\gamma}{\theta} x^{\gamma-1} \exp\left(-\frac{x^\gamma}{\theta}\right) \qquad x \geq 0$$
$$= 0 \qquad\qquad\qquad otherwise \qquad (1)$$

where, θ determines the spread of the function, γ determines the shape of the function and A modulates the peak value of the function. For $\gamma = 1$, the function reduces to an exponential form, while for $\gamma = 2$, we obtain the *Rayleigh* form.

Target regions in the image are detected by determining all locations in the image where the gray scale values have a "low" probability of belonging to the background. The probability of each gray scale value of belonging to the background is determined by the background cluster that it is closest to, and is computed in a maximum likelihood sense as follows. Let x represent a gray scale value, where $0 \leq x \leq 255$, and $L(x \in c_k)$ be the *likelihood* of x belonging to cluster c_k, where c_k, $k = 1 \ldots n$, represents the n background clusters. $L(x \in c_k)$ is computed as:

$$L(x \in c_k) = \frac{\frac{A_k \gamma_k}{\theta_k} x^{\gamma_k - 1} \exp\left(-\frac{x^{\gamma_k}}{\theta_k}\right)}{\frac{A_k \gamma_k}{\theta_k} x_{peak,k}^{\gamma_k - 1} \exp\left(-\frac{x_{peak,k}^{\gamma_k}}{\theta_k}\right)}, \qquad (2)$$

where, $x_{peak,k}$ represents the gray level value at the peak of cluster k, and the probability that x belongs to the background is obtained as:

$$P(B/x) = max\{L(x \in c_k), k = 1 \ldots n\}, \qquad (3)$$

where, B represents the background class, and the probability of x belonging to the target is $(1 - P(B/x))$. The false alarm rate at this stage is directly proportional to the threshold probability chosen as "low". Typically, at this stage all possible target locations are detected, at the cost of a high false alarm rate. Figures 2 and 3 show examples of typical FLIR images, their histograms and the Weibull approximations.

3 Target Segmentation Using Focused Analysis

Once all possible target locations in the image have been identified, a focused analysis approach is used to improve the segmentation around each candidate target location. A region-growing procedure is used to initially segment the target, which is then followed by a *refinement* stage, where the accuracy of the boundary of the segmented target region is improved using the salient edge information (obtained from the *salient edge image*) in and around the target region.

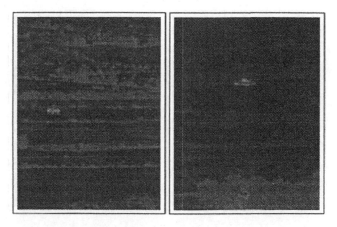

Fig. 2. FLIR images of a truck and a tank.

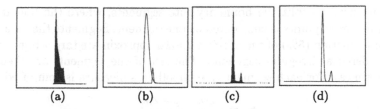

(a)   (b)   (c)   (d)

Fig. 3. (a), (c) Histograms of the images in Figure 2. (b), (d) Weibull approximations of the histograms.

3.1 A Local Shape Driven Region Growing Paradigm

A region-growing paradigm is used to segment each candidate target detected. To grow the region around the detected areas, surrounding pixels with a low probability of belonging to the background are added to the region. The process is iteratively continued by adding pixels to the region with higher probabilities of belonging to the background, while the size of the segmented region does not exceed the *a priori* known object size, the compactness of the region does not exceed a predefined maximum, and the boundary of the region does not get too irregular.

The last constraint uses the fact that the targets to be segmented are man-made and hence show some kind of regularity. On the other hand, boundaries of objects that belong to the background (like vegetation, clouds, etc.) are usually characterized by irregular boundaries which exhibit frequent changes in curvature. To determine the percentage of the boundary that is irregular, we first extract the boundary (contour) of the region. The corner points of the boundary are detected. For man-made objects, these corner points can be regarded as the end points of the linear segments that make up its boundary. The contour (boundary) between these corner points should be relatively smooth. The corner points are detected using a cubic B-spline fitting method [2]. Next, the corner

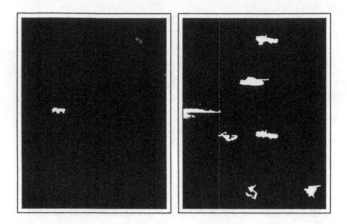

Fig. 4. Region based segmentation of the images in Figure 2.

points are used to split the boundary into segments, where the boundary between two corner points is considered as one segment. Segments that are smaller than some fraction (5%) of the *a priori* known approximate target boundary size are considered as irregular segments. The rest of the segments are then tested for smoothness. For each segment i, a smoothness measure is computed as:

$$S_s^i = \frac{\Delta\kappa_{max} - \Delta\kappa^i}{\Delta\kappa_{max} - \Delta\kappa_{min}}, \tag{4}$$

where, $\Delta\kappa$ is the average change of curvature of the segment, $\Delta\kappa_{min}$ is the smallest average change of curvature for all the segments of the target boundary and $\Delta\kappa_{max}$ is the largest average change of curvature for all the segments. The measure S_s^i satisfies $0 \le S_s^i \le 1$.

The smoothness measures are used to order the segments, and segments with a smoothness measure greater than a certain threshold value are considered to be smooth. This threshold value is computed as follows:

$$\Theta = 1 - e^{\frac{-A\mu_{\Delta\kappa}}{\rho_{\Delta\kappa}}}, \tag{5}$$

where, $\mu_{\Delta\kappa}$ and $\rho_{\Delta\kappa}$ are the mean and standard deviation of the average change in curvature of all the segments, and A is a constant. By making Θ a function of the average smoothness of the segments and the deviation of the smoothness of the segments, the threshold value is indicative of the overall smoothness of the boundary. Figure 4 shows the final region-based segmentation for the images in figure 2. A closer look at the segmentation for one of the examples is shown in figure 5.

3.2 Integrating Salient Image Contours with Segmented Regions for Refining Target Boundaries

A new paradigm for the refinement of region-based segmentation results using edge information is presented next. Since segmentation using region-based tech-

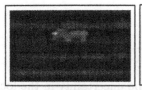

Fig. 5. A closer look at the segmentation of the truck in Figure 2.

niques are inefficient in locating exact target boundaries and tend to miss small parts of the target, the boundaries of the segmented targets are refined using the salient edge segments in the image. An *edge contour* is obtained from an edge as a linked list of edge pixels. An *edge segment* is defined as that part of an *edge contour* between its corner points.

Given the target boundary obtained from the region growing process and the salient edge segments in and around the target region, the boundary refinement problem is stated as:

For every point s on the region boundary, find its new location as a selection from a set of candidate edge element locations $\bar{z} = \{z_j, j = 0 \ldots n\}$, where $z_0 = s$.

This problem is formulated as a classification problem, where point s will take one of the labels given by $\bar{z}$. Using Bayes decision rule, choose z_j as the new location if

$$p(s|z_j)P(z_j) \geq p(s|z_k)P(z_k) \qquad \forall k \neq j, \tag{6}$$

where $p(s|z_j)$ represents the conditional density function of (s, z_j) and $P(z_j)$ is the *prior* probability of z_j. The *prior* probability of each candidate location z_j is estimated as the proximity of the salient edge segment to which z_j belongs to the boundary of the target region. Proximity of an edge segment is defined as the percentage of its segment that is *close* to the region boundary. The *closeness* of a segment pixel to a region boundary is determined by a *Refinement Search Circle* (RSC) of radius λ. If an edge segment pixel lies within a RSC placed on any region boundary point, then the edge segment pixel is considered as *close* to the region boundary. Each salient edge segment $(SE_i, 0 \leq i \leq q$, where q is the number of salient edge segments and $SE_0 = s$) is assigned a proximity weight determined by the number of its pixels that are close to the target region boundary, and is given by:

$$Prox(SE_i) = \frac{\#\ edge\ elements\ of\ segment\ SE_i\ that\ is\ close\ to\ region\ boundary}{\#\ edge\ elements\ in\ SE_i}$$
$$\tag{7}$$

Hence, under suitable assumptions [5], the *priors* are computed as:

$$P(z_j) \approx Prox(SE_i^j). \tag{8}$$

All points lying on the same salient edge segment will have the same *prior* probability. The density function $p(s|z_j)$ is assumed to be Gaussian distributed around z_j with variance σ_j^2 as:

$$p(s|z_j) = \frac{1}{\sqrt{2\pi}\sigma_j} \exp \frac{-(s - z_j)^2}{2\sigma_j^2}. \tag{9}$$

The standard deviation σ_j represents the uncertainty in the location of the edge element z_j and is approximated as:

$$\sigma_j = \frac{Maximum\ strength\ of\ a\ salient\ edge\ element\ in\ the\ image}{Strength\ of\ salient\ edge\ element\ at\ z_j}, \tag{10}$$

where the strength of an edge element is proportional to gradient of the image at the edge element location. Therefore, the edge element with the highest strength has unit variance. Assume that the standard deviation at z_0, $\sigma_0 = \sigma_{max}$, where σ_{max} represents the largest standard deviation among all the salient edge elements in the image. Using the criteria in equation (6) and the density function given in equation (9), z_j is chosen as the candidate location if the following is satisfied:

$$(s - z_j)^2 \leq 2\sigma_j^2 [\log \frac{P(z_j)\sigma_k}{P(z_k)\sigma_j} + \frac{(s - z_k)^2}{2\sigma_k^2}] \qquad \forall\, k \neq j. \tag{11}$$

The computational burden associated with the above criterion is reduced by assigning a *prior* probability $P(z_0) = Prox(SE_0) = \Gamma$, where $0 < \Gamma \leq 1$. Substituting $k = 0$ and $\sigma_0 = \sigma_{max}$ in equation (11) and noting that $z_0 = s$, the winning candidate has to satisfy the condition that:

$$(s - z_j)^2 \leq 2\sigma_j^2 [\log \frac{P(z_j)\sigma_{max}}{\Gamma\sigma_j}]. \tag{12}$$

Thus, by assigning a user defined *prior* probability Γ, the search space is restricted to a circle of radius λ given by the largest value of right hand side in equation (12). The largest value of $|s - z_j|$ in equation (12) is obtained when $P(z_j)$ is at its maximum P_{max}, and $\sigma_j = (P_{max}\sigma_{max}/\Gamma)e^{-0.5}$, and is given by:

$$\lambda = |s - z_j|_{max} = \frac{P_{max}\sigma_{max}}{\Gamma}e^{-0.5}. \tag{13}$$

Since P_{max} is the maximum value $P(z_j)$ can take, which in turn is the maximum proximity weight that is possible (equation (8)), $P_{max} = 1$. Therefore the search radius becomes $\lambda = e^{-0.5}\sigma_{max}/\Gamma$. The radius has been deliberately denoted by λ to emphasize the equivalence of this search radius to the RSC used for finding the *closeness* of the edge segments to the region boundary. From equation (11), it is seen that when choosing from edge locations that belong to the same salient segment, the problem reduces to finding the edge location that is closest to the boundary point.

At the end of the boundary refinement stage, an edge image is obtained which represents the location of the refined target region boundary. To complete breaks in the boundary, with an emphasis on incorporating small parts of the

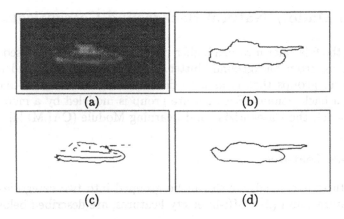

Fig. 6. (a) Target (b) Boundary of target obtained from region-based segmentation (c) Salient edges around the target. (Note: salient edges that belong to a long contour are separated by crosses on the contour.) (d) The refined boundary of the target.

target missed by the region segmentation, the following algorithm is used: (1) Remove isolated edge pixels. These are edge pixels with no neighbors in the refined boundary. (2) For each edge pixel at the site of a break in the boundary, determine if it is part of a salient edge. If it is, incorporate the complete edge segment into the boundary. Repeat this process iteratively until no more salient edge information can be incorporated. (3) Enforce linear connectivity at all the remaining breaks. Figures 6 and 7 show examples of the refinement process.

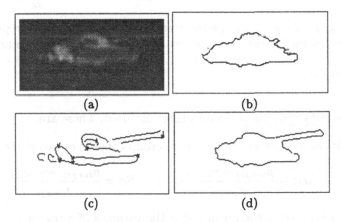

Fig. 7. (a) Target (b) Boundary of target obtained from region-based segmentation (c) Salient edges around the target. (Note: salient edges that belong to a long contour are separated by crosses on the contour.) (d) The refined boundary of the target.

4 Man-Made / Natural Background Categorization

To reduce the false alarm rate, a higher interpretation module is used to identify target regions from background clutter. Image statistics are used to derive a set of feature groups that is used to discriminate between man-made objects and natural backgrounds. Each feature group is modeled by a modular neural network model, the Categorizing and Learning Module (CALM) [4].

4.1 Image Features

Image features extracted, which can be grouped into two categories; (1) Geometric Features, and (2) FLIR-intensity Features, are described below.

Geometric Features Two geometric features are used. These are:

- **Saliency:** This feature group is based on the number of *salient segments* in the target region and consists of two elements:

$$S1 = \frac{\sum salient\ edge\ segments\ in\ target\ region}{\sum edge\ segments\ in\ target\ region}$$

$$S2 = \frac{\sum salient\ edge\ segments\ in\ target\ region}{\sum salient\ segments\ in\ entire\ image} \tag{14}$$

- **Boundary Regularity:** A feature group of three elements is used to characterize the smoothness of the target boundary.

$$R1 = \mu_{\Delta K}, \qquad R2 = \sigma_{\Delta K}, \qquad R3 = \frac{\sum_i L_i}{L}, \tag{15}$$

where, $\mu_{\Delta K}$ and σ_{DeltaK} are the same as in equation (5), L_i represents the ith segment on the target boundary that is very small (chosen as 5% of the total target boundary length) and L represents the number of segments on the boundary.

FLIR-Intensity Features Two features are used. These are:

- **Gray level statistics:** This feature group consists of two elements:

$$Mn = \frac{\mu_{target} - \mu_{image}}{\mu_{image}}, \qquad Sd = \frac{\sigma_{target} - \sigma_{image}}{\sigma_{image}}, \tag{16}$$

where, μ represents the mean and σ the variance of a region.
- **Gray level clusters:**

$$CLUS = No.\ of\ distinct\ gray\ level\ clusters\ in\ target\ region$$

$$SPREAD = Distance\ between\ furthest\ cluster\ centroids \tag{17}$$

4.2 Man-made / Natural Background Classification

Features extracted from the target regions are then classified as man-made or natural using modified CALM networks. Modifications to the CALM were done to incorporate a confidence value with its decisions and to speed up its convergence. Each feature group was modeled by a CALM module and the decision from each of the feature groups was combined to get the final classification result. Details of this can be found in [5].

5 Summary

In this paper we have presented new target detection/segmentation algorithms for use in automatic target recognition systems. By modeling the background by Weibull distributions, a good initial detection of target locations is obtained. A focused analysis of each target location is then performed by a region-growing procedure which uses the underlying background probabilities for checking region homogeneity and local shape characteristics for determining the convergence/stopping criteria. To get better representations of the target boundary, salient edge information in the image is used to refine the boundary obtained by the region-growing method. A novel method to do this using a Bayes' minimum risk classification approach is presented. Finally, the false alarm rate is reduced by using a higher level interpretation module built using modified CALM neural networks, that uses geometric and intensity based features from each detected/segmented region to classify it as man-made or natural. The target detection/segmentation algorithms have been successfully used on second generation FLIR images. A 100% detection rate with a false alarm rate of 5% was obtained when the segmentation method was tested on 200 images from the HULI9306_SIG subset of the COMANCHE data set provided to us by the Night Vision Laboratory.

References

1. B. Bhanu, "Automatic target recognition: State of the art survey," *IEEE Transactions on Aerospace and Electronic Systems*, vol. AES-22, no. 4, pp. 364–379, 1986.
2. G. Medioni and Y. Yasumoto, "Corner detection and curve representation using cubic B-splines," *Computer Vision, Graphics and Image Processing*, vol. 39, pp. 267–278, 1987.
3. K. Rao and J. Liou, "Salient contour extraction for target recognition," *SPIE: Acquisition, Tracking and Pointing*, vol. 1482, pp. 293–306, 1991.
4. J. M. J. Murre, *Learning and Categorization in Modular Neural Networks*. Lawrence Erlbaum Associates, Publishers, 1992.
5. D. Nair and J. K. Aggarwal, "Robust automatic detection/segmentation in 2nd generation FLIR images," *Technical Report, Computer & Vision Research Center, University of Texas at Austin*, TR-96-03-103, 1996.

Generalized Image Matching: Statistical Learning of Physically-Based Deformations

Chahab Nastar[1], Baback Moghaddam[2] and Alex Pentland[2]

[1] INRIA Rocquencourt, B.P. 105, 78153 Le Chesnay Cédex, France
[2] MIT Media Laboratory, 20 Ames St., Cambridge, MA 02139, USA

Abstract. We describe a novel approach for image matching based on deformable intensity surfaces. In this approach, the intensity surface of the image is modeled as a deformable 3D mesh in the $(x, y, I(x, y))$ space. Each surface point has 3 degrees of freedom, thus capturing fine surface changes. A set of representative deformations within a class of objects (e.g. faces) are statistically learned through a Principal Components Analysis, thus providing a priori knowledge about object-specific deformations. We demonstrate the power of the approach by examples such as image matching and interpolation of missing data. Moreover this approach dramatically reduces the computational cost of solving the governing equation for the physically based system by approximately three orders of magnitude.

1 Introduction

In recent years, computer vision research has witnessed a growing interest in eigenvector analysis and subspace decomposition methods [15]. This general analysis framework lends itself to several closely related formulations in object modeling and recognition which employ the *principal modes* or characteristic *degrees-of-freedom* for description. The identification and parametric representation of a system in terms of these principal modes is at the core of recent advances in physically-based modeling [20, 17] and parametric descriptions of shape [6, 2, 10]. On the other hand, *view-based* eigentechniques have recently provided some of the best results in object recognition [21, 19].

In this paper, we propose a new method which combines both the physically-based modes of vibration and the statistically-based modes of variation. In view of some recent critiques of physical modeling (e.g. [4]) our motivation here is to ground physically-based models in actual real-world statistics in order to obtain a more realistic and data-driven model for the underlying phenomenon [13, 5].

Furthermore, we seek to unify the shape and texture components of an image in a single compact mathematical framework. Current work in the area of image-based object modeling deals with the shape (2D) and texture (grayscale) components of an image in an independant manner [3, 11]. Our novel representation combines both the spatial (XY) and grayscale (I) components of the image into a 3D surface (or manifold) and then efficiently solves for a dense correspondance map in the XYI space. This "manifold matching" technique can be viewed as a more general formulation for image correspondance which, unlike optical flow, does *not* require a constant brightness assumption.

In principal, any two image manifolds can be matched in this way (though sometimes erroneously), therefore we must further constrain the space of allowable manifold deformations to specific object classes (eg., frontal views of faces). These characteristic deformations (or "principal warps") are learned through a statistical Principal Components Analysis (PCA) [9] which identifies the principal subspace in which the final correspondance field must lie. Since the Karhunen-Loeve Transform (KLT) [12] in PCA corresponds to a unitary linear change of basis, which can be appended to the modal transform used in solving the physical system, we can ultimately derive a compact reduced-order form of the governing equation which combines both the dynamics of the physical system and the "learned" deformations which were observed in actual training data.

2 Deformable intensity surfaces for image matching

Our idea of using intensity surfaces for matching and recognition comes from the observation that the transformation of shape to intensity is quasi-linear under controlled lighting conditions ; in other terms, *the intensity of the 2D image reflects the actual 3D shape.* Our system focuses on matching and recognition in the 3D space defined by $(x, y, I(x, y))$, that we will call the XYI space (see [18] for details).

In our formulation, deforming the intensity surface of *image*1 into the one of *image*2 in XYI takes place in 5 steps :

1. Reduce, if necessary, the number of graylevels in *image*1 and *image*2 down to the same number g of graylevels (typically $g = 32$).
2. Initialize the deformable surface S as a subsampling of the intensity surface of *image*1.
3. Convert *image*2 to its 3D binary representation, *image*3.
4. Compute Euclidean distance maps at each voxel of *image*3 [7, 22].
5. Let S deform dynamically in *image*3 with the external force derived from the distance maps created at step 2.

Note that steps 1 to 4 are pre-processing steps. Steps 1 and 2 provide respectively intensity and spatial smoothing of the image. The dynamic process of step 5 is described in [17] ; to sum up, the intensity surface S is modeled as a deformable mesh of size $N = n \times n'$ nodes, ruled by Lagrangian dynamics :

$$\mathbf{M\ddot{U}} + \mathbf{C\dot{U}} + \mathbf{KU} = \mathbf{F}(t) \qquad (1)$$

where $\mathbf{U} = [\ldots, \Delta x_i, \Delta y_i, \Delta z_i, \ldots]^T$ is a vector storing nodal displacements, $\mathbf{M}$, $\mathbf{C}$ and $\mathbf{K}$ are respectively the mass, damping and stiffness matrices of the system, and $\mathbf{F}$ is the external (or "image") force. The above equation is of order $3N$.

At each node M_i of the mesh, the image force points to *the closest point* P_i in the 3D binary image *image*3 [17]. Figure 1 shows a representation of the deformation process. Note that the external forces (dashed arrows) *do not* necessarily correspond to the final displacement field of the surface since the closest point P_i is updated at each time iteration. The elasticity of the surface provides an intrinsic smoothness constraint for computing the final displacement field. Note that our formulation provides an interesting alternative to optical flow

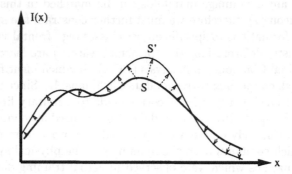

Fig. 1. Intensity surface S being pulled towards S' by image forces

methods, without the classical *brightness constraint* [8]. Indeed, the brightness constraint corresponds to a particular case of our formulation[3] where the closest point P_i has to have the same intensity as M_i ($\overrightarrow{M_i P_i}$ is parallel to the XY plane). We do not make that assumption here.

The vibration modes $\phi(i)$ of the previous deformable surface are the vector solutions of the eigenproblem [1] :

$$\mathbf{K}\phi = \omega^2 \mathbf{M}\phi \qquad (2)$$

where $\omega(i)$ is the i-th eigenfrequency of the system. Solving the governing equations in the modal basis leads to scalar equations where the unknown $\tilde{u}(i)$ is the amplitude of mode i :

$$\ddot{\tilde{u}}(i) + \tilde{c}_i\dot{\tilde{u}}(i) + \omega(i)^2\tilde{u}(i) = \tilde{f}_i(t) \qquad i = 1,\ldots,3N. \qquad (3)$$

The closed-form expression of the displacement field is now :

$$\mathbf{U} \approx \sum_{i=1}^{P} \tilde{u}(i)\phi(i) \qquad (4)$$

with $P \ll 3N$, which means that only P scalar equations of the type of (3) need to be solved. The modal superposition equation (4) can be seen as a Fourier expansion with high-frequencies neglected [16].

We make use of *the analytic expressions of the modes* which are known sine and cosine functions for specific surface topologies. For quadrilateral surface meshes that have plane topology (which is the case of the intensity surfaces), the eigenfrequencies of the system are [16] :

$$\omega^2(p,p') = 4K/M\left(\sin^2\frac{p\pi}{2n} + \sin^2\frac{p'\pi}{2n'}\right) \qquad (5)$$

---

[3] In fact, by simply disabling the I component of our deformations we can obtain a standard 2D deformable mesh which yields correspondances similar to an optical flow technique with thin-plate regularizers.

K is the stiffness of each spring, M the mass of each node, p and p' are the mode parameters. The modes of vibration are :

$$\phi(p, p') = [\ldots, \cos \frac{p\pi(2i-1)}{2n} \cos \frac{p'\pi(2j-1)}{2n'}, \ldots]^T \qquad (6)$$

where $i = 1, \ldots, n$ and $j = 1, \ldots, n'$. These analytic expressions avoid the call to costly eigenvector-extraction routines ; moreover, they allow the total number of modes to be easily adjusted.

3 Statistical Modeling

In theory, our deformable intensity surface can undergo any possible deformation. Thus, it seems interesting to *learn* the deformations of a specific class of objects and add them as *constraints* into our system. This is an important step for guiding the deformations of our mesh when performed within a specific object class and also allows us to deal with occlusions and missing data, as we shall see later.

Our approach to learning the space of allowable manifold deformations particular to a specific object class Ω (eg., frontal faces) is that of *unsupervised learning*. Particularly, we perform a PCA on a selected training set of deformations in order to recover the principal components of the warps. This approach is actually part of a more complete statistical formulation for estimating the probability density function of these warps in the high-dimensional vector space $\tilde{U} \in \mathcal{R}^P$ (see [14]). The estimated class-conditional density $P(\tilde{U}|\Omega)$ can be ultimately used in a Bayesian framework for a variety of tasks such as regression, interpolation, inference and classification. However, in this paper, we have concentrated mainly on the dimensionality-reduction aspect of PCA in order to obtain a lower-dimensional subspace in which to solve for the manifold correspondance field.

Given a training set of suitable warp vectors $\{\tilde{U}^t\}$ for $t = 1...N_T$, the principal warps are obtained by solving the eigenvalue problem

$$\Lambda = E^T \Sigma E \qquad (7)$$

where Σ is the covariance matrix of the training set, E is the eigenvector matrix of Σ and Λ is the corresponding diagonal matrix of eigenvalues. The unitary matrix E defines a coordinate transform (rotation) which *decorrelates* the data and makes explicit the *invariant subspaces* of the matrix operator Σ. In PCA, a partial KLT is performed to identify the largest-eigenvalue eigenvectors and obtain a principal component feature vector $\hat{U} = E_M^T (\tilde{U} - \tilde{U}_0)$, where $\tilde{U}_0$ the mean warp vector and E_M is a submatrix of E containing the principal eigenvectors. This KLT can be seen as a linear transformation $\hat{U} = \mathcal{T}(\tilde{U}) : \mathcal{R}^P \rightarrow \mathcal{R}^L$ which extracts a lower-dimensional subspace of the KL basis corresponding to the maximal eigenvalues. These principal components preserve the major linear correlations in the data and discard the minor ones.[4]

---

[4] In practice the number of training images N_T is far less than the dimensionality of the data, P, consequently the covariance matrix Σ is singular. However, the first $L < N_T$ eigenvectors can always be computed (estimated) from N_T samples using, for example, a Singular Value Decomposition.

By ranking the eigenvectors of the KL expansion with respect to their eigenvalues and selecting the first L principal components we form an orthogonal decomposition of the vector space $\mathcal{R}^P$ into two mutually exclusive and complementary subspaces: the principal subspace (or feature space) $\{\mathbf{E}_i\}_{i=1}^L$ containing the principal components and its orthogonal complement $\bar{F} = \{\mathbf{E}_i\}_{i=L+1}^P$. In this paper, we simply discard the orthogonal subspace and work entirely within the principal subspace $\{\mathbf{E}_i\}_{i=1}^L$, hereafter referred to simply by the matrix $\mathbf{E}$.

4 Combining physics and statistics

Instead of solving the unconstrained governing equation (1), we compute the projection of the unknown $\mathbf{U}$ (dimension : 3N = 3nn'), first into a modal subbasis (dimension P), then into a KL subspace (dimension L) :

$$\mathbf{U} \xrightarrow{\Phi} \tilde{\mathbf{U}} \xrightarrow{\mathbf{E}} \hat{\mathbf{U}} \tag{8}$$

The first transform is the projection into the modal subspace :

$$\mathbf{U} = \Phi \tilde{\mathbf{U}} \tag{9}$$

The second transform is the projection of the modal amplitudes into the PCA subspace :

$$\tilde{\mathbf{U}} = \mathbf{E}\hat{\mathbf{U}} + \tilde{\mathbf{U}}_0 \tag{10}$$

Equations (9) and (10) yield the global transform :

$$\mathbf{U} = \Psi\hat{\mathbf{U}} + \mathbf{U}_0 \tag{11}$$

where the global transformation matrix Ψ is simply : $\Psi = \Phi \mathbf{E}$ and $\mathbf{U}_0 = \Phi \tilde{\mathbf{U}}_0$. Note that Ψ is a rectangular orthogonal matrix.

By premultiplying equation (1) by Ψ^T and changing unknowns (equation (11)), we obtain :

$$\Psi^T \mathbf{M} \Psi \ddot{\hat{\mathbf{U}}} + \Psi^T \mathbf{C} \Psi \dot{\hat{\mathbf{U}}} + \Psi^T \mathbf{K} \Psi \hat{\mathbf{U}} = \Psi^T \mathbf{F}(t) - \Psi^T \mathbf{K} \mathbf{U}_0 \tag{12}$$

Let :

$$\hat{\mathbf{M}} = \Psi^T \mathbf{M} \Psi \tag{13}$$

$$\hat{\mathbf{C}} = \Psi^T \mathbf{C} \Psi \tag{14}$$

$$\hat{\mathbf{K}} = \Psi^T \mathbf{K} \Psi = \mathbf{E}^T \Omega^2 \mathbf{E} \tag{15}$$

$$\hat{\mathbf{F}}(t) = \Psi^T \mathbf{F}(t) - \Psi^T \mathbf{K} \mathbf{U}_0 = \Psi^T \mathbf{F}(t) - \mathbf{E}^T \Omega^2 \tilde{\mathbf{U}}_0 \tag{16}$$

Note that the new mass, damping and stiffness matrices, as well as the new external force, do not involve heavy computations because : (i) we make the common assumption that $\mathbf{M}$ and $\mathbf{C}$ are scalar matrices ($\mathbf{M} = M\mathbf{I}$, $\mathbf{C} = C\mathbf{I}$ where M and C are mass and damping scalars) , and (ii) Ω^2 is a diagonal matrix. We now end up with the standard Lagrangian equation of unknown $\hat{\mathbf{U}}$.

$$\hat{\mathbf{M}}\ddot{\hat{\mathbf{U}}} + \hat{\mathbf{C}}\dot{\hat{\mathbf{U}}} + \hat{\mathbf{K}}\hat{\mathbf{U}} = \hat{\mathbf{F}}(t) \tag{17}$$

Solving this equation for $\hat{\mathbf{U}}$ and then changing basis back to the canonical basis (equation (11)) provides the estimated displacement $\mathbf{U}$. By using this method, the resulting displacement $\mathbf{U}$ is constrained to lie along those learned deformation modes that are characteristic of the object class.

5   Experimental Results

We conduct our experiments with facial imagery. The manifold matching technique described in this paper requires rough alignment of the two input images in order to function properly. In our experiments, this alignment was obtained using an automatic face-processing system which extracts faces from the input image and normalizes for translation, scale and slight rotations (both in-plane and out-of-plane). This system is described in detail in [14].

For the learning phase of our technique, we choose a set of 50 faces to be warped into a reference face. Each of these faces has a $N = 128 \times 128$ resolution, and the manifolds are matched in a modal subspace whose dimension is suitably chosen $P = 3 \times 128^2/4^2 = 3072$ [18]. We then perform a Principal Components Analysis on the spectra of these warps.

Figure 2 shows the modes of variation along individual KL-eigenvectors extracted from the learning set. For example, we can see that $\vec{E_1}$ represents change in global headshape (as well as the size of the eyes). Eigenvectors $\vec{E_2}$ and $\vec{E_3}$ represent a change in the chin size and forehead, respectively. Higher-order eigenvectors, for example $\vec{E_{10}}$ represent subtler variations in facial appearance (e.g. eye shape). By looking at the KL-eigenvalues, it is easy to draw the percentage of the data variance that is captured versus the number of eigenvalues. Figure 3 shows that 90% of the data is adequately captured by $L = 25$ principal eigenvectors.

5.1   Subspace Warps

Figure 4 shows an example of matching a test image to that of the reference using both the unconstrained and constrained warps. This basic example illustrates how a dense correspondence field can be obtained between two images from different objects. Figure 5 displays the modal spectrum and its reconstruction in the KL space. The total reconstruction error is on the order of 4%, demonstrating that by solving the reduced-order physical system (equation (12)), we have not significantly sacrificed accuracy. In addition, solving this equation requires considerably less computation. The degrees of freedom in the original mesh were $3N = 3 \times 128 \times 128 \approx 50,000$. In the modal subspace, the degrees of freedom were reduced to $P = 3 \times 32 \times 32 \approx 3,000$, and finally in the KL subspace, the degrees of freedom were further reduced to $L = 25$, thus achieving a compression factor of approximately $5000 : 1$.

5.2   Interpolation of Missing data

One of the advantages of learned warps is that, during the matching process, the deformations are constrained for a specific object. Consequently, invalid deformations arising out of missing data (e.g. object occlusion) are automatically disallowed.

The first example illustrates an experiment where regions of the face were occluded with a black bar (to simulate occlusion or incomplete data), as shown in figure 6 (top row). If we attempt an unconstrained warp in the modal space, an invalid reconstruction will be obtained (figure 6 bottom left and center). On the other hand, if the deformation is constrained by the learned modes, we obtain a better reconstruction of the missing data as shown in figure 6 (bottom right).

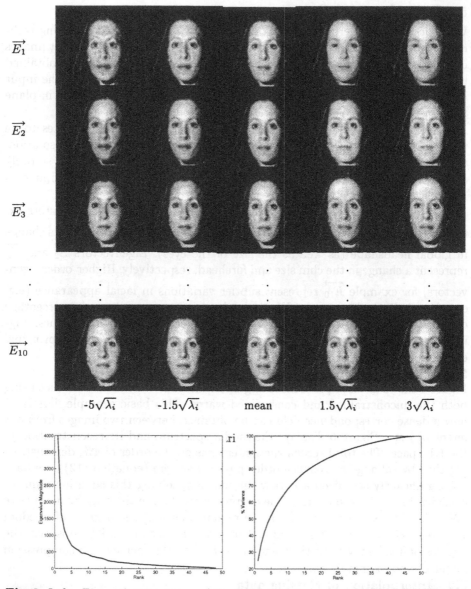

$$-5\sqrt{\lambda_i} \qquad -1.5\sqrt{\lambda_i} \qquad \text{mean} \qquad 1.5\sqrt{\lambda_i} \qquad 3\sqrt{\lambda_i}$$

Fig. 3. Left : Eigenvalue spectrum of the PCA transform. Right : Cumulative eigenvalue spectrum of the PCA transform

Fig. 4. From left to right : rest image ; reference image ; unconstrained warp in modal space ; constrained warp in KL-space

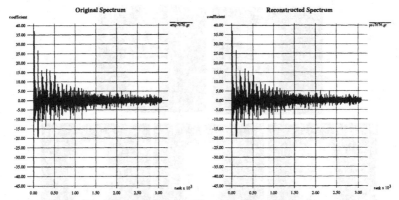

Fig. 5. Left : the original spectrum of the deformation. Right : reconstruction of the spectrum in the KL-subspace.

This example illustrates how our principal warp formulation effectively functions as a model-based image interpolant for a given class of objects.

The second example is similar in spirit to the first, except where the missing data is replaced by an arbitrary image region (in this case a texture), for example when one object partially occludes another. Here once again we see how the learned principal warps can yield a much better reconstruction and interpolation of non-matching image regions (figures 7).

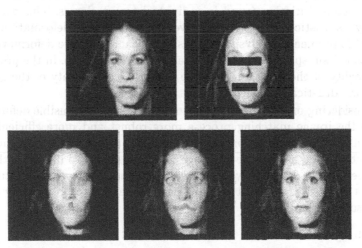

Fig. 6. Top : we wish to warp the left image into the right image **Bottom** : image warps ; left : in the real space. center : in the unconstrained modal subspace. right : in the constrained principal subspace

6 Conclusions

We have described a novel approach for image matching based on deformable intensity surfaces. In this approach, the intensity surface of the image is modeled as a deformable surface embedded in XYI space. Our approach is thus a *generalization* of optical flow and deformable shape matching methods (which consider

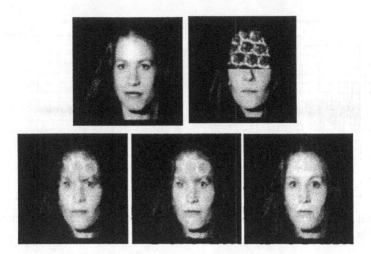

Fig. 7. See caption of figure 6

only changes in XY), of statistical texture models such as "eigenfaces" (which consider only changes in I an assume an already existing XY correspondance), and of hybrid methods with treat shape and texture separately and sequentially.

We have further shown how to tailor the space of allowable XYI deformations to fit the actual variation found in individual target classes. This was accomplished by a statistical analysis of observed image-to-image deformations using a Principal Components Analysis. The result is that the image deformation is restricted to the subspace of physically-plausible deformations. In the process, the dimensionality of the matching and the numerical complexity of the governing equation are drastically reduced.

By considering only the low-dimensional subspace of plausible deformations, we make the image matching process more robust and more efficient. We in effect "build in" statistical *a priori* knowledge about how the object can vary in order to obtain the best image-to-image match possible. To illustrate the power of this method we have shown that we can interpolate missing data despite occlusions and noise, and that we can use this method to obtain very compact image descriptions.

References

1. K. J. Bathe. *Finite Element Procedures in Engineering Analysis.* Prentice-Hall, 1982.
2. A. Baumberg and D. Hogg. Learning flexible models from image sequences. In *Proceedings of the Third European Conference on Computer Vision 1994 (ECCV'94)*, Stockholm, Sweden, May 1994.
3. David Beymer. Vectorizing face images by interleaving shape and texture computations. A.I. Memo No. 1537, Artificial Intelligence Laboratory, Massachusetts Institute of Technology, 1995.

4. T. Boult. Physics in a fantasy world vs. robust statistical estimation. In *NSF/ARPA workshop on 3D Object Representation in Computer Vision*, New York, USA, December 1994.

5. T.F. Cootes and C.J. Taylor. Combining point distribution models with shape models based on finite element analysis. *Image and Vision Computing*, 13(5), 1995.

6. T.F. Cootes, C.J. Taylor, D.H. Cooper, and J. Graham. Training models of shape from sets of examples. In *Proceedings of the British Machine Vision Conference*, Leeds, 1992.

7. P. E. Danielsson. Euclidean distance mapping. *Computer Vision, Graphics, and Image Processing*, 14:227–248, 1980.

8. B.K.P. Horn and G. Schunck. Determining optical flow. *Artificial Intelligence*, 17:185–203, 1981.

9. I.T. Jolliffe. *Principal Component Analysis*. Springer-Verlag, New York, 1986.

10. C. Kervrann and F. Heitz. Learning structure and deformation modes of nonrigid objects in long image sequences. In *Proceedings of the International Workshop on Automatic Face- and Gesture-Recognition*, 1995.

11. A. Lanitis, C. J. Taylor, and T. F. Cootes. A unified approach to coding and interpreting face images. In *Proceedings of the International Conference on Computer Vision 1995 (ICCV'95)*, Cambridge, MA, June 1995.

12. M.M. Loeve. *Probability Theory*. Van Nostrand, Princeton, 1955.

13. J. Martin, A. Pentland, and R. Kikinis. Shape analysis of brain structures using physical and experimental modes. In *IEEE Conference on Computer Vision and Pattern Recognition*, Seattle, USA, June 1994.

14. B. Moghaddam and A. Pentland. Probabilistic visual learning for object detection. In *IEEE Proceedings of the Fifth International Conference on Computer Vision*, Cambridge, USA, june 1995.

15. H. Murase and S. K. Nayar. Visual learning and recognition of 3D objects from appearance. *International Journal of Computer Vision*, 14(5), 1995.

16. C. Nastar. Vibration modes for nonrigid motion analysis in 3D images. In *Proceedings of the Third European Conference on Computer Vision (ECCV '94)*, Stockholm, May 1994.

17. C. Nastar and N. Ayache. Fast segmentation, tracking, and analysis of deformable objects. In *Proceedings of the Fourth International Conference on Computer Vision (ICCV '93)*, Berlin, May 1993.

18. C. Nastar and A. Pentland. Matching and recognition using deformable intensity surfaces. In *IEEE International Symposium on Computer Vision*, Coral Gables, USA, November 1995.

19. A. Pentland, B. Moghaddam, T. Starner, and M. Turk. View based and modular eigenspaces for face recognition. In *IEEE Proceedings of Computer Vision and Pattern Recognition*, 1994.

20. A. Pentland and S. Sclaroff. Closed-form solutions for physically based shape modelling and recognition. *IEEE Transactions on Pattern Analysis and Machine Intelligence*, PAMI-13(7):715–729, July 1991.

21. M. Turk and A. Pentland. Face recognition using eigenfaces. In *IEEE Proceedings of Computer Vision and Pattern Recognition*, pages 586–591, 1991.

22. Q.Z. Ye. The signed euclidean distance transform and its applications. In *International Conference on Pattern Recognition*, pages 495–499, 1988.

Reasoning about Occlusions during Hypothesis Verification

Charlie Rothwell

INRIA, 2004, Route des Lucioles, Sophia Antipolis, 06902 CEDEX, France

Abstract. In this paper we study the limitations of current verification strategies in object recognition and suggest how they may be enhanced. On the whole *object topology* is exploited little during verification. In practice, understanding the connectivity relationships between features in the image, or on the object, can lead to significantly more accurate evaluations of recognition hypotheses. We study how topology reasoning allows us to hypothesize the presence of occlusions in the image. Analysis of these hypotheses provides information which turns out to be crucial to the quality of our overall verification results.

1 Introduction

In an object recognition system, the process which is most used to differentiate between identification hypotheses is the final stage, called *verification*. The typical processing path for a recognition system is to start off by extracting features from the image, then to do feature grouping and model selection (indexing), and penultimately a correspondence stage is used to pair together sets of model and image features. Finally, in mature systems a verification step is then included which ultimately determines the degree of correctness of the model-image correspondences. Of course not all recognition systems follow this path, and often we find that verification is not performed as a separate process to correspondence. Nevertheless, a fairly general list of recognition algorithms which record the variations in the methods contains the efforts of Ayache and Faugeras [1], Pollard, *et al.* [9], Grimson and Lozano-Pérez [6], Bolles and Horaud [2], Faugeras and Hébert [5], Thompson and Mundy [14], Huttenlocher and Ullman [7], Lamdan and Wolfson [8], Stein and Medioni [13], and Califano and Mohan [3].

In this article we study how verification based on understanding model and image *topology* leads to better verification results. We use the term topology to represent the connectivity relationships between features. The importance of the use of topology in vision is discussed more completely in [10], here it is sufficient to state that *none* of the processing techniques discussed in this paper would be possible without a suitable understanding of topology. Typically, verification is based entirely on *geometric* methods.

Locally, topology conveys whether two model features are adjacent (connected) and should be seen as such in an image. Globally it reveals whether a continuous chain of features is present between any pair of primitives, and so defines notions of global connectivity. Topology thus enables neighbourhood based inferences. The observance of a particular model feature in an image would most likely indicate that all features adjacent to it should also be visible. The failure to find the adjacent features in the image would thus indicate either of two things: perhaps an occlusion is present and so the features are hidden; or conversely that the hypothesis is wrong and should either be discarded or at least have its importance diminished. Discrimination between these two types of incident can only be achieved by simultaneous topological and occlusion analysis.

Commonly verification is done as follows: the final conclusion of a recognition algorithm is that a set of individual model features matches a set of individual image features (rather than just saying that the model matches the image). The precise set of correspondences implies a geometric mapping from the model to the image. Additionally, one would also recover a measure of the number of matching features as a percentage of the whole; this number would form the basis of a verification score on which a criterion for the acceptance of the hypothesis would be built. The actual score would be computed by projecting all of the model features (perhaps a set of line segments, or even edgels from an acquisition image) to the image, and then counting the number of projected features which find *image support*. The hypothesis is accepted if a certain proportion of the features are matched. A test for image support of a projected model feature could be whether the feature lies close to an observed image feature which has a similar orientation. A complete description of such a verification strategy can be found in [14]. Note especially that approach is entirely *geometric*.

In the rest of this article we build on the work of [12], but include the following considerations in order to improve the verification method of the previous paragraph:

- Negative evidence must be explained. Primarily we should be able to locate occlusion events which justify the lack of measurement of a scene-model match. Failure to find occlusion events reduces the likelihood that a hypothesis is correct.
- Unless there is positive evidence of occlusion, some notion of object topology must be preserved. Therefore two image features should not be marked as coming from adjacent features on an object's boundary unless they are either connected in the image, or unless there exists an occlusion event between them.
- Under generic viewing conditions there must be uniqueness of solution. A single feature cannot belong to more than one object.

This last point requires development. The nature of the results given by different recognition systems varies dramatically depending on the application. Often the recognition problem is posed as the task of finding a specific object in a scene, and then immediately terminating the processing [6]. This problem is significantly different from, and easier, than that of identifying all objects in a scene which might correspond to any of a number of objects in a model base (that is a complete evaluation of the scene). In this situation, a single mistake can lead to catastrophic failures in interpretation because any one decision influences all subsequent processing. Thus, we would be likely to find that accepting a particular hypothesis might in some way prevent the formation of another hypothesis (particularly if the former is incorrect). It is perhaps therefore wise to be conservative and to allow multiple interpretations so that no truly correct hypothesis is discarded. Certainly, in light of the results given in [12], where it was stated that a significant number of false positives are likely to be recovered in a scene, we should at first follow this line of thinking.

However, most applications might be expected to provide unique and accurate scene interpretations. We are therefore interested in moving towards the notion of single feature interpretations whilst still working in relatively unstructured and unknown environments. Unfortunately, the immaturity of most verification schemes makes this difficult. It is for this reason that we have taken another look at the competences of these schemes, and we see that combining a topological representation with our previous geometric measures provides an initial step towards improvement.

We now describe how topological reasoning can be used as an aid to verification. First we show how a system such as that described in [12] produces incorrect recognition results due to the frailty of geometric verification methods, and then we show how topology reasoning introduces opportunities for occlusion analysis leading to more robust recognition. A more complete version of this paper is available as [11].

Fig. 1. *A model from the library having an outer boundary similar to that of the bracket visible in the image is hypothesized as being present in the scene.*

2 A sample recognition system

The recognition system we base our work on is called LEWIS, which uses invariant-indexing to form recognition hypotheses. A complete review of LEWIS can be found in [12], as well as examples of the system recognizing objects and performance statistics. Here we concentrate on examples failures of the system in identifying objects.

A characteristic example of recognition failure is shown in Fig. 1. After edgel detection and extraction of lines and conics, features are grouped together to form indexes based on plane projective invariants. Five lines around the boundary of the bracket just above centre-left in the image happens to have an index value which matches the invariants for a particular object which is distinct from the bracket in the model base.

Under verification the model for the incorrect hypothesis is projected to the image. Within LEWIS a model is represented as a set of edgel data recovered from an acquisition view of the isolated object, and a combination of the fitted lines, conics, and computed invariants for the edgel data. In essence, only the edgel data of the model are projected to the image. These are matched to image edgels using distance and orientation criteria: their orientations must differ by less than fifteen degrees and separations by less than five pixels. In the right-hand image of Fig. 1 the edges which found complete support are drawn in white, those which found matching image edgels within five pixels but whose orientation differed too much are drawn in grey (these may not be visible), and those failing both matching criteria are marked in black. In this example 55.0% of the model features were found to have complete image support. Any score over 50% is considered by LEWIS to be sufficiently high, and so the hypothesis is marked as accepted. Obviously there is an error as the wrong object has been identified. However, using geometric verification measures it cannot be ruled out.

A different failure is shown in Fig. 2. Here a false positive is created due to the presence of spurious scene features (linear texture). The problem arises because unconnected image features provide support for the model hypothesis over a large area. In this case 55.2% of the model edgels found matches in the scene even though only a few of them actually projected onto the features used for indexing.

3 Enhanced verification methods

Figures 1 and 2 highlight two areas where the insufficiency of current reasoning causes verification to fail. We cannot correct these errors using purely geometrical processes but rather must exploit integrated topological structure. Advances can be made rapidly

Fig. 2. *In this case an invariant configuration causes the estimation of the incorrect pose of an object. This is due to symmetry on part of the model.*

when we start to diffuse the local acceptance or rejection of a hypothesis around an object boundary. In short, in considering the acceptance of any one model element, we must analyse the outcome of the verification procedure for its neighbouring elements.

The two driving notions in verification thus become *topology* and consequently the understanding of *occlusion*. The topology of the image features which are assigned correspondences to model features must match the topology of the model features exactly. Exception can only be permitted due to occlusion, or when we are faced with the time-old problem of extracting reliable segmentations from images. We thus need to develop algorithms for analyzing the topology of features and for estimating the presence of occlusions. Now, in cluttered scenes we are very seldom able to recover image support for the entire boundary of the projection of a model hypothesized through indexing. Often this is because the objects are occluded in scenes, or perhaps they may suffer partial self-occlusion if they possess their own three-dimensional structure. The key issue is that the projection of the model into the image indicates where occlusions might arise (due to the loss of image support) and so our task becomes that of finding independent evidence for the occlusions. If we cannot, then it is likely that the original recognition hypothesis is incorrect and so we might do better by considering a different interpretation.

Occlusion events are typically marked by the presence of 'T' junctions in the image edgel structure. Generically, an object feature which undergoes occlusion will be cut and terminated by a locally straight transverse line segment. Thus when occlusions are hypothesized by the sudden loss of image support for the projection of the model we should look for 'T' junctions. Hypothesizing occlusions is roughly done as follows (see [11] for a more thorough treatment): find all of the projected model edgels which have image support, these form connected sub-sets. Then, at the boundaries of each of these sets (where image support is first totally lost), hypothesize an occlusion event.

We have evaluated occlusion hypotheses in two ways. The first uses the edgel data computed as the initial step of image processing. Initially we recover all of the junctions in the original edgel image which have an order greater or equal to three. Junctions of order three are those where three edgel chain curves meet at a single point. We declare the presence of a 'T' junction when the angle between any pair of the edges meeting at an order 3 junction are within twenty degrees of 180 degrees. The large tolerance reflects the fact that edgel contours are frequently displaced by significant amounts near junctions, and it also allows for occlusion by curved objects. Then, if the edgel-'T' junction lies sufficiently close to where the recognition hypothesis deduced that there should be an occlusion event, then we can add confidence to the original recognition hypothesis.

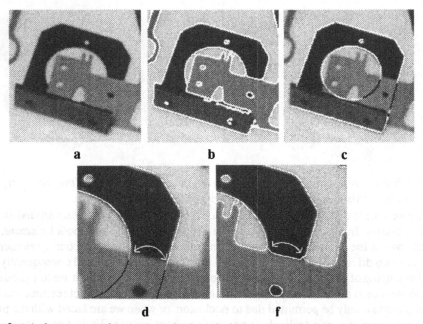

Fig. 3. *(a) shows part of an image which includes an object from the model base. In (b) are edgels from the edgel detector and (c) the (correct) projection of the model. This suggests occlusions at the points marked by the arrows in (d). Near to these points are triple junctions in the edgel description which have the forms of 'T's, as shown in (f).*

In practice this type of processing does not resolve all of the occlusion events we find in an image. This is because edgel detectors are notoriously poor at recovering meaningful connectivity at junctions. Consequently, it will seldom recover all of the 'T' junctions. We therefore resort to a second test which examines the overall structure of the image intensity data near to the locations of each hypothesized occlusion. This is done by parametric model fitting to junctions such as suggested by Deriche and Blaszka [4], though there are many other approaches to model fitting, some of which are mentioned in [11]. The type of parametric model fitted by [4] assumes that the surface is composed of a number of constant intensity plateaux which meet at the junction and are separated by straight edges. Each plateau represents an image region and smoothing is accounted for between the image regions through an approximate parametrization of Gaussian smoothing. The algorithm of [4] fits such a model over a specific window size at a given location in the image (where we suspect that there is an occlusion), and returns a number of different parameters which represent the interpretation of the intensity surface. The key measures which are returned are a fitting cost, the grey level values of the plateaux, and the angles at which the edges come into the junction. We can then estimate whether the junction is a real 'T' junction by looking at the angles between the edges and by making sure that the plateaux have sufficiently different grey levels.

Examples - edgel contour junctions

In Fig. 3 we show how data from the edgel detector hypothesizes 'T' junctions near to where occlusion events should be found on the strength of a specific recognition hypothesis. This is an example of positive support for an object hypothesis, with the measurement of the low-level junction description enhancing the confidence in a hypothesis. In

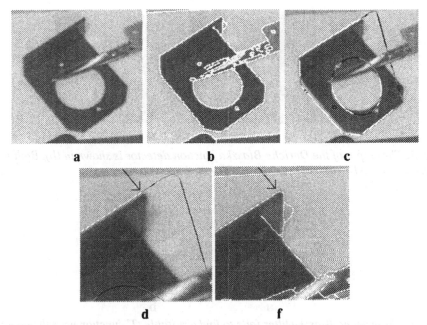

Fig. 4. *The edgels for sub-image (a) are shown in (b). LEWIS deduces the incorrect hypothesis (c). Near the white-grey transition of the projected outline in (d), we expect to find a junction in the edgel description (that is if the hypothesis were correct). However, as shown in (f) the failure of the edgel detector to record a junction in the right place suggests that the hypothesis is false.*

Fig. 4 we demonstrate an example of negative evidence which renders a hypothesis unlikely (or perhaps may even cause it to be rejected outright). The projected model curves are shown initially following a contour in the image, but at one stage (due to an incorrect model being hypothesized), the projected model curve parts quite clearly from the image curve. The edgel detector fails to find any junctions near the point of departure and so one can start to doubt whether the original hypothesis is correct.

This type of reasoning appears very attractive on the weight of the two examples given. However, most edgel detectors have difficulty recovering correct edgel contour connectivity near junctions. This means that failure to record the presence of a 'T' junction can be taken as only partial evidence against a hypothesis, and should not be used too strongly. Conversely, if we are able to find a 'T' junction where predicted by a recognition hypothesis then we can add considerable weight to the hypothesis. The progress we make in marking 'T' junctions is apparent. Henceforth, the projected edgel data from the hypothesis which originally failed to gain image support (and so was treated neutrally) can be treated as *positive support* if associated with a confirmed occlusion event.

Examples - parametric junction models

The alternative approach to evaluating 'T' junctions fits parametric junction models. In Figs 5 and 6 we show similar examples to those for the edgel junction reasoning, but this time 'T' junctions are evaluated using the Deriche-Blaszka operator. In the first case a pair of suitable 'T' junctions are recovered and shown superimposed in the figure, and in the second case no such fit could be found and so the hypothesis is marked as being unlikely. Thus we see that parametric model fitting can enhance the understanding of

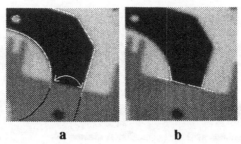

a   **b**

Fig. 5. *The output of the Deriche-Blaszka junction detector is shown in (b). Both junctions appear to be sufficiently close to 'T' junctions.*

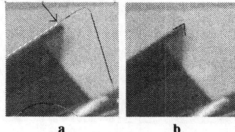

a   **b**

Fig. 6. *The Deriche-Blaszka filter fails to find a suitable 'T' junction near the potential occlusion event already discussed in Fig. 4. Instead it finds the 'L' junction shown in (b). Such a feature has no relation to an occlusion event, and so the original hypothesis is likely to be false.*

recognition hypotheses when directed by prior topological reasoning.

Again, these examples of the use of the parametric junction model approach do not describe the whole truth. Over repeated trials we have found that the Deriche-Blaszka model does not actually represent the image intensity surface correctly. In [11] we discuss to some extent why this is the case; overall more work is required in developing junction detectors which model the image intensity surface correctly.

In summary, we have so far demonstrated how occlusion events can be hypothesized by studying the model topology information contained in hypotheses produced by a typical recognition system. We have also seen that hypothesized occlusions can be evaluated in different ways. Two methods have been discussed: the first using bottom-up information recovered from the original output of an edgel filter; and the second derives top-down data resulting from the application of parametric junction model fitting. For certain cases the second method is more accurate and more reliable, but in general it relies on making incorrect assumptions about the shape of the intensity surface (which is seldom composed of smoothed constant-intensity plateaux). Nevertheless we can employ both methods with caution. Whenever either approach suggests the presence of a 'T' junction, we can be fairly confident that it is right. However, they both frequently reject junctions which do actually correspond to occlusion events.

3.1 How to update hypotheses

Now, whenever we find an occluded region which is terminated at one end by a verified 'T' junction, it can be marked as making a positive contribution to the hypothesis. This is to say that the score of *visible model edgels* used to compute the overall verification score should be incremented by the number of edgels in the occluded region. Consequently we can transform a hypothesis such as that in Fig. 7 from a 70.5% score to

a   **b**

Fig. 7. *The original verification method produced only a 70.5% score for the hypothesis shown in (a). However, after the prediction and verification of the various 'T' junctions bounding the occluded parts of the hypothesized object, we increase the verification score to 93.6%. Edgels which were previously unmatched, but are now marked as positively occluded, are depicted in white in (b). The overall verification score is incremented by the number of edgels in the positively identified occluded regions.*

a   **b**   **c**

Fig. 8. *The original verification score for the object in (b) was 70.7%. After occlusion reasoning this rises to 83.6% which is a clear indication that the hypothesis is correct.*

a re-evaluated score of 93.6%, and hence have little doubt that the hypothesis is correct. Ideally we would hope that the new score would tend towards 100%, but there are always short sections of projected model curve which project near to image features with the wrong orientation (and so are not marked as being caused by occlusion). Taking these into account would make the hypothesis in Fig. 7 take on a score of little less than 100% (in fact 98.2% when we ignore small section of less than five pixels in length). The dominance of a good hypothesis such as this one significantly enhances our understanding of the scene. From a number of experiments we are able to conclude that a final matching score of over 90% leaves little doubt as to the identity of an object. Another example is in Fig. 8 where the score of an occluded object rises from 70.7% to 83.6% after occlusion reasoning (and 91.3% after removal of short unmatched chains).

3.2 Correctness of image topology

So far in this paper the effects we have been interested in have been dominated by model topology and cause-and-effect reasoning for connected parts of the model to be either occluded or visible. It is quite understandable that we can make reciprocal considerations with regard to image topology, or more properly between the consistency of both of the model and image descriptions. In an ideal world (where we would of course cease to be frustrated by problems in segmentation), the image topology should match the projection of the model topology exactly.

Even with our current segmentation abilities we can develop simple tests which

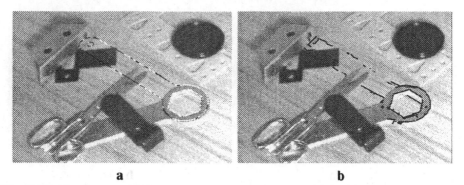

a b

Fig. 9. *The poorly oriented hypothesis in Fig. 2 required support from unconnected texture features. These have an image topology which is inconsistent with that of the model. As there are no indications of occlusion events at the ends of each image curve, we infer that the correspondences are incorrect. Subsequently we can doubt the hypothesis.*

show some robustness to errors in the extraction of the low-level image description, and which indicate whether the grounds for believing a hypothesis should be reduced. For instance, we can start off by seeing which sets of image features have been given a correspondence with a single topological feature from a model. Due to the trivial connectivity constraints which exist on a lone model feature, one would also expect the corresponding image features to be connected. There are a number of different reasons for the image features to become distinct, though still to remain topologically linked. Perhaps the segmentation and fitting procedures separated the features at the geometric level by attributing each one to different geometric objects, or perhaps the edgel detector erroneously placed a junction between them to provide connectivity with other features. However, in both of these cases the image topology is consistent with a single model feature, and so we need not doubt the integrity of a particular hypothesis.

Conversely, we should mark cases in which connectivity has been lost. Of course there might be a perfectly reasonable explanation such as the presence of an occlusion event, but if not, we should add further doubt to the interpretation. Thus, our way of reasoning is again led back to the detection of occlusion events.

We show the success of this line of reasoning in Fig. 9. Model features have been projected into the image and have found sufficient image support along their lengths. However, the support has actually been provided by sets of unconnected image features. We thus test for the presence of occlusion events at the ends of the image features, and if they are not found, we mark the hypothesis as being unreasonable.

3.3 Uniqueness of description

By this stage of the proceedings the hypotheses have undergone a detailed level of topological analysis. Those which have been attributed near perfect scores are very likely to be correct, whilst those with poorer verification tallies may either be erroneous, or might just be suffering due to difficulties in segmentation or occlusion event detection.

We now return to the fact that a single feature in an image is caused by a single scene feature. Therefore, a consequence of recognition should be that the correspondence between model and image features is at most one-to-one. Should two hypotheses match a single image feature we can be sure that *at least one* of the hypotheses is incorrect, and so should try and eliminate the least likely. This process is risky should the confidence levels in the hypotheses be poorly defined, but as our abilities at verification improve,

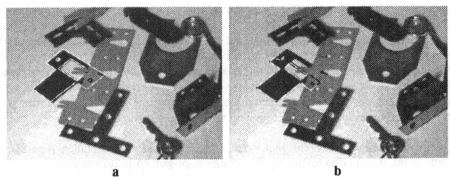

a   b

Fig. 10. *In Fig. 8 we were able to find two hypotheses which matched to common scene features. After all of the topological processing the hypothesis in (a) scored 91.3% and that in (b) 68.2%. Any score over 90% provides very strong confidence, and so we can eliminate any other hypotheses which match to the same image features.*

we can start to attribute error measures with reasonable accuracy.

Remembering that once occlusion analysis has been performed, we can mark hypotheses more clearly for acceptance or rejection. Thus we can proceed by accepting the single hypothesis which has gained the highest recognition score. Then, all of the image features which have been given a correspondence to any of the features in this model are marked as being explained. All other hypotheses which have correspondences with these image features are marked as being inconsistent, and rejected. We then take the next best hypothesis, and proceed by examining its image features. In short, this process ensures a uniqueness of description of the image features. An example of this type of reasoning is given in Fig. 10 where we are able to rule out the hypothesis in (b) because it shares scene features with the very highly scored hypothesis in (a).

4 Conclusions

In this article we have demonstrated how reasoning about model and image topology enhances our object recognition verification abilities. A typical recognition system such as that of [14] computes the final match score for any hypothesis by determining whether a set of independent model features finds support in an image. [12] demonstrated that such a geometric strategy does not produce conclusive recognition results. We have found that diffusing verification information around connected components of a model means that a lack of image support can actually be turned into *positive evidence*. This in turn means that verification thresholds can perhaps be raised from 50% up to somewhere in excess of a 90% level of image support.

The development of our verification algorithm involves reasoning about discrepancies between the model and image topologies. The differences are used to hypothesize where occlusion events should lie in the image. The presence of such events strengthens a recognition hypothesis, and the lack of one suggests that a hypothesis might be false. The detection of the junctions is done via both edgel detector output, and the Deriche-Blaszka feature detector [4]. Neither of these filters function perfectly, though when either hypothesizes the presence of a 'T' junction we can be relatively sure that it is correct.

Of course our results are not entirely complete. Whilst working with single images, we need to analyse the effects of other feature detectors. There are a large number of other filters which require evaluation in either of the domains of bottom-up or top-down processing. We also need to test the algorithms on a more varied range of objects of which a three-dimensional model base is the ultimate goal.

On a different level, we have only made use of the boundary information contained within the models. Such geometric primitives obviously provide very easy access to object descriptions. However, a full verification scheme should include analysis about surface properties such as texture, and also even colour. Certainly with the aid of top-down segmentation based on the recognition hypotheses one would be able to test out other object properties in conjunction with the more geometric aspects.

A more complete version of this paper is available through ftp as a technical report [11] from: <URL ftp://ftp.inria.fr/INRIA/publication/publi-ps-gz/RR/RR-2673.ps.gz>.

Acknowledgments

A number of people were involved with discussions during the original development of LEWIS and its re-implementation within the TargetJr software system initially developed at GE CRD (on which this paper is based). As such I would like to thank David Forsyth, Bill Hoffman, Joe Mundy, and Andrew Zisserman. Thierry Blaszka provided his implementation of the Deriche-Blaszka filter. I also acknowledge discussions with Theo Papadopoulo and Nour-Eddine Deriche, and Pippa Hook helped to ensure the correctness of the English in this article. This research was funded by a Human Capital and Mobility grant from the European Community.

References

[1] N. Ayache and O. Faugeras. HYPER: A New Approach for the Recognition and Positioning of Two-Dimensional Objects. *PAMI*, 8(1):44–54, 1986.

[2] R. Bolles and R. Horaud. 3DPO: A Three-dimensional Part Orientation System. *IJRR*, 5(3):3–26, 1986.

[3] A. Califano and R. Mohan. Systematic design of indexing strategies for object recognition. *Proc. CVPR*, p.709–710, 1993.

[4] R. Deriche and T. Blaszka. Recovering and characterizing image features using an efficient model based approach. *Proc. CVPR*, p.530–535, 1993.

[5] O. Faugeras and M. Hébert. The representation, recognition, and locating of 3d shapes from range data. *IJRR*, 5:27–52, 1986.

[6] W.E.L. Grimson and T. Lozano-Pérez. Localizing overlapping parts by searching the interpretation tree. *PAMI*, 9(4):469–482, 1987.

[7] D. Huttenlocher and S. Ullman. Recognizing Solid Objects by Alignment with an Image. *IJCV*, 5(2):195–212, 1990.

[8] Y. Lamdan and H. Wolfson. Geometric Hashing: A General and Efficient Model-Based Recognition Scheme. *Proc. ICCV*, p.238–249, 1988.

[9] S. Pollard, J. Porrill, J. Mayhew, and J. Frisby. Matching geometrical descriptions in three-space. *IVC*, 5(2):73–78, 1987.

[10] C. Rothwell, J. Mundy, and W. Hoffman. Representing objects using topology. In preparation, 1996.

[11] C. Rothwell. The importance of reasoning about occlusions during hypothesis verification in object recognition. TR 2673, INRIA, 1995.

[12] C. Rothwell. *Object recognition through invariant indexing*. Oxford University Press, 1995.

[13] F. Stein and G. Medioni. Structural Indexing: Efficient 3-D Object Recognition. *PAMI*, 14(2):125–145, 1992.

[14] D. Thompson and J. Mundy. Three-dimensional model matching from an unconstrained viewpoint. *Proc. ICRA*, p.208–220, 1987.

Object Recognition Using Multidimensional Receptive Field Histograms

Bernt Schiele and James L. Crowley

LIFIA/GRAVIR, 46 Ave Félix Viallet, 38031 Grenoble, France

Abstract. This paper presents a technique to determine the identity of objects in a scene using histograms of the responses of a vector of local linear neighborhood operators (receptive fields). This technique can be used to determine the most probable objects in a scene, independent of the object's position, image-plane orientation and scale. In this paper we describe the mathematical foundations of the technique and present the results of experiments which compare robustness and recognition rates for different local neighborhood operators and histogram similarity measurements.

1 Introduction and Motivation

Swain and Ballard [10] have developed a technique which identifies objects in an image by matching a color histogram from a region of the image with a color histogram from a sample of the object. Their technique has been shown to be remarkably robust to changes in the object's orientation, changes of the scale of the object, partial occlusion or changes of the viewing position. However, the major drawback of their method is its sensitivity to the color and intensity of the light source and color of the object to be detected. Several authors have improved the performance of the color histogram approach by introducing measures which are less sensitive to illumination changes (see i.e. [5, 6, 2]).

The color histogram approach is an attractive method for object recognition, because of its simplicity, speed and robustness. However, its reliance on object color and (to a lesser degree) light source intensity make it inappropriate for many recognition problems. The focus of our work has been to develop a similar technique using local descriptions of an object's shape provided by a vector of linear receptive fields. For the Swain and Ballard algorithm, it can be seen that robustness to scale and rotation are provided by the use of color. Robustness to changes in viewing angle and to partial occlusion are due to the use of *histogram matching*. Thus it is natural to exploit the power of histogram matching to perform recognition based on histograms of local shape properties. The most general method to measure such properties is the use of a vector of linear local neighborhood operations, or receptive fields. We have compared sensitivity and recognition reliability for a variety of local neighborhood operations, and present the results of the most successful functions below.

The first part of the paper presents our generalization of the color histogram method (section 2–4). Section 5 shows the robustness of different local neighborhood operations to additive Gaussian noise. In the second part we show the use of the histogram matching of receptive field vectors for object recognition (section 6) and experimental results (section 7).

2 Multidimensional Receptive Field Histograms

One can identify the following parameters for the *multidimensional receptive field histogram* approach:

- The choice of local property measurements (section 3),
- Measurement for the comparison of the histograms (section 4),
- Design parameters of the histograms: number of dimensions of the histogram and resolution of each axis.

The local properties should be chosen so that they are either invariant or equivariant to scale and 2D–rotation [1]. Invariant means that the local characteristics does not change with scale or 2D–rotation, while equivariant means that they vary in a uniform manner which is represented by a translation in a parameter space. Unfortunately most of the available characteristics are only scale invariant *or* 2D–rotation invariant. Therefore we use equivariant local characteristics which allow us to select an arbitrary scale and rotation (see e.g. [4, 3]). Section 3 describes the filters and normalizations which can be used.

The comparison measurement determines the separability between histograms, as we will see in the experiments described below. Different measures for the histogram comparison are introduced in section 4.

The design parameters of the histograms determine the separability between the histograms of different objects. In [8] we concluded that reducing of the resolution (number of bins per histogram axis) results in an improvement of the stability of the histograms with respect to view point changes, but also diminishes the discrimination between objects. From the experiments of [8] we concluded also that discrimination can be recovered by improving the number of histograms dimensions provided by independent local properties.

3 The local characteristics

In this section we briefly describe receptive field functions which can be used for object recognition. The calculation of local properties can be divided into the local linear point-spread function (formula (1)), and the normalization function used during measurements of local properties.

$$Img_{Mask}(x, y) = \sum_{i,j=-m,-n}^{m,n} Img(x+i, y+j) Mask(i, j) \qquad (1)$$

3.1 Filter

The first results we present are with non–equivariant filters. We have used these simple filters in our first experiments to test the power of our approach. This is followed by the description of two equivariant filter classes, Gabor filters and Gaussian derivatives.

---

[1] Recent results have shown that the technique is quite robust to 3D rotation. These results have been submitted to the *International Workshop on Object Representation for Computer Vision* at this conference [8]

Gradient and Laplacian Operators Our first experiments were performed with first derivative and Laplacian operators given by:

$$M_{dx} = \begin{pmatrix} -1 & 0 & 1 \\ -1 & 0 & 1 \\ -1 & 0 & 1 \end{pmatrix} \qquad M_{dy} = \begin{pmatrix} -1 & -1 & -1 \\ 0 & 0 & 0 \\ 1 & 1 & 1 \end{pmatrix} \qquad M_{lap} = \begin{pmatrix} -1 & -2 & -1 \\ -2 & 12 & -2 \\ -1 & -2 & -1 \end{pmatrix}$$

Gabor filter Gabor filters are local compact filters tuned to a spatial frequency band. Gabor filters are defined by modulating a Gaussian window with a cosine and an imaginary sine giving an even and odd filter pair. The main advantage of the Gabor filters is that one can freely choose the frequency (and therefore the scale) as well as the bandwidth of the filter.

A Gabor filter pair is compact in both space and frequency. In our experiments we have used a two–dimensional formulation of the Gabor functions proposed by Daugman [1] (in the Fourier domain):

$$G(u, v) = e^{-\pi\left((u-u_0)^2\alpha^2 + (v-v_0)^2\beta^2\right)} e^{-2\pi i(x_0(u-u_0)+y_0(v-v_0))} \qquad (2)$$

where (x_0, y_0) are the center coordinates of the filter, (α, β) define the width and the length, and (u_0, v_0) specify the modulation in x and y direction, which has the spatial frequency $\omega_0 = \sqrt{u_0^2 + v_0^2}$ and direction $\theta_0 = \arctan(v_0/u_0)$.

To design a Gabor filter, we follow a method proposed by Westelius [11] to choose the standard deviation α and the spatial frequency ω_0. These two parameters determine the size and bandwidth of the filter.

Gaussian derivatives By using the Gaussian derivatives one can explicitly select the scale. This is achieved by adapting the variance σ of the derivative. Given the Gaussian distribution $f(x, y)$ we obtain the first derivative in x-direction:

$$f(x, y) = e^{-\frac{x^2+y^2}{2\sigma^2}} \qquad \frac{\partial f(x, y)}{\partial x} = -\frac{x}{\sigma^2} f(x, y)$$

3.2 Normalization

The effects of variation in signal intensity can be removed by normalizing the inner product of a filter with a signal during convolution. Normalization should be considered from at least two points of view. The first point concerns how well the normalized convolution behaves in the presence of additive noise (see experiments in section 5). The second point concerns how the normalized convolution responds to variations in signal intensity due to differences in ambient light intensity, aperture setting or digitizer gain. We have compared the robustness of correlation with *no* normalization and with two other forms of normalization.

Normalization by energy Dividing by neighborhood energy removes variations in signal strength which may be due to light source intensity variation, and thus provide a filter output vector histogram which is invariant to illumination intensity. Energy normalization also turns out to be the most robust in respect to additive Gaussian noise. Therefore we have used energy normalization in most of our experiments.

$$Img_{ene}(x, y) = \frac{\sum_{i,j} Img(x+i, y+j) Mask(i, j)}{\sqrt{\sum_{i,j} Img(x+i, y+j)^2} \sqrt{\sum_{i,j} Mask(i, j)^2}}$$

Normalization by mean and variance By Variance normalization we refer to subtracting the mean of each neighborhood and then dividing by the variance of the neighborhood. Variance normalization is relatively sensitive to additive Gaussian noise. This makes *Variance*-normalization unusable in our context.

$$Img_{var}(x,y) = \frac{\sum_{i,j}(Img(x+i,y+j) - \overline{Img(x,y)})Mask(i,j)}{\sqrt{\sum_{i,j}(Img(x+i,y+j) - \overline{Img(x,y)})^2}\sqrt{\sum_{i,j}Mask(i,j)^2}}$$

with $\overline{Img(x,y)} = \frac{1}{(2m+1)(2n+1)}\sum_{i,j=-m,-n}^{m,n} Img(x+i,y+j)$.

4 Histogram Comparison

This section describes possible measurements for comparing histograms. The analysis of these measurements is important, since the "intersection"–measurement, used by Swain and Ballard [10], has limitations for the use for multidimensional receptive field histograms. For object recognition using receptive field histograms we compare a histogram T from a database to a newly observed histogram H.

Sum of squared distances The sum of squared differences (SSD) is commonly used in signal processing:

$$SSD(H,T) = \sum_{i,j}(H(i,j) - T(i,j))^2 \tag{3}$$

χ^2 **– test** The proper method proposed by mathematical statistics for the comparison of two histograms is the χ^2–test. χ^2 is used here to calculate the "distance" between two histograms. We have used two different calculations for χ^2 [7]: χ_T^2 is defined, when the theoretical distribution (here T) is known exactly. Although we do not know the theoretical distribution in the general case, we have found that χ_T^2 works well in practice:

$$\chi_T^2(H,T) = \sum_{i,j} \frac{(H(i,j) - T(i,j))^2}{T(i,j)} \tag{4}$$

The second calculation χ_{TH}^2 compares two real histograms. χ_{TH}^2 also gives good results. For the moment it is not clear which of the two χ^2 measurements is more reliable:

$$\chi_{TH}^2(H,T) = \sum_{i,j} \frac{(H(i,j) - T(i,j))^2}{H(i,j) + T(i,j)} \tag{5}$$

Intersection Swain and Ballard [10] used the following intersection value to compare two color–histograms:

$$\cap(H,T) = \sum_{i,j} \min(H(i,j), T(i,j)) \tag{6}$$

The advantage of this measurement is, that background pixels are explicitly neglected when they don't occur in the Model histogram $T(i,j)$. In their original work they reported the need for a sparse distribution of the colors in the histogram in order to be able to distinguish between different objects. Our experiments have verified this requirement. Unfortunately, multidimensional receptive field histograms are not generally sparse, and a more sophisticated comparison measure is required.

Bayes Rule The last section below considers the use of Bayes rule to determine for each pixel or set of pixels, the probability that it is the projection of a part of a specified object. In [9] we have introduced the following formula:

$$p(O_n | \bigwedge_k M_k) = \frac{\prod_k p(M_k|O_n)p(O_n)}{\sum_n \prod_k p(M_k|O_n)p(O_n)} \tag{7}$$

with $p(O_n)$ the a priori probability of the object O_n, and $p(M_k|O_n)$ is the probability density function of object O_n, which can be directly derived from the histogram of object O_n. This formula can be used to determine for each subregion of an image the probability of the occurrence of each object O_n only based on the multidimensional receptive field histograms of each object (see for details and recognition results [9]).

5 Robustness to additive Gaussian noise

In this section we report the results of an experiment which was designed to determine how sensitive the different combinations of filter and normalization are in respect to additive Gaussian noise. For this experiment we used 8 artificial images. We will summarize the results for one image, which we call *Sin* which contains a sine–curve with the wavelength of 45 pixels.

Figure 1 show the results. To the *Sin*–image we added Gaussian noise with variance $\sigma = 1, 2, 3, \ldots, 20$ (abscissa in the diagrams). We store the histogram of the initial image (which is equivalent to $\sigma = 0$). This histogram is then compared (by using χ^2_{TH} as distance measurement) to the histograms with additive Gaussian noise. This distance correspond to the ordinate of the diagrams.

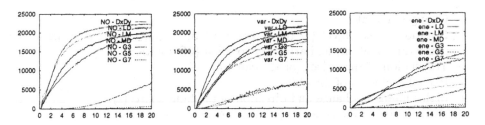

Fig. 1. *Sin:* Left: Robustness with no normalization. Middle: Robustness with Variance normalization. Right Robustness with Energy normalization

In these diagrams we look mainly at the relative behavior of the different filter normalization combinations, rather than at the absolute value of the χ^2_{TH} distance between the images (which depends strongly on the design parameter of the histograms). In this experiment we used seven different pairs of filters (see section 3): $DxDy$ for M_{dx} and M_{dy}, LD for M_{lap} and Direction of the first derivative, LM for M_{lap} and Magnitude of the first derivative, MD for Magnitude and Direction of the first derivative, $G3$ for Gabor filter with wavelength of 2.8 pixel (7×7 window) in x and in y direction, $G5$ for Gabor filter with wavelength of 5.7 pixel (15 × 15 window) in x and in y direction, $G7$ for Gabor filter with wavelength of 11.3 pixel (30 × 30 window) in x and in y direction.

The first statement we can make is, that the Gabor filters are much more robust to additive Gaussian noise than the other filters (e.g. figure 1). This is not surprising, since the Gabor filters are known to be robust to additive Gaussian noise (one part of the Gabor function is a Gaussian smoothing function). Only in the case of the *Variance* normalization do Gabor filters fail to behave properly (see figure 1). The second statement that we can make is in relation to the different normalizations: *no* normalization behave rather nicely (figure 1). The *Variance* normalization on the other hand disturbs the nice behavior of the Gabor filters (figure 1). But the best normalization for all of the filters is the *Energy* normalization (figure 1).

In the following sections we will use only *Energy* normalization since it seems to be the most robust normalization for the considered filters in respect to additive noise. The following section shows quite satisfactory results with this normalization in the recognition experiments (see section 7).

6 Using Multidimensional Receptive Field Histograms for Object Recognition

The first part of this section defines the object recognition task by the analysis of the "degrees of freedom". The second part describes the use of multidimensional receptive field histograms for this object recognition task. Section 7 gives experimental results of this approach.

Degrees of freedom within the object recognition task Possible changes of the object's appearance must be considered in the object recognition task. Possible changes include:

- Changes in scale
- Rotation of the object (or the camera): we distinguish rotation in the image plane (2D rotation) and arbitrary rotation (3D rotation)
- Translation of the object (or the camera)
- Partial occlusion of the object
- Light: intensity change and direction of the light source(s)
- Noise (noise of the camera, quantization noise, blur, ...)

In our approach, changes in to scale and 2D rotation are handled by the use of steerable filters [4, 3]. Therefore we will have only one image for one object and will generalize from this image to all considered scales and 2D rotations (see experiments in section 7).

The histograms themselves are invariant with respect to translation of the image or the object, since position information is completely removed. Furthermore the histogram matching is relatively immune to minor occlusions. This was demonstrated by Swain and Ballard in the original work on color histograms [10].

Signal intensity variations are accommodated by the use of energy normalized convolution with robust filters such as Gabor filters. For simplicity, our first experiments were based on simple mask operators as introduced in section 3 which are not necessarily invariant to light intensity changes.

To test robustness in relation to noise we completed a series of experiments with artificial and real images, where we added Gaussian noise. The impact on the histograms (measured with an appropriate distance measure) are shown in section 5.

In this article we do not consider the other degrees of freedom mentioned above: 3D rotation and light direction. In [8] we examined the robustness of the approach to image–plane rotation and view point changes (3D-rotation).

Application for Object Recognition The system we describe here is only an initial experiment to demonstrate the capabilities of the approach for object recognition. Further investigation must be performed in the use the multidimensional receptive filed histograms in a more thorough manner.

In this experimental version of the system, the database consists of histograms of each object at a set of scales and 2D orientations. A new histogram of an observed object is compared to each histogram of the database to find the closest match.

7 Experimental results

This section describes three experiments with the use of multidimensional receptive field histogram for object recognition: in the first experiments we consider scale, in the second we consider scale and image–plane rotation. In the last experiment we generalize from one single view of an object to 5 different scales.

7.1 Scale Experiment

In this section we report results from a recognition experiment with different scales of objects. We employed two series of images of 31 objects (see figure 2) at 6 different scales (approximate difference between each scale is 10%, see figure 2). The total number of images is therefore $2 \times 31 \times 6 = 372$. The first series have been used to calculate the histogram database and the second series have been used as test–set.

As mentioned above (section 6) we have different parameters in the *multidimensional receptive field histogram* approach. In this experiment we varied the local properties and the histogram comparison measurement The design parameter of the histograms have been fixed (2–dimensional with resolution of 32 cells per axis, for variation of the design parameters see [8]). For local properties, we used the same pairs of filters as in section 5. All experiments were performed with only two filters, as a minimal limiting case (in [8] we showed that recognition rates can be improved by increasing the number of local properties measured at each pixel).

Table 1 shows the recognition rate for different filter–pairs and different histogram comparison measurements. The first column of table 1 shows the filter–pairs. The first row shows the histogram comparison measurement as introduced in section 4: the two χ^2 measurements χ_T^2 and χ_{TH}^2, sum of squared differences (SSD) and the intersection measurement. The table shows a recognition rate of 100%, when we choose the filter pair magnitude and direction of gradient, and the comparison measurement χ_T^2.

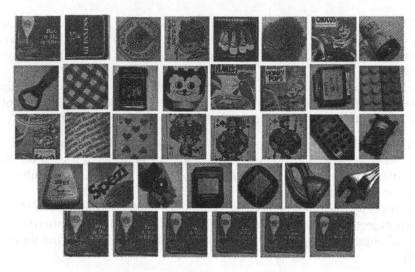

Fig. 2. Top: The 31 objects of the scale experiment. Bottom: The 6 different scales

Filter	χ^2_T	χ^2_{TH}	SSD	intersection
MD	100.0	98.9	89.8	91.4
DxDy	97.8	97.8	90.9	62.9
LD	97.3	97.3	88.7	86.0
LM	94.1	94.6	82.8	26.9
G3	86.6	86.0	64.0	43.5
G5	93.5	91.4	81.7	57.5
G7	97.8	97.8	92.4	34.4

Table 1. Recognition results with 31 Objects at 6 different scales

Following the results of table 1 we can analyze the different histogram comparison measurements: χ^2_T almost always gives the best results. χ^2_{TH} works nearly as well as χ^2_T. The SSD also gave quite good results nearly all of the time. The intersection (originally used by Swain and Ballard) give good results in some particular cases. Nevertheless the average performance over all of the filter pairs is not satisfactory. To summarize the table we can conclude that the χ^2 are the best, followed by SSD and intersection. In other experiments (e.g. section 7.3, 7.2 and [8]), we did make similar observation in relation to the histogram comparison measurements. Therefore we state that the χ^2 measurements are the best to compare *multidimensional receptive field histograms*.

7.2 2D Rotation experiment

This section presents results of an experiment where we considered the effects of 2D (image plane) rotation of objects at different changes in scale.

In this particular experiment we had 10 objects at 8 different orientations. The difference between the orientations was roughly $45°$. Furthermore we took images of each object at 5 different scales, where the difference between each scale was approximately 10%. Therefore the whole image–set contains $10 \times 8 \times 5 = 400$ images.

For the experiment we divided the image set into database and the test–set. The database consists of three different scales, respectively the first, the third and the fifth scale. Therefore $3 \times 8 \times 10 = 240$ histograms are in the database. The remaining 2 scales are then tested against the database (test–set is therefore $2 \times 8 \times 10 = 160$ histograms of images).

We can report here the effects of 2D rotation and scale changes on recognition rates (see table 2) of three filter–pairs (for the description of the abbreviations see section 7.1).

Filter	χ^2_T	χ^2_{TH}	SSD	prod	intersection
DxDy	99.4	99.4	81.3	10.0	66.9
G3	86.9	85.0	53.1	10.6	56.9
G5	88.8	87.5	54.4	18.8	19.4

Table 2. Recognition results with 10 Objects at 5 different scales and 8 different orientations

As we already concluded from the scale experiment, the χ^2 measurements give the best results (χ^2_T slightly better than χ^2_{TH}). SSD gives good results for DxDy and intersection does not give satisfactory results for any of the reported filters.

7.3 Experiment: generalizing scales from one single view

Up to now we always took images of the same object at different scales. Since this is not always practical we want to take only one image of an object and to generalize to a range of scales. This is demonstrated in a second scale experiment.

This second scale experiment uses only one image of each object at one particular scale (of the first series). Starting from this single image we calculate 5 histograms, each corresponding to a different scale of the object. Therefore we have to use "steerable" filters as Gabor filters or Gaussian derivatives. In this particular experiment we used first order Gaussian derivatives (in x and in y direction = dxdy) and the magnitude and direction of the first Gaussian derivative (= magdir) with $\sigma = 0.8, 0.9, 1.0, 1.1$ and 1.2. This was done with all 31 objects of the first experiment (see section 7.1) so that the histogram database contains $5 \times 31 = 155$ histograms. As a test–set we used the images of the 31 objects of the second series at 5 different scales. For each of those images we calculated the histogram with $\sigma = 1.0$. These histograms are then compared to the histogram database.

Filter	χ^2_T	χ^2_{TH}	SSD	intersection
magdir	99.4	100	19.4	99.4
dxdy	98.7	98.1	9.0	91.6

Table 3. Recognition results of the second scale experiment: 31 Objects at 5 different scales, where we generalized the scales from one single view of each object

Table 3 shows the results of the experiment. Once again the χ^2 give the best results. The intersection measurement gives quite good results too. This time SSD doesn't give good results at all.

This experiment shows that we can "steer" the scale, so that it is possible to calculate all considered scales from one single image of an object.

8 Conclusion and Perspective

In this paper we have shown how the color histogram matching technique of Swain and Ballard can be generalized to use vectors of local image properties measured by normalized convolution with local receptive fields. We have found that this technique present a fast and robust method to determine if a specified object is present in an image of a scene. This method can be used with very local filters for gradient and Laplacian, as well as with more noise resistant filters such as Gabor filters and Gaussian derivatives. We have demonstrated that the method is most reliable and robust when the inner product of the receptive field at each neighborhood is normalized by the energy of the neighborhood. Our experiments have also demonstrated that the χ^2 test provides the most reliable form of histogram comparison for this method.

Relatively high recognition rates have been demonstrated with vectors composed of only two receptive field. We showed in [8] that these rates can be made even higher by increasing the number of filters included in the vector. The increase in memory required can be off-set by decreasing the quantization of the histograms.

References

1. J. G. Daugman. High confidence visual recognition of persons by test of statistical independance. *IEEE PAMI*, 15(11):1148–1161, November 1993.
2. F. Ennesser and G. Medioni. Finding waldo, or focus of attention using local color information. *IEEE PAMI*, 17(8):805–809, 1995.
3. L. M. J. Florack, B. M. ter Haar Romeny, J. J. Koenderink, and M. A. Viergever. General intensity tranformations and second order invariants. In *SCIA '91*, 1991.
4. W. T. Freeman and E. H. Adelson. The design and use of steerable filters. *IEEE PAMI*, 13(9):891–906, 1991.
5. B. V. Funt and G. D. Finlayson. Color constant color indexing. *IEEE PAMI*, 17(5):522–529, 1995.
6. G. Healey and D. Slater. Using illumination invariant color histogram descriptors for recognition. In *CVPR*, pages 355–360, 1994.
7. W. H. Press, S. A. Teukolsky, W. T. Vetterling, and B. P. Flannery. *Numerical Recipes in C*. Cambridge University Press, 2nd edition, 1992.
8. B. Schiele and J. L. Crowley. The robustness of object recognition to rotation using multidimensional receptive field histograms. submitted to International Workshop on Object Representation for Computer Vision, April 1996.
9. B. Schiele and J. L. Crowley. Probabilistic object recognition using multidimensional receptive field histograms. submitted to ICPR'96, August 1996.
10. M.J. Swain and D.H. Ballard. Color indexing. *IJCV*, 7(1):11–32, 1991.
11. C.-J. Westelius. *Preattentive Gaze Control for Robot Vision*. PhD thesis, Department of Electrical Engineering, Linköping University, 1992.

Normalization by Optimization

Ralph Schiller

Technische Universität Hamburg-Harburg, 21075 Hamburg, Germany

Abstract. An approach to normalization is presented for both the affine and the projective case. The approach is based on group factorization as well as on optimizing parameter invariant integrals, in order to overcome the difficult problem of parameterization. Related work has been carried out by [6] and by [4] for affine transformations and by [5] for projective transformations. To avoid some drawbacks inherent to projective transformations it is suitable to integrate point information or explore 'thick' curves.

1 Introduction

An approach to normalization is presented mainly based on minimizing some fundamental properties of a subgroup in order to normalize objects up to this subgroup. In the projective case it is important that the normalization of the object is not a transformation to an abstract canonical frame but rather a reconstruction of the physical test pattern. This is relevant because small distortions of the object will be large distortions of the normalized object if the line which is mapped on the line at infinity is close to the object. Especially [5] had to face this problem but [2] proved the problem being inherent to projective transformations. We therefore propose centered curves which are already optimal in some way so that the normalization will be as extreme as the projective transformation of the object.

In case of affine transformations we will generalize some results of [6] and [4] in $\mathbb{R}^2$ to curves in $\mathbb{R}^n$ and surfaces in $\mathbb{R}^3$, which is necessary for stereo-graphic reconstruction. [6] were rather interested in the interpretation of 2d-images from 3d-objects and [4] emphasis was on texture so that they need not care so much for the parameterization of curves. But exactly the parameterization of curves is a specific problem for affine and projective invariant pattern recognition as the digitalization grid is rigid so that the amount of pixel coordinates of two equivalent but digitalized image contours may extremely vary. It is the very advantage of extremising low order parameter invariant integrals or sums that the parameterization problem can be neglected.

2 From Projective Transformations to Affine Transformations

The congruences and affine transformations are subgroups of the projective transformations. The first task is to normalize the non-affine part of the transformation in case we want to describe objects position invariant due to the pin-hole camera model. Therefore we remark the following factorization of projective transformations: A projective transformation can always be factorized in an affine part and a nonlinear part.

$$P : \mathbb{R}^2 \to \mathbb{R}^2 \ (x, y) \mapsto \left(\frac{x}{gx + hy + 1}, \frac{y}{gx + hy + 1} \right)$$

$$A : \mathbb{R}^2 \to \mathbb{R}^2 \ (x, y) \mapsto ((a - cg)x + (b - ch)y + c, (d - fg)x + (e - hf)y + f)$$

$$A(P(x, y)) = \left(\frac{ax + by + c}{gx + hy + 1}, \frac{ex + dy + f}{gx + hy + 1} \right)$$

Now let **P** be a convex polygon in $\mathbb{R}^2$. Due to our principle we want to extremize here some fundamental property of the affine group in order to normalize the object under projective transformations up to affine transformations. The first approach was to choose the ratio of two distances on a line. Using this fundamental invariant it is possible to define an affine arclength for a convex polygon based on intersections of tangents. The analog for continuous convex smooth curves would be to extremize the affine arclength $\int \sqrt[3]{|\, x'(s)x''(s)\,|}\,ds$. But Åström [2] has proved that projective transformations are quite powerful so that the infimum for the smooth curves would be ellipses as Blaschke [3] has proved the extremal property of ellipses with respect to affine arclength. Therefore it does not seem to be useful just to normalize a single curve. We will look at two classes of objects:

- a convex curve with a point in the interior of its bounded domain
- a convex curve with another convex curve in the interior of its bounded domain

Another fundamental property of affine transformations is that the center of mass (with respect to area) transforms consistently under affine transformations. Now it is shown that this never happens under 'pure' projective transformations and thus we can normalize curves up to affine transformations.

Theorem 1. *Let P be a convex polygon with nonempty interior. Then the only projective transforms preserving the convexity and mass center of the polygon are the affine ones (i.e. those for which the non-linear part of the projective transform is reduced to identity).*

Proof. Due to the factorization of the projective transformation in an affine and a nonlinear part, it is sufficient to verify the theorem for the nonlinear part.

So we are facing the following projective transformations:

$$P(x, y) = \left(\frac{x}{gx + hy + 1}, \frac{y}{gx + hy + 1} \right) \qquad gx_i + hy_i + 1 \geq 0 \qquad i = 0, \ldots, n$$

if (x_i, y_i) are the vertices of the polygon. The inequalities define the feasible region as a convex closed set, which has to be bounded. In this way the lines are described which do not meet the interior of the closed polygon. Particularly if we fix some parameter (g_1, h_1), then the corresponding line $G_1 = \{(x, y) : g_1 x + h_1 y + 1 = 0\}$ is mapped on the line at infinity. Let G_0 be the line through the origin parallel to G_1, then G_0 divides the convex polygon in two half-spaces, H^-, H^+, as the center of mass lies in the origin. Let H^+ be the half with points between G_0 and G_1. The line G_0 is mapped identically on itself. The polygon is convex and convex sets within the polygon are mapped on convex sets. To each point h^- on the border of the polygon in H^- corresponds a unique point h^+ on the border of the polygon in H^+ lying on a line through the origin. While the Euclidean distance of such a point h^- shrinks, the Euclidean distance of one such corresponding point h^+ expands. Therefore the origin cannot be center of mass of the transformed polygon. For that reason look without loss of generality at a line parallel to the y-axis intersecting the positive x-axis: $\{(x, y) : gx + 1 = 0\}$ $g < 0$. Then we find for a vector $(r \cos \phi, r \sin \phi)$ with $r \geq 0$ $\quad \phi \in [0, \frac{\pi}{2}] \cup [\frac{3\pi}{2}, 2\pi]$

$$\sqrt{(\frac{r \cos \phi}{gr \cos \phi + 1})^2 + (\frac{r \sin \phi}{gr \cos \phi + 1})^2} = \sqrt{(\frac{r^2}{gr \cos \phi + 1})^2} = \frac{r}{gr \cos \phi + 1}$$

As $r \cos \phi \geq 0$ and $g < 0$ and the line is admissible, the denominator lies between zero and one. If ϕ lies in the opposite angle region, the denominator will be greater than one.

If the center of mass of a convex test pattern is marked, then we are capable of computing a normalized equivalent pattern which characterizes the test pattern up to an affine transformation; for there has to be one position in which the marked point is the center of mass and so there exists one such projective transformation. If there were another transformation being not an affine equivalent one, then the above theorem would be wrong.

Frequently we know one point in the interior of a convex test pattern but it need not be the center of mass. The question is whether this point is the center of mass of a projective equivalent pattern.

Theorem 2. *Let P be a convex polygon. Then there exists at least one projective transform preserving the convexity of the polygon such that the origin will be transformed in the mass center of the transformed polygon.*

Proof. Again, only the nonlinear part of a projective transformation has to be taken into consideration, and therefore one may assume that the origin has to be mapped on itself. The single pieces of area transform in the following way:

$$\begin{vmatrix} \frac{x_k}{gx_k + hy_k + 1} & \frac{x_{k+1}}{gx_{k+1} + hy_{k+1} + 1} \\ \frac{y_k}{gx_k + hy_k + 1} & \frac{y_{k+1}}{gx_{k+1} + hy_{k+1} + 1} \end{vmatrix} = \frac{\begin{vmatrix} x_k & x_{k+1} \\ y_k & y_{k+1} \end{vmatrix}}{(gx_k + hy_k + 1)(gx_{k+1} + hy_{k+1} + 1)}$$

In order that the center of mass of the polygon and the origin coincide, two equations have to be met: (the polygon is assumed to be positively orientated)

$$0 = x_s = \sum_{k=0}^{n} \frac{\begin{vmatrix} x_k & x_{k+1} \\ y_k & y_{k+1} \end{vmatrix} (x_k(gx_{k+1} + hy_{k+1} + 1) + x_{k+1}(gx_k + hy_k + 1))}{(gx_k + hy_k + 1)^2(gx_{k+1} + hy_{k+1} + 1)^2}$$

$$0 = y_s = \sum_{k=0}^{n} \frac{\begin{vmatrix} x_k & x_{k+1} \\ y_k & y_{k+1} \end{vmatrix} (y_k(gx_{k+1} + hy_{k+1} + 1) + y_{k+1}(gx_k + hy_k + 1))}{(gx_k + hy_k + 1)^2(gx_{k+1} + hy_{k+1} + 1)^2}$$

and further the inequalities:

$$gx_i + hy_i + 1 \geq 0 \qquad i = 0, \ldots, n.$$

Both equalities may be regarded as conditions for the gradient of a function to be zero; for g we have

$$\frac{\partial}{\partial g} \frac{1}{(gx_k + hy_k + 1)(gx_{k+1} + hy_{k+1} + 1)} = \frac{-1(x_k(gx_{k+1} + hy_{k+1} + 1) + x_{k+1}(gx_k + hy_k + 1))}{(gx_k + hy_k + 1)^2(gx_{k+1} + hy_{k+1} + 1)^2}$$

and alike for h. So we find

$$F(g, h) = \sum_{k=0}^{n} \frac{-1 \begin{vmatrix} x_k & x_{k+1} \\ y_k & y_{k+1} \end{vmatrix}}{(gx_k + hy_k + 1)(gx_{k+1} + hy_{k+1} + 1)}$$

This function has a minimum and a maximum on the compact region defined by the inequalities. The single summands are negative as the areas are all positive and the denominators are positive because of the constraint inequalities. The function will decrease without limit towards the border of the compact domain. Therefore a maximum will not be attained on the border. But as there has to be a maximum, the corresponding gradient must have a zero point.

Subsequently if the affine part of the decomposition is fixed the transformation is unique because of theorem 1. However, a single point is principally hard to detect. Therefore we look at objects which may be normalized in a more robust way and which are still of some practical significance. We consider two closed convex curves: $(x(t), y(t))$ lies in the bounded open domain of $(\tilde{x}(t), \tilde{y}(t))$. Particularly we will look at curves of the following type: $(\tilde{x}(t), \tilde{y}(t)) = (lx(t), ly(t))$ $l > 1$ whose center of mass coincide. Another fundamental property of affine transformations is that area ratios are preserved. So we want to extremize the ratio of the area of the two curves in order to normalize the curves up to affine transformations.

The task is to describe the object consisting of the two convex curves projectively invariant (up to affine transformations).

$$F(g, h) = \frac{\int \frac{\tilde{x}(t)\dot{\tilde{y}}(t) - \tilde{y}(t)\dot{\tilde{x}}(t)}{(g\tilde{x}(t) + h\tilde{y}(t) + 1)^2} dt}{\int \frac{x(t)\dot{y}(t) - y(t)\dot{x}(t)}{(gx(t) + hy(t) + 1)^2} dt}$$

Consequently we face the following optimization problem with the restriction that the convexity of the objects is preserved.

$$F(g,h) \longrightarrow min \qquad g\tilde{x}(t) + h\tilde{y}(t) + 1 \geq 0$$

The feasible region is again a compact set and towards the border of the set the values of the function grow without a limit. So a minimum of the function has to lie in the interior of the set and the gradient of the function is of the following form:

$$\frac{\partial F(g,h)}{\partial g} = \frac{\int -2\tilde{x}\frac{\tilde{x}\dot{\tilde{y}}-\dot{\tilde{x}}\tilde{y}}{(g\tilde{x}+h\tilde{y}+1)^3}dt \int \frac{x\dot{y}-\dot{x}y}{(gx+hy+1)^2}dt - \int \frac{\tilde{x}\dot{\tilde{y}}-\dot{\tilde{x}}\tilde{y}}{(g\tilde{x}+h\tilde{y}+1)^2}dt \int -2x\frac{x\dot{y}-\dot{x}y}{(gx+hy+1)^3}dt}{(\int \frac{x\dot{y}-y\dot{x}}{(gx+hy+1)^2}dt)^2}$$

This implies, that the gradient of the function is zero if the centers of mass of the two curves coincide.

$$\frac{\int \tilde{x}\frac{\tilde{x}\dot{\tilde{y}}-\tilde{y}\dot{\tilde{x}}}{(g\tilde{x}+h\tilde{y}+1)^3}dt}{\int \frac{\tilde{x}\dot{\tilde{y}}-\tilde{y}\dot{\tilde{x}}}{(g\tilde{x}+h\tilde{y}+1)^2}dt} = \frac{\int x\frac{x\dot{y}-y\dot{x}}{(gx+hy+1)^3}dt}{\int \frac{x\dot{y}-y\dot{x}}{(gx+hy+1)^2}dt}$$

So if our test object consists of two curves which are centered, then the test pattern will already be an extremal object. In order to demonstrate uniqueness of the optimization problem we restrict to the particular class of objects already mentioned above: $(\tilde{x}(t), \tilde{y}(t)) = (lx(t), ly(t))$ $l > 1$ where the center of mass of $(x(t), y(t))$ lies in the origin. If we parameterize our curve $(x(t), y(t))$ with the area-parameterization [1] we will derive the following simplified function which is to be optimized:

$$F(g,h) = \frac{\int \frac{l^2}{(glx(t)+hly(t)+1)^2}dt}{\int \frac{1}{(gx(t)+hy(t)+1)^2}dt} \longrightarrow min \quad glx(t) + hly(t) + 1 \geq 0$$

We know that the parameter $(g,h) = (0,0)$ represents a local minimum as the centers of mass of our curves coincide due to their construction. Further our two curves are apparently projective equivalent. Due to theorem 1 we know that the center of mass of our curve cannot remain in the origin under the above 'pure' projective transformations unless $(g,h) = (0,0)$. The relation of our objects is $(x(t), y(t)) \sim (lx(t), ly(t))$ and the relation of the projective transformed objects is

$$\left(\frac{x(t)}{gx(t)+hy(t)+1}, \frac{y(t)}{gx(t)+hy(t)+1}\right) \sim \left(\frac{lx(t)}{glx(t)+hly(t)+1}, \frac{ly(t)}{glx(t)+hly(t)+1}\right)$$

and if there were a second minimum then the centers of mass would have to coincide. But as $l > 1$ the line $\{(x,y) : glx + hly + 1 = 0\}$ is parallel to $\{(x,y) : gx + hy + 1 = 0\}$ and closer to the object. Thus similarly as in theorem 1 the centers of mass of the transformed objects cannot coincide. So there has to be a unique minimum.

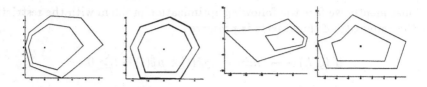

Fig. 1. Projective transformed object, normalized object up to affine transformations after some iterations. (The normalization of the non-convex curves is based on the corresponding convex curves)

3   From Affine Transformations to Congruences

Due to our principle we want to extremize some fundamental property of the congruences in order to normalize our objects under equiaffine transformations up to congruences. We therefore chose the Euclidean distance and found later on that there had been performed some research in this respect.

Already Brady and Yuille [6] have pointed out for a large class of curves $\gamma(t) = (x(t), y(t))$ that the following minimization problem has a unique solution in $\mathbb{R}^2$ up to congruences:

$$\int \| A\dot{\gamma}(t) + b \| \, dt \longrightarrow \quad min$$

under the restriction that the determinant of A is equal one. The particular problem is in this case that it is not possible to determine the parameters explicitly in general so that we are forced to use iterative algorithms. The advantage is that we need not care for the parameterization problem as the integral is parameter independent. We also used the squared distances and derived the same explicit formula as [4]. But [4] were not that interested in contours and therefore could neglect the parameterization problem which raises in this case as the integral is no longer parameter independent. Nevertheless a complete explicit formula was derived in $\mathbb{R}^2$ [9] by using Arbter's [1] area parameterization. The whole algorithm was tested with real images [9] and the object recognition results were very satisfying.

In $\mathbb{R}^3$ we prefer the ordinary Euclidean distance, as it is not so easy to parameterize our curve in this case. Therefore it will be shown, that the problem - to find among all equiaffine equivalent curves that one, with minimal Euclidean arclength - has a unique solution for a large class of point sets in $\mathbb{R}^n$ unless the object is degenerate. We want to mention that it is not so difficult to demonstrate the existence of a solution. Consider three points in $\mathbb{R}^2$ which are not collinear: $(x_1, y_1), \cdots, (x_3, y_3)$. The linear transformation with determinant one minimizing the following term $\sum_{k=1}^{n} \sqrt{(ax_k + by_k)^2 + (cx_k + dy_k)^2}$ will transform the three points in such a way that they represent the vertices of an equilateral triangle [9]. So if there are more points then there will always be a lower bound build up by an equilateral triangle. In $\mathbb{R}^3$ the same is true for a regular tetrahedron and so on.

Theorem 3. *The optimization problem has a unique solution up to congruences, unless the polygon can be embedded in a hyperplane.*

Proof. Let P_1 be a polygon: $(x_1, y_1), \cdots, (x_n, y_n)$ in $\mathbb{R}^n$ and $P_2 = AP + z$ an affine equivalent polygon. Suppose that both these polygons are optimal due to our minimization problem. According to the matrix decomposition theorem [7] we can factorize A in the following way: $A = O_1 D O_2$ where O_1 and O_2 are orthogonal matrices and D is a diagonal matrix with positive diagonal elements. Thus we redefine our polygons in the following way: $P_1 := O_2 P_1$ and $P_2 := D P_1$. P_1 and P_2 are again optimal as an orthogonal matrix does not change the Euclidean distances. So the transformation matrix between the two optimal polygons is a diagonal matrix. Now we can prove uniqueness by proving that the following restricted optimization problem has a unique solution.

$$\sum_{k=1}^{n} \sqrt{\sum_{i=1}^{m} \lambda_i^2 x_{ik}^2} \longrightarrow min \qquad \prod_{i=1}^{m} \lambda_i = 1 \qquad \lambda_i > 0$$

- The objective function:
  In order to verify that there is a unique solution we consider one single summand as a function: $\sqrt{\sum_{i=1}^{m} \lambda_i^2 x_i^2}$. This function is convex, because, if λ and μ are two vectors and $0 \leq \alpha \leq 1$ the convexity condition can be expressed in the following relation:

$$\sqrt{\sum_{i=1}^{m}(\alpha\lambda_i + (1-\alpha)\mu_i)^2 x_i^2} \leq \alpha\sqrt{\sum_{i=1}^{m}\lambda_i^2 x_i^2} + (1-\alpha)\sqrt{\sum_{i=1}^{n}\mu_i^2 x_i^2}$$

The relation is true due to the Minkowski inequality. As each single summand is a convex function and as a sum of convex functions leads to a convex function, our function has to be convex too.
If F is a convex function than the level-sets $N_c := \{x : F(x) \leq c\}$ are convex closed sets, and our special level-sets are restricted too in case none of the components k of the vectors x are identically zero; for if you assume that the level-sets are not restricted, then we could find a $\lambda \neq O$, because of the convexity of the level sets, such that for all $\alpha \geq 0$ we find $\alpha\lambda \in N_c$, i.e.

$$\sum_{k=1}^{n}\sqrt{\sum_{i=1}^{m}(\alpha\lambda_i)^2 x_{ik}^2} \leq c \Longleftrightarrow \alpha\sum_{k=1}^{n}\sqrt{\sum_{i=1}^{m}\lambda_i^2 x_{ik}^2} \leq c$$

But this can only happen if the whole expression is equal zero.
- The feasible region:
  The set $\prod_{i=1}^{m}\lambda_i \geq 1$ $\quad \lambda_i > 0$ is a convex set; for if

$$\prod_{i=1}^{m}\lambda_i \geq 1 \qquad \prod_{i=1}^{m}\mu_i \geq 1 \qquad \lambda_i > 0 \; \mu_i > 0$$

we calculate due to the concavity of the logarithm for $1 \geq \alpha \geq 0$

$$\log\left(\prod_{i=1}^{m}(\alpha\lambda_i + (1-\alpha)\mu_i)\right) = \sum_{i}^{m}\log(\alpha\lambda_i + (1-\alpha)\mu_i)$$

$$\geq \alpha\sum_{i}^{m}\log\lambda_i + (1-\alpha)\sum_{i}^{m}\log\mu_i$$

Now taking the exponential of the two sides, one gets:

$$\prod_{i=1}^{m}(\alpha\lambda_i + (1-\alpha)\mu_i) \geq (\prod_{i=1}^{m}\lambda_i)^{\alpha}(\prod_{i=1}^{m}\mu_i)^{(1-\alpha)} \geq 1$$

if the two products are greater than one. And especially due to the last inequality we find that the constraint set $\prod_{i=1}^{m}\lambda_i = 1$ $\lambda_i > 0$ is the border of a strictly convex set.

There exists a neighbourhood of the origin which does not contain a point of the feasible region as the Euclidean norm of an admissible point is always greater than one. Therefore one can find a factor such that the unit-level set, stretched by this factor, will intersect the constraint set in a unique point due to the strict convexity of the constraint set.

3.1 Line objects in $\mathbb{R}^3$

The algorithm was tested on some objects in $\mathbb{R}^3$ and each time it led to a 'unique' solution, see figure 2. The algorithm may be as well applied on objects which are build up by several curves.

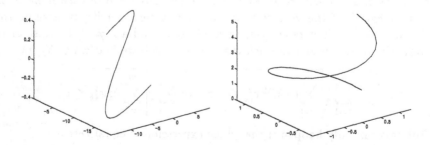

Fig. 2. Affine transformed and normalized helix

3.2 Normalization of Surfaces in $\mathbb{R}^3$

Closely related to the problem of minimizing the arclength under equiaffine transformations is the problem of minimizing the surface of an object in $\mathbb{R}^3$ under equiaffine transformations:

Let $x = (x_1, x_2, x_3)$ and $y = (y_1, y_2, y_3)$ be two vectors in $\mathbb{R}^3$. Then the crossproduct $Z = x \times y$ is the vector whose components consists of the minors of the matrix which rows are x and y. There is an important relation among a vector z, the crossproduct Z and the affine transformed vectors z^*, Z^*. For example Blaschke [3] has proved that: $z^* Z^* = zZ$. So we can easily prove that the crossproduct acts as a contravariant tensor under affine transformations: $Z^* = A^{t^{-1}} Z$. Let $\Phi(u, v) := X(u, v)i + Y(u, v)j + Z(u, v)k$ be a surface with a compact parameter domain K. The area $I(\Phi)$ of the surface Φ on K is defined by

$$\int_K |N(u, v)| d(u, v) = \int_K |\frac{\partial \Phi}{\partial u} \times \frac{\partial \Phi}{\partial v}| d(u, v).$$

Particularly we find

$$I(\Phi) = \int_K \sqrt{(\frac{\partial(Y, Z)}{\partial(u, v)})^2 + (\frac{\partial(Z, X)}{\partial(u, v)})^2 + (\frac{\partial(X, Y)}{\partial(u, v)})^2} d(u, v) \text{ with } \frac{\partial(Y, Z)}{\partial(u, v)} = \begin{vmatrix} \frac{\partial Y}{\partial u} & \frac{\partial Y}{\partial v} \\ \frac{\partial Z}{\partial u} & \frac{\partial Z}{\partial v} \end{vmatrix}$$

In [8] it is proved that the surface area as defined above is invariant under 'admissible' parameter transformations. $N(u, v)$ is the normal vector. So if we consider the problem to find one surface among all equiaffine equivalent surfaces with minimal surface area, we find due to the contravariance of the crossproduct the following problem:

$$\int_K \| A^{t^{-1}}((\frac{\partial(Y, Z)}{\partial(u, v)}), (\frac{\partial(Z, X)}{\partial(u, v)}), (\frac{\partial(X, Y)}{\partial(u, v)}))d(u, v) \| \longrightarrow min$$

with the restriction that the determinant of the matrix A is one. But as the matrices form a group we may replace the term $A^{t^{-1}}$ with A. The structure of the problem, to minimize Euclidean distances under equiaffine transformations is the same as above. Therefore this new optimization problem for surfaces will again have a unique solution.

4 Conclusion

The emphasis in our paper is on parameter invariant integrals - area in the projective case and perimeter in the affine case - in order to overcome the difficult problem of parameterization. To normalize the rotation due to our principle we used such a normalization scheme as: $\sum_{k=1}^{n}(ax_k + by_k)^2 \longrightarrow min \quad a^2 + b^2 = 1$. But we omit this passage here as the problem of parameterization is not so difficult in the case of congruences and finish with some normalized objects:

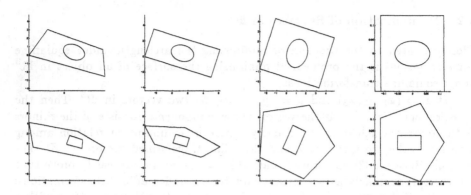

Fig. 3. a) Projective transformed object b) projective normalized object c) affine normalized object d) congruent normalized object

5 Acknowledgments

I would like to thank my supervisor Hans Burkhardt for inspiration and guidance.

References

1. Arbter, K., Snyder, W., Burkhardt, H. and Hirzinger, G.: Application of Affine-Invariant Fourier Descriptors to 3-D Objects. IEEE Trans. on Pattern Analysis and Machine Intelligence, PAMI-12(7):640-647, Juli 1990
2. Åström, K.: Fundamental difficulties with projective normalization of planar curves. Second Joint European - US Workshop Ponta Delgada, Azores, Portugal, October 1993 Proceedings
3. Blaschke, W.: Vorlesungen über Differentialgeometrie und goemetrische Grundlagen von Einsteins Relativitätstheorie. Berlin Verlag von Julius Springer 1923
4. Blake, A., Marinos, C.: Shape form texture: Estimation, isotropy and moments. Artificial Intell., vol. 45, pp. 323-380, 1990
5. Sinclair, D., Blake, A.: Isoperimetric Normalization of Planar Curves. IEEE Transactions on Pattern Analysis and Machine Intelligence, Vol. 16, No. 8, August 1994
6. Brady, M., Yuille, A.: An extremum principle for shape from contour. IEEE Transactions on Pattern Analysis and Machine Intelligence, vol PAMI-6, 1984
7. Fischer, G.: Analytische Geometrie. Wiesbaden: Vieweg, 1985
8. Heuser, H.: Lehrbuch der Analysis, Teil 2, Teubner
9. Schiller, R.: Distance geometric approach to recognize a set of finitely many points. Interner Bericht 10/94, Technische Informatik I, TU-HH, Juli 1994.

Extracting Curvilinear Structures:
A Differential Geometric Approach

Carsten Steger

Forschungsgruppe Bildverstehen (FG BV)
Informatik IX, Technische Universität München
Orleansstr. 34, 81667 München, Germany
E-mail: stegerc@informatik.tu-muenchen.de

Abstract. In this paper a method to extract curvilinear structures from digital images is presented. The approach is based on differential geometric properties of the image function. For each pixel, the second order Taylor polynomial is computed by convolving the image with the derivatives of a Gaussian smoothing kernel. Line points are required to have a vanishing gradient and a high curvature in the direction perpendicular to the line. The use of the Taylor polynomial and the Gaussian kernels leads to a single response of the filter to each line. Furthermore, the line position can be determined with sub-pixel accuracy. Finally, the algorithm scales to lines of arbitrary width. An analysis about the scale-space behaviour of two typical line types (parabolic and bar-shaped) is given. From this analysis, requirements and useful values for the parameters of the filter can be derived. Additionally, an algorithm to link the individual line points into lines and junctions that preserves the maximum number of line points is given. Examples on aerial images of different resolution illustrate the versatility of the presented approach.

1   Introduction

Extracting lines in digital images is an important low-level operation in computer vision that has many applications, especially in photogrammetric and remote sensing tasks. There it can be used to extract linear features, like roads, railroads, or rivers, from satellite or low resolution aerial imagery.

The published schemes to line detection can be classified into three categories. The first approach detects lines by only considering the gray values of the image [4, 8]. Line points are extracted by using purely local criteria, e.g., local gray value differences. Since this will generate a lot of false hypotheses for line points, elaborate and computationally expensive perceptual grouping schemes have to be used to select salient lines in the image [5, 8]. Furthermore, lines cannot be extracted with sub-pixel accuracy.

The second approach is to regard lines as objects having parallel edges [9, 11]. In a first step, the local direction of a line is determined for each pixel. Then two edge detection filters are applied in the direction perpendicular to the line. Each edge detection filter is tuned to detect either the left or right edge of the line. The responses of each filter are combined in a non-linear way to yield the final response of the operator [9]. The advantage of this approach is that since the edge detection filters are based on the derivatives of Gaussian kernels, the procedure can be iterated over the scale-space parameter σ to detect lines of arbitrary widths. However, because special directional edge detection filters have to be constructed that are not separable, the approach is computationally expensive.

In the third approach, the image is regarded as a function $z(x, y)$ and lines are detected as ridges and ravines in this function by locally approximating the image function by its second or third order Taylor polynomial. The coefficients of this polynomial are usually determined by using the facet model, i.e., by a least squares fit of the polynomial to the image data over a window of a certain size [6, 1, 7]. The direction of the line is determined from the Hessian matrix of the Taylor polynomial. Line points are then found by selecting pixels that have a high second directional derivative, i.e., a high curvature, perpendicular to the line direction. The advantage of this approach is that lines can be detected with sub-pixel accuracy without having to construct specialized directional filters. However, because the convolution masks that are used to determine the coefficients of the Taylor polynomial are rather poor estimators for the first and second partial derivatives this approach usually leads to multiple responses to a single line, especially when masks larger than 5×5 are used to suppress noise. Therefore, the approach does not scale well and cannot be used to detect lines that are wider than about 5 pixels.

In this paper an approach to line detection that uses the differential geometric approach of the third category of operators will be presented. In contrast to those, the coefficients of a second order Taylor polynomial are determined by convolving the image with the derivatives of a Gaussian smoothing kernel. Because of this, the algorithm can be scaled to lines of arbitrary width. Furthermore, the behaviour of the algorithm in scale space is investigated for various types of lines. Finally, an algorithm to link the detected line points into a topologically sound data structure of lines and junctions is presented.

2 Detection of Line Points

2.1 Models for Lines in 1D

Many approaches to line detection consider lines in 1D to be bar-shaped, i.e., the ideal line of width $2w$ and height h is assumed to have a profile given by

$$f_b(x) = \begin{cases} h, & |x| \le w \\ 0, & |x| > w \end{cases} . \tag{1}$$

However, due to sampling effects of the sensor lines usually do not have this profile. Figure 1 shows a profile of a line in an aerial image. As can be seen, no flat bar profile is apparent. Therefore, in this paper lines are assumed to have an approximately parabolic profile. The ideal line of width $2w$ and height h is then given by

$$f_p(x) = \begin{cases} h(1 - (x/w)^2), & |x| \le w \\ 0, & |x| > w \end{cases} . \tag{2}$$

The line detection algorithm will be developed for this type of profile, but the implications of applying it to bar-shaped lines will be considered later on.

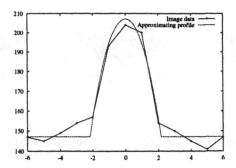

Fig. 1. Profile of a line in an aerial image and approximating ideal line profile

2.2 Detection of Lines in 1D

In order to detect lines with a profile given by (2) in an image $z(x)$ without noise, it is sufficient to determine the points where $z'(x)$ vanishes. However, it is usually convenient to select only salient lines. A useful criterion for salient lines is the magnitude of the second derivative $z''(x)$ in the point where $z'(x) = 0$. Bright lines on a dark background will have $z''(x) \ll 0$ while dark lines on a bright background will have $z''(x) \gg 0$.

Real images will contain a significant amount of noise. Therefore, the scheme described above is not sufficient. In this case, the first and second derivatives of $z(x)$ should be estimated by convolving the image with the derivatives of the Gaussian smoothing kernel

$$g_\sigma(x) = \frac{1}{\sqrt{2\pi}\sigma} e^{-\frac{x^2}{2\sigma^2}} \ . \tag{3}$$

The responses, i.e., the estimated derivatives, will be:

$$
\begin{aligned}
r_p(x, \sigma, w, h) &= g_\sigma(x) * f_p(x) \\
&= \frac{h}{w^2} \big((w^2 - x^2 - \sigma^2)(\phi_\sigma(x + w) - \phi_\sigma(x - w)) - \\
&\quad 2\sigma^2 x (g_\sigma(x + w) - g_\sigma(x - w)) - \\
&\quad \sigma^4 (g'_\sigma(x + w) - g'_\sigma(x - w)) \big)
\end{aligned}
\tag{4}
$$

$$
\begin{aligned}
r'_p(x, \sigma, w, h) &= g'_\sigma(x) * f_p(x) \\
&= \frac{h}{w^2} \big(-2x(\phi_\sigma(x + w) - \phi_\sigma(x - w)) + \\
&\quad (w^2 - x^2 - 3\sigma^2)(g_\sigma(x + w) - g_\sigma(x - w)) - \\
&\quad 2\sigma^2 x (g'_\sigma(x + w) - g'_\sigma(x - w)) - \\
&\quad \sigma^4 (g''_\sigma(x + w) - g''_\sigma(x - w)) \big)
\end{aligned}
\tag{5}
$$

$$
\begin{aligned}
r''_p(x, \sigma, w, h) &= g''_\sigma(x) * f_p(x) \\
&= \frac{h}{w^2} \big(-2(\phi_\sigma(x + w) - \phi_\sigma(x - w)) -
\end{aligned}
$$

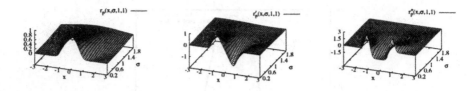

Fig. 2. Scale-space behaviour of the ideal line f_p when convolved with the derivatives of Gaussian kernels for $x \in [-3, 3]$ and $\sigma \in [0.2, 2]$

$$4x(g_\sigma(x+w) - g_\sigma(x-w)) +$$
$$(w^2 - x^2 - 5\sigma^2)(g'_\sigma(x+w) - g'_\sigma(x-w)) -$$
$$2\sigma^2 x(g''_\sigma(x+w) - g''_\sigma(x-w)) -$$
$$\sigma^4(g'''_\sigma(x+w) - g'''_\sigma(x-w))) \qquad (6)$$

where

$$\phi_\sigma(x) = \int_{-\infty}^{x} e^{-\frac{t^2}{2\sigma^2}} dt . \qquad (7)$$

Equations (4)–(6) give a complete scale-space description of how the ideal line profile f_p will look like when it is convolved with the derivatives of Gaussian kernels. Figure 2 shows the responses for an ideal line with $w = 1$ and $h = 1$ (i.e., a bright line on a dark background) for $x \in [-3, 3]$ and $\sigma \in [0.2, 2]$. As can be seen from this figure, $r'_p(x, \sigma, w, h) = 0 \Leftrightarrow x = 0$ for all σ. Furthermore, $r''_p(x, \sigma, w, h)$ takes on its maximum negative value at $x = 0$ for all σ. Hence it is possible to determine the precise location of the line for all σ. Furthermore, it can be seen that because of the smoothing the ideal line will be flattened out as σ increases. This means that if large values for σ are used, the threshold to select salient lines will have to be set to an accordingly smaller value. Section 4 will give an example of how this can be used in practice to select appropriate thresholds.

For a bar profile without noise no simple criterion that depends only on $z'(x)$ and $z''(x)$ can be given since $z'(x)$ and $z''(x)$ vanish in the interval $[-w, w]$. However, if the bar profile is convolved with the derivatives of the Gaussian kernel, a smooth function is obtained in each case. The responses will be:

$$r_b(x, \sigma, w, h) = h(\phi_\sigma(x+w) - \phi_\sigma(x-w)) \qquad (8)$$
$$r'_b(x, \sigma, w, h) = h(g_\sigma(x+w) - g_\sigma(x-w)) \qquad (9)$$
$$r''_b(x, \sigma, w, h) = h(g'_\sigma(x+w) - g'_\sigma(x-w)) . \qquad (10)$$

Figure 3 shows the scale-space behaviour of a bar profile with $w = 1$ and $h = 1$ when it is convolved with the derivatives of a Gaussian. It can be seen that the bar profile gradually becomes "round" at its corners. The first derivative will vanish only at $x = 0$

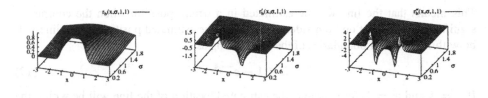

Fig. 3. Scale-space behaviour of the bar-shaped line f_b when convolved with the derivatives of Gaussian kernels for $x \in [-3, 3]$ and $\sigma \in [0.2, 2]$

for all $\sigma > 0$ because of the infinite support of $g_\sigma(x)$. However, the second derivative $r_b''(x, \sigma, w, h)$ will not take on its maximum negative value for small σ. In fact, for $\sigma \leq 0.2$ it will be approximately zero. Furthermore, there will be two distinct minima in the interval $[-w, w]$. It is, however, desirable for $r_b''(x, \sigma, w, h)$ to exhibit a clearly defined minimum at $x = 0$. After some lengthy calculations it can be shown that

$$\sigma \geq w/\sqrt{3} \tag{11}$$

has to hold for this. Furthermore, it can be shown that $r_b''(x, \sigma, w, h)$ will have its maximum negative response in scale-space for $\sigma = w/\sqrt{3}$. This means that the same scheme as described above can be used to detect bar-shaped lines as well. However, the restriction on σ must be observed. The same analysis could be carried out for other types of lines as well, e.g., roof-shaped lines. However, it is expected that no fundamentally different results will be obtained. For all σ above a certain value that depends on the line type the responses will show the desired behaviour of $z'(0) = 0$ and $z''(0) \ll 0$ with $z''(x)$ having a distinct minimum.

The discussion so far has assumed that lines have the same contrast on both sides of the line. This is rarely true in real images, however. For simplicity, only asymetrical bar-shaped lines

$$f_a(x) = \begin{cases} 0, & x < -w \\ 1, & |x| \leq w \\ h, & x > w \end{cases} \tag{12}$$

are considered ($h \in [0, 1]$). The corresponding responses will be:

$$r_a(x, \sigma, w, h) = \phi_\sigma(x + w) - (h - 1)\phi_\sigma(x - w) \tag{13}$$
$$r_a'(x, \sigma, w, h) = g_\sigma(x + w) - (h - 1)g_\sigma(x - w) \tag{14}$$
$$r_a''(x, \sigma, w, h) = g_\sigma'(x + w) - (h - 1)g_\sigma'(x - w) . \tag{15}$$

The location where $r_a'(x, \sigma, w, h) = 0$, i.e., the position of the line, is given by

$$x = -\frac{\sigma^2}{2} \ln(1 - h) . \tag{16}$$

This means that the line will be estimated in a wrong position when the contrast is significantly different on both sides of the line. The estimated position of the line will be within the actual boundaries of the line as long as

$$h \leq 1 - e^{-\frac{2w}{\sigma^2}} .$$ (17)

If $\sigma = 1$ and $w = 1$, for example, the estimated location of the line will be within the actual line if $h \leq 0.86466$. This means that relatively large contrast differences can be handled. Note, however, that as $h \to 1$, i.e., as the bar line profile is gradually transformed into a step edge profile, the location of the line $x \to \infty$. Fortunately, $r_a''(x, \sigma, w, h)$ will have a small value as $h \to 1$, so by simple thresholding these erroneously located lines can be eliminated.

2.3 Lines in 1D, Discrete Case

The analysis so far has been carried out for analytical functions $z(x)$. For discrete signals only two modifications have to be made. The first one is the choice of how to implement the convolution in discrete space. Integrated Gaussian kernels were chosen as convolutions masks, mainly because they give automatic normalization of the masks and a direct criterion on how many coefficients are needed for a given approximation error. The integrated Gaussian is obtained if one regards the discrete image z_n as a piecewise constant function $z(x) = z_n$ for $x \in (n - \frac{1}{2}, n + \frac{1}{2}]$ and integrating the continuous Gaussian kernel over this area. The convolution masks will be given by:

$$g_{n,\sigma} = \phi_\sigma(n + \tfrac{1}{2}) - \phi_\sigma(n - \tfrac{1}{2})$$ (18)

$$g_{n,\sigma}' = g_\sigma(n + \tfrac{1}{2}) - g_\sigma(n - \tfrac{1}{2})$$ (19)

$$g_{n,\sigma}'' = g_\sigma'(n + \tfrac{1}{2}) - g_\sigma'(n - \tfrac{1}{2}) .$$ (20)

The approximation error is set to 10^{-4} in each case. Of course, other schemes, like Lindeberg's discrete Gaussian derivative approximations [10] or a recursive computation [3], are suitable for the implementation as well.

The second problem that has to be solved is how to determine the location of a line in the discrete case. In principle, one could use a zero crossing detector for this task. However, this would yield the position of the line only with pixel accuracy. In order to overcome this, the second order Taylor polynomial of z_n is examined. Let r, r', and r'' be the locally estimated derivatives at point n of the image that are obtained by convolving the image with g_n, g_n', and g_n''. Then the Taylor polynomial is given by

$$p(x) = r + r'x + \frac{1}{2}r''x^2 .$$ (21)

The position of the line, i.e., the point where $p'(x) = 0$ is

$$x = -\frac{r'}{r''} .$$ (22)

The point n is declared a line point if this position falls within the pixel's boundaries, i.e., if $x \in [-\frac{1}{2}, \frac{1}{2}]$ and the second derivative r'' is larger than a user-specified threshold. Please note that in order to extract lines, the response r, which is the smoothed local image intensity, is unnecessary and therefore does not need to be computed.

2.4 Detection of Lines in 2D

Curvilinear structures in 2D can be modeled as curves $s(t)$ that exhibit a characteristic 1D line profile (e.g., f_p or f_b) in the direction perpendicular to the line, i.e., perpendicular to $s'(t)$. Let this direction be $n(t)$. This means that the first directional derivative in the direction $n(t)$ should vanish and the second directional derivative should be of large absolute value. No assumption can be made about the derivatives in the direction of $s'(t)$. For example, let $z(x,y)$ be an image that results from sweeping the profile f_p along a circle $s(t)$ of radius r. The second directional derivative perpendicular to $s'(t)$ will have a large negative value, as desired. However, the second directional derivative along $s'(t)$ will also be non-zero.

The only problem that remains is to compute the direction of the line locally for each image point. In order to do this, the partial derivatives r_x, r_y, r_{xx}, r_{xy}, and r_{yy} of the image will have to be estimated. This can be done by convolving the image with the appropriate 2D Gaussian kernels. The direction in which the second directional derivative of $z(x,y)$ takes on its maximum absolute value will be used as the direction $n(t)$. This direction can be determined by calculating the eigenvalues and eigenvectors of the Hessian matrix

$$H(x,y) = \begin{pmatrix} r_{xx} & r_{xy} \\ r_{xy} & r_{yy} \end{pmatrix} . \tag{23}$$

The calculation can be done in a numerically stable and efficient way by using one Jacobi rotation to annihilate the r_{xy} term. Let the eigenvector corresponding to the eigenvalue of maximum absolute value, i.e., the direction perpendicular to the line, be given by (n_x, n_y) with $\|(n_x, n_y)\|_2 = 1$. As in the 1D case, a quadratic polynomial will be used to determine whether the first directional derivative along (n_x, n_y) vanishes within the current pixel. This point will be given by

$$(p_x, p_y) = (tn_x, tn_y) , \tag{24}$$

where

$$t = -\frac{r_x n_x + r_y n_y}{r_{xx} n_x^2 + 2r_{xy} n_x n_y + r_{yy} n_y^2} . \tag{25}$$

Again, $(p_x, p_y) \in [-\frac{1}{2}, \frac{1}{2}] \times [-\frac{1}{2}, \frac{1}{2}]$ is required in order for a point to be declared a line point. As in the 1D case, the second directional derivative along (n_x, n_y), i.e., the maximum eigenvalue, can be used to select salient lines.

2.5 Examples

Figure 4(b) gives an example of the results obtainable with the presented approach. Here, bright line points were extracted from the input image given in Fig. 4(a). This image is part of an aerial image with a ground resolution of 2 m. Figure 4(c) shows the results that were obtained using the facet model. In both cases the sub-pixel location (p_x, p_y) of the line points and the direction (n_x, n_y) perpendicular to the line are symbolized by vectors. The strength of the line, i.e., the absolute value of the second directional derivative along (n_x, n_y) is symbolized by gray values. Line points with high saliency have dark gray values.

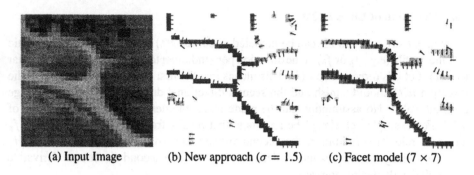

| (a) Input Image | (b) New approach ($\sigma = 1.5$) | (c) Facet model (7×7) |

Fig. 4. Line points detected in image (a) using the new approach (b) and using the facet model (c)

From Fig. 4 it can be seen that in the approach presented here there will always be a single response to a given line. When the facet model is used, multiple responses are quite common. Note, for example, the line that enters the image in the middle of the left hand side. This makes linking the individual line points into lines rather complicated. In [1] the response of the operator is thinned before linking to get around this problem. However, this operation throws away useful information since diagonal lines will be thinned unnecessarily. In the new approach the linking will be considerably easier and no thinning operation is needed.

3 Linking Line Points into Lines

After individual line pixels have been extracted, they must be linked into lines. In order to facilitate later mid-level vision processes, e.g., perceptual grouping, the resulting data structure should contain explicit information about the lines as well as the junctions between them. This data structure should be topologically sound in the sense that junctions are represented by points and not by extended areas as in [1]. Furthermore, since the presented approach yields only single responses to each line, no thinning operation needs to be performed prior to linking. This assures that the maximum information about the line points will be present in the data structure.

Since there is no suitable criterion to classify the line points into junctions and normal line points in advance without having to resort to extended junction areas, another approach has been adopted. From the algorithm in Sect. 2 the following data are obtained for each pixel: the orientation of the line $(n_x, n_y) = (\cos \alpha, \sin \alpha)$, a measure of strength of the line (the second directional derivative in the direction of α), and the sub-pixel location of the line (p_x, p_y).

Starting from the pixel with maximum second derivative, lines will be constructed by adding the appropriate neighbour to the current line. Since it can be assumed that the line point detection algorithm will yield a fairly accurate estimate for the local direction of the line, only three neighbouring pixels that are compatible with this direction are examined. For example, if the current pixel is (c_x, c_y) and the current orientation of the line is in the interval $[-22.5°, 22.5°]$, these points will be $(c_x + 1, c_y - 1)$,

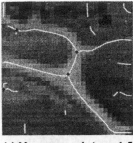 

(a) New approach ($\sigma = 1.5$)   (b) Facet model (7×7)

Fig. 5. Linked lines detected using the new approach (a) and using the facet model (b). Lines are drawn in white while junctions are displayed as black crosses.

$(c_x + 1, c_y)$, and $(c_x + 1, c_y + 1)$. The choice about the appropriate neighbour to add to the line is based on the distance between the respective sub-pixel locations and the angle difference of the two points. Let $d = \|p_2 - p_1\|_2$ be the distance between the two points and $\beta = |\alpha_2 - \alpha_1|$, $\beta \in [0, \pi/2]$, be the angle difference between those points. The neighbour that is added to the line is the one that minimizes $d + w\beta$. In the current implementation, $w = 1$ is used. This algorithm will select each line point in the correct order. At junction points, it will select one branch to follow without detecting the junction. This will be detected later on. The algorithm of adding line points is continued until no more line points are found in the current neighbourhood or until the best matching candidate is a point that has already been added to another line. If this happens, the point is marked as a junction, and the line that contains the point is split into two lines at the junction point.

New lines will be created as long as the starting point has a second directional derivative that lies above a certain, user-selectable upper threshold. Points are added to the current line as long as their second directional derivative is greater than another user-selectable lower threshold. This is similar to a hysteresis threshold operation [2].

The contour linking approach presented here is similar to that given in [6]. However, there the best neighbour is determined from a neighbourhood that does not depend on the current direction of the line. Furthermore, the author does not mention whether explicit junction information is generated by the algorithm.

With a slight modification the algorithm is able to deal with multiple responses if it is assumed that with the facet model approach no more than three parallel responses are generated. No such case has been encountered for mask sizes of up to 13×13. Under this assumption, the algorithm can proceed as above. Additionally, if there are multiple responses to the line in the direction perpendicular to the line (e.g., the pixels $(c_x, c_y - 1)$ and $(c_x, c_y + 1)$ in the example above), they are marked as processed if they have roughly the same orientation as (c_x, c_y). The termination criterion for lines has to be modified to stop at processed line points instead of line points that are contained in another line.

Figure 5 shows the result of linking the line points in Fig. 4 into lines. The results

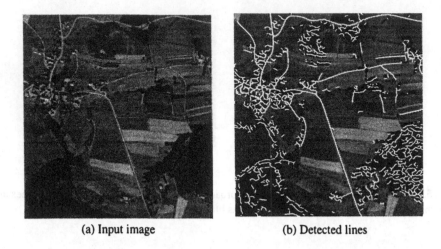

(a) Input image   (b) Detected lines

Fig. 6. Salient lines detected in the complete aerial image

are overlaid onto the original image. In this case, the upper threshold was set to zero, i.e., all lines, no matter how faint, were selected. If an upper threshold of 5 were used only the salient lines would be selected. It is apparent that the lines obtained with the new approach are much smoother than the lines obtained with the facet model. Furthermore, the geometric accuracy in case of unequal contrast is better with the new approach. Note, for example, the line that enters the image at the bottom right corner. This line has quite a different contrast on both sides. With the new approach the line is within half a pixel of the true location of the line while with the facet model it lies more than one pixel from the true line.

4 Further Examples

In this section some more examples of the versatility of the proposed approach will be given. Figure 6(a) shows the complete aerial image from which the image in Fig. 4 was taken. In this example, $\sigma = 1.5$ and only bright lines that had a second derivative with an absolute value larger than 8 were selected. The lower threshold for the hysteresis was set to 3. It can be seen from Fig. 6(b) that the algorithm is able to extract most of the salient lines from the image.

Figure 7 shows that the presented approach scales very well. In Fig. 7(a) an aerial image with a ground resolution of ≈ 25 cm is displayed. The lines in this image are approximately bar-shaped. If 50 pixel wide lines are to be detected, i.e., if $w = 25$, according to (11), a $\sigma \geq 14.4337$ should be selected. In fact, $\sigma = 15$ was used for this image. If lines with a contrast of $h \geq 70$ are to be selected, (10) shows that these lines will have a second derivative of ≈ -0.10316. Therefore, the threshold for the absolute value of the second derivative was set to 0.1. The lower threshold was set to 0.025. Figure 7(b) displays the lines that were detected with these parameters. As can be seen,

640

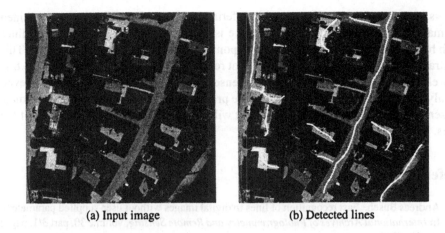

(a) Input image   (b) Detected lines

Fig. 7. Salient lines detected in a high resolution aerial image

all of the roads were detected. Most of the lines in this image have different contrasts on both sides of the line. Therefore it is not surprising that the detected lines deviate slightly from the true centers of the lines. This is especially true for the line in the bottom right corner of the image. However, even this line is detected within the boundaries of the actual line.

5 Conclusions

In this paper a low-level approach to the extraction of curvilinear structures from images was presented. An analysis of the scale-space behaviour of two distinct line types was carried out. The results of this analysis help tremendously in the selection of the appropriate parameters for the algorithm. The advantages of this approach are that line extraction is done using only the first and second directional derivatives of the image. No specialized directional filters are needed. This makes the approach computationally efficient. For instance, the 520×560 image in Fig. 6 was processed in 8 seconds on a HP 735 workstation. Furthermore, since the derivatives are estimated by convolving the image with the derivatives of a Gaussian smoothing kernel, only a single response is generated for each line. The algorithm has no problems extracting line points where three or more lines meet.

An algorithm has been presented that links the extracted line points into a data structure containing lines and junctions. Although the algorithm itself does not attempt any perceptual grouping, the data structure that is generated will facilitate this in a higher-level step.

The presented approach shows two fundamental limitations. Firstly, if a line has highly different contrasts on each side oft the line, the position of the line will be estimated in a different position than the actual center of the line. This is a fundamental limitation of other approaches as well [9, 1]. In this paper, an analysis was carried out that shows how the position will vary with differing contrasts. Secondly, only a combined estimate of the width and height of the line is returned. This means, that narrow lines with high contrast will result in similar responses as broad lines with low contrast. This contrasts with the approach given in [9] that returns an estimate of the width of the line as well as the height of the line at the expense of computational complexity. However, if only lines of a certain range of widths are present in an image, the combined estimate presents no fundamental limitation since it will then depend only on the contrast of the lines.

References

1. Andreas Busch. Fast recognition of lines in digital images without user-supplied parameters. In *International Archives of Photogrammetry and Remote Sensing*, volume 30, part 3/1, pages 91–97, 1994.
2. John Canny. A computational approach to edge detection. *IEEE Transactions on Pattern Analysis and Machine Intelligence*, 8(6):679–698, 1986.
3. Rachid Deriche. Recursively implementing the gaussian and its derivatives. Rapport de Recherche 1893, INRIA, Sophia Antipolis, France, April 1993.
4. M. A. Fischler, J. M. Tenenbaum, and H. C. Wolf. Detection of roads and linear structures in low-resolution aerial imagery using a multisource knowledge integration technique. *Computer Graphics and Image Processing*, 15:201–223, 1981.
5. Martin A. Fischler. The perception of linear structure: A generic linker. In *Image Understanding Workshop*, pages 1565–1579, San Francisco, CA, USA, 1994. Morgan Kaufmann Publishers.
6. Frank Glazer. Curve finding by ridge detection and grouping. In W. Kropatsch and H. Bischof, editors, *Mustererkennung*, Informatik Xpress 5, pages 109–116. Deutsche Arbeitsgemeinschaft für Mustererkennung, 1994.
7. Robert M. Haralick, Layne T. Watson, and Thomas J. Laffey. The topographic primal sketch. *International Journal of Robotics Research*, 2(1):50–72, 1983.
8. Bruno Jedynak and Jean-Philippe Rozé. Tracking roads in satellite images by playing twenty questions. In A. Gruen, O. Kuebler, and P. Agouris, editors, *Automatic Extraction of Man-Made Objects from Aerial and Space Images*, pages 243–253, Basel, Switzerland, 1995. Birkhäuser Verlag.
9. T. M. Koller, G. Gerig, G. Székely, and D. Dettwiler. Multiscale detection of curvilinear structures in 2-D and 3-D image data. In *Fifth International Conference on Computer Vision*, pages 864–869. IEEE Computer Society Press, 1995.
10. Tony Lindeberg. Discrete derivative approximations with scale-space properties: A basis for low-level feature extraction. *Journal of Mathematical Imaging and Vision*, 3(4):349–376, 1993.
11. J. Brian Subirana-Vilanova and Kah Kay Sung. Multi-scale vector-ridge-detection for perceptual organization without edges. A.I. Memo 1318, MIT Artificial Intelligence Laboratory, Cambridge, MA, USA, December 1992.

Affine / Photometric Invariants
for Planar Intensity Patterns

Luc Van Gool[1], Theo Moons[1]* and Dorin Ungureanu[1]

Katholieke Universiteit Leuven, ESAT – MI2,
K. Mercierlaan 94, B–3001 Heverlee, BELGIUM

Abstract. The paper contributes to the viewpoint invariant recognition of planar patterns, especially labels and signs under affine deformations. By their nature, the information of such 'eye-catchers' is not contained in the outline or frame — they often are affinely equivalent like parallelograms and ellipses — but in the intensity content within. Moment invariants are well suited for their recognition. They need a closed bounding contour, but this is comparatively easy to provide for the simple shapes considered. On the other hand, they characterize the intensity patterns without the need for error prone feature extraction. This paper uses moments as the basic features, but extends the literature in two respects: (1) deliberate mixes of different types of moments to keep the order of the moments (and hence also the sensitivity to noise) low and yet have a sufficiently large number to safeguard discriminant power; and (2) invariance with respect to photometric changes is incorporated in order to find the simplest moment invariants that can cope with changing lighting conditions which can hardly be avoided when changing viewpoint. The paper gives complete classifications of such affine / photometric moment invariants. Experiments are described that illustrate the use of some of them.

1 Introduction

A lot of research has been put into the extraction of invariants for planar shapes under geometrical deformations [5, 6]. Most work has focused on the shapes' contours. For certain applications, however, it would be more effective if one could use invariants derived from the intensity patterns bounded by the contour. For example, if one is to recognize labels or traffic signs, the contours will typically contain little information. If affine distortions are considered, many of them will be affinely equivalent (e.g. all parallelograms or all ellipses). Contour invariants will be difficult to apply to many of the patterns and therefore moment invariants are considered. Affine distortions are the most general type of geometric transformations that can be considered in this case, because expressions of moments that are projectively invariant do not exist [12]. Note the natural complementarity: with complex outlines contour invariants can be used and it would be

---

* Postdoctoral Research Fellow of the Belgian National Fund for Scientific Research (N.F.W.O.).

difficult to extract closed contours as required by the moments; in the case of non-discriminant outlines of the forementioned, simple parametric types closed contours can be fitted rather easily for the calculation of the moments, whereas contour invariants are of little use.

The goal being invariant characterisation of the intensity patterns, care has to be taken of photometric changes as well. When the camera changes its position relative to the pattern, its intensities will in general change. This paper contains a complete and systematic classification of invariants for the combined effect of affine deformations and photometric changes. A second concern is to keep the order of the moments low, since high orders introduce more noise. To that effect, the mixing of different kinds of moments and the combination of coplanar patterns are considered. Different kinds of moments have been defined (geometric, Legendre, Zernike, rotational, etc.). The reported work is based on the traditional "geometric" moments and only moments up to the second order are considered. These choices are in keeping with the results of noise sensitivity tests [11].

Of course, the paper adds to a large body of literature on moments and moment invariants. It is impossible to give a complete overview here (for a partial review, see Prokop and Reeves [8]). Nevertheless, there are a number of contributions which are directly related to the presented work. Maitra [4] and Abo-Zaid et al. [1] discussed variations of Hu's metric and scaling moment invariants [3] that are also invariant under global scaling of the intensity. Another strand of research has concentrated on deriving moment invariants under affine transformations [2, 9, 10]. Reiss [9] combined affine and photometric invariance and his work comes closest to that reported. Nevertheless, most of the moment invariants that are given here are novel.

The paper is organized as follows: first, notation and terminology is established in Section 2. In Section 3 the affine / photometric invariants are systematically classified according to the highest order of the moments involved. For each case, invariants under affine transformations, photometric invariants, as well as combined affine / photometric invariants are given. Section 4 then reports on ongoing experiments with these moment invariants. Section 5 summarizes the results.

2 Affine / Photometric Transformations

2.1 Definitions and Notations

If x and y stand for the row and column coordinates of an image with intensities $i(x, y)$ and a planar object's closed contour C circumscribes the region Ω in the image, then

$$MS_{Cpq} = \iint_\Omega x^p y^q \, dx \, dy \qquad \text{and} \qquad MI_{Cpq} = \iint_\Omega i(x, y) x^p y^q \, dx \, dy$$

are the *shape (p, q)-moment* and *intensity (p, q)-moment* resp.. Both are said to be of order $p + q$. For brevity, in the sequel *"nth-order moments"* stand for the set of moments up to and including the nth-order moments. If M_{Cpq} rather than MS_{Cpq} or MI_{Cpq} is specified, this is to mean that the corresponding expression can be used with either of them. Notation often will be simplified to MS_{pq} and MI_{pq} if there is only one contour involved.

The *moment invariants* discussed in the paper are deliberately made function of both shape moments and intensity moments. This mix lowers the required order of the moments and thus contributes to their robustness. With "*n th-order moment invariants*" will be meant moment invariants that combine moments up to order n.

2.2 Geometric and Photometric Changes

The geometric deformations considered are affine transformations

$$\begin{pmatrix} x' \\ y' \end{pmatrix} = \begin{pmatrix} a_{11} & a_{12} \\ a_{21} & a_{22} \end{pmatrix} \begin{pmatrix} x \\ y \end{pmatrix} + \begin{pmatrix} b_1 \\ b_2 \end{pmatrix} \tag{1}$$

with $|A| = a_{11}a_{22} - a_{12}a_{21} \neq 0$. This implies that the camera is relatively far from the object.

Two kinds of photometric changes are considered: *pure scaling*

$$I'(x,y) = sI(x,y) \;\; ;$$

and *scaling combined with an offset*

$$I'(x,y) = sI(x,y) + o \;.$$

The assumptions that the camera is relatively far from the object and that the object is planar can greatly simplify the analysis of the photometric changes. Typically light sources are far from the objects as well. The geometry of light reflection is the same for all points in that case, i.e. they share the same angles of light incidence and camera viewing direction. Also for the more sophisticated models of diffuse reflection the change in camera or light position will in that case result in an overall scaling of intensity [7]. The offset allows to better model the combined effect of diffuse and specular reflection [14] and has been found to give better performance [9].

The actions of the photometric and affine changes on the moments come out to commute. Hence, one might first normalize against one type of transformation and then against the other. Alternatively, one may normalize against one and switch to the use of invariants for the other. To some extent, this latter strategy typically is what has happened in the literature. The photometric offset can e.g. be eliminated through the use of intensity minus average intensity and the photometric scale parameter can be eliminated by normalizing the resulting intensity's variance [9]. After these normalisations one then has to deal with affine deformations exclusively. The resulting affine invariants may look simpler than the ones given here, but the inherent complexity is at least comparable. However, the normalisation steps are quite expensive computationally, since they require a pixel-wise modification. Not normalising means that one has to deal with larger numbers (original intensities and coordinates instead of deviations from average values), but this is outweighed by far by only using lower order moments.

2.3 Geometric and Photometric Effects on Moments

As already mentioned before, only moments up to 2nd order will be considered. How do these transformations affect the value of the moments? The affine transformation (1) changes the vector

$$\mathbf{m}^t = (M_{20} \; M_{11} \; M_{02} \; M_{10} \; M_{01} \; M_{00})$$

of second-order moments as

$$\mathbf{m}' = \mathbf{T}_m \mathbf{m} \tag{2}$$

where

$$\mathbf{T}_m = |A| \begin{pmatrix} a_{11}^2 & 2a_{11}a_{12} & a_{12}^2 & 2a_{11}b_1 & 2a_{12}b_1 & b_1^2 \\ a_{11}a_{21} & a_{11}a_{22} + a_{12}a_{21} & a_{12}a_{22} & a_{11}b_2 + a_{21}b_1 & a_{12}b_2 + a_{22}b_1 & b_1b_2 \\ a_{21}^2 & 2a_{21}a_{22} & a_{22}^2 & 2a_{21}b_2 & 2a_{22}b_2 & b_2^2 \\ 0 & 0 & 0 & a_{11} & a_{12} & b_1 \\ 0 & 0 & 0 & a_{21} & a_{22} & b_2 \\ 0 & 0 & 0 & 0 & 0 & 1 \end{pmatrix}$$

with $|A| = a_{11}a_{22} - a_{12}a_{21}$ as before. As can be seen from the transformation matrix, the 0th-order moment M_{00}, as well as the 1st-order moments (including M_{00}) can also be considered in isolation. Observe that shape and intensity moments transform in exactly the same way under affine transformations.

If the model of the photometric changes is restricted to *pure scaling*, then

$$M I'_{ij} = s M I_{ij} \qquad \text{and} \qquad M S'_{ij} = M S_{ij} \; .$$

The shape moments are trivially invariant, because they do not involve intensities. To obtain photometric intensity moment invariants, it suffices to take the ratio of two intensity moment invariants. Shape and intensity moments can thus be used separately.

If the intensity changes by *a scale factor s and an offset o*, then

$$M'_{Iij} = s M_{Iij} + o M_{Sij} \qquad \text{and} \qquad M'_{Sij} = M_{Sij} \; .$$

The result is a 2-dimensional group of transformations acting on the vectors $(M I_{ij}, M S_{ij})^T$. Again the shape moments are trivially invariant. However, the intensity moments can no longer be used in isolation. Observe that the photometric changes act identically on all intensity moments, irrespective of their order.

The actions of the photometric and the geometric transformations commute. As a consequence, the overall group of affine / photometric transformations is a direct product of the affine group and the group of intensity transformations. This implies that combined affine / photometric invariants exist if the number of moments surpasses the sum of the orbit dimensions of both actions taken separately, and that the invariants are found as common expressions in the sets of affine and photometric moment invariants separately. A Lie group theoretical strategy was used to classify all independent affine / photometric moment invariants up to the second order. This classification is given in the next section. Space restrictions do not allow to discuss the generation of these invariants. The interested reader is referred to [13] for details on the classification strategy.

3 Affine / Photometric Moment Invariants

This section classifies the independent moment invariants up to the second order. For each order, affine, photometric (for both models), and combined affine / photometric invariants are given, which are mixtures of both shape and intensity moments. If the invariants only involve shape moments, combinations of different patterns (different bounding contours) are considered, until an invariant involving an intensity moment is found, because the assumption is that the useful information is contained in the intensity pattern. In practice, additional contours can be generated by making an invariant subdivision of the pattern under consideration. Two examples are given in Fig. 1. An elliptical pattern can be divided into the original ellipse and an ellipse with the same center of gravity, orientation and eccentricity, but having half the size along each axis. Similarly, a parallelogram shaped pattern can be subdivided into e.g. the whole pattern and the pattern inside the diamond that emerges by connecting the midpoints on each side.

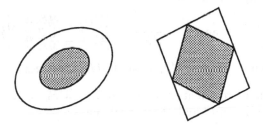

Fig. 1. *Invariant subdivisions of an ellipse and a parallelogram.*

3.1 Zeroth-order Moment Invariants

Affine Invariants : MI_{00} / MS_{00}'

Photometric Invariants.

Photometric scaling : MS_{C00}; MS_{D00} and MI_{C00}/MI_{D00}

Photometric scaling + offset : MS_{C00}; MS_{D00}; MS_{E00},

and $\frac{MI_{C00}MS_{D00}-MI_{D00}MS_{C00}}{MI_{C00}MS_{E00}-MI_{E00}MS_{C00}}$

Affine / Photometric Invariants.

Photometric scaling : MI_{C00}/MI_{D00} and MS_{C00}/MS_{D00}

Photometric scaling + offset : MS_{C00}/MS_{D00}; MS_{C00}/MS_{E00},

and $\frac{MI_{C00}MS_{D00}-MI_{D00}MS_{C00}}{MI_{C00}MS_{E00}-MI_{E00}MS_{C00}}$

3.2   First-order Moment Invariants

Affine Invariants : $MI_{C00}\,/\,MS_{C00}$; $MS_{D00}\,/\,MS_{C00}$,

$$\text{and} \qquad \frac{1}{MS_{C00}} \begin{vmatrix} \frac{MI_{C10}}{MI_{C00}} - \frac{MS_{C10}}{MS_{C00}} & \frac{MI_{C10}}{MI_{C00}} - \frac{MS_{D10}}{MS_{D00}} \\ \frac{MI_{C01}}{MI_{C00}} - \frac{MS_{C01}}{MS_{C00}} & \frac{MI_{C01}}{MI_{C00}} - \frac{MS_{D01}}{MS_{D00}} \end{vmatrix} \qquad (3)$$

Photometric Invariants.

Photometric scaling : MI_{10}/MI_{00}, MI_{01}/MI_{00}, MS_{10}, MS_{01}, MS_{00}

Photometric scaling + offset : MS_{10}, MS_{01}, MS_{00}, and $\frac{MI_{10}MS_{00}-MI_{00}MS_{10}}{MI_{01}MS_{00}-MI_{00}MS_{01}}$

Affine / Photometric Invariants.

Photometric scaling :

3 shape moments : MS_{C00}/MS_{D00}, and MS_{C00}/MS_{E00}

1 intensity + 2 shape moments : MS_{C00}/MS_{D00}, and (3)

2 intensity + 1 shape moments :

$$MI_{C00}/MI_{D00} \qquad \text{and} \qquad \frac{1}{MS_{C00}} \begin{vmatrix} \frac{MI_{C10}}{MI_{C00}} - \frac{MS_{C10}}{MS_{C00}} & \frac{MI_{C10}}{MI_{C00}} - \frac{MI_{D10}}{MI_{D00}} \\ \frac{MI_{C01}}{MI_{C00}} - \frac{MS_{C01}}{MS_{C00}} & \frac{MI_{C01}}{MI_{C00}} - \frac{MI_{D01}}{MI_{D00}} \end{vmatrix}$$

3 intensity moments : MI_{C00}/MI_{D00}, and MI_{C00}/MI_{E00}

2 intensity + 2 shape moments : MI_{C00}/MI_{D00}, MS_{C00}/MS_{D00}, (3),
   (3) with the rôles of C and D reversed, and

$$\begin{vmatrix} \frac{MI_{C10}}{MI_{C00}} - \frac{MS_{C10}}{MS_{C00}} & \frac{MI_{D10}}{MI_{D00}} - \frac{MS_{C10}}{MS_{C00}} \\ \frac{MI_{C01}}{MI_{C00}} - \frac{MS_{C01}}{MS_{C00}} & \frac{MI_{D01}}{MI_{D00}} - \frac{MS_{C01}}{MS_{C00}} \end{vmatrix} : \begin{vmatrix} \frac{MI_{C10}}{MI_{C00}} - \frac{MS_{C10}}{MS_{C00}} & \frac{MS_{D10}}{MS_{D00}} - \frac{MS_{C10}}{MS_{C00}} \\ \frac{MI_{C01}}{MI_{C00}} - \frac{MS_{C01}}{MS_{C00}} & \frac{MS_{D01}}{MS_{D00}} - \frac{MS_{C01}}{MS_{C00}} \end{vmatrix}$$

Photometric scaling + offset : MS_{C00}/MS_{D00},

$$\frac{MI_{C00}}{(MI_{C00}MS_{D00} - MI_{D00}MS_{C00})} \begin{vmatrix} \frac{MI_{C10}}{MI_{C00}} - \frac{MS_{C10}}{MS_{C00}} & \frac{MS_{C10}}{MS_{C00}} - \frac{MS_{D10}}{MS_{D00}} \\ \frac{MI_{C01}}{MI_{C00}} - \frac{MS_{C01}}{MS_{C00}} & \frac{MS_{C01}}{MS_{C00}} - \frac{MS_{D01}}{MS_{D00}} \end{vmatrix} \qquad (4)$$

(4) with the rôles of C and D reversed, and

$$\frac{MI_{C00}MI_{D00}MS_{D00}}{(MI_{C00}MS_{D00} - MI_{D00}MS_{C00})^2} \begin{vmatrix} \frac{MI_{C10}}{MI_{C00}} - \frac{MS_{C10}}{MS_{C00}} & \frac{MI_{D10}}{MI_{D00}} - \frac{MS_{D10}}{MS_{D00}} \\ \frac{MI_{C01}}{MI_{C00}} - \frac{MS_{C01}}{MS_{C00}} & \frac{MI_{D01}}{MI_{D00}} - \frac{MS_{D01}}{MS_{D00}} \end{vmatrix}$$

3.3 Second-order Moment Invariants

Affine Invariants :

2nd order moments of 1 type only :

$$\frac{(M_{20}M_{00} - M_{10}^2)(M_{02}M_{00} - M_{01}^2) - (M_{11}M_{00} - M_{10}M_{01})^2}{M_{00}^6} \tag{5}$$

2nd order moments + 0th order moment of other type : $M I_{00}/M S_{00}$, and (5)

2nd order intensity moments + 1th order shape moments :
$M I_{00}/M S_{00}$, (5) for intensity moments, and

$$\frac{\beta\epsilon^2 - 2\alpha\delta\epsilon + \gamma\delta^2}{M S_{00}^8} \tag{6}$$

where

$$\alpha = (M I_{11} M I_{00} - M I_{10} M I_{01}) \ , \qquad \delta = (M S_{00} M I_{10} - M S_{10} M I_{00}) \ ,$$
$$\beta = (M I_{20} M I_{00} - M I_{10}^2) \ , \qquad \epsilon = (M S_{00} M I_{01} - M S_{01} M I_{00}) \ ,$$
$$\gamma = (M I_{02} M I_{00} - M I_{01}^2) \ .$$

2nd order intensity moments + 2nd order shape moments :
$M I_{00}/M S_{00}$, (5) for intensity moments, (5) for shape moments,
(6), (6) with the rôles of intensity and shape moments reversed, and

$$\frac{(\alpha A - \gamma B)^2 + (\alpha A - \beta C)^2 + 2(\beta A - \alpha B)(\gamma A - \alpha C)}{M I_{00}^6 M S_{00}^6} \tag{7}$$

where

$$A = (M S_{11} M S_{00} - M S_{10} M S_{01}) \ , \quad B = (M S_{20} M S_{00} - M S_{10}^2) \ ,$$
$$\text{and} \quad C = (M S_{02} M S_{00} - M S_{01}^2) \ .$$

Photometric Invariants.

Photometric scaling : $M I_{10}/M_{00}, M I_{01}/M_{00}, M I_{20}/M_{00}, M I_{11}/M_{00}, M I_{02}/M_{00}$

Photometric scaling + offset : $PI(00, 10, 01), PI(00, 10, 20), PI(00, 10, 11),$
$PI(00, 10, 02), PI(10, 01, 20), PI(10, 01, 11), PI(10, 01, 02), PI(01, 20, 11),$
$PI(01, 20, 02),$ and $PI(20, 11, 02),$ where

$$PI(ij, kl, mn) = \begin{vmatrix} M_{Iij} & M_{Ikl} \\ M_{Sij} & M_{Skl} \end{vmatrix} : \begin{vmatrix} M_{Iij} & M_{Imn} \\ M_{Sij} & M_{Smn} \end{vmatrix}$$

Affine / Photometric Invariants.

Photometric scaling :

2nd order intensity moments + 0th order shape moment :

$$\frac{(M I_{20} M I_{00} - M I_{10}^2)(M I_{02} M I_{00} - M I_{01}^2) - (M I_{11} M I_{00} - M I_{10} M I_{01})^2}{M I_{00}^4 M S_{00}^2} \tag{8}$$

2nd order intensity moments + 1th order shape moment : (8), and

$$\frac{\beta\epsilon^2 - 2\alpha\delta\epsilon + \gamma\delta^2}{M I_{00}^4 M S_{00}^2} \tag{9}$$

with the same notations as in (6).

2nd order intensity moments + 2nd order shape moment : (8), (9),
  (5) for shape moments, (9) with the rôles of intensity and shape moments
  reversed, and

$$\frac{(\alpha A - \gamma B)^2 + (\alpha A - \beta C)^2 + 2(\beta A - \alpha B)(\gamma A - \alpha C)}{M I_{00}^4 M S_{00}^8}$$

with the same notations as in (7).

Photometric scaling + offset : the simplest (of the 4 existing) invariant(s) is

$$\frac{M S_{00}^4 \left(\bar{\beta}\bar{\epsilon}^2 - 2\bar{\alpha}\bar{\delta}\bar{\epsilon} + \bar{\gamma}\bar{\delta}^2\right)}{(\bar{\alpha} A - \bar{\gamma} B)^2 + (\bar{\alpha} A - \bar{\beta} C)^2 + 2(\bar{\beta} A - \bar{\alpha} B)(\bar{\gamma} A - \bar{\alpha} C)}$$

with A, B, C as defined earlier for eq. (7) and

$$\bar{\alpha} = (M I_{10} M S_{00} - M S_{10} M I_{00})(M I_{01} M S_{00} - M S_{01} M I_{00}) \ ,$$
$$\bar{\beta} = (M I_{10} M S_{00} - M S_{10} M I_{00})^2 \qquad \bar{\delta} = M S_{00}(M I_{10} M S_{00} - M S_{10} M I_{00})$$
$$\bar{\gamma} = (M I_{01} M S_{00} - M S_{01} M I_{00})^2 \qquad \bar{\epsilon} = M S_{00}(M I_{01} M S_{00} - M S_{01} M I_{00})$$

4 Experiments

Due to space restrictions, only one experiment will be discussed here. Other
experiments are given in [13]. The goal of this experiment is to test the discriminatory power of invariant (4). To this end, its value is computed for images
of postcards. Fig. 2 shows four of the postcards used in the test. To test the
invariance of the expression, each of the images is transformed mathematically
for several combinations of affine transfomations as well as photometric scaling
and offset. The original image and three of its transformed versions are shown in
Fig. 3. The intensity pattern of the postcard is delineated by an parallellogram
fitting program. In the experiment, the invariant (4) is calculated for the original
image and its transformed versions. Since this invariant needs two contours C
and D, the postcards's outline is taken as one contour and a parallellogram corresponding to one half of the postcards is used for the second contour. As there
are 4 possible choices for this second part, which cannot be distinguished in an
affine context. Therefore, (4) is computed for each of the four possibilities and
their values are summed. The results of this computation are given in Table 1
for the four postcards shown in Fig. 2 and their corresponding transformed versions as the one shown in Fig. 3. Clearly, the values of this symmetric invariant
expression are quite stable over the different transformations, whereas the value
is significantly different for the different postcards.

Fig. 2. *Four postcards used in the test.*

Fig. 3. *The original and three transformed versions of the first postcard in Fig. 2.*

5 Conclusions

The central theme of the paper is the generation and classification of simple and robust moment invariants by (1) the deliberate mixing of intensity and shape moments, and (2) the provision for joint invariance under geometric (affine) deformations and photometric changes,

The mixing of intensity and shape moments has a number of advantages. First, it is a *necessary* condition for the use of intensity moments in combination with offsets in the photometric changes, a fact that does not seem to have been made explicit in the literature. Without the shape moments no group action would result. Second, mixing both moment types *as a deliberate strategy* leads to their optimal use and simpler moment invariants, whereas their occasional mixing in the past seems to have been rather *ad hoc* and thereby not exploited to the full.

Table 1. *Values of the invariant (4) for the postcards in Fig. 2 and their transformed versions as in Fig. 3.*

	original	transform 1	transform 2	transform 3
postcard 1	0.224	0.230	0.225	0.210
postcard 2	-1.818	-1.814	-1.850	-1.840
postcard 3	0.590	0.587	0.592	0.560
postcard 4	0.125	0.130	0.121	0.133

It has been observed that the more discriminating moment invariants may be of higher order than the second. Thus, there is a conflict between robustness against noise and discriminant power. As the experiments have shown, the lower order invariants proposed here are effective, especially since there are more of them available now.

References

1. A. Abo-Zaid, O. Hinton, and E. Horne, About moment normalisation and complex moment descriptors, *Proc. 4th International Conference on Pattern Recognition,* pp.399–407, 1988.
2. J. Flusser and T. Suk, Pattern recognition by affine moment invariants, *Pattern Recognition,* Vol. 26, No. 1, pp.167–174, 1993.
3. M. Hu, Visual pattern recognition by moment invariants, *IRE Trans. Information Theory,* IT-8, pp.179–187, 1962.
4. S. Maitra, Moment invariants, *Proc. IEEE,* Vol.67, pp.697-699, 1979.
5. J.L. Mundy and A. Zisserman (eds.), *Geometric Invariance in Computer Vision,* MIT Press, Cambridge, Massachusetts, 1992.
6. J.L. Mundy, A. Zisserman, and D. Forsyth (eds.), *Applications of Invariance in Computer Vision,* Lecture Notes in Computer Science, Vol. **825,** pp. 89–106, Springer-Verlag, Berlin / Heidelberg / New York / Tokyo, 1994.
7. M. Oren and S. Nayar, Seeing beyond Lambert's law, *Proc. European Conference on Computer Vision,* pp. 269–280, 1994.
8. R. Prokop and A. Reeves, A survey of moment-based techniques for unoccluded object representation and recognition, *CVGIP: Models and Image Processing,* Vol.54, No.5, pp. 438–460, 1992.
9. T. Reiss, *Recognizing planar objects using invariant image features,* Lecture Notes in Computer Science, Vol. **676,** pp. 89–106, Springer-Verlag, Berlin / Heidelberg / New York / Tokyo, 1993.
10. H. Schulz-Mirbach, *Anwendung von Invarianzprinzipien zur Merkmalgewinnung in der Mustererkennung,* VDI Verlag, 1995.
11. C.-H. Teh and R. Chin, On image analysis by the methods of moments, *IEEE Transactions on Pattern Analysis and Machine Intelligence,* Vol. 10, No. 4, pp. 496–513, 1988.
12. L. Van Gool, T. Moons, E. Pauwels and A. Oosterlinck, Vision and Lie's approach to invariance, *Image and Vision Computing,* Vol. 13 (1995), no. 4, pp. 259–277.
13. L. Van Gool, T. Moons, and D. Ungureanu, *Affine/photometric invariants for planar intensity patterns,* K.U.Leuven Technical Report KUL/ESAT/MI2/9511, Katholieke Universiteit Leuven, Belgium, 1995.
14. L. Wolff, On the relative brightness of specular and diffuse reflection, *Proc. Computer Vision and Pattern Recognition,* pp.369–376, 1994.

Image Synthesis from a Single Example Image

Thomas Vetter[1] and Tomaso Poggio[2]

[1] Max–Planck–Institut für biologische Kybernetik,
Spemannstr.38 D–72076 Tübingen, Germany
Email: *vetter@mpik-tueb.mpg.de*
[2] Center for Biological and Computational Learning
M.I.T. Cambridge, MA 02139 U.S.A.
Email: *poggio@ai.mit.edu*

Abstract. The need to generate new views of a 3D object from a single real image arises in several fields, including graphics and object recognition. While the traditional approach relies on the use of 3D models, we exploit 2D image transformations that are specific to the relevant object class and learnable from example views of other "prototypical" objects of the same class.

For *linear object classes* we show that linear transformations can be learned exactly from a basis set of 2D prototypical views. We demonstrate the approach on artificial objects and then show preliminary evidence that the technique can effectively "rotate" high-resolution face images from a single 2D view.

Index Items – 3D Object recognition, rotation invariance, deformable models, image synthesis

1 Introduction

View-based approaches to 3D object recognition and graphics may avoid the explicit use of 3D models by exploiting the memory of several views of the object and the ability to interpolate or generalize among them. In many situations however a sufficient number of views may not be available. In an extreme case we may have to do this with only one real view. Consider for instance the problem of recognizing a specific human face under a different pose or expression when only one example picture is given. Our visual system is certainly able to perform this task [1]. The obvious explanation is that we exploit prior information about how face images transform, learned through extensive experience with other faces. Thus the idea (see [2]), is to learn class-specific image-plane transformations from examples of objects of the same class and then to apply them to the real image of the new object in order to synthesize virtual views.

The work described in this paper is based on the idea of *linear object classes*. These are 3D objects whose 3D shape can be represented as a linear combination of a sufficiently small number of prototypical objects. Linear object classes have the properties that new orthographic views of any object of the class under *uniform affine 3D transformations*, and in particular rigid transformations in

3D, can be generated exactly if the corresponding transformed views are known for the set of prototypes. Thus if the training set consist of frontal and rotated views of a set of prototype faces, any rotated view of a new face can be generated from a single frontal view – provided that the linear class assumption holds.

Key to our approach is a representation of an object view in terms of a *shape vector* and a *texture vector* (see also Jones and Poggio [3] and Beymer and Poggio [4]). The first gives the image-plane coordinates of feature points of the object surface; the second provides their color or grey-level. Assuming correspondence, we will represent an image as follows: we code its 2D-shape as the deformation field of selected feature points – each corresponding in the limit to a pixel – from a reference image in the same pose, which serves as the origin of our coordinate system. The texture is coded as the intensity map of the image with feature points set in correspondence with the reference image. Thus each component of the shape and the texture vector refers to the same feature point e.g. pixel. Notice, we do *not* need correspondence between different poses as in the parallel deformation technique of Poggio and Brunelli [5] and Beymer et al.[6].

2 Linear Object Classes

2.1 Shape of 3D objects

Consider a 3D view of an three-dimensional object, which is defined in terms of pointwise features [2]. A 3D view can be represented by a vector $\mathbf{X} = (x_1, y_1, z_1, x_2, \ldots, y_n, z_n)^T$, that is, by the x, y, z-coordinates of its n feature points (we factor out translation). Assume that $\mathbf{X} \in \Re^{3n}$ is the linear combination of q 3D views $\mathbf{X}_i$ of *other* objects of the same dimensionality, such that:

$$\mathbf{X} = \sum_{i=1}^{q} \alpha_i \mathbf{X}_i. \tag{1}$$

$\mathbf{X}$ is then the linear combination of q vectors in a $3n$ dimensional space, each vector representing an object of n pointwise features. Consider now the linear operator L associated with a desired uniform transformation such as for instance a specific rotation in 3D. Let us define $\mathbf{X}^r = L\mathbf{X}$ the rotated 3D view of object $\mathbf{X}$. Because of the linearity of the group of uniform linear transformations $\mathcal{L}$, it follows that

$$\mathbf{X}^r = \sum_{i=1}^{q} \alpha_i \mathbf{X}_i^r \tag{2}$$

Thus, *if a 3D view of an object can be represented as the weighted sum of views of other objects, its rotated view is a linear combination of the rotated views of the other objects with the same weights.* Of course for an arbitrary 2D view that is a projection of a 3D view, a decomposition like (1) does not in general imply a decomposition of the rotated 2D views (it is a necessary but not a sufficient condition).

2D projections of 3D objects

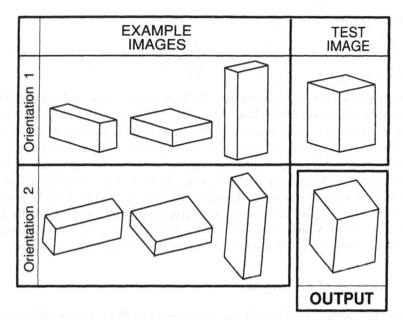

Fig. 1. *Learning an image transformation according to a rotation of three-dimensional cuboids from one orientation (upper row) to a new orientation (lower row). The 'test' cuboid (upper row right) can be represented as a linear combination of the two-dimensional coordinates of the three example cuboids in the upper row. The linear combination of the three example views in the lower row, using the coefficients evaluated in the upper row, results in the correct transformed view of the test cuboid as output (lower row right). Notice that correspondence between views in the two different orientations is not needed and different points of the object may be occluded in the different orientations.*

The question we want to answer here is, "Under which conditions the 2D projections of 3D objects satisfy equation (1) to (2)?" The answer will clearly depend on the types of objects we use and also on the projections we allow. We define:

A set of 3D views (of objects) $\{\mathbf{X}_i\}$ *is a* **linear object class** *under a linear projection P if* $dim\{\mathbf{X}_i\} = dim\{\mathbf{PX}_i\}$ *with* $\mathbf{X}_i \in \Re^{3n}$ *and* $\mathbf{PX}_i \in \Re^p$ *and* $p < 3n$

Note that the linear projection P is not restricted to projections from $3D$ to $2D$, but may also "drop" occluded points. Now assume $\mathbf{x} = P\mathbf{X}$ and $\mathbf{x}_i = P\mathbf{X}_i$ being the projections of elements of a linear object class with

$$\mathbf{x} = \sum_{i=1}^{q} \alpha_i \mathbf{x}_i. \qquad (3)$$

Then $\mathbf{x}^r = P\mathbf{X}^r$ can be constructed without knowing $\mathbf{X}^r$ using α_i of equation

(3) and the given $\mathbf{x}_i^r = P\mathbf{X}_i^r$ of the other objects

$$\mathbf{x}^r = \sum_{i=1}^{q} \alpha_i \mathbf{x}_i^r. \tag{4}$$

Figure 1 shows a very simple example of a linear object class and the construction of a new view of an object. Taking the 8 corners of a cuboid as features, a 3D view $\mathbf{X}$, as defined above, is an element of $\Re^{24}$; however, the dimension of the class of all cuboids is only 3, so any cuboid can be represented as a linear combination of three cuboids. For any projection, that preserve these 3 dimensions, we can apply equations (3) and (4). The projection in figure 1 projects all non occluded corners orthographically onto the image-plane ($\mathbf{x} = P\mathbf{X} \in \Re^{14}$) preserving the dimensionality. Notice that the orthographic projection of an exactly frontal view of a cuboid, which would result in a rectangle as image, would preserve 2 dimensions only, so equation (4) could not guarantee the correct result.

2.2 Texture of 3D objects

In this section we extend our linear space model from a representation based on the location of feature points only to textured feature points. The texture of a 3D object can be represented by a vector $\mathbf{t} \in \Re^n$, that is, by the texture values of its n feature points. Since the texture or image irradiance of an object is in general a complex function of albedo, surface orientation and the direction of illumination, we have to distinguish different situations.

Let us first consider the easy case of objects all with the same identical texture: corresponding feature points of different objects have the same intensity or color. In this situation a single texture map (e.g. the reference image) is sufficient for a whole object class.

Next consider the situation in which the texture is a function of *albedo* only, that is independent of the surface normal. Then a linear texture class can be formulated in a way equivalent to equations (1) through (4). Assuming $\mathbf{t}_i$ being the texture maps of corresponding feature points of a class of objects. Then equation

$$\mathbf{t} = \sum_{i=1}^{q} \beta_i \mathbf{t}_i \tag{5}$$

with β_i (different to α_i in equation (3)) implies for the "rotated" textures

$$\mathbf{t}^r = \sum_{i=1}^{q} \beta_i \mathbf{t}_i^r, \tag{6}$$

assuming that the texture is independent of the surface orientation and the projection does not change the dimensionality of the texture space. In an application the coefficients α_i for the shape and coefficients β_i for the texture can be computed separately. In face recognition experiments [4] the coefficients β_i were already used as a rotation invariant representation of textures of faces.

3 An Implementation for Grey-Level Pixel Images

We applied the linear class idea to grey-level images of human faces, each given in two orientations (22.5° and 0°) in a resolution of 256-by-256 pixels and 8 bit.

3.1 Computation of the correspondence

Our method assumes correspondence, and thus we have to find corresponding feature points in the two images and the associated spatial difference vector. For each prototype image separately, we computed the correspondence relative to a reference image, that is we found for every pixel location in the reference image, e.g. a pixel at the tip of the nose, the corresponding pixel location in the other images. This is in general a hard problem but here we need correspondence only between prototypical objects shown in the same pose, that is only between quite similar images. This makes it feasible to compute correspondence with automatic techniques such as optical flow algorithms. We use a coarse-to-fine gradient-based gradient method [7] and follow an implementation described in [8]. For every point x, y in an image I, the error term $E = \sum(I_x \delta x + I_y \delta y - \delta I)^2$ is minimized for $\delta x, \delta y$, with I_x, I_y being the spatial image derivatives and δI the difference of intensity of the two compared images. The final result of this computation $(\delta x, \delta y)$ is used as an approximation of the spatial displacement of a pixel from one image to the other. For an image of n-by-n pixels these displacements represent a correspondence field $\Delta \mathbf{x}$ of the image to a reference image with $\Delta \mathbf{x} \in \Re^{2n^2}$.

3.2 Projection of a test image into a linear shape and texture space

First we compute the correspondence field $\Delta \mathbf{x}$ of the new test image to the reference image (which is in the same pose). Using the correspondence fields $\Delta \mathbf{x}_i$ of the other example images to the reference image and following equation (3), we computed the coefficients α_i by minimizing $||\Delta \mathbf{x} - \sum_{i=1}^{q} \alpha_i \Delta \mathbf{x}_i||^2$ with a SVD algorithm. We solved equation (5) for the coefficients β_i assuming that the texture is a function of the albedo only. As a preliminary step the textures of all prototypical objects were mapped along the correspondence fields onto the reference image, so that all corresponding pixels in the prototypical images were mapped to the same pixel location in the reference image.

3.3 Synthesis of the new image.

The final step is image rendering. We compute the new texture and the new correspondence fields applying the computed coefficients β_i and α_i to the examples given in the second orientation (equations (6) and (4)). The new image is then generated combining this texture and correspondence field. This is possible because both are given in the coordinates of the reference image. That means that for every pixel in the reference image we know the pixel texture value and

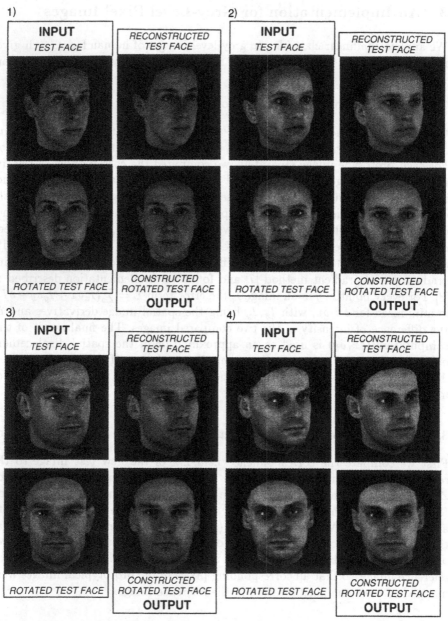

Fig. 2. *Four examples of artificially rotated human faces. Each test face (upper left) is "rotated" using 49 different faces (not shown) as examples, the results are marked as output. For comparison, the "true" rotated test face is shown on the lower left (this face was not used in the computation). The difference between synthetic and real rotated face is due to the incomplete example set, since the same difference can already be seen in the reconstruction of the input test face using the 49 example faces (upper right).*

the vector pointing to the new location. The new location generally does not coincide with the equally spaced grid of pixels of the destination image. A commonly used solution of this problem is known as forward warping [9], which for every new pixel, uses the nearest three points to linearly approximate the pixel intensity.

4 Is the Linear Class Assumption Valid for Human Faces?

For man made objects, which often consist of cuboids, cylinders or other geometric primitives, the assumption of linear object classes seems natural. However, in the case of faces it is not clear how many example faces are necessary to synthesize any other face and in fact, it is unclear if the assumption of a linear class is appropriate at all. The key test for the linear class hypothesis in this case is how well the synthesized rotated face approximates the "true" rotated face. We tested our approach on a small set of 50 faces, each given in two orientations (22.5° and 0°). Figure 2 shows four tests using the technique described earlier. In each case one face was selected as test face and the 49 remaining faces were used as prototypical examples. Each test face is shown on the upper left and the output image produced by our technique on the lower right, showing a rotated test face. The true rotated test face from the data base (not used in the computation) is shown on the lower left. We also show in the upper right the synthesis of the test face through the 49 example faces in the test orientation. This reconstruction of the test face should be understood as the projection of the test face into the shape and texture space of the other 49 example faces. A perfect reconstruction of the test face would be a necessary (not sufficient!) requirement for the claim that the 50 faces are a linear object class. Overall, the results are not perfect but, considering the small size of the sample set, the reconstruction is quite good. In our experiments, even in the cases were human observers judge the synthetic faces most different from the original face images, the synthetic images look like human faces and show clear characteristics of the target faces. The similarity of the reconstruction to the input test face suggests that an example set size of the order of hundred faces may be sufficient to construct a huge variety of different faces. We conclude that the linear object class approach may be a satisfactory approximation even for complex objects such as faces. On the other hand, it is also obvious that the reconstruction of every specific mole or wrinkle in a face requires an almost infinite number of examples. To overcome this problem, correspondence between images taken from different viewpoints could be used to map the specific texture on the new pose [3, 4].

5 Discussion

Linear combinations of linear drawing images of a single object have been already successfully used to create a new image of the same object [10]. Here we created a new grey-level image of an object using linear combinations of images of different objects of the same class. Given only a single image of an object, we are able to generate additional synthetic images of this object under the assumption that the "linear class" property holds. This is demonstrated not only for objects purely defined through their shape but also for smooth objects with texture.

This approach based on two-dimensional models does not need any depth information, so the sometime difficult step of generating three-dimensional models from two-dimensional images is superfluous. Since no correspondence is necessary between images representing objects in different poses and orientations, fully automated algorithms can be applied for the correspondence finding step. For object recognition tasks our approach has several implications. Our technique can provide additional artificial example images of an object when only a single image is given. Alternatively, the coefficients, which result from a decomposition of shape and texture into the prototypical shapes and textures, give us directly a representation of the object which is invariant under any 3D affine transformation and which may be used for recognition.

References

1. N. Troje and H. Bülthoff, "Face recognition under varying pose: The role of texture and shape," *Vision Research in press.*
2. T. Poggio and T. Vetter, "Recognition and structure from one 2D model view: observations on prototypes, object classes, and symmetries," A.I. Memo No. 1347, Artificial Intelligence Laboratory, Massachusetts Institute of Technology, 1992.
3. M. Jones and T. Poggio, "Model-based matching of line drawings by linear combination of prototypes," in *ICCV'95*, 1995.
4. D. Beymer and T. Poggio, "Face recognition from one model view," in *ICCV'95*, 1995.
5. T. Poggio and R. Brunelli, "A novel approach to graphics," Technical report 1354, MIT Media Laboratory Perceptual Computing Section, 1992.
6. D. Beymer, A. Shashua, and T. Poggio, "Example-based image analysis and synthesis," A.I. Memo No. 1431, Artificial Intelligence Laboratory, Massachusetts Institute of Technology, 1993.
7. J. Bergen, P. Anandan, K. Hanna, and R. Hingorani, "Hierarchical model-based motion estimation," in *Proceedings of the European Conference on Computer Vision*, (Santa Margherita Ligure, Italy), pp. 237–252, 1992.
8. J. Bergen and R. Hingorani, "Hierarchical motion-based frame rate conversion," technical report, David Sarnoff Research Center Princeton NJ 08540, 1990.
9. G. Wolberg, *Image Warping.* Los Alamitos CA: IEEE Computer Society Press, 1990.
10. S. Ullman and R. Basri, "Recognition by linear combinations of models.," *IEEE Transactions on Pattern Analysis and Machine Intelligence*, vol. 13, pp. 992–1006, 1991.

Complexity of Indexing: Efficient and Learnable Large Database Indexing [*]

Michael Werman and Daphna Weinshall

Institute of Computer Science, The Hebrew University of Jerusalem
91904, Jerusalem, Israel
{werman,daphna}@cs.huji.ac.il

Abstract. Object recognition starts from a set of image measurements (including locations of points, lines, surfaces, color, and shading), which provides access into a database where representations of objects are stored. We describe a *complexity theory of indexing*, a meta-analysis which identifies the best set of measurements (up to algebraic transformations) such that: (1) the representation of objects are linear subspaces and thus easy to **learn**; (2) direct indexing is **efficient** since the linear subspaces are of minimal rank. The index complexity is determined via a simple process, equivalent to computing the rank of a matrix. We readily re-derive the index complexity of the few previously analyzed cases. We then compute the best index for new cases: 6 points in one perspective image, and 6 directions in one para-perspective image; the most efficient representation of a color is a plane in $3D$ space. For future applications with any vision problem where the relations between shape and image measurements can be written down in an algebraic form, we give an automatic process to construct the most efficient database that can be directly obtained by learning from examples.

1 Introduction

Any recognition system starts with a finite list of n measurements computed on the image. The vector of n measurements defines the indexing space $\mathcal{R}^n$, and each image corresponds to a point in $\mathcal{R}^n$. Since objects appear in many different forms in different images (due to changes in viewing position, illumination, etc), each object is mapped to a collection of points. This collection, the set of all points corresponding to a feasible image of the object, is the object's *picture manifold* in the *indexing space*.

The particular form of an object's *picture manifold* is very important for object recognition, whether index-based or search-based:

Direct indexing: in the simplest approach each object's *picture manifold* is stored in the database. Now given a set of measurements computed from an image, which is a point in $\mathcal{R}^n$, we use the point as an index into a table,

---

[*] This research was supported the Israeli Ministry of Science under Grant 032.7568, and by the U.S. Office of Naval Research under Grant N00014-93-1-1202, R&T Project Code 4424341—01.

pointing to a cell where the identity of the observed object is written (e.g., geometric hashing [4]).

Search: If we have parametric representations of the picture manifolds, we can efficiently store those representations. Given a set of measurements computed from an image, we search the database to find the picture manifold which covers the measurements point in $\mathcal{R}^n$ (e.g., alignment [2]).

The first approach is the most time-efficient since no search is required. However, clearly the form of the *picture manifold* is critical here: if it is too complex, it may not be feasible to represent this picture manifold in memory. On the other extreme, the second approach is space-efficient since only a parametric representation of the picture manifold is stored. However, recognition requires search, and may therefore take too long to be useful.

For an optimal system we probably need to consider a hybrid of the 2 approaches to object recognition. In order to do that, we need to better understand the form of objects' *picture manifold*. Finally and most importantly, we need to be able to manipulate the measurements defining the indexing space, optimizing for simplicity and efficiency of the picture manifold at the same time. That is, we seek a transformation of the measurements such that:

Simplicity- the picture manifold is a linear subspace, a manifold which is both easy to represent and easy to learn from examples.

Efficiency- the picture manifold is of a low rank.

Indexing spaces where objects' picture manifolds are both simple and low dimensional can be used in recognition systems which are based on either one of the approaches described above. Defining and computing good indexing spaces is the subject of the present paper.

1.1 Previous work:

The concepts we study here are closely related to the concept of invariants, which is a subject of much interest in computer vision and pattern recognition. To analyze the picture manifold, we need to start with a relation between the vector of measurements and the object. This relation should not depend on variables of the imaging process, such as the camera orientation (viewing position) or the color of the illumination source; thus we need an invariant relation between image measurement and objects.

Invariant relations were divided in the literature into 2 types:

Model-free invariant: there exist image measurements which identify the object uniquely, so that their value is completely determined by the object regardless of the details of the imaging process. This is what is usually called an invariant in the literature. Such invariant relations do not exist for $2D$ images of general $3D$ objects, but for special classes of objects, such as planar or symmetrical objects, invariant relations may be found [6, 8].

Model-based invariant: a relation which includes mixed terms, representing image measurements or model parameters. Model-based invariant relations exist for many interesting vision problems [9].

If a model-free invariant is used to construct the database (the indexing space), then the picture manifold of an object is a point. For example, geometric hashing [4] uses the class of planar objects composed of features, where model-free invariants exist and therefore the picture manifolds are points. If on the other hand a model-based invariant is used, the picture manifold may be a complex hypersurface in the indexing space. For example, Murase & Nayar [7] use a very general class of objects and illumination conditions, and therefore they have to represent picture manifolds which take on rather complex shapes.

Clearly, for object recognition using either direct indexing or search, it is better to use an indexing space where the picture manifolds are as simple as possible (for example, hyperplanes rather than quadrics), and have as low a rank as possible. Jacobs [1, 3] defined the concept of indexing space, and studied the complexity of picture manifolds. However, Jacobs did not address the manipulation of the indexing space. Achieving this goal, obtaining linear manifolds that are easy to learn from examples while maintaining low rank (efficiency), is a major contribution of our present study.

1.2 Description of our approach:

We start from a general vision problem, with one or more cameras, where the image measurements include points, lines, or surfaces, graylevels, color and texture. For each such problem we find an optimal set of transformed measurements, to define an indexing space where the picture manifolds are linear (hyperplanes) and of the lowest possible rank (among all linear picture manifolds). The lowest rank of the picture manifold is called the **index complexity** of the vision problem.

At the end we get the best linear relation between objects and image measurements. Objects are then represented by low rank hyperplanes. We advocate the use of linear representations of objects even at a modest cost of space efficiency for the following reasons: (1) Hyperplanes can be learned (or interpolated) robustly and efficiently from a small number of examples; this is NOT the case for general manifolds. Moreover, parametric representations of general manifolds may not even be computable (cf. [7]). (2) Using noisy measurements, it is easy to find the distance of a measurement point from a hyperplane (unlike general manifolds), which is the problem that an object recognition algorithm needs to solve.

Our approach allows us to find class constraints, on the permitted set of objects, such that the index efficiency is increased (and the rank of the picture manifold is decreased). With sufficient class constraints we can reduce the rank of the picture manifold to 0, achieving the most efficient indexing possible via model-free invariants. This was the goal of a few studies [6, 8]; these studies gave examples of class constraints but did not offer a general way to generate them, like we do here.

The rest of this paper is organizes as follows: In Section 2 we define the concept of index complexity, which is the rank of the object's picture manifold in the indexing space. We show how to compute the lowest complexity given a particular vision problem. The lowest complexity of an index is simply the rank of a matrix, which we define and call the complexity-matrix. Many examples are discussed in Section 3. In Section 4 the addition of class constraints, for the purpose of reducing index complexity, is explained.

2 Complexity of indexing

A vision problem, describing a set of images, is defined by 3 types of variables:

1. **T** denotes the set of parameters describing the unknown **imaging variables**, including camera transformation and color of illumination.
2. **S** denotes the set of parameters describing the **object shape**, where shape denotes both geometrical and physical features (including color and shading).
3. **D** denotes the data - a set of **image measurements**.

Each image provides a number of relations between these variables, of the form $F(\mathbf{T}, \mathbf{S}, \mathbf{D}) = 0$; for example, an image of 1 point provides 2 such relations, one for the x coordinate and one for the y. The vision problem is fully described by the equations, relating these variables over the given images:

$$\mathcal{V} = \{F_l(\mathbf{T}, \mathbf{S}, \mathbf{D}) = 0\}_{l=1}^{M}$$

2.1 Database indexing:

Each vision problem $\mathcal{V}$ produces an ideal $\mathcal{I}$, which is the set of all the algebraic relations between the image measurables and the model shape. $\mathcal{I}$ is generated by a set of relations, obtained from the relations in $\mathcal{V}$ by eliminating the imaging variables **T**. Thus:

$$\mathcal{I} = \{f_l(\mathbf{S}, \mathbf{D}) = 0\}_{l=1}^{L}$$

$L \leq M$. The ideal $\mathcal{I}$ includes precisely all true relationships between the model parameters **S** and the image measurements **D**. The polynomials in $\mathcal{I}$ do not include the imaging variables **T**. We will often not distinguish between the ideal and its set of generators. The ideal is the smallest set including the generators closed under addition and multiplication by polynomials in **S** and **D**.

[11] describes a general elimination method to obtain the set of model-image relations from the vision problem using Gröbner bases (which is equivalent to Gauss elimination in linear systems).

2.2 Complexity of indexing:

We start by analyzing the simple case where the set of possible model-image relations $\mathcal{I}$ of our vision problem $\mathcal{V}$ includes a single relation $I : f(\mathbf{S}, \mathbf{D}) = 0$. The number of relations is often chosen to be one, as this requires the smallest number of constraints in the vision problem so that the imaging variables can be eliminated. Moreover, taking a larger set of equations can result in a combinatorial blowup. We define the complexity of I, a measure of the space and time needed to implement the database index obtained from it. As explained in the introduction, the complexity of an index reflects the size of the smallest hyperplane that needs to be stored in the database, so that the object can be retrieved by the index. In the appendix we define and compute joint index complexity when multiple indices are available.

To begin with, we rewrite the invariant relation I in the following compact way, explicitly separating image variables $\mathbf{D}$ from shape variables $\mathbf{S}$:

$$f(\mathbf{S}, \mathbf{D}) = \sum_{k=1}^{r+1} g_k(\mathbf{S}) * h_k(\mathbf{D}) = 0 \tag{1}$$

where g_k and h_k are polynomial functions of the shape $\mathbf{S}$ and the image $\mathbf{D}$ respectively. We call (1) the canonical representation of $f(\mathbf{S}, \mathbf{D})$. Note that if $f(\mathbf{S}, \mathbf{D})$ is algebraic, as we assume here, this decomposition always exists.

Definition 1 Index. The index, which includes all the image measurables in the invariant relation (1), is the r-dimensional vector:

$$[\frac{h_1(\mathbf{D})}{h_{r+1}(\mathbf{D})}, \ldots, \frac{h_r(\mathbf{D})}{h_{r+1}(\mathbf{D})}] \in \mathcal{R}^r \tag{2}$$

The indexing space is therefore $\mathcal{R}^r$. The manifold of all the possible images of object $\mathbf{S}$ is the hyperplane of rank $(r-1)$ defined by $\sum \frac{h_i(\mathbf{D})}{h_{r+1}(\mathbf{D})} \frac{g_i(\mathbf{S})}{g_{r+1}(\mathbf{S})} = -1$

In the simplest case where $r = 1$, we have:

$$f(\mathbf{S}, \mathbf{D}) = h_1(\mathbf{D})g_1(\mathbf{S}) + h_2(\mathbf{D})g_2(\mathbf{S}) = 0 \quad \Rightarrow \quad \frac{h_1(\mathbf{D})}{h_2(\mathbf{D})} = -\frac{g_2(\mathbf{S})}{g_1(\mathbf{S})}$$

This means that the object $\mathbf{S}$ has a model-free invariant: the database is a 1-dimensional table, where each object is represented by a point (cell) in the table, and the cell's value is $-g_2(\mathbf{S})/g_1(\mathbf{S})$. The image provides a direct access to the table via $h_1(\mathbf{D})/h_2(\mathbf{D})$.

More generally, the index given in (2) is an element of $\mathcal{R}^r$. If $r = 1$ it is a model-free invariant; if $r > 1$ it is a model-based invariant as defined in [9]. The database now is an r-dimensional table, and a pointer to object $\mathbf{S}$ is stored in all the cells of the table that satisfy (1). In order to achieve efficient recognition, we would like to find a representation of the form (1) with the smallest r.

Definition 2 Index Complexity. The index complexity $\mathcal{C}$ of a relation $f(\mathbf{S}, \mathbf{D})$ is the rank of the smallest hyperplane that the relation defines. Thus if r is the smallest number of terms such that:

$$f(\mathbf{S}, \mathbf{D}) = \sum_{k=1}^{r+1} g_k(\mathbf{S}) * h_k(\mathbf{D}) = 0 \tag{3}$$

then $\mathcal{C} = r - 1$.

When the index ideal $\mathcal{I}$ of a vision problem has a single relation in it, $\mathcal{C}$ is also the index complexity of the vision problem.

2.3 Computing index complexity:

We can always write the algebraic expression $f(\mathbf{S}, \mathbf{D}) = 0$ as a sum of multiplications; let

$$f(\mathbf{S}, \mathbf{D}) = \sum_{i=1}^{n} \sum_{j=1}^{m} m_{ij} s_i d_j = \mathbf{s} \cdot M \cdot \mathbf{d}^T = 0 \qquad (4)$$

where s_i and d_j are distinct products of element of $\mathbf{S}$ and $\mathbf{D}$ respectively, $\mathbf{s} = [s_1, \ldots, s_n]$, $\mathbf{d} = [d_1, \ldots, d_m]$, and M is the $n \times m$ matrix whose elements are m_{ij}.

Definition 3. M is the **complexity-matrix** of a relation $f(\mathbf{S}, \mathbf{D}) = 0$.

Theorem 4. *The minimal representation of* $f(\mathbf{S}, \mathbf{D}) = \sum_{k=1}^{r+1} g_k(\mathbf{S}) * h_k(\mathbf{D})$ *has* $r + 1$ *terms where* $r + 1$ *is equal to the rank of the complexity-matrix* M.

Proof. The theorem follows from (4), the fact that elementary operations on the rows and columns of a matrix are algebraic operations, and that the rank of a matrix is the minimal number of outer products of vectors that sum to the matrix.

Algorithm to compute the complexity $\mathcal{C}$ and the corresponding index:

1. Compute the SVD decomposition of $M = U \Sigma V^T$; the rank of M is equal to the number of non-0 elements in the diagonal matrix Σ, and the complexity is $\mathcal{C} = \text{rank}(M) - 2$.
2. Since by construction $f(\mathbf{S}, \mathbf{D}) = \mathbf{s} U \Sigma V^T \mathbf{d}^T = \mathbf{g}(\mathbf{S}) \cdot \mathbf{h}^T(\mathbf{D})$ where $\mathbf{g}(\mathbf{S}) = [g_1(\mathbf{S}), \ldots, g_{r+1}(\mathbf{S})]$, $\mathbf{h}(\mathbf{D}) = [h_1(\mathbf{D}), \ldots, h_{r+1}(\mathbf{D})]$

   shape parameterization - $\mathbf{g}(\mathbf{S}) = \mathbf{s} U \sqrt{\Sigma}$

   most efficient index - $\mathbf{h}(\mathbf{D}) = \mathbf{d} V \sqrt{\Sigma}$

3 Examples

3.1 Reformulation of known results

We first reformulate the following known results:

1. The index complexity of an affine or perspective image of symmetrical points is -1, namely, the most efficient index does not depend on the model - it only verifies symmetry in the image.
2. The index complexity of a perspective image of 5 coplanar points is 0, in other words, there is a model-free invariant.
3. The index complexity of 5 points, projected with an affine camera, is 1.

3.2 New results

$\mathcal{C} = 3$ 6 points and a perspective camera:

We work out this example in detail because it shows the power and relative simplicity of our method. We start by using homogeneous coordinates to represent the $3D$ coordinates of the 6 points. Since we are working in $\mathcal{P}^3$, 5 points define a basis; we select the first 5 points to be a particularly convenient projective basis, leading to the following representation of the $3D$ shape of the points:

$$
P_1 = \begin{pmatrix} 1 \\ 0 \\ 0 \\ 1 \end{pmatrix} \quad
P_2 = \begin{pmatrix} 0 \\ 1 \\ 0 \\ 1 \end{pmatrix} \quad
P_3 = \begin{pmatrix} 0 \\ 0 \\ 1 \\ 1 \end{pmatrix} \quad
P_4 = \begin{pmatrix} 0 \\ 0 \\ 0 \\ 1 \end{pmatrix} \quad
P_5 = \begin{pmatrix} 1 \\ 1 \\ 1 \\ 1 \end{pmatrix} \quad
P_6 = \begin{pmatrix} X_1 \\ Y_1 \\ Z_1 \\ W_1 \end{pmatrix}
$$

Similarly we use homogeneous coordinates to represent the projected $2D$ coordinates of the 6 points. Since we are working in $\mathcal{P}^2$, 4 points define a basis; we select the first 4 points to be the projective basis, leading to the following representation of the image of the points:

$$
p_1 = \begin{pmatrix} 1 \\ 0 \\ 1 \end{pmatrix} \quad
p_2 = \begin{pmatrix} 0 \\ 1 \\ 1 \end{pmatrix} \quad
p_3 = \begin{pmatrix} 0 \\ 0 \\ 1 \end{pmatrix} \quad
p_4 = \begin{pmatrix} 1 \\ 1 \\ 1 \end{pmatrix} \quad
p_5 = \begin{pmatrix} a_0 \\ b_0 \\ c_0 \end{pmatrix} \quad
p_6 = \begin{pmatrix} a_1 \\ b_1 \\ c_1 \end{pmatrix}
$$

Given any image of the 6 points, we can always compute the $2D$ projective transformation which will transform the points to the representation given above.

With this choice of coordinates, the relation between the $3D$ shape and the image measurements is the following:

$$
\begin{pmatrix} \alpha & -\delta & -\delta & \delta \\ -\delta & \beta & -\delta & \delta \\ \alpha & \beta & \gamma & \delta \end{pmatrix}
\begin{pmatrix} 1 & 0 & 0 & 0 & 1 & X_1 \\ 0 & 1 & 0 & 0 & 1 & Y_1 \\ 0 & 0 & 1 & 0 & 1 & Z_1 \\ 1 & 1 & 1 & 1 & 1 & W_1 \end{pmatrix}
= \begin{pmatrix} 1 & 0 & 0 & 1 & a_0 & a_1 \\ 0 & 1 & 0 & 1 & b_0 & b_1 \\ 0 & 0 & 1 & 1 & c_0 & c_1 \end{pmatrix}
$$

Using the terminology of Section 2:

1. **T** denotes the set of 4 variables α, β, γ, δ, which represent the camera unknowns that we would like to eliminate.
2. **S** denotes the set of 4 variables X_1, Y_1, Z_1, W_1, - the $3D$ projective coordinates of the 6th point.
3. **D** denotes the set of 6 variables a_0, b_0, c_0, a_1, b_1, c_1; these are the image measurements - the projective coordinates of the points.

Using elimination we get 1 constraint in $\mathcal{I}$:

$$
\begin{aligned}
f(\mathbf{S}, \mathbf{D}) = {} & a_0 b_1 Z_1 Y_1 + c_1 Y_1{}^2 b_0 - c_1 X_1{}^2 a_0 - a_0 a_1 Z_1{}^2 + a_0 b_1 Z_1{}^2 - b_1 Y_1{}^2 b_0 + c_0 a_1 Z_1{}^2 - \\
& c_0 b_1 Z_1{}^2 - b_0 a_1 Z_1{}^2 + b_0 b_1 Z_1{}^2 + a_1 X_1{}^2 a_0 - a_1 Y_1 a_0 X_1 + a_1 Y_1 a_0 Z_1 + c_1 Y_1 b_0 Z_1 - Y_1 a_1 Z_1 c_0 + \\
& c_0 a_1 Z_1 X_1 - c_0 b_1 Z_1 Y_1 - b_0 a_1 Z_1 X_1 - a_1 Y_1 b_0 Z_1 - c_1 X_1 Y_1 b_0 + X_1 b_1 Y_1 b_0 + c_1 X_1 Y_1 a_0 + \\
& X_1 a_0 b_1 Z_1 - c_1 X_1 a_0 Z_1 + c_0 X_1 b_1 Z_1 - X_1 b_0 b_1 Z_1 + c_1 X_1 a_0 W_1 - c_0 a_1 Z_1 W_1 + c_0 b_1 Z_1 W_1 + \\
& a_0 a_1 Z_1 W_1 - a_0 b_1 Z_1 W_1 + b_1 Y_1 b_0 W_1 + b_0 a_1 Z_1 W_1 - b_0 b_1 Z_1 W_1 - c_1 Y_1 b_0 W_1 - a_1 X_1 a_0 W_1 - \\
& 2 X_1 b_1 Y_1 a_0 + 2 a_1 Y_1 b_0 X_1 = 0
\end{aligned}
$$

Representing $f(\mathbf{S}, \mathbf{D})$ as a sum of multiplication as required in (4), we see that there are 10 shape monomials s_i, thus $n = 10$ and $\mathbf{s} = [X_1^2, X_1Y_1, X_1Z_1, X_1W_1, Y_1^2, Y_1Z_1, Y_1W_1, Z_1^2, Z_1W_1, W_1^2]$. There are 9 image monomials d_j, thus $m = 9$ and $\mathbf{d} = [a_0a_1, a_0b_1, a_0c_1, b_0a_1, b_0b_1, b_0c_1, c_0a_1, c_0b_1, c_0c_1]$.

We can now construct the 10×9 complexity matrix M. We use Gaussian elimination to decompose M as $M = UW$, where U is 10×5, and W is 5×9. Many matrices U satisfy these conditions, and we choose the "simplest":

$$
M =
\begin{bmatrix}
1 & -1 & 0 & -1 & 0 & 1 & 0 & -1 & 1 & 0 \\
0 & -2 & 1 & 0 & 0 & 1 & 0 & 1 & -1 & 0 \\
-1 & 1 & -1 & 1 & 0 & 0 & 0 & 0 & 0 & 0 \\
0 & 2 & -1 & 0 & 0 & -1 & 0 & -1 & 1 & 0 \\
0 & 1 & -1 & 0 & -1 & 0 & 1 & 1 & -1 & 0 \\
0 & -1 & 0 & 0 & 1 & 1 & -1 & 0 & 0 & 0 \\
0 & 0 & 1 & 0 & 0 & -1 & 0 & 1 & -1 & 0 \\
0 & 0 & 1 & 0 & 0 & -1 & 0 & -1 & 1 & 0 \\
0 & 0 & 0 & 0 & 0 & 0 & 0 & 0 & 0 & 0
\end{bmatrix}
=
\begin{bmatrix}
1 & 1/2 & 0 & 0 & -1/2 \\
0 & 1 & -1 & 0 & 1/2 \\
-1 & -1/2 & 1 & 0 & 0 \\
0 & -1 & 1 & 0 & -1/2 \\
0 & -1/2 & 1 & 1 & 1/2 \\
0 & 1/2 & 0 & -1 & 0 \\
0 & 0 & -1 & 0 & 1/2 \\
0 & 0 & -1 & 0 & -1/2 \\
0 & 0 & 0 & 0 & 0
\end{bmatrix}
\begin{bmatrix}
1 & 0 & 0 & -1 & 0 & 0 & 0 & 0 & 0 \\
0 & -2 & 0 & 0 & 0 & 2 & 0 & 0 & 0 \\
0 & 0 & -1 & 0 & 0 & 1 & 0 & 0 & 0 \\
0 & 0 & 0 & 0 & -1 & 0 & 1 & 0 & 0 \\
0 & 0 & 0 & 0 & 0 & 0 & 2 & -2 & 0
\end{bmatrix}
$$

Since the rank of M is 5, the complexity of the index is 3. From the above decomposition we get

shape parameterization: $\mathbf{g}(\mathbf{S}) = \mathbf{s} \cdot U = [X_1^2 - X_1W_1, -2X_1Y_1 + 2Y_1Z_1, -X_1Z_1 + Y_1Z_1, -Y_1^2 + Y_1W_1, 2Z_1^2 - 2Z_1W_1]$

most efficient index: $\mathbf{h}(\mathbf{D}) = \mathbf{d} \cdot V = [a_1a_0 - c_1a_0, 0.5a_1a_0 + a_0b_1 - 0.5c_1a_0 - b_0a_1 - 0.5b_0b_1 + 0.5b_0c_1, -a_0b_1 + c_1a_0 + b_0a_1 + b_0b_1 - c_0a_1 - c_0b_1, b_0b_1 - b_0c_1, -0.5a_1a_0 + 0.5a_0b_1 - 0.5b_0a_1 + 0.5b_0b_1 + 0.5c_0a_1 - 0.5c_0b_1]$

It is interesting to compare this result with the literature. Jacobs [3] showed recently that for 6 points in a single perspective image, the lowest rank picture manifold is a non-linear manifold of rank 3 in $\mathcal{R}^4$. We obtained the same rank 3, while using a different 4-dimensional measurement vector (a different indexing space). In this transformed space, the picture manifold is of the same low rank of 3 (provenly the lowest rank possible), and yet it is linear with all the benefits that come from linearity. In addition, our result was obtained automatically and relatively effortlessly, as we are using general and not problem-specific tools.

$\mathcal{C} = 3$: 4 points and a weak perspective camera:

The vision problem is a single image of 4 points, obtained by a weak perspective camera. The index complexity is 3. Thus the most efficient linear picture manifold is a $3D$ hyperplane in $\mathcal{R}^4$.

$\mathcal{C} = 3$ 6 lines and a para-perspective camera:

We took a vision problem of 6 lines projected para-perspectively (using an affine camera model) to a single image. We wrote down the equations relating the 6 unknown camera parameters, the 24 unknown parameters of the lines, and the 6 known projected directions in the image. Using elimination we obtained a single index $f(\mathbf{S}, \mathbf{D})$. The rank of the complexity matrix of the index is 5, thus the index complexity is 3.

For convenience of notation, we define the function $\Theta(a, b, c, d, e, f)$ as:

$$\Theta(a, b, c, d, e, f) = \begin{vmatrix} a & b & b \\ d & d & c \\ e & f & e \end{vmatrix} - \begin{vmatrix} a & a & b \\ c & d & c \\ e & f & f \end{vmatrix}$$

Using the analysis described above we computed the most efficient index:

$$\mathbf{h}(\mathbf{D}) = [\Theta(sl_1, sl_2, sl_3, sl_4, sl_5, sl_6), \ \Theta(sl_1, sl_2, sl_3, sl_5, sl_4, sl_6), \ \Theta(sl_1, sl_3, sl_5, sl_4, sl_6, sl_2), \ \Theta(sl_1, sl_3, sl_4, sl_2, sl_6, sl_5), \ \Theta(sl_1, sl_5, sl_4, sl_2, sl_6, sl_3)]$$

where sl_i is the projected slope of the i-th line.

$C = 2$: **The color at a point:**

Using the linear combination color model [5], the color of an object or a light source can be described as a linear combination of a set of basis color functions of wavelength λ. It has been argued that 3 basis functions can approximate well the color of natural objects and light sources. We use a model where the color spaces of objects and light sources are approximated by the span of 3 basis functions each. Our camera model has 4 different color sensors. Thus we have 4 measurements at each point (the output of the 4 sensors), and 6 unknowns - the 3 coefficients of the object color space and the 3 coefficients of the illuminant color space. With 4 equations we can eliminate the 3 illumination parameters, leaving us with a relationship between the measured colors (the sensors' output) and the reflectance properties (the color coefficients) of the object at a given point.

The index complexity of this relationship is 2, meaning that all the possible measured colors of a point (regardless of illumination color) sit in a plane in a 3-dimensional color space. The analysis also gives that there is no smaller subspace of the colors, which is insensitive to the illuminant color.

4   Adding class constraints

Once it has been shown that model-free invariants do not exist for unconstrained objects, attention had turned to characterizing the constraints (or classes of objects) which would lead to model-free invariants [6, 8], or in our language, lead to index complexity 0. The present analysis allows us to do so directly, as part of a more general question: what class constraints on objects reduce the index complexity (and thus make their recognition more efficient)? We determine sufficient and necessary conditions on class constraints to reduce index complexity, in particular to reduce it to 0 (implying the existence of model-free invariants). We start from a relation in the indices set

$$f(\mathbf{S}, \mathbf{D}) = \sum_{k=1}^{C+2} g_k(\mathbf{S}) * h_k(\mathbf{D}) = 0$$

where $g_k(\mathbf{S})$ and $h_k(\mathbf{D})$ are polynomial functions of the shape and image measurements respectively. Every class constraint of the form $\lambda(\mathbf{S}) = 0$, where $\lambda(\mathbf{S})$ divides some $\sum \alpha_i g_k(\mathbf{S})$, reduces the dimension of the best index by at least 1. Thus:

Theorem 5 class constraints:. *To reduce the index complexity of a vision problem from $\mathcal{C}$ to $p < \mathcal{C}$, the class constraints should provide at most $(\mathcal{C}-p)$ independent constraints of the form $\lambda_i(\mathbf{S}) = 0$, where each $\lambda_i(\mathbf{S})$ divides some $\sum \alpha_i g_k$ modulo the $\lambda_j(\mathbf{S})$, $j < i$.*

Clearly there is a trade off between index complexity, which is higher for more general (and less constrained) classes, and the density of the database, which is smaller for more general classes (as there are fewer types of such general objects).

Example: given 6 points and a perspective camera, we showed in Section 3.2 that the index complexity is 3, and we computed $\mathbf{g}(\mathbf{S})$ and $\mathbf{h}(\mathbf{D})$. It immediately follows that:

- If any of the parameters of the 6th point X_1, Y_1, Z_1 equals 0, the complexity goes down from 3 to 1; if any 2 of these parameters are equal, the complexity also goes down from 3 to 1. Thus *if 4 of the 6 points are coplanar, the index complexity of the system is 1.*
- If $X_1 = Y_1 = W_1$, the complexity is 0, namely, there is a model-free invariant.

Appendix: Handling multiple indices

Up to now we considered the case where the ideal of indices $\mathcal{I}$, obtained via elimination from the vision problem $\mathcal{V}$, has a single index function in it. Typically, however, the rank of the ideal is $L > 1$, namely, L independent indices can be computed from the image. We can compute the rank complexity of each index independently; thus object representation and recognition will require using L different (but minimal) indexing tables. Instead, we look for a single transformation of the measurement space into R^Φ, where each of the L relations defines a hyperplane of rank $\Phi - 1$. Now the index complexity (or the rank of the object manifold) is $\mathcal{C} = \Phi - L$. Even if $\mathcal{C}$ is larger than the individual index complexity of each of the L relations, we see two advantages to the joint approach:

- The table is sparser and therefore indexing into it is more robust.
- Object representation and recognition requires a single table.

We propose the following algorithm to compute the joint index complexity.

1. Obtain the **joint complexity matrix**.
   - Write the L relations as sums of multiplications, as in (4), with n_l distinct products of elements of $\mathbf{S}$ (shape) and m_l distinct products of elements of $\mathbf{D}$ (image measurements), for $1 \leq l \leq L$.
   - Let $\tilde{s}_i$ denote all the distinct products of elements of $\mathbf{S}$ (shape), which appear in any of the L relations, $i = 0..N$ and $N \leq \sum_l^L n_l$.

- For each relation, rewrite (4) using $\tilde{s}_i$ (we can always do this by adding terms whose coefficients are 0).
- Construct the individual complexity matrix M_l for each relation. The size of M_l, the complexity matrix of the l-th relation, is $N \times m_l$.
- Concatenate the L matrices M_l from left to right, giving us the joint complexity matrix M of size $N \times \sum_1^L m_l$.

Note the asymmetrical role of rows and columns here: the row variables define the elements of the joint indexing table, and thus should be the same for all indices; the column variables define the elements of the index, and thus can (and should) vary for different indices.

2. Compute the SVD decomposition of the joint complexity matrix $M = U\Sigma V^T$. Let Φ denote the rank of M, then $\Phi - L$ is the joint complexity of the indices.
3. The shape parametrization is $\mathbf{g(S)} = \tilde{s}U\sqrt{\Sigma}$, where $\tilde{s} = [\tilde{s}_1, \ldots, \tilde{s}_N]$.
4. Because of the way M was constructed, we can find a matrix V_l such that the individual complexity matrix of the l-th relation can be decomposed as $M_l = U\Sigma V_l^T$. The most efficient index is $\mathbf{h}^l(\mathbf{D}) = \mathbf{d}_l V_l \sqrt{\Sigma}$.

The result of using this algorithm on 7 points in a perspective image can be seen in this volume [10].

References

1. D. T. Clemens and D. W. Jacobs. Space and time bounds on indexing 3-D models from 2-D images. *IEEE T-PAMI*, 13(10):1007–1017, 1991.
2. D. P. Huttenlocher and S. Ullman. Recognizing solid objects by alignment with an image. *IJCV*, 5:195–212, 1990.
3. D. W. Jacobs. Matching 3-D models to 2-D images. *IJCV*, 1995. in press.
4. Y. Lamdan and H. Wolfson. Geometric hashing: a general and efficient recognition scheme. In *Proc. 2nd ICCV*, pages 238–251, Tarpon Springs, FL, 1988. IEEE, Washington, DC.
5. L. T. Maloney and B. Wandell. A computational model of color constancy. *JOSA*, 1:29–33, 1986.
6. Y. Moses and S. Ullman. Limitations of non model-based recognition schemes. In G. Sandini, editor, *Proc. 2nd ECCV, Lecture Notes in Computer Science*, volume 588, pages 820–828. Springer Verlag, 1992.
7. H. Murase and S. K. Nayar. Visual learning and recognition of 3-D objects from appearance. *IJCV*, 1995. in press.
8. C.A. Rothwell, D.A. Forsyth, A. Zisserman, and J.L. Mundy. Extracting projective structure from single perspective views of 3d point sets. In *Proc. 4th ICCV*, pages 573–582, Berlin, Germany, 1993. IEEE, Washington, DC.
9. D. Weinshall. Model-based invariants for 3D vision. *IJCV*, 10(1):27–42, 1993.
10. D. Weinshall, M. Werman, and A. Shashua. Duality of multi-point and multi-frame Geometry: Fundamental Shape Matrices and Tensors. In *Proc. 4th ECCV*, Cambridge, UK, 1996.
11. M. Werman and A. Shashua. Elimination: An approach to the study of 3D-from-2D. In *Proc. ICCV*, June 1995.
12. M. Werman and D. Weinshall. Complexity of indexing: Efficient and learnable large database indexing. TR 95-7, Hebrew University, 1995.

Structure from Motion (2)

Spatiotemporal Representations
for Visual Navigation

LoongFah Cheong, Cornelia Fermüller, and Yiannis Aloimonos

Computer Vision Laboratory, Center for Automation Research, University of
Maryland, College Park, MD 20742-3275

Abstract. The study of visual navigation problems requires the integration of visual processes with motor control. Most essential in approaching this integration is the study of appropriate spatio-temporal representations which the system computes from the imagery and which serve as interfaces to all motor activities. Since representations resulting from exact quantitative reconstruction have turned out to be very hard to obtain, we argue here for the necessity of representations which can be computed easily, reliably and in real time and which recover only the information about the 3D world which is really needed in order to solve the navigational problems at hand. In this paper we introduce a number of such representations capturing aspects of 3D motion and scene structure which are used for the solution of navigational problems implemented in visual servo systems. In particular, the following three problems are addressed: (a) to change the robot's direction of motion towards a fixed direction, (b) to pursue a moving target while keeping a certain distance from the target, and (c) to follow a wall-like perimeter. The importance of the introduced representations lies in the following:

- They can be extracted using minimal visual information, in particular the sign of flow measurements or the the first order spatiotemporal derivatives of the image intensity function. In that sense they are direct representations needing no intermediate level of computation such as correspondence.
- They are global in the sense that they represent how three-dimensional information is globally encoded in them. Thus, they are robust representations since local errors do not affect them.
- Usually, from sequences of images, three-dimensional quantities such as motion and shape are computed and used as input to control processes. The representations discussed here are given directly as input to the control procedures, thus resulting in a real time solution.

1 Introduction

A continuous interplay between visual processing and motor activity is a characteristic of most existing systems that interact with their environments. Initial

The support of the Office of Naval Research under Contract N00014-93-1-0257, National Science Foundation under Grant IRI-90-57934 and the Austrian "Fonds zur Förderung der wissenschaftlichen Forschung" project No S7003 and a postgraduate scholarship from Tan Kah Khee Foundation is gratefully acknowledged.

attempts at these difficult problems followed a modular approach. The goal of Computational Vision was defined as the reconstruction of an accurate description of the system's spatiotemporal environment. Assuming that this information can be acquired exactly, sensory feedback robotics was concerned with the planning and execution of the robot's activities. The problem with such separation of perception from action was that both computational goals turned out to be intractable [1, 2, 4, 5]. Recently a number of studies have been published which argue for a closer coupling by means of achieving solutions to a number of specialized visuo-motor control problems [13, 14]. We also encounter the so-called approach of image-based control [3, 6, 16, 18]. The principle behind this approach is that instead of computing intermediate representations, directly available image measurements are used as feedback for the control loop. Most commonly, a number of feature points extracted from the image are tracked over time. The idea behind this is that the chosen image features can be directly related to the parameters of the robot's joints. However, image features of this kind, in general, are not easily extractable, nor easily tracked. Furthermore, even simple kinematic maps are no longer simple when it comes to inverting them.

Here we argue that from the viewpoint of computational perception the essence of understanding the coupling of perception and action will come from understanding the appropriate spatiotemporal representations which the system computes from the imagery. By considering three visuo-servo motor control tasks, we present representations for 3D rigid motion and shape of increasing levels of complexity. In all three tasks we consider a robot system consisting of a body and a camera which can move independently of the body. The first and simplest task consists of changing the robot's direction of motion towards a fixed direction using visual information. For this task we need only partial egomotion information about the robot. In the second task the robot must pursue a moving target while keeping a certain distance from the target. The additional visual representation computed is the time to collision. In the third and last task, the robot has to follow a perimeter. This task requires the system to obtain partial depth information about the perimeter.

2 Task One: Moving Towards a Fixed Direction

The robotic system considered in all three tasks consists, as illustrated in Fig. 1, of a body on wheels with a camera on top of the body. To describe the system's degrees of freedom we define two coordinate systems, attached to the camera and the robot. The robot moves on a surface and is constrained in its movement to a forward translation T_R (along z_R) and a rotation ω_R around the vertical axis, i.e., the y_R-axis. The camera is positioned along a vertical axis that passes through the center of rotation of the robot and it has two independent rotational degrees of freedom. Its orientation measured with respect to the coordinate frame of the robot is given by its tilt, θ_x, and its pan, θ_y; there is no roll, i.e. $\theta_z = 0$.

We first consider the simple task of moving towards a new direction. Referring to the mobile robot illustrated in Fig. 2, the problem can be stated as follows: a

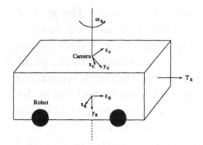

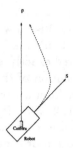

Fig. 1. Camera and robot local coordinate systems.

Fig. 2. The robot, currently moving forward with speed S, aims to veer towards **p**. The dotted path represents the trajectory generated in-flight by the servo system.

robot, moving forward with speed S, is required to head towards a new direction, along which some feature **p** lies; **p** is selected beforehand by some higher level process. The robot first directs the camera at **p**, so that the line of sight is now positioned along **p**. We assume such gaze shifts are accomplished by fast saccadic movements.

The robot must now make a series of steering decisions so that the forward motion and the direction of the heading have the same projection on the xz-plane of the camera coordinate system, i.e., θ_y has to become zero. These steering movements are controlled by a servomechanism, which derives information from images captured by the camera.

2.1 Global Motion Patterns

Traditional studies on visual motion rely on the optical flow field, whose computation requires optimization techniques to be invoked. For real-time systems, we have to consider time constraints imposed on the actions the system performs. Here we do not utilize the value of the optic flow but a more robust measurement, namely the sign of the projection of the flow **u** along a set of directions **n** on the image. That is for a direction **n** on the image at a point (x, y) we measure the sign of $u_n = \mathbf{u} \cdot \mathbf{n}$, if it can be computed reliably there.

In [9, 10] it has been shown that when these measurements are considered along a set of appropriately chosen directions they give rise to a global structure encoding 3D motion information in form of patterns in the image plane. In the following sections we will make use of some of these classes of vectors and their corresponding patterns, namely the "copoint patterns" and the "coaxis patterns".

The coaxis vectors are obtained as follows: Consider an infinite class of cones with apex at the nodal point of the eye and axis (A, B, C). The intersections of the cones with the image plane give rise to a set of conic sections. The (A, B, C) coaxis vectors are the vectors perpendicular to these conic sections. In order to speak of positive and negative values we define an orientation on these vectors as shown in Fig. 3.

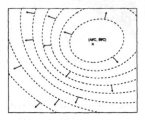

Fig. 3. Field lines corresponding to an axis (A, B, C) and positive coaxis vectors (A, B, C).

A camera of focal length f is undergoing a rigid motion with translation (U, V, W) (and thus Focus of Expansion or FOE $= (x_0, y_0)$) and rotation (α, β, γ). If we consider for a class of (A, B, C) coaxis vectors the orientation of the translational components, we find that a second order curve $h(A, B, C, x_0, y_0; x, y) = 0$ through the FOE separates the positive from the negative components (Fig. 4a), where

$$h(A, B, C, x_0, y_0; x, y) = x^2(Cf + By_0) + y^2(Cf + Ax_0) - xy(Ay_0 + Bx_0)$$
$$-xf(Af + Cx_0) - yf(Bf + Cy_0) + f^2(Ax_0 + By_0)(1)$$

Similarly, the positive and negative components of the (A, B, C) coaxis vectors due to rotation are separated by a straight line $g(A, B, C, \alpha, \beta, \gamma; x, y) = 0$, where

$$g(A, B, C, \alpha, \beta, \gamma; x, y) = y(\alpha C - \gamma A) - x(\beta C - \gamma B) + \beta Af - \alpha Bf. \quad (2)$$

The line passes through the point $(\frac{\alpha f}{\gamma}, \frac{\beta f}{\gamma})$ where the rotation axis pierces the image plane, the Axis of Rotation point (AOR) (Fig. 4b).

Combining the constraints due to translation and rotation, we obtain a pattern as shown in Fig. 4c, the coaxis pattern.

For a second kind of classification, the copoint vectors, which are the motion vectors perpendicular to straight lines through the point (r, s) similar patterns are obtained.

2.2 Using Visual Patterns in the Servomechanism of a Moving System

The retinal motion field perceived on the robot's camera is due to translation and rotation. The direction of translational motion is defined by the angle between the direction in which the robot is moving and the direction in which the camera is pointing. The rotation originates from body motion and is mainly due to the robot's turning around the y-axis. There could also be some rotation around the

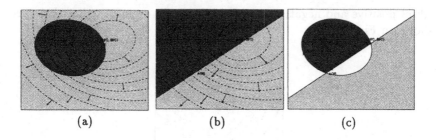

(a)   (b)   (c)

Fig. 4. (a) The (A, B, C) coaxis vectors due to translation are negative if they lie within a second-order curve defined by the FOE, and are positive at all other locations. (b) The coaxis vectors due to rotation separate the image plane into a half-plane of positive values and a half-plane of negative values. (c) A general rigid motion defines an area of positive coaxis vectors and an area of negative coaxis vectors. The rest of the image plane is not considered.

x-axis because the surface on which the robot is moving might be uneven, but there will be no or only very small rotation around the z-axis.

For the visual servoing task considered we examine those coaxis vectors which correspond to axes in the xy-plane, the $(A, B, 0)$ coaxis patterns. During the process of steering while the FOE is not aligned with the y-axis, the hyperbola (1) separating the positive from the negative translational vectors becomes

$$h(A, B, 0, x_0, y_0; x, y):$$
$$By_0 x^2 + Ax_0 y^2 - (Ay_0 + Bx_0)xy - Af^2 x - Bf^2 y + f^2 Ax_0 + Bf^2 y_0 = 0 \tag{3}$$

Since within the field of view f^2 is much larger than the quadratic terms in the image coordinates (x^2, y^2, xy), (3) can be approximated by $\hat{h}$ as

$$\hat{h}(A, B, 0, x_0, y_0; x, y) : y = \tfrac{A}{B}x_0 + y_0 - \tfrac{A}{B}x$$

which describes a line with slope $\frac{-A}{B}$ and intercept $\frac{A}{B}x_0 + y_0$. If the FOE is at the y-axis, the intercept becomes y_0.

The rotation around the y-axis is controlled by the robot's steering mechanism and thus the robot has knowledge about this rotation at least to a good approximation and can compensate for the resulting flow component by subtracting it from the visual motion field. This additional knowledge makes one of the patterns, namely the pattern corresponding to the axis $(1, 0, 0)$, especially suitable for deriving x_0 directly. We call this pattern the α-pattern, the corresponding vectors the α-vectors, and the corresponding hyperbola the α-hyperbola, since they do not contain any rotation around the x-axis (denoted in equations as α). Equation (3) becomes:

$$h(1, 0, 0, x_0, y_0; x, y) : x_0 y^2 - y_0 xy - xf^2 + x_0 f^2 = 0$$

which simplifies to $x = x_0$. An additional rotation around the y-axis causes a small nearly constant component added to every flow vector and thus the line in the translational pattern will be shifted. Nevertheless, such approximation will not affect the successful accomplishment of the task. As the robot approaches its goal, the steering motion it has to apply becomes smaller and smaller. Thus the additional rotational flow field decreases which in turn allows the FOE to be estimated more accurately.

2.3 Servo System

To set up the control loop equation, we must relate the robot's motion to the image motion. Referring to the coordinate systems defined in Fig. 1, we denote the velocity of the robot's forward translation by S and the velocity of its rotation around the y-axis by β. θ_x and θ_y are the pan and tilt of the orientation of the camera with respect to the coordinate frame of the robot.

Using T to denote translation, and ω to denote rotation, we can then express the motions of the robot and that of the camera in their respective coordinate frames denoted by subscripts R and C as follows:

$$
\begin{aligned}
\mathbf{T_R} &= (0, 0, S)^T & \omega_{\mathbf{R}} &= (0, \beta, 0)^T \\
\mathbf{T_C} &= {}^C R_R(\mathbf{T_R} + \omega_{\mathbf{R}} \times \mathbf{P}) = (S \sin\theta_y, -S \cos\theta_y \sin\theta_x, S \cos\theta_y \cos\theta_x)^T \\
\omega_{\mathbf{C}} &= {}^C R_R \, \omega_{\mathbf{R}} & &= (0, \beta \cos\theta_x, \beta \sin\theta_x)^T
\end{aligned}
$$

Thus the coordinates of the FOE (x_0, y_0) that we computed for the camera motion are related to pan and tilt as follows:

$$
x_0 = \frac{\tan\theta_y}{\cos\theta_x} f \quad y_0 = -\tan\theta_x \, f \tag{4}
$$

The position of x_0 is used as the input to the servo system to control the amount of steering the robot has to perform. If the servo system is operated with a proportional controller, the rotational speed, β, of the robot will be given by $\beta = Kx_0$. Writing β as $\dfrac{d\theta_y}{dt}$, and substituting (4) for x_0, we obtain $\frac{d\theta_y}{dt} = \frac{\tan\theta_y}{\cos\theta_x} f$. Approximating $\tan\theta_y$ by θ_y and $\cos\theta_x$ by 1 we obtain a linear control equation $\beta = \frac{d\theta_y}{dt} = K\theta_y$.

3 Task Two: Pursuing a Moving Target

We next consider the case where the servo system is called upon to follow a time-varying input. We encounter such a situation when the system is involved in the pursuit of a moving target. For addressing this task we need more elaborate control mechanisms, such as PID controler and gain scheduling. In addition, we impose a second constraint on the control system: while tracking a target, we want the system to maintain a constant distance away from the target. Such a task requires more complex visual information processing. In particular, we compute from the images the time to contact, from the observer to the target, and we use it in the control of the robot.

3.1 Estimating Time to Contact when the Observer and the Target are Moving

Time to contact between the camera and a scene point is defined as the value $\frac{Z}{W}$ with Z being the depth and W the relative forward translational speed of the camera with respect to the point. However, computing the time to contact of a point on the target is very much complicated by the difficulty in computing the target's relative motion if the image of the target only covers a small part of the image plane. In the literature a number of methods have been proposed which are based on the utilization of divergence [12, 15, 17] or tracking [8], suffering from the problems of requiring on the one hand knowledge about the motion and shape of the target, and on the other hand that the robot keeps track of its own motion during tracking.

The way to circumvent these problems is to use visual information not from the area covered by the target, but from the area surrounding the target. In particular, if the target is moving on the ground we can utilize the area at the bottom of the target which is at about the same distance as the target. From this we can obtain a measure $\frac{Z}{W}$, where W is the forward translational velocity of the robot with regard to the static scene and Z is the depth of points close to the target. Using the methods described in Section 2.1 and 2.2 we can estimate the motion of the observer. In general, knowing the 3D motion and knowing the normal flow at a point allows us to derive $\frac{Z}{W}$ at the corresponding point as

$$\frac{Z}{W} = \frac{(x - x_0)n_x + (y - y_0)n_y}{u_n - (u_{\rot}n_x + v_{\rot}n_y)}.$$

For the copoint vectors with gradient $(1,0)$ and the α-vectors, the value of the flow along the gradient can be approximated as $u_n = \frac{W}{Z}(x - x_0)$ and thus $\frac{Z}{W} = \frac{x - x_0}{u_n}$. Since we have estimated x_0 we can estimate the time to contact from this relationship.

4 Task Three: Perimeter Following

4.1 Estimating Functions of Depth

For many visual tasks requiring some depth or shape information, instead of computing exact depth measurements, it may be sufficient to compute less informative descriptions of shape and depth, such as functions of depth and shape where the functions are such that they can be computed easily from well-defined image information. This idea is demonstrated here by means of the task of perimeter or wall following.

Perimeter following in our application is described as follows: A robot (car) is moving on a road which is bounded on one side by a wall-like perimeter. On the basis of visual information the robot has to control its steering in order to keep its distance from the perimeter at a constant value and maintain its forward direction as nearly parallel to the perimeter as possible. The perimeter is defined as a planar textured structure in the scene (connected or not) perpendicular to the plane of the road.

Usually perimeter following is addressed either through general motion and depth reconstruction or by computing the slopes of lines parallel to the road (boundary lines on highways), which means that the boundary first has to be detected and thus the segmentation problem has to be solved.

The strategy applied here to the perimeter following task is as follows. While the robot is moving forward it has its camera directed at some point on the perimeter. As it continues moving it maintains the relative orientation of the camera with regard to its forward translation. It compares distance information derived from flow fields obtained during its motion with distance information computed from a flow field obtained when it was moving parallel to the road. This distance information will tell what the robot's steering direction is with respect to the perimeter.

The distance information we use is the scaled directional derivative of inverse depth along (imaginary) lines on the perimeter. From the observed flow field, normal flow measurements along (imaginary) lines through the image center are selected and compared to normal flow measurements along (imaginary) lines of equal slope in the reference flow field. The details of the computations are outlined below.

4.2 Direct Visual Depth Cue

For comparison reasons, we assume that the angle θ_y and the angle θ_x between the forward direction and the camera direction (determining x_0 and y_0) remain constant. The robot's rotational velocity (around the x-axis and y-axis) can change in any way.

The motion perceived in the images of the camera is due to a translation (U, V, W) and a rotation (α, β). Let us consider the normalized normal flow values $f_n(\mathbf{x}, \mathbf{n}) = \frac{u_n}{n_x}$ along $(0,0)$ copoint vectors, that is perpendicular to imaginary lines through the image center. We obtain:

$$f_n(\mathbf{x}, \mathbf{n}) = \frac{u_n}{n_x} = \frac{(-U - V\frac{n_y}{n_x})}{Z} + \alpha f \frac{n_y}{n_x} - \beta f \qquad (5)$$

Along each of the lines through the center $\frac{n_y}{n_x}$ is constant and thus $(-U - V\frac{n_y}{n_x})$ and $(\alpha f \frac{n_y}{n_x} - \beta f)$ are also constant and (5) describes $f_n(\mathbf{x}, \mathbf{n})$ as a function which is linear in the inverse depth. For any two points P_1 and P_2 with coordinates $\mathbf{x}_1$ and $\mathbf{x}_2$ along such a line the difference $(f_n(\mathbf{x}_1, \mathbf{n}) - f_n(\mathbf{x}_2, \mathbf{n}))$ is independent of the rotation. We thus compute the directional derivative $D(f_n(\mathbf{x}, \mathbf{n}))_{\mathbf{n}^\perp}$ of $f_n(\mathbf{x}, \mathbf{n})$ at points on lines with slope $k = -\frac{n_x}{n_y}$ in the direction of a unit vector $\mathbf{n}^\perp = (-n_y, n_x)$ parallel to the image lines. Dropping, in the notation of the directional derivative, the dependence of f_n on $\mathbf{x}$ and $\mathbf{n}$ we obtain

$$D(f_n)_{\mathbf{n}^\perp} = (-U - V\frac{n_y}{n_x})D(\frac{1}{Z})_{\mathbf{n}^\perp} \qquad (6)$$

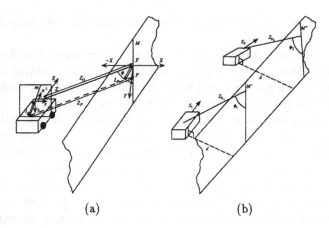

Fig. 5. (a)Geometric configuration during perimeter following. (b) Comparing the values from two configurations.

Let Z_0 be the depth at the fixation point and Φ the angle between the Z-axis and a line M on the perimeter whose image m is the the line in the image parallel to $\mathbf{n}^\perp$ along which we compute depth (see Fig. 5a). Equation (6) for any point along the line m becomes

$$D(f_n)_{\mathbf{n}^\perp} = (U + V\frac{n_y}{n_x})\frac{1}{\tan \Phi Z_0 f} \qquad (7)$$

It can be shown that the value of $|D(f_n)_{\mathbf{n}^\perp}|$ along certain directions $\mathbf{n}^\perp$ decreases as the robot steers towards the perimeter for directions (see Fig. 5b). Lines with such direction $\mathbf{n}^\perp$ have slopes which are of opposite sign to the slope of the image of a line parallel to the road of the perimeter passing through the image center. [7]

5 Experiments

Figures 6 and 7 show experiments on the first task. The mobile platform on which the camera was mounted was a conventionally steered vehicle, the camera's field of view was approximately $30°$ and the servo system was operated with a proportionality constant of $K = 0.1$. Figure 7 shows some images taken by a camera mounted on a mobile platform, as the latter is making steering movements. The feature $\mathbf{p}$ corresponds to the star in the center of the image, mounted on a tripod and initially located at a distance of 5 m from the camera. Figure 6 displays the configuration of this setting including the robot's trajectory. Figure 7a, c, and e show images taken by the system at three time instants (as marked in Fig. 6) with the normal flow fields superimposed. Figures 7b, d, and f show the positive and negative α-vectors as computed from the normal flow fields in black and white and the line approximating the α-hyperbola which has been fitted to the data.

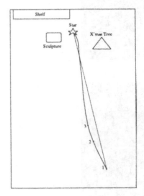

Fig. 6. 3D configuration as studied in task 1.

(a)   (b)

(c)   (d)

(e)   (f)

Fig. 7. Task 1: Some scenes along the trajectory.

In a second experiment concerning Task 3 a mobile platform with a camera mounted on it moved along an alley-like perimeter. The servomechanism was implemented as a simple proportional control. The scene contained a highly textured perimeter. We thus derived image measurements along a number of lines and used the mean of the computed estimates as input to the servo system. Figure 8a shows one of the reference images, which was taken when the robot was moving parallel to the perimeter. The lines along which image measurements were taken are overlaid on the image in white. Figure 8b shows the normal flow field computed for this same image. The successfulness of the technique was evaluated by studying whether and how the system corrected its path when we moved it either closer or further away from the perimeter. For the experiment shown the system always recovered to a movement parallel to the perimeter. Figure 8c displays an image taken when the robot steered towards the perimeter and Figure 8d shows when it again moved away from the perimeter.

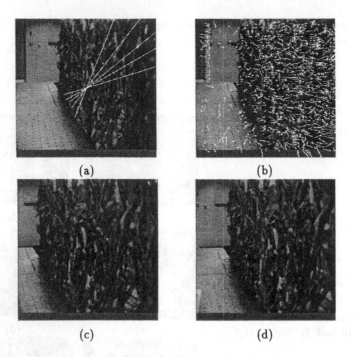

(a) (b)

(c) (d)

Fig. 8. Task 3: The robot moves along an "alley".

6 Conclusions

A new way of making use of visual information for autonomous behavior has been presented. Visual representations which are manifested through geometrical constraints defined on the flow in various directions and on the normal flow were used as input to the servomechanism. 3D motion and structure representations derived from these constraints were applied to the solution of a number of navigational problems involving the control of a system's 3D motion with respect to its environment and to other moving objects. Some of the constraints described, however, are of a general nature, and thus might be utilized in various modified forms for the solution of other navigational problems.

References

1. Aloimonos, J. (Y.): Purposive and qualitative active vision. Proc. DARPA Image Understanding Workshop (1990) 816–828
2. J. Aloimonos, I. Weiss, and A. Bandopadhay, Active vision. *International Journal of Computer Vision*, 2:333–356, 1988.
3. R.C. Arkin, R. Murphy, M. Pearson and D. Vaughn, Mobile robot docking operations in a manufacturing environment: Progress in visual perceptual strategies. In *Proc. IEEE International Workshop on Intelligent Robots and Systems*, pages 147–154, 1989.

684

4. R. Bajcsy, Active perception. *Proc. of the IEEE*, 76:996–1005, 1988.
5. D. Ballard and C. Brown, Principles of animate vision. *CVGIP: Image Understanding*, 45:3–21, Special Issue on Purposive, Qualitative, Active Vision, Y. Aloimonos (Ed.), 1992.
6. B. Espiau, F. Chaumette, and P. Rives, A new approach to visual servoing in robotics. *IEEE Trans. on Robotics and Automation*, 8:313–326, 1992.
7. C. Fermüller, L.F. Cheong and Y. Aloimonos, 3D Motion and Shape Representations in Visual Servo Control. Technical Report, Center for Automation Research, University of Maryland, CAR-TR-799, July 1995.
8. C. Fermüller and Y. Aloimonos, Tracking facilitates 3-D motion estimation. *Biological Cybernetics*, 67:147–158, 1992.
9. C. Fermüller and Y. Aloimonos, Direct perception of three-dimensional motion through patterns of visual motion. *Science*, 270:1973–1976, 1995.
10. C. Fermüller and Y. Aloimonos, On the geometry of visual correspondence. *International Journal of Computer Vision*, to appear, 1995.
11. E. Francois and P. Bouthemy, Derivation of qualitative information in motion analysis. *Image and Vision Computing*, 8:279–288, 1990.
12. R.C. Nelson and Y. Aloimonos, Obstacle avoidance using flow field divergence. *IEEE Trans. on Pattern Analysis and Machine Intelligence*, 11:1102–1106, 1989.
13. D. Raviv and M. Herman, Visual Servoing from 2D image cues. In Y. Aloimonos (Ed.), *Active Perception*, Advances in Computer Vision, pages 191–229, Lawrence Erlbaum, Hillsdale, NJ, 1993.
14. J. Santos-Victor, G. Sandini, F. Curotto and S. Garibaldi, Divergent stereo for robot navigation: Learning from bees. In *Proc. IEEE Conference on Computer Vision and Pattern Recognition*, pages 434–439, 1993.
15. G. Sandini, F. Gandolfo, E. Grosso and M. Tistarelli, Vision during action. In Y. Aloimonos (Ed.), *Active Perception*, Advances in Computer Vision, pages 151–190. Lawrence Erlbaum, Hillsdale, NJ, 1993.
16. S.B. Skaar, W.H. Brockman, and R. Hanson, Camera-space manipulation. *International Journal of Robotics Research*, 6:20–32, 1987.
17. M. Subbarao, Bounds on time-to-collision and rotational component from first-order derivatives of image flow. *Computer Vision, Graphics, and Image Processing*, 50:329–341, 1990.
18. L.E Weiss and A.C. Sanderson, Dynamic sensor-based control of robots with visual feedback. *IEEE Trans. on Robotics and Automation*, 3:404–417, 1987.

Decoupling the 3D Motion Space by Fixation

Konstantinos Daniilidis[1*] and Inigo Thomas[2]

[1] Computer Science Institute, University of Kiel
[2] GRASP Laboratory, University of Pennsylvania

Abstract. Fixation is defined as the ability of an active visual system to keep the projection of an environmental point stationary in the image. We show in this paper that fixation enables the decoupling of the 3D-motion parameters by projecting appropriately the spherical motion field in two latitudinal directions with respect to two different poles of the image sphere. Both computational steps are based on one-dimensional searches along meridians of the image sphere. We do not use the efference copy of the fixational rotation of the camera. Performance of the algorithm is tested on real world sequences with fixation accomplished either off-line or during the recording using an active camera.

1 Introduction

The ability to perceive the three-dimensional motion relative to the environment is crucial for every robot acting in a dynamically changing world. The estimation of 3D motion parameters has been addressed in the past as a reconstruction problem: Given a monocular image sequence the goal was to obtain the relative 3D motion to every scene component as well as a relative depth map of the environment. Solutions given suffer under instability problems and require an immense computational effort which excludes a real time reactive behavior. In this study, we will show the computational advantages with respect to 3D motion estimation of the fixation on an environmental point. There is a large amount of work in biological and computer vision research on how fixation is achieved [1, 2]. The evident advantages of overcoming the field of view, foveal sensing, and reducing the motion blur have been considered sufficiently justifying the fixational movements so that only sporadic approaches delved into the computational advantages of fixation.

We show in this paper that the ability to fixate on a stationary point combined with the appropriate representation of the motion field enables the decoupling of the 3D-motion parameters. We use a spherical image surface which can be mapped 1:1 to the image plane. We do not use any information from the motor encoders or from the input in the fixation feedback loop (called the efference copy in biology). Fixation is formulated only as a constraint on the

---

* Preusserstr. 1-9 24105 Kiel, Germany, kd@informatik.uni-kiel.de. We heavily acknowledge the constructive discussions with Gerald Sommer, Ruzena Bajcsy, Yiannis Aloimonos, and Hans-Peter Mallot.

motion field. This constraint reduces the number of unknowns from five to four. The translation direction remains unknown (two parameters) but instead of the angular velocity (three unknowns) we obtain only the torsion - rotation about the target direction– and the time to collision to the fixated scene point. The new representation for the fixated motion field is based on two projections. Assuming that the fixated target point is the pole of the sphere we show that the latitudinal projection of the motion field has the property of being constant along a meridian. The constant value is equal to the torsion and the meridian contains the heading direction. Taking as a new pole the normal to this meridian we again project the flow field in the latitudinal direction and obtain a similar pattern: A meridian with respect to the new pole where the new latitudinal projection is constant and equal to the time to impact to the target. This new meridian fully constrains the heading direction. We are, thus, able to compute the heading direction by applying only two onedimensional searches. In case of a heading direction outside the field of view we replace the second projection with the solution of an equation in the two remaining unknowns.

2 Problem Statement

We assume that the imaging surface is a sphere with unit radius. We denote by $\hat{p}$ the points on this sphere resulting from the projection $\hat{p} = P/\|P\|$. The mapping of the planar imaging surface to a spherical surface is one to one. Let $x = P/\hat{z}^T P$ be a point on the image plane $Z = 1$ with the optical axis parallel to the Z-axis with unit vector $\hat{z}$. If $\dot{x}$ is the motion field on that plane then it can be easily proved that the spherical motion field reads

$$\dot{p} = \frac{1}{\|x\|}(\hat{p} \times (\dot{x} \times \hat{p})).$$ (1)

Most of the authors assume that for a small field of view the two fields are approximately equal. However, for a large field of view the above equation should be used. Special care should be taken in the mapping of the planar discretization noise onto the sphere. We assume that the observer is moving with instantaneous linear velocity v and angular velocity ω relative to the environment so that the velocity of a scene point P can be written as $\dot{P} = v + \omega \times P$. In case of pure ego motion all equations are valid with the opposite sign for the velocities v and ω. The spherical motion field reads

$$\dot{p} = \frac{1}{\|P\|}(\hat{p} \times (v \times \hat{p})) + \omega \times \hat{p}$$ (2)

where we can observe the classical decomposition into a translational component depending on the environment ($\|P\|$) and the rotational term depending only on the image position. The spherical motion field vector lies on the tangential plane at point $\hat{p}$ so that $\dot{p}^T \hat{p} = 0$. As we mentioned at the beginning we suppose that a control algorithm exists that makes a target point $\hat{t}$ on the sphere be fixated which means $\dot{t} = 0$. From (2) follows that

$$-\frac{v \times \hat{t}}{\|T\|} + \omega \quad \text{is parallel to} \quad \hat{t}$$

where $\|T\|$ is the distance to the target scene point. Hence, the angular velocity in case of fixation reads

$$\omega = \gamma \hat{t} + \frac{v \times \hat{t}}{\|T\|} \tag{3}$$

It is constrained to be a function of the linear velocity and possesses only one degree of freedom γ: the torsion around the target point $\hat{t}$. Thus, after fixation the flow field contains three components (Fig. 1): A translational one due to v, a fixational equal to the second term $v \times \hat{t}/\|T\|$ of (3), and a torsional component $\gamma \hat{t}$.

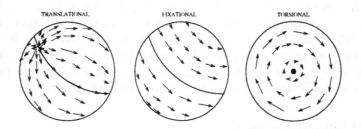

TRANSLATIONAL. FIXATIONAL. TORSIONAL.

Fig. 1. The three components of a fixated motion field.

After inserting the fixation angular velocity (3) into (2) the spherical motion field of a point $\hat{p}$ different from the target reads

$$\dot{p} = \hat{p} \times (v \times (\frac{\hat{p}}{\|P\|} - \frac{\hat{t}}{\|T\|})) + \gamma(\hat{t} \times \hat{p}). \tag{4}$$

After eliminating the structure information $\|P\|$ by taking the scalar product with $v \times \hat{p}$ we obtain the "epipolar" equation for the fixated motion field

$$(v \times \hat{p})^T (\dot{p} - \frac{\hat{t}}{\|T\|} \hat{p}^T v - \gamma(\hat{t} \times \hat{p})) = 0 \tag{5}$$

which corresponds to the instantaneous version of epipolar equation for general motion

$$(v \times \hat{p})^T (\dot{p} - \omega \times \hat{p}) = 0. \tag{6}$$

We see that the depth-free equation (5) contains three unknowns for the scaled linear velocity $v/\|T\|$ plus one unknown for the torsion γ around the target. Furthermore, the equation (5) is quadratic in the components of v and bilinear in (v, γ).

The first and most important result obtained by Bandopadhay and Ballard [3] and by Aloimonos et al. [4] was that fixation reduces the number of unknowns from five to four. Their fixation constraint was that $\omega = (v_y, -v_x, \gamma)$ which is direct implication of (3) if we set the target parallel to the optical axis: $\hat{t} = \hat{z}$. In the work of Fermüller and Aloimonos [5] fixation is exploited to compute the line on the image which passes through the FOE. Using only normal flow the location of the FOE on this line is found by matching patterns to the repeatedly detranslated flow. Then the FOE is localized using the rotational control signals of the camera movements in two subsequent time points and making the assumption that translation direction is almost constant despite fixation. The equation of fixational motion field (4) is used by Taalebinezhaad [6]. The flow field in the Brightness Change Constraint Equation (BCCE) is substituted by the fixational motion field. As the BCCE at every pixel introduces a new unknown (depth) an additional assumption of minimal variation of depth near the fixation point is added. To convert the resulting minimization into an eigenvalue problem it is further assumed that the torsion γ is already computed in a preceding step. This step is solved assuming local frontoparallel patches. However, this assumption enables a local and linear computation of rotation and translation without fixation. Raviv and Herman [7] study the surfaces in the world that produce constant flow in the image. They show that the level sets of equal latitudinal flow are cylinders and that the longitudinal flow is zero along two planes. The intersection of these planes with the cylinder corresponds to the points in the world that produce zero flow. The first part in [8] is identical to the work by Raviv and Herman [7]. They derive the equal flow cylinders and planes. However, Thomas et. al. [8] apply their findings of zero longitudinal flow to determine the angle between the target and the velocity v. This plane always appears in the image as a line, provided that the FOV is 180 degrees. These results are tested using a novel 180 degrees field of view camera. In [9] fixation is combined with the log-polar transformation. Using the second order spatial derivatives of the fixated log-polar field it is shown that the time to collision can be computed using only the radial component of the velocity. Advantages of the polar transformation in case of fixation are also shown in [10] where the heading direction is computed using two specific lines through the center of the image. The work of Barth and Tsuji [11] addresses the issue of how to fixate in the direction of the translation. Their technique is based on the following heuristic. They group the flow vectors near the point of fixation into two groups: positive and negative flows. The difference in the average of the flow values at these groups indicates the direction of translation with respect to the current fixation direction. Based on this value the robot is controlled to turn towards the direction of translation. The same issue is addressed in [12] using an affine model for the optical flow field. Servoing towards the heading direction is achieved by minimizing the lateral translational components by means of a task function.

3 Projections of the fixated motion field

We proceed by projecting the fixated spherical motion field (4) into two different orthogonal basis systems of the tangential plane at an arbitrary point on the sphere. The first projection assumes that the target direction $\hat{t}$ is the pole of the sphere defining thus a latitudinal and a longitudinal unit vector

$$\hat{\phi}_1 = \frac{\hat{t} \times \hat{p}}{\|\hat{t} \times \hat{p}\|} \quad \text{and} \quad \hat{\theta}_1 = \hat{\phi}_1 \times \hat{p},$$

respectively, lying in the tangential plane of point $\hat{p}$.

The second projection assumes as a pole the unit vector in the direction of $v \times \hat{t}$ yielding a latitudinal and a longitudinal unit vector

$$\hat{\phi}_2 = \frac{(v \times \hat{t}) \times \hat{p}}{\|(v \times \hat{t}) \times \hat{p}\|} \quad \text{and} \quad \hat{\theta}_2 = \hat{\phi}_2 \times \hat{p},$$

respectively. Through the course of exposition the reader may consult Figure 2 where the projections are illustrated.

Meridian with constant $\frac{\dot{p}^T \hat{\phi}_1}{\|\hat{t} \times \hat{p}\|}$ Meridian with constant $\frac{\dot{p}'^T \hat{\phi}_2}{\|\hat{t} \times \hat{n}_1\|}$

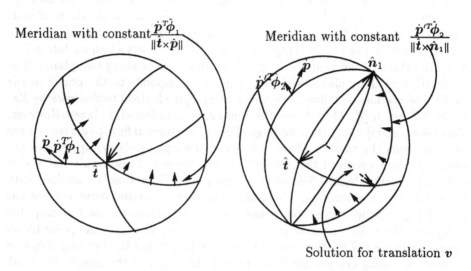

Solution for translation v

Fig. 2. The meridians with respect to the target pole $\hat{t}$ are drawn on the left sphere. The spherical flow $\dot{p}$ is projected on the latitudinal direction. The first step of the algorithm is a 1D search for the meridian with constant $\dot{p}^T \hat{\phi}_1 / \|\hat{t} \times \hat{p}\|$. In the second step (see the right sphere) the pole is $\hat{n}_1$ perpendicular to the meridian found in the first step. The flow without torsion $\dot{p}' = \dot{p} - \gamma(\hat{t} \times \hat{p})$ is projected on the new latitudinal directions. A 1D search among the meridians with respect to pole $\hat{n}_1$ for the meridian with constant $\dot{p}'^T \hat{\phi}_2 / \|\hat{t} \times \hat{n}_1\|$ yields a second big circle. The intersection of the big circles found in the two steps gives the solution for the desired translation direction v.

The latitudinal projection using the target direction $\hat{t}$ as a pole reads

$$\dot{p}^T \hat{\phi}_1 = \frac{1}{\|\hat{t} \times \hat{p}\|} v^T (\hat{t} \times \hat{p}) \left(\frac{1}{\|P\|} - \frac{\hat{p}^T \hat{t}}{\|T\|} \right) + \gamma \|\hat{t} \times \hat{p}\|.$$

Because the angle between the target $\hat{t}$ and the considered point is known we divide by its sine which is equal to $\|\hat{t} \times \hat{p}\|$:

$$\frac{\dot{p}^T \hat{\phi}_1}{\|\hat{t} \times \hat{p}\|} - \gamma = \frac{1}{\|\hat{t} \times \hat{p}\|^2} v^T (\hat{t} \times \hat{p}) \left(\frac{1}{\|P\|} - \frac{\hat{p}^T \hat{t}}{\|T\|} \right). \tag{7}$$

We see that the latitudinal component minus the torsion vanishes if the considered point lies on the plane spanned by the target and the translation direction.

Thus, we are able to constrain the translation direction if we find the meridian with longitude η where the term $\frac{\dot{p}^T \hat{\phi}_1}{\|\hat{t} \times \hat{p}\|}$ is constant independent of the latitude $\|\hat{t} \times \hat{p}\|$. Unfortunately this is not the only case where this term becomes constant. Suppose that a part of the environment is planar. Let the equation of the plane be $\hat{N}^T X = d$ and assume that the target is on the optical axis. If the plane normal reads $N = (\cos \alpha \sin \beta, \sin \alpha \sin \beta, \cos \beta)$ then it can be easily proved that

$$\frac{1}{\|\hat{t} \times \hat{p}\|^2} v^T (\hat{t} \times \hat{p})(\frac{1}{\|P\|} - \frac{\hat{p}^T \hat{t}}{\|T\|}) = \frac{1}{d} v^T \hat{\phi}_1 \sin \beta \cos(\alpha - \eta), \qquad (8)$$

which is independent of the latitude $\|\hat{t} \times \hat{p}\|$. Hence, all meridians that are projections of lines on planes in the scene will have a constant latitudinal projection independent of the colatitude angle. Furthermore, the right hand side of (7) will vanish on the meridians that are projections of infinite depths ($1/\|P\| = 0$) and on the entire field of view if the translation is parallel to the target direction: $v \times \hat{t} = 0$.

To summarize the defeating configurations:

1. There may exist meridians with constant latitudinal projection if these meridians are projections of planar parts of the environment or of scene points at infinity.
2. The latitudinal projection is everywhere constant if we fixate on the translation direction or if translation does not exist at all.

Suppose now that the unit vector $\hat{n}_1$ in the direction $v \times \hat{t}$ is given and let it be the new pole. The new pole introduces new meridians and latitudes. Since torsion can be computed in the first projection above we consider the latitudinal projection of the torsion-free flow

$$(\dot{p} - \gamma(\hat{t} \times \hat{p}))^T \hat{\phi}_2 = \frac{1}{\|P\| \|(v \times \hat{t}) \times \hat{p}\|} (\hat{p} \times (v \times \hat{p}))^T ((v \times \hat{t}) \times \hat{p}) + \frac{\|v \times \hat{t}\|}{\|T\|} \|\hat{p} \times \hat{n}_1\|,$$

where $\hat{n}_1$ is the unit vector $\frac{v \times \hat{t}}{\|v \times \hat{t}\|}$ known from the first projection. Hence, we can divide the left hand side and rewrite the right hand side as following:

$$\frac{(\dot{p} - \gamma(\hat{t} \times \hat{p}))^T \hat{\phi}_2}{\|\hat{p} \times \hat{n}_1\|} = \frac{1}{\|P\| \|(v \times \hat{t}) \times \hat{p}\|} \hat{p}^T (v \times (v \times \hat{t})) + \frac{\|v\| \|\hat{v} \times \hat{t}\|}{\|T\|}.$$

Considering now meridians through the pole $v \times \hat{t}$ we obtain following cases where the torsion-free latitudinal component will be constant.

1. On the meridian with normal $v \times (v \times \hat{t})$.
2. On the meridians containing points with infinite depth.

The detection of the meridian with normal

$$\hat{n}_2 = \frac{v \times (\hat{v} \times \hat{t})}{\|\hat{v} \times \hat{t}\|}$$

allows the full computation of the translation direction

$$\hat{v} = \hat{n}_1 \times \hat{n}_2.$$

Having obtained the heading direction we know the sine of the angle between the heading direction and the target $\|\hat{v} \times \hat{t}\|$. The remaining constant after vanishing of the first term in (3) yields $\lambda = \|v\|/\|T\|$ which is the fourth and last unknown of the motion problem in case of fixation. The inverse of it can be interpreted as the time to collision to an object at the same distance as the target in the motion direction.

To find meridians of constant value in the first and the second latitudinal projections we compute for every meridian the mean and the variance over the latitude. Then, we search for the meridians on which this variance is minimized. The means on these meridians yield the torsion and the inverse of the time to collision, in the first and second projection respectively.

Although in the first projection all meridians - or sectors of them - were contained in the field of view this is not the case in the second projection where the meridians are with respect to the new pole $\hat{n}_1$. It is very easy to imagine this case if for example $\hat{n}_1 = (0, 1, 0)$. We will see in the experiments that in such a case the variance of the second latitudinal projection gets its minimum at the border of the field of view. A corrective saccade can then shift the focus of expansion inside the field of view and the process can be continued with a refixation on a new point. If we want to avoid a corrective saccade we must replace the second search with a procedure as follows. The first step constrains the translational velocity to the plane with normal $\hat{n}_1$. Thus, we can write

$$\hat{v} = \cos \chi t + \sin \chi (\hat{n}_1 \times \hat{t}), \tag{9}$$

where χ is the remaining degree of freedom of the translation direction or, in the terms of the formulation above, the longitude of the searched meridian in the second step. Let rewrite (5) as

$$(\hat{v} \times \hat{p})^T (\dot{p}' - \lambda \hat{t} \, \hat{p}^T \hat{v}) = 0, \tag{10}$$

where $\dot{p}' = \dot{p} - \gamma(\hat{t} \times \hat{p})$ is known from the second step and $\lambda = \|v\|/\|T\|$ is the inverse of the time to collision. If we insert $\hat{v}$ from (9) in (10) in the above equation we obtain

$$\cos \chi (\hat{t} \times \hat{p})^T \dot{p}' + \sin \chi ((\hat{n}_1 \times \hat{t}) \times \hat{p})^T \dot{p}' = \lambda \sin \chi \hat{p}^T \hat{n}_1 (\cos \chi \hat{p}^T \hat{t} + \sin \chi (\hat{n}_1 \times \hat{t})^T \hat{p}). \tag{11}$$

This is a nonlinear equation in the two unknowns χ and λ which can be solved numerically with nonlinear minimization.

To summarize, we present the algorithmic steps of our method:

1. Choose a sampling step for the longitude angle η with respect to pole $\hat{t}$ - in reality being always the optical axis if we fixate on the center. Divide the optical flow field in groups with the same longitude η corresponding to meridians. Compute for every group the mean and the variance of

$$\frac{\dot{p}^T \hat{\phi}_1}{\|\hat{t} \times \hat{p}\|}.$$

Carry out an 1D-search for the minimum η_{min} of the variance. The new pole $\hat{n}_1$ reads $(\sin \eta_{min}, -\cos \eta_{min}, 0)$ if $\hat{t}$ is the optical axis.

2. Compute for all points the longitude angle χ with respect to the new pole $\hat{n}_1$ and group the vectors with the same χ. Compute for every group the mean and the variance of

$$\frac{(\dot{p} - \gamma(\hat{t} \times \hat{p}))^T \hat{\phi}_2}{\|\hat{p} \times \hat{n}_1\|}$$

and search for the minimum χ_{min} of the variance. Divide the mean by $\|\hat{v} \times \hat{t}\|$ in order to obtain the inverse of the time to collision $\|v\|/\|T\|$. If χ_{min} is near the border of the field of view then either carry out a saccade towards χ_{min} or apply the nonlinear minimization described above.

4 Experimental Results

We tested the proposed algorithms with synthetic as well as real data. Real data experiments were carried out using sequences recorded by passive as well as active cameras. In the non-fixated sequence we emulated the fixation by appropriately rotating the optical flow field. In all the experiments, the 1D-search of the first step runs over 45 samples of the 180 degrees η-range. The sampling interval for χ in the 1D search of the second step is one degree. If the focus of expansion lies outside the field of view we replace the second step with the alternative nonlinear minimization method applying the Levenberg-Marquardt method.

We produce synthetic motion fields assuming a scene looking like a corridor. In the first experiment we assume a wide field of view of 90 degrees and we apply translations $v = (\sin \chi_{gt}, 0, \cos \chi_{gt})$ where χ_{gt} is the ground truth angle between translation and target direction. The latter is assumed to coincide with the optical axis. In this as well as all subsequent simulations it turns out that the error in the azimuthal angle η of the translation direction was under 2 deg and the relative error in the torsion γ under 3%. Therefore we will plot only the error in the χ-angle and the inverse of the time to collision λ. In Fig. 3 we show the error in the angle χ for translation directions deviating from 5 to 40 degrees from the target direction. The motion field is corrupted by gaussian noise with relative standard deviation of 10% and 20%. We tested for two torsion values 0 and 0.005, shown in the left and right of Fig. 3 respectively. We observe that the error increases with the deviation of the translation from the target direction and its behavior is not smooth in presence of torsion. The second synthetic experiment concerns a smaller field of view (45 deg) in presence of torsion and relative optical flow error of 10%. Since the second step can be applied only for $\chi < 20$ deg we applied in all steps the nonlinear minimization with respect to χ and λ. The results (Fig. 4) are significantly better than the 1D-search for even a larger field of view (see above) but with the additional cost of an iterative method. The same initial values were used in the nonlinear minimization for all translation directions. In the following image sequences we computed the optical flow with

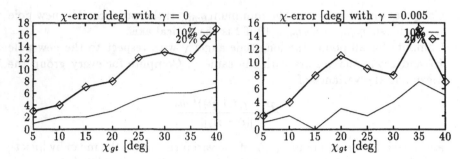

Fig. 3. The error in the χ-angle as a function of the translation direction for a field of view of 90 degrees and two values of torsion: 0 (left) and 0.005 (right). The motion field is corrupted by gaussian relative error with standard deviation of 10% and 20%.

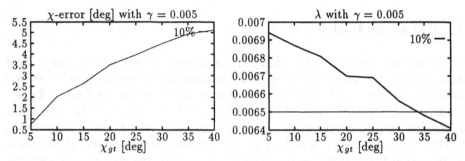

Fig. 4. The error in the χ-angle (left) and the inverse of the time to collision $\lambda = \|v\|/\|T\|$ as computed by the alternative to the second step. The field of view is 45 deg and the relative error in the optical flow is 10%.

a standard differential method which assumes a constant flow field in the local neighborhood of every pixel. The spatiotemporal derivatives are computed with binomial kernels which are approximations of the first derivative of a Gaussian. The computed flow field is first mapped to the plane $Z = 1$ using the intrinsic parameters and then transformed to a spherical flow field using (1). The first sequence is the well known synthetic Yosemite sequence (Courtesy of Lynn Quam at SRI) which contains both translation with ground truth ($\eta = 90$ deg, $\chi = -9.84$ deg) and rotation with ground truth $\omega = (0.00023, 0.00162, 0.00028)$. The original and the fixated flow fields (Fig. 5 left) are computed only for the part of the image that contains ego motion (the clouds area is excluded). The minimum of the variance of the first latitudinal projection (Fig. 5, top right) gives an η estimate of 97.37 deg and a torsion estimate of -0.00063 (the opposite sign is due to our formulation of scene motion). Since the minimum of the variance of the second latitudinal projection (Fig. 5, bottom right) is at the limit of the field of view we again apply the nonlinear minimization for the second step and obtain $\chi = -5.96$ and $\lambda = 0.00145$. The second sequence is already fixated during its recording with an active camera (Fig. 6). Up to the fixational movement the motion of the observer is pure translational with ground truth measured manually ($\eta_{gt} = 0$ deg and $\chi_{gt} = 9.2$ deg). Because the focus of expansion is

694

inside the field of view solutions are obtained by applying both steps of our algorithm yielding the estimates $\eta = -2$ deg and $\chi_{gt} = 5$ deg (Fig. 7). As already observed in the simulations the main error is in the deviation of the translation from the target direction in the second step. The observed robustness of the first step is consistent with the theoretical results by Maybank and Jepson [13, 14] in case of general motion who proved that if the observed surface is irregular the line through the center and the focus of expansion can be robustly estimated.

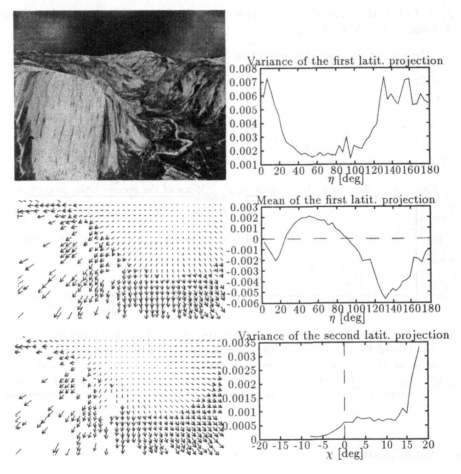

Fig. 5. The 1st and the 14th image of the Yosemite sequence (above), the original flow field (bottom left), and the fixated flow field (bottom right). The variance (top left) and the mean (top right) of the first latitudinal projection for the Yosemite sequence. The minimum of the variance gives the angle η and the mean for this η gives the torsion. The variance of the second latitudinal projection has its minimum at the right bound of χ indicating a focus of expansion outside the field of view.

Fig. 6. The 1st (left) and the 10th (middle) of the real fixated sequence and the computed optical flow field (right)

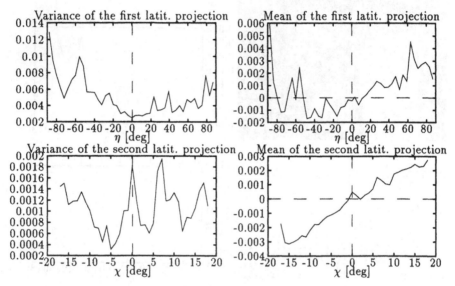

Fig. 7. The variance (top left) and the mean (top right) of the first latitudinal projection for the real fixated sequence. The minimum of the variance gives the angle η and the mean for this η gives the torsion. The variance (bottom left) and the mean (top right) of the second latitudinal projection for the real fixated sequence giving the angle χ at the minimum of the variance and the inverse of the time to collision, respectively.

5 Conclusion

It was proven in the past that fixation reduces the number of unknowns in the structure from motion problem from five to four. We showed in this paper that fixation can further simplify the computation of 3D-motion parameters from a monocular sequence. Appropriate projections of the spherical flow field enable the decoupling of the motion parameters in two groups: The first contains the azimuthal angle of the translation and the torsion. The second parameter group contains the polar angle of the translation direction and the time to collision to the fixated target. Two 1D searches in each of the two projections yield all

696

four unknowns. In contrast to other algorithms, we do not make any use of the measurements of camera movements necessary for fixation. We assume for the second search that the focus of expansion is inside the field of view. If this is not the case we can apply a two-unknowns nonlinear minimization or even better carry out a correcting saccade that will bring the focus of expansion inside the field of view. The algorithm was tested in three real world sequences fixated off-line or actively. Without applying any special method for accurate computation of the flow we obtained very promising results.

References

1. R.H.S. Carpenter. *Movements of the Eyes*. Pion Press, London, 1988.
2. K.J. Bradshaw, P.F. McLauchlan, I.D. Reid, and D.W. Murray. Saccade and pursuit on an active head-eye platform. *Image and Vision Computing*, 12:155–163., 1994.
3. A. Bandopadhay and D.H. Ballard. Egomotion perception using visual tracking. *Computational Intelligence*, 7:39–47, 1990.
4. Y. Aloimonos, I. Weiss, and A. Bandyopadhyay. Active Vision. In *Proc. Int. Conf. on Computer Vision*, pages 35–54, London, UK, June 8-11, 1987.
5. C. Fermüller and Y. Aloimonos. The role of fixation in visual motion anaylsis. *International Journal of Computer Vision*, 11:165–186, 1993.
6. M.A. Taalebinezhaad. Direct recovery of motion and shape in the general case by fixation. *IEEE Trans. Pattern Analysis and Machine Intelligence*, 14:847–853, 1992.
7. D. Raviv and M. Herman. A unified approach to camera fixation and vision-based road following. *IEEE Trans. Systems, Man, and Cybernetics*, 24:1125–1141, 1994.
8. I. Thomas, E. Simoncelli, and R. Bajcsy. Spherical retinal flow for a fixating observer. In *Proc. IEEE Workshop on Visual Behaviors*, pages 37–41, 1994.
9. M. Tistarelli and G. Sandini. Dynamic aspects in active vision. *CVGIP: Image Understanding*, 56:108–129, 1992.
10. K. Daniilidis. Computation of 3D-motion parameters using the log-polar transform. In *V. Hlavac et al. (Ed.), Proc. Int. Conf. Computer Analysis of Images and Patterns CAIP, Prag*, pages 82–89, 1995.
11. M.J. Barth and S. Tsuji. Egomotion determination through an intelligent gaze control strategy. *IEEE Trans. Systems, Man, and Cybernetics*, 23:1424–1432, 1993.
12. V. Sundareswaran, P. Bouthemy, and F. Chaumette. Active camera self-orientation using dynamic image parameters. In *Proc. Third European Conference on Computer Vision*, pages 111–115. Stockholm, Sweden, May 2-6, J.O. Eklundh (Ed.), Springer LNCS 800, 1994.
13. S. Maybank. *Theory of Reconstruction from Image Motion*. Springer-Verlag, Berlin et al., 1993.
14. A.D. Jepson and D.J. Heeger. Subspace methods for recovering rigid motion II: Theory. Technical Report RBCV-TR-90-36, University of Toronto, 1990.

Automatic Singularity Test for Motion Analysis by an Information Criterion

Kenichi Kanatani*

Department of Computer Science
Gunma University, Kiryu, Gunma 376, Japan

Abstract. The structure-from-motion algorithm from two views fails
if the object is a planar surface or the camera motion is a pure rotation. This paper presents a new scheme for automatically detecting these
anomalies without using any knowledge about the noise in the images.
This judgment does not involve any empirically adjustable thresholds,
either. The basic principle of our scheme is to choose a model that has
"higher predicting capability" measured by the *geometric information
criterion* (*geometric AIC*).

1 Introduction

The *structure-from-motion* algorithm from two views has been studied by many
researchers in the past [3, 5, 11]. However, the algorithm fails if all the feature
points are coplanar in the scene, because a planar surface is a degenerate *critical
surface* that gives rise to ambiguity of 3-D interpretation [4, 9]. Hence, a different
algorithm is necessary for a planar surface. The planar surface algorithm has also
been studied by many researchers in the past [7, 8, 10]. However, both the general
and the planar surface algorithms assume that the translation of the camera is
not zero; if the camera motion is a pure rotation around the center of the lens,
no 3-D information can be obtained. If follows that the structure-from-motion
analysis must take the following steps:

1. Test if the translation is 0—we call this the *rotation test*. If so, output a
   warning message and stop.
2. Test if the object is a planar surface—we call this the *planarity test*. If so,
   apply the planar surface algorithm.
3. Else, apply the general algorithm.

In practice, however, the images have noise, and the general algorithm applied in
the presence of noise produces some (unreliable) solution even when the camera
motion is a pure rotation or the object is a planar surface. In the past, the
above tests have been done by introducing an ad-hoc criterion and an empirically
adjustable threshold. For example, based on the fact that the smallest eigenvalue

* This work was in part supported by the Ministry of Education, Science, Sports and
Culture, Japan under a Grant in Aid for Scientific Research B (No. 07458067) and
the Okawa Institute of Information and Telecommunication.

of a matrix involved in the analysis should be a multiple root in the absence of noise if the object is a planar surface, the object is judged as planar if its smallest two eigenvalues are close enough. However, *how can we determine the threshold for such a judgement?*

- First of all, we need to know the accuracy of the detected feature points, because the threshold should be set high if the accuracy is low while it should be set low if the accuracy is high. However, the accuracy of the feature points detected by an image processing operation depends on not only the operation itself but also various imaging conditions such as the focus, the resolution, the lighting, the shape of the object, and its orientation and position. Hence, the accuracy is different from image to image, and it is almost impossible to predict it in advance.
- Even if the accuracy can be predicted, what can be obtained is a *probability* of the noise, since the noise is a random phenomenon. Following the statistical theory of testing of hypotheses, we can set the threshold in such a way that the probability (the *significance level*) that a planar surface is judged as non-planar is $a\%$. However, how can we set that significance level? The result of the judgment differs if the significance level is set different.

In the past, little attention has been paid to this problem. Sometimes, thresholds are adjusted so that the experiment in question works well. In this paper, we present a theoretical framework for doing the planarity and rotation tests *without knowing the magnitude of the image noise* and *without introducing any empirically adjustable thresholds*. What makes this possible is the introduction of the *geometric information criterion (geometric AIC)*, which measures the *predictive capacity* of the model in statistical terms.

2   3-D Reconstruction from Two Views

Define an XYZ camera coordinate system in such a way that the origin O is at the center of the lens and the Z-axis is along the optical axis. If the distance between the origin O and the photo-cell array is taken as the unit of length, the image plane can be identified with $Z = 1$; the imaging geometry can be regarded as perspective projection onto it. Define an xy image coordinate system on the image plane $Z = 1$ in such a way that the origin o is on the optical axis and the x- and y-axes are parallel to the X- and Y-axes, respectively. Then, a point (x, y) on the image plane can be represented by vector $x = (x, y, 1)^\top$ (the superscript $\top$ designates transpose).

Suppose the camera is rotated around the center of the lens by R and translated by h (Fig. 1(a)). We call $\{h, R\}$ the *motion parameters*. If we define the $X'Y'Z'$ camera coordinate system and the $x'y'$ image coordinate system with respect to the camera position after this motion, a point (x', y') on the image plane, which can be represented by vector $x' = (x', y', 1)$ with respect to the $X'Y'Z'$ coordinate system, can be represented by vector Rx' with respect to the XYZ coordinate system. If follows that vectors x and x' can be images of

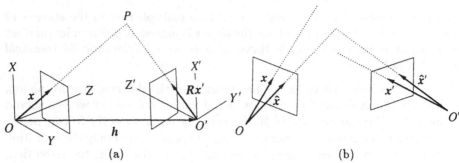

Fig. 1. (a) Geometry of camera motion. (b) Vectors **x** and **x**′ are corrected so that their lines of sight meet in the scene.

the same point in the scene if and only if the following *epipolar equation* holds[2] [2, 4, 9, 12]:

$$|x, h, Rx'| = 0. \tag{1}$$

This equation states that the *lines of sight* defined by vectors x and x' should meet in the scene. In the presence of image noise, vectors x and x' may not satisfy eq. (1) even if the motion parameters $\{h, R\}$ are correct. In order to determine a unique point in the scene, we must correct vectors x and x' so that their lines of sight meet in the scene (Fig. 1(b)).

3 General Model

Given N corresponding points (x_α, y_α), (x'_α, y'_α), $\alpha = 1, ..., N$, between the two images, we decompose the vectors x_α and x'_α that represent them into the form

$$x_\alpha = \bar{x}_\alpha + \Delta x_\alpha, \qquad x'_\alpha = \bar{x}'_\alpha + \Delta x'_\alpha, \tag{2}$$

where $\bar{x}_\alpha$ and $\bar{x}'_\alpha$ represent the "true" positions supposedly observed if noise did not exist (i.e., if the feature points detected from the gray-level images were accurate). We regard the noise terms Δx_α and $\Delta x'_\alpha$ as independent Gaussian random variables of mean **0** and respective covariance matrices[3] $V[x_\alpha] \ (= E[\Delta x_\alpha \Delta x_\alpha^\top])$ and $V[x'_\alpha] \ (= E[\Delta x'_\alpha \Delta x'^\top_\alpha])$ ($E[\cdot]$ denotes expectation). Although it is in general very difficult to predict the accuracy of feature point detection in advance, it is often possible to predict qualitative characteristics such as uniformity and isotropy. So, we assume that the covariance matrices $V[x_\alpha]$ and $V[x'_\alpha]$ are known *up to scale*. In other words, we assume that there exists an unknown constant ϵ, which we call the *noise level*, such that

$$V[x_\alpha] = \epsilon^2 V_0[x_\alpha], \qquad V[x'_\alpha] = \epsilon^2 V_0[x'_\alpha] \tag{3}$$

---

[2] $|\mathbf{a}, \mathbf{b}, \mathbf{c}|$ denotes the scalar triple product of vectors **a**, **b**, and **c**.

[3] Since the third components of vectors $\mathbf{x}_\alpha$ and $\mathbf{x}'_\alpha$ are both 1, the covariance matrices $V[\mathbf{x}_\alpha]$ and $V[\mathbf{x}'_\alpha]$ are singular matrices of rank 2 whose ranges are the XY and $X'Y'$ planes, respectively.

for known matrices $V_0[x_\alpha]$ and $V_0[x_\alpha]$, which we call the *normalized covariance matrices*.

In the presence of noise, the corresponding points x_α and x'_α are not accurate, so they must be corrected so as to satisfy the epipolar equation (1). Infinitely many solutions exist for this correction. From among them, we choose the one that is statistically the most likely. This can be done by choosing the vectors $\hat{x}_\alpha$ and $\hat{x}'_\alpha$ that minimize the sum of the squared *Mahalanobis distances* [6]:

$$J = \sum_{\alpha=1}^{N}(x_\alpha - \hat{x}_\alpha, V_0[x_\alpha]^-(x_\alpha - \hat{x}_\alpha)) + \sum_{\alpha=1}^{N}(x'_\alpha - \hat{x}'_\alpha, V_0[x'_\alpha]^-(x'_\alpha - \hat{x}'_\alpha)). \quad (4)$$

Throughout this paper, we denote the inner product of vectors by $(\cdot, \cdot)$ and the (*Moore-Penrose*) *generalized inverse*[4] by $(\cdot)^-$. The first order solution of the above minimization is given as follows [6]:

$$\hat{x}_\alpha = x_\alpha - \frac{(x_\alpha, Gx'_\alpha)V_0[x_\alpha]Gx'_\alpha}{(\hat{x}'_\alpha, G^\top V_0[x_\alpha]G\hat{x}'_\alpha) + (\hat{x}_\alpha, GV_0[x'_\alpha]G^\top \hat{x}_\alpha)},$$

$$\hat{x}'_\alpha = x'_\alpha - \frac{(x_\alpha, Gx'_\alpha)V_0[x'_\alpha]G^\top x_\alpha}{(\hat{x}'_\alpha, G^\top V_0[x_\alpha]G\hat{x}'_\alpha) + (\hat{x}_\alpha, GV_0[x'_\alpha]G^\top \hat{x}_\alpha)}. \quad (5)$$

Here, the matrix G is defined by

$$G = h \times R, \quad (6)$$

and called the *essential matrix* [2, 4, 9, 12]; we define the product $a \times U$ of vector $a = (a_i)$ and matrix $U = (U_{ij})$ as a matrix whose (ij) element is[5] $\sum_{k,l=1}^{3} \varepsilon_{ikl}a_kU_{lj}$.

The minimum of the function J defined by eq. (4) is a function of the motion parameters $\{h, R\}$, so we write it as $J[h, R]$. Substituting eqs. (5) into eq. (4), we obtain the following expression [6]:

$$J[h, R] = \sum_{\alpha=1}^{N} \frac{(x_\alpha, Gx'_\alpha)^2}{(x'_\alpha, G^\top V_0[x_\alpha]Gx'_\alpha) + (x_\alpha, GV_0[x'_\alpha]G^\top x_\alpha)}. \quad (7)$$

The statistically most likely values of the motion parameters $\{h, R\}$ are those that minimize this function. From eq. (7), we can immediately see that the scale of the translation h is indeterminate[6], so we normalize it into $\|h\| = 1$ (we denote the norm of a vector by $\|\cdot\|$). The minimization of eq. (7) can be conducted accurately and effectively by a numerical technique called

---

[4] It is computed by applying the *spectral decomposition* [6] and replacing nonzero eigenvalues by their reciprocals.

[5] The symbol ε_{ijk} is the *Eddington epsilon*, taking values 1 and -1 if (iji) is an even and odd permutations of (123), respectively, and value 0 otherwise.

[6] This corresponds to the well known fact that a large camera motion relative to a large object in the distance is indistinguishable from a small camera motion relative to a small object near the camera as long as images are the only source of information.

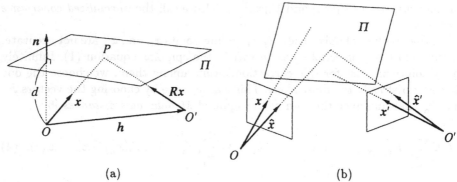

(a)   (b)

Fig. 2. (a) Geometry of camera motion relative to a planar surface. (b) Vectors **x** and **x**$'$ are corrected so that their lines of sight meet on the surface Π.

renormalization [5]. According to our statistical model, the motion parameters $\{\hat{h}, \hat{R}\}$ that minimize eq. (7) and the corrected positions $\hat{x}_\alpha$ and $\hat{x}'_\alpha$ are the *maximum likelihood estimators* of $\{h, R\}$, $\bar{x}_\alpha$, and $\bar{x}'_\alpha$, respectively.

4   Planar Surface Model

Suppose the object is a planar surface Π. Let n be its unit surface normal with respect to the XYZ coordinate system, and d its distance (positive in the direction n) from the origin O; we call $\{n, d\}$ the *surface parameters* (Fig. 2(a)). It can be easily shown that vectors x and x' can be images of the same point on the surface Π if and only if the following equation holds [2, 4, 8, 9, 10, 12]:

$$x' \times Ax = 0. \tag{8}$$

Here, A is a matrix defined by

$$A = R^\top(hn^\top - dI), \tag{9}$$

where I is the unit matrix. Eq. (8) states that the lines of sights defined by x and x' should meet on the surface Π; this is a stronger condition than the epipolar equation (1), which is automatically implied. However, eq. (8) may not exactly hold in the presence of noise, so we must correct x and x' so as to satisfy it (Fig. 2(b)).

Given N corresponding points x_α, x'_α, $\alpha = 1, ..., N$, the maximum likelihood estimators $\hat{x}_\alpha$ and $\hat{x}'_\alpha$ of $\bar{x}_\alpha$ and $\bar{x}'_\alpha$ for fixed surface and motion parameters $\{n, d\}$ and $\{h, R\}$ are the vectors that minimize the function J given by eq. (4) under the constraint that they satisfy eq. (8). The first order solution is given as follows [6]:

$$\hat{x}_\alpha = x_\alpha - (x'_\alpha \times AV_0[x_\alpha])^\top W_\alpha(x'_\alpha \times Ax_\alpha),$$
$$\hat{x}'_\alpha = x'_\alpha + ((Ax_\alpha) \times V_0[x'_\alpha])^\top W_\alpha(x'_\alpha \times Ax_\alpha), \tag{10}$$

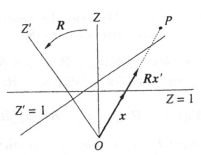

Fig. 3. Pure rotation of the camera.

$$W_\alpha = \left(x'_\alpha \times AV_0[x_\alpha]A^\top \times x'_\alpha + (Ax_\alpha) \times V_0[x'_\alpha] \times (Ax_\alpha) \right)^-_2. \qquad (11)$$

The product $a \times A \times b$ of vectors $a = (a_i)$ and $b = (b_i)$ and a matrix $A = (A_{ij})$ is defined to be a matrix whose (ij) element is $\sum^3_{k,l,m,n=1} \varepsilon_{ikm}\varepsilon_{jln}a_{kb_l}A_{mn}$. The symbol $(\cdot)^-_r$ denotes the *rank-constrained generalized inverse*[7] [6]. The minimum of the function J is a function of the surface and motion parameters $\{n, d\}$ and $\{h, R\}$, so we write it as $J[n, d, h, R]$. Substituting eqs. (10) into eq. (4), we obtain the following expression [6]:

$$J[n, d, h, R] = \sum^N_{\alpha=1}(x'_\alpha \times Ax_\alpha, W_\alpha(x'_\alpha \times Ax_\alpha)). \qquad (12)$$

The maximum likelihood estimators $\{\hat{n}, \hat{d}\}$ and $\{\hat{h}, \hat{R}\}$ of the surface and motion parameters $\{n, d\}$ and $\{h, R\}$ are the values that minimize eq. (12). This minimization can be conducted accurately and effectively by a numerical technique called *renormalization* [7].

5 Pure Rotation Model

Let x and x' be the images of the same point in the scene. From Fig. 3, it is easily seen that the camera motion is a pure rotation if and only if there exists a rotation matrix R such that

$$x \times Rx' = 0. \qquad (13)$$

Given N corresponding points x_α, x'_α, $\alpha = 1, ..., N$, the maximum likelihood estimators $\hat{x}_\alpha$ and $\hat{x}'_\alpha$ of $\bar{x}_\alpha$ and $\bar{x}'_\alpha$ for a fixed rotation R are the vectors that

---

[7] The matrix obtained by applying the spectral decomposition, replacing all but the largest r eigenvalues by 0, and computing the generalized inverse. This operation is necessary to prevent numerical instability of the computation; the expression inside the parentheses becomes singular in the limit as x_α and x'_α approach $\bar{x}_\alpha$ and $\bar{x}'_\alpha$, respectively.

minimize the function J given by eq. (4) under the constraint that they satisfy eq. (13). The first order solution is given as follows [6]:

$$\hat{x}_\alpha = x_\alpha + ((Rx'_\alpha) \times V_0[x_\alpha])^\top W_\alpha (x_\alpha \times Rx'_\alpha),$$
$$\hat{x}'_\alpha = x'_\alpha - (x_\alpha \times RV_0[x'_\alpha])^\top W_\alpha (x_\alpha \times Rx'_\alpha), \qquad (14)$$

$$W_\alpha = \Big((Rx'_\alpha) \times V_0[x_\alpha] \times (Rx'_\alpha) + x_\alpha \times RV_0[x'_\alpha]R^\top \times x_\alpha\Big)_2^-. \qquad (15)$$

The minimum of the function J is a function of the rotation R, so we write it as $J[R]$. Substituting eqs. (14) into eq. (4), we obtain the following expression [6]:

$$J[R] = \sum_{\alpha=1}^{N} (x_\alpha \times Rx'_\alpha, W_\alpha(x_\alpha \times Rx'_\alpha)). \qquad (16)$$

The maximum likelihood estimators $\hat{R}$ of the rotation R is the matrix that minimizes eq. (16). This minimization can be conducted numerically by steepest descent. A good approximation[8] of $\hat{R}$ is analytically obtained by applying the singular value decomposition [4].

6 Geometric Model

Regardless of the shape of the object and the motion of the camera, the problem can be generalized in abstract terms as follows. A pair of corresponding vectors x and x' can be identified with a *six*-dimensional direct sum vector $x \oplus x'$. The third components of x and x' are both 1, so vector $x \oplus x'$ is constrained to be in the four-dimensional affine subspace

$$\mathcal{X} = \{(x, y, 1, x', y', 1)^\top | x, y, x', y' \in \mathcal{R}\}, \qquad (17)$$

which we call the *data space* ($\mathcal{R}$ denotes the set of all real numbers). Observing N corresponding points x_α and x'_α is equivalent to observing N vectors $x_\alpha \oplus x'_\alpha$ in the data space $\mathcal{X}$.

Suppose there exists a constraint on the shape of the object and/or the motion of the camera that can be expressed as L equations parameterized by an n-dimensional vector u in the form

$$F^{(k)}(x, x'; u) = 0, \qquad k = 1, ..., L. \qquad (18)$$

These L equations need not be algebraically independent[9] as equations of x and x'. We call the number r of independent equations the *rank* of the constraint.

---

[8] As shown shortly, what we actually need is not the estimate $\hat{R}$ itself but the residual $J[\hat{R}]$. Since $\hat{R}$ minimizes $J[R]$, the residual $J[\hat{R}]$ can be accurately computed even from an approximate value of $\hat{R}$.

[9] In order to avoid pathological cases, we assume that each of the L equations defines a manifold of *codimension 1* in the data space $\mathcal{X}$ such that the L manifolds intersect each other *transversally* [6].

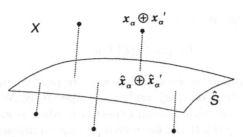

Fig. 4. The model S is optimally fitted to the data points, and the data points are optimally projected onto it.

Eq. (18) then defines a manifold S of *codimension*[10] r in the data space $\mathcal{X}$. We call S the *(geometric) model*. The domain $\mathcal{U}$ of the vector u that parameterizes the constraint is called the *parameter space*. If it is an n'-dimensional manifold in $\mathcal{R}^n$, the model S is said to have n' *degrees of freedom*.

Given N corresponding points x_α, x'_α, $\alpha = 1, ..., N$, in the presence of noise, there may not exist an instance of S that exactly passes through all the N points $x \oplus x'_\alpha \in \mathcal{X}$. So, we optimally *fit* the model S by adjusting the parameter $u \in \mathcal{U}$ in the sense of *maximum likelihood estimation*. Let $\hat{S}$ be the resulting optimal fit. We then optimally *project* each point $x_\alpha \oplus x'_\alpha \in \mathcal{X}$ onto $\hat{S}$ in the sense of *maximum likelihood estimation* (Fig. 4). Let $\hat{x}_\alpha \oplus \hat{x}'_\alpha$ be the resulting optimal projection; $\hat{x}_\alpha$ and $\hat{x}'_\alpha$ are the maximum likelihood estimators of $\bar{x}_\alpha$ and $\bar{x}'_\alpha$, respectively. Let $\hat{J}$ be the residual of the function J for the maximum likelihood estimators $\hat{x}_\alpha$ and $\hat{x}'_\alpha$. It can be proved that $\hat{J}/\epsilon^2$ is subject to a χ^2 distribution with $rN - n'$ degrees of freedom in the first order [6]. Hence, an unbiased estimator of the squared noise level ϵ^2 is obtained in the following form:

$$\hat{\epsilon}^2 = \frac{\hat{J}}{rN - n'} \tag{19}$$

7 Geometric Information Criterion

The "goodness" of a model can be measured by its "predicting capability" [1]. Let x_α^* and $x_\alpha^{*'}$ be *future data* that have the same probability distribution as the current data x_α and x'_α and are independent of x_α and x'_α. The residual of model S for the maximum likelihood estimators $\hat{x}_\alpha$ and $\hat{x}'_\alpha$, which are computed from the current data x_α and x'_α, *with respect to the future data* x_α^* *and* $x_\alpha^{*'}$ is

$$\hat{J}^* = \sum_{\alpha=1}^{N}(x_\alpha^* - \hat{x}_\alpha, V_0[x_\alpha]^-(x_\alpha^* - \hat{x}_\alpha)) + \sum_{\alpha=1}^{N}(x_\alpha^{*'} - \hat{x}'_\alpha, V_0[x'_\alpha]^-(x_\alpha^{*'} - \hat{x}'_\alpha)). \tag{20}$$

---

[10] The difference between the dimension of the space (in this case $\mathcal{X}$) and the dimension of the manifold.

This is a random variable; its expectation is

$$I(S) = E^*[E[\hat{J}^*]], \tag{21}$$

where $E[\cdot]$ and $E^*[\cdot]$ denotes expectation with respect to the current data x_α and x'_α and the future data x^*_α and $x^{*'}_\alpha$, respectively. We call $I(S)$ the *expected residual*. A model with a small expected residual is expected to have high predicting capability [1]. It can be proved [6] that an unbiased estimator of the expected residual is given by

$$AIC(S) = \hat{J} + 2(pN + n')\epsilon^2, \tag{22}$$

where $p = 4 - r$ is the dimension of the manifold S. We call eq. (22) the *geometric information criterion*, or *geometric AIC* for short.

8 Automatic Model Selection

Let S_1 be a model of dimension p_1 and codimension r_1 with n'_1 degrees of freedom, and S_2 a model of dimension p_2 and codimension r_2 with n'_2 degrees of freedom. Suppose model S_2 is obtained by adding an additional constraint to model S_2. We say that model S_2 is *stronger* than model S_1, or model S_1 is *weaker* than model S_2, and write

$$S_2 \succ S_1. \tag{23}$$

Let $\hat{J}_1$ and $\hat{J}_2$ be the residuals of S_1 and S_2, respectively. If model S_1 is correct, the squared noise level ϵ^2 is estimated by eq. (19). Substituting it into the expression for the geometric AIC, we obtain

$$AIC(S_1) = \hat{J}_1 + \frac{2(p_1 N + n'_1)}{r_1 N - n'_1}\hat{J}_1, \quad AIC(S_2) = \hat{J}_2 + \frac{2(p_2 N + n'_2)}{r_1 N - n'_1}\hat{J}_1. \tag{24}$$

Recalling that the geometric AIC is an estimator of the expected residual (see eq. (20)), we put

$$K = \sqrt{\frac{AIC(S_2)}{AIC(S_1)}} = \sqrt{\frac{r_1 N - n'_1}{(2p_1 + r_1)N + n'_1}\left(\frac{\hat{J}_2}{\hat{J}_1} + \frac{2(p_2 N + n'_2)}{r_1 N - n'_1}\right)}. \tag{25}$$

This quantity describes the ratio of the expected deviation from model S_2 to the expected deviation from model S_1. It follows that if

$$K < 1 \tag{26}$$

model S_2 is *preferable* to S_1 with regard to the predicting capability. This criterion requires no knowledge about the noise magnitude and involves no empirically adjustable thresholds.

9   Planarity and Rotation Tests

The epipolar equation (1) defines a model S of codimension 1 in the four-dimensional data space $\mathcal{X}$ parameterized by the motion parameters $\{h, R\}$, which have five degrees of freedom. Hence, its geometric AIC is

$$AIC(S) = \hat{J} + (6N + 10)\epsilon^2. \tag{27}$$

The three components of eq. (8) are algebraically independent, so the rank of the constraint that the object is a planar surface is 2. Hence, it defines a model S_Π of codimension 2 in $\mathcal{X}$ parameterized by the surface and motion parameters $\{n, d\}$ and $\{h, R\}$, which have eight degrees of freedom. Its geometric AIC is

$$AIC(S_\Pi) = \hat{J}_\Pi + (4N + 16)\epsilon^2. \tag{28}$$

The three components of eq. (13) are algebraically independent, so the rank of the constraint that the camera motion is a pure rotation is 2. Hence, it defines a model S_R of codimension 2 in $\mathcal{X}$ parameterized by the rotation R, which has three degrees of freedom. Its geometric AIC is

$$AIC(S_R) = \hat{J}_\Pi + (4N + 6)\epsilon^2. \tag{29}$$

From eqs. (1), (8), and (13), we observe the following order of strength:

$$S_R \succ S_\Pi \succ S. \tag{30}$$

Since the general model S is also valid for the planar surface model S_Π and the pure rotation model S_R, the object is judged to be a planar surface if

$$K_\Pi = \sqrt{\frac{N-5}{7N+5}\left(\frac{\hat{J}_\Pi}{\hat{J}} + \frac{4N+16}{N-5}\right)} < 1, \tag{31}$$

and the camera motion is judged to be a pure rotation if

$$K_R = \sqrt{\frac{N-5}{7N+5}\left(\frac{\hat{J}_R}{\hat{J}} + \frac{4N+6}{N-5}\right)} < 1. \tag{32}$$

10   Examples

Two planar grids hinged together with angle $\pi - \theta$ were defined in the scene, and images that simulate two views from different camera positions were generated. The image size and the focal length were assumed to be 512×512 (pixels) and 600 (pixels), respectively. Fig. 5(a) shows the images for $\theta = 50°$. The x- and y-coordinates of each grid point were perturbed by independent random Gaussian noise of mean 0 and standard deviation σ (pixels). Using the grid points as feature points, we conducted the planarity test 100 times, each time using

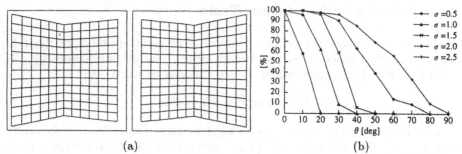

(a)   (b)

Fig. 5. (a) Images of two planar grids hinged together in the scene ($\theta = 50°$). (b) The percentage of the instances judged as planar.

different noise. Fig. 5(b) shows, for various values of θ and for $\sigma = 0.5 \sim 3.0$, the percentage of the instances for which the object is judged as planar. We see that the threshold for the test is automatically adjusted to the noise.

If $\sigma = 1.0$, the percentage is approximately 50% for $\theta = 22°$. Fig. 6(a) shows one instance for which the object is judged as planar ($K_\Pi = 0.87$), and Fig. 6(b) shows the 3-D shapes reconstructed from the general and planar surface models. The true shape is superimposed in dashed lines. Fig. 7(a) shows a one instance for which the object is judged as non-planar ($K_\Pi = 1.13$). From these, we see that although the perturbed images look almost the same, the reconstructed shape from the general model has little sense if the object is judged as planar, while the non-planar shape can be reconstructed fairly well if the object is judged as non-planar.

11 Concluding Remarks

We have presented a scheme for automatically testing if the object can be regarded as a planar surface or if the camera motion is a pure rotation without using any knowledge about the noise in the images. This judgment does not involve any empirically adjustable thresholds, either. The basic principle of our scheme is to choose a model that has "higher predicting capability" measured by the *geometric information criterion (geometric AIC)*. Our approach presents a new paradigm for geometric model selection in a wide range of problems of robotics and computer vision.

References

1. Akaike, H.: A new look at the statistical model identification. IEEE Trans. Automation Control **19**-6 (1974) 776–723
2. Faugeras, O.: Three-Dimensional Computer Vision: A Geometric Viewpoint. MIT Press, Cambridge, MA, 1993
3. Faugeras, O. D. and S. Maybank, S.: Motion from point matches: Multiplicity of solutions. Int. J. Comput. Vision **4**-3 (1990) 225–246

708

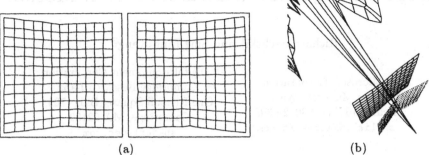

Fig. 6. (a) An instance for which the object is judged as planar. (b) Reconstructed 3-D shapes and the true shape.

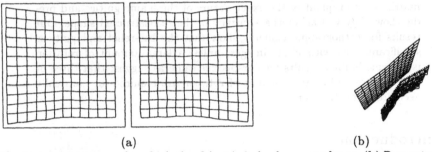

Fig. 7. (a) An instance for which the object is judged as non-planar. (b) Reconstructed 3-D shapes and the true shape.

4. Kanatani, K.: Geometric Computation for Machine Vision. Oxford University Press, Oxford, 1993
5. Kanatani, K.: Renormalization for motion analysis: Statistically optimal algorithm. IEICE Trans. Inf. & Sys. **E77-D**-11 (1994) 1233–1239
6. Kanatani, K.: Statistical Optimization for Geometric Computation: Theory and Practice. Elsevier Science, Amsterdam, 1996
7. Kanatani, K. and Takeda, S.: 3-D motion analysis of a planar surface by renormalization. IEICE Trans. Inf. & Syst. **E78-D**-8 (1995) 1074–1079
8. Longuet-Higgins, L. C.: The reconstruction of a plane surface from two perspective projections. Proc. Roy. Soc. Lond. **B227** (1986) 399–410
9. Maybank, S.: Theory of Reconstruction from Image Motion. Springer, Berlin, 1993
10. Weng, J., Ahuja, N. and Huang, T. S.: Motion and structure from point correspondences with error estimation: Planar surfaces. IEEE Trans. Sig. Proc. **39**-12 (1991) 2691–2717
11. Weng, J., Ahuja, N. and Huang, T. S.: Optimal motion and structure estimation. IEEE Trans. Patt. Anal. Mach. Intell. **15**-9 (1993) 864–884
12. Weng, J., Huang, T. S. and Ahuja, N.: Motion and Structure from Image Sequences. Springer, Berlin, 1993

Shape Ambiguities in Structure from Motion

Richard Szeliski[1] and Sing Bing Kang[2]

[1] Microsoft Corporation,
One Microsoft Way,
Redmond, WA 98052-6399
szeliski@microsoft.com

[2] Digital Equipment Corporation,
Cambridge Research Lab,
One Kendall Square, Bldg. 700,
Cambridge, MA 02139
sbk@crl.dec.com

Abstract. This paper examines the fundamental ambiguities and uncertainties inherent in recovering structure from motion. By examining the eigenvectors associated with null or small eigenvalues of the Hessian matrix, we can quantify the exact nature of these ambiguities and predict how they will affect the accuracy of the reconstructed shape. Our results for orthographic cameras show that the bas-relief ambiguity is significant even with many images, unless a large amount of rotation is present. Similar results for perspective cameras suggest that three or more frames and a large amount of rotation are required for metrically accurate reconstruction.

1 Introduction

Structure from motion is one of the classic problems in computer vision and has received a great deal of attention over the last decade. It has wide-ranging applications, including robot vehicle guidance and obstacle avoidance, and the reconstruction of 3-D models from imagery. Unfortunately, the quality of results available using this approach is still often very disappointing. More precisely, while the qualitative estimates of structure and motion look reasonable, the actual quantitative (*metric*) estimates can be significantly distorted.

Much progress has been made recently in identifying the sources of errors and instabilities in the structure from motion process. It is now widely understood that the arbitrary algebraic manipulation of the imaging equations to derive closed-form solutions (e.g., [1]) can lead to algorithms that are numerically ill-conditioned or unstable in the presence of measurement errors. To overcome this, statistically optimal algorithms for estimating structure and motion have been developed [2, 3, 4]. It is also understood that using more feature points and images results in better estimates, and that certain configurations of points (at least in the two frame case) are pathological and cannot be reconstructed.

An example of an algorithm which generates very good results is the factorization approach of Tomasi and Kanade [5]. This algorithm assumes orthography and is implemented using an object-centered representation and singular value decomposition. It uses many points and frames, and for most sequences, a large amount of object rotation (usually 360°). However, when only a small range of viewpoints is present (e.g., the "House" sequence in [5], Fig. 7), the reconstruction no longer appears metric (the house walls are not perpendicular).

In this paper, we demonstrate that it is precisely this last factor, i.e., the overall rotation of the object, or equivalently, the variation in viewpoints, which critically determines the quality of the reconstruction. The ambiguity in object shape due to small viewpoint variation often looks like it might be a *projective* deformation of the Euclidean shape, which is interesting—several researchers have argued recently in favor of trying to recover only this projective structure [6, 7, 8]. In fact, we show that the major ambiguity in the reconstruction is a simple depth scale uncertainty, i.e., the classic *bas-relief* ambiguity which exists for two-frame structure from motion under orthographic projection [9].

To derive our results, we use eigenvalue analysis of the covariance matrix for the structure and motion estimates. Our results are significant for two reasons. First, we show how to theoretically derive the expected ambiguity in a reconstruction, and also derive some intuitive guidelines for selecting imaging situations which can be expected to produce reasonable results. Second, since the primary ambiguities are very well characterized by a small number of modes, this information can be used to construct better on-line (recursive) estimation algorithms.

Our paper is structured as follows. After reviewing previous work, we present our formulation of the structure from motion problem and develop our technique for analyzing ambiguities using eigenvector analysis of the information (Hessian) matrix. We then present the results of our analysis for two different camera models: 1-D orthographic cameras and 2-D perspective cameras (more examples and results are presented in [10]). We conclude with a discussion of the main sources of errors and ambiguities, and directions for possible future work.

2 Previous work

Structure from motion has been extensively studied in computer vision. Early papers on this subject develop algorithms to compute the structure and motion from a small set of points matched in two frames using an *essential parameter* approach [1]. The performance of this approach can be significantly improved using non-linear least squares (*optimal estimation*) techniques [2, 3]. More recent research focuses on extraction of shape and motion from longer image sequences using both batch and recursive (Kalman filter) formulations [4, 5, 11, 12, 13]. Another line of research has addressed recovering affine [14] or projective [6, 7, 8] structure estimates. For a more detailed review of related work, please see [4, 10].

The nature of structure and motion errors, which is the main focus of this paper, has also previously been studied. Weng *et al.* perform some of the earliest and most detailed error analyses of the two-frame essential parameter approach [3]. Adiv [15] and Young and Chellappa [16] analyze continuous-time (optical flow) based algorithms using the concept of the Cramer-Rao lower bound. Oliensis and Thomas [17] show how modeling the motion error can significantly improve the performance of recursive algorithms.

In this paper, we extend these previous results using an eigenvalue analysis of the covariance matrix. This analysis can pinpoint the exact nature of structure from motion ambiguities and the largest sources of reconstruction error. We also focus on multi-frame optimal structure from motion algorithms, which have not been studied in great detail.

3 Problem formulation and uncertainty analysis

The equation which projects the ith 3-D point $\mathbf{p}_i$ (given time-varying motion parameters $\mathbf{m}_j$) into the jth frame at location $\mathbf{u}_{ij}$ is

$$\mathbf{u}_{ij} = \mathcal{P}\left(T(\mathbf{p}_i, \mathbf{m}_j)\right). \tag{1}$$

The perspective projection $\mathcal{P}$ (defined below) is applied to a rigid transformation

$$T(\mathbf{p}_i, \mathbf{m}_j) = \mathbf{R}_j \mathbf{p}_i + \mathbf{t}_j, \tag{2}$$

where $\mathbf{R}_j$ is a rotation matrix and $\mathbf{t}_j$ is a translation applied after the rotation. A variety of alternative representations are possible for the rotation matrix [18]. In this paper, we represent the rotation matrix as a function of a quaternion, since this representation has no singularities.

The standard perspective projection equation used in computer vision is

$$\begin{pmatrix} u \\ v \end{pmatrix} = \mathcal{P}_1 \begin{pmatrix} x \\ y \\ z \end{pmatrix} \equiv \begin{pmatrix} f\frac{x}{z} \\ f\frac{y}{z} \end{pmatrix}, \tag{3}$$

where f is a product of the focal length of the camera and the pixel scale factor (assuming that pixels are square). An alternative object-centered formulation, which we introduced in [4] is

$$\begin{pmatrix} u \\ v \end{pmatrix} = \mathcal{P}_2 \begin{pmatrix} x \\ y \\ z \end{pmatrix} \equiv \begin{pmatrix} s\frac{x}{1+\eta z} \\ s\frac{y}{1+\eta z} \end{pmatrix}. \tag{4}$$

Here, we assume that the (x, y, z) coordinates before projection are with respect to a reference frame Π_j that has been displaced away from the camera by a distance t_z along the optical axis, with $s = f/t_z$ and $\eta = 1/t_z$ (Fig. 1). The projection parameter s can be interpreted as a *scale factor* and η as a *perspective distortion factor*. Our alternative perspective formulation allows us to model both orthographic and perspective cameras using the same model.

In our previous work, we used the iterative Levenberg-Marquardt algorithm to estimate $\{\mathbf{p}_i, \mathbf{m}_j\}$ from $\{\mathbf{u}_{ij}\}$, since it provides a statistically optimal solution [2, 3, 4, 12]. The Levenberg-Marquardt method is a standard non-linear least squares technique [19] which minimizes

$$C(\mathbf{a}) = \sum_i \sum_j c_{ij} |\tilde{\mathbf{u}}_{ij} - \mathbf{f}_{ij}(\mathbf{a})|^2, \tag{5}$$

where $\tilde{\mathbf{u}}_{ij}$ is the observed image measurement, $\mathbf{f}_{ij}(\mathbf{a}) = \mathbf{u}(\mathbf{p}_i, \mathbf{m}_j)$ is given in (1), and $\mathbf{a}$ contains the 3-D points $\mathbf{p}_i$, the motion parameters $\mathbf{m}_j$, and any additional unknown calibration parameters. The weight c_{ij} in (5) describes the confidence in measurement $\mathbf{u}_{ij}$.

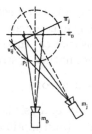

Fig. 1. Sample configuration of cameras ($\mathbf{m}_j$), 3-D points ($\mathbf{p}_i$), image planes(Π_j), and screen locations ($\mathbf{u}_{ij}$)

3.1 Uncertainty analysis

Regardless of the solution technique, the uncertainty in the recovered parameters—assuming that image measurements are corrupted by small Gaussian noise errors—can be determined by computing the inverse covariance or *information* matrix **A**. This matrix is formed by computing outer products of the *Jacobians* of the measurement equations

$$\mathbf{A} = \sum_i \sum_j c_{ij} \frac{\partial \mathbf{f}_{ij}^T}{\partial \mathbf{a}} \frac{\partial \mathbf{f}_{ij}}{\partial \mathbf{a}^T}. \tag{6}$$

For notational succinctness, we use the symbol

$$\mathbf{H}_{ij} = \begin{bmatrix} \frac{\partial \mathbf{f}_{ij}^T}{\partial \mathbf{p}_i} \\ \frac{\partial \mathbf{f}_{ij}^T}{\partial \mathbf{m}_j} \end{bmatrix}$$

to denote the non-zero portion of the full Jacobian $\frac{\partial \mathbf{f}_{ij}^T}{\partial \mathbf{a}}$.

The **A** matrix has the structure

$$\mathbf{A} = \begin{bmatrix} \mathbf{Ap} & \mathbf{Apm} \\ \mathbf{Apm}^T & \mathbf{Am} \end{bmatrix}. \tag{7}$$

The matrices **Ap** and **Am** are block diagonal, with diagonal entries

$$\mathbf{Ap}_i = \sum_j \frac{\partial \mathbf{f}_{ij}^T}{\partial \mathbf{p}_i} \frac{\partial \mathbf{f}_{ij}}{\partial \mathbf{p}_i^T} \quad \text{and} \quad \mathbf{Am}_j = \sum_i \frac{\partial \mathbf{f}_{ij}^T}{\partial \mathbf{m}_j} \frac{\partial \mathbf{f}_{ij}}{\partial \mathbf{m}_j^T}, \tag{8}$$

respectively (assuming $c_{ij} = 1$), while **Apm** is dense, with entries

$$\mathbf{Ap}_i \mathbf{m}_j = \frac{\partial \mathbf{f}_{ij}^T}{\partial \mathbf{p}_i} \frac{\partial \mathbf{f}_{ij}}{\partial \mathbf{m}_j^T}. \tag{9}$$

The information matrix has previously been used in the context of structure from motion to determine *Cramer-Rao lower bounds* on the parameter uncertainties by taking the inverse of the diagonal entries [15, 16]. The Cramer-Rao bounds, however, can be arbitrarily weak, especially when $\mathbf{A}$ is singular or near-singular. In this paper, we use eigenvector analysis of $\mathbf{A}$ to find the dominant directions in the uncertainty (covariance) matrix and their magnitudes, which gives us more insight into the exact nature of structure from motion ambiguities.

3.2 Estimating reconstruction errors

An important benefit of uncertainty analysis is that we can easily quantify the expected amount of reconstruction (and motion) error for an optimal structure from motion algorithm. In the case of RMS reconstruction error, the positional uncertainty matrix $\mathbf{C_{p_i}}$ can be computed by inverting $\mathbf{A}$ and looking at its upper left block (the block corresponding to the $\mathbf{p}_i$ variables).[1] The RMS reconstruction and motion error can also be computed directly by summing the *inverse* eigenvalues of the information $\mathbf{A}$.

What is the advantage of the second approach, if computing eigenvalues is just as expensive as inverting matrices? First, we can compute the first few eigenvalues more cheaply (and in less space) than the matrix inverse, and these tend to dominate the overall reconstruction error. Second, it justifies the approach in the paper, which is to look at the minimum eigenvalue as the prime indicator of reconstruction error.

3.3 Ambiguities in structure from motion

Because structure from motion attempts to recover both the structure of the world and the camera motion without any external (prior) knowledge, it is subject to certain ambiguities. The most fundamental (but most innocuous) of these is the coordinate frame (also known as pose, or Euclidean) ambiguity, i.e., we can move the origin of the coordinate system to an arbitrary place and pose and still obtain an equally valid solution.

The next most common ambiguity is the scale ambiguity (for a perspective camera) or the depth ambiguity (for an orthographic camera). This ambiguity can be removed with a small amount of additional knowledge, e.g., the absolute distance between camera positions.

A third ambiguity, and the one we focus on in this paper, is the *bas-relief ambiguity*. In its pure form, this ambiguity occurs for a two frame problem with an orthographic camera, and is a confusion between the *relative depth* of the object and the amount of object rotation. In this paper, we focus on the *weak* form of this ambiguity, i.e., the very large *bas-relief uncertainty* which occurs with imperfect measurements even when we use more than two frames and/or perspective cameras. A central result of this paper is that the bas-relief ambiguity captures the largest uncertainties arising in structure from motion. However, when examined in detail, it appears that a larger class of deformations (i.e., projective) more fully characterizes the errors which occur in structure from motion.

To characterize these ambiguities, we will use eigenvector analysis of the information matrix, as explained in Section 3.1. Absolute ambiguities will show

---

[1] Note that this is *not* the same as simply inverting $\mathbf{A_p}$.

up as zero eigenvalues (unless we add additional constraints or knowledge to remove them), whereas weak ambiguities will show up as small eigenvalues.

4 Orthography: single scanline

Let us begin our analysis with an orthographic scanline camera, where the unknowns are the 2-D point positions $\mathbf{p}_i = (x_i, z_i)$ and the rotation angles θ_j.[2] The imaging equations are

$$u_{ij} = c_j x_i - s_j z_i \tag{10}$$

with $c_j = \cos\theta_j$ and $s_j = \sin\theta_j$.

The Jacobian for the 1-D orthographic camera is

$$\mathbf{H}_{ij} = \left[\frac{\partial u_{ij}}{\partial x_i}\ \frac{\partial u_{ij}}{\partial z_i} \Big| \frac{\partial u_{ij}}{\partial \theta_j} \right]^T = \left[c_j\ -s_j \big| -(c_j z_i + s_j x_i) \right]^T , \tag{11}$$

and the entries in the information matrix are

$$\mathbf{A}_{\mathbf{p}_i} = \left[\begin{matrix} \sum_j c_j^2 & -\sum_j c_j s_j \\ -\sum_j c_j s_j & \sum_j s_j^2 \end{matrix} \right] = \left[\begin{matrix} C & -D \\ -D & S \end{matrix} \right], \tag{12}$$

$$\mathbf{A}_{\mathbf{p}_i \mathbf{m}_j} = \left[\begin{matrix} -c_j^2 z_i - c_j s_j x_i \\ c_j s_j z_i + s_j^2 x_i \end{matrix} \right], \tag{13}$$

$$\mathbf{A}_{\mathbf{m}_j} = \left[\sum_i (c_j z_i + s_j x_i)^2 \right] = \left[c_j^2 Z + 2 c_j s_j W + s_j^2 X \right], \tag{14}$$

with $C = \sum_j c_j^2$, $D = \sum_j c_j s_j$, $S = \sum_j s_j^2$, $Z = \sum_i z_i^2$, $W = \sum_i z_i x_i$, and $X = \sum_i z_i^2$.

Before analyzing the complete information matrix, let us look at the two subblocks $\mathbf{A}_\mathbf{p}$ and $\mathbf{A}_\mathbf{m}$. If we know the motion, the structure uncertainty is determined by $\mathbf{A}_{\mathbf{p}_i}$ and is simply the triangulation error, i.e., $\sigma_x^2 \propto C^{-1}$ and $\sigma_z^2 \propto S^{-1}$ (note that for small rotations, σ_x^2 is generally much smaller than σ_z^2). If we know the structure, the motion accuracy is determined by $\mathbf{A}_{\mathbf{m}_j}$ and is inversely proportional to the variance in depth along the viewing direction (s_j, c_j).

What about ambiguities in the solution? Under orthography, the traditional scale ambiguity does not exist. However, translations along the optical axis cannot be estimated, and an overall pose (coordinate frame) ambiguity still exists. This manifests itself as the null (zero eigenvalue) eigenvector

$$\mathbf{e}_0 = \left[z_0\ -x_0\ \cdots\ z_N\ -x_N \big| 1\ \cdots\ 1 \right]^T .$$

4.1 Two frames: the bas-relief ambiguity

Let us say we only have two frames, and we have fixed $\theta_0 = 0, c_0 = 1, s_0 = 0, \theta_1 = \theta, c_1 = c, s_1 = s$ (Fig. 2). Then

$$\mathbf{A}_{\mathbf{p}_i} = \left[\begin{matrix} 1 + c^2 & -cs \\ -cs & s^2 \end{matrix} \right], \mathbf{A}_{\mathbf{p}_i \mathbf{m}} = \left[\begin{matrix} -c^2 z_i - cs x_i \\ cs z_i + s^2 x_i \end{matrix} \right], \mathbf{A}_\mathbf{m} = \left[c^2 Z + 2 cs W + s^2 X \right]. \tag{15}$$

---

[2] We do not estimate the horizontal translation since it can be determined from the motion of the centroid of the image points [5].

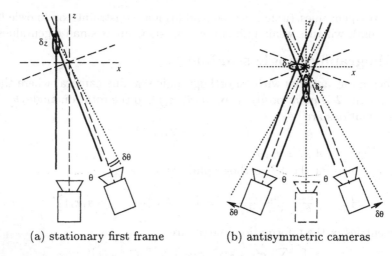

| (a) stationary first frame | (b) antisymmetric cameras |

Fig. 2. Orthographic projection, two frames.

The solid lines indicate the viewing rays, while the thin lines indicate the optical axes and image planes. The diagonal dashed lines are the displaced viewing rays, while the ellipses indicate the positional uncertainty in the reconstruction due to uncertainty in motion (indicated as $\delta\theta$).

The bas-relief ambiguity manifests itself as a null eigenvector

$$\mathbf{e}_0 = \begin{bmatrix} 0 & cz_0 + sx_0 & 0 & \cdots & cz_N + sx_N \big| -s \end{bmatrix}^T.$$

as can be verified by inspection. This is as we expected, i.e., the primary uncertainty in the structure is entirely in the depth (z) direction, and is a scale uncertainty (proportional to z). Note however that this uncertainty is proportional to $cz + sx$ rather than z, as can be seen by inspecting Fig. 2a.

An alternative parameterization of the two-frame problem is to set $\theta_0 = -\theta_1$ (Fig. 2b), in which case the null eigenvector is

$$\mathbf{e}_0 = \begin{bmatrix} s^2x_0 & -c^2z_i & \cdots & s^2x_N & -c^2z_N \mid cs \end{bmatrix}^T. \tag{16}$$

It shows that the primary effect of the bas-relief ambiguity is a "squashing" of the z values for a small increase in motion, with a much smaller "bulging" in the x values.[3] This squashing and bulging is an affine deformation of the true structure.

4.2 More than two frames, equi-angular motion constraint

To simplify the analysis, we assume for the moment that we know we have an equi-angular image sequence, i.e., that the rotation angles are given by $\theta_j = j\Delta\theta$, $j \in \{-J, \ldots, J\}$, $J = \frac{F+1}{2}$, where F is the total number of frames (imagine

---

[3] Note that the total interframe rotation is now 2θ.

Fig. 2b with more cameras). In this case, we have

$$\mathbf{H}_{ij}^T = \left[\, c_j \ -s_j \, \middle| \, -j(c_j z_i + s_j x_i) \,\right] \tag{17}$$

$$\mathbf{A_{p_i}} = \begin{bmatrix} \sum_j c_j^2 & 0 \\ 0 & \sum_j s_j^2 \end{bmatrix} = \begin{bmatrix} C & 0 \\ 0 & S \end{bmatrix}, \tag{18}$$

$$\mathbf{A_{p_i m}} = \begin{bmatrix} -\sum_j jc_j s_j x_i \\ \sum_j jc_j s_j z_i \end{bmatrix} = \begin{bmatrix} -Ex_i \\ Ez_i \end{bmatrix}, \tag{19}$$

$$\mathbf{A_m} = \left[\sum_j j^2 c_j^2 Z + \sum_j j^2 s_j^2 X \right] = \left[C'Z + S'X \right], \tag{20}$$

with $E = \sum_j jc_j s_j$, $C' = \sum_j j^2 c_j^2$, $S' = \sum_j j^2 s_j^2$, and C, D, S, Z, W, X defined as in (13–14). In this case, the smallest eigenvalue eigenvector has the form

$$\mathbf{e}_0 = \left[\, \alpha x_0 \ -\beta z_0 \ \cdots \ \alpha x_N \ -\beta z_N \, \middle| \, 1 \,\right]^T. \tag{21}$$

This will be an eigenvector if we can satisfy the matrix equation $\mathbf{Ae} = \lambda \mathbf{e}$, i.e.,

$$\left[\begin{array}{c|c} \mathbf{A_p} & \mathbf{A_{pm}} \\ \hline \mathbf{A_{pm}^T} & \mathbf{A_m} \end{array} \right] \begin{bmatrix} \alpha x_0 \\ -\beta z_0 \\ \vdots \\ -\beta z_N \\ \hline 1 \end{bmatrix} = \lambda \begin{bmatrix} \alpha x_0 \\ -\beta z_0 \\ \vdots \\ -\beta z_N \\ \hline 1 \end{bmatrix},$$

which reduces to the following three equations:

$$\alpha C - E = \alpha \lambda$$
$$\beta S - E = \beta \lambda$$
$$(S' - \alpha E)X + (C' - \beta E)Z = \lambda.$$

Substituting $\alpha = \frac{E}{C-\lambda}$ and $\beta = \frac{E}{S-\lambda}$ into the third equation, we obtain a cubic in λ,

$$(S-\lambda)(S'(C-\lambda) - E^2)X + (C-\lambda)(C'(S-\lambda) - E^2)Z - (S-\lambda)(C-\lambda)\lambda = 0, \tag{22}$$

which can be solved analytically using a package such as *Mathematica*® [20].

Assuming that the smallest eigenvalue is very small, we can use the approximation $\alpha \approx \frac{E}{C}$ to obtain a quadratic in λ,

$$(S - \lambda)(S'C - E^2)X + C(C'(S - \lambda) - E^2)Z - (S - \lambda)C\lambda = 0. \tag{23}$$

Furthermore, using the small angle approximations, $C \approx \sum_j 1 \equiv J_0$, $S \approx \Delta\theta^2 J_2$, $E \approx \Delta\theta J_2$, $C' \approx J_2$, and $S' \approx \Delta\theta^2 J_4$, where $J_2 = \sum_j j^2$ and $J_4 = \sum_j j^4$, we obtain after some manipulation

$$\lambda_{\min} \approx \frac{\Delta\theta^4 X J_2 (J_0 J_4 - J_2^2)}{J_0 J_2 Z + \Delta\theta^2 [X(J_0 J_4 - J_2^2) + J_0 J_2]}. \tag{24}$$

Notice that the minimum eigenvalue is related to the fourth power of $\Delta\theta$, i.e., doubling the inter-frame rotation reduces the RMS error by a factor of 4

λ_{min}	$F = 2$	$F = 3$	$F = 4$	$F = 5$	$F = 6$	$F = 7$	$F = 8$
$\theta_{tot} = 11.5°$	0.000000	0.000067	0.000079	0.000088	0.000096	0.000104	0.000112
$\theta_{tot} = 22.9°$	0.000000	0.001087	0.001283	0.001418	0.001547	0.001677	0.001810
$\theta_{tot} = 34.4°$	0.000000	0.005618	0.006597	0.007277	0.007931	0.008594	0.009269
$\theta_{tot} = 45°$	0.000000	0.016854	0.019688	0.021673	0.023596	0.025552	0.027547
$\theta_{tot} = 60°$	0.000000	0.054679	0.063442	0.069678	0.075782	0.082017	0.088389
$\theta_{tot} = 90°$	0.000000	0.272977	0.316453	0.348500	0.380039	0.412200	0.444997

Table 1. Minimum eigenvalues for 1-D orthographic known equi-angular motion

(assuming that $Z \gg \Delta\theta^2$). Increasing the extent of the x_i compared to the z_i directly increases the minimum eigenvalue, i.e., it decreases the structure uncertainty. This result is somewhat surprising, and suggests that flatter objects can be reconstructed better.

We can numerically compute the values of λ for a range of J and $\Delta\theta$ values. For example, with $J = 1$, $\Delta\theta = 0.1$ rad $\approx 6°$, and $X = Z = 1$, we have $\lambda = \{0.0000664436, 1.98064, 3.0193\}$. For the smallest eigenvalue, $\lambda = 0.0000664436$, we have a corresponding $\alpha = 0.0666676$ and $\beta = 10.0001$.

Once the smallest eigenvalue and eigenvector have been computed, we can easily determine some additional eigenvectors. Any vector which consists purely of x_i or z_i values which is also orthogonal to $\mathbf{Apm}$ is an eigenvector, e.g.,

$$\mathbf{e} = \begin{bmatrix} x_1 & 0 & -x_0 & 0 & \cdots & 0 & | & 0 \end{bmatrix}.$$

The eigenvalues corresponding to the pure x eigenvectors are C, while the z eigenvalues are S. In other words, once the global bas-relief uncertainty has been accounted for (squashing in z and smaller bulging in x), the variance in x position estimates is proportional to C^{-1} and in z positions is proportional to S^{-1}, i.e., exactly the expected triangulation error for known camera positions.

For the above example with $J = 1$ (3 frames), $\Delta\theta = 0.1$ rad $\approx 6°$, and $X = Z = 1$, the values for C and S are 2.98 and 0.0199, respectively. From this, we see that the correlated depth uncertainty due to the motion uncertainty is a factor of $0.0199/0.00006644 = 300$ times greater than the individual depth uncertainties. A full table of λ_{min} as a function of $F = 2J + 1$ (the number of frames) and $\theta_{tot} = (F - 1)\Delta\theta$ (the total rotation angle) is shown in Table 1.

4.3 More than two frames, without motion constraint

If we take the same data set as above, but remove the additional knowledge of equi-angular steps, we end up solving for each motion (angle) estimate separately. The equations for $\mathbf{Ap_i}$, $\mathbf{Ap_i m_j}$, and $\mathbf{Am_j}$ are given in (13–14), with $D = 0$. In this case, we do not have a closed form solution. However, performing a numerical eigenvalue analysis of the $\mathbf{A}$ matrix using a set of 9 points sampled on the unit square, i.e., $\{(x,z), x,z \in \{-1,0,1\}\}$, we obtain results that are very close to those shown in Table 1 (see [10] for details).

$\lambda_{\min}$	$F = 2$	$F = 3$	$F = 4$	$F = 5$	$F = 6$	$F = 7$	$F = 8$
$\theta_{\text{tot}} = 11.5°$	0.000175	0.000214	0.000239	0.000269	0.000299	0.000331	0.000364
$\theta_{\text{tot}} = 22.9°$	0.000690	0.001289	0.001462	0.001633	0.001803	0.001981	0.002158
$\theta_{\text{tot}} = 34.4°$	0.001512	0.004372	0.004972	0.005491	0.006009	0.006510	0.007024
$\theta_{\text{tot}} = 45°$	0.002512	0.009905	0.011282	0.012020	0.012959	0.013460	0.014070
$\theta_{\text{tot}} = 60°$	0.004234	0.020246	0.022853	0.021650	0.021870	0.020495	0.019727
$\theta_{\text{tot}} = 90°$	0.008381	0.032074	0.032623	0.027976	0.026149	0.023367	0.021596

Table 2. Minimum eigenvalues for 3-D perspective projection, equi-angular rotation around y axis, $\eta = 0.1$.

5 Perspective in 3-D

Our full-length paper contains analyses of orthography in 3-D and the perspective scanline model [10]. Due to space limitations, we now jump directly to the full 3-D perspective model. Here, we know that the two-frame problem has a solution, although our results on the simpler camera models suggest that the reconstructions may be particularly sensitive to noise.

In this section, we briefly discuss results of numerical eigenvalue analysis of pure object-centered rotation (which in camera-centered coordinates is actually both rotation and translation), and pure forward translation. Ignoring the effects of motion across the retina, these two cases capture the basic motion cues available to structure from motion. In our experiments, we used a 15-point data set consisting of the 8 corners of a unit cube, the 6 cube faces, and the origin.

5.1 Mostly rotations

The computed eigenvalues for pure rotation are shown in Table 2. Compared to the orthographic case (Table 1), we see some striking differences. First, the two-frame problem is now soluble (up to a scale ambiguity, of course). Second, for small viewing angles, there is marked improvement even for multiple frames. Third, the results for large viewing angles with small η's are significantly inferior to the orthographic results. This appears to be caused by ambiguities in camera motion along the optical axis (t_z), which are neglected in the orthographic case.

The tables of $\lambda_{\min}$ with varying η are presented in [10] for the two and three frame problems. For the two-frame case, doubling the amount of perspective distortion η results in a fourfold increase in $\lambda_{\min}$ (and hence a halving of the RMS error). For the three-frame case, the results are less sensitive to η.

For a typical minimum eigenvector (e.g., a three-frame problem with $\eta = 0.1$ and $\theta_{tot} = 11.5°$), the majority of the ambiguity is depth scaling. However, the eigenvector is not a pure affine transform of the 3-D coordinates (this has been verified numerically). Our conjecture is that the minimum eigenvector may be a *projective* transformation of the 3-D points, i.e., that the main ambiguity is projective, but we have not yet found a proof for this conjecture.

5.2 Looming

The motion of a camera forward in a 3-D world creates a different kind of parallax, which can also be exploited to compute structure from motion. To compute the ambiguities in this kind of motion, we used the same approach as before, except with no rotation and pure forward motion ($t_z \neq 0$).

Using our usual 15-point data set results in some unexpected behavior: four of the eigenvalues are zero. This is because the z coordinates of the three points on the optical axis cannot be recovered as they lie on the focus of expansion. This is a severe limitation of recovering structure from looming: points near the focus of expansion are recovered with extremely poor accuracy. In our experiments, we use a 12-point data set instead, i.e., the 15-point set with the three points $(x, y) = (0, 0)$ removed. The numerical results can be found in [10].

In one set of experiments, we calculate $\lambda_{\min}$ as a function of the number of frames F and the total extent of forward motion t_z (the object being viewed is a unit cube with coordinates $[-1, 1]^3$). The two-frame results are almost as good as the three frame results with the same extent of motion. The value of $\lambda_{\min}$ appears to depend quadratically on the total extent of motion. Overall, however, these results are much worse than those available with object-centered rotation.

In another set of experiments, we calculate $\lambda_{\min}$ as a function of η, i.e., the amount of perspective distortion. It appears that $\lambda_{\min}$ depends cubically on η, at least for small t_zs. To obtain reasonable estimates, therefore, it is necessary to both use a wide field of view and a large amount of motion relative to the scene depth.

6 Discussion

The results presented in this paper suggest that in many situations where structure from motion might be applied, the solutions are extremely sensitive to noise. In fact, very few results of convincing quality are available. Those cases where metrically accurate results have been demonstrated almost always use a large amount of rotation [5].

This raises the obvious question: are current structure from motion algorithms of practical significance? The situation is perhaps not that bad. For large object rotations, we can indeed recover accurate reconstructions. Furthermore, for scene reconstruction, using cameras with large fields of view, several camera mounted in different directions, or even panoramic images, should remove most of the ambiguities.

The general approach developed in this paper, i.e., eigenvalue analysis of the Hessian (information) matrix appears to explain most of the known ambiguities in structure from motion. However, there are certain ambiguities (e.g., depth reversals under orthography, or multiplicities of solutions with few points and frames) which will not be detected by this analysis because they correspond to multiple local minima of the cost function in the parameter space. Furthermore, analysis of the information matrix can only predict the sensitivity of the results to *small* amounts of image noise. Further study using empirical methods is required to determine the limitations of our approach.

Using the minimum eigenvalue to predict the overall reconstruction error may fail when the dominant ambiguities are in the motion parameters (e.g., what appears to be happening under perspective for large motions). Computing the RMS_{pos} error directly from the covariance matrix $\mathbf{A}^{-1}$ would be more useful in these cases, and we plan to carry out this analysis.

6.1 Future work

We are currently performing an error analysis on the results of an optimal structure from motion algorithm [4] with noisy data to see if they agree with the errors predicted by our analysis. In particular, we are estimating the (scaled) metric, affine, and projective reconstruction errors to determine which kinds of errors dominate.

In future work, we plan to compare results available with object-centered and camera-centered representations (Equations 3–4). Our guess is that the former will produce estimates of better quality. Similarly, we would like to analyze the effects of mis-estimating internal calibration parameters such as focal length, and to study the feasibility of estimating them as part of the reconstruction process. The results presented here have assumed for now that feature points are visible in all images. Our approach generalizes naturally to missing data points. In particular, we would like to study the effects feature tracks with relatively short lifetimes.

Finally, it appears that the portion of the uncertainty matrix which is correlated can be accounted for by a small number of modes. This suggest that an efficient recursive structure from motion algorithm could be developed which avoids the need for using full covariance matrices [17] but which performs significantly better than algorithms which ignore such correlations.

7 Conclusions

This paper has developed new techniques for analyzing the fundamental ambiguities and uncertainties inherent in structure from motion. Our approach is based on examining the eigenvalues and eigenvectors of the Hessian matrix in order to quantify the nature of these ambiguities. The eigenvalues can also be used to predict the overall accuracy of the reconstruction.

Under orthography, the bas-relief ambiguity dominates the reconstruction error, even with large numbers of frames. This ambiguity disappears, however, for large object-centered rotations. For perspective cameras, two-frame solutions are possible, but there must still be a large amount of object rotation for best performance. Using three of more frames avoids some of the sensitivities associated with two-frame reconstructions. Translations towards the object are an alternative source of shape information, but these appear to be quite weak unless large fields of views and large motions are involved.

When available, prior information about the structure or motion (e.g., absolute distances, perpendicularities) can be used to improve the accuracy of the reconstructions. Whether 3-D reconstruction errors (for modeling) or motion estimation errors (for navigation) are most significant for a given application determines the conditions which produce acceptable results. In any case, careful error analysis is essential in ensuring that the results of structure from motion algorithms are sufficiently reliable to be used in practice.

References

1. H. C. Longuet-Higgins. A computer algorithm for reconstructing a scene from two projections. *Nature*, 293:133–135, 1981.
2. M. E. Spetsakis and J. Y. Aloimonos. Optimal motion estimation. In *IEEE Workshop on Visual Motion*, pp. 229–237, 1989.
3. J. Weng, N. Ahuja, and T. S. Huang. Optimal motion and structure estimation. *IEEE Trans. Pattern Analysis and Machine Intelligence*, 15(9):864–884, 1993.
4. R. Szeliski and S. B. Kang. Recovering 3D shape and motion from image streams using nonlinear least squares. *J. Vis. Commun. and Image Repr.*, 5(1):10–28, 1994.
5. C. Tomasi and T. Kanade. Shape and motion from image streams under orthography: A factorization method. *Int'l J. of Computer Vision*, 9(2):137–154, 1992.
6. O. D. Faugeras. What can be seen in three dimensions with an uncalibrated stereo rig? In *Second European Conf. Computer Vision (ECCV'92)*, pp. 563–578, 1992
7. R. Hartley, R. Gupta, and T. Chang. Stereo from uncalibrated cameras. In *IEEE Conf. Computer Vision and Pattern Recognition (CVPR'92)*, pp. 761–764, 1992.
8. R. Mohr, L. Veillon, and L. Quan. Relative 3D reconstruction using multiple uncalibrated images. In *IEEE Conf. Computer Vision and Pattern Recognition (CVPR'93)*, pp. 543–548, 1993.
9. H. C. Longuet-Higgins. Visual motion ambiguity. *Vision Research*, 26(1):181–183, 1986.
10. R. Szeliski and S. B. Kang. Shape ambiguities in structure from motion. Technical Report 96/1, Digital Equipment Corporation, Cambridge Research Lab, Cambridge, MA, 1996.
11. N. Cui, J. Weng, and P. Cohen. Extended structure and motion analysis from monocular image sequences. In *Third Int'l Conf. Computer Vision (ICCV'90)*, pp. 222–229, 1990.
12. C. J. Taylor, D. J. Kriegman, and P. Anandan. Structure and motion in two dimensions from multiple images: A least squares approach. In *IEEE Workshop on Visual Motion*, pp. 242–248, 1991.
13. A. Azarbayejani and A. P. Pentland. Recursive estimation of motion, structure, and focal length. *IEEE Trans. Pattern Analysis and Machine Intelligence*, 17(6):562–575, 1995.
14. J. J. Koenderink and A. J. van Doorn. Affine structure from motion. *J. of the Optical Society of America A*, 8:377–385538, 1991.
15. G. Adiv. Inherent ambiguities in recovering 3-D motion and structure from a noisy flow field. *IEEE Trans. Pattern Analysis and Machine Intelligence*, 11(5):477–490, 1989.
16. G.-S. Y. Young and R. Chellappa. Statistical analysis of inherent ambiguities in recovering 3-d motion from a noisy flow field. *IEEE Trans. Pattern Analysis and Machine Intelligence*, 14(10):995–1013, 1992.
17. J. Oliensis and J. I. Thomas. Incorporating motion error in multi-frame structure from motion. In *IEEE Workshop on Visual Motion*, pp. 8–13, 1991.
18. N. Ayache. *Artificial Vision for Mobile Robots: Stereo Vision and Multisensory Perception*. MIT Press, Cambridge, MA, 1991.
19. W. H. Press, B. P. Flannery, S. A. Teukolsky, and W. T. Vetterling. *Numerical Recipes in C: The Art of Scientific Computing*. Cambridge University Press, Cambridge, England, 1992.
20. Stephen Wolfram. *Mathematica: A System for Doing Mathematics by Computer*. Addison-Wesley, Redwood City, CA, 1991.

Author Index

Author Index

Springer-Verlag
and the Environment

We at Springer-Verlag firmly believe that an international science publisher has a special obligation to the environment, and our corporate policies consistently reflect this conviction.

We also expect our business partners – paper mills, printers, packaging manufacturers, etc. – to commit themselves to using environmentally friendly materials and production processes.

The paper in this book is made from low- or no-chlorine pulp and is acid free, in conformance with international standards for paper permanency.

Lecture Notes in Computer Science

For information about Vols. 1–992

please contact your bookseller or Springer-Verlag